Broadband RF
and
Microwave
Amplifiers

Broadband RF
and
Microwave
Amplifiers

Andrei Grebennikov • Narendra Kumar
Binboga S. Yarman

CRC Press
Taylor & Francis Group
Boca Raton London New York

CRC Press is an imprint of the
Taylor & Francis Group, an **informa** business

CRC Press
Taylor & Francis Group
6000 Broken Sound Parkway NW, Suite 300
Boca Raton, FL 33487-2742

First issued in paperback 2017

© 2016 by Taylor & Francis Group, LLC
CRC Press is an imprint of Taylor & Francis Group, an Informa business

No claim to original U.S. Government works

ISBN-13: 978-1-4665-5738-3 (hbk)
ISBN-13: 978-1-138-80020-5 (pbk)

Visit the Taylor & Francis Web site at
http://www.taylorandfrancis.com

and the CRC Press Web site at
http://www.crcpress.com

Contents

Preface

The main objective of this book is to present all relevant information required for radio frequency (RF) and microwave broadband power amplifier design, including well-known historical and recent novel schematic configurations, theoretical approaches, circuit simulation results, and practical implementation techniques and technologies. This book can be very useful for lecturing to promote the systematic way of thinking with analytical calculations, circuit simulation, and practical verification. The demonstration of not only new results based on new technologies or circuit schematics is given, but some sufficiently old ideas or approaches are also introduced that could be very useful in modern design practice or could contribute to the appearance of new general design ideas and specific circuit design techniques. As a result, this book is intended for and can be recommended to *university-level professors* as a comprehensive reference material to help in lecturing for graduate and postgraduate students, to *researchers and scientists* to combine the theoretical analysis with practical design and to provide a sufficient basis for innovative ideas and circuit design techniques, and to *practicing designers and engineers* as the book contains numerous well-known and novel practical circuits and theoretical approaches with a detailed description of their operational principles and applications.

In Chapter 1, the basic two-port networks are introduced to describe the behavior of linear and nonlinear circuits. To characterize the nonlinear properties of the bipolar or field-effect transistors (FETs), their equivalent circuit elements are expressed through the impedance Z-parameters, admittance Y-parameters, or hybrid H-parameters. On the other hand, the transmission ABCD-parameters are very important in the design of the distributed circuits such as a transmission line or cascaded elements, whereas the scattering S-parameters are widely used to simplify a measurement procedure. Monolithic implementation of lumped inductors and capacitors is usually required at microwave frequencies and for portable devices.

Chapter 2 describes the basic principles of power amplifier design procedure. On the basis of spectral-domain analysis, the concept of a conduction angle is introduced with simple and clear analyses of the basic Classes A, AB, B, and C of the power amplifier operation. Accuracy of nonlinear modeling for metal–oxide–semiconductor field-effect transistor (MOSFET), metal–semiconductor field-effect transistor (MESFET), high-electron-mobility transistor (HEMT), and bipolar devices, including heterojunction bipolar transistors (HBTs), is a very important part of the power amplifier design, especially having a great demand in modern microwave monolithic-integrated circuits. The possibility of the maximum power gain for a stable power amplifier is discussed and analytically derived. The concept and design of push–pull amplifiers using balanced transistors and different types of transmission-line transformers and combiners are given.

Chapter 3 begins with the basic principles of the impedance matching and broadband power amplifier design using lumped and distributed parameters. Generally, the matching design procedure is based on the methods of circuit analysis, optimization, and synthesis. According to the first method, the circuit parameters are calculated at one frequency chosen in advance (usually the center or high-bandwidth frequency), and then, the power amplifier performance is analyzed across the entire frequency bandwidth. To synthesize the broadband matching/compensation network, it is necessary to choose the maximum attenuation level or reflection coefficient magnitude inside the operating frequency bandwidth and then to obtain the parameters of matching networks by using special tables and formulas to convert the lumped elements into distributed ones. The matching technique with a prescribed amplitude–frequency response and practical examples of broadband power amplifiers designed to operate from RF to millimeter-wave frequencies are discussed.

Dissipative or lossy gain-compensation-matching circuits can provide an important trade-off between power gain, reflection coefficient, and the operating frequency bandwidth. Moreover, the resistive nature of such a simple matching circuit may also improve amplifier stability and reduce its size and cost. Chapter 4 describes the power amplifier design based on a broadband concept that provides some advantages when there is no need to tune the resonant-circuit parameters. In this case, it is sufficiently easy to provide multioctave amplification from very low frequencies up to ultrahigh frequencies using the power MOSFET devices when lossy gain compensation is provided. This can be possible due to some margin in a power gain at lower frequencies for these devices, because its value decreases with frequency by approximately 6 dB per octave. Besides, lossy gain-compensating networks and resistive-feedback power amplifiers can provide lower input reflection coefficients, smaller gain ripple, a more predictable amplifier design, and can contribute to the amplifier stability factors that are superior to those of lossless-matching networks. Several design techniques including graphical and decomposition synthesis methods are briefly described.

Chapter 5 explains the design of broadband RF and microwave amplifiers using real frequency techniques (RFTs). Different network termination arrangements, such as the single-match case with complex load and resistive source and the double-match case where both load and source impedances are complex, are discussed using matching networks with lumped elements and transmission lines. RFT provides several advantages over most of the usual techniques: it does not require any active component model and no predetermined matching network topology is necessary. The automated design procedure based on the RFT is introduced to resolve the input- and output-matching problems. A simplified real frequency technique (SRFT) utilizes the measured data obtained from the generator and the load networks, and neither an *a priori* choice of an interstage equalizer topology, nor an analytic form of the system transfer function is assumed. The optimization process of the design procedure is carried out directly in terms of a physically realizable, unit-normalized reflection coefficient that describes the equalizer alone. Numerous design examples based on MATLAB® programming process clearly illustrate all aspects of the RFT design techniques.

In modern wireless communication systems, it is required that the power amplifier could operate with high efficiency over a wide frequency range to simultaneously provide multiband and multistandard signal transmission. Chapter 6 starts with a detailed description of the reactance compensation technique and the conventional design of a high-efficiency switchmode Class-E power amplifier requiring a high value of the loaded quality factor Q_L to satisfy the necessary harmonic impedance conditions at the output device terminal is described. However, if a sufficiently small value of Q_L is selected, a high-efficiency broadband operation of the Class-E power amplifier can be realized by applying reactance compensation technique. Usually, the bandwidth limitation in power amplifiers comes from the low value of the device transition frequency and large output capacitance; therefore, silicon laterally diffused metal–oxide–semiconductor field-effect transistor (LDMOSFET) technology has been the preferred choice up to 2.2 GHz. As an alternative, gallium nitride (GaN) HEMT technology enables high efficiency, large breakdown voltage, high-power density, and significantly higher broadband performance due to higher transition frequency and smaller periphery, resulting in the smaller input and output capacitances and less parasitics. Different types of broadband Class-E power amplifiers based on different implementation technologies are given.

Chapter 7 describes the historical aspect, basic structures, and main principles of the Doherty amplifiers. The Doherty amplifier with a series-connected load and inverted Doherty architectures are also described and discussed. To increase efficiency over the power-backoff range, the switchmode broadband Class-E mode can be used in the load network. Examples of the lumped Doherty amplifier implemented in monolithic microwave-integrated circuits, digitally driven Doherty technique, and broadband capability of the two-stage Doherty amplifier are given.

Chapter 8 begins with the historical aspect and basic principles of the low-noise amplifier (LNA) design including basic topologies, minimum noise figure, and linearization techniques. When it is necessary to achieve high-gain and low-noise performance over a sufficiently wide frequency range, the LNAs are designed using lossless-matching circuits. However, the lossy feedback LNAs have been shown to be capable of the flat gain over a very wide bandwidth with a sufficiently low-noise figure, small size, and convenience in practical implementation. Several design techniques including an iterative optimization or interactive graphical design approach are very useful in view of many variables and conflicting objectives of high gain, flat and broadband gain, and low-noise figure. Finally, some practical circuit schematics of the broadband millimeter-wave LNAs are given and discussed.

The potential of the traveling-wave or distributed amplification for obtaining power gains over very wide frequency bands has been recognized yet in the mid-1930s when it was found that the gain–bandwidth performance is greatly affected by the capacitance and transconductance of the conventional vacuum tube. In Chapter 9, the basic principles of distributed amplification and circuit implementation of microwave GaAs FET-distributed amplifiers are first introduced and described. Different architectures such as cascaded and matrix-distributed power amplifiers, different techniques and technologies including extended resonant and complementary metal–oxide–semiconductor (CMOS) implementations are given. Finally, the noise properties of distributed amplifiers using MOSFET, MESFET, and HBT devices are described.

Ultrawideband (UWB) transmission technology is very attractive for its low-cost and low-power communication applications, occupying a very wide frequency range. Initially, it was mainly used for radar-based applications because of the wideband nature of the signal that results in very accurate-timing information. However, due to further developments in high-speed switching and narrowband pulse generation technology, UWB has become more attractive for low-cost communication applications where large-frequency bandwidths are achieved by using very narrow time-duration baseband pulses of appropriate shape and duration. Larger-transmission bandwidths are preferred to achieve higher data rates without the need to increase transmitting power, resulting in the ability for increasingly fine resolution for multipath arrivals. This leads to reduced fading per resolved path since the impulsive nature of the transmitted waveforms prevents significant overlap and, hence, reduces the possibility of destructive combining. Market considerations require that UWB-based products be implemented using CMOS technology to achieve low-power and low-cost integration. Chapter 10 describes the basic circuit schematics of the CMOS amplifiers for UWB applications including distributed feedback, common-gate, lossy-matched configurations, and noise-canceling technique.

MATLAB is a registered trademark of The MathWorks, Inc. For product information, please contact:

The MathWorks, Inc.
3 Apple Hill Drive
Natick, MA 01760-2098, USA
Tel: 508 647 7000
Fax: 508-647-7001
E-mail: info@mathworks.com
Web: www.mathworks.com

Acknowledgments

Dr. Andrei Grebennikov especially wishes to thank his wife, Galina Grebennikova, for performing an important numerical calculation and computer artwork design, and for her constant encouragement, inspiration, support, and assistance.

Dr. Narendra Kumar would like to acknowledge his wife, Dr. Phuvaneswary Sangeran, for the great support and motivation.

Dr. Binboga S. Yarman would like to acknowledge his wife, Dr. Sema Yarman, for her eternal support without expectation.

Introduction

The power transmission from a generator to a load constitutes one of the fundamental problems in the design of broadband communication systems that generally involves the design of an impedance-transforming or matching network to transform a given load impedance into another specified impedance over some frequency range. The history of broadband impedance matching is very long and goes back to the end of the 1920s when the broadband frequency-selective filtering circuit of a recurrent structure was applied to a vacuum-tube amplifier [1]. In this case, the circuit of a bandpass type was arranged to match the anode resistance of one vacuum-tube amplifier with high-input impedance of the succeeding amplifier, with three shunt branches consisting of parallel combinations of inductors and capacitors, and three series branches consisting of series-resonant circuits [2]. The proper impedance matching between amplifying stages was also obtained with the impedance-correcting networks of a ladder type connected in series with broadband wave filters [3,4]. The filtering circuit having a proper terminating resistor was able to present a substantially constant pure resistance over a considerable range of RFs [5].

At that time, it was discovered that the amplification in a broadband vacuum-tube amplifier was limited not only by the amplifying ability of the vacuum tube at low frequencies, but also by its shunt capacitance because it limits the broadband-coupling impedance that can be built up across the input and output circuits of a vacuum tube. Many forms of networks can be employed to maintain nearly uniform impedance across the shunt capacitor where this capacitor is regarded as one element. The greatest shunt capacitance across which the uniform impedance can be developed over a frequency bandwidth is defined as $C = 2/R\omega_c$, where R is the terminal shunt resistance and ω_c is the angular cutoff frequency. This relation leads to the ultimate theoretical limitation on the broadband performance of the coupling impedance

$$\Delta\omega CR \leq 2 \tag{I.1}$$

in which $\Delta\omega$ is the bandwidth in terms of angular frequency [6]. This formula is valid not only for low-pass filters, but also for bandpass filters of the same total bandwidth. Note that the theoretical limit is based on an infinite number of circuit elements, and cannot be exceeded in any passive network.

Figure I.1 shows the practical interstage-coupling circuits from the simple to the complex, where the addition of a few circuit elements in a preferred arrangement allows a very close approximation of the theoretical performance to be obtained [6]. Here, the relative bandwidths and low-pass filter components are shown in the first column. The first example (a) includes no filter sections, but only the shunt capacitance with a resistor in parallel. The second example (b) includes a constant-k filter half-section representing the coupling impedance with series inductance and resistance in the parallel path. The constant-k half-section provides for maximum capacitance directly across the impedance terminals on the left-hand side. The third example (c) has an m-derived filter half-section, which provides for matching the image impedance with the resistor at the end on the right-hand side. The second column of Figure I.1 shows the practical low-pass circuits where the resistor and reactive elements are rearranged in a ladder network in the most convenient order. The third column shows the bandpass networks exactly analogous to the low-pass networks

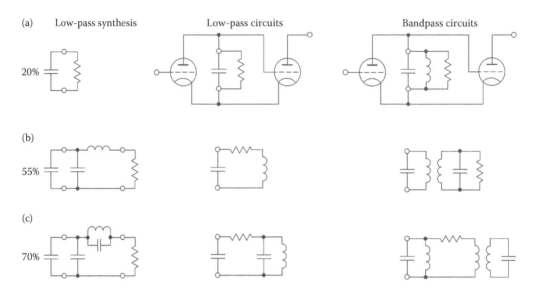

(a)   Low-pass synthesis   Low-pass circuits   Bandpass circuits

20%

(b)

55%

(c)

70%

FIGURE I.1
Two-terminal low-pass and bandpass-coupling circuits.

of the second column when each reactance element of the low-pass filter becomes a tuned circuit in the bandpass analog.

Figure I.2 shows the generalized block diagram of a matching network to transform the source impedance Z_S to the load impedance Z_L by means of a two-port matching network. Here, the source is assumed to consist of an ideal voltage source connected in series with the source impedance, and maximum power delivery to the load is then obtained when the proper impedance is presented to the source. For a lossless impedance-transforming network and a given load impedance Z_L, the impedance Z_{in} presented by the matching network must be conjugate of Z_S [4]. Generally, there are four different network termination arrangements, with the most general broadband *double-match* case, where both the load and source impedances are complex. The simpler broadband *single-match* case involves a complex load and a resistive source with $X_S = 0$. Filter networks are either *doubly terminated* with resistances R_L and R_S or *singly terminated* with $Z_L = R_L$ and $Z_S = 0$. In the latter case, either an ideal voltage or current source may provide the excitation.

With a ladder-type impedance-transforming network, since the series impedances and shunt admittances are frequency dependent, all the series branches will have a given physical configuration and all the shunt branches will have the inverse configuration. For example, if the series branches are inductances, the shunt branches will be capacitances. By using other series arm configurations, a considerable variety of networks can be obtained. The property of this impedance-transforming network configuration with the load impedance described by a certain mathematical function can be expressed as a single polynomial function

$$R(\text{or } G) = \frac{F(x)}{A_0 + A_1 x + A_2 x^2 + \ldots + A_n x^n} \tag{I.2}$$

where R(or G) is the resistance (or conductance) of the corrected structure, $A_0, A_1, \ldots A_n$ are the polynomial coefficients that specify the values of the elements in the network

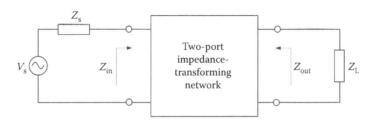

FIGURE I.2
Generalized block diagram of a two-port impedance-transforming network.

implicitly, $F(x)$ is either the resistance or conductance component of the impedance, and x is the function of frequency [4]. As it follows from Equation I.2, to secure the proper resistance and conductance from the corrected structure, it is simply necessary to choose such values of the constants $A_0, A_1, \ldots A_n$ that the polynomial in the denominator satisfies the equation with sufficient accuracy. Note that such a problem of approximating a given curve by a polynomial of a given degree is a well-known one in mathematics.

However, it is not possible to match arbitrary impedance to a pure resistance over the whole frequency spectrum, or even at all frequencies within a finite frequency band. On the other hand, it is evidently possible to obtain a match at any desired number of frequencies, provided the given impedance has a finite-resistive component at those frequencies. Such a matching, however, has little practical value because it is incorrect to assume that one can obtain a reasonable match over a frequency band by correctly matching at a sufficiently large number of frequencies within the desired band. In this case, the statement of any matching problem must include the maximum tolerance on the match, as well as the minimum bandwidth within which the match is to be obtained. Therefore, it is reasonable to expect that, for a given load impedance and a given maximum tolerance, there is an upper limit to the bandwidth that can be obtained by means of a physically realizable impedance-matching network. With reference to Figure I.2 where Z_L is a given impedance function of frequency, a nondissipative impedance-transforming network must be designed such that, when terminated in Z_L, the magnitude of the input reflection coefficient $\Gamma = (Z_{in} - 1)/(Z_{in} + 1)$ is smaller than, or equal to, a specified value $|\Gamma|_{max}$ at all frequencies within a specified band.

As opposed to a step-by-step procedure leading to a ladder structure used previously, the first systematic investigation of matching networks was made by Bode for a class of very useful load impedance consisting of a resistor R shunted by a capacitor C. He showed that the fundamental gain–bandwidth limitation on the matching network in this case takes the form

$$\int_0^\infty \ln \frac{1}{|\Gamma(\omega)|} \, d\omega \leq \frac{\pi}{RC} \tag{I.3}$$

where $\Gamma(\omega)$ is the input frequency-dependent reflection coefficient corresponding to the input impedance Z_{in} in Figure I.2 [7]. If $|\Gamma|$ is kept constant and equal to $|\Gamma|_{max}$ over a frequency band of width $\Delta\omega$ and is made equal to unity over the rest of the frequency spectrum, Equation I.3 yields

$$\Delta\omega \ln \frac{1}{|\Gamma(\omega)|_{max}} \leq \frac{\pi}{RC} \tag{I.4}$$

indicating that the product of the bandwidth by the minimum passband value of $\ln(1/|\Gamma|_{\max})$ has a maximum limit fixed by the product RC. From Equation I.3 also follows that approaching a perfect match by making $|\Gamma|$ very small at any frequency results in an unnecessary waste of the area represented by the integral, and therefore in reduction of the bandwidth. This limitation found by Bode can be applied to any impedance consisting of a reactive two-port network terminated in a parallel RC configuration.

The Bode gain–bandwidth theory for broadbanding the RC single match was then extended by Fano by utilizing the established doubly terminated filter theory [8]. Fano technique was based on Darlington theorem stating that any physically realizable impedance function can be considered as the input impedance of a reactive two-port network terminated in a pure resistance, and this resistance can be made equal to 1 Ω in all cases by incorporating an appropriate ideal transformer in the reactive network [9]. In this case, the load impedance Z_L in Figure I.2 can be replaced with a Darlington resistively terminated LC two-port network, so that the doubly terminated filter with unity resistors on both ends and two LC-cascaded two-port networks can be obtained as the equalizer. The overall problem was to design a Chebyshev or elliptic equal-ripple doubly terminated filter having a specified number of elements, the solution of which resulted in a set of integral constraints with proper weighting functions depending on the load impedance [8]. The method was illustrated by the design of simple matching networks for two cases of the load impedances, one consisting of a resistance in series with an inductance and the other consisting of a capacitance shunting a series RL combination. The lossless reactive two-port network (or equalizer) with a finite number of elements can also be designed to approximate a transfer characteristic that is a prescribed function of frequency, rather than a constant, over the useful frequency band [10]. Matthaei modified and extended Fano design procedure and presented a practical means for selecting the optimum design parameters [11]. In the mid-1960s, Youla developed a new theory based on the principle of complex normalization that avoids some difficulties encountered in Fano approach and which can be generalized to the design of equalizers with an active load [12]. Finally, explicit formulas for computing the optimum design parameters of the low-pass and bandpass impedance-matching networks having Butterworth and Chebyshev responses of arbitrary order for a class of the most practical RLC load were then derived [13,14].

Analytic theory applied to the impedance-matching network design is essential to understand the gain–bandwidth limitations of the given loads to be matched, with explicit formulas available only for certain simple cases. However, it is limited to simple single or double matching in which the generator and load networks include at most two reactive elements, namely, either a capacitor or an inductor, as shown in Figure I.3. If the number of reactive elements increases on the load side, either theory becomes inaccessible or the resulting gain performance turns out to be nonoptimal, and the equalizer structures become unnecessarily complicated and even sometimes completely unrealizable. These restrictions take either the form of a simultaneous set of difficult algebraic equations or a set of integral expressions that must be satisfied simultaneously. Even if the load model is known, the procedure presents great numerical difficulties that in principle are resolvable but in practice become almost intractable, especially if the load includes more than two reactive elements. Proceeding by direct numerical optimization of the elements of the equalizer is also a difficult matter because these elements relate to the system gain in highly nonlinear combinations. Furthermore, such direct optimization requires a specific assumption of the equalizer topology.

In 1977, a new numerical approach known as the *RFT* was introduced by Carlin for the solution of the broadband matching of an arbitrary load to a resistive generator that leads

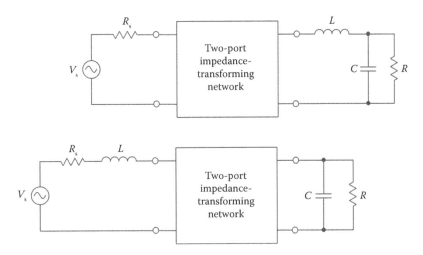

FIGURE I.3
Two-port impedance-transforming network with simple source and load networks.

to a simple technique for the design of a lossless two-port equalizer [15]. It is based on the measured data obtained from the generator and the load networks, and neither a preliminary choice of an equalizer topology, nor an analytic form of the system transfer function is assumed. The optimization process of the design procedure is carried out directly in terms of a physically realizable, unit-normalized reflection coefficient that describes the equalizer alone. Among the variety of objective functions, it can be maximizing the minimum passband gain, or minimizing the maximum noise figure. On the basis of a generalized form of the double-matching design procedure by assuming the complex generator and load networks when one can represent generator and load networks as lossless Darlington two-port networks with resistive termination resulting in a cascaded structure, an example of the design of the broadband multistage microwave lumped-element FET amplifier was demonstrated [16]. The SRFT, in which the active two-port network was treated on the basis of its measured S-parameters, including the feedback parameter S_{12}, was then later applied to the design of the broadband three-stage microwave microstrip HEMT amplifiers using an efficient optimization procedure to minimize the objective function in terms of the transducer power gain, input and output reflection coefficients, and noise figure [17].

References

1. *History of Broadband Impedance Matching*, GHN: IEEE Global History Network, 2012.
2. E. L. Norton, Filtering circuits, U.S. Patent 1,788,538, January 1931 (filed Apr. 1929).
3. H. W. Bode, Wave filter, U.S. Patent 1,814,238, July 1931 (filed Apr. 1930).
4. H. W. Bode, A method of impedance correction, *Bell Syst. Tech. J.*, 9, 794–838, 1930.
5. W. B. van Roberts, Resistance coupled amplifier, U.S. Patent 1,925,340, September 1933 (filed Mar. 1929).
6. H. A. Wheeler, Wide-band amplifiers for television, *Proc. IRE*, 27, 429–438, 1939.
7. H. W. Bode, *Network Analysis and Feedback Amplifier Design*, New York: Van Nostrand, 1945.

8. R. M. Fano, *Theoretical Limitations on the Broadband Matching of Arbitrary Impedances*, Technical Report No. 41, MIT: Research Laboratory of Electronics, 1948.

9. S. Darlington, Synthesis of reactance 4-poles, *J. Math. Phys.*, 18, 275–353, 1939.

10. A. P. Brogle Jr., The design of reactive equalizers, *Bell Syst. Tech. J.*, 28, 716–750, 1949.

11. G. L. Matthaei, Synthesis of Chebyshev impedance-matching networks, filters, and interstages, *IRE Trans. Circuit Theory*, CT-3, 163–172, 1956.

12. D. C. Youla, A new theory of broad-band matching, *IEEE Trans. Circuit Theory*, CT-11, 30–50, 1964.

13. W.-K. Chen, Explicit formulas for the synthesis of optimum broad-band impedance-matching networks, *IEEE Trans. Circuits Syst.*, CAS-24, 157–169, 1977.

14. W.-K. Chen and T. Chaisrakeo, Explicit formulas for the synthesis of optimum bandpass Butterworth and Chebyshev impedance-matching networks, *IEEE Trans. Circuits Syst.*, CAS-27, 928–942, 1980.

15. H. J. Carlin, A new approach to gain–bandwidth problems, *IEEE Trans. Circuits Syst.*, CAS-24, 170–175, 1977.

16. B. S. Yarman and H. J. Carlin, A simplified real frequency technique applied to broad-band multistage microwave amplifiers, *IEEE Trans. Microw. Theory Tech.*, MTT-30, 2216–2222, 1982.

17. R. Soares, A. Perennec, A. Olomo Ngongo, and P. Jarry, Application of a simplified real-frequency synthesis method to distributed-element amplifier design, *Int. J. Microw. Millim.-Wave Comput.-Aided Eng.*, 1, 365–378, 1991.

Authors

Andrei Grebennikov (senior member of IEEE) earned his engineering diploma in radio electronics from the Moscow Institute of Physics and Technology and a PhD in radio engineering from the Moscow Technical University of Communications and Informatics, in 1980 and 1991, respectively. He obtained long-term academic and industrial experience working with Moscow Technical University of Communications and Informatics (Russia), Institute of Microelectronics (Singapore), M/A-COM (Ireland), Infineon Technologies (Germany/Austria), Bell Labs, Alcatel-Lucent (Ireland) and Microsemi (USA) as an engineer, researcher, lecturer and educator. He read lectures as a guest professor in University of Linz (Austria) and presented short courses and tutorials as an invited speaker at International Microwave Symposia, European and Asia-Pacific Microwave Conferences, Institute of Micro-electronics, Singapore, Motorola Design Centre, Malaysia, Tomsk State University of Control Systems and Radioelectronics, Russia and Aachen Technical University, Germany. He is an author and co-author of more than 100 papers, has 25 European and U.S. patents and patent applications, and has written eight books dedicated to radiofrequency and microwave circuit design.

Narendra Kumar (senior member of IEEE and Fellow of IET) earned his PhD in electrical engineering from RWTH Technical University Aachen, Germany. He was in the R&D of Motorola Solutions, starting in early 1999, where he held a position of principal staff engineer. He has several U.S. patents, all assigned to Motorola Solutions, in the area of radiofrequency (RF) and microwave amplifier circuitry. Currently, he is with University of Malaya as associate professor in the Department of Electrical Engineering. Since 2011, he has been appointed as visiting professor to Istanbul University, Turkey. He has authored and coauthored more than 50 papers in technical journals and conferences, and 2 international books. He has conducted seminars related to RF and microwave power amplifiers in Europe and Asia Pacific. He was appointed to the IEEE Industry Relations Team of Asia Pacific, in 2015, to support the industry–academic collaboration effort.

Binboga S. Yarman (fellow of IEEE) earned his BSc in electrical engineering from Technical University of Istanbul in 1974, a MSc from Stevens Institute of Technology, New Jersey, in 1978 and a PhD from Cornell University, Ithaca, New York, in 1982. He was a member of the technical staff at Microwave Technology Center of David Sarnoff Research Center, Princeton, New Jersey. He served as professor at Anatolia University-Eskisehir, Middle East Technical University-Ankara, Technical University of Istanbul and Istanbul University. He was one of the founders of I-ERDEC Maryland (1983), STFA SAVRONIK, a defence electronics company in Turkey (1986) and ARES Security Systems Inc. (1990). He was the chief technical adviser to the Turkish Prime Ministry Office and director of Electronic and Technical Security of Turkey (1993–1999). He was the founding president of Isik University, Istanbul, Turkey (1996–2004). He was a visiting professor at Ruhr University, Bochum, Germany (1987–1994), and Tokyo Institute of Technology, Tokyo, Japan (2006–2008). Dr. B. S. Yarman has published more than 200 scientific and technical papers in the fields of electrical/electronic engineering, microwave engineering, computer engineering, mathematics and management. He holds four U.S. patents assigned to the U.S. Air Force. He has served in various technical and scientific committees, since 1980, in the United States and Turkey. He received the Young Turkish Scientist Award in 1986 and the National Research and Technology Counsel of Turkey Technology Award in 1987. He is an Alexander Von Humboldt Research Fellow (1987), Bonn, Germany, a member of New York Academy of Science (1994) and "Man of the Year in Science and Technology (1988)" of Cambridge Biography Center, U.K.

1

Two-Port Network Parameters

Two-port equivalent circuits are widely used in RF and microwave circuit design to describe the electrical behavior of both active devices and passive networks [1,2]. The two-port network impedance Z-parameters, admittance Y-parameters, or hybrid H-parameters are very important to characterize the nonlinear properties of any type of bipolar or field-effect transistors used as an active device of the power amplifier. The transmission $ABCD$-parameters of a two-port network are very convenient for designing the distributed circuits such as transmission lines or cascaded active or passive elements. The scattering S-parameters are useful to characterize linear circuits and are required to simplify the measurement procedure. Transmission lines are widely used in matching networks of high-power or low-noise amplifiers, directional couplers, power combiners, and power dividers. Monolithic implementation of lumped inductors and capacitors is usually required at microwave frequencies and for portable devices.

1.1 Traditional Network Parameters

The basic diagram of a two-port transmission system can be represented by the equivalent circuit shown in Figure 1.1, where V_S is the voltage source, Z_S is the source impedance, Z_L is the load impedance, and linear network is a time-invariant two-port network without independent source. The two independent phasor currents I_1 and I_2 (flowing across input and output terminals) and phasor voltages V_1 and V_2 characterize such a two-port network. For autonomous oscillator systems, in order to provide an appropriate analysis in the frequency domain of the two-port network in the negative one-port representation, it is sufficient to set the source impedance to infinity. For a power amplifier or oscillator design, the elements of the matching or resonant circuits, which are assumed to be linear or appropriately linearized, can be found among the linear network elements, or additional two-port linear networks can be used to describe their frequency domain behavior.

For a two-port network, the following equations can be considered as boundary conditions:

$$V_1 + Z_S I_1 = V_S \tag{1.1}$$

$$V_2 + Z_L I_2 = V_L \tag{1.2}$$

Suppose that it is possible to obtain a unique solution for the linear time-invariant circuit shown in Figure 1.1. Then the two linearly independent equations, which describe the general two-port network in terms of circuit variables V_1, V_2, I_1, and I_2, can be expressed in a matrix form as

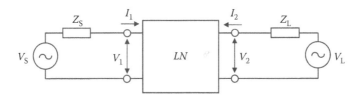

FIGURE 1.1
Basic diagram of two-port nonautonomous transmission system.

$$[M][V] + [N][I] = 0 \tag{1.3}$$

or

$$\left.\begin{aligned} m_{11}V_1 + m_{12}V_2 + n_{11}I_1 + n_{12}I_2 &= 0 \\ m_{21}V_1 + m_{22}V_2 + n_{21}I_1 + n_{22}I_2 &= 0 \end{aligned}\right\} \tag{1.4}$$

The complex 2×2 matrices $[M]$ and $[N]$ in Equation 1.3 are independent of the source and load impedances Z_S and Z_L and voltages V_S and V_L, respectively. They depend only on the circuit elements inside the linear network.

If matrix $[M]$ in Equation 1.3 is nonsingular when $|M| \neq 0$, then this matrix equation can be rewritten in terms of $[I]$ as

$$[V] = -[M]^{-1}[N][I] = [Z][I] \tag{1.5}$$

where $[Z]$ is the open-circuit impedance two-port network matrix. In a scalar form, matrix Equation 1.5 is given by

$$V_1 = Z_{11} I_1 + Z_{12} I_2 \tag{1.6}$$

$$V_2 = Z_{21} I_1 + Z_{22} I_2 \tag{1.7}$$

where Z_{11} and Z_{22} are the open-circuit driving-point impedances, and Z_{12} and Z_{21} are the open-circuit transfer impedances of the two-port network. The voltage components V_1 and V_2 due to the input current I_1 can be found by setting $I_2 = 0$ in Equations 1.6 and 1.7, resulting in an open-circuited output terminal. Similarly, the same voltage components V_1 and V_2 are determined by setting $I_1 = 0$ when the input terminal becomes open-circuited. The resulting driving-point impedances can be written as

$$Z_{11} = \left.\frac{V_1}{I_1}\right|_{I_2=0} \qquad Z_{22} = \left.\frac{V_2}{I_2}\right|_{I_1=0} \tag{1.8}$$

whereas the two transfer impedances are

$$Z_{21} = \left.\frac{V_2}{I_1}\right|_{I_2=0} \qquad Z_{12} = \left.\frac{V_1}{I_2}\right|_{I_1=0} \tag{1.9}$$

Dual analysis can be used to derive the short-circuit admittance matrix when the current components I_1 and I_2 are considered as outputs caused by V_1 and V_2. If matrix $[N]$ in Equation 1.3 is nonsingular when $|N| \neq 0$, this matrix equation can be rewritten in terms of $[V]$ as

$$[I] = -[N]^{-1}[M][V] = [Y][V] \tag{1.10}$$

where $[Y]$ is the short-circuit admittance two-port network matrix. In a scalar form, matrix Equation 1.10 is written as

$$I_1 = Y_{11} V_1 + Y_{12} V_2 \tag{1.11}$$

$$I_2 = Y_{21} V_1 + Y_{22} V_2 \tag{1.12}$$

where Y_{11} and Y_{22} are the short-circuit driving-point admittances, and Y_{12} and Y_{21} are the short-circuit transfer admittances of the two-port network. In this case, the current components I_1 and I_2 due to the input voltage source V_1 are determined by setting $V_2 = 0$ in Equations 1.11 and 1.12, thus creating a short-circuited output terminal. Similarly, the same current components I_1 and I_2 are determined by setting $V_1 = 0$ when the input terminal becomes short-circuited. As a result, the two driving-point admittances are

$$Y_{11} = \frac{I_1}{V_1}\bigg|_{V_2=0} \qquad Y_{22} = \frac{I_2}{V_2}\bigg|_{V_1=0} \tag{1.13}$$

whereas the two transfer admittances are

$$Y_{21} = \frac{I_2}{V_1}\bigg|_{V_2=0} \qquad Y_{12} = \frac{I_1}{V_2}\bigg|_{V_1=0} \tag{1.14}$$

In some cases, an equivalent two-port network representation can be redefined in order to express the voltage source V_1 and output current I_2 in terms of the input current I_1 and output voltage V_2. If the submatrix

$$\begin{bmatrix} m_{11} & n_{12} \\ m_{21} & n_{22} \end{bmatrix}$$

given in Equation 1.4 is nonsingular, then

$$\begin{bmatrix} V_1 \\ I_2 \end{bmatrix} = -\begin{bmatrix} m_{11} & n_{12} \\ m_{21} & n_{22} \end{bmatrix}^{-1} \begin{bmatrix} n_{11} & m_{12} \\ n_{21} & m_{22} \end{bmatrix} \begin{bmatrix} I_1 \\ V_2 \end{bmatrix} = [H] \begin{bmatrix} I_1 \\ V_2 \end{bmatrix} \tag{1.15}$$

where $[H]$ is the hybrid two-port network matrix. In a scalar form, matrix Equation 1.15 is expressed as

$$V_1 = h_{11} I_1 + h_{12} V_2 \tag{1.16}$$

$$I_2 = h_{21} I_1 + h_{22} V_2 \tag{1.17}$$

where h_{11}, h_{12}, h_{21}, and h_{22} are the hybrid H-parameters. The voltage source V_1 and current component I_2 are determined by setting $V_2 = 0$ for the short-circuited output terminal in Equations 1.16 and 1.17 as

$$h_{11} = \left. \frac{V_1}{I_1} \right|_{V_2=0} \qquad h_{21} = \left. \frac{I_2}{I_1} \right|_{V_2=0} \tag{1.18}$$

where h_{11} is the driving-point input impedance and h_{21} is the forward current transfer function. Similarly, the input voltage source V_1 and output current I_2 are determined by setting $I_1 = 0$ when input terminal becomes open-circuited as

$$h_{12} = \left. \frac{V_1}{V_2} \right|_{I_1=0} \qquad h_{22} = \left. \frac{I_2}{V_2} \right|_{I_1=0} \tag{1.19}$$

where h_{12} is the reverse voltage transfer function and h_{22} is the driving-point output admittance.

The transmission parameters, often used for passive device analysis, are determined for the independent input voltage source V_1 and input current I_1 in terms of the output voltage V_2 and output current I_2. In this case, if the submatrix

$$\begin{bmatrix} m_{11} & n_{11} \\ m_{21} & n_{21} \end{bmatrix}$$

given in Equation 1.4 is nonsingular, we obtain

$$\begin{bmatrix} V_1 \\ I_1 \end{bmatrix} = -\begin{bmatrix} m_{11} & n_{11} \\ m_{21} & n_{21} \end{bmatrix}^{-1} \begin{bmatrix} m_{12} & n_{12} \\ m_{22} & n_{22} \end{bmatrix} \begin{bmatrix} V_2 \\ -I_2 \end{bmatrix} = [ABCD] \begin{bmatrix} V_2 \\ -I_2 \end{bmatrix} \tag{1.20}$$

where $[ABCD]$ is the forward transmission two-port network matrix. In a scalar form, we can write

$$V_1 = AV_2 - BI_2 \tag{1.21}$$

$$I_1 = CV_2 - DI_2 \tag{1.22}$$

where A, B, C, and D are the transmission parameters. The voltage source V_1 and current component I_1 are determined by setting $I_2 = 0$ for the open-circuited output terminal in Equations 1.21 and 1.22 as

$$A = \left. \frac{V_1}{V_2} \right|_{I_2=0} \qquad C = \left. \frac{I_1}{V_2} \right|_{I_2=0} \tag{1.23}$$

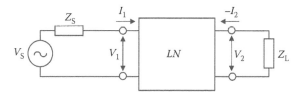

FIGURE 1.2
Basic diagram of loaded two-port transmission system.

where A is the reverse voltage transfer function and C is the reverse transfer admittance. Similarly, the input independent variables V_1 and I_1 are determined by setting $V_2 = 0$ when the output terminal is short-circuited as

$$B = -\frac{V_1}{I_2}\bigg|_{V_2=0} \qquad D = -\frac{I_1}{I_2}\bigg|_{V_2=0} \tag{1.24}$$

where B is the reverse transfer impedance and D is the reverse current transfer function. The reason a minus sign is associated with I_2 in Equations 1.20 through 1.22 is that historically, for transmission networks, the input signal is considered as flowing to the input port, whereas the output current is flowing to the load. The direction of the current $-I_2$ entering the load is shown in Figure 1.2.

1.2 Scattering Parameters

The concept of incident and reflected voltage and current parameters can be illustrated by a single-port network shown in Figure 1.3, where the network impedance Z is connected to the signal source V_S with the internal impedance Z_S. In a common case, the terminal current I and voltage V consist of incident and reflected components (assume their root-mean-square values). When the load impedance Z is equal to the conjugate of the source impedance expressed as $Z = Z_S^*$, the terminal current I becomes the incident current I_i, which is written as

$$I_i = \frac{V_S}{Z_S^* + Z_S} = \frac{V_S}{2\,\mathrm{Re}\,Z_S} \tag{1.25}$$

The terminal voltage V, defined as the incident voltage V_i, can be determined from

$$V_i = \frac{Z_S^*\,V_S}{Z_S^* + Z_S} = \frac{Z_S^*\,V_S}{2\,\mathrm{Re}\,Z_S} \tag{1.26}$$

Consequently, the incident power, which is equal to the maximum available power from the source, can be obtained by

$$P_i = \mathrm{Re}\,(V_i I_i^*) = \frac{|V_S|^2}{4\,\mathrm{Re}\,Z_S} \tag{1.27}$$

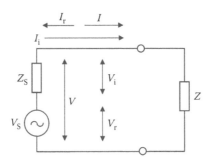

FIGURE 1.3
Incident and reflected voltages and currents.

The incident power can be rewritten in a normalized form using Equation 1.26 as

$$P_i = \frac{|V_i|^2 \, \mathrm{Re}\, Z_S}{|Z_S^*|^2} \tag{1.28}$$

This allows the normalized incident voltage wave a to be defined as the square root of the incident power P_i by

$$a = \sqrt{P_i} = \frac{V_i \sqrt{\mathrm{Re}\, Z_S}}{Z_S^*} \tag{1.29}$$

Similarly, the normalized reflected voltage wave b defined as the square root of the reflected power P_r can be written as

$$b = \sqrt{P_r} = \frac{V_r \sqrt{\mathrm{Re}\, Z_S}}{Z_S} \tag{1.30}$$

The incident power P_i can be expressed through the incident current I_i and the reflected power P_r can be expressed through the reflected current I_r, respectively, as

$$P_i = |I_i|^2 \, \mathrm{Re}\, Z_S \tag{1.31}$$

$$P_r = |I_r|^2 \, \mathrm{Re}\, Z_S \tag{1.32}$$

As a result, the normalized incident voltage wave a and reflected voltage wave b can be given by

$$a = \sqrt{P_i} = I_i \sqrt{\mathrm{Re}\, Z_S} \tag{1.33}$$

$$b = \sqrt{P_r} = I_r \sqrt{\mathrm{Re}\, Z_S} \tag{1.34}$$

The parameters a and b can also be called the *normalized incident* and *reflected current waves* or simply *normalized incident* and *reflected waves*, respectively, since the normalized current waves and the normalized voltage waves represent the same parameters.

The voltage V and current I related to the normalized incident and reflected waves a and b can be written as

$$V = V_i + V_r = \frac{Z_S^*}{\sqrt{\operatorname{Re} Z_S}} a + \frac{Z_S}{\sqrt{\operatorname{Re} Z_S}} b \tag{1.35}$$

$$I = I_i - I_r = \frac{1}{\sqrt{\operatorname{Re} Z_S}} a - \frac{1}{1\sqrt{\operatorname{Re} Z_S}} b \tag{1.36}$$

where

$$a = \frac{V + Z_S I}{2\sqrt{\operatorname{Re} Z_S}} \quad b = \frac{V - Z_S^* I}{2\sqrt{\operatorname{Re} Z_S}} \tag{1.37}$$

The source impedance Z_S is often purely real and is therefore used as the normalized impedance. In microwave design technique, the characteristic impedance of the passive two-port networks, including transmission lines and connectors, is considered as real and equal to 50 Ω. This is very important for measuring S-parameters when all transmission lines, source, and load should have the same real impedance. For $Z_S = Z_S^* = Z_0$, where Z_0 is the characteristic impedance, the ratio of the normalized reflected wave and the normalized incident wave for a single-port network is called the *reflection coefficient* Γ defined as

$$\Gamma = \frac{b}{a} = \frac{V - Z_S^* I}{V + Z_S I} = \frac{V - Z_S I}{V + Z_S I} = \frac{Z - Z_S}{Z + Z_S} \tag{1.38}$$

where $Z = V/I$.

For a two-port network shown in Figure 1.4, the normalized reflected waves b_1 and b_2 can also be represented by the normalized incident waves a_1 and a_2, respectively, as

$$b_1 = S_{11} a_1 + S_{12} a_2 \tag{1.39}$$

$$b_2 = S_{21} a_1 + S_{22} a_2 \tag{1.40}$$

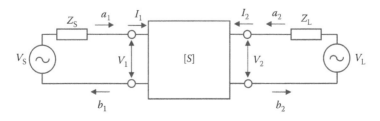

FIGURE 1.4
Basic diagram of S-parameter two-port network.

or, in a matrix form,

$$\begin{bmatrix} b_1 \\ b_2 \end{bmatrix} = \begin{bmatrix} S_{11} & S_{11} \\ S_{21} & S_{21} \end{bmatrix} \begin{bmatrix} a_1 \\ a_2 \end{bmatrix} \tag{1.41}$$

where the incident waves a_1 and a_2 and the reflected waves b_1 and b_2 for complex source and load impedances Z_S and Z_L are given by

$$a_1 = \frac{V_1 + Z_S I_1}{2\sqrt{\mathrm{Re}\,Z_S}} \quad a_2 = \frac{V_2 + Z_L I_2}{2\sqrt{\mathrm{Re}\,Z_L}} \tag{1.42}$$

$$b_1 = \frac{V_1 - Z_S^* I_1}{2\sqrt{\mathrm{Re}\,Z_S}} \quad b_2 = \frac{V_2 - Z_L^* I_2}{2\sqrt{\mathrm{Re}\,Z_L}} \tag{1.43}$$

where S_{11}, S_{12}, S_{21}, and S_{22} are the S-parameters of the two-port network.

From Equation 1.41, it follows that if $a_2 = 0$, then

$$S_{11} = \frac{b_1}{a_1}\bigg|_{a_2=0} \quad S_{21} = \frac{b_2}{a_1}\bigg|_{a_2=0} \tag{1.44}$$

where S_{11} is the reflection coefficient and S_{21} is the transmission coefficient for ideal matching conditions at the output terminal when there is no incident power reflected from the load. Similarly,

$$S_{12} = \frac{b_1}{a_2}\bigg|_{a_1=0} \quad S_{22} = \frac{b_2}{a_2}\bigg|_{a_1=0} \tag{1.45}$$

where S_{12} is the transmission coefficient and S_{22} is the reflection coefficient for ideal matching conditions at the input terminal.

1.3 Conversions between Two-Port Parameters

The parameters describing the same two-port network through different two-port matrices (impedance, admittance, hybrid, or transmission) can be cross-converted, and the elements of each matrix can be expressed by the elements of other matrices. For example, Equations 1.11 and 1.12 for the Y-parameters can be easily solved for the independent input voltage source V_1 and input current I_1 as

$$V_1 = -\frac{Y_{22}}{Y_{21}} V_2 - \frac{1}{Y_{21}} I_2 \tag{1.46}$$

$$I_1 = -\frac{Y_{11} Y_{22} - Y_{12} Y_{21}}{Y_{21}} V_2 - \frac{Y_{11}}{Y_{21}} I_2 \tag{1.47}$$

By comparing the equivalent Equations 1.21 and 1.22 and Equations 1.46 and 1.47, the direct relationships between the transmission *ABCD*-parameters and admittance *Y*-parameters are written as

$$A = -\frac{Y_{22}}{Y_{21}} \quad B = -\frac{1}{Y_{21}} \tag{1.48}$$

$$C = -\frac{\Delta Y}{Y_{21}} \quad D = -\frac{Y_{11}}{Y_{21}} \tag{1.49}$$

where $\Delta Y = Y_{11}Y_{22} - Y_{12}Y_{21}$.

A summary of the relationships between the impedance *Z*-parameters, admittance *Y*-parameters, hybrid *H*-parameters, and transmission *ABCD*-parameters is shown in Table 1.1, where $\Delta Z = Z_{11}Z_{22} - Z_{12}Z_{21}$ and $\Delta H = h_{11}h_{22} - h_{12}h_{21}$.

To convert *S*-parameters to the admittance *Y*-parameters, it is convenient to represent Equations 1.42 and 1.43 as

$$I_1 = (a_1 - b_1)\frac{1}{\sqrt{Z_0}} \quad I_2 = (a_2 - b_2)\frac{1}{\sqrt{Z_0}} \tag{1.50}$$

$$V_1 = (a_1 + b_1)\sqrt{Z_0} \quad V_1 = (a_2 + b_2)\sqrt{Z_0} \tag{1.51}$$

TABLE 1.1

Relationships between *Z*-, *Y*-, *H*-, and *ABCD*-Parameters

	[Z]		[Y]		[H]		[ABCD]	
[Z]	Z_{11}	Z_{12}	$\dfrac{Y_{22}}{\Delta Y}$	$-\dfrac{Y_{12}}{\Delta Y}$	$\dfrac{\Delta H}{h_{22}}$	$\dfrac{h_{12}}{h_{22}}$	$\dfrac{A}{C}$	$\dfrac{AD-BC}{C}$
	Z_{21}	Z_{22}	$-\dfrac{Y_{21}}{\Delta Y}$	$\dfrac{Y_{11}}{\Delta Y}$	$-\dfrac{h_{21}}{h_{22}}$	$\dfrac{1}{h_{22}}$	$\dfrac{1}{C}$	$\dfrac{D}{C}$
[Y]	$\dfrac{Z_{22}}{\Delta Z}$	$-\dfrac{Z_{12}}{\Delta Z}$	Y_{11}	Y_{12}	$\dfrac{1}{h_{11}}$	$-\dfrac{h_{12}}{h_{11}}$	$\dfrac{D}{B}$	$-\dfrac{AD-BC}{B}$
	$-\dfrac{Z_{21}}{\Delta Z}$	$\dfrac{Z_{11}}{\Delta Z}$	Y_{21}	Y_{22}	$\dfrac{h_{21}}{h_{11}}$	$\dfrac{\Delta H}{h_{11}}$	$-\dfrac{1}{B}$	$\dfrac{A}{B}$
[H]	$\dfrac{\Delta Z}{Z_{22}}$	$\dfrac{Z_{12}}{Z_{22}}$	$\dfrac{1}{Y_{11}}$	$-\dfrac{Y_{12}}{Y_{11}}$	h_{11}	h_{12}	$\dfrac{B}{D}$	$\dfrac{AD-BC}{D}$
	$-\dfrac{Z_{21}}{Z_{22}}$	$\dfrac{1}{Z_{22}}$	$\dfrac{Y_{21}}{Y_{11}}$	$\dfrac{\Delta Y}{Y_{11}}$	h_{21}	h_{22}	$-\dfrac{1}{D}$	$\dfrac{C}{D}$
[ABCD]	$\dfrac{Z_{11}}{Z_{21}}$	$\dfrac{\Delta Z}{Z_{21}}$	$-\dfrac{Y_{22}}{Y_{21}}$	$-\dfrac{1}{Y_{21}}$	$-\dfrac{\Delta H}{h_{21}}$	$-\dfrac{h_{11}}{h_{21}}$	A	B
	$\dfrac{1}{Z_{21}}$	$\dfrac{Z_{22}}{Z_{21}}$	$-\dfrac{\Delta Y}{Y_{21}}$	$-\dfrac{Y_{11}}{Y_{21}}$	$-\dfrac{h_{22}}{h_{21}}$	$-\dfrac{1}{h_{21}}$	C	D

where it is assumed that the source and load impedances are real and equal to $Z_S = Z_L = Z_0$. Substituting Equations 1.50 and 1.51 into Equations 1.11 and 1.12 results in

$$\frac{a_1 - b_1}{\sqrt{Z_0}} = Y_{11}(a_1 + b_1)\sqrt{Z_0} + Y_{12}(a_2 + b_2)\sqrt{Z_0} \tag{1.52}$$

$$\frac{a_2 - b_2}{\sqrt{Z_0}} = Y_{21}(a_1 + b_1)\sqrt{Z_0} + Y_{22}(a_2 + b_2)\sqrt{Z_0} \tag{1.53}$$

which can then be respectively converted to

$$-b_1(1 + Y_{11}Z_0) - b_2 Y_{12}Z_0 = -a_1(1 - Y_{11}\, Z_0) + a_2 Y_{12}Z_0 \tag{1.54}$$

$$-b_1\, Y_{21}Z_0 - b_2(1 + Y_{22}Z_0) = a_1\, Y_{21}Z_0 - a_2(1 - Y_{22}\, Z_0) \tag{1.55}$$

In this case, Equations 1.54 and 1.55 can be solved for the reflected waves b_1 and b_2 as

$$\begin{aligned} &b_1\left[(1 + Y_{11}Z_0)(1 + Y_{22}Z_0) - Y_{12}Y_{21}Z_0^2\right] \\ &= a_1\left[(1 - Y_{11}Z_0)(1 + Y_{22}Z_0) + Y_{12}Y_{21}Z_0^2\right] - 2\, a_2\, Y_{12}Z_0 \end{aligned} \tag{1.56}$$

$$\begin{aligned} &b_2\left[(1 + Y_{11}Z_0)(1 + Y_{22}Z_0) - Y_{12}Y_{21}Z_0^2\right] \\ &= -2\, a_1\, Y_{21}Z_0 + a_2\left[(1 + Y_{11}Z_0)(1 - Y_{22}Z_0) + Y_{12}Y_{21}Z_0^2\right] \end{aligned} \tag{1.57}$$

Comparing equivalent Equations 1.44 and 1.45 and Equations 1.56 and 1.57 gives the following relationships between the scattering S-parameters and admittance Y-parameters:

$$S_{11} = \frac{(1 - Y_{11}Z_0)(1 + Y_{22}Z_0) + Y_{12}Y_{21}Z_0^2}{(1 + Y_{11}Z_0)(1 + Y_{22}Z_0) - Y_{12}Y_{21}Z_0^2} \tag{1.58}$$

$$S_{12} = \frac{-2\, Y_{12}Z_0}{(1 + Y_{11}Z_0)(1 + Y_{22}Z_0) - Y_{12}Y_{21}Z_0^2} \tag{1.59}$$

$$S_{21} = \frac{-2\, Y_{21}Z_0}{(1 + Y_{11}Z_0)(1 + Y_{22}Z_0) - Y_{12}Y_{21}Z_0^2} \tag{1.60}$$

$$S_{22} = \frac{(1 + Y_{11}Z_0)(1 - Y_{22}Z_0) + Y_{12}Y_{21}Z_0^2}{(1 + Y_{11}Z_0)(1 + Y_{22}Z_0) - Y_{12}Y_{21}Z_0^2} \tag{1.61}$$

The relationships between S-parameters with Z-, H-, and $ABCD$-parameters can be obtained in a similar fashion. Table 1.2 shows the conversions between S-parameters and

TABLE 1.2

Conversions between *S*-Parameters and *Z*-, *Y*-, *H*-, and *ABCD*-Parameters

S-Parameters through *Z*-, *Y*-, *H*-, and *ABCD*-Parameters	*Z*-, *Y*-, *H*-, and *ABCD*-Parameters through *S*-Parameters

$$S_{11} = \frac{(Z_{11} - Z_0)(Z_{22} + Z_0) - Z_{12}Z_{21}}{(Z_{11} + Z_0)(Z_{22} + Z_0) - Z_{12}Z_{21}}$$

$$Z_{11} = Z_0 \frac{(1 + S_{11})(1 - S_{22}) + S_{12}S_{21}}{(1 - S_{11})(1 - S_{22}) - S_{12}S_{21}}$$

$$S_{12} = \frac{2\,Z_{12}Z_0}{(Z_{11} + Z_0)(Z_{22} + Z_0) - Z_{12}Z_{21}}$$

$$Z_{12} = Z_0 \frac{2\,S_{12}}{(1 - S_{11})(1 - S_{22}) - S_{12}S_{21}}$$

$$S_{21} = \frac{2\,Z_{21}Z_0}{(Z_{11} + Z_0)(Z_{22} + Z_0) - Z_{12}Z_{21}}$$

$$Z_{21} = Z_0 \frac{2\,S_{21}}{(1 - S_{11})(1 - S_{22}) - S_{12}S_{21}}$$

$$S_{22} = \frac{(Z_{11} + Z_0)(Z_{22} - Z_0) - Z_{12}Z_{21}}{(Z_{11} + Z_0)(Z_{22} + Z_0) - Z_{12}Z_{21}}$$

$$Z_{22} = Z_0 \frac{(1 - S_{11})(1 + S_{22}) + S_{12}S_{21}}{(1 - S_{11})(1 - S_{22}) - S_{12}S_{21}}$$

$$S_{11} = \frac{(1 - Y_{11}Z_0)(1 + Y_{22}Z_0) + Y_{12}Y_{21}Z_0^2}{(1 + Y_{11}Z_0)(1 + Y_{22}Z_0) - Y_{12}Y_{21}Z_0^2}$$

$$Y_{11} = \frac{1}{Z_0} \frac{(1 - S_{11})(1 + S_{22}) + S_{12}S_{21}}{(1 + S_{11})(1 + S_{22}) - S_{12}S_{21}}$$

$$S_{12} = \frac{- 2\,Y_{12}Z_0}{(1 + Y_{11}Z_0)(1 + Y_{22}Z_0) - Y_{12}Y_{21}Z_0^2}$$

$$Y_{12} = \frac{1}{Z_0} \frac{-2\,S_{12}}{(1 + S_{11})(1 + S_{22}) - S_{12}S_{21}}$$

$$S_{21} = \frac{- 2\,Y_{21}Z_0}{(1 + Y_{11}Z_0)(1 + Y_{22}Z_0) - Y_{12}Y_{21}Z_0^2}$$

$$Y_{21} = \frac{1}{Z_0} \frac{-2\,S_{21}}{(1 + S_{11})(1 + S_{22}) - S_{12}S_{21}}$$

$$S_{22} = \frac{(1 + Y_{11}Z_0)(1 - Y_{22}Z_0) + Y_{12}Y_{21}Z_0^2}{(1 + Y_{11}Z_0)(1 + Y_{22}Z_0) - Y_{12}Y_{21}Z_0^2}$$

$$Y_{22} = \frac{1}{Z_0} \frac{(1 + S_{11})(1 - S_{22}) + S_{12}S_{21}}{(1 + S_{11})(1 + S_{22}) - S_{12}S_{21}}$$

$$S_{11} = \frac{(h_{11} - Z_0)(1 + h_{22}Z_0) - h_{12}h_{21}Z_0}{(h_{11} + Z_0)(1 + h_{22}Z_0) - h_{12}h_{21}Z_0}$$

$$h_{11} = Z_0 \frac{(1 + S_{11})(1 + S_{22}) - S_{12}S_{21}}{(1 - S_{11})(1 + S_{22}) + S_{12}S_{21}}$$

$$S_{12} = \frac{2\,h_{12}Z_0}{(h_{11} + Z_0)(1 + h_{22}Z_0) - h_{12}h_{21}Z_0}$$

$$h_{12} = \frac{2\,S_{12}}{(1 - S_{11})(1 + S_{22}) + S_{12}S_{21}}$$

$$S_{21} = \frac{-2\,h_{21}Z_0}{(h_{11} + Z_0)(1 + h_{22}Z_0) - h_{12}h_{21}Z_0}$$

$$h_{21} = \frac{-2\,S_{21}}{(1 - S_{11})(1 + S_{22}) + S_{12}S_{21}}$$

$$S_{22} = \frac{(h_{11} + Z_0)(1 - h_{22}Z_0) + h_{12}h_{21}Z_0}{(h_{11} + Z_0)(1 + h_{22}Z_0) - h_{12}h_{21}Z_0}$$

$$h_{22} = \frac{1}{Z_0} \frac{(1 - S_{11})(1 - S_{22}) - S_{12}S_{21}}{(1 - S_{11})(1 + S_{22}) + S_{12}S_{21}}$$

$$S_{11} = \frac{AZ_0 + B - CZ_0^2 - DZ_0}{AZ_0 + B + CZ_0^2 + DZ_0}$$

$$A = \frac{(1 + S_{11})(1 - S_{22}) + S_{12}S_{21}}{2\,S_{21}}$$

$$S_{12} = \frac{2(AD - BC)Z_0}{AZ_0 + B + CZ_0^2 + DZ_0}$$

$$B = Z_0 \frac{(1 + S_{11})(1 + S_{22}) - S_{12}S_{21}}{2\,S_{21}}$$

$$S_{21} = \frac{2\,Z_0}{AZ_0 + B + CZ_0^2 + DZ_0}$$

$$C = \frac{1}{Z_0} \frac{(1 - S_{11})(1 - S_{22}) - S_{12}S_{21}}{2\,S_{21}}$$

$$S_{22} = \frac{-AZ_0 + B - CZ_0^2 + DZ_0}{AZ_0 + B + CZ_0^2 + DZ_0}$$

$$D = \frac{(1 - S_{11})(1 + S_{22}) + S_{12}S_{21}}{2\,S_{21}}$$

Z-, Y-, H-, and $ABCD$-parameters for the simplified case when the source impedance Z_S and the load impedance Z_L are equal to the characteristic impedance Z_0 [3].

1.4 Interconnections of Two-Port Networks

When analyzing the behavior of a particular electrical circuit in terms of the two-port network parameters, it is often necessary to define the parameters of a combination of the two or more internal two-port networks. For example, the general feedback amplifier circuit consists of an active two-port network representing the amplifier stage, which is connected in parallel with a passive feedback two-port network. In general, the two-port networks can be interconnected using parallel, series, series–parallel, or cascade connections.

To characterize the resulting two-port networks, it is necessary to take into account which currents and voltages are common for individual two-port networks. The most convenient set of parameters is one for which the common currents and voltages represent a simple linear combination of the independent variables. For the interconnection shown in Figure 1.5a, the two-port networks characterized by the impedance matrices $[Z_a]$ and $[Z_b]$ are connected in series for both the input and output terminals. Therefore, the currents flowing through these terminals are equal when

$$I_1 = I_{1a} = I_{1b} \quad I_2 = I_{2a} = I_{2b} \tag{1.62}$$

or, in a matrix form,

$$[I] = [I_a] = [I_b] \tag{1.63}$$

The terminal voltages of the resulting two-port network represent the corresponding sums of the terminal voltages of the individual two-port networks when

$$V_1 = V_{1a} + V_{1b} \quad V_2 = V_{2a} + V_{2b} \tag{1.64}$$

or, in a matrix form,

$$[V] = [V_a] + [V_b] \tag{1.65}$$

In this case, the currents are common components both for the resulting and individual two-port networks. Consequently, to describe the properties of entire circuit, it is most convenient to use the impedance matrices, and we can write for each two-port network using Equation 1.62, respectively,

$$[V_a] = [Z_a][I_a] = [Z_a][I] \tag{1.66}$$

$$[V_b] = [Z_b][I_b] = [Z_b][I] \tag{1.67}$$

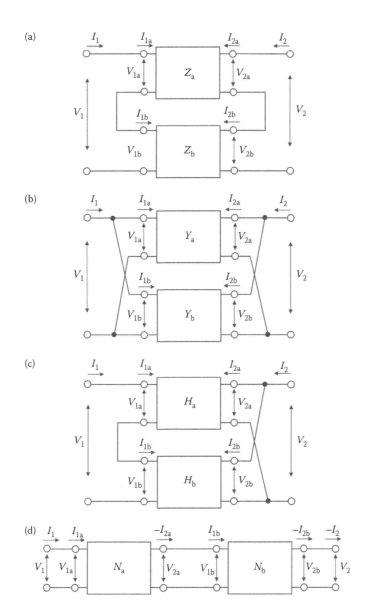

FIGURE 1.5
Different interconnections of two-port networks.

Adding both sides of Equations 1.66 and 1.67 yields

$$[V] = [Z][I] \tag{1.68}$$

where

$$[Z] = [Z_a] + [Z_b] = \begin{bmatrix} Z_{11a} + Z_{11b} & Z_{12a} + Z_{12b} \\ Z_{21a} + Z_{21b} & Z_{22a} + Z_{22b} \end{bmatrix} \tag{1.69}$$

The circuit shown in Figure 1.5b is composed of the two-port networks characterized by the admittance matrices $[Y_a]$ and $[Y_b]$ and connected in parallel, where the common components for both resulting and individual two-port networks are input and output voltages, respectively,

$$V_1 = V_{1a} = V_{1b} \quad V_2 = V_{2a} = V_{2b} \tag{1.70}$$

or, in a matrix form,

$$[V] = [V_a] = [V_b] \tag{1.71}$$

Consequently, to describe the entire circuit properties in this case, it is convenient to use the admittance matrices that give the resulting matrix equation in the form

$$[I] = [Y][V] \tag{1.72}$$

where

$$[Y] = [Y_a] + [Y_b] = \begin{bmatrix} Y_{11a} + Y_{11b} & Y_{12a} + Y_{12b} \\ Y_{21a} + Y_{21b} & Y_{22a} + Y_{22b} \end{bmatrix} \tag{1.73}$$

The series connection of the input terminals and parallel connection of the output terminals are characterized by the circuit in Figure 1.5c, which shows a series–parallel connection of two-port networks. The common components for this circuit are the input currents and the output voltages. As a result, it is most convenient to analyze the entire circuit properties using hybrid matrices. The resulting two-port hybrid matrix is equal to the sum of the two individual hybrid matrices, which is written as

$$[H] = [H_a] + [H_b] = \begin{bmatrix} h_{11a} + h_{11b} & h_{12a} + h_{12b} \\ h_{21a} + h_{21b} & h_{22a} + h_{22b} \end{bmatrix} \tag{1.74}$$

Figure 1.5d shows the cascade connection of the two individual two-port networks. For such an approach using the one-by-one interconnection of the two-port networks, the output voltage and the output current of the first network is equal to the input voltage and the input current of the second one, respectively, when

$$V_1 = V_{1a} \quad I_1 = I_{1a} \tag{1.75}$$

$$V_{2a} = V_1 \quad - I_{2a} = I_{1b} \tag{1.76}$$

$$V_{2b} = V_2 \quad - I_{2b} = -I_2 \tag{1.77}$$

In this case, it is convenient to use a system of *ABCD*-parameters given by Equations 1.21 and 1.22. As a result, for the first individual two-port network shown in Figure 1.5d,

$$\begin{bmatrix} V_{1a} \\ I_{1a} \end{bmatrix} = \begin{bmatrix} A_a & B_a \\ C_a & D_a \end{bmatrix} \begin{bmatrix} V_{2a} \\ -I_{2a} \end{bmatrix} \tag{1.78}$$

or, using Equations 1.75 and 1.76,

$$\begin{bmatrix} V_1 \\ I_1 \end{bmatrix} = \begin{bmatrix} A_a & B_a \\ C_a & D_a \end{bmatrix} \begin{bmatrix} V_{1b} \\ I_{1b} \end{bmatrix} \tag{1.79}$$

Similarly, for the second individual two-port network,

$$\begin{bmatrix} V_{1b} \\ I_{1b} \end{bmatrix} = \begin{bmatrix} A_b & B_b \\ C_b & D_b \end{bmatrix} \begin{bmatrix} V_{2b} \\ -I_{2b} \end{bmatrix} = \begin{bmatrix} A_b & B_b \\ C_b & D_b \end{bmatrix} \begin{bmatrix} V_2 \\ -I_2 \end{bmatrix} \tag{1.80}$$

Then, substituting matrix Equation 1.80 to matrix Equation 1.79 yields

$$\begin{bmatrix} V_1 \\ I_1 \end{bmatrix} = \begin{bmatrix} A_a & B_a \\ C_a & D_a \end{bmatrix} \begin{bmatrix} A_b & B_b \\ C_b & D_b \end{bmatrix} \begin{bmatrix} V_2 \\ -I_2 \end{bmatrix} = \begin{bmatrix} A & B \\ C & D \end{bmatrix} \begin{bmatrix} V_2 \\ -I_2 \end{bmatrix} \tag{1.81}$$

Consequently, the transmission matrix of the resulting two-port network obtained by the cascade connection of the two or more individual two-port networks is determined by multiplying the transmission matrices of the individual networks. This important property is widely used in the analysis and design of transmission networks and systems.

1.5 Practical Two-Port Networks

1.5.1 Single-Element Networks

The simplest networks, which include only one element, can be constructed by a series-connected admittance Y, as shown in Figure 1.6a, or by a parallel-connected impedance Z, as shown in Figure 1.6b.

The two-port network consisting of the single series admittance Y can be described in a system of Y-parameters as

$$I_1 = YV_1 - YV_2 \tag{1.82}$$

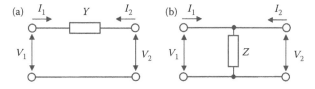

FIGURE 1.6
Single-element networks.

$$I_2 = -Y V_1 + Y V_2 \tag{1.83}$$

or, in a matrix form,

$$[Y] = \begin{bmatrix} Y & -Y \\ -Y & Y \end{bmatrix} \tag{1.84}$$

which means that $Y_{11} = Y_{22} = Y$ and $Y_{12} = Y_{21} = -Y$. The resulting matrix is a singular matrix with $|Y| = 0$. Consequently, it is impossible to determine such a two-port network with the series admittance Y-parameters through a system of Z-parameters. However, by using H- and $ABCD$-parameters, it can be described, respectively, by

$$[H] = \begin{bmatrix} 1/Y & 1 \\ -1 & 0 \end{bmatrix} \quad [ABCD] = \begin{bmatrix} 1 & 1/Y \\ 0 & 1 \end{bmatrix} \tag{1.85}$$

Similarly, for a two-port network with the single parallel impedance Z,

$$[Z] = \begin{bmatrix} Z & Z \\ Z & Z \end{bmatrix} \tag{1.86}$$

which means that $Z_{11} = Z_{12} = Z_{21} = Z_{22} = Z$. The resulting matrix is a singular matrix with $|Z| = 0$. In this case, it is impossible to determine such a two-port network with the parallel impedance Z-parameters through a system of Y-parameters. By using H- and $ABCD$-parameters, this two-port network can be described by

$$[H] = \begin{bmatrix} 0 & 1 \\ -1 & 1/Z \end{bmatrix} \quad [ABCD] = \begin{bmatrix} 1 & 0 \\ 1/Z & 1 \end{bmatrix} \tag{1.87}$$

1.5.2 π- and *T*-Type Networks

The basic configurations of a two-port network that usually describe the electrical properties of the active devices can be represented in the form of a π-circuit shown in Figure 1.7a and in the form of a *T*-circuit shown in Figure 1.7b. Here, the π-circuit includes the current source $g_m V_1$ and the *T*-circuit includes the voltage source $r_m I_1$.

By deriving the two-loop equations using Kirchhoff's current law or applying Equations 1.13 and 1.14 for the π-circuit, one can obtain

$$I_1 - (Y_1 + Y_3) V_1 + Y_3 V_2 = 0 \tag{1.88}$$

$$I_2 + (g_m - Y_3) V_1 + (Y_2 + Y_3) V_2 = 0 \tag{1.89}$$

Equations 1.88 and 1.89 can be rewritten as matrix Equation 1.3 with

$$[M] = \begin{bmatrix} 1 & 0 \\ 0 & 1 \end{bmatrix} \text{ and } [N] = \begin{bmatrix} -(Y_1 + Y_3) & Y_3 \\ -g_m + Y_3 & -(Y_2 + Y_3) \end{bmatrix}$$

Since matrix $[M]$ is nonsingular, such a two-port network can be described by a system of Y-parameters as

$$[Y] = -[M]^{-1}[N] = \begin{bmatrix} Y_1 + Y_3 & -Y_3 \\ g_m - Y_3 & Y_2 + Y_3 \end{bmatrix} \tag{1.90}$$

Similarly, for a two-port network in the form of a T-circuit using Kirchhoff's voltage law or applying Equations 1.8 and 1.9, one can obtain

$$[Z] = -[M]^{-1}[N] = \begin{bmatrix} Z_1 + Z_3 & Z_3 \\ r_m + Z_3 & Z_2 + Z_3 \end{bmatrix} \tag{1.91}$$

If $g_m = 0$ for a π-circuit and $r_m = 0$ for a T-circuit, their corresponding matrices in a system of $ABCD$-parameters can be written as, for π-circuit,

$$[ABCD] = \begin{bmatrix} 1 + \dfrac{Y_2}{Y_3} & \dfrac{1}{Y_3} \\ Y_1 + Y_2 + \dfrac{Y_1 Y_2}{Y_3} & 1 + \dfrac{Y_1}{Y_3} \end{bmatrix} \tag{1.92}$$

and, for T-circuit,

$$[ABCD] = \begin{bmatrix} 1 + \dfrac{Z_2}{Z_3} & Z_1 + Z_2 + \dfrac{Z_1 Z_2}{Z_3} \\ \dfrac{1}{Z_3} & 1 + \dfrac{Z_1}{Z_3} \end{bmatrix} \tag{1.93}$$

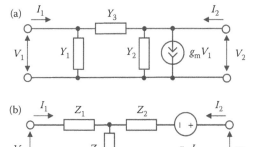

FIGURE 1.7
Basic diagrams of π- and T-networks.

Based on the appropriate relationships between impedances of a *T*-circuit and admittances of a π-circuit, these two circuits become equivalent with respect to the effect on any other two-port network. For a π-circuit shown in Figure 1.8a,

$$
\begin{aligned}
I_1 &= Y_1\ V_{13} + Y_3\ V_{12} = Y_1\ V_{13} + Y_3(V_{13} - V_{23}) \\
&= (Y_1 + Y_3)\ V_{13} - Y_3\ V_{23}
\end{aligned}
\tag{1.94}
$$

$$
\begin{aligned}
I_2 &= Y_2\ V_{23} - Y_3\ V_{12} = Y_2\ V_{23} - Y_3(V_{13} - V_{23}) \\
&= -Y_3 V_{13} + (Y_2 + Y_3)\ V_{23}
\end{aligned}
\tag{1.95}
$$

Solving Equations 1.94 and 1.95 for voltages V_{13} and V_{23} yields

$$
V_{13} = \frac{Y_2 + Y_3}{Y_1 Y_2 + Y_1 Y_2 + Y_1 Y_2} I_1 + \frac{Y_3}{Y_1 Y_2 + Y_1 Y_2 + Y_1 Y_2} I_2
\tag{1.96}
$$

$$
V_{23} = \frac{Y_3}{Y_1 Y_2 + Y_1 Y_2 + Y_1 Y_2} I_1 + \frac{Y_1 + Y_3}{Y_1 Y_2 + Y_1 Y_2 + Y_1 Y_2} I_2
\tag{1.97}
$$

Similarly, for a *T*-circuit shown in Figure 1.8b,

$$
\begin{aligned}
V_{13} &= Z_1 I_1 + Z_3 I_3 = Z_1 I_1 + Z_3(I_1 + I_2) \\
&= (Z_1 + Z_3) I_1 + Z_3 I_2
\end{aligned}
\tag{1.98}
$$

$$
\begin{aligned}
V_{23} &= Z_2 I_2 + Z_3 I_3 = Z_2 I_2 + Z_3(I_1 + I_2) \\
&= Z_3 I_1 + (Z_2 + Z_3) I_2
\end{aligned}
\tag{1.99}
$$

and the equations for currents I_1 and I_2 can be obtained by

$$
I_1 = \frac{Z_2 + Z_3}{Z_1 Z_2 + Z_1 Z_2 + Z_1 Z_2} V_{13} - \frac{Z_3}{Z_1 Z_2 + Z_1 Z_2 + Z_1 Z_2} V_{23}
\tag{1.100}
$$

$$
I_2 = -\frac{Z_3}{Z_1 Z_2 + Z_1 Z_2 + Z_1 Z_2} V_{13} + \frac{Z_1 + Z_3}{Z_1 Z_2 + Z_1 Z_2 + Z_1 Z_2} V_{23}
\tag{1.101}
$$

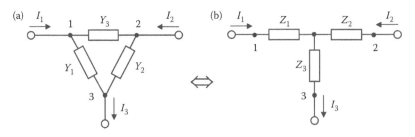

FIGURE 1.8
Equivalence of π- and *T*-circuits.

TABLE 1.3

Relationships between π- and T-Circuit Parameters

T- to π-Transformation	π- to T-Transformation
$Y_1 = \dfrac{Z_2}{Z_1 Z_2 + Z_2 Z_3 + Z_1 Z_3}$	$Z_1 = \dfrac{Y_2}{Y_1 Y_2 + Y_2 Y_3 + Y_1 Y_3}$
$Y_2 = \dfrac{Z_1}{Z_1 Z_2 + Z_2 Z_3 + Z_1 Z_3}$	$Z_2 = \dfrac{Y_1}{Y_1 Y_2 + Y_2 Y_3 + Y_1 Y_3}$
$Y_3 = \dfrac{Z_3}{Z_1 Z_2 + Z_2 Z_3 + Z_1 Z_3}$	$Z_3 = \dfrac{Y_3}{Y_1 Y_2 + Y_2 Y_3 + Y_1 Y_3}$

To establish a T- to π-transformation, it is necessary to equate the coefficients for V_{13} and V_{23} in Equations 1.100 and 1.101 to the corresponding coefficients in Equations 1.94 and 1.95. Similarly, to establish a π- to T-transformation, it is necessary to equate the coefficients for I_1 and I_2 in Equations 1.98 and 1.99 to the corresponding coefficients in Equations 1.96 and 1.97. The resulting relationships between admittances for a π-circuit and impedances for a T-circuit are given in Table 1.3.

1.6 Three-Port Network with Common Terminal

The concept of a two-port network with two independent sources can generally be extended to any multiport networks. Figure 1.9 shows the three-port network where all three independent sources are connected to a common point. The three-port network matrix Equation 1.3 can be described in a scalar form as

$$\left.\begin{array}{l} m_{11}V_1 + m_{12}V_2 + m_{13}V_3 + n_{11}I_1 + n_{12}I_2 + n_{13}I_3 = 0 \\ m_{21}V_1 + m_{22}V_2 + m_{23}V_3 + n_{21}I_1 + n_{22}I_2 + n_{23}I_3 = 0 \\ m_{31}V_1 + m_{32}V_2 + m_{33}V_3 + n_{31}I_1 + n_{32}I_2 + n_{33}I_3 = 0 \end{array}\right\} \qquad (1.102)$$

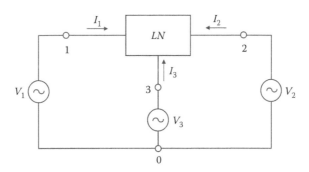

FIGURE 1.9
Basic diagram of three-port network with common terminal.

If matrix [N] in Equation 1.102 is nonsingular when $|N| \neq 0$, this system of three equations can be rewritten in admittance matrix representation in terms of the voltage matrix [V], similarly to a two-port network, as

$$\begin{bmatrix} I_1 \\ I_2 \\ I_3 \end{bmatrix} = \begin{bmatrix} Y_{11} & Y_{12} & Y_{13} \\ Y_{21} & Y_{22} & Y_{23} \\ Y_{31} & Y_{32} & Y_{33} \end{bmatrix} \begin{bmatrix} V_1 \\ V_2 \\ V_3 \end{bmatrix} \qquad (1.103)$$

The matrix [Y] in Equation 1.103 is the indefinite admittance matrix of the three-port network and represents a singular matrix with two important properties: the sum of all terminal currents entering the circuit is equal to zero, that is, $I_1 + I_2 + I_3 = 0$, and all terminal currents entering the circuit depend on the voltages between circuit terminals, which makes the sum of all terminal voltages equal to zero, that is, $V_{13} + V_{32} + V_{21} = 0$.

According to the first property, adding the left and right parts of matrix Equation 1.103 results in

$$(Y_{11} + Y_{21} + Y_{31})V_1 + (Y_{12} + Y_{22} + Y_{32})V_2$$
$$+(Y_{13} + Y_{23} + Y_{33})V_3 = 0 \qquad (1.104)$$

Since all terminal voltages (V_1, V_2, and V_3) can be set independently from each other, Equation 1.104 can be satisfied only if any column sum is identically zero,

$$\left. \begin{aligned} Y_{11} + Y_{21} + Y_{31} &= 0 \\ Y_{12} + Y_{22} + Y_{32} &= 0 \\ Y_{13} + Y_{23} + Y_{33} &= 0 \end{aligned} \right\} \qquad (1.105)$$

Terminal currents will neither decrease nor increase with the simultaneous change of all terminal voltages by the same magnitude. Consequently, if all terminal voltages are equal to a nonzero value when $V_1 = V_2 = V_3 = V_0$, a lack of the terminal currents occurs when $I_1 = I_2 = I_3 = 0$. For example, since from the first row of the matrix Equation 1.103 follows that $I_1 = Y_{11}V_1 + Y_{12}V_2 + Y_{13}V_3$, then we can write

$$0 = (Y_{11} + Y_{12} + Y_{13})V_0 \qquad (1.106)$$

which results, due to the nonzero value V_0, in

$$Y_{11} + Y_{12} + Y_{13} = 0 \qquad (1.107)$$

Applying the same approach to other two rows results in

$$\left. \begin{aligned} Y_{11} + Y_{12} + Y_{13} &= 0 \\ Y_{21} + Y_{22} + Y_{23} &= 0 \\ Y_{31} + Y_{32} + Y_{33} &= 0 \end{aligned} \right\} \qquad (1.108)$$

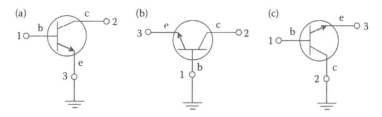

FIGURE 1.10
Bipolar transistors with different common terminals.

Consequently, by using Equations 1.105 through 1.108, the indefinite admittance Y-matrix of a three-port network can be rewritten by

$$[Y] = \begin{bmatrix} Y_{11} & Y_{12} & -(Y_{11} + Y_{12}) \\ Y_{21} & Y_{22} & -(Y_{21} + Y_{22}) \\ -(Y_{11} + Y_{21}) & -(Y_{12} + Y_{22}) & Y_{11} + Y_{12} + Y_{21} + Y_{22} \end{bmatrix} \qquad (1.109)$$

By selecting successively terminal 1, 2, and 3 as the datum terminal, the corresponding three two-port admittance matrices of the initial three-port network can be obtained. In this case, the admittance matrices will correspond to a common-emitter configuration shown in Figure 1.10a, a common-base configuration shown in Figure 1.10b, and a common-collector configuration of the bipolar transistor shown in Figure 1.10c, respectively. If the common-emitter device is treated as a two-port network characterized by four Y-parameters (Y_{11}, Y_{12}, Y_{21}, and Y_{22}), the two-port matrix of the common-collector configuration with grounded collector terminal is simply obtained by deleting the second row and the second column in matrix Equation 1.109. For the common-base configuration with grounded base terminal, the first row and the first column should be deleted because the emitter terminal is considered the input terminal.

A similar approach can be applied to the indefinite three-port impedance network. This allows the Z-parameters of the impedance matrices of the common-base and the common-collector configurations through known impedance Z-parameters of the common-emitter configuration of the transistor to be determined. Parameters of the three-port network, which can describe the electrical behavior of the three-port bipolar or field-effect transistor configured with different common terminals, are given in Table 1.4.

TABLE 1.4
Y- and Z-Parameters of Active Device with Different Common Terminal

	Y-Parameters		Z-Parameters	
Common emitter (source)	Y_{11} Y_{12}		Z_{11} Z_{12}	
	Y_{21} Y_{22}		Z_{21} Z_{22}	
Common base (gate)	$Y_{11} + Y_{12} + Y_{21} + Y_{22}$	$-(Y_{12} + Y_{22})$	$Z_{11} + Z_{12} + Z_{21} + Z_{22}$	$-(Z_{12} + Z_{22})$
	$-(Y_{21} + Y_{22})$	Y_{22}	$-(Z_{21} + Z_{22})$	Z_{22}
Common collector (drain)	Y_{11}	$-(Y_{11} + Y_{12})$	Z_{11}	$-(Z_{11} + Z_{12})$
	$-(Y_{11} + Y_{21})$	$Y_{11} + Y_{12} + Y_{21} + Y_{22}$	$-(Z_{11} + Z_{21})$	$Z_{11} + Z_{12} + Z_{21} + Z_{22}$

1.7 Lumped Elements

Generally, passive hybrid or integrated circuits are designed based on lumped elements, distributed elements, or combination of both types of elements. Distributed elements represent any sections of the transmission lines of different lengths, types, and characteristic impedances. The basic lumped elements are inductors and capacitors that are small in size in comparison with the transmission-line wavelength λ, and usually their linear dimensions are less than λ/10 or even λ/16. In applications where lumped elements are used, their basic advantages are small physical size and low production cost. However, their main drawbacks are lower quality and lower power-handling capability compared with distributed elements.

1.7.1 Inductors

Inductors are lumped elements that store energy in a magnetic field. The lumped inductors can be implemented using several different configurations such as a short section of a strip conductor or wire, a single loop, or a spiral. The printed high-impedance microstrip-section inductor is usually used for low inductance values, typically less than 2 nH, and often meandered to reduce the component size. Printed microstrip single-loop inductors are not very popular due to their limited inductance per unit area. The approximate expression for the microstrip short-section inductance in free space is given by

$$L(\text{nH}) = 0.2 \times 10^{-3} l \left[\ln\left(\frac{l}{W+t} \right) + 1.193 + \frac{W+t}{3l} \right] K_g \tag{1.110}$$

where the conductor length l, conductor width W, and conductor thickness t are in microns, and the term K_g accounts for the presence of a ground plane defined as

$$K_g = 0.57 - 0.145 \ln \frac{W}{h}, \quad \text{for} \quad \frac{W}{h} > 0.05 \tag{1.111}$$

where h is the spacing from ground plane [4,5].

Spiral inductors can have a circular configuration, a rectangular (square) configuration as shown in Figure 1.11a, or an octagonal configuration as shown in Figure 1.11b, if the technology allows 45° routing. The circular geometry is superior in electrical performance, whereas the rectangular shapes are easy to lay out and fabricate. Printed inductors are based on using thin-film or thick-film Si or GaAs fabrication processes, and the inner conductor is pulled out to connect with other circuitry through a bondwire or an air bridge, or by using multilevel crossover metal. The general expression for a spiral inductor, which is also valid for its planar integration within accuracy of around 3%, is based on a Wheeler formula and can be obtained as

$$L(\text{nH}) = \frac{K_1 n^2 d_{\text{avg}}}{1 + K_2 \rho} \tag{1.112}$$

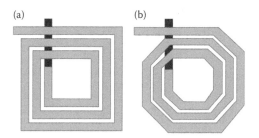

FIGURE 1.11
Spiral inductor layouts.

where n is the number of turns, $d_{avg} = (d_{out} + d_{in})/2$ is the average diameter, $\rho = (d_{out} + d_{in})/(d_{out} - d_{in})$ is the fill ratio, d_{out} is the outer diameter in μm, d_{in} is the inner diameter in μm, and the coefficients K_1 and K_2 are layout dependent: square—$K_1 = 2.34$, $K_2 = 2.75$, hexagonal—$K_1 = 2.33$, $K_2 = 3.82$, and octagonal—$K_1 = 2.25$, $K_2 = 3.55$ [6,7].

In contrast to the capacitors, high-quality inductors are not readily available in a standard complementary metal-oxide-semiconductor (CMOS) technology. Therefore, it is necessary to use special techniques to improve the inductor electrical performance. By using a standard CMOS technology with only two metal layers and a heavily doped substrate, the spiral inductor will have a large series resistance compared with three–four metal layer technologies, and the substrate losses become a very important factor due to the relatively low resistivity of silicon. A major source of substrate losses is the capacitive coupling when current is flowing not only through the metal strip but also through the silicon substrate. Another important source of substrate losses is the inductive coupling when, due to the planar inductor structure, the magnetic field penetrates deeply into the silicon substrate, inducing current loops and related losses. However, the latter effects are particularly important for large-area inductors and can be overcome by using silicon micromachining techniques [8].

The simplified equivalent circuit for the CMOS spiral microstrip inductor is shown in Figure 1.12, where L_s models the self and mutual inductances, R_s is the series coil resistance, C_{ox} is the parasitic oxide capacitance from the metal layer to the substrate, R_{si} is the resistance of the conductive silicon substrate, C_{si} is the silicon substrate parasitic capacitance, and C_c is the parasitic coupling capacitance [9]. The parasitic silicon substrate capacitance C_{si} is sufficiently small and in most cases can be neglected. Such a model shows an accurate agreement between simulated and measured data within 10% across a variety of inductor geometries and substrate dopings up to 20 GHz [10]. At frequencies well below the inductor self-resonant radian frequency ω_{SRF}, the coupling capacitance C_c between metal segments due to fringing fields in both the dielectric and air regions can also be neglected since the relative dielectric constant of the oxide is small enough [11]. In this case, if one side of the inductor is grounded, the self-resonant radian frequency of the spiral inductor can approximately be calculated from

$$\omega_{SRF} = \frac{1}{\sqrt{L_s C_{ox}}} \sqrt{\frac{L_s - R_s^2 C_{ox}}{L_s - R_{si}^2 C_{ox}}} \tag{1.113}$$

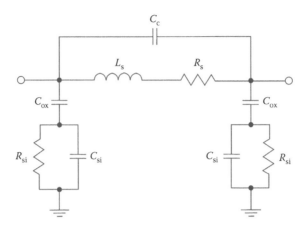

FIGURE 1.12
Equivalent circuit of a square spiral inductor.

At frequencies higher than self-resonant frequency ω_{SRF}, the inductor exhibits a capacitive behavior. The self-resonant frequency ω_{SRF} is limited mainly by the parasitic oxide capacitance C_{ox}, which is inversely proportional to the oxide thickness between the metal layer and substrate. The frequency at which the inductor quality factor Q is maximal can be obtained as

$$\omega_Q = \frac{1}{\sqrt{L_s C_{ox}}} \sqrt{\frac{R_s}{2R_{si}}} \left(\sqrt{1 + \frac{4R_{si}}{3R_s}} - 1 \right)^{0.5} \tag{1.114}$$

The inductor metal conductor series resistance R_s can be easily calculated at low frequencies as the product of the sheet resistance and the number of squares of the metal trace. However, at high frequencies, the skin effect and other magnetic field effects will cause a nonuniform current distribution in the inductor profile. In this case, a simple increase in the diameter of the inductor metal turn does not necessarily correspondingly reduce the inductor series resistance. For example, for the same inductance value, the difference in resistance between the two inductors, when one of them has a metal strip twice as wide, is only a factor of 1.35 [12]. Moreover, at very high frequencies, the largest contribution to the series resistance does not come from the longer outer turns, but from the inner turns. This phenomenon is a result of the generation of circular eddy currents in the inner conductors, whose direction is such that they oppose the original change in magnetic field. On the inner side of the inner turn, coil current and eddy current flow in the same direction, so the current density is larger than average. On the outer side, both currents cancel, and the current density is smaller than average. As a result, the current in the inner turn is pushed to the inside of the conductor.

In hybrid or monolithic applications, bondwires are used to interconnect different components such as lumped elements, planar transmission lines, solid-state devices, and integrated circuits. These bondwires, which are usually made of gold or aluminum, have 0.5- to 1.0-mil diameters, and their lengths are electrically shorter compared with the operating wavelength. To characterize the electrical behavior of the bondwires, simple formulas in terms of their inductances and series resistances can be used. As a first-order approximation, the parasitic capacitance associated with bondwires can be neglected. When $l \gg d$, where l is the bondwire length in μm and d is the bondwire diameter in μm,

$$L(\text{nH}) = 0.2 \times 10^{-3}\, l\left(\ln\frac{4l}{d} + 0.5\frac{d}{l} - 1 + C \right) \tag{1.115}$$

where $C = \tanh(4\delta/d)/4$ is the frequency-dependent correction factor, which is a function of the bondwire diameter and its material's skin depth δ [6,13].

1.7.2 Capacitors

Capacitors are lumped elements that store energy due to an electric field between two electrodes (or plates) when a voltage is applied across them. In this case, a charge of equal magnitude but of opposite sign accumulates on the opposing capacitor plates. The capacitance depends on the area of the plates, separation, and dielectric material between them. The basic structure of a chip capacitor shown in Figure 1.13a consists of two parallel plates, each of area $A = W \times l$ and separated by a dielectric material of thickness d and permittivity $\varepsilon_0\varepsilon_r$, where ε_0 is the free-space permittivity (8.85×10^{-12} farads/m) and ε_r is the relative dielectric constant.

Chip capacitors are usually used in hybrid integrated circuits when relatively high capacitance values are required. In the parallel-plate configuration, the capacitance is commonly expressed as

$$C\,(\text{pF}) = 8.85 \times 10^{-3}\varepsilon_r\frac{Wl}{d} \tag{1.116}$$

where W, l, and d are dimensions in millimeters. Generally, the low-frequency bypass capacitor values are expressed in microfarads and nanofarads, high-frequency blocking and tuning capacitors are expressed in picofarads, and parasitic or fringing capacitances are written in femtofarads. This basic formula given by Equation 1.116 can also be applied to capacitors based on a multilayer technique [5]. The lumped-element equivalent circuit of a capacitor is shown in Figure 1.13b, where L_s is the series plate inductance, R_s is the series

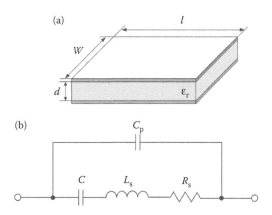

FIGURE 1.13
Parallel capacitor topology and its equivalent circuit.

contact and plate resistance, and C_p is the parasitic parallel capacitance. When $C \gg C_p$, the radian frequency ω_{SRF}, at which the reactances of series elements C and L_s become equal, is called the *capacitor self-resonant frequency*, and the capacitor impedance is equal to the resistance R_s.

For monolithic applications where relatively low capacitances (typically less than 0.5 pF) are required, planar series capacitances in the form of microstrip or interdigital configurations can be used. These capacitors are simply formed by gaps in the center conductor of the microstrip lines, and they do not require any dielectric films. The gap capacitor shown in Figure 1.14a can be equivalently represented by a series coupling capacitance and two parallel fringing capacitances [14]. The interdigital capacitor is a multifinger periodic structure, as shown in Figure 1.14b, where the capacitance occurs across a narrow gap between thin-film transmission-line conductors [15]. These gaps are essentially very long and folded to use a small amount of area. In this case, it is important to keep the size of the capacitor very small relative to the wavelength, so that it can be treated as a lumped element. A larger total width-to-length ratio results in the desired higher shunt capacitance and lower series inductance. An approximate expression for the total capacitance of interdigital structure with $s = W$ and length l less than a quarter wavelength can be given by

$$C \text{ (pF)} = (\varepsilon_r + 1) \, l \, [(N-3)A_1 + A_2] \tag{1.117}$$

where N is the number of fingers and

$$A_1 \left(\frac{\text{pF}}{\mu\text{m}} \right) = 4.409 \tanh \left[0.55 \left(\frac{h}{W} \right)^{0.45} \right] \times 10^{-6} \tag{1.118}$$

$$A_2 \left(\frac{\text{pF}}{\mu\text{m}} \right) = 9.92 \tanh \left[0.52 \left(\frac{h}{W} \right)^{0.5} \right] \times 10^{-6} \tag{1.119}$$

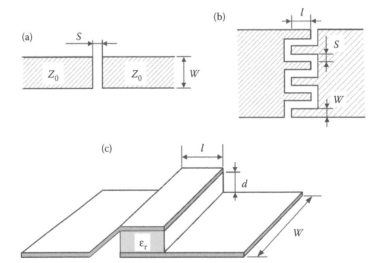

FIGURE 1.14
Different series capacitor topologies.

where h is the spacing from the ground plane.

Series planar capacitors with larger values, which are called metal–insulator–metal (MIM) capacitors, can be realized by using an additional thin dielectric layer (typically less than 0.5 μm) between two metal plates, as shown in Figure 1.14c [5]. The bottom plate of the capacitor uses a thin unplated metal, and the dielectric material is typically silicon nitride (Si_3N_4) for integrated circuits on GaAs and SiO_2 for integrated circuits on Si. The top plate uses a thick plated conductor to reduce the loss in the capacitor. These capacitors are used to achieve higher capacitance values in small areas (10 pF and greater), with typical tolerances from 10% to 15%. The capacitance can be calculated according to Equation 1.116.

1.8 Transmission Line

Transmission lines are widely used in matching circuits of power amplifiers, in hybrid couplers, or power combiners and dividers. When the propagated signal wavelength is compared to its physical dimension, the transmission line can be considered as a two-port network with distributed parameters, where the voltages and currents vary in magnitude and phase over length.

1.8.1 Basic Parameters

Schematically, a transmission line is often represented as a two-wire line, as shown in Figure 1.15a, where its electrical parameters are distributed along its length. The physical properties of a transmission line are determined by four basic parameters:

1. The series inductance L due to the self-inductive phenomena of two conductors
2. The shunt capacitance C in view of the proximity between two conductors
3. The series resistance R due to the finite conductivity of the conductors
4. The shunt conductance G, which is related to the dielectric losses in the material

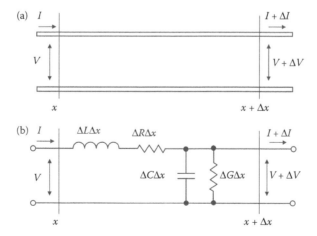

FIGURE 1.15
Transmission line schematics.

As a result, a transmission line of length Δx represents a lumped-element circuit shown in Figure 1.15b, where ΔL, ΔC, ΔR, and ΔG are the series inductance, shunt capacitance, series resistance, and shunt conductance per unit length, respectively. If all these elements are distributed uniformly along the transmission line, and their values do not depend on the chosen position of Δx, this transmission line is called the *uniform transmission line*. Any finite length of the uniform transmission line can be viewed as a cascade of section length Δx.

To define the distribution of the voltages and currents along the uniform transmission line, it is necessary to write the differential equations using Kirchhoff's voltage law for instantaneous values of the voltages and currents in the line section of length Δx, distant x from its beginning. For the sinusoidal steady-state condition, the telegrapher equations for $V(x)$ and $I(x)$ are given by

$$\frac{d^2V(x)}{dx^2} - \gamma^2 V(x) = 0 \qquad (1.120)$$

$$\frac{d^2I(x)}{dx^2} - \gamma^2 I(x) = 0 \qquad (1.121)$$

where $\gamma = \alpha + j\beta = \sqrt{(\Delta R + j\omega\Delta L)(\Delta G + j\omega\Delta C)}$ is the complex propagation constant (which is a function of frequency), α is the attenuation constant, and β is the phase constant. The general solutions of Equations 1.120 and 1.121 for voltage and current of the traveling wave in the transmission line can be written as

$$V(x) = A_1 \exp(-\gamma x) + A_2 \exp(\gamma x) \qquad (1.122)$$

$$I(x) = \frac{A_1}{Z_0} \exp(-\gamma x) - \frac{A_2}{Z_0} \exp(\gamma x) \qquad (1.123)$$

where $Z_0 = \sqrt{(\Delta R + j\omega\Delta L) / (\Delta G + j\omega\Delta C)}$ is the characteristic impedance of the transmission line, $V_i = A_1\exp(-\gamma x)$ and $V_r = A_2\exp(\gamma x)$ represent the incident voltage and the reflected voltage, respectively, and $I_i = A_1\exp(-\gamma x)/Z_0$ and $I_r = A_2\exp(\gamma x)/Z_0$ are the incident current and the reflected current, respectively. From Equations 1.122 and 1.123, it follows that the characteristic impedance of the transmission line Z_0 represents the ratio of the incident (reflected) voltage to the incident (reflected) current at any position on the line as

$$Z_0 = \frac{V_i(x)}{I_i(x)} = \frac{V_r(x)}{I_r(x)} \qquad (1.124)$$

For a lossless transmission line, when $R = G = 0$ and the voltage and current do not change with position, the attenuation constant $\alpha = 0$, the propagation constant $\gamma = j\beta = j\omega\sqrt{\Delta L\Delta C}$, and the phase constant $\beta = \omega\sqrt{\Delta L\Delta C}$. Consequently, the characteristic impedance is reduced to $Z_0 = \sqrt{L/C}$ and represents a real number. The wavelength is defined as $\lambda = 2\pi/\beta = 2\pi / \omega\sqrt{\Delta L\Delta C}$, and the phase velocity as $v_p = \omega/\beta = 1 / \sqrt{\Delta L\Delta C}$.

Figure 1.16 represents a transmission line of the characteristic impedance Z_0 terminated with a load Z_L. In this case, the constants A_1 and A_2 are determined at the position $x = l$ by

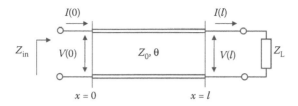

FIGURE 1.16
Loaded transmission line.

$$V(l) = A_1 \exp(-\gamma\, l) + A_2 \exp(\gamma\, l) \tag{1.125}$$

$$I(l) = \frac{A_1}{Z_0}\exp(-\gamma\, l) - \frac{A_2}{Z_0}\exp(\gamma\, l) \tag{1.126}$$

and equal to

$$A_1 = \frac{V(l) + Z_0 I(l)}{2}\exp(\gamma\, l) \tag{1.127}$$

$$A_2 = \frac{V(l) - Z_0 I(l)}{2}\exp(-\gamma\, l) \tag{1.128}$$

As a result, wave equations for voltage $V(x)$ and current $I(x)$ can be rewritten as

$$V(x) = \frac{V(l) + Z_0 I(l)}{2}\exp[\gamma(l - x)] + \frac{V(l) - Z_0 I(l)}{2}\exp[-\gamma(l - x)] \tag{1.129}$$

$$I(x) = \frac{V(l) + Z_0 I(l)}{2Z_0}\exp[\gamma(l - x)] - \frac{V(l) - Z_0 I(l)}{2Z_0}\exp[-\gamma(l - x)] \tag{1.130}$$

which allows their determination at any position on the transmission line.

The voltage and current amplitudes at $x = 0$ as functions of the voltage and current amplitudes at $x = l$ can be determined from Equations 1.129 and 1.130 as

$$V(0) = \frac{V(l) + Z_0 I(l)}{2}\exp(\gamma\, l) + \frac{V(l) - Z_0 I(l)}{2}\exp(-\gamma\, l) \tag{1.131}$$

$$I(0) = \frac{V(l) + Z_0 I(l)}{2Z_0}\exp(\gamma\, l) - \frac{V(l) - Z_0 I(l)}{2Z_0}\exp(-\gamma\, l) \tag{1.132}$$

By using the ratios $\cosh x = [\exp(x) + \exp(-x)]/2$ and $\sinh x = [\exp(x) - \exp(-x)]/2$, Equations 1.131 and 1.132 can be rewritten in the form

$$V(0) = V(l)\cosh(\gamma\, l) + Z_0 I(l)\sinh(\gamma\, l) \tag{1.133}$$

$$I(0) = \frac{V(l)}{Z_0} \sinh(\gamma\, l) + I(l)\cosh(\gamma\, l) \tag{1.134}$$

which represents the transmission equations of the symmetrical reciprocal two-port network expressed through the *ABCD*-parameters when $AD - BC = 1$ and $A = D$. Consequently, the transmission *ABCD*-matrix of the lossless transmission line with $\alpha = 0$ can be defined as

$$[ABCD] = \begin{bmatrix} \cos\theta & jZ_0\sin\theta \\ \dfrac{j\sin\theta}{Z_0} & \cos\theta \end{bmatrix} \tag{1.135}$$

Using the formulas to transform *ABCD*-parameters into *S*-parameters yields

$$[S] = \begin{bmatrix} 0 & \exp(-j\theta) \\ \exp(-j\theta) & 0 \end{bmatrix} \tag{1.136}$$

where $\theta = \beta l$ is the electrical length of the transmission line.

In the case of the loaded lossless transmission line, the reflection coefficient Γ is defined as the ratio between the reflected voltage wave and the incident voltage wave given at a position x as

$$\Gamma(x) = \frac{V_r}{V_i} = \frac{A_2}{A_1}\exp(2j\beta x) \tag{1.137}$$

By taking into account Equations 1.127 and 1.128, the reflection coefficient for $x = l$ can be defined as

$$\Gamma = \frac{Z - Z_0}{Z + Z_0} \tag{1.138}$$

where Γ represents the load reflection coefficient and $Z = Z_L = V(l)/I(l)$. If the load is mismatched, only part of the available power from the source is delivered to the load. This power loss is called the *return loss (RL)*, and is calculated in decibels as

$$RL = -20\log_{10}|\Gamma| \tag{1.139}$$

For a matched load when $\Gamma = 0$, a return loss is of ∞ dB. A total reflection with $\Gamma = 1$ means a return loss of 0 dB when all incident power is reflected.

According to the general solution for voltage at a position x in the transmission line,

$$V(x) = V_i(x) + V_r(x) = V_i[1 + \Gamma(x)] \tag{1.140}$$

Hence, the maximum amplitude (when the incident and reflected waves are in phase) is

$$V_{max}(x) = |V_i|[1 + |\Gamma(x)|] \tag{1.141}$$

and the minimum amplitude (when these two waves are 180° out of phase) is

$$V_{min}(x) = |V_i|[1 - |\Gamma(x)|] \tag{1.142}$$

The ratio of V_{max} to V_{min}, which is a function of the reflection coefficient Γ, represents the *voltage standing wave ratio* (*VSWR*). The *VSWR* is a measure of mismatch and can be written as

$$VSWR = \frac{V_{max}}{V_{min}} = \frac{1 + |\Gamma|}{1 - |\Gamma|} \tag{1.143}$$

which can change from 1 to ∞ (where $VSWR = 1$ implies a matched load). For a load impedance with zero imaginary part when $Z_L = R_L$, the *VSWR* can be calculated as $VSWR = R_L/Z_0$ when $R_L \geq Z_0$ and $VSWR = Z_0/R_L$ when $Z_0 \geq R_L$.

From Equations 1.133 and 1.134, it follows that the input impedance of the loaded lossless transmission line can be obtained as

$$Z_{in} = \frac{V(0)}{I(0)} = Z_0 \frac{Z_L + jZ_0 \tan(\theta)}{Z_0 + jZ_L \tan(\theta)} \tag{1.144}$$

which gives an important dependence between the input impedance, the transmission-line parameters (electrical length and characteristic impedance), and the arbitrary load impedance.

1.8.2 Microstrip Line

Planar transmission lines as an evolution of the coaxial and parallel-wire lines are compact and readily adaptable to hybrid and monolithic integrated circuit fabrication technologies at RF and microwave frequencies [16]. In a microstrip line, the grounded metallization surface covers only one side of dielectric substrate, as shown in Figure 1.17. Such a configuration is equivalent to a pair-wire system for the image of the conductor in the ground plane, which produces the required symmetry [17]. In this case, the electric and magnetic field lines are located in both the dielectric region between the strip conductor and the ground plane and in the air region above the substrate. As a result, the electromagnetic

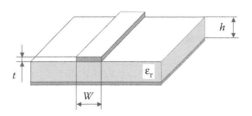

FIGURE 1.17
Microstrip-line structure.

wave propagated along a microstrip line is not a pure transverse electromagnetic (TEM), since the phase velocities in these two regions are not the same. However, in a quasistatic approximation, which gives sufficiently accurate results as long as the height of the dielectric substrate is very small compared with the wavelength, it is possible to obtain the explicit analytical expressions for its electrical characteristics. Since a microstrip line is an open structure, it has a major fabrication advantage over the stripline due to simplicity of practical realization, interconnection, and adjustments.

The exact expression for the characteristic impedance of a lossless microstrip line with finite strip thickness is given by [18,19]

$$
Z_0 = \begin{cases} \dfrac{60}{\sqrt{\varepsilon_{re}}} \ln\left(\dfrac{8h}{W_e} + \dfrac{W_e}{4h} \right) & \text{for } \dfrac{W}{h} \le 1 \\[4mm] \dfrac{120\pi}{\sqrt{\varepsilon_{re}}} \left[\dfrac{W_e}{h} + 1.393 + 0.667 \ln\left(\dfrac{W_e}{h} + 1.444 \right) \right]^{-1} & \text{for } \dfrac{W}{h} \ge 1 \end{cases}
\tag{1.145}
$$

where

$$
\frac{W_e}{h} = \frac{W}{h} + \frac{\Delta W}{h}
\tag{1.146}
$$

$$
\frac{\Delta W}{h} = \begin{cases} \dfrac{1.25}{\pi} \dfrac{t}{h}\left(1 + \ln\dfrac{4\pi\, W}{t} \right) & \text{for } \dfrac{W}{h} \le \dfrac{1}{2}\pi \\[4mm] \dfrac{1.25}{\pi} \dfrac{t}{h}\left(1 + \ln\dfrac{2h}{t} \right) & \text{for } \dfrac{W}{h} \ge \dfrac{1}{2}\pi \end{cases}
\tag{1.147}
$$

$$
\varepsilon_{re} = \frac{\varepsilon_r + 1}{2} + \frac{\varepsilon_r - 1}{2} \frac{1}{\sqrt{1 + 12h/W}} - \frac{\varepsilon_r - 1}{4.6} \frac{t}{h} \sqrt{\frac{h}{W}}
\tag{1.148}
$$

Figure 1.18 shows the characteristic impedance Z_0 of a microstrip line with zero strip thickness as a function of the normalized strip width W/h for various ε_r according to Equations 1.145 to 1.148.

In practice, it is possible to use a sufficiently simple formula to estimate the characteristic impedance Z_0 of a microstrip line with zero strip thickness written as [20]

$$
Z_0 = \frac{120\pi}{\sqrt{\varepsilon_r}} \frac{h}{W} \frac{1}{1 + 1.735\varepsilon_r^{-0.0724}(W/h)^{-0.836}}
\tag{1.149}
$$

For a microstrip line in a quasi-TEM approximation, the conductor loss factor α_c (in Np/m) as a function of the microstrip-line geometry can be obtained by

$$
\alpha_c = \begin{cases} 1.38A \dfrac{R_s}{hZ_0} \dfrac{32 - (W_e/h)^2}{32 + (W_e/h)^2} & \text{for } \dfrac{W}{h} \le 1 \\[4mm] 6.1 \times 10^{-5} A \dfrac{R_s Z_0 \varepsilon_{re}}{h} \left(\dfrac{W_e}{h} + \dfrac{0.667\, W_e/h}{1.444 + W_e/h} \right) & \text{for } \dfrac{W}{h} \ge 1 \end{cases}
\tag{1.150}
$$

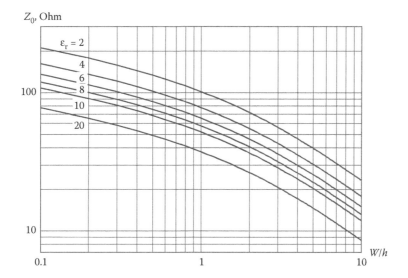

FIGURE 1.18
Microstrip-line characteristic impedance versus W/h.

with

$$A = 1 + \frac{h}{W_e}\left(1 + \frac{1}{\pi}\ln\frac{2B}{t}\right) \tag{1.151}$$

$$B = \begin{cases} 2\pi\, W & \text{for } \dfrac{W}{h} \leq 1/2\pi \\[2mm] h & \text{for } \dfrac{W}{h} \geq 1/2\pi \end{cases} \tag{1.152}$$

where W_e/h is given by Equations 1.146 and 1.147 [21].

The dielectric loss factor α_d (in Np/m) can be calculated by

$$\alpha_d = 27.3\frac{\varepsilon_r}{\varepsilon_r - 1}\frac{\varepsilon_{re} - 1}{\sqrt{\varepsilon_{re}}}\frac{\tan\delta}{\lambda} \tag{1.153}$$

Conductor loss is a result of several factors related to the metallic material composing the ground plane and walls, among which are conductivity, skin effect, and surface ruggedness. For most microstrip lines (except for some kinds of semiconductor substrate such as silicon), the conductor loss is much more significant than the dielectric loss. The conductor losses increase with increasing characteristic impedance due to greater resistance of narrow strips.

1.8.3 Coplanar Waveguide

A coplanar waveguide (CPW) is similar in structure to a slotline, the only difference being a third conductor centered in the slot region. The center strip conductor and two outer grounded conductors lie in the same plane on substrate surface, as shown in Figure 1.19 [22,23]. A coplanar configuration has some advantages such as low dispersion, ease of attaching shunt and series circuit components, no need for via holes, and simple realization of short-circuited ends, which makes a CPW very suitable for hybrid and monolithic integrated circuits. In contrast to the microstrip and stripline, the CPW has shielding between adjacent lines that creates a better isolation between them. However, like the microstrip and stripline, the CPW can be also described by a quasi-TEM approximation for both numerical and analytical calculations. Because of high dielectric constant of the substrate, most of the RF energy is stored in the dielectric and the loading effect of the grounded cover is negligible if it is more than two slot widths away from the surface. Similarly, the thickness of the dielectric substrate with higher relative dielectric constants is not so critical, and should practically be once or twice the width W of the slots.

The approximate expression of the characteristic impedance Z_0 for zero metal thickness, which is satisfactorily accurate in a wide range of substrate thicknesses, can be written as

$$Z_0 = \frac{30\pi}{\sqrt{\varepsilon_{re}}} \frac{K(k')}{K(k)} \tag{1.154}$$

where

$$\varepsilon_{re} = 1 + \frac{\varepsilon_r - 1}{2} \frac{K(k')}{K(k)} \frac{K(k_1)}{K(k_1')} \tag{1.155}$$

$k = (s/s + 2W)$, $k_1 = \sinh(\pi s/4h)/\sinh(\pi(s + 2W)/4h)$, $k' = \sqrt{1 - k^2}$, $k_1' = \sqrt{1 - k_1^2}$, and K is the complete elliptic integral of the first kind [24].

The values of ratios $K(k)/K(k')$ and $K(k_1)/K(k_1')$ can be defined from

$$\frac{K(k)}{K(k')} = \begin{cases} \pi / \ln\left(2\dfrac{1 + \sqrt{k'}}{1 - \sqrt{k'}} \right) & \text{for } 0 \le k \le \dfrac{1}{\sqrt{2}} \\[4mm] \dfrac{1}{\pi}\ln\left(2\dfrac{1 + \sqrt{k}}{1 - \sqrt{k}} \right) & \text{for } \dfrac{1}{\sqrt{2}} \le k \le 1 \end{cases} \tag{1.156}$$

which provides the relative error lower than 3×10^{-6} [25]. Figure 1.20 shows the characteristic impedance Z_0 of a CPW as a function of the parameter $s/(s + 2W)$ for various ε_r according to Equations 1.154 and 1.155.

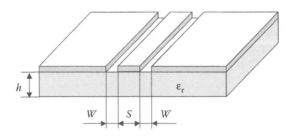

FIGURE 1.19
Coplanar waveguide structure.

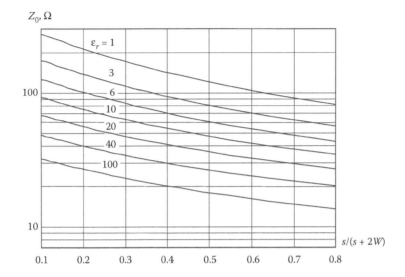

FIGURE 1.20
Coplanar waveguide characteristic impedance versus $s/(s + 2W)$.

1.9 Noise Figure

There are several primary noise sources in the electrical circuit. *Thermal* or *white noise* is created by the random motion of charge carriers due to thermal excitation, being always found in any conducting medium whose temperature is above absolute zero whatever the nature of the conduction process or the nature of the mobile charge carriers [26]. This random motion of carriers creates a fluctuating voltage on the terminals of each resistive element which increases with temperature. However, if the average value of such a voltage is zero, then the noise power on its terminal is not zero being proportional to the resistance of the conductor and to its absolute temperature. The resistor as a thermal noise source can be represented by either of the noise sources shown in Figure 1.21.

The noise voltage source and noise current source can be respectively described by Nyquist equations through their mean-square noise voltage and noise current values as

$$\overline{e_n^2} = 4kTR\Delta f \tag{1.157}$$

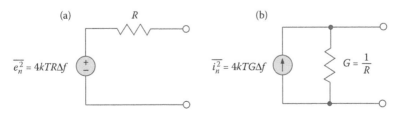

FIGURE 1.21
Equivalent circuits to represent thermal noise sources.

$$\overline{i_n^2} = \frac{4kT\Delta f}{R} \tag{1.158}$$

where $k = 1.38 \times 10^{-23}$ J/K is the Boltzmann constant, T is the absolute temperature, and $kT = 4 \times 10^{-21}$ W/Hz $= -174$ dBm/Hz at ambient temperature $T = 290$ K. The thermal noise is proportional to the frequency bandwidth Δf, and it can be represented by the voltage source in series with resistor R, or by the current source in parallel to the resistor R. The maximum noise power can be delivered to the load when $R = R_L$, where R_L is the load resistance, being equal to $kT\Delta f$. Hence, the noise power density when the noise power is normalized by Δf is independent of frequency and is considered as white noise. The root-mean-square noise voltage and current are proportional to the square root of the frequency bandwidth Δf.

Shot noise is associated with the carrier injection through the device p–n junction, being generated by the movement of individual electrons within the current flow. In each forward biased junction, there is a potential barrier that can be overcome by the carriers with higher thermal energy. Such a process is random and mean-square noise current can be given by

$$\overline{i_n^2} = 2qI\Delta f \tag{1.159}$$

where q is the electron charge and I is the direct current flowing through the p–n junction. The shot noise depends on the thermal energy of the carriers near the potential barrier and its power density is independent of frequency. It has essentially a flat spectral distribution and can be treated as the thermal or white type of noise with current source $\overline{i_n^2}$ connected in parallel to the small-signal junction resistance. In a voltage noise representation, when the noise voltage source is connected in series with such a resistor, it can be written as

$$\overline{e_n^2} = 2kTr\Delta f \tag{1.160}$$

where $r = kT/qI$ is the junction resistance.

It is well-known that any linear noisy two-port network can be represented as a noise-free two-port part with noise sources at the input and the output connected in a different way [27,28]. For example, the noisy linear two-port network with internal noise sources shown in Figure 1.22a can be redrawn, either in the impedance form with external series voltage noise sources shown in Figure 1.22b or in the admittance form with external parallel current noise sources shown in Figure 1.22c.

However, to fully describe the noise properties of the two-port network at fixed frequency, it is sometimes convenient to represent it through the noise-free two-port part and the noise sources equivalently located at the input. Such a circuit is equivalent to the configurations with noise sources located at the input and the output [29]. In this case, it is enough to use four parameters: the noise spectral densities of both noise sources and the real and imaginary parts of its correlation spectral density. These four parameters can be defined by measurements at the two-port network terminals. The two-port network current and voltage amplitudes are related to each other through a system of two linear algebraic equations. By taking into account the noise sources at the input and the output, these equations in the impedance and admittance forms can be respectively written as

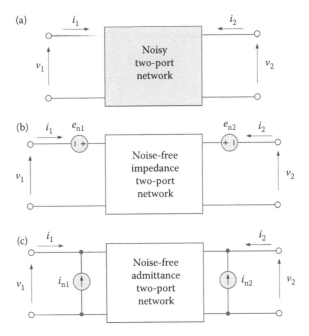

FIGURE 1.22
Linear two-port network with noise sources.

$$V_1 = Z_{11}I_1 + Z_{12}I_2 - V_{n1} \tag{1.161}$$

$$V_2 = Z_{21}I_1 + Z_{22}I_2 - V_{n2} \tag{1.162}$$

and

$$I_1 = Y_{11}V_1 + Y_{12}V_2 - I_{n1} \tag{1.163}$$

$$I_2 = Y_{21}V_1 + Y_{22}V_2 - I_{n2} \tag{1.164}$$

where the voltage and current noise amplitudes represent the Fourier transforms of noise fluctuations.

The equivalent two-port network with voltage and current noise sources located at its input is shown in Figure 1.23a, where [Y] is the two-port network admittance matrix and ratios between current and voltage amplitudes can be written as

$$I_1 = Y_{11}(V_1 + V_{ni}) + Y_{12}V_2 - I_{ni} \tag{1.165}$$

$$I_2 = Y_{21}(V_1 + V_{ni}) + Y_{22}V_2 \tag{1.166}$$

From comparison of Equations 1.163 and 1.164 with Equations 1.165 and 1.166, respectively, it follows that

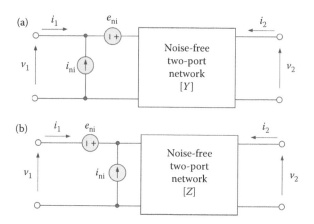

FIGURE 1.23
Linear two-port network with noise sources at input.

$$V_{ni} = -\frac{I_{n2}}{Y_{21}}$$ (1.167)

$$I_{ni} = I_{n1} - \frac{Y_{11}}{Y_{21}} I_{n2}$$ (1.168)

representing the relationships between the current noise sources at the input and the output corresponding to the circuit shown in Figure 1.22c and the voltage and current noise sources at the input only corresponding to the circuit shown in Figure 1.23a. In this case, Equations 1.167 and 1.168 are valid only if $Y_{21} \neq 0$ that always takes place in practice. Similar equations can be written for the circuit with the series noise voltage source followed by a parallel noise current source shown in Figure 1.23b in terms of impedance Z-parameters to represent the relationships between the voltage noise sources at the input and the output corresponding to the circuit shown in Figure 1.22b. The use of voltage and current noise sources at the input enables the combination of all internal two-port network noise sources.

To evaluate a quality of the two-port network, it is important to know the amount of noise added to a signal passing through it. Usually this can be done by introducing an important parameter such as a *noise figure* or *noise factor*. The noise figure of the two-port network is intended as an indication of its noisiness. The lower the noise figure, the less noise is contributed by the two-port network. The noise figure is defined as

$$F = \frac{S_{in}/N_{in}}{S_{out}/N_{out}}$$ (1.169)

where S_{in}/N_{in} is the signal-to-noise ratio available at the input and S_{out}/N_{out} is the signal-to-noise ratio available at the output.

For a two-port network characterizing by the available power gain G_A, the noise figure can be rewritten as

$$F = \frac{S_{in}/N_{in}}{G_A S_{in}/G_A (N_{in} + N_{add})} = 1 + \frac{N_{add}}{N_{in}}$$ (1.170)

where N_{add} is the additional noise power added by the two-port network referred to the input. From Equation 1.170, it follows that the noise figure depends on the source impedance Z_S shown in Figure 1.24a, but not on the circuit connected to the output of the two-port network.

Hence, if the two-port network is driven from the source with impedance $Z_S = R_S + jX_S$, the noise figure F of this two-port network in terms of the model shown in Figure 1.24b with input voltage and current noise sources and noise-free two-port network can be obtained by

$$F = 1 + \frac{\overline{|e_n + Z_S i_n|^2}}{4kTR_S \Delta f}$$

$$= 1 + \frac{R_n + |Z_S|^2 G_n + 2\sqrt{R_n G_n}\,\mathrm{Re}(CZ_S)}{R_S} \tag{1.171}$$

where

$$R_n = \frac{\overline{e_n^2}}{4kT\Delta f} \tag{1.172}$$

is the equivalent input-referred noise resistance corresponding to the noise voltage source, where $\overline{e_n^2} = \overline{e_{nS}^2} + \overline{e_{ni}^2}$

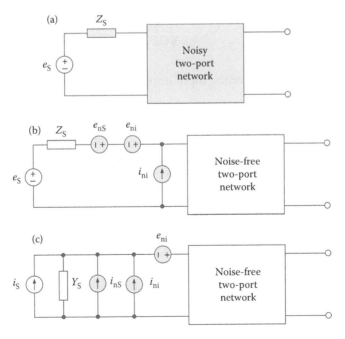

FIGURE 1.24
Linear two-port networks to calculate noise figure.

$$G_n = \frac{\overline{i_n^2}}{4kT\Delta f} \qquad (1.173)$$

is the equivalent input-referred noise conductance corresponding to the noise current source, where $\overline{i_n^2} = \overline{i_{ni}^2}$, and

$$C = \frac{\overline{i_n e_n^*}}{\sqrt{\overline{i_n^2}\ \overline{e_n^2}}} \qquad (1.174)$$

is the correlation coefficient representing a complex number less than or equal to unity in magnitude [28]. Here, G_n and R_n generally do not represent the particular circuit immittances but depend on the bias level resulting in a dependence of the noise figure on the operating bias point of the active device.

As the source impedance Z_S is varied over all values with positive R_S, the noise figure F has a minimum value of

$$F_{min} = 1 + 2\sqrt{R_n G_n}\left[\sqrt{1 - (\operatorname{Im}C)^2} + \operatorname{Re}C\right] \qquad (1.175)$$

which occurs for the optimum source impedance $Z_{Sopt} = R_{Sopt} + jX_{Sopt}$ given by

$$|Z_{Sopt}|^2 = \frac{R_n}{G_n} \qquad (1.176)$$

$$X_{Sopt} = \sqrt{\frac{R_n}{G_n}}\operatorname{Im}C \qquad (1.177)$$

As a result, the noise figure F for the input impedance Z_S, which is not optimum, can be expressed in terms of F_{min} as

$$\begin{aligned} F &= F_{min} + |Z_S - Z_{Sopt}|^2\frac{G_n}{R_S}\\ &= F_{min} + [(R_S - R_{Sopt})^2 + (X_S - X_{Sopt})^2]\frac{G_n}{R_S} \end{aligned} \qquad (1.178)$$

Similarly, the noise figure F can be equivalently expressed using a model shown in Figure 1.24c with source admittance $Y_S = G_S + jB_S$ as

$$\begin{aligned} F &= F_{min} + |Y_S - Y_{Sopt}|^2\frac{R_n}{G_S}\\ &= F_{min} + [(G_S - G_{Sopt})^2 + (B_S - B_{Sopt})^2]\frac{R_n}{G_S} \end{aligned} \qquad (1.179)$$

where F_{min} is the minimum noise figure of the two-port network, which can be realized with respect to the source admittance Y_S, $Y_{Sopt} = G_{Sopt} + jB_{Sopt}$ is the optimal source admittance, and R_n is the equivalent noise resistance that measures how rapidly the noise figure degrades when the source admittance Y_S deviates from its optimum value Y_{Sopt} [30]. Since the admittance Y_S is generally complex, then its real and imaginary parts can be controlled independently. To obtain the minimum value of the noise figure, the two matching conditions of $G_S = G_{Sopt}$ and $B_S = B_{Sopt}$ must be satisfied.

In a multistage transmitter system, the input signal travels through a cascade of many different components, each of which may degrade the signal-to-noise ratio to some degree. For a cascade of two stages having available gains G_{A1} and G_{A2} and noise figures F_1 and F_2, using Equation 1.170 results in the output-to-input noise power ratio N_{out}/N_{in} written as

$$\frac{N_{out}}{N_{in}} = G_{A2}\left[G_{A1}\left(1 + \frac{N_{add1}}{N_{in}}\right) + \frac{N_{add2}}{N_{in}}\right] = G_{A1}G_{A2}\left(F_1 + \frac{F_2 - 1}{G_{A1}}\right) \qquad (1.180)$$

where N_{add1} and N_{add2} are the additional noise powers added by the first and second stages, respectively. Consequently, an overall noise figure $F_{1,2}$ for a two-stage system based on Equation 1.169 can be given by

$$F_{1,2} = F_1 + \frac{1}{G_{A1}}(F_2 - 1) \qquad (1.181)$$

Equation 1.181 can be generalized to a multistage transmitter system with n stages as

$$F_{1,n} = F_1 + \frac{F_2 - 1}{G_{A1}} + \cdots + \frac{F_n - 1}{G_{A1}G_{A2}\ldots G_{A(n-1)}} \qquad (1.182)$$

which means that the noise figure of the first stage has the predominant effect on the overall noise figure, unless G_{A1} is small or F_2 is large [31].

References

1. L. O. Chua, C. A. Desoer, and E. S. Kuh, *Linear and Nonlinear Circuits*, New York: McGraw-Hill, 1987.
2. D. M. Pozar, *Microwave Engineering*, New York: John Wiley & Sons, 2004.
3. D. A. Frickey, Conversions between S, Z, Y, h, ABCD, and T parameters which are valid for complex source and load impedances, *IEEE Trans. Microwave Theory Tech.*, MTT-42, 205–211, 1994.
4. E. F. Terman, *Radio Engineer's Handbook*, New York: McGraw-Hill, 1945.
5. I. J. Bahl, *Lumped Elements for RF and Microwave Circuits*, Boston: Artech House, 2003.
6. H. A. Wheeler, Simple inductance formulas for radio coils, *Proc. IRE*, 16, 1398–1400, 1928.
7. S. S. Mohan, M. del Mar Hershenson, S. P. Boyd, and T. H. Lee, Simple accurate expressions for planar spiral inductances, *IEEE J. Solid-State Circuits*, SC-34, 1419–1424, 1999.

8. J. M. Lopez-Villegas, J. Samitier, C. Cane, P. Losantos, and J. Bausells, Improvement of the quality factor of RF integrated inductors by layout optimization, *IEEE Trans. Microwave Theory Tech.*, MTT-48, 76–83, 2000.

9. J. R. Long and M. A. Copeland, The modeling, characterization, and design of monolithic inductors for silicon RF IC's, *IEEE J. Solid-State Circuits*, SC-32, 357–369, 1997.

10. N. A. Talwalkar, C. P. Yue, and S. S. Wong, Analysis and synthesis of on-chip spiral inductors, *IEEE Trans. Electron Devices*, ED-52, 176–182, 2005.

11. N. M. Nguyen and R. G. Meyer, Si IC-compatible inductors and *LC* passive filters, *IEEE J. Solid-State Circuits*, SC-25, 1028–1031, 1990.

12. J. Craninckx and M. S. J. Steyaert, A 1.8-GHz low-phase-noise CMOS VCO using optimized hollow spiral inductors, *IEEE J. Solid-State Circuits*, SC-32, 736–744, 1997.

13. S. L. March, Simple equations characterize bond wires, *Microwaves RF*, 30, 105–110, 1991.

14. P. Benedek and P. Silvester, Equivalent capacitances of microstrip gaps and steps, *IEEE Trans. Microwave Theory Tech.*, MTT-20, 729–733, 1972.

15. G. D. Alley, Interdigital capacitors and their applications in lumped element microwave integrated circuits, *IEEE Trans. Microwave Theory Tech.*, MTT-18, 1028–1033, 1970.

16. R. M. Barrett, Microwave printed circuits—The early years, *IEEE Trans. Microwave Theory Tech.*, MTT-32, 983–990, 1984.

17. D. D. Grieg and H. F. Engelmann, Microstrip—A new transmission technique for the kilo-megacycle range, *Proc. IRE*, 40, 1644–1650, 1952.

18. E. O. Hammerstad, Equations for microstrip circuit design, *Proc. 5th Eur. Microwave Conf.*, Hamburg, Germany, pp. 268–272, 1975.

19. I. J. Bahl and R. Garg, Simple and accurate formulas for microstrip with finite strip thickness, *Proc. IEEE*, 65, 1611–1612, 1977.

20. R. S. Carson, *High-Frequency Amplifiers*, New York: John Wiley & Sons, 1975.

21. K. C. Gupta, R. Garg, and R. Chadha, *Computer-Aided Design of Microwave Circuits*, Dedham: Artech House, 1981.

22. C. P. Weng, Coplanar waveguide: A surface strip transmission line suitable for nonreciprocal gyromagnetic device applications, *IEEE Trans. Microwave Theory Tech.*, MTT-17, 1087–1090, 1969.

23. R. N. Simons, *Coplanar Waveguide Circuits, Components, and Systems*, New York: John Wiley & Sons, 2001.

24. G. Ghione and C. Naldli, Analytical formulas for coplanar lines in hybrid and monolithic, *Electron. Lett.*, 20, 179–181, 1984.

25. W. Hilberg, From approximations to exact relations for characteristic impedances, *IEEE Trans. Microwave Theory Tech.*, MTT-17, 259–265, 1969.

26. A. van der Ziel, *Noise*, Englewood Cliffs: Prentice-Hall, 1954.

27. H. C. Montgomery, Transistor noise in circuit applications, *Proc. IRE*, 40, 1461–1471, 1952.

28. H. Rothe and W. Dahlke, Theory of noise fourpoles, *Proc. IRE*, 44, 811–818, 1956.

29. A. G. T. Becking, H. Groendijk, and K. S. Knol, The noise factor of four-terminal networks, *Philips Res. Rep.*, 10, 349–357, 1955.

30. IRE Subcommittee 7.9 on Noise (H. A. Haus, Chairman), Representation of noise in linear two-ports, *Proc. IRE*, 48, 69–74, 1960.

31. H. T. Friis, Noise figures of radio receivers, *Proc. IRE*, 32, 419–422, 1944.

2

Power Amplifier Design Principles

Power amplifier design procedure requires accurate active device modeling, effective impedance matching (depending on the technical requirements and operating conditions), stability in operation, and ease of practical implementation. The quality of the power amplifier design is evaluated by achieving maximum power gain across the required frequency bandwidth under stable operating conditions with minimum amplifier stages, and the requirements for linearity or high efficiency can be considered where they are needed. For stable operation, it is necessary to evaluate the operating frequency domains where the active device may be potentially unstable.

2.1 Basic Classes of Operation: A, AB, B, and C

As established in the 1920s, power amplifiers can generally be classified into three classes according to their mode of operation: *linear mode* when its operation is confined to the substantially linear portion of the vacuum-tube characteristic curve; *critical mode* when the anode current ceases to flow, but operation extends beyond the linear portion up to the saturation and cutoff (or pinch-off) regions; and *nonlinear mode* when the anode current ceases to flow during a portion of each cycle, with a duration that depends on the grid bias [1]. When high efficiency is required, power amplifiers of the third class are employed since the presence of harmonics contributes to the attainment of high efficiencies. In order to suppress harmonics of the fundamental frequency to deliver a sinusoidal signal to the load, a parallel resonant circuit can be used in the load network, which bypasses harmonics through a low-impedance path, and, by virtue of its resonance to the fundamental, receives energy at that frequency. At the very beginning of the 1930s, power amplifiers operating in the first two classes with 100% duty ratio were called the Class-A power amplifiers, whereas the power amplifiers operating in the third class with 50% duty ratio were assigned to Class-B power amplifiers [2].

The best way to understand the electrical behavior of a power amplifier and the fastest way to calculate its basic electrical characteristics such as output power, power gain, efficiency, stability, or harmonic suppression is to use a spectral-domain analysis. Generally, such an analysis is based on the determination of the output response of the nonlinear active device when applying the multiharmonic signal to its input port, which analytically can be written as

$$i(t) = f[v(t)] \tag{2.1}$$

where $i(t)$ is the output current, $v(t)$ is the input voltage, and $f(v)$ is the nonlinear transfer function of the device. Unlike the spectral-domain analysis, time-domain analysis establishes the relationships between voltage and current in each circuit element in the time domain when a system of equations is obtained applying Kirchhoff's law to the circuit to be analyzed. As a result, such a system will be composed of nonlinear integro-differential equations describing a nonlinear circuit. The solution to this system can be found by applying the numerical-integration methods.

The voltage $v(t)$ in the frequency domain generally represents the multiple-frequency signal at the device input, which is written as

$$v(t) = V_0 + \sum_{k=1}^{N} V_k \cos(\omega_k t + \phi_k) \tag{2.2}$$

where V_0 is the constant voltage, V_k is the voltage amplitude, ϕ_k is the phase of the k-order harmonic component ω_k, $k = 1, 2, \ldots, N$, and N is the number of harmonics.

The spectral-domain analysis, based on substituting Equation 2.2 into Equation 2.1 for a particular nonlinear transfer function of the active device, determines the output spectrum as a sum of the fundamental-frequency and higher-order harmonic components, the amplitudes and phases of which will determine the output signal spectrum. Generally, it is a complicated procedure that requires a harmonic-balance technique to numerically calculate an accurate nonlinear circuit response. However, the solution can be found analytically in a simple way when it is necessary to only estimate the basic performance of a power amplifier in terms of the output power and efficiency. In this case, a technique based on a piecewise-linear approximation of the device transfer function can provide a clear insight to the basic behavior of a power amplifier and its operation modes. It can also serve as a good starting point for a final computer-aided design and optimization procedure.

The piecewise-linear approximation of the active device current–voltage transfer characteristic is a result of replacing the actual nonlinear dependence $i = f(v_{in})$, where v_{in} is the voltage applied to the device input, by an approximated one that consists of the straight lines tangent to the actual dependence at the specified points. Such a piecewise-linear approximation for the case of two straight lines is shown in Figure 2.1a.

The output current waveforms for the actual current–voltage dependence (dashed curve) and its piecewise-linear approximation by two straight lines (solid curve) are plotted in Figure 2.1b. Under large-signal operation mode, the waveforms corresponding to these two dependences are practically the same for the most part, with negligible deviation for small values of the output current close to the pinch-off region of the device operation and significant deviation close to the saturation region of the device operation. However, the latter case results in a significant nonlinear distortion and is used only for high-efficiency operation modes when the active period of the device operation is minimized. Hence, at least two first output-current components, dc and fundamental, can be calculated through a Fourier-series expansion with sufficient accuracy. Therefore, such a piecewise-linear approximation with two straight lines can be effective for a quick estimate of the output power and efficiency of the linear power amplifier.

The piecewise-linear active device current–voltage characteristic is defined as

$$i = \begin{cases} 0 & v_{in} \leq V_p \\ g_m(v_{in} - V_p) & v_{in} \geq V_p \end{cases} \tag{2.3}$$

where g_m is the device transconductance and V_p is the pinch-off voltage.

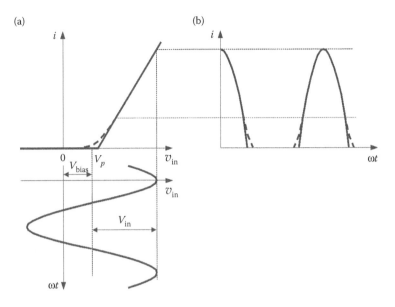

FIGURE 2.1
Piecewise-linear approximation technique.

Let us assume the input signal to be in a cosine form:

$$v_{in} = V_{bias} + V_{in} \cos \omega t \qquad (2.4)$$

where V_{bias} is the input dc bias voltage.

At the point on the plot when the voltage $v_{in}(\omega t)$ becomes equal to a pinch-off voltage V_p and where $\omega t = \theta$, the output current $i(\theta)$ takes a zero value. At this moment,

$$V_p = V_{bias} + V_{in} \cos \theta \qquad (2.5)$$

and the angle θ can be calculated from

$$\cos \theta = -\frac{V_{bias} - V_p}{V_{in}} \qquad (2.6)$$

As a result, the output current represents a periodic pulsed waveform described by the cosine pulses with maximum amplitude I_{max} and width 2θ as

$$i = \begin{cases} I_q + I \cos \omega t & -\theta \le \omega t < \theta \\ 0 & \theta \le \omega t < 2\pi - \theta \end{cases} \qquad (2.7)$$

where the conduction angle 2θ indicates the part of the RF current cycle, during which a device conduction occurs, as shown in Figure 2.2. When the output current $i(\omega t)$ takes a zero value, one can write

$$i = I_q + I \cos \theta = 0 \qquad (2.8)$$

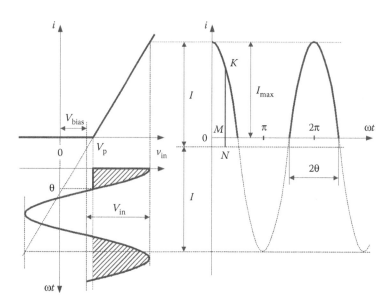

FIGURE 2.2
Schematic definition of conduction angle.

Taking into account that $I = g_m V_{in}$ for a piecewise-linear approximation, Equation 2.7 can be rewritten for $i > 0$ by

$$i = g_m V_{in}(\cos \omega t - \cos \theta) \tag{2.9}$$

When $\omega t = 0$, then $i = I_{max}$ and

$$I_{max} = I(1 - \cos \theta) \tag{2.10}$$

The Fourier-series expansion of the even function when $i(\omega t) = i(-\omega t)$ contains only even components of this function and can be written as

$$i(\omega t) = I_0 + I_1 \cos \omega t + I_2 \cos 2\omega t + \cdots + I_N \cos N\omega t \tag{2.11}$$

where the dc, fundamental-frequency, and any nth-harmonic components are calculated by

$$I_0 = \frac{1}{2\pi} \int_{-\theta}^{\theta} g_m V_{in}(\cos \omega t - \cos \theta)\, d\omega t = I\gamma_0(\theta) \tag{2.12}$$

$$I_1 = \frac{1}{\pi} \int_{-\theta}^{\theta} g_m V_{in}\, (\cos \omega t - \cos \theta)\cos \omega t\, d\omega t = I\gamma_1(\theta) \tag{2.13}$$

and

$$I_n = \frac{1}{\pi} \int_{-\theta}^{\theta} g_m V_{in}(\cos \omega t - \cos \theta) \cos n\omega t \, d\omega t = I\gamma_n(\theta) \tag{2.14}$$

where $\gamma_n(\theta)$ are called the coefficients of expansion of the output-current cosine waveform or the current coefficients [3,4]. They can be analytically defined for the dc and fundamental components as

$$\gamma_0(\theta) = \frac{1}{\pi}(\sin \theta - \theta \cos \theta) \tag{2.15}$$

$$\gamma_1(\theta) = \frac{1}{\pi}\left(\theta - \frac{\sin 2\theta}{2}\right) \tag{2.16}$$

and for the second- and higher-order harmonic components as

$$\gamma_n(\theta) = \frac{1}{\pi}\left[\frac{\sin(n-1)\theta}{n(n-1)} - \frac{\sin(n+1)\theta}{n(n+1)}\right] \tag{2.17}$$

where $n = 2, 3, \dots$.

The dependences of $\gamma_n(\theta)$ for the dc, fundamental-frequency, second-, and higher-order current components are shown in Figure 2.3. The maximum value of $\gamma_n(\theta)$ is achieved when $\theta = 180°/n$. A special case is $\theta = 90°$, when odd current coefficients are equal to zero, that is, $\gamma_3(\theta) = \gamma_5(\theta) = \dots = 0$. The ratio between the fundamental-frequency and dc components $\gamma_1(\theta)/\gamma_0(\theta)$ varies from 1 to 2 for any values of the conduction angle, with a minimum value of 1 for $\theta = 180°$ and a maximum value of 2 for $\theta = 0°$, as shown in Figure 2.3a. Besides, it is necessary to pay attention to the fact that the current coefficient $\gamma_3(\theta)$ becomes negative within the interval of $90° < \theta < 180°$, as shown in Figure 2.3b. This implies the proper phase changes of the third current harmonic component when its values are negative. Consequently, if the harmonic components with $\gamma_n(\theta) > 0$ achieve positive maximum values at the time moments corresponding to the middle points of the current waveform, the harmonic components with $\gamma_n(\theta) < 0$ can achieve negative maximum values at these same time moments. As a result, the combination of different harmonic components with proper loading will result in flattening of the current or voltage waveforms, thus improving efficiency of the power amplifier. The amplitude of corresponding current harmonic component can be obtained by

$$I_n = \gamma_n(\theta) g_m V_{in} = \gamma_n(\theta) I \tag{2.18}$$

In some cases, it is necessary for an active device to provide a constant value of I_{max} at any values of θ that requires an appropriate variation of the input voltage amplitude V_{in}. In this case, it is more convenient to use the coefficients α_n defined as a ratio of the nth current harmonic amplitude I_n to the maximum current waveform amplitude I_{max},

$$\alpha_n = \frac{I_n}{I_{max}} \tag{2.19}$$

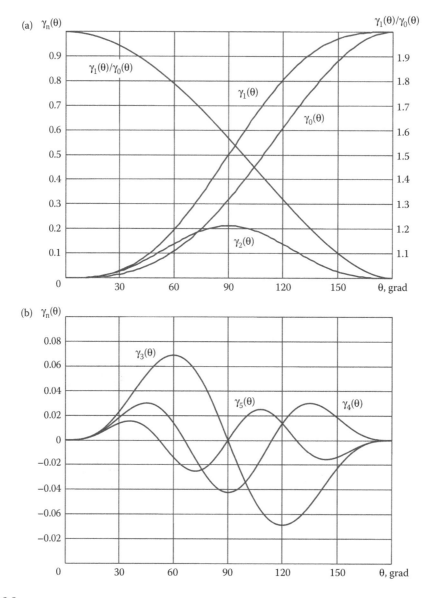

FIGURE 2.3
Dependences of $\gamma_n(\theta)$ for dc, fundamental, and higher-order current components.

From Equations 2.10, 2.18, and 2.19, it follows that

$$\alpha_n = \frac{\gamma_n(\theta)}{1 - \cos\theta} \tag{2.20}$$

and the maximum value of $\alpha_n(\theta)$ is achieved when $\theta = 120°/n$.

To analytically determine the operation classes of the power amplifier, consider a simple resistive stage shown in Figure 2.4, where L_{ch} is the ideal choke inductor with zero series resistance and infinite reactance at the operating frequency, C_b is the dc-blocking capacitor

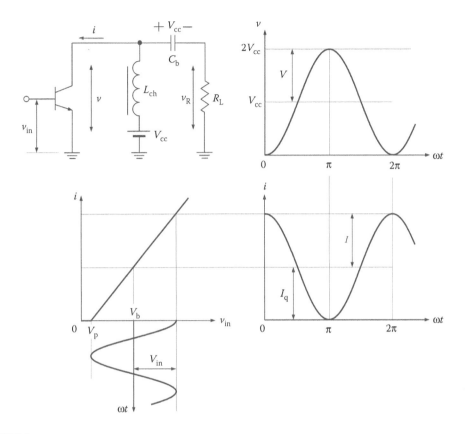

FIGURE 2.4
Voltage and current waveforms in Class-A operation.

with infinite value having zero reactance at the operating frequency, and R_L is the load resistor. The dc supply voltage V_{cc} is applied to both plates of the dc-blocking capacitor, being constant during the entire signal period. The active device behaves as an ideal voltage- or current-controlled current source having zero saturation resistance.

For an input cosine voltage given by Equation 2.4, the operating point must be fixed at the middle point of the linear part of the device transfer characteristic with $V_{in} \le V_{bias} - V_p$. Normally, in order to simplify an analysis of the power amplifier operation, the device transfer characteristic is represented by a piecewise-linear approximation. As a result, the output current is cosinusoidal,

$$i = I_q + I \cos \omega t \tag{2.21}$$

with the quiescent current I_q greater or equal to the collector current amplitude I. In this case, the output collector current contains only two components—dc and cosine—and the averaged current amplitude is equal to a quiescent current I_q.

The output voltage v across the device collector represents a sum of the dc supply voltage V_{cc} and cosine voltage v_R across the load resistor R_L. Consequently, the greater output current i, the greater voltage v_R across the load resistor R_L and the smaller output voltage v. Thus, for a purely real load impedance when $Z_L = R_L$, the collector voltage v is shifted by 180° relatively to the input voltage v_{in} and can be written as

$$v = V_{cc} + V \cos(\omega t + 180^\circ) = V_{cc} - V \cos \omega t \qquad (2.22)$$

where V is the output voltage amplitude.

Substituting Equation 2.21 into Equation 2.22 yields

$$v = V_{cc} - (i - I_q)R_L \qquad (2.23)$$

where $R_L = V/I$, and Equation 2.23 can be rewritten as

$$i = \left(I_q + \frac{V_{cc}}{R_L}\right) - \frac{v}{R_L} \qquad (2.24)$$

which determines a linear dependence of the collector current versus collector voltage. Such a combination of the cosine collector voltage and current waveforms is known as a Class-A operation mode. In practice, because of the device nonlinearities, it is necessary to connect a parallel LC circuit with resonant frequency equal to the operating frequency to significantly suppress any possible harmonic components.

Circuit theory prescribes that the collector efficiency η can be written as

$$\eta = \frac{P}{P_0} = \frac{1}{2}\frac{I}{I_q}\frac{V}{V_{cc}} = \frac{1}{2}\frac{I}{I_q}\xi \qquad (2.25)$$

where

$$P_0 = I_q V_{cc} \qquad (2.26)$$

is the dc output power,

$$P = \frac{IV}{2} \qquad (2.27)$$

is the power delivered to the load resistance R_L at the fundamental frequency f_0, and

$$\xi = \frac{V}{V_{cc}} \qquad (2.28)$$

is the collector voltage peak factor.

Then, by assuming the ideal conditions of zero saturation voltage when $\xi = 1$ and maximum output current amplitude when $I/I_q = 1$, from Equation 2.25, it follows that the maximum collector efficiency in a Class-A operation mode is equal to

$$\eta = 50\% \qquad (2.29)$$

However, as it also follows from Equation 2.25, increasing the value of I/I_q can further increase the collector efficiency. This leads to a step-by-step nonlinear transformation of the current cosine waveform to its pulsed waveform when the amplitude of the collector current exceeds zero value during only a part of the entire signal period. In this case, an active device is operated in the active region followed by the operation in the pinch-off region when the collector current is zero, as shown in Figure 2.5. As a result, the frequency spectrum at the device output will generally contain the second, third, and higher-order harmonics of the fundamental frequency. However, owing to the high quality of the parallel resonant LC circuit, only the fundamental-frequency signal flows into the load, while the short-circuit conditions are fulfilled for higher-order harmonic components. Therefore, ideally, the collector voltage represents a purely sinusoidal waveform with the voltage amplitude $V \leq V_{cc}$.

Equation 2.8 for the output current can be rewritten through the ratio between a quiescent current I_q and a current amplitude I as

$$\cos\theta = -\frac{I_q}{I} \tag{2.30}$$

As a result, the basic definitions for nonlinear operation modes of a power amplifier through half the conduction angle θ can be introduced as

- When $\theta > 90°$, then $\cos\theta < 0$ and $I_q > 0$, corresponding to Class-AB operation.

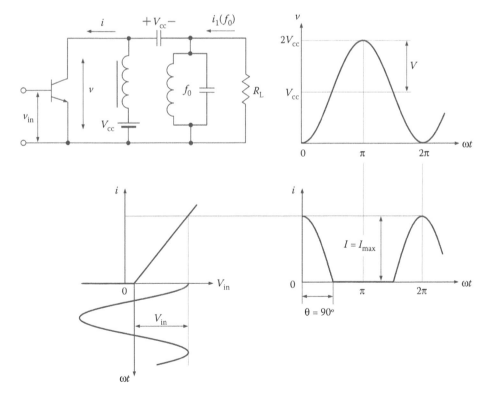

FIGURE 2.5
Voltage and current waveforms in Class-B operation.

- When $\theta = 90°$, then $\cos\theta = 0$ and $I_q = 0$, corresponding to Class-B operation.
- When $\theta < 90°$, then $\cos\theta > 0$ and $I_q < 0$, corresponding to Class-C operation.

The periodic pulsed output current $i(\omega t)$ is represented as a Fourier-series expansion by Equation 2.11, where the dc current component is a function of θ in the operation modes with $\theta < 180°$, in contrast to a Class-A operation mode where $\theta = 180°$ and the dc current is equal to the quiescent current during the entire period.

The collector efficiency of a power amplifier with parallel resonant circuit, biased to operate in a nonlinear mode with certain conduction angle, can be obtained by

$$\eta = \frac{P_1}{P_0} = \frac{1}{2}\frac{I_1}{I_0}\xi = \frac{1}{2}\frac{\gamma_1}{\gamma_0}\xi \tag{2.31}$$

which is a function of θ only, where P_1 is the output power at fundamental frequency and

$$\frac{\gamma_1}{\gamma_0} = \frac{\theta - \sin\theta\cos\theta}{\sin\theta - \theta\cos\theta} \tag{2.32}$$

The vacuum-tube Class-B power amplifiers had been defined as those which operate with a negative grid bias such that the anode current is practically zero with no excitation grid voltage, and in which the output power is proportional to the square of the excitation voltage [5]. If $\xi = 1$ and $\theta = 90°$, then from Equations 2.15 and 2.16, it follows that the maximum collector efficiency in a Class-B operation mode is equal to

$$\eta = \frac{\pi}{4} \cong 78.5\% \tag{2.33}$$

The fundamental-frequency power delivered to the load $P_L = P_1$ is defined as

$$P_1 = \frac{VI_1}{2} = \frac{VI\gamma_1(\theta)}{2} \tag{2.34}$$

showing its direct dependence on the conduction angle 2θ. This means that reduction in θ results in lower γ_1, and, in order to increase the fundamental-frequency power P_1, it is necessary to increase the current amplitude I. Since the current amplitude I is determined by the input voltage amplitude V_{in}, the input power P_{in} must be increased. The collector efficiency increases with reduced value of θ as well and becomes maximum when $\theta = 0°$, where the ratio γ_1/γ_0 is maximal, as follows from Figure 2.3a. For instance, the collector efficiency η increases from 78.5% to 92% when θ reduces from 90° to 60°. However, it requires increasing the input voltage amplitude V_{in} by 2.5 times, resulting in lower values of the power-added efficiency (*PAE*), which is defined as

$$PAE = \frac{P_1 - P_{in}}{P_0} = \frac{P_1}{P_0}\left(1 - \frac{1}{G_p}\right) \tag{2.35}$$

where

$$G_{\mathrm{p}} = \frac{P_1}{P_{\mathrm{in}}} \qquad (2.36)$$

is the operating power gain.

The vacuum-tube Class-C power amplifiers had been defined as those that operate with a negative grid bias more than sufficient to reduce the anode current to zero with no excitation grid voltage, and in which the output power varies as the square of the anode voltage between limits [5]. The main distinction between Class-B and Class-C is in the duration of the output current pulses, which are shorter for Class-C when the active device is biased beyond the cutoff point. It should be noted that, for the device transfer characteristic ideally represented by a square-law approximation, the odd-harmonic current coefficients $\gamma_n(\theta)$ are not equal to zero in this case, although there is no significant difference between the square-law and linear cases [6]. To achieve the maximum anode (collector) efficiency in Class-C, the active device should be biased (negative) considerably past the cutoff (pinch-off) point to provide the sufficiently low conduction angles [7].

In order to obtain an acceptable trade-off between a high power gain and a high power-added efficiency in different situations, the conduction angle should be chosen within the range of $120° \leq 2\theta \leq 190°$. If it is necessary to provide high collector efficiency of the active device having a high-gain capability, it is necessary to choose a Class-C operation mode with θ close to $60°$. However, when the input power is limited and power gain is not sufficient, a Class-AB operation mode is recommended with small quiescent current when θ is slightly greater than $90°$. In the latter case, the linearity of the power amplifier can be significantly improved. From Equation 2.32, it follows that the ratio of the fundamental-frequency component of the anode (collector) current to the dc current is a function of θ only, which means that, if the operating angle is maintained constant, the fundamental component of the anode (collector) current will replicate linearly to the variation of the dc current, thus providing the linear operation of the Class-C power amplifier when dc current is directly proportional to the grid (base) voltage [8].

2.2 Load Line and Output Impedance

The graphical method of laying down a load line on the family of the static curves representing anode current against anode voltage for various grid potentials was already well known in the 1920s [9]. If an active device is connected in a circuit in which the anode load is a pure resistance, the performance may be analyzed by drawing the load line where the lower end of the line represents the anode supply voltage and the slope of the line is established by the load resistance, that is, the load resistance is equal to the value of the intercept on the voltage axis divided by the value of the intercept on the current axis.

In a Class-A operation mode, the output voltage v across the device anode (collector or drain) represents a sum of the dc supply voltage V_{cc} and cosine voltage across the load resistance R_L, and can be defined by Equation 2.22. In this case, the power dissipated in the load and the power dissipated in the device is equal when $V_{\mathrm{cc}} = V$, and the load resistance $R_L = V/I$ is equal to the device output resistance R_{out} [7]. In a pulsed operation mode (Class-AB, B, or C), since the parallel LC circuit is tuned to the fundamental frequency, ideally, the voltage across the load resistor R_L represents a cosine waveform. By using

Equations 2.7, 2.13, and 2.22, the relationship between the collector current i and the collector voltage v during a time period of $-\theta \leq \omega t < \theta$ can be expressed by

$$i = \left(I_q + \frac{V_{cc}}{\gamma_1 R_L} \right) - \frac{v}{\gamma_1 R_L} \tag{2.37}$$

where the fundamental current coefficient γ_1 as a function of θ is determined by Equation 2.16, and the load resistance is defined by $R_L = V/I_1$, where I_1 is the fundamental current amplitude. Equation 2.37 determining the dependence of the collector current on the collector voltage for any values of conduction angle in the form of a straight-line function is called the *load line* of the active device. For a Class-A operation mode with $\theta = 180°$ when $\gamma_1 = 1$, the load line defined by Equation 2.37 is identical to the load line defined by Equation 2.24.

Figure 2.6 shows the idealized active device output I–V curves and load lines for different conduction angles according to Equation 2.37 with the corresponding collector and current waveforms. From Figure 2.6, it follows that the maximum collector current amplitude I_{max} corresponds to the minimum collector voltage V_{sat} when $\omega t = 0$, and is the same for any conduction angle. The slope of the load line defined by its slope angle β is different for different conduction angles and values of the load resistance, and can be obtained by

$$\tan \beta = \frac{I_{max}}{V(1 - \cos\theta)} = \frac{1}{\gamma_1 R_L} \tag{2.38}$$

from which it follows that greater slope angle β of the load line results in smaller value of the load resistance R_L for the same θ.

The load resistance R_L for the active device as a function of θ, which is required to terminate the device output to deliver the maximum output power to the load, can be written in a general form as

$$R_L(\theta) = \frac{V}{\gamma_1(\theta)I} \tag{2.39}$$

which is equal to the device equivalent output resistance R_{out} at the fundamental frequency [5]. The term "equivalent" means that this is not a real physical device resistance as in a Class-A mode, but its equivalent output resistance, the value of which determines the optimum load, which should terminate the device output to deliver maximum fundamental-frequency output power. The equivalent output resistance is calculated as a ratio between the amplitudes of the collector cosine voltage and fundamental-frequency collector current component, which depends on the angle θ.

In a Class-B mode when $\theta = 90°$ and $\gamma_1 = 0.5$, the load resistance R_L^B is defined as $R_L^B = 2V/I_{max}$. Alternatively, taking into account that $V_{cc} = V$ and $P_{out} = I_1 V/2$ for the fundamental-frequency output power, the load resistance $R_L^B = V/I_1$ can be written in a simple idealized analytical form with zero saturation voltage V_{sat} as

$$R_L^B = \frac{V_{cc}^2}{2P_{out}} \tag{2.40}$$

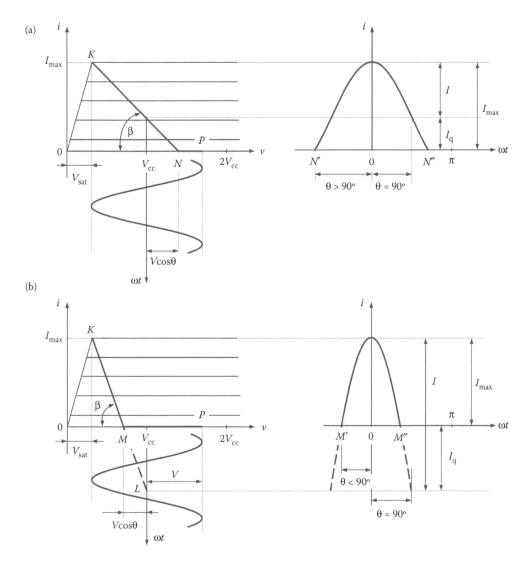

FIGURE 2.6
Collector current waveforms in Class-AB (a) and Class-C (b) operations.

In general, the entire load line represents a broken line PK including a horizontal part, as shown in Figure 2.6. Figure 2.6a represents a load line PNK corresponding to a Class-AB mode with $\theta > 90°$, $I_q > 0$, and $I < I_{max}$. Such a load line moves from point K corresponding to the maximum output current amplitude I_{max} at $\omega t = 0$ and determining the device saturation voltage V_{sat} through the point N located at the horizontal axis v where $i = 0$ and $\omega t = \theta$. For a Class-AB operation, the conduction angle for the output current pulse between points N' and N'' is greater than 180°. Figure 2.6b represents a load line PMK corresponding to a Class-C mode with $\theta < 90°$, $I_q < 0$, and $I > I_{max}$. For a Class-C operation, the load line intersects a horizontal axis v in a point M, and the conduction angle for the output current pulse between points M' and M'' is smaller than 180°. Hence, generally the load line represents a broken line with the first section having a slope angle β and the other horizontal section with zero current i. In a Class-B mode, the collector current represents half-cosine pulses with the conduction angle of $2\theta = 180°$ and $I_q = 0$.

Now, let us consider a Class-B operation with increased amplitude of the cosine collector voltage. In this case, as shown in Figure 2.7, an active device is operated in the saturation, active, and pinch-off regions, and the load line represents a broken line *LKMP* with three linear sections (*LK*, *KM*, and *MP*). The new section *KL* corresponds to the saturation region, resulting in a half-cosine output current waveform with depression in the top part. With further increase of the output voltage amplitude, the output current pulse can be split into two symmetrical pulses containing a significant level of the higher-order harmonic components. The same result can be achieved by increasing a value of the load resistance R_L when the load line is characterized by smaller slope angle β.

The collector current waveform becomes asymmetrical for the complex load, the impedance of which represents the load resistance and capacitive or inductive reactance. In this case, the Fourier-series expansion of the output current given by Equation 2.11 includes a particular phase for each harmonic component. Then, the output voltage at the device collector is written as

$$v = V_{cc} - \sum_{n=1}^{\infty} I_n \left| Z_n \right| \cos\left(n\omega t + \phi_n\right) \tag{2.41}$$

where I_n is the amplitude of the nth output current harmonic component, $\left| Z_n \right|$ is the magnitude of the load-network impedance at the nth output current harmonic component, and ϕ_n is the phase of the nth output current harmonic component. Assuming that Z_n is zero for $n = 2, 3, \ldots$, which is possible for a resonant load network having negligible impedance at any harmonic component except the fundamental, Equation 2.41 can be rewritten as

$$v = V_{cc} - I_1 \left| Z_1 \right| \cos\left(\omega t + \phi_1\right) \tag{2.42}$$

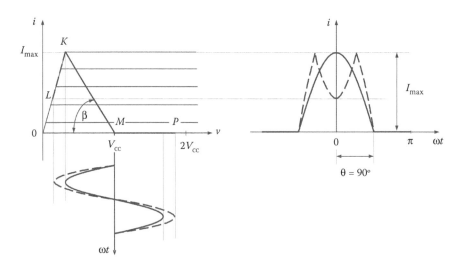

FIGURE 2.7
Collector current waveforms for the device operating in saturation, active, and pinch-off regions.

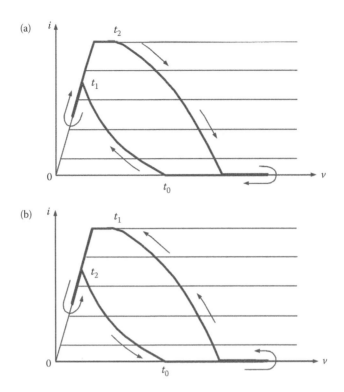

FIGURE 2.8
Load lines for (a) inductive and (b) capacitive load impedances.

As a result, for the inductive load impedance, the depression in the collector current waveform reduces and moves to the left-hand side of the waveform, whereas the capacitive load impedance causes the depression to deepen and shift to the right-hand side of the collector current waveform [10]. This effect can simply be explained by the different sign for the phase angle ϕ_1 in Equation 2.42, as well as generally by the different phase conditions for fundamental and higher-order harmonic components composing the collector current waveform, and is illustrated by the different load lines for (a) inductive and (b) capacitive load impedances shown in Figure 2.8. Note that now the load line represents a two-dimensional curve with a complicated behavior.

2.3 Nonlinear Active Device Models

Generally, for an accurate power amplifier simulation and matching circuit design for different operating frequencies and output-power levels, it is necessary to represent an active device in the form of a nonlinear equivalent circuit, which can adequately describe the small- and large-signal electrical behavior of the power amplifier up to the device transition frequency f_T and higher to its maximum frequency f_{max} that allows a sufficient number of harmonic components to be taken into account. Accurate device modeling is extremely important to develop monolithic integrated circuits. Better approximations of the final design can only be achieved if the nonlinear device behavior is described accurately.

2.3.1 LDMOSFETs

Figure 2.9a shows the cross-section of the physical structure of an LDMOSFET (lateral diffusion metal-oxide semiconductor field-effect transistor) device where a heavily doped p^+-sinker is inserted between top source and p^+-substrate (source grounding) for low resistivity to provide high-current flow between the drain and source terminals [11]. The lightly doped p-epilayer and n-drift layer are required to provide sufficient distance between regions to prevent latchup (forward-biased p–n diodes) and for the drain-source breakdown protection. The parasitic gate-drain capacitance is directly related to the overlap of the gate oxide onto the heavily doped n^+-source region. To describe accurately the nonlinear properties of the large-size MOSFET (metal-oxide semiconductor field-effect transistor) device, it is necessary to consider its two-dimensional gate-distributed nature along both the channel length and channel width, resulting in lower values of the intrinsic series gate and shunt gate-source resistances. Figure 2.9b shows the nonlinear MOSFET equivalent circuit with extrinsic parasitic elements, which can properly describe the nonlinear behavior of both VDMOSFET (vertical diffusion metal-oxide semiconductor field-effect transistor) and LDMOSFET devices [12,13].

The nonlinear current source $i(v_{gs}, v_{ds}, \tau)$ as a function of the input gate-source and output drain-source voltages incorporating self-heating effect can be described sufficiently simple and accurate using hyperbolic functions [13,14]. Careful analytical description of

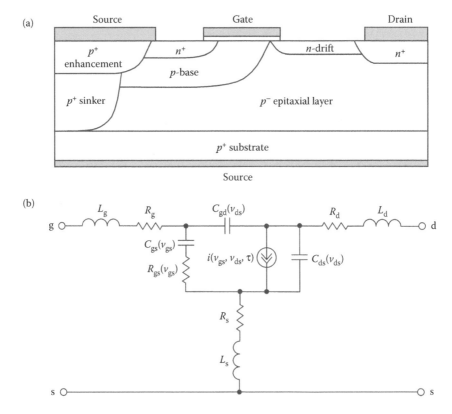

FIGURE 2.9
Nonlinear LDMOSFET model and its physical structure.

the transition from quadratic to linear regions of the device transfer characteristic enables the more accurate prediction of the intermodulation distortion [15]. The overall channel carrier transit time τ also includes an effect of the transcapacitance required for charge conservation. The drain-source capacitance C_{ds} and gate-drain capacitance C_{gd} are considered as the junction capacitances that strongly depend on the drain-source voltage. The extrinsic parasitic elements are represented by the gate and drain bondwire inductances L_g and L_d, source inductance L_s, source and drain bulk and ohmic resistances R_s and R_d, and gate contact and ohmic resistance R_g. The effect of the gate-source channel resistance R_{gs} becomes significant at higher frequencies close to the transition frequency $f_T = g_m/2\pi C_{gs}$, where g_m is the device transconductance. To account for self-heating effect and substrate losses, a special four-port thermal circuit and a series combination of the resistance and capacitance between the external drain and source terminals can be included [16].

An empirical nonlinear model developed for silicon LDMOS transistors (or LDMOSFET devices), which is single-piece and continuously differentiable, can be written as

$$I_{ds} = \beta V_{gst}^{VGexp}(1 + \lambda V_{ds})\tanh\left(\frac{\alpha V_{ds}}{V_{gst}}\right)[1 + K_1\exp(V_{BReff1})]$$

$$+ I_{ss}\exp\left(\frac{V_{ds} - V_{BR}}{V_T}\right) \tag{2.43}$$

where

$$V_{gst} = V_{st}\ln\left[1 + \exp\left(\frac{V_{gst2}}{V_{st}}\right)\right]$$

$$V_{gst2} = V_{gst1} - \frac{1}{2}\left(V_{gst1} + \sqrt{\left(V_{gst1} - V_K\right)^2 + \Delta^2} - \sqrt{V_K^2 + \Delta^2}\right)$$

$$V_{gst1} = V_{gs} - V_{th0} - \gamma V_{ds}$$

$$V_{BReff1} = \frac{V_{ds} - V_{BReff}}{K_2} + M_3\frac{V_{ds}}{V_{BReff}}$$

$$V_{BReff} = \frac{V_{BR}}{2}\left[1 + \tanh(M_1 - V_{gst}M_2)\right]$$

where λ is the drain current slope parameter, β is the transconductance parameter, V_{th0} is the forward threshold voltage, V_{st} is the subthreshold slope coefficient, V_T is the temperature voltage, I_{ss} is the forward diode leakage current, V_{BR} is the breakdown voltage, K_1, K_2, M_1, M_2, and M_3 are the breakdown parameters, and V_K, V_{Gexp}, Δ, and γ are the gate-source voltage parameters [14].

The gate-source capacitance C_{gs} can be analytically described as a function of the gate-source voltage since it is practically independent of the drain-source voltage. It is equal to the oxide capacitance C_{ox} in the accumulation region, significantly reduces in the depletion region, slightly decreases and reaches its minimum in the weak-inversion region, and then significantly increases in the moderate-inversion region and becomes practically constant in the strong-inversion or saturation region, as shown in Figure 2.10 [17]. The approximation function for the gate-source capacitance C_{gs} as the dependence of V_{gs} can be derived by using two components containing the hyperbolic functions as

$$
\begin{aligned}
C_{gs} = C_{gs1} + C_{gs2}\{1 + \tanh[C_{gs6}(V_{gs} + C_{gs3})]\} \\
+ C_{gs4}[1 - \tanh(C_{gs5}V_{gs})]
\end{aligned}
\tag{2.44}
$$

where C_{gs1}, C_{gs2}, C_{gs3}, C_{gs4}, C_{gs5}, and C_{gs6} are the approximation parameters [14].

The gate-source resistance R_{gs} is determined by the effect of the channel inertia in responding to rapid changes of the time-varying gate-source voltage, and varies in such a manner that the charging time $\tau_g = R_{gs}C_{gs}$ remains approximately constant. Thus, the increase of R_{gs} in the velocity saturation region, when the channel conductivity decreases, is partially compensated by the decrease of C_{gs} due to nonuniform channel charge distribution [18]. The effect of R_{gs} becomes significant at higher frequencies close to the transition frequency f_T of the MOSFET and cannot be taken into consideration when designing RF circuits that operate below 2 GHz, as used for commercial wireless applications [19,20].

2.3.2 GaAs MESFETs and GaN HEMTs

Adequate representation for MESFETs (metal-semiconductor field-effect transistors) and HEMTs (high-electron-mobility transistors) in a frequency range up to at least 25 GHz can be provided using a nonlinear model shown in Figure 2.11a, which is very similar to a nonlinear MOSFET model [21,22]. The intrinsic model is described by the channel charging resistance R_{gs}, which represents the resistive path for the charging of the gate-source

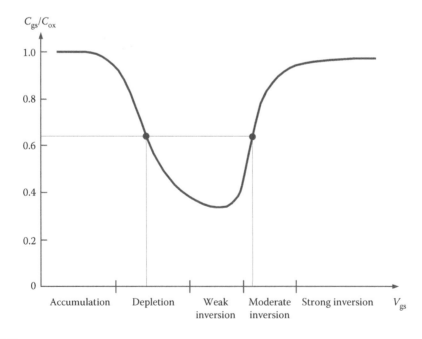

FIGURE 2.10
Gate-source capacitance versus gate-source voltage.

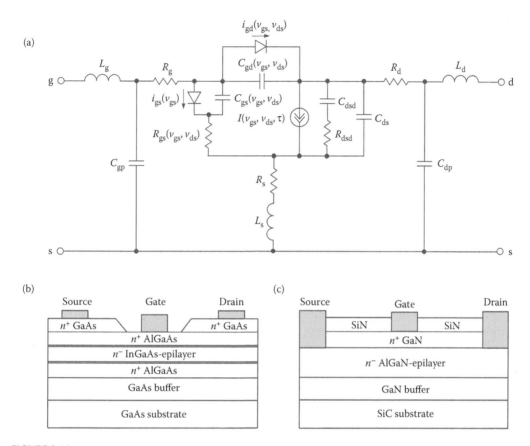

FIGURE 2.11
Nonlinear MESFET and HEMT model with HEMT physical structures.

capacitance C_{gs}, feedback gate-drain capacitance C_{gd}, and drain-source capacitance C_{ds} with the gate-source diode to model the forward conduction current $i_{gs}(v_{gs})$ and gate-drain diode to account for the gate-drain avalanche current $i_{gd}(v_{gs}, v_{ds})$, which can occur at large-signal operation conditions. The gate-source capacitance C_{gs} and gate-drain capacitance C_{gd} represent the charge depletion region and can be treated as the voltage-dependent Schottky-barrier diode capacitances, being the nonlinear functions of the gate-source voltage v_{gs} and drain-source voltage v_{ds}. For negative gate-source voltage and small drain-source voltage, these capacitances are practically equal. However, when the drain-source voltage is increased beyond the current saturation point, the gate-drain capacitance C_{gd} is much more heavily back-biased than the gate-source capacitance C_{gs}. Therefore, the gate-source capacitance C_{gs} is significantly more important and usually dominates the input impedance of the MESFET or HEMT device. The influence of the drain-source capacitance C_{ds} on the device behavior is insignificant and its value is bias independent. The capacitance C_{dsd} and resistance R_{dsd} model the dispersion of the MESFET or HEMT current–voltage characteristics due to trapping effect in the device channel, which leads to discrepancy between the dc and S-parameter measurements at higher frequencies [23,24]. A large-signal model for monolithic power amplifier design must be accurate for all operating conditions. In addition, the model parameters should be easily extractable and the model must be as simple as possible. Various nonlinear MESFET and HEMT models with different complexity are available, and each one can be considered sufficiently accurate for a particular application. For example, although the Materka model does not fulfill charge conservation, it seems to be an acceptable compromise between accuracy and model simplicity for MESFETs, but not for HEMTs, where it is preferable to use the Angelov model [25,26]. For example, it can be used to predict the large-signal behavior of the pHEMT (pseudomorphic HEMT) devices using in high-power, high-efficiency 60-GHz MMICs (microwave monolithic integrated circuits) [27]. By using three additional terms of the gate power-series function in the Angelov model, the better accuracy can be achieved in a large-signal modeling of AlGaN/GaN HEMT devices on SiC substrate [28]. This model can also be improved by incorporating two additional analytical expressions to model the device behavior in saturation region [29].

Figure 2.11b shows the cross-section of the physical structure of an InGaAs/AlGaAs HEMT device, where an undoped InGaAs *n*-epilayer is used as a channel and two heavily *n*-doped AlGaAs layers with a high energetic barrier for holes are necessary to maximize high electron mobility in the channel. In this case, spacing between AlGaAs layer and InGaAs channel is optimized to achieve high breakdown voltage. An example of the physical structure of a AlGaN/GaN HEMT device is shown in Figure 2.11c, where an undoped AlGaN *n*-epilayer is used as a channel, an *n*-type doped GaN layer can suppress dispersion in the device current–voltage characteristics, and a SiN passivation layer with optimized parameters contributes to a lower-trap device structure [30]. Thermal conductivity of a GaN HEMT device is improved by using a SiC substrate. Note that the GaN-based technology can provide higher breakdown voltage, wider bandwidth, and higher efficiency of the power amplifier due to high charge density and the ability to operate at higher voltages for GaN HEMT devices, which are characterized by lower output capacitance and on-resistance [31,32].

The basic electrical properties of the MESFET or HEMT device can be characterized by the admittance Y-parameters expressed through the device intrinsic small-signal equivalent circuit as

$$Y_{11} = \frac{j\omega C_{gs}}{1 + j\omega C_{gs}R_{gs}} + j\omega C_{gd} \tag{2.45}$$

$$Y_{12} = -j\omega C_{gd} \tag{2.46}$$

$$Y_{21} = \frac{g_m \exp(-j\omega\tau)}{1 + j\omega C_{gs}R_{gs}} - j\omega C_{gd} \tag{2.47}$$

$$Y_{22} = \frac{1}{R_{ds}} + j\omega(C_{ds} + C_{gd}) \tag{2.48}$$

where g_m is the device transconductance and R_{ds} is the differential drain-source resistance [13]. In this case, the dispersion effect, which is important at higher frequencies and modeled by C_{dsd} and R_{dsd}, cannot be taken into account.

By separating Equations 2.45 through 2.48 into their real and imaginary parts, the elements of the small-signal equivalent circuit can be analytically determined as

$$C_{gd} = -\frac{\text{Im } Y_{12}}{\omega} \tag{2.49}$$

$$C_{gs} = -\frac{\operatorname{Im} Y_{11} - \omega C_{gd}}{\omega}\left[1+\left(\frac{\operatorname{Re} Y_{11}}{\operatorname{Im} Y_{11} - \omega C_{gd}}\right)^2\right] \tag{2.50}$$

$$R_{gs} = \frac{\operatorname{Re} Y_{11}}{(\operatorname{Im} Y_{11} - \omega C_{gd})^2 + (\operatorname{Re} Y_{11})^2} \tag{2.51}$$

$$g_m = \sqrt{(\operatorname{Re} Y_{21})^2 + (\operatorname{Im} Y_{21} + \omega C_{gd})^2}\ \sqrt{1+(\omega C_{gs}R_{gs})^2} \tag{2.52}$$

$$\tau = \frac{1}{\omega}\sin^{-1}\left(\frac{-\omega C_{gd} - \operatorname{Im} Y_{21} - \omega C_{gs}R_{gs}\operatorname{Re} Y_{21}}{g_m}\right) \tag{2.53}$$

$$C_{ds} = \frac{\operatorname{Im} Y_{22} - \omega C_{gd}}{\omega} \tag{2.54}$$

$$R_{ds} = \frac{1}{\operatorname{Re} Y_{22}} \tag{2.55}$$

which are valid for a wide frequency range up to the transition frequency f_T [33]. Assuming that all extrinsic parasitic elements are known, the only problem is then to determine the admittance Y-parameters of the intrinsic two-port network from on-bias experimental data [34]. Consecutive steps shown in Figure 2.12 can represent such a determination procedure [35]:

- Measurement of the S-parameters of the extrinsic device
- Transformation of the S-parameters to the impedance Z-parameters with subtraction of the series inductances L_g and L_d
- Transformation of the impedance Z-parameters to the admittance Y-parameters with subtraction of the parallel capacitances C_{gp} and C_{dp}
- Transformation of the admittance Y-parameters to the impedance Z-parameters with subtraction of series resistances R_g, R_s, R_d, and inductance L_s
- Transformation of the impedance Z-parameters to the admittance Y-parameters of the intrinsic device two-port network

A simple and accurate nonlinear Angelov model is capable of modeling the drain current–voltage characteristics and its derivatives, as well as the gate-source and gate-drain capacitances, for different submicron gate-length HEMT devices and commercially available MESFETs. The drain current source is described by using the hyperbolic functions as

$$I_{ds} = I_{pk}(1 + \tanh\psi)(1 + \lambda V_{ds})\tanh\alpha V_{ds} \tag{2.56}$$

where I_{pk} is the drain current at maximum transconductance with the contribution from the output conductance subtracted, λ is the channel-length modulation parameter, and $\alpha = \alpha_0 + \alpha_1\tanh\psi$ is the saturation voltage parameter, where α_0 is the saturation voltage

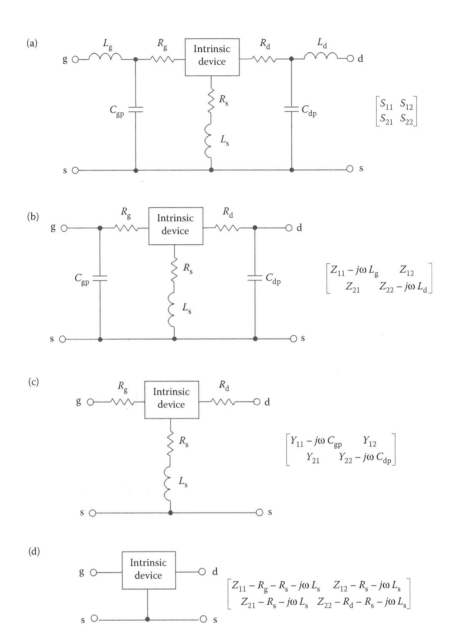

FIGURE 2.12
Method for extracting device intrinsic Z-parameters.

parameter at pinch-off and α_1 is the saturation voltage parameter at $V_{gs} > 0$. The parameter ψ is a power-series function centered at V_{pk} with the bias voltage V_{gs} as a variable,

$$\psi = P_1 (V_{gs} - V_{pk}) + P_2 (V_{gs} - V_{pk})^2 + P_3 (V_{gs} - V_{pk})^3 + \cdots \tag{2.57}$$

where V_{pk} is the gate voltage for maximum transconductance g_{mpk}. The model parameters as a first approximation can be easily obtained from the experimental $I_{ds}(V_{gs}, V_{ds})$ curves at a saturated channel condition when all higher terms in ψ are assumed to be zero and λ is the slope of the $I_{ds} - V_{ds}$ characteristic.

The same hyperbolic functions can be used to model the intrinsic device capacitances C_{gs} and C_{ds}. When an accuracy of 5%–10% is sufficient, the gate-source capacitance C_{gs} and gate-drain capacitance C_{gd} can be described by

$$C_{gs} = C_{gs0}[1 + \tanh(P_{1gsg}V_{gs})]\,[1 + \tanh(P_{1gsd}V_{ds})] \tag{2.58}$$

$$C_{gd} = C_{gd0}[1 + \tanh(P_{1gdg}V_{gs})][1 - \tanh(P_{1gdd}V_{ds} + P_{1cc}V_{gs}V_{ds})] \tag{2.59}$$

where the product $P_{1cc}V_{gs}V_{ds}$ reflects the cross-coupling of V_{gs} and V_{ds} on C_{gd} and the coefficients P_{1gsg}, P_{1gsd}, P_{1gdg}, and P_{1gdd} are the fitting parameters.

2.3.3 Low- and High-Voltage HBTs

Figure 2.13a shows the modified Gummel–Poon nonlinear model of the bipolar transistor with extrinsic parasitic elements [36,37]. Such a hybrid-π equivalent circuit can model the nonlinear electrical behavior of bipolar transistors, in particularly HBT (heterojunction bipolar transistor) devices, with sufficient accuracy up to about 20 GHz. The intrinsic model is described by the dynamic diode resistance r_π, the total base–emitter junction capacitance and base charging diffusion capacitance C_π, the base–collector diode required to account for the nonlinear effects at the saturation, the internal collector–base junction capacitance C_{ci}, the external distributed collector–base capacitance C_{co}, the collector–emitter capacitance C_{ce}, and the nonlinear current source $i(v_{be}, v_{ce})$. The lateral and base semiconductor resistances underneath the base contact and the base semiconductor resistance underneath the emitter are combined into a base-spreading resistance r_b. The extrinsic parasitic elements are represented by the base bondwire inductance L_b, emitter ohmic resistance r_e, emitter inductance L_e, collector ohmic resistance r_c, and collector bondwire inductance L_c. To increase the usable operating frequency range of the device up to 50 GHz, it is necessary to include the collector current delay time τ in the collector current source as $g_m\exp(-j\omega\tau)$. The more complicated models, such as VBIC, HICUM, or MEXTRAM, include the effects of self-heating of a bipolar transistor, take into account the parasitic p–n–p transistor formed by the base, collector, and substrate regions, provide an improved description of depletion capacitances at large forward bias, and take into account avalanche and tunneling currents and other nonlinear effects corresponding to distributed high-frequency effects [38].

Figure 2.13b shows the modified version of a bipolar transistor equivalent circuit, where $C_c = C_{co} + C_{ci}$, $r_{b1} = r_bC_{ci}/C_c$, and $r_{b2} = r_bC_{co}/C_c$ [39]. Such an equivalent circuit becomes possible due to an equivalent π- to T-transformation of the elements r_b, C_{co}, and C_{ci} and a condition $r_b \ll (C_{ci} + C_{co})/\omega C_{ci}C_{co}$, which is usually fulfilled over a frequency range close to the device maximum frequency f_{max}. Then, from a comparison of the transistor nonlinear models, for a bipolar transistor in Figure 2.13b, for a MOSFET device in Figure 2.9b, and for a MESFET device in Figure 2.11a, it is easy to detect the circuit similarity of all these equivalent circuits, which means that the basic circuit design procedure is very similar for any type of the bipolar or field-effect transistors (FETs). The main difference is in the device physics and values of the model parameters. However, techniques for representation of the input and output impedances, stability analysis based on feedback effect, and derivation of power gain and efficiency are very similar.

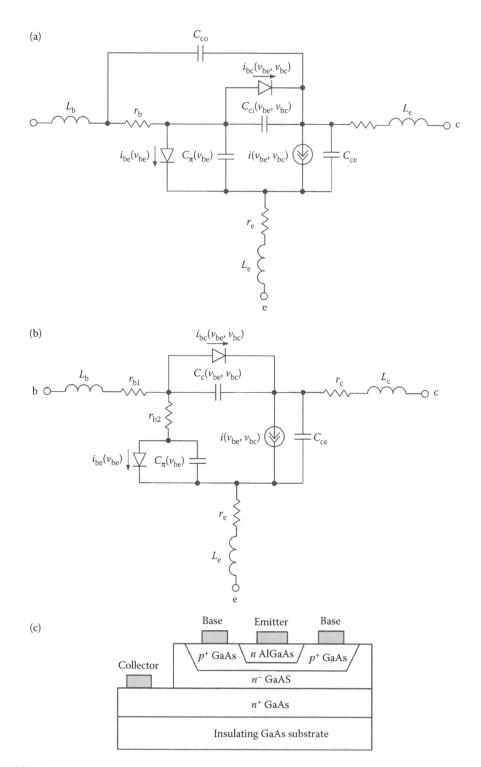

FIGURE 2.13
Nonlinear BJT and HBT models and HBT physical structure.

The cross-section of a physical structure of an AlGaAs/GaAs HBT device is shown in Figure 2.13c, with a heavily *p*-doped base to reduce base resistance and a lightly *n*-doped emitter to minimize emitter capacitance [40]. The lightly *n*-doped collector region allows collector–base junction to sustain relatively high voltages without breaking down. The forward-bias emitter-injection efficiency is very high since the wider-bandgap AlGaAs emitter injects electrons into the GaAs *p*-base at lower energy level, but the holes are prevented from flowing into the emitter by a high energy barrier, thus resulting in the ability to decrease base length, base-width modulation, and increase frequency response. By using a wide bandgap InGaP layer instead of an AlGaAs one, the device performance over temperature can be improved [41]. The high-linearity power performance in Class-AB condition at the backoff power level, the ruggedness under mismatch and overdrive condition, and the long lifetime of the InGaP/GaAs HBT technology makes it very attractive for the 28-V power amplifier applications [42]. The growth process used for a high-voltage HBT device is identical to the process used for the conventional low-voltage HBT device, which is widely used in handset power amplifiers, except for changes to the collector because of the higher voltage operating requirements. The epitaxial growth process starts with a highly doped *n*-type collector layer and a lightly *n*-doped collector drift region, then followed by a heavily doped *p*-type base layer and an InGaP emitter layer, and finishes with an InGaAs cap layer [43]. As a result, the high-voltage HBT devices exhibit collector–base breakdown voltages higher than 70 V.

The bipolar transistor intrinsic Y-parameters can be written as

$$Y_{11} = \frac{1}{r_\pi} + j\omega(C_\pi + C_{ci}) \tag{2.60}$$

$$Y_{12} = -j\omega C_{ci} \tag{2.61}$$

$$Y_{21} = g_m\exp(-j\omega\tau) - j\omega C_{ci} \tag{2.62}$$

$$Y_{22} = \frac{1}{r_{ce}} + j\omega C_{ci} \tag{2.63}$$

where r_{ce} is the output Early resistance that models the effect of the base-width modulation on the transistor characteristics due to variations in the collector–base depletion region.

After separating Equations 2.60 through 2.63 into their real and imaginary parts, the elements of the intrinsic small-signal equivalent circuit can be determined analytically as [44]

$$C_\pi = \frac{\text{Im}(Y_{11} + Y_{12})}{\omega} \tag{2.64}$$

$$r_\pi = \frac{1}{\text{Re}\,Y_{11}} \tag{2.65}$$

$$C_{ci} = -\frac{\text{Im}\,Y_{12}}{\omega} \tag{2.66}$$

$$g_m = \sqrt{(\mathrm{Re}\, Y_{21})^2 + (\mathrm{Im}\ Y_{21} + \mathrm{Im}\, Y_{12})^2} \tag{2.67}$$

$$\tau_\pi = \frac{1}{\omega}\cos^{-1}\frac{\mathrm{Re}\, Y_{21} + \mathrm{Re}\, Y_{12}}{\sqrt{(\mathrm{Re}\, Y_{21})^2 + (\mathrm{Im}\, Y_{21} + \mathrm{Im}\, Y_{12})^2}} \tag{2.68}$$

$$r_{ce} = \frac{1}{\mathrm{Re}\, Y_{22}} \tag{2.69}$$

A simple nonlinear HBT model for computer-aided simulations can be based on representation of the collector current source through the power series and diffusion capacitances through the hyperbolic functions [45]. To equivalently represent the input impedance of a bipolar transistor, it needs to take into account that C_{ce} is usually much smaller than C_c. As a result, the equivalent output capacitance can be defined as $C_{out} \cong C_c$. The input equivalent resistance R_{in} can approximately be represented by the base resistance r_b, while the input equivalent capacitance can be defined as $C_{in} \cong C_\pi + C_c$. The feedback effect of the collector capacitance C_c through C_{co} and C_{ci} is sufficiently high when load variations are directly transferred to the device input with a significant extent.

2.4 Power Gain and Stability

Power amplifier design aims for maximum power gain and efficiency for a given value of output power with a predictable degree of stability. In order to extract the maximum power from a generator, it is a well-known fact that the external load should have a vector value that is a conjugate of the internal impedance of the source [46]. The power delivered from a generator to a load, when matched on this basis, will be called the available power of the generator [47]. In this case, the power gain of the four-terminal network is defined as the ratio of power delivered to the load impedance connected to the output terminals to power available from the generator connected to the input terminals, usually measured in decibels, and this ratio is called the *power gain* irrespective of whether it is greater or less than one [48,49].

Figure 2.14 shows the basic block schematic of a single-stage power amplifier circuit, which includes an active device, an input matching circuit to match with the source impedance, and an output matching circuit to match with the load impedance. Generally,

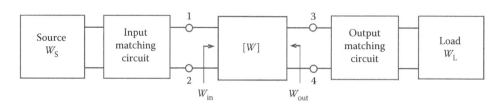

FIGURE 2.14
Block schematic of single-stage power amplifier.

the two-port active device is characterized by a system of the immittance W-parameters, that is, any system of impedance Z-parameters, hybrid H-parameters, or admittance Y-parameters [50,51]. The input and output matching circuits transform the source and load immittances W_S and W_L into specified values between points 1-2 and 3-4, respectively, by means of which the optimal design operation mode of the power amplifier is realized.

The operating power gain G_P, which represents the ratio of power dissipated in the active load $\mathrm{Re}\,W_L$ to the power delivered to the input port of the active device, can be expressed in terms of the immittance W-parameters as

$$G_P = \frac{|W_{21}|^2 \,\mathrm{Re}\,W_L}{|W_{22} + W_L|^2 \,\mathrm{Re}\,W_{in}} \tag{2.70}$$

where

$$W_{in} = W_{11} - \frac{W_{12}\,W_{21}}{W_{22} + W_L} \tag{2.71}$$

is the input immittance and W_{ij} ($i, j = 1, 2$) are the immittance two-port parameters of the active device equivalent circuit.

The transducer power gain G_T, which represents the ratio of power dissipated in the active load $\mathrm{Re}\,W_L$ to the power available from the source, can be expressed in terms of the immittance W-parameters as

$$G_T = \frac{4|W_{21}|^2 \,\mathrm{Re}\,W_S \;\mathrm{Re}\,W_L}{|(W_{11} + W_S)(W_{22} + W_L) - W_{12}W_{21}|^2} \tag{2.72}$$

The operating power gain G_P does not depend on the source parameters and characterizes only the effectiveness of the power delivery from the input port of the active device to the load. This power gain helps to evaluate the gain property of a multistage amplifier when the overall operating power gain $G_{P(total)}$ is equal to the product of each stage G_P. The transducer power gain G_T includes an assumption of conjugate matching of the load and the source.

The bipolar transistor simplified small-signal π-hybrid equivalent circuit shown in Figure 2.15 provides an example for a conjugate-matched bipolar power amplifier. The

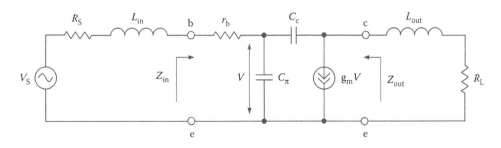

FIGURE 2.15
Simplified equivalent circuit of matched bipolar power amplifier.

impedance Z-parameters of the equivalent circuit of the bipolar transistor in a common-emitter configuration can be written as

$$Z_{11} = r_b + \frac{1}{g_m + j\omega C_\pi} \qquad Z_{12} = \frac{1}{g_m + j\omega C_\pi}$$

$$Z_{21} = -\frac{1}{j\omega C_c}\frac{g_m - j\omega C_c}{g_m + j\omega C_\pi} \qquad Z_{22} = \left(1 + \frac{C_\pi}{C_c}\right)\frac{1}{g_m + j\omega C_\pi} \tag{2.73}$$

where g_m is the transconductance, r_b is the series base resistance, C_π is the base–emitter capacitance including both diffusion and junction components, and C_c is the feedback collector capacitance.

By setting the device feedback impedance Z_{12} to zero and complex conjugate-matching conditions at the input of $R_S = \mathrm{Re}Z_{in}$ and $L_{in} = -\mathrm{Im}Z_{in}/\omega$ and at the output of $R_L = \mathrm{Re}Z_{out}$ and $L_{out} = -\mathrm{Im}Z_{out}/\omega$, the small-signal transducer power gain G_T can be calculated from

$$G_T = \left(\frac{f_T}{f}\right)^2 \frac{1}{8\pi f_T r_b C_c} \tag{2.74}$$

where $f_T = g_m/2\pi C_\pi$ is the device transition frequency.

Figure 2.16 shows the simplified circuit schematic for a conjugate-matched FET power amplifier. The admittance Y-parameters of the small-signal equivalent circuit of any FET device in a common-source configuration can be written as

$$Y_{11} = \frac{j\omega C_{gs}}{1 + j\omega C_{gs}R_{gs}} + j\omega C_{gd} \qquad Y_{12} = -j\omega C_{gd}$$

$$Y_{21} = \frac{g_m}{1 + j\omega C_{gs}R_{gs}} - j\omega C_{gd} \qquad Y_{22} = \frac{1}{R_{ds}} + j\omega(C_{ds} + C_{gd}) \tag{2.75}$$

where g_m is the transconductance, R_{gs} is the gate-source resistance, C_{gs} is the gate-source capacitance, C_{gd} is the feedback gate-drain capacitance, C_{ds} is the drain-source capacitance, and R_{ds} is the differential drain-source resistance.

Since the value of the gate-drain capacitance C_{gd} is usually relatively small, the effect of the feedback admittance Y_{12} can be neglected in a simplified case. Then, it is necessary

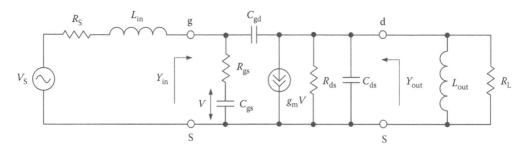

FIGURE 2.16
Simplified equivalent circuit of matched FET power amplifier.

to set $R_S = R_{gs}$ and $L_{in} = 1/\omega^2 C_{gs}$ for input matching, while $R_L = R_{ds}$ and $L_{out} = 1/\omega^2 C_{ds}$ for output matching. Hence, the small-signal transducer power gain G_T can be approximately calculated from

$$G_T(C_{gd} = 0) = MAG = \left(\frac{f_T}{f}\right)^2 \frac{R_{ds}}{4R_{gs}} \tag{2.76}$$

where $f_T = g_m/2\pi C_{gs}$ is the device transition frequency and *MAG* is the maximum available gain representing a theoretical limit on the power gain that can be achieved under complex conjugate-matching conditions.

From Equations 2.74 and 2.76, it follows that the small-signal power gain of a conjugate-matched power amplifier for any type of the active device drops off as $1/f^2$ or 6 dB per octave. Therefore, $G_T(f)$ can be readily predicted at a certain frequency f, if a power gain is known at the transition frequency f_T, by

$$G_T(f) = G_T(f_T)\left(\frac{f_T}{f}\right)^2 \tag{2.77}$$

It should be noted that previous analysis is based upon the linear small-signal consideration when generally nonlinear device current source as a function of both input and output voltages can be characterized by the linear transconductance g_m as a function of the input voltage and the output differential resistance R_{ds} as a function of the output voltage. This is a result of a Taylor-series expansion of the output current as a function of the input and output voltages with maintaining only the dc and linear components. Such an approach helps to understand and derive the maximum achievable power-amplifier parameters in a linear approximation. In this case, an active device is operated in a Class-A mode when one-half of the dc power is dissipated in the device, while the other half is transformed to the fundamental-frequency output power flowing into the load, resulting in a maximum ideal collector efficiency of 50%. The device output resistance R_{out} remains constant and can be calculated as a ratio of the dc supply voltage to the dc current flowing through the active device. In a common case, for a complex conjugate-matching procedure, the device output immittance under large-signal consideration should be calculated using a Fourier-series analysis of the output current and voltage fundamental components. This means that, unlike a linear Class-A mode, an active device is operated in a device linear region for only part of the entire period, and its output resistance is defined as a ratio of the fundamental-frequency output voltage to the fundamental-frequency output current. This is not a physical resistance resulting in a power loss inside the device, but an equivalent resistance required to use for a conjugate matching procedure. In this case, the complex conjugate matching concept is valid when it is necessary first to compensate for the reactive part of the device output impedance and second to provide a proper load resistance resulting in a maximum power gain for a given supply voltage and required output power delivered to the load. Note that this is not a maximum available small-signal power gain, which can be achieved in a linear operation mode, but a maximum achievable large-signal power gain that can be achieved for a particular operation mode with a certain conduction angle. Of course, the maximum large-signal power gain is smaller than the small-signal power gain for the same input power, since the output power in a nonlinear operation mode also includes the powers at the harmonic components of the fundamental frequency.

Therefore, it makes more practical sense not to separately introduce the concepts of the gain match with respect to the linear power amplifiers and the power match in nonlinear power amplifier circuits since the maximum large-signal power gain, being a function of the angle θ, corresponds to the maximum fundamental-frequency output power delivered to the load due to large-signal conjugate output matching. It is very important to provide a conjugate matching at both input and output device ports to achieve maximum power gain in a large-signal mode. In a Class-A mode, the maximum small-signal power gain ideally remains constant regardless of the output power level.

The transistor characterization in a large-signal mode can be done based on equivalent quasi-harmonic nonlinear approximation under the condition of sinusoidal port voltages [52]. In this case, the large-signal impedances are generally determined in the following manner. The designer tunes the load network (often by trial and error) to maximize the output power to the required level using a particular transistor at a specified frequency and supply voltage. Then, the transistor is removed from the circuit and the impedance seen by the collector is measured at the carrier frequency. The complex-conjugate of the measured impedance then represents the equivalent large-signal output impedance of the transistor at that frequency, supply voltage, and output power. Similar design process is used to measure the input impedance of the transistor in order to maximize power-added efficiency of the power amplifier.

In early radiofrequency vacuum-tube transmitters, it was observed that the tubes and associated circuits may have damped or undamped oscillations depending upon the circuit losses, the feedback coupling, the grid and anode potentials, and the reactance or tuning of the parasitic circuits [53,54]. Various parasitic oscillator circuits such as the tuned-grid–tuned-anode circuit with capacitive feedback, Hartley, Colpitts, or Meissner oscillators can be realized at high frequencies, which potentially can be eliminated by adding a small resistor close to the grid or anode connections of the tubes for damping the circuits. Inductively coupled rather than capacitively coupled input and output circuits should be used wherever possible.

According to the immittance approach applied to the stability analysis of the active non-reciprocal two-port network, it is necessary and sufficient for its unconditional stability if the following system of equations can be satisfied for the given active device:

$$\text{Re}\,[W_S(\omega) + W_{in}(\omega)] > 0 \tag{2.78}$$

$$\text{Im}\,[W_S(\omega) + W_{in}(\omega)] = 0 \tag{2.79}$$

or

$$\text{Re}\,[W_L(\omega) + W_{out}(\omega)] > 0 \tag{2.80}$$

$$\text{Im}\,[W_L(\omega) + W_{out}(\omega)] = 0 \tag{2.81}$$

where $\text{Re}W_S$ and $\text{Re}W_L$ are considered to be greater than zero [55,56]. The active two-port network can be treated as unstable or potentially unstable in the case of the opposite signs in Equations 2.78 and 2.80.

Analysis of Equation 2.78 or Equation 2.80 on extremum results in a special relationship between the device immittance parameters called the device stability factor

$$K = \frac{2 \operatorname{Re} W_{11} \operatorname{Re} W_{22} - \operatorname{Re}(W_{12} W_{21})}{|W_{12} W_{21}|} \tag{2.82}$$

which shows a stability margin indicating how far from zero value are the real parts in Equations 2.78 and 2.80 if they are positive [56]. An active device is unconditionally stable if $K \geq 1$ and potentially unstable if $K < 1$.

When the active device is potentially unstable, an improvement of the power amplifier stability can be provided with the appropriate choice of the source and load immittances W_S and W_L. In this case, the circuit stability factor K_T is defined in the same way as the device stability factor K, taking into account $\operatorname{Re} W_S$ and $\operatorname{Re} W_L$ along with the device W-parameters, and written as

$$K_T = \frac{2 \operatorname{Re}(W_{11} + W_S) \operatorname{Re}(W_{22} + W_L) - \operatorname{Re}(W_{12} W_{21})}{|W_{12} W_{21}|} \tag{2.83}$$

If the circuit stability factor $K_T \geq 1$, the power amplifier is unconditionally stable. However, the power amplifier becomes potentially unstable if $K_T < 1$. The value of $K_T = 1$ corresponds to the border of the circuit unconditional stability. The values of the circuit stability factor K_T and device stability factor K become equal if $\operatorname{Re} W_S = \operatorname{Re} W_L = 0$.

For the active device stability factor $K > 1$, the operating power gain G_P has to be maximized. By analyzing Equation 2.70 on extremum, it is possible to find optimum values $\operatorname{Re} W_L^o$ and $\operatorname{Im} W_L^o$ when the operating power gain G_P is maximal [57,58]. As a result

$$G_{Pmax} = \left| \frac{W_{21}}{W_{12}} \right| \bigg/ \left(K + \sqrt{K^2 - 1} \right) \tag{2.84}$$

The power amplifier with an unconditionally stable active device provides a maximum power gain operation only if the input and output of the active device are conjugate-matched with the source and load impedances, respectively. For the lossless input matching circuit when the power available at the source is equal to the power delivered to the input port of the active device, that is, $P_S = P_{in}$, the maximum operating power gain is equal to the maximum transducer power gain, that is, $G_{Pmax} = G_{Tmax}$.

The domains of the device's potential instability include the operating frequency ranges where the active device stability factor is equal to $K < 1$. Within the bandwidth of such a frequency domain, parasitic oscillations can occur, defined by internal positive feedback and operating conditions of the active device. The instabilities may not be self-sustaining, induced by the RF drive power but remaining on its removal. One of the most serious cases of the power amplifier instability can occur when there is a variation of the load impedance. Under these conditions, the transistor may be destroyed almost instantaneously. However, even it is not destroyed, the instability can result in an increased level of the spurious emissions in the output spectrum of the power amplifier tremendously. Generally, the following classification for linear instabilities can be made [59]:

- Low-frequency oscillations produced by thermal feedback effects
- Oscillations due to internal feedback

- Negative resistance or conductance-induced instabilities due to transit-time effects, avalanche multiplication, etc.
- Oscillations due to external feedback as a result of insufficient decoupling of the dc supply, etc.

Therefore, it is very important to determine the effect of the device feedback parameters on the origin of the parasitic self-oscillations and to establish possible circuit configurations of the parasitic oscillators. Based on the simplified bipolar equivalent circuit shown in Figure 2.15, the device stability factor can be expressed through the parameters of the transistor equivalent circuit as

$$K = 2r_b g_m \frac{1 + \dfrac{g_m}{\omega_T C_c}}{\sqrt{1 + \left(\dfrac{g_m}{\omega C_c}\right)^2}} \tag{2.85}$$

where $\omega_T = 2\pi f_T$ [13,39].

At very low frequencies, the bipolar transistors are potentially stable and the fact that $K \to 0$ when $f \to 0$ in Equation 2.85 can be explained by simplifying the bipolar equivalent circuit. In practice, at low frequencies, it is necessary to take into account the dynamic base–emitter resistance r_π and Early collector–emitter resistance r_{ce}, the presence of which substantially increases the value of the device stability factor. This gives only one unstable frequency domain with $K < 1$ and low-boundary frequency f_{p1}. However, an additional region of possible low-frequency oscillations can occur due to thermal feedback where the collector junction temperature becomes frequently dependent, and the common-base configuration is especially affected by this [60].

Equating the device stability factor K with unity allows us to determine the high-boundary frequency of a frequency domain of the bipolar transistor potential instability as

$$f_{p2} = \frac{g_m}{2\pi C_c} \Bigg/ \sqrt{(2r_b g_m)^2 \left(1 + \frac{g_m}{\omega_T C_c}\right)^2 - 1} \tag{2.86}$$

When $r_b g_m > 1$ and $g_m \gg \omega_T C_c$, Equation 2.86 is simplified to

$$f_{p2} \approx \frac{1}{4\pi\, r_b C_\pi} \tag{2.87}$$

At higher frequencies, a presence of the parasitic reactive intrinsic transistor parameters and package parasitics can be of great importance in view of power amplifier stability. The parasitic series emitter lead inductance L_e shown in Figure 2.17 has a major effect on the device stability factor. The presence of L_e leads to the appearance of the second frequency domain of potential instability at higher frequencies. The circuit analysis shows that the second frequency domain of potential instability can be realized only under the particular ratios between the normalized parameters $\omega_T L_e / r_b$ and $\omega_T r_b C_c$ [13,39]. For example, the second domain does not occur for any values of L_e when $\omega_T r_b C_c \geq 0.25$.

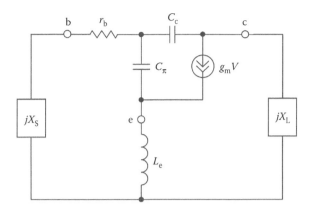

FIGURE 2.17
Simplified bipolar π-hybrid equivalent circuit with emitter lead inductance.

An appearance of the second frequency domain of the device potential instability is the result of the corresponding changes in the device feedback phase conditions and takes place only under a simultaneous effect of the collector capacitance C_c and emitter lead inductance L_e. If the effect of one of these factors is lacking, the active device is characterized by only the first domain of its potential instability.

Figure 2.18 shows the potentially realizable equivalent circuits of the parasitic oscillators. If the value of a series-emitter inductance L_e is negligible, the parasitic oscillations can occur only when the values of the source and load reactances are positive, that is $\mathrm{Im}Z_S = jX_S > 0$ and $\mathrm{Im}Z_L = jX_L > 0$. In this case, the parasitic oscillator shown in Figure 2.18a

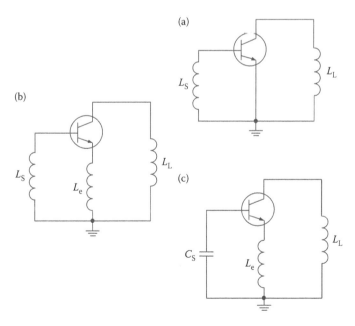

FIGURE 2.18
Equivalent circuits of parasitic bipolar oscillators.

represents the inductive three-point circuit, where the inductive elements L_S and L_L in combination with the collector capacitance C_c form a Hartley oscillator. From a practical point of view, the more the value of the collector dc-feed inductance exceeds the value of the base-bias inductance, the more likely low-frequency parasitic oscillators can be created. It was observed that a very low inductance, even a short between the emitter and the base, can produce very strong and dangerous oscillations, which may easily destroy a transistor [59]. Therefore, it is recommended to increase a value of the base choke inductance and to decrease a value of the collector choke inductance.

The presence of L_e leads to narrowing of the first frequency domain of the potential instability, which is limited to the high-boundary frequency f_{p2}, and can contribute to appearance of the second frequency domain of the potential instability at higher frequencies. The parasitic oscillator that corresponds to the first frequency domain of the device potential instability can be realized only if the source and load reactances are inductive, that is, $\text{Im}Z_S = jX_S > 0$ and $\text{Im}Z_L = jX_L > 0$, with the equivalent circuit of such a parasitic oscillator shown in Figure 2.18b. The parasitic oscillator corresponding to the second frequency domain of the device potential instability can be realized only if the source reactance is capacitive and the load reactance is inductive, that is, $\text{Im}Z_S = -jX_S < 0$ and $\text{Im}Z_L = jX_L > 0$, with the equivalent circuit shown in Figure 2.18c. The series-emitter inductance L_e is an element of fundamental importance for the parasitic oscillator that corresponds to the second frequency domain of the device potential instability. It changes the circuit phase conditions so it becomes possible to establish the oscillation phase-balance condition at high frequencies. However, if it is possible to eliminate the parasitic oscillations at high frequencies by other means, increasing of L_e will result to narrowing of a low-frequency domain of potential instability, thus making the power amplifier potentially more stable, although at the expense of reduced power gain.

Similar analysis of the MOSFET power amplifier also shows two frequency domains of MOSFET potential instability due to the internal feedback gate-drain capacitance C_{gd} and series source inductance L_s [13]. Because of the very high gate-leakage resistance, the value of the low-boundary frequency f_{p1} is sufficiently small. For usually available conditions for power MOSFET devices when $g_m R_{ds} = 10 \div 30$ and $C_{gd}/C_{gs} = 0.1 \div 0.2$, the high boundary frequency f_{p2} can approximately be calculated from

$$f_{p2} \approx \frac{1}{4\pi\, R_{gs} C_{gs}} \tag{2.88}$$

It should be noted that power MOSFET devices have a substantially higher value of $g_m R_{ds}$ at small values of the drain current than at its high values. Consequently, for small drain current, the MOSFET device is characterized by a wider domain of potential instability. This domain is significantly wider than the same first domain of the potential instability of the bipolar transistor. The series source inductance L_s contributes to the appearance of the second frequency domain of the device potential instability. The potentially realizable equivalent circuits of the MOSFET parasitic oscillators are the same as for the bipolar transistor, as shown in Figure 2.18 [13].

Thus, to prevent the parasitic oscillations and to provide a stable operation mode of any power amplifier, it is necessary to take into consideration the following common requirements:

- Use an active device with stability factor $K > 1$

- If it is impossible to choose an active device with $K > 1$, it is necessary to provide the circuit stability factor $K_T > 1$ by the appropriate choice of the real parts of the source and load immittances
- Disrupt the equivalent circuits of the possible parasitic oscillators
- Choose proper reactive parameters of the matching circuit elements adjacent to the input and output ports of the active device, which are necessary to avoid the self-oscillation conditions

Generally, the parasitic oscillations can arise at any frequency within the potential instability domains for particular values of the source and load immittances W_S and W_L. The frequency dependences of W_S and W_L are very complicated and very often cannot be predicted exactly, especially in multistage power amplifiers. Therefore, it is very difficult to propose a unified approach to provide a stable operation mode of the power amplifiers with different circuit configurations and operation frequencies. In practice, the parasitic oscillations can arise close to the operating frequencies due to the internal positive feedback inside the transistor and at the frequencies sufficiently far from the operating frequencies due to the external positive feedback created by the surface-mounted elements. As a result, the stability analysis of the power amplifier must include the methods to prevent the parasitic oscillations in different frequency ranges.

It should be noted that expressions in Equations 2.78 through 2.84 are given by using the device immittance Z- or Y-parameters that allow the power gain and stability to be calculated using the parameters of the device equivalent circuit and to physically understand the corresponding effect of each circuit parameter, but not through the scattering S-parameters, which are very convenient during the measurement procedure required for device modeling. Moreover, by using modern simulation tools, there is no need to even draw stability circles on a Smith chart or analyze the stability factor across the wide frequency range since the K-factor is just a derivation from the basic stability conditions and usually is a function of linear parameters, which can only reveal linear instabilities. Besides, it is difficult to predict unconditional stability for a multistage power amplifier because parasitic oscillations can be caused by the interstage circuits. In this case, the easiest and most effective way to provide stable operation of the multistage power amplifier (or single-stage power amplifier) is to simulate the real part of the device input impedance $Z_{in} = V_{in}/I_{in}$ at the input terminal of each transistor as a ratio between the input voltage and current by placing a voltage node and a current meter, as shown in Figure 2.19a. If $\mathrm{Re}Z_{in} < 0$, then either a low-value series resistor must be added to the device base terminal as a part of the input matching circuit or a load-network configuration can be properly chosen to provide the resulting positive value of $\mathrm{Re}Z_{in}$. In this case, not only linear instabilities with small-signal soft startup oscillation conditions but also nonlinear instabilities with large-signal hard startup oscillation conditions or parametric oscillations can be identified around the operating region. Figure 2.19b shows the parallel RC stabilizing circuit with a bypass capacitor C_{bypass} connected in series to the input port of a GaN HEMT device [61]. In this case, using a stabilizing resistor R_{gate} and a low-value gate-bias resistor R_{bias} improves the stability factor considerably at low frequencies without affecting the device performance at higher frequencies.

Figure 2.20 shows the example of a stabilized bipolar VHF (very high frequency) power amplifier configured to operate in a zero-bias Class-C mode. Conductive input and output loading due to resistances R_1 and R_2 eliminate a low-frequency instability domain. The series inductors L_3 and L_4 contribute to higher power gain if the resistance values are too

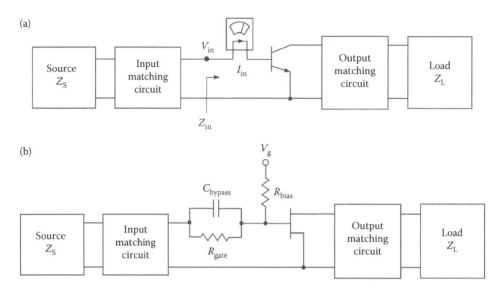

FIGURE 2.19
Single-stage power amplifiers with measured device input impedance.

small, and can compensate for the capacitive input and output device impedances. To provide a negative-bias Class-C mode, the shunt inductor L_2 can be removed. The equivalent circuit of the potential parasitic oscillator at higher frequencies is realized by means of the parasitic reactive parameters of the transistor and external circuitry. The only possible equivalent circuit of such a parasitic oscillator at these frequencies is shown in Figure 2.18c. It can only be realized if the series-emitter lead inductance is present. Consequently, the electrical length of the emitter lead should be reduced as much as possible, or, alternatively, the appropriate reactive immittances at the input and output transistor ports are provided. For example, it is possible to avoid the parasitic oscillations at these frequencies if the inductive immittance is provided at the input of the transistor and capacitive

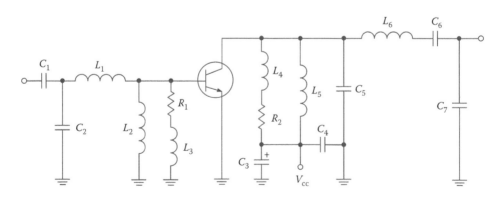

FIGURE 2.20
Stabilized bipolar Class-C VHF power amplifier.

reactance is provided at the output of the transistor. This is realized by an input series inductance L_1 and an output shunt capacitance C_5.

2.5 Push–Pull and Balanced Power Amplifiers

Generally, if it is necessary to increase an overall output power of the power amplifier, several active devices can be used in parallel or push–pull configurations. In a parallel configuration, the active devices are not isolated from each other, which requires a very good circuit symmetry, and output impedance becomes too small in the case of high output power. The latter drawback can be eliminated in a push–pull configuration, which provides increased values of the input and output impedances. In this case, for the same output power level, the input impedance Z_{in} and output impedance Z_{out} are approximately four times as high as that of in a parallel connection of the active devices since a push–pull arrangement is essentially a series connection. At the same time, the loaded quality factors of the input and output matching circuits remain unchanged because both the real and reactive parts of these impedances are increased by the factor of four. Very good circuit symmetry can be provided using balanced active devices with common emitters (or sources) in a single package. The basic concept of a push–pull operation can be analyzed by using the corresponding circuit schematic shown in Figure 2.21 [62].

2.5.1 Basic Push–Pull Configuration

It is most convenient to consider an ideal Class-B operation, which means that each transistor conducts exactly half a cycle (equal to 180°) with zero quiescent current. Let us also assume that the number of turns of both primary and secondary windings of the output transformer T_2 is equal when $n_1 = n_2$ and the collector current of each transistor can be represented in the following half-sinusoidal form:

For the first transistor

$$i_{c1} = \begin{cases} +I_c \sin \omega t & 0 \le \omega t < \pi \\ 0 & \pi \le \omega t < 2\pi \end{cases} \tag{2.89}$$

For the second transistor

$$i_{c2} = \begin{cases} 0 & 0 \le \omega t < \pi \\ -I_c \sin \omega t & \pi \le \omega t < 2\pi \end{cases} \tag{2.90}$$

where I_c is the output current amplitude.

Being transformed through the output transformer T_2 with the appropriate phase conditions, the total current flowing through the load R_L is obtained as

$$i_R(\omega t) = i_{c1}(\omega t) - i_{c2}(\omega t) = I_c \sin \omega t \tag{2.91}$$

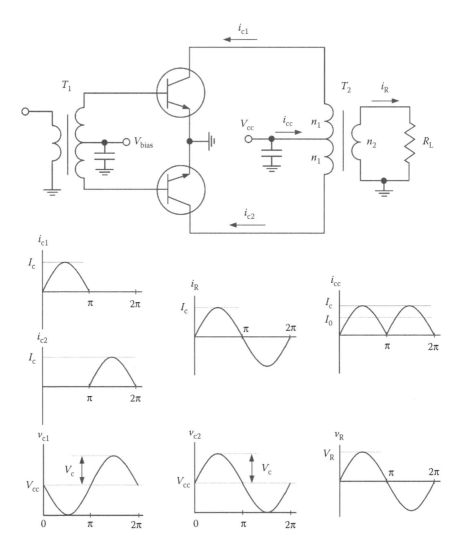

FIGURE 2.21
Basic concept of push–pull operation.

The current flowing into the center tap of the primary windings of the output transformer T_2 is the sum of the collector currents, resulting in

$$i_{cc}(\omega t) = i_{c1}(\omega t) + i_{c2}(\omega t) = I_c |\sin \omega t| \tag{2.92}$$

Ideally, the even-order harmonics being in phase are canceled out and should not appear at the load. In practice, a level of the second-harmonic component of 30–40 dB below the fundamental is allowable. However, it is necessary to connect a bypass capacitor to the center tap of the primary winding to exclude power losses due to even-order harmonics. The current $i_R(\omega t)$ produces the load voltage $v_R(\omega t)$ onto the load R_L as

$$v_R(\omega t) = I_c R_L \sin(\omega t) = V_R \sin(\omega t) \tag{2.93}$$

where V_R is the load voltage amplitude.

The total dc collector current is defined as the average value of $i_{cc}(\omega t)$, which yields

$$I_0 = \frac{1}{2\pi} \int_0^{2\pi} i_{cc}(\omega t)d\omega t = \frac{2}{\pi} I_c \tag{2.94}$$

The total dc power P_0 and fundamental-frequency output power P_{out} for the ideal case of zero saturation voltage of both transistors when $V_c = V_{cc}$ and taking into account that $V_R = V_c$ for equal turns of windings when $n_1 = n_2$ are calculated respectively from

$$P_0 = \frac{2}{\pi} I_c V_{cc} \tag{2.95}$$

$$P_{out} = \frac{I_c V_{cc}}{2} \tag{2.96}$$

Consequently, the maximum theoretical collector efficiency that can be achieved in a push–pull Class-B operation is equal to

$$\eta = \frac{P_{out}}{P_0} = \frac{\pi}{4} \cong 78.5\% \tag{2.97}$$

In a balanced circuit, identical sides carry 90° quadrature or 180° out-of-phase signals of equal amplitude. In the latter case, if perfect balance is maintained on both sides of the circuit, the difference between signal amplitudes becomes equal to zero in each midpoint of the circuit, as shown in Figure 2.22. This effect is called the *virtual grounding*, and this midpoint line is referred to as the *virtual ground*. The virtual ground, being actually inside the balanced transistor package having two identical transistor chips, reduces a common-mode inductance and results in better stability and usually higher power gain [63].

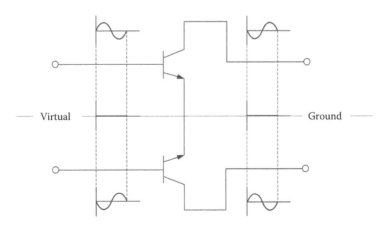

Virtual   Ground

FIGURE 2.22
Basic concept of balanced transistor.

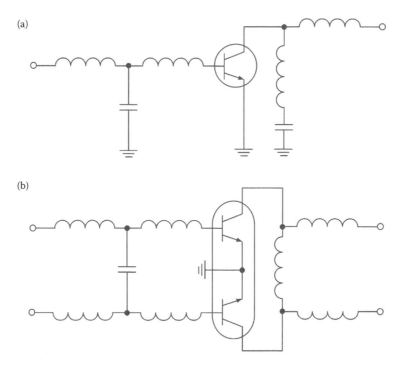

FIGURE 2.23
Matching technique for single-ended (a) and balanced (b) transistors.

When using a balanced transistor, new possibilities for both internal and external imped-ance matching procedure emerge. For instance, for a push–pull operation mode of two single-ended transistors, it is necessary to provide reliable grounding for input and output matching circuits for each device, as shown in Figure 2.23a. Using the balanced transistors significantly simplifies the matching circuit topologies, with the series inductors and shunt capacitors connected between amplifying paths, as shown in Figure 2.23b, where the dc-blocking capacitors are not needed [64]. Such an approach can provide additional design flexibility when, for example, a two-stage monolithic push–pull X-band GaAs MESFET power amplifier can be optimized for either small-signal, high-gain operation, or for large-signal power saturated operation by changing the lengths of the bondwires that form the shunt inductance at the drain circuits of each stage [65].

2.5.2 Baluns

For a push–pull operation of the power amplifier with a balanced transistor, it is also nec-essary to provide the unbalanced-to-balanced transformation referenced to the ground both at the input and at the output of the power amplifier. The most suitable approach to solve this problem in the best possible manner at high frequencies and microwaves is to use the transmission-line baluns (balanced-to-unbalanced transmission-line transform-ers). The first transmission-line balun for coupling a single coaxial line having a quarter wavelength at the center bandwidth frequency to a push–pull coaxial line (or a pair of coaxial lines), which maintains perfect balance over a wide frequency range was intro-duced and described by Lindenblad in 1939 [66,67].

Figure 2.24a shows the basic structure and equivalent circuit of a simple coaxial balun, where port A is the unbalanced port and port B is the balanced port. To be a perfect balun when a balanced load is connected to port terminals B, the shield return current $I_2 - I_3$ would equal to I_1, which would ideally represent the delayed input current, and the output terminal voltages would be equal and opposite with respect to the ground. In this case, if the characteristic impedance Z_0 of the coaxial transmission line is equal to the input impedance at the unbalanced end of the transformer, the total impedance from both outputs at the balanced end of the transformer will be equal to the input impedance. Hence, such a transmission-line transformer can be used as a 1:1 balun. The equivalent circuit for this coaxial balun demonstrates the basic drawback of this balun, when its inner conductor is shielded from ground having practically infinite impedance to ground, whereas the outer shield does have a finite impedance to ground when a balun is placed above a printed circuit board. The presence of the lower ground plane creates a shunt short-circuit stub with the characteristic impedance Z_1 across one of the load and this converts the high-pass balun structure into a bandpass one. As a result, this stub has a dramatic effect on the balun performance, with the bandwidth being reduced to about an octave based on phase imbalance. One of the solutions is simply to raise the transmission line as high as possible above the printed circuit board and make both conductors symmetrical with respect to lower ground plane. The other solution is to attach a compensating stub to the other load, as shown in Figure 2.24b, which results in perfect amplitude and phase balance above the low-frequency cutoff region providing less than 1-dB insertion loss achieved from 5 MHz to 2.5 GHz [68].

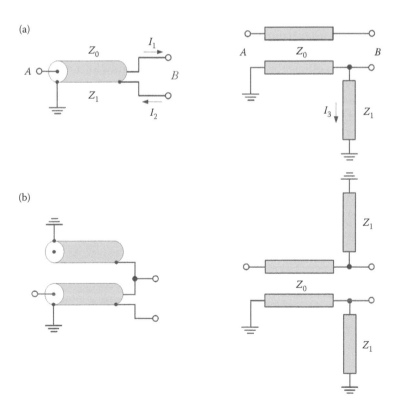

FIGURE 2.24
Basic structures and equivalent circuits of coaxial baluns.

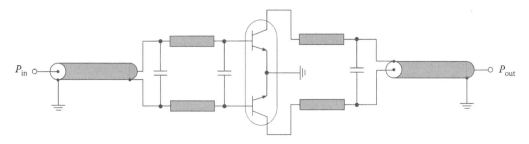

FIGURE 2.25
Push–pull bipolar power amplifier with input and output baluns.

Figure 2.25 shows the basic structure of a push–pull power amplifier with a balanced bipolar transistor including the input and output matching circuits. To extend the operating frequency range to lower frequencies, the outside of the coaxial line of the balun can be loaded with a low-loss ferrite core, which acts as a choke to force equal and opposite currents in the inner and outer conductor and isolate the 180° output from the input ground terminal by creating a high and lossy impedance for Z_1. In this case, the measured S-parameters of the back-to-back connected baluns showed an insertion loss of about 0.5–0.6 dB and better than 20-dB return loss over 50–1000 MHz [69]. As a result, four broadband GaN HEMT power amplifier units were combined using such a low-loss coaxial balun that transforms an unbalanced 50-Ω load into two 25-Ω impedances that are 180° out of phase and each of the 25-Ω end is driven by a pair of the power amplifier units connected in parallel. A similar balun is used at the input to create the 180° out-of-phase input to the two pairs of the power amplifier units, resulting in over 100-W output power and higher than 60% drain efficiency across the frequency bandwidth of 100–1000 MHz. The lower cutoff bandwidth frequency can be provided to cover down to 10 MHz by adding lower-frequency ferrites, but it may affect performance at high bandwidth frequencies. Generally, since ferrite has limited bandwidth, it is possible to use several ferrite cores to broaden the frequency bandwidth. For example, by using a low-frequency ferrite core covering 1–10 MHz, a medium-frequency ferrite core covering 10–200 MHz, and a high-frequency ferrite core covering high frequencies above 200 MHz, the balun can cover 1 MHz to 2.5 GHz with the loss of 0.25 dB at low frequency and 1.3 dB at 2.5 GHz [70].

Figure 2.26 shows the load-network arrangement with two coaxial baluns combined to provide a push–pull operation of the balanced power amplifier by creating a

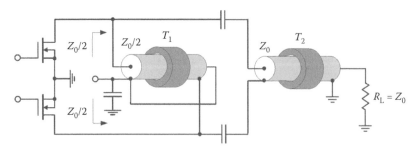

FIGURE 2.26
Load-network arrangement with two coaxial baluns for push–pull operation.

balanced-to-unbalanced impedance transformation with higher spectral purity. Ideally, the 180° out-of-phase currents from both active devices will have pure half-sinusoidal waveforms in a Class-B operation mode, which contain (according to the Fourier-series expansion) only fundamental and even harmonic components. This implies a 180° shift between the fundamental components from both active devices and in-phase condition for remaining even harmonic components. In this case, the transformer T_1 representing a phase inverter is operated as a filter for even harmonics because currents flow through its inner and outer conductors in opposite directions. For each fundamental component flowing through its inner and outer conductors in the same directions, it works as an RF choke, the impedance of which depends on the ferrite core permeability, and power supply can be conveniently connected to its middle point bypassed by a large-value capacitor. Consequently, since the transformer T_2 represents a 1:1 coaxial balun, in order to provide maximum power delivery to the load R_L, the output equivalent impedance of each active device should be twice as small, being equal to the characteristic impedance of the transformer T_1. As an example, four broadband MOSFET power amplifier units, each having such a load-network configuration with two coaxial baluns, were combined in a single module achieving an output power of 100 W and over 40% drain efficiency across the frequency range of 2–76 MHz with the second- and third-harmonic suppression better than 28 dB [71].

The miniaturized compact input unbalanced-to-balanced transformer shown in Figure 2.27 covers the frequency bandwidth up to an octave with well-defined rejection-mode impedances [72]. To avoid the parasitic capacitance between the outer conductor and the ground, the coaxial semirigid transformer T_1 is mounted atop microstrip shorted stub l_1 and soldered continuously along its length. The electrical length of this stub is usually chosen from the condition of $\theta \leq \pi/2$ on the high bandwidth frequency depending on the matching requirements. To maintain circuit symmetry on the balanced side of the transformer network, another semirigid coaxial section T_2 with unconnected center conductor is soldered continuously along microstrip shorted stub l_2. The lengths of T_2 and l_2 are equal to the lengths of T_1 and l_1, respectively. Because the input short-circuited microstrip stubs provide inductive impedances, the two series capacitors C_1 and C_2 of the

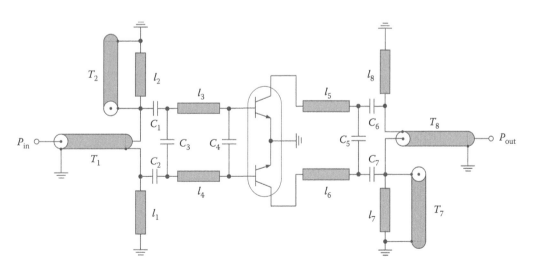

FIGURE 2.27
Push–pull power amplifier with compact balanced-to-unbalanced transformers.

same value are used for matching purposes, thereby forming the first high-pass matching section and providing dc blocking at the same time. The practical circuit realization of the output matching circuit and balanced-to-unbalanced transformer can be the same as for the input matching circuit.

Figure 2.28a shows the basic structure of a compensated coaxial balun, including a series open-circuit line to compensate for the short-circuited line reactance over a wide frequency range, where Z_a and Z_b represent the characteristic impedances of coaxial lines a and b, respectively, and Z_{ab} is the characteristic impedance of the balanced line ab composed of the outer conductors of coaxial lines a and b [73]. In this case, the compensating line provides control of the reactance around center bandwidth frequency. It should be noted that this balun has exactly the same equivalent circuit as the Marchand balun [74]. However, the Roberts balun is easier to implement, and both baluns are inherently bandpass networks.

An equivalent circuit of the compensated balun is shown in Figure 2.28b, from which it follows that it is desirable for broader bandwidth to make Z_{ab} as large as possible relative to R_L, where R_L is a balanced load. The input impedance Z_{in} seen by the series coaxial line is written as

$$Z_{in} = \frac{jR_L Z_{ab} \tan \theta_{ab}}{R_L + jZ_{ab} \tan \theta_{ab}} - jZ_b \cot \theta_b$$

$$= \frac{R_L}{\dfrac{R_L^2}{Z_{ab}^2 \tan^2 \theta_{ab}} + 1} + \frac{jR_L^2 Z_{ab} \tan \theta_{ab}}{R_L^2 + Z_{ab}^2 \tan^2 \theta_{ab}} - jZ_b \cot \theta_b \qquad (2.98)$$

From Equation 2.98, it follows that the resistive term of the input impedance is independent of the compensating reactance Z_b. Let the electrical lengths of line b and ab be equal, making $\theta = \theta_{ab} = \theta_b$, and let the characteristic impedances $Z_{ab} = R_L$ and $Z_a = Z_b$. Then, Equation 2.98 is simplified to

$$Z_{in} = R_L \sin^2 \theta + j \cot \theta \, (R_L \sin^2 \theta - Z_a) \qquad (2.99)$$

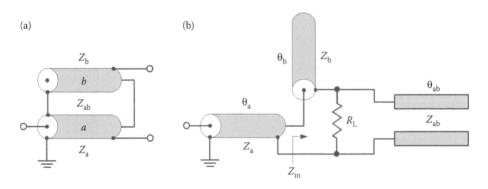

FIGURE 2.28
Structure (a) and equivalent circuit (b) of broadband compensated bandpass balun.

and the input impedance becomes perfectly matched at two widely separated frequencies (symmetrically disposed about a center bandwidth frequency corresponding to $\theta = 90°$ when $\cot\theta = 0$) according to

$$\sin^2\theta = \frac{Z_a}{R_L} \tag{2.100}$$

As an example, for matching a 70-Ω balanced load to a 50-Ω unbalanced load when $Z_a = Z_b = 50 \ \Omega$, $Z_{ab} = 50 \ \Omega$, and $\theta_{ab} = \theta_b$, the values of θ for which a perfect match is expected are calculated from Equation 2.100 as equal to 58° and 122°. Since θ is linearly proportional to frequency, the frequency band between points of perfect match has a ratio of 2.1. The mismatch at the worst point in this band (namely, at the center bandwidth frequency) corresponds to a *VSWR* of 1.4 [73]. Such a compensated balun utilizing a multidielectric structure achieved a broadband performance of up to 3:1 in a simple coplanar configuration with an input return loss better than 15 dB, amplitude imbalance better than 0.35 dB, and phase imbalance better than 1.5° over the frequency range from 15 to 45 GHz [75].

Figure 2.29 shows the circuit schematic of a broadband microstrip balun with normalized parameters using a three-section Wilkinson divider for power splitting followed by Lange coupled-line directional couplers for phase shifting [76]. This planar balun structure can be easily fabricated on alumina substrate using a conventional monolithic process, resulting in good broadband amplitude and phase balance performance. In order to achieve the required tight coupling over 3:1 bandwidth, interdigitated Lange couplers in an unfolded configuration to minimize bondwire connections were employed. The balun fabricated on a 10-mil alumina substrate achieved an amplitude imbalance of ±0.6 dB, average phase imbalance of 7° (with worst case of 11°) and maximum insertion loss of 1.2 dB from 6 to 20 GHz. With a single-section Wilkinson divider and additional short-length correction line on the noninverting arm, the amplitude response within ±1 dB, phase difference of 180 ± 4°, and insertion loss of the order of 1 dB over 6–18 GHz were measured [77].

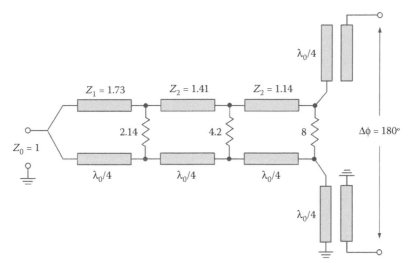

FIGURE 2.29
Circuit schematic of broadband planar balun.

2.5.3 Balanced Power Amplifiers

The balanced amplifier technique using the quadrature 3-dB couplers for power dividing and combining represents an alternative approach to the push–pull operation. Figure 2.30a shows the basic circuit schematic of a balanced amplifier where two power amplifier units of the same performance are arranged between the input splitter and output combiner, each having a 90° phase difference between coupled and through ports. The fourth port of each quadrature coupler must be terminated with a ballast resistor R_{bal}, which is equal to 50 Ω for a 50-Ω system impedance. The input signal is split into two equal-amplitude components by the first 90° hybrid coupler with 0° and 90° paths, then amplified, and finally recombined by the second 90° hybrid coupler. Owing to proper phase shifting, both signals in the load of the isolated port of the combiner are canceled, and the load connected at the combiner output port sees the sum of these two signals. The theory of balanced amplifiers has been given by Kurokawa when the operating frequency bandwidth over 1.2 octaves can be obtained with one-section distributed quarterwave 3-dB directional couplers [78]. For a wide frequency range, the main advantages of the balanced design are the improved input and output impedance matching, gain flatness, intermodulation distortion, and potential design simultaneously for minimum noise figure and good input match. As an example, a four-stage balanced bipolar amplifier achieved a power gain of 20 ± 0.5 dB and an input *VSWR* less than 1.2 across the octave frequency bandwidth from 0.8 to 1.6 GHz [79]. By extending the wide operating frequency range to higher frequencies, an output power of around 23 dBm with gain variations close to 1 dB over 4.5–6.5 GHz and 8–12 GHz was achieved for the balanced microstrip GaAs MESFET amplifiers [80,81].

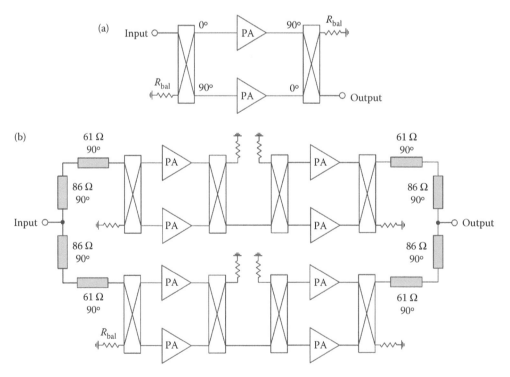

FIGURE 2.30
Schematics of balanced power amplifiers with quadrature hybrid couplers.

If the individual amplifiers with equal performance in the balanced pair are not perfectly matched at certain frequencies, then a signal in the 0° path of the coupler will be reflected from the corresponding amplifier and a signal in the 90° path of the coupler will be similarly reflected from the other amplifier. The reflected signals will again be phased with 90° and 0°, respectively, and the total reflected power as a sum of the in-phase reflected signals flows into the isolated port and dissipates on the ballast resistor R_{bal}. As a result, an input *VSWR* of the quadrature coupler does not depend on the equal load mismatch level. This gives a constant well-defined load to the driver stage, improving amplifier stability and driver power flatness across the operating frequency range. Generally, the stability factor of a balanced stage can be an order of magnitude higher than its single-ended equivalent, depending on the *VSWR* and isolation of the quadrature couplers. If one of the amplifiers fails or is turned off, the balanced configuration provides a gain reduction of −6 dB only. Besides, the balanced structure ideally provides the cancellation in the load of the third-order products such as $2f_1 + f_2$, $2f_2 + f_1$, $3f_1$, $3f_2$, ..., and attenuation by 3 dB of the second-order products such as $f_1 \pm f_2$, $2f_1$, $2f_2$, ... In a microstrip implementation for octave-band power amplifiers, one of the most popular couplers for power dividing and combining is a 3-dB Lange hybrid coupler.

Figure 2.30b shows the circuit schematic of a two-way balanced module consisting of two pairs of cascaded balanced amplifier stages, where the respective output powers are combined using simple two-element power combiners, which are composed of two quarterwave transmission lines with different characteristic impedances [82]. Based on this architecture and GaAs MESFET devices with the gate periphery of 1×1000 μm^2, an output power of 1 W across 7.25–12 GHz was achieved.

Figure 2.31a shows the microstrip single-section topology of a coupled-line directional coupler, which can be used for broadband power dividing and combining. Its electrical properties are described using a concept of two types of excitations for the coupled lines in TEM approximation. In this case, for the even mode, the currents flowing in the strip conductors are equal in amplitude and flow in the same direction. The electric field has

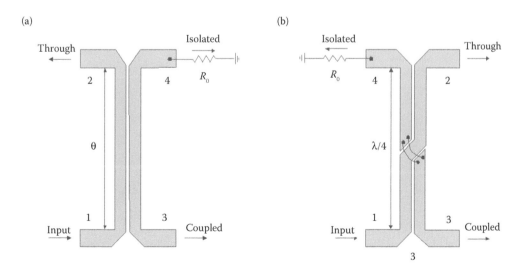

FIGURE 2.31
Coupled-line directional couplers.

even symmetry about the center line, and no current flows between two strip conductors. For the odd mode, the currents flowing in the strip conductors are equal in amplitude, but flow in opposite directions. The electric field lines have an odd symmetry about the center line, and a voltage null exists between these two strip conductors. An arbitrary excitation of the coupled lines can always be treated as a superposition of appropriate amplitudes of even and odd modes.

Therefore, the characteristic impedance for even excitation mode Z_{0e} and the characteristic impedance for the odd excitation mode Z_{0o} characterize the coupled lines. For two coupled equal-strip lines used in a standard system with the characteristic impedance of Z_0, $Z_0^2 = Z_{0e}Z_{0o}$ and

$$Z_{0e} = Z_0\sqrt{\frac{1+C}{1-C}} \tag{2.101}$$

$$Z_{0o} = Z_0\sqrt{\frac{1-C}{1+C}} \tag{2.102}$$

where the midband voltage coupling coefficient C of the directional coupler is defined as

$$C = \frac{Z_{0e} - Z_{0o}}{Z_{0e} + Z_{0o}} \tag{2.103}$$

where $C = 0$ for zero coupling and $C = 1$ for completely superposed transmission lines [83,84].

An analysis in terms of the scattering S-parameters results in a condition of $S_{11} = S_{14} = 0$ for any electrical lengths of the coupled lines, which means that the output port 4 is isolated from the matched input port 1. Varying the coupling between the lines and their widths can change the characteristic impedances Z_{0e} and Z_{0o}. In this case,

$$S_{12} = \frac{\sqrt{1-C^2}}{\sqrt{1-C^2}\cos\theta + j\sin\theta} \tag{2.104}$$

$$S_{13} = \frac{jC\sin\theta}{\sqrt{1-C^2}\cos\theta + j\sin\theta} \tag{2.105}$$

where θ is the electrical length of the coupled-line section.

The voltage-split ratio K is defined as the ratio between voltages at port 2 and port 3 as

$$K = \left|\frac{S_{12}}{S_{13}}\right| = \frac{\sqrt{1-C^2}}{C\sin\theta} \tag{2.106}$$

where K can be controlled by varying the coupling coefficient C and electrical length θ.

For a quarter-wavelength-long coupler when $\theta = 90°$, Equations 2.104 and 2.105 reduce to

$$S_{12} = -j\sqrt{1 - C^2} \qquad (2.107)$$

$$S_{13} = C \qquad (2.108)$$

from which it follows that equal voltage split between the output ports 2 and 3 can be provided with $C = 1/\sqrt{2}$ (or 3 dB).

If it is necessary to provide the output ports 2 and 3 at one side, it is possible to use a construction of a microstrip directional coupler with crossed bondwires, as shown in Figure 2.31b. The strip crossover for a stripline directional coupler can be easily achieved with the three-layer sandwich. The microstrip 3-dB directional coupler fabricated on alumina substrate for idealized zero strip thickness should have the calculated strip spacing of less than 10 μm. Such a narrow value easily explains the great interest to the constructions of the directional couplers with larger spacing.

A popular way to increase the coupling between two edge-coupled microstrip lines is to use several parallel narrow microstrip lines interconnected with each other by the bondwires, as shown in Figure 2.32. For an interdigitated Lange coupler shown in Figure 2.32a, four coupled microstrip lines are used, resulting in a 3-dB coupling over an octave or more bandwidth [85]. In this case, the signal flowing to the input port 1 is distributed between the output ports 2 and 3 with the phase difference of 90°. However, this structure is quite complicated for practical implementation when, for alumina substrate with $\epsilon_r = 9.6$, the dimensions of a 3-dB interdigitated Lange coupler are $W/h = 0.107$ and $S/h = 0.071$, where W is the width of each strip and S is the spacing between adjacent strips.

Figure 2.32b shows the unfolded Lange coupler with four strips of equal length offering the same electrical performance but easier for circuit modeling [86]. The even-mode characteristic impedance Z_{e4} and odd-mode characteristic impedance Z_{o4} of the unfolded Lange coupler with $Z_0^2 = Z_{e4}Z_{o4}$ in terms of the characteristic impedances of a two-conductor line (which is identical to any pair of adjacent lines in the coupler) can be obtained as

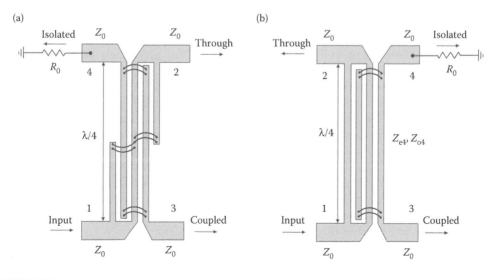

FIGURE 2.32
Lange directional couplers.

$$Z_{e4} = \frac{Z_{0o} + Z_{0e}}{3Z_{0o} + Z_{0e}} Z_{0e} \tag{2.109}$$

$$Z_{o4} = \frac{Z_{0e} + Z_{0o}}{3Z_{0e} + Z_{0o}} Z_{0o} \tag{2.110}$$

where Z_{0e} and Z_{0o} are the even- and odd-mode characteristic impedances of the two-conductor pair [87].

The midband voltage coupling coefficient C is given by

$$C = \frac{Z_{e4} - Z_{o4}}{Z_{e4} + Z_{o4}} = \frac{3(Z_{0e}^2 - Z_{0o}^2)}{3(Z_{0e}^2 + Z_{0o}^2) + 2Z_{0e}Z_{0o}} \tag{2.111}$$

The even- and odd-mode characteristic impedances Z_{0e} and Z_{0o}, as functions of the characteristic impedance Z_0 and coupling coefficient C, are determined by

$$Z_{0e} = Z_0 \sqrt{\frac{1+C}{1-C}} \; \frac{4C - 3 + \sqrt{9 - 8C^2}}{2C} \tag{2.112}$$

$$Z_{0o} = Z_0 \sqrt{\frac{1-C}{1+C}} \; \frac{4C + 3 - \sqrt{9 - 8C^2}}{2C} \tag{2.113}$$

For alumina substrate with $\varepsilon_r = 9.6$, the dimensions of such a 3-dB unfolded Lange coupler are $W/h = 0.112$ and $S/h = 0.08$, where W is the width of each strip and S is the spacing between the strips.

2.6 Transmission-Line Transformers and Combiners

The transmission-line transformers and combiners can provide very wide operating bandwidths up to frequencies of 3 GHz and higher [62,88]. They are widely used in matching networks for antennas and power amplifiers in the HF and VHF bands, and their low losses make them especially useful in high-power circuits [89,90]. Typical structures for transmission-line transformers and combiners consist of parallel wires, coaxial cables, or bifilar twisted wire pairs. In the latter case, the characteristic impedance can be easily determined by the wire diameter, the insulation thickness, and, to some extent, the twisting pitch [91,92]. For coaxial cable transformers with correctly chosen characteristic impedance, the theoretical high-frequency bandwidth limit is reached when the cable length comes in order of a half wavelength, with the overall achievable bandwidth being about a decade. By introducing the low-loss high-permeability ferrites alongside a good–quality, semi-rigid coaxial or symmetrical strip cable, the low-frequency limit can be significantly improved providing bandwidths of several or more decades.

The concept of a broadband impedance transformer consisting of a pair of interconnected transmission lines was first disclosed and described by Guanella [93,94]. Figure 2.33a shows a Guanella transformer system with transmission-line character achieved by an arrangement comprising one pair of cylindrical coils, which are wound in the same sense and are

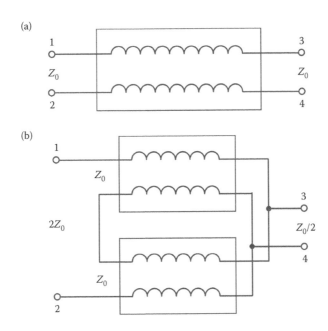

FIGURE 2.33
Schematic configurations of Guanella 1:1 (a) and 4:1 (b) transformers.

spaced a certain distance apart by an intervening dielectric. In this case, one cylindrical coil is located inside the insulating cylinder and the other coil is located on the outside of this cylinder. For the currents flowing through both windings in opposite directions, the corresponding flux in the coil axis is negligibly small. However, for the currents flowing in the same direction through both coils, the latter may be assumed to be connected in parallel, and a coil pair represents a considerable inductance for such currents and acts like a choke coil. With terminal 4 being grounded, such a 1:1 transformer provides matching of the balanced source to unbalanced load. In this case, if terminal 2 is grounded, it simply represents a delay line. In a particular case, when terminals 2 and 3 are grounded, the transformer performs as a phase inverter.

A series–parallel connection of a plurality of coil pairs can produce a match between unequal source and load resistances. Figure 2.33b shows a 4:1 impedance (2:1 voltage) transmission-line transformer, where the two pairs of cylindrical transmission-line coils are connected in series at the input and in parallel at the output. For the characteristic impedance Z_0 of each transmission line, this results in twice higher impedance $2Z_0$ at the input and twice lower impedance $Z_0/2$ at the output. By grounding terminal 4, such a 4:1 impedance transformer provides impedance matching of the balanced source to the unbalanced load. In this case, when terminal 2 is grounded, it performs as a 4:1 *unun* (*un*balanced-to-*un*balanced transformer). With a series–parallel connection of n coil pairs with the characteristic impedance Z_0 each, the input impedance is equal to nZ_0 and the output impedance is equal to Z_0/n. Since voltages that have equal delays through the transmission lines are added, such a technique results in the so-called *equal-delay* transmission-line transformers.

The simplest transmission-line transformer is a quarterwave transmission line, whose characteristic impedance is chosen to give the correct impedance transformation. However, this transformer provides a narrowband performance valid only around frequencies, for which the transmission line is odd multiples of a quarter wavelength. If a ferrite sleeve is added to the transmission line, common-mode currents flowing in both transmission-line inner and outer conductors in phase and in the same direction are suppressed and the load may be balanced and floating above ground [95,96]. If the characteristic impedance of the transmission line is equal to the terminating impedances, the transmission is inherently broadband. If not, there will be a dip in the response at the frequency, at which the transmission line is a quarter-wavelength long.

A coaxial cable transformer, whose physical configuration is shown in Figure 2.34a and equivalent circuit representation with polarity reversing is shown in Figure 2.34b, consists of the coaxial line arranged inside the ferrite core, or wound around the ferrite core. Either end of the load resistor can be grounded, depending on the desired output polarity. The larger the core permeability, the fewer the turns required for a given low-frequency response and the larger the overall bandwidth. Owing to its practical configuration, the coaxial cable transformer takes a position between the lumped and distributed systems. Therefore, at lower frequencies, its equivalent circuit represents a conventional low-frequency transformer, as shown in Figure 2.34c, while at higher frequency, it is a transmission line with the characteristic impedance Z_0, as shown in Figure 2.34d. The advantage of such a transformer is that the parasitic interturn capacitance determines its characteristic impedance, whereas this parasitic capacitance negatively contributes to the transformer frequency performance of the conventional wire-wound transformer with discrete windings by resonating with the leakage inductance that produces a loss peak.

When $R_S = R_L = Z_0$, the transmission line can be considered a transformer with a 1:1 impedance transformation. To avoid any resonant phenomena, especially for complex loads, which can contribute to the significant output power variations, the length l of the

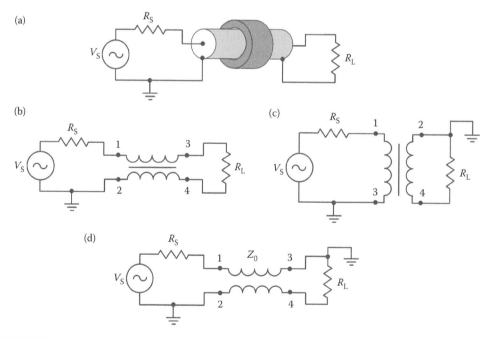

FIGURE 2.34
Schematic configurations of coaxial cable transformer.

transmission line, as a rule of thumb, is kept to no more than an eighth of a wavelength $\lambda_{\min}$ at the highest operating frequency,

$$l \leq \frac{\lambda_{\min}}{8} \tag{2.114}$$

where $\lambda_{\min}$ is the minimum wavelength in the transmission line corresponding to the high operating frequency $f_{\max}$.

The low-frequency bandwidth limit of a coaxial cable transformer is determined by the effect of the magnetizing inductance L_m of the outer surface of the outer conductor, according to the equivalent low-frequency transformer model shown in Figure 2.35a, where the transmission line is represented by the ideal 1:1 transformer [90]. The resistance R_0 represents the losses of the transmission line. An approximation to the magnetizing inductance can be made by considering the outer surface of the coaxial cable to be the same as that of a straight wire (or linear conductor), which, at higher frequencies where the skin effect causes the current to be concentrated on the outer surface, would have the self-inductance of

$$L_m(\text{nH}) = 2l\left[\ln\left(\frac{2l}{r}\right) - 1 \right] \tag{2.115}$$

where l is the length of the coaxial cable in centimeters and r is the radius of the outer surface of the outer conductor in centimeters [90].

High permeability of core materials results in shorter transmission lines. If a toroid is used for the core, the magnetizing inductance L_m is obtained by

$$L_m(\text{nH}) = 4\pi \, n^2 \mu \frac{A_e}{L_e} \tag{2.116}$$

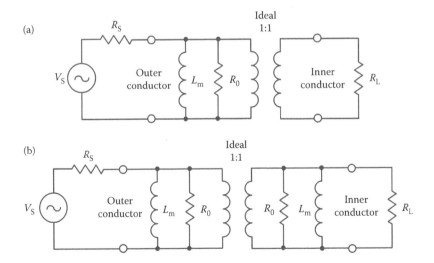

FIGURE 2.35
Low-frequency models of 1:1 coaxial cable transformer.

where n is the number of turns, μ is the core permeability, A_e is the effective cross-sectional area of the core in cm^2, and L_e is the average magnetic path length in cm [97].

Considering the transformer equivalent circuit with a lossless transmission line shown in Figure 2.35a, the ratio between the power delivered to the load P_L and power available at the source $P_S = V_S^2/8R_S$, when $R_S = R_L$, can be obtained from

$$\frac{P_L}{P_S} = \frac{(2\omega L_m)^2}{R_S^2 + (2\omega L_m)^2}$$

(2.117)

which gives the minimum operating frequency f_{min} for a given magnetizing inductance L_m, when taking into account the maximum decrease of the output power by 3 dB, as

$$f_{min} \geq \frac{R_S}{4\pi \, L_m}$$

(2.118)

A similar low-frequency model for a coaxial cable transformer using twisted or parallel wires is shown in Figure 2.35b [90]. Here, the model is symmetrical as both conductors are exposed to magnetic material and therefore contribute identically to the losses and low-frequency performance of the transformer.

An approach using a transmission line based on a single bifilar wound coil to realize a broadband 1:4 impedance transformation was introduced by Ruthroff [98,99]. In this case, by using a core material of a sufficiently high permeability, the number of turns can be significantly reduced. Figure 2.36a shows the circuit schematic of an unbalanced-to-unbalanced

(a)

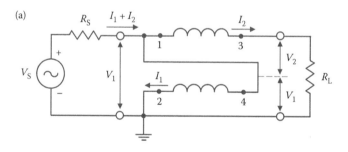

(b)

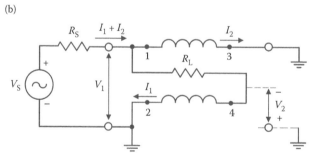

FIGURE 2.36
Schematic configurations of Ruthroff 1:4 impedance transformer.

1:4 transmission line transformer, where terminal 4 is connected to the input terminal 1. As a result, for $V = V_1 = V_2$, the output voltage is twice the input voltage, and the transformer has a 1:2 voltage step-up ratio. As the ratio of input voltage to input current is one-fourth the load voltage to load current, the transformer is fully matched for maximum power transfer when $R_L = 4R_S$, and the transmission-line characteristic impedance Z_0 is equal to the geometric mean of the source and load impedances,

$$Z_0 = \sqrt{R_S R_L} \qquad (2.119)$$

where R_S is the source resistance and R_L is the load resistance. Figure 2.36b shows an impedance transformer acting as a phase inverter, where the load resistance is included between terminals 1 and 4 to become a 1:4 balun. This technique is called the *bootstrap effect*, which does not have the same high-frequency response as the Guanella equal-delay approach because it adds a delayed voltage to a direct one [100]. The delay becomes excessive when the transmission line reaches a significant fraction of a wavelength.

Figure 2.37a shows the physical implementation of a Ruthroff 4:1 impedance transformer using a coaxial cable arranged inside the ferrite core. At lower frequencies, such a transformer can be considered an ordinary 2:1 voltage autotransformer. To improve the performance at higher frequencies, it is necessary to add an additional phase-compensating line of the same length, as shown in Figure 2.37b, resulting in a Guanella ferrite-based 4:1 impedance transformer. In this case, a ferrite core is necessary only for the upper line because the outer conductor of the lower line is grounded at both ends, and no current flows through it. A current I driven into the inner conductor of the upper line produces a current I that flows in the outer conductor of the upper line, resulting in a current $2I$ flowing into the load R_L. Because the voltage $2V$ from the transformer input

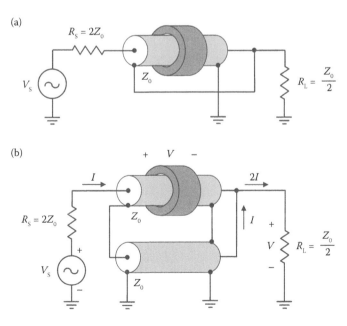

FIGURE 2.37
Schematic configurations of 4:1 coaxial cable transformer.

is divided in two equal parts between the coaxial line and the load, such a transformer provides impedance transformation from $R_S = 2Z_0$ into $R_L = Z_0/2$, where Z_0 is the characteristic impedance of each coaxial line. The bandwidth extension for the Ruthroff transformers can also be achieved by using the transmission lines with step-function and exponential changes in their characteristic impedances [101,102]. To adopt this transmission line transformer for microwave planar applications, the coaxial line can be replaced by a pair of stacked strip conductors or coupled microstrip lines [103,104]. For asymmetric broadside-coupled microstrip lines, the coupling coefficient is much stronger because both electric and magnetic coupling are present, resulting in wider bandwidth and smaller size [105].

Figure 2.38 shows similar arrangements for the 3:1 voltage coaxial cable transformers, which produce 9:1 impedance transformation. A current I driven into the inner conductor of the upper line in Figure 2.38a will cause a current I to flow in the outer conductor of the upper line. This current then produces a current I in the outer conductor of the lower line, resulting in a current $3I$ flowing into the load R_L. The lowest coaxial line can be removed, resulting in a 9:1 impedance coaxial cable transformer shown in Figure 2.38b. The characteristic impedance of each transmission line is specified by the voltage applied to the end

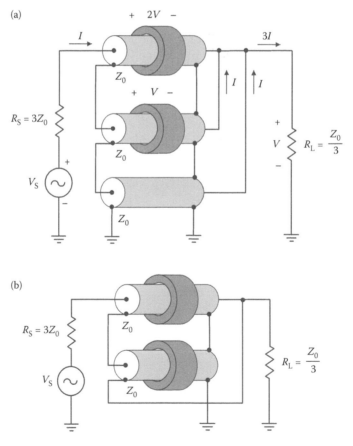

FIGURE 2.38
Schematic configurations of 9:1 coaxial cable transformer.

of the line and the current flowing through the line and is equal to Z_0. In a 0.18-μm CMOS process with six metal layers, a 1:9 transmission-line transformer with broadside-coupled and multiple-metal stacked transmission lines achieved a broadband impedance transformation from 5.0 ± 0.1 Ω optimal load impedance of the power cell to 50-Ω load with a bandwidth of 4.4–6.6 GHz and an insertion loss of about 1 dB [106].

By using the transmission-line baluns with different integer-transformation ratios in certain connection, it is possible to obtain the fractional-ratio baluns and ununs [88,107,108]. Figure 2.39 shows a transformer configuration for obtaining an impedance ratio of 2.25:1, which consists of a 1:1 Guanella balun on the top combined with a 1:4 Guanella balun where voltages on the left-hand side are in series and on the right-hand side are in parallel [107]. In this case, the left-hand side has the higher impedance. In a matched condition, this transformer should have a high-frequency response similar to a single transmission line. By grounding the corresponding terminals (shown by a dashed line), it becomes a broadband unun. Different ratios can be obtained with other configurations. For example, using a 1:9 Guanella balun below the 1:1 unit results in a 1.78:1 impedance ratio, whereas the impedance ratio becomes 1.56:1 with a 1:16 balun.

On the other hand, the overall 1:1.5 voltage transformer configuration can be achieved by using the cascade connection of a 1:3 voltage transformer to increase the impedance 9 times, and a 2:1 voltage transformer to decrease the impedance 4 times, the block schematic of which is shown in Figure 2.40a [108]. The practical configuration of this transformer using coaxial cables and ferrite cores is shown in Figure 2.40b. Here, the currents $I/3$ in the inner conductors of two lower lines cause an overall current $2I/3$ in the outer conductor of the upper line, resulting in a current $2I/3$ flowing into the load R_L. A load voltage $3V/2$ is 180° out of phase with a longitudinal voltage $V/2$ along the upper line, resulting in a voltage V at the transformer input. The lowest line can also be eliminated with direct connection of the points at both ends of its inner conductor, as in the case of the 2:1 and 3:1 Ruthroff voltage transformers shown in Figures 2.37a and 2.38b, respectively. If the source impedance is 50 Ω, then the characteristic impedance of all three transmission lines should be 75 Ω. In this case, the matched condition corresponds to a load impedance of 112.5 Ω.

By using the coaxial cable transformers, the output powers from two or more power amplifiers can be combined. Figure 2.41 shows an example of such a transformer, which

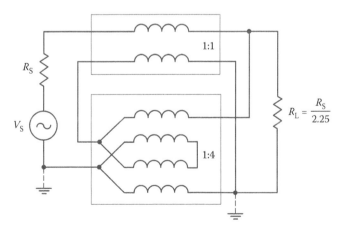

FIGURE 2.39
Schematic configuration of equal-delay 2.25:1 unun.

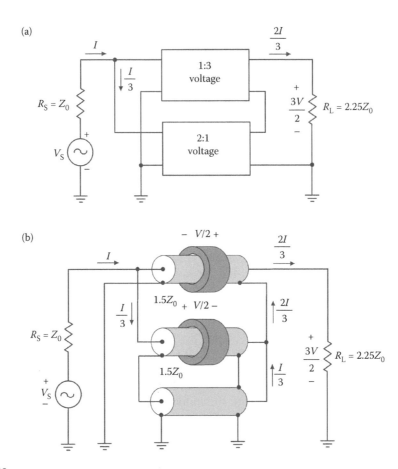

FIGURE 2.40
Schematic configurations of fractional 1:2.25 impedance transformer.

combines two in-phase signals when both signals are delivered to the load R_L and no signal will be dissipated in the ballast resistor R_0 if their amplitudes are equal [98,109]. The main advantage of this transformer is zero longitudinal voltage along the line for equal input powers; as a result, no losses occur in the ferrite core. When one power amplifier defaults or disconnects, the longitudinal voltage becomes equal to half a voltage of another power amplifier. For this transformer, it is possible to combine two 180° out-of-phase signals when the ballast resistor is considered the load, and the load resistor in turn is considered the ballast resistor.

The schematic of another hybrid coaxial cable transformer using as a combiner is shown in Figure 2.42. The advantage of this combiner is that both the load R_L and the ballast resistor R_0 are grounded. In this case, the optimum characteristic impedance of each coaxial cable is defined as $R_S \sqrt{2}$. These hybrid transformer-based combiners can also be used for power dividing when the output power from a single power amplifier is divided and delivered into two independent loads. In this case, the original load and the two signal sources should be switched. It should be noted that the term "hybrid" comes not from the fact that the transformer might be constructed of two different entities (e.g., cable and resistor), but just because it is being driven by two signals as opposed to only one. Consequently, the

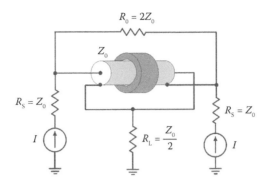

FIGURE 2.41
Coaxial cable combiner.

hybrid transformer represents a four-port passive device having two input ports, one sum port, and one difference port. The unique characteristic of the hybrid transformer is ability to isolate the two input signal sources.

Figure 2.43a shows a coaxial cable two-way combiner where the input signals have the same amplitudes and phases at ports 2 and 3 are matched at higher frequencies when all lines are of the same lengths and $R_S = Z_0 = R_L/2 = R_0/2$ [88]. In this case, the isolation between these input ports can be calculated in decibels by

$$C_{23} = 10\log_{10}[4(1 + 4\cot^2\theta)] \tag{2.120}$$

where θ is the electrical length of each transmission line. In order to improve the isolation, the symmetrical ballast resistor R_0 should be connected through two additional lines (coaxial cables), as shown in Figure 2.43b, where all transmission lines have the same electrical lengths.

Figure 2.44 shows a coaxial cable two-way combiner, which is fully matched and isolated in pairs [88]. Such combiners can effectively be used in high-power broadcasting VHF FM and VHF-UHF TV transmitters. In this case, for power amplifiers with the identical output

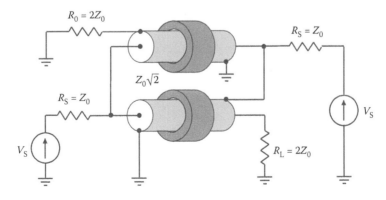

FIGURE 2.42
Two-cable hybrid combiner.

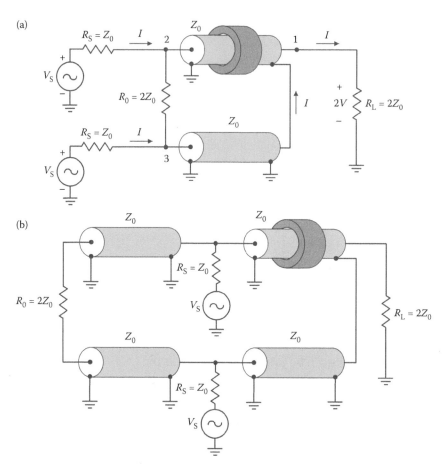

FIGURE 2.43
Coaxial cable combiners with increased isolation.

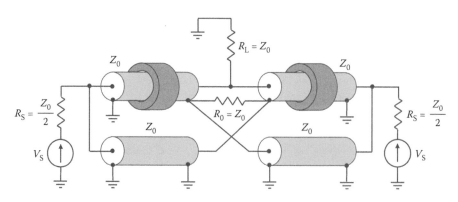

FIGURE 2.44
Fully matched and isolated coaxial cable combiner.

impedances $R_{S1} = R_{S2} = Z_0/2$, it is necessary to choose the ballast resistor R_0 and the load R_L of equal values as $R_0 = R_L = Z_0$, where Z_0 is the characteristic impedance of each transmission line of the same length.

References

1. A. A. Oswald, Power amplifiers in trans-Atlantic radio telephony, *Proc. IRE*, 13, 313–324, 1925.
2. L. E. Barton, High audio power from relatively small tubes, *Proc. IRE*, 19, 1131–1149, 1931.
3. A. I. Berg, *Theory and Design of Vacuum-Tube Generators* (in Russian), Moskva: GEI, 1932.
4. P. H. Osborn, A study of Class B and C amplifier tank circuits, *Proc. IRE*, 20, 813–834, 1932.
5. C. E. Fay, The operation of vacuum tubes as Class B and Class C amplifiers, *Proc. IRE*, 20, 548–568, 1932.
6. F. E. Terman and J. H. Ferns, The calculation of Class C amplifier and harmonic generator performance of screen-grid and similar tubes, *Proc. IRE*, 22, 359–373, 1934.
7. L. B. Hallman, A Fourier analysis of radio-frequency power amplifier wave forms, *Proc. IRE*, 20, 1640–1659, 1932.
8. W. L. Everitt, Optimum operating conditions for Class C amplifiers, *Proc. IRE*, 22, 152–176, 1934.
9. C. E. Kilgour, Graphical analysis of output tube performance, *Proc. IRE*, 19, 42–50, 1931.
10. V. I. Kaganov, *Transistor Radio Transmitters* (in Russian), Moskva: Energiya, 1976.
11. N. Chevaux and M. M. De Souza, Comparative analysis of VDMOS/LDMOS power transistors for RF amplifiers, *IEEE Trans. Microwave Theory Tech.*, MTT-57, 2643–2651, 2009.
12. G. A. Holle and H. C. Reader, Nonlinear MOSFET model for the design of RF power amplifiers, *IEE Proc. Circuits Devices Syst.*, 139, 574–580, 1992.
13. A. Grebennikov, *RF and Microwave Power Amplifier Design*, New York: McGraw-Hill, 2004.
14. W. R. Curtice, J. A. Pla, D. Bridges, T. Liang, and E. E. Shumate, A new dynamic electro-thermal nonlinear model for silicon RF LDMOS FETs, *1999 IEEE MTT-S Int. Microwave Symp. Dig.*, Anaheim, CA, vol. 2, pp. 419–422, 1999.
15. C. Fager, J. C. Pedro, N. B. Carvalho, and H. Zirath, Prediction of IMD in LDMOS transistor amplifiers using a new large-signal model, *IEEE Trans. Microwave Theory Tech.*, MTT-50, 2834–2842, 2002.
16. W. R. Curtice, L. Dunleavy, W. Claussen, and R. Pengelly, New LDMOS model delivers powerful transistor library: The CMC model, *High Frequency Electron.*, 3, 18–25, 2004.
17. Y. P. Tsividis, *Operation and Modeling of the MOS Transistor*, New York: McGraw-Hill, 1987.
18. R. Sung, P. Bendix, and M. B. Das, Extraction of high-frequency equivalent circuit parameters of submicron gate-length MOSFET's, *IEEE Trans. Electron Devices*, ED-45, 1769–1775, 1998.
19. M. C. Ho, K. Green, R. Culbertson, J. Y. Yang, D. Ladwig, and P. Ehnis, A physical large signal Si MOSFET model for RF circuit design, *1997 IEEE MTT-S Int. Microwave Symp. Dig.*, Denver, CO, pp. 391–394, 1997.
20. B. J. Cheu and P. K. Ko, Measurement and modeling of short-channel MOS transistor gate capacitances, *IEEE J. Solid-State Circuits*, SC-22, 464–472, 1987.
21. J. -M. Dortu, J. -E. Muller, M. Pirola, and G. Ghione, Accurate large-signal GaAs MESFET and HEMT modeling for power MMIC amplifier design, *Int. J. Microwave Millimeter-Wave Computer-Aided Eng.*, 5, 195–208, 1995.
22. L.-S. Liu, J. -G. Ma, and G. -I. Ng, Electrothermal large-signal model for III-V FETs including frequency dispersion and charge conservation, *IEEE Trans. Microwave Theory Tech.*, MTT-38, 822–824, 1990.
23. C. -J. Wei, Y. Tkachenko and D. Bartle, An accurate large-signal model of GaAs MESFET which accounts for charge conservation, dispersion, and self-heating, *IEEE Trans. Microwave Theory Tech.*, MTT-46, 1638–1644, 1998.

24. A. Jarndal and G. Kompa, Large-signal model for AlGaN/GaN HEMTs accurately predicts trapping- and self-heating-induced dispersion and intermodulation distortion, *IEEE Trans. Electron Devices*, ED-54, 2830–2836, 2007.
25. T. Kacprzak and A. Materka, Compact DC model of GaAs FET's for large-signal computer calculation, *IEEE J. Solid-State Circuits*, SC-18, 211–213, 1983.
26. I. Angelov, H. Zirath, and N. Rorsman, A new empirical nonlinear model for HEMT and MESFET devices, *IEEE Trans. Microwave Theory Tech.*, MTT-40, 2258–2266, 1992.
27. O. S. A. Tang, K. H. G. Duh, S. M. J. Liu, P. M. Smith, W. F. Kopp, T. J. Rogers, and D. J. Pritchard, Design of high-power, high-efficiency 60-GHz MMIC's using an improved nonlinear PHEMT model. *IEEE J. Solid-State Circuits*, SC-32, 1326–1333, 1997.
28. I. Angelov, V. Desmaris, K. Dynefors, P. A. Nilsson, N. Rorsman, and H. Zirath, On the large-signal modelling of AlGaN/GaN HEMTs and SiC MESFETs, *13th Eur. GAAS Symp. Dig.*, Paris, pp. 309–312, 2005.
29. A. Garcia-Osorio, J. R. Loo-Yau, J. A. Reynoso-Hernandez, S. Ortega, and J. L. del Valle-Padilla, An empirical *I-V* nonlinear model suitable for GaN FET Class F PA design, *Microwave Optical Technol. Lett.*, 53, 1256–1259, 2011.
30. K. Joshin and T. Kikkawa, High-power and high-efficiency GaN HEMT amplifiers, *2008 IEEE Radio Wireless Symp. Dig.*, Orlando, FL, pp. 65–68, 2008.
31. T. Ishida, GaN HEMT technologies for space and radio applications, *Microwave J.*, 54, 56–66, 2011.
32. U. K. Mishra, L. Chen, T. E. Kazior, and Y. F. Wu, GaN-based RF power devices and amplifiers, *Proc. IEEE*, 96, 287–305, 2008.
33. M. Berroth and R. Bosch, Broad-band determination of the FET small-signal equivalent circuit, *IEEE Trans. Microwave Theory Tech.*, MTT-38, 891–895, 1990.
34. Q. Fan, J. H. Leach, and H. Morkoc, Small signal equivalent circuit modeling for AlGaN/GaN HFET: Hybrid extraction method for determining circuit elements of AlGaN/GaN HFET, *Proc. IEEE*, 98, 1140–1150, 2010.
35. G. Dambrine, A. Cappy, F. Heliodore, and E. Playez, A new method for determining the FET small-signal equivalent circuit, *IEEE Trans. Microwave Theory Tech.*, MTT-36, 1151–1159, 1988.
36. N. M. Rohringer and P. Kreuzgruber, Parameter extraction for large-signal modeling of bipolar junction transistors, *Int. J. Microwave Millimeter-Wave Computer-Aided Eng.*, 5, 161–272, 1995.
37. J. P. Fraysee, D. Floriot, P. Auxemery, M. Campovecchio, R. Quere, and J. Obregon, A non-quasi-static model of GaInP/AlGaAs HBT for power applications, *1997 IEEE MTT-S Int. Microwave Symp. Dig.*, Denver, CO, vol. 2, pp. 377–382, 1997.
38. M. Reisch, *High-Frequency Bipolar Transistors*, Berlin: Springer, 2003.
39. V. M. Bogachev and V. V. Nikiforov, *Transistor Power Amplifiers* (in Russian), Moskva: Energiya, 1978.
40. P. M. Asbeck, M. F. Chang, K. -C. Wang, D. L. Miller, G. J. Sullivan, N. H. Sheng, E. A. Sovero, and J. A. Higgins, Heterojunction bipolar transistors for microwave and millimeter-wave integrated circuits, *IEEE Trans. Microwave Theory Tech.*, MTT-35, 1462–1468, 1987.
41. Y. -S. Lin and J. -J. Jiang, Temperature dependence of current gain, ideality factor, and offset voltage of AlGaAs/GaAs and InGaP/GaAs HBTs, *IEEE Trans. Electron Devices*, ED-56, 2945–2951, 2009.
42. N. L. Wang, W. Ma, S. Xu, E. Camargo, X. Sun, P. Hu, Z. Tang, H. F. Chau, A. Chen, and C. P. Lee, 28-V high-linearity and rugged InGaP/GaAs HBT, *2006 IEEE MTT-S Int. Microwave Symp. Dig.*, San Francisco, CA, pp. 881–884, 2006.
43. C. Steinberger, T. Landon, C. Suckling, J. Nelson, J. Delaney, J. Hitt, L. Witkowski, G. Burgin, R. Hajji, and O. Krutko, 250 W HVHBT Doherty with 57% WCDMA efficiency linearized to −55 dBc for 2c11 6.5 dB PAR, *IEEE J. Solid-State Circuits*, SC-43, 2218–2228, 2008.
44. D. Costa, W. U. Liu, and J. S. Harris, Direct extraction of the AlGaAs/GaAs heterojunction bipolar transistor small-signal equivalent circuit, *IEEE Trans. Electron Devices*, ED-38, 2018–2024, 1991.
45. I. Angelov, K. Choumei, and A. Inoue, An empirical HBT large-signal model for CAD, *Int. J. RF Microwave Computer-Aided Eng.*, 13, 518–533, 2003.

46. W. L. Everitt, Output networks for radio-frequency power amplifiers, *Proc. IRE*, 19, 725–737, 1931.
47. H. T. Friis, Noise figure of radio receivers, *Proc. IRE*, 32, 419–422, 1944.
48. S. Roberts, Conjugate-image impedances, *Proc. IRE*, 34, 198–204, 1946.
49. S. J. Haefner, Amplifier-gain formulas and measurements, *Proc. IRE*, 34, 500–505, 1946.
50. R. L. Pritchard, High-frequency power gain of junction transistors, *Proc. IRE*, 43, 1075–1085, 1955.
51. A. R. Stern, Stability and power gain of tuned power amplifiers, *Proc. IRE*, 45, 335–343, 1957.
52. L. S. Houselander, H. Y. Chow, and R. Spense, Transistor characterization by effective large-signal two-port parameters, *IEEE J. Solid-State Circuits*, SC-5, 77–79, 1970.
53. B. J. Thompson, Oscillations in tuned radio-frequency amplifiers, *Proc. IRE*, 19, 421–437, 1931.
54. G. W. Fyler, Parasites and instability in radio transmitters, *Proc. IRE*, 23, 985–1012, 1935.
55. F. B. Llewellyn, Some fundamental properties of transmission systems, *Proc. IRE*, 40, 271–283, 1952.
56. D. F. Page and A. R. Boothroyd, Instability in two-port active networks, *IRE Trans. Circuit Theory*, CT-5, 133–139, 1958.
57. J. M. Rollett, Stability and power gain invariants of linear two-ports, *IRE Trans. Circuit Theory Appl.*, CT-9, 29–32, 1962.
58. J. G. Linvill and L. G. Schimpf, The design of tetrode transistor amplifiers, *Bell Syst. Tech. J.*, 35, 813–840, 1956.
59. O. Muller and W. G. Figel, Stability problems in transistor power amplifiers, *Proc. IEEE*, 55, 1458–1466, 1967.
60. O. Muller, Internal thermal feedback in fourpoles, especially in transistors, *Proc. IEEE*, 52, 924–930, 1964.
61. Application Note AN-010, *GaN for LDMOS Users*, Nitronex Corp., 2008.
62. H. L. Krauss, C. W. Bostian, and F. H. Raab, *Solid State Radio Engineering*, New York: John Wiley & Sons, 1980.
63. L. B. Max, Balanced transistors: A new option for RF design, *Microwaves*, 16, 42–46, 1977.
64. J. Johnson, A look inside those integrated two-chip amps, *Microwaves*, 19, 54–59, 1980.
65. V. Sokolov and R. E. Williams, Development of GaAs monolithic power amplifiers in X-band, *IEEE Trans. Electron Devices*, ED-27, 1164–1171, 1980.
66. N. E. Lindenblad, Television transmitting antenna for empire state building, *RCA Rev.*, 3, 387–408, 1939.
67. N. E. Lindenblad, Junction between single and push-pull lines, U.S. Patent 2,231,839, February 1941 (filed May 1939).
68. A. Riddle, Ferrite and wire baluns with under 1dB loss to 2.5 GHz, *1998 IEEE MTT-S Int. Microwave Symp. Dig.*, Baltimore, MD, pp. 617–620, 1998.
69. K. Krishnamurthy, T. Driver, R. Vetury, and J. Martin, 100 W GaN HEMT power amplifier module with >60% efficiency over 100–1000 MHz bandwidth, *2010 IEEE MTT-S Int. Microwave Symp. Dig.*, Anaheim, CA, pp. 940–943, 2010.
70. A. K. Ezzedine and H. C. Huang, 10W ultra-broadband power amplifier, *2008 IEEE MTT-S Int. Microwave Symp. Dig.*, Atlanta, GA, pp. 643–646, 2008.
71. V. V. Nikiforov, T. T. Kulish, and I. V. Shevnin, Broadband HF-VHF MOSFET power amplifier design (in Russian), *Poluprovodnikovaya Elektronika v Teknike Svyazi*, 23, 27–36, 1983.
72. L. B. Max, Apply wideband techniques to balanced amplifiers, *Microwaves*, 19, 83–88, 1980.
73. W. K. Roberts, A new wide-band balun, *Proc. IRE*, 45, 1628–1631, 1957.
74. N. Marchand, Transmission-line conversion transformers, *Electronics*, 17, 142–145, 1944.
75. K. W. Hamed, A. P. Freundorfer, and Y. M. M. Antar, A novel 15 to 45 GHz monolithic passive balun for MMICs applications, *2003 IEEE MTT-S Int. Microwave Symp. Dig.*, Philadelphia, PA, pp. 31–34, 2003.
76. J. Rogers and R. Bhatia, A 6 to 20 GHz planar balun using a Wilkinson divider and Lange couplers, *1991 IEEE MTT-S Int. Microwave Symp. Dig.*, Fort Worth, TX, pp. 865–868, 1991.
77. B. J. Minnis and M. Healy, New broadband balun structures for monolithic microwave integrated circuits, *1991 IEEE MTT-S Int. Microwave Symp. Dig.*, Fort Worth, TX, pp. 425–428, 1991.

78. K. Kurokawa, Design theory of balanced transistor amplifiers, *Bell Syst. Tech. J.*, 44, 1675–1798, 1965.

79. R. S. Engelbrecht and K. Kurokawa, A wideband low noise L-band balanced transistor amplifier, *Proc. IEEE*, 53, 237–247, 1965.

80. R. E. Neidert and H. A. Willing, Wide-band gallium arsenide power MESFET amplifiers, *IEEE Trans. Microwave Theory Tech.*, MTT-24, 342–350, 1976.

81. K. B. Niklas, R. B. Gold, W. T. Wilser, and W. R. Hitchens, A 12–18 GHz medium-power GaAs MESFET amplifier, *IEEE J. Solid-State Circuits*, SC-13, 520–527, 1978.

82. K. B. Niklas, W. T. Wilser, R. B. Gold, and W. R. Hitchens, Application of the two-way balanced amplifier concept to wide-band power amplification using GaAs MESFET's, *IEEE Trans. Microwave Theory Tech.*, MTT-28, 172–179, 1980.

83. B. M. Oliver, Directional electromagnetic couplers, *Proc. IRE*, 42, 1686–1692, 1954.

84. E. M. T. Jones and J. T. Bolljahn, Coupled-strip-transmission-line filters and directional couplers, *IRE Trans. Microwave Theory Tech.*, MTT-4, 75–81, 1956.

85. J. Lange, Interdigitated stripline quadrature hybrid, *IEEE Trans. Microwave Theory Tech.*, MTT-17, 1150–1151, 1969.

86. R. Waugh and D. LaCombe, Unfolding the Lange coupler, *IEEE Trans. Microwave Theory Tech.*, MTT-20, 777–779, 1972.

87. W. P. Ou, Design equations for an interdigitated directional coupler, *IEEE Trans. Microwave Theory Tech.*, MTT-23, 253–255, 1975.

88. Z. I. Model, *Networks for Combining and Distribution of High Frequency Power Sources* (in Russian), Moskva: Sov. Radio, 1980.

89. J. Sevick, *Transmission Line Transformers*, Norcross: Noble Publishing, 2001.

90. C. Trask, Transmission line transformers: Theory, design and applications, *High Frequency Electron.*, 4, 46–53, 2005, 5, 26–33, 2006.

91. E. Rotholz, Transmission-line transformers, *IEEE Trans. Microwave Theory Tech.*, MTT-29, 148–154, 1981.

92. J. Horn and G. Boeck, Design and modeling of transmission line transformers, *Proc. 2003 IEEE SBMO/MTT-S Int. Microwave Optoelectron. Conf.*, Foz do Iguacu, Brazil, vol. 1, pp. 421–424.

93. G. Guanella, New method of impedance matching in radio-frequency circuits, *Brown Boveri Rev.*, 31, 327–329, 1944.

94. G. Guanella, High-frequency matching transformer, U.S. Patent 2,470,307, May 1949 (filed Apr. 1945).

95. R. K. Blocksome, Practical wideband RF power transformers, combiners, and splitters, *Proc. RF Technol. Expo 86*, pp. 207–227, 1986.

96. J. L. B. Walker, D. P. Myer, F. H. Raab, and C. Trask, *Classic Works in RF Engineering: Combiners, Couplers, Transformers, and Magnetic Materials*, Norwood: Artech House, 2005.

97. J. Sevick, Magnetic materials for broadband transmission line transformers, *High Frequency Electron.*, 4, 46–52, 2005.

98. C. L. Ruthroff, Some broad-band transformers, *Proc. IRE*, 47, 1337–1342, 1959.

99. C. L. Ruthroff, Broadband transformers, U.S. Patent 3,037,175, May 1962 (filed May 1958).

100. J. Sevick, A simplified analysis of the broadband transmission line transformer, *High Frequency Electron.*, 3, 48–53, 2004.

101. R. T. Irish, Method of bandwidth extension for the Ruthroff transformer, *Electron. Lett.*, 15, 790–791, 1979.

102. S. C. Dutta Roy, Optimum design of an exponential line transformer for wide-band matching at low frequencies, *Proc. IEEE*, 67, 1563–1564, 1979.

103. M. Engels, M. R Jansen, W. Daumann, R. M. Bertenburg, and F. J. Tegude, Design methodology, measurement and application of MMIC transmission line transformers, *1995 IEEE MTT-S Int. Microwave Symp. Dig.*, Orlando, FL, vol. 3, pp. 1635–1638, 1995.

104. S. P. Liu, Planar transmission line transformer using coupled microstrip lines, *1998 IEEE MTT-S Int. Microwave Symp. Dig.*, Baltimore, MD, vol. 2, pp. 789–792, 1998.

105. I. J. Bahl, Broadband and compact impedance transformers for microwave circuits, *IEEE Microwave Mag.*, 7, 56–62, 2006.

106. H. K. Chiou and H. Y. Liao, Broadband and low-loss 1:9 transmission-line transformer in 0.18-μm CMOS process, *IEEE Electron Device Lett.*, 31, 921–923, 2010.
107. J. Sevick, Design of broadband ununs with impedance ratios less than 1:4, *High Frequency Electron.*, 3, 44–50, 2004.
108. S. E. London and S. V. Tomashevich, Line transformers with fractional transformation factor, *Telecommunications Radio Eng.*, 28/29, 129–130, 1974.
109. H. Granberg, *Broadband Transformers and Power Combining Techniques for RF*, Application Note AN747, Motorola Semiconductor, Inc., 1993.

3

Lossless Matched Broadband Power Amplifiers

Generally, the matching design procedure is based on the methods of circuit analysis, optimization, and synthesis. According to the first method, the circuit parameters are calculated at one frequency chosen in advance (usually the center or high-bandwidth frequency), and then, the power amplifier performance is analyzed across the entire frequency bandwidth. To synthesize the broadband matching/compensation network, it is necessary to choose the maximum attenuation level or reflection coefficient magnitude inside the operating frequency bandwidth and then to obtain the parameters of matching networks by using special tables and formulas to convert the lumped elements into distributed ones. For push–pull power amplifiers, it is very convenient to use both lumped and distributed parameters when the lumped capacitors are connected in parallel to the microstrip lines due to the effect of virtual ground.

3.1 Impedance Matching

In a common case, an optimum solution depends on the circuit requirements, such as the simplicity in practical realization, the frequency bandwidth and minimum power ripple, design implementation and adjustability, stable operation conditions, and sufficient harmonic suppression. As a result, many types of the matching networks are available that are based on the lumped elements and transmission lines. To simplify and visualize the matching design procedure, an analytical approach when all parameters of the matching circuits are calculated using simple analytical equations alongside with their Smith chart visualization can be used.

3.1.1 Basic Principles

Impedance matching is necessary to provide maximum delivery to the load of the RF power available from the source by using some impedance-matching network that can modify the load as viewed from the generator [1]. This means that generally, when the electrical signal propagates in the circuit, a portion of this signal might be reflected at the interface between the sections with different impedances. Therefore, it is necessary to establish the conditions that allow to fully transmitting the entire RF signal without any reflection. To determine an optimum value of the load impedance Z_L, at which the power delivered to the load is maximal, the equivalent circuit shown in Figure 3.1a can be considered.

In this case, the power delivered to the load can be defined as

$$P = \frac{1}{2} V_{in}^2 \operatorname{Re}\left(\frac{1}{Z_L}\right) = \frac{1}{2} V_S^2 \left|\frac{Z_L}{Z_S + Z_L}\right|^2 \operatorname{Re}\left(\frac{1}{Z_L}\right) \tag{3.1}$$

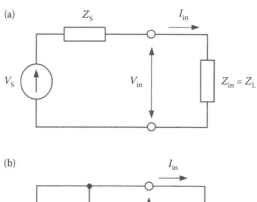

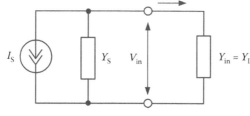

FIGURE 3.1
Equivalent circuits with voltage and current sources.

where $Z_S = R_S + jX_S$ is the source impedance, $Z_L = R_L + jX_L$ is the load impedance, V_S is the source voltage amplitude, and V_{in} is the load voltage amplitude. Substituting the real and imaginary parts of the source and load impedances Z_S and Z_L into Equation 3.1 yields

$$P = \frac{1}{2} V_S^2 \frac{R_L}{(R_S + R_L)^2 + (X_S + X_L)^2} \tag{3.2}$$

If the source impedance Z_S is fixed, then, it is necessary to vary the real and imaginary parts of the load impedance Z_L until maximum power is delivered to the load. To maximize the output power, the following analytical conditions in the form of derivatives with respect to the output power are written:

$$\frac{\partial P}{\partial R_L} = 0 \quad \frac{\partial P}{\partial X_L} = 0 \tag{3.3}$$

Applying these conditions and taking Equation 3.2 into consideration, the system of two equations can be obtained as

$$\frac{1}{(R_L + R_S)^2 + (X_L + X_S)^2} - \frac{2R_L(R_L + R_S)}{[(R_L + R_S)^2 + (X_L + X_S)^2]^2} = 0 \tag{3.4}$$

$$\frac{2X_L(X_L + X_S)}{[(R_L + R_S)^2 + (X_L + X_S)^2]^2} = 0 \tag{3.5}$$

Simplifying Equations 3.4 and 3.5 results in

$$R_S^2 - R_L^2 + (X_L + X_S)^2 = 0 \tag{3.6}$$

$$X_L(X_L + X_S) = 0 \tag{3.7}$$

By solving Equations 3.6 and 3.7 simultaneously for R_S and X_S, one can obtain

$$R_S = R_L \tag{3.8}$$

$$X_L = -X_S \tag{3.9}$$

or, in an impedance form,

$$Z_L = Z_S^* \tag{3.10}$$

where (*) denotes the complex-conjugate value [2].

Equation 3.10 is called an *impedance conjugate-matching condition*, and its fulfillment results in maximum power delivered to the load for fixed source impedance. Maximum power delivered to the load must be equal to

$$P = \frac{V_S^2}{8R_S} \tag{3.11}$$

The *admittance conjugate-matching condition*, which is applied to the equivalent circuit shown in Figure 3.1b, is given as

$$Y_L = Y_S^* \tag{3.12}$$

and can be readily obtained in the same way. Maximum power delivered to the load in this case can be defined as

$$P = \frac{I_S^2}{8G_S} \tag{3.13}$$

where $G_S = \mathrm{Re}Y_S$ is the source conductance and I_S is the source current amplitude.

Thus, the conjugate-matching conditions in a common case can be determined through the immittance parameters, representing any system of the impedance Z-parameters or admittance Y-parameters in the form

$$W_L = W_S^* \tag{3.14}$$

The matching circuit is connected between the source and the input of an active device, as shown in Figure 3.2a, and between the output of an active device and the load, as shown in

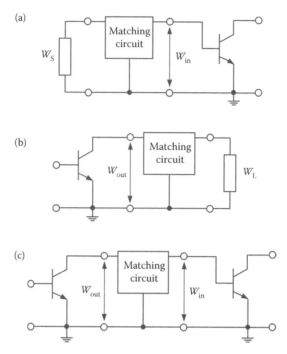

FIGURE 3.2
Matching circuit arrangements.

Figure 3.2b. For a multistage amplifier, the load represents an input circuit of the next stage. Therefore, the matching circuit in this case is connected between the output of the active device of the preceding amplifier stage and the input of the active device of the succeeding stage of the power amplifier, as shown in Figure 3.2c. The main objective is to properly transform the load immittance W_L to the optimum device output immittance W_{out}, the value of which is properly determined by the supply voltage, output power, device saturation voltage, and selected operation class to maximize the amplifier efficiency and output power. It should be noted that Equation 3.14 is given in a general immittance form without an indication of whether it is used in a small-signal or large-signal application. In the latter case, this only means that the device immittance W-parameters are fundamentally averaged over a large-signal swing across the device-equivalent circuit parameters and that the conjugate-matching principle is valid in both the small-signal application and the large-signal application, where the optimum equivalent device output resistance (or conductance) at the fundamental frequency is matched to the load resistance (or conductance) and the effect of the device output reactive elements is compensated by the conjugate reactance of the load network. In addition, the matching circuits should be designed to provide the required voltage and current waveforms at the device output and the stability of operation conditions. The losses in the output-matching circuits must be as small as possible to deliver the output power to the load with maximum efficiency. Finally, it is desirable that the matching circuit must be easy to tune.

3.1.2 Matching with Lumped Elements

Generally, there is a variety of configurations for matching networks to efficiently deliver the signal from the source to the load, and application of any of these matching networks in the power amplifier depends on its class of operation, level of output power, operating frequency, frequency bandwidth, or level of harmonic suppression. The lumped matching networks in the form of (a) L-transformer, (b) π-transformer, or (c) T-transformer shown in Figure 3.3, respectively, have proved to be effective for a long time for power amplifier design [1]. The simplest and most popular matching network is the circuit in the form of the L-transformer. The transforming properties of this matching circuit can be analyzed by using the equivalent transformation of a parallel into a series representation of the RX network.

Consider the parallel RX network shown in Figure 3.4a, where R_1 is the real (resistive) part and X_1 is the imaginary (reactive) part of the network impedance $Z_1 = jX_1R_1/(R_1 + jX_1)$, and the series RX network shown in Figure 3.4b, where R_2 is the resistive part and X_2 is the reactive part of the circuit impedance $Z_2 = R_2 + jX_2$. These two impedance networks (series and parallel) can be considered equivalent at some frequency if $Z_1 = Z_2$, which results in

$$R_2 + jX_2 = \frac{R_1 X_1^2}{R_1^2 + X_1^2} + j\frac{R_1^2 X_1}{R_1^2 + X_1^2} \tag{3.15}$$

Equation 3.15 can be rearranged into two separate equations for real and imaginary parts as

$$R_1 = R_2(1 + Q^2) \tag{3.16}$$

$$X_1 = X_2(1 + Q^{-2}) \tag{3.17}$$

where $Q = R_1/|X_1| = |X_2|/R_2$ is the network quality factor, which is equal for both the series and parallel RX networks.

Consequently, if the reactive impedance or reactance $X_1 = -X_2(1 + Q^{-2})$ is connected in parallel to the series circuit composed of the resistance R_2 and reactance X_2, it allows the reactance of the series circuit to be compensated. In this case, the input impedance of such

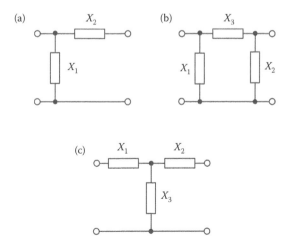

FIGURE 3.3
Matching networks in the form of L-, π-, and T-transformers.

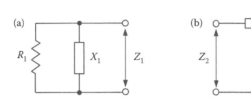

FIGURE 3.4
Impedance parallel and series-equivalent networks.

a two-port network, which is shown in Figure 3.5, will be only resistive and equal to R_1. Consequently, to transform the resistance R_1 into the other resistance R_2 at a certain frequency, it is sufficient to connect a two-port L-transformer between them with the opposite signs of the reactances X_1 and X_2, the parameters of which can be easily calculated from the following simple equations:

$$|X_1| = \frac{R_1}{Q} \tag{3.18}$$

$$|X_2| = R_2 Q \tag{3.19}$$

where

$$Q = \sqrt{\frac{R_1}{R_2} - 1} \tag{3.20}$$

is the circuit-loaded quality factor expressed through the resistances to be matched. Thus, to design a matching circuit with fixed resistances to be matched, first, we need to calculate the quality factor Q according to Equation 3.20 and then to define the reactive elements, according to Equations 3.18 and 3.19.

Owing to the opposite signs of the reactances X_1 and X_2, the two possible circuit configurations (one in the form of a low-pass filter section and another in the form of a high-pass filter section) with the same transforming properties can be realized, which are shown in Figure 3.6 together with the design equations. In practice, the single two-port L-transformers can be used as the input or interstage-matching circuits in power

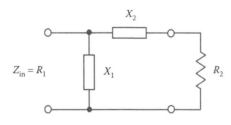

FIGURE 3.5
Input impedance of a two-port network.

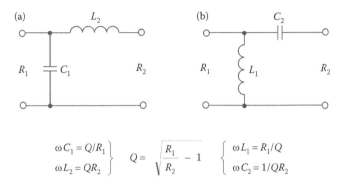

$$\left.\begin{array}{l} \omega C_1 = Q/R_1 \\ \omega L_2 = QR_2 \end{array}\right\} \quad Q = \sqrt{\dfrac{R_1}{R_2} - 1} \quad \left\{\begin{array}{l} \omega L_1 = R_1/Q \\ \omega C_2 = 1/QR_2 \end{array}\right.$$

FIGURE 3.6
L-type matching circuits and relevant equations.

amplifiers, where the requirements for the out-of-band suppression and harmonic control required for higher efficiency are not as high as for the output-matching circuits. In this case, the main advantage of such an *L*-transformer is its simplicity when the only two reactive elements with fast tuning are needed. For larger values of $Q \geq 10$, it is possible to use a cascade connection of *L*-transformers, which allows wider frequency bandwidth and transformer efficiency to be realized.

The matching circuits in the form of (a) π-transformer and (b) *T*-transformer can be realized by the appropriate connection of two *L*-transformers, as shown in Figure 3.7. For each *L*-transformer, the resistance R_1 and the resistance R_2 are transformed to some intermediate resistance R_0 with the value of $R_0 < (R_1, R_2)$ for a π-transformer and the value of $R_0 > (R_1, R_2)$ for a *T*-transformer. The value of R_0 is not fixed and can be chosen to be arbitrary depending on the frequency bandwidth. This means that, compared to the simple *L*-transformer with fixed parameters for the same ratio of R_2/R_1, the circuit parameters of the π- or *T*-transformer can be different. However, they provide narrower frequency bandwidths due to higher quality factors because the intermediate resistance R_0 is either greater or smaller than each of the resistances R_1 and R_2. By taking into account the two possible

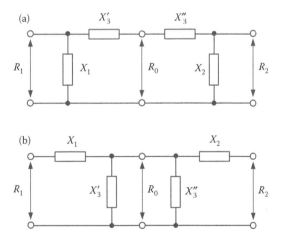

FIGURE 3.7
Matching circuits developed by connecting two *L*-transformers.

circuit configurations of the *L*-transformer shown in Figure 3.6, it is possible to develop the different circuit configurations of the two-port transformers shown in Figure 3.7a, where $X_3 = X_3' + X_3''$, and in Figure 3.7b, where $X_3 = X_3' X_3'' / (X_3' + X_3'')$.

Several of the most widely used two-port π-transformers, together with the design equations, are shown in Figure 3.8 [3,4]. The π-transformers are usually used as output-matching circuits of the high-power amplifiers in a Class-B operation when it is necessary to achieve a sinusoidal drain (or collector) voltage waveform by appropriate harmonic suppression. In addition, it is convenient to use some of them as interstage-matching circuits in low-power and medium-power amplifiers when it is necessary to provide sinusoidal voltage waveforms both at the drain (or collector) of the driver-stage transistor and at the

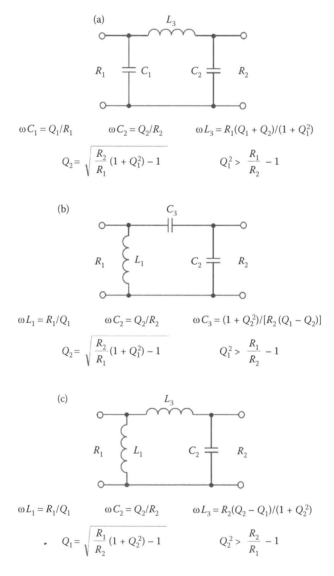

(a)

$$\omega C_1 = Q_1/R_1 \qquad \omega C_2 = Q_2/R_2 \qquad \omega L_3 = R_1(Q_1 + Q_2)/(1 + Q_1^2)$$

$$Q_2 = \sqrt{\frac{R_2}{R_1}(1 + Q_1^2) - 1} \qquad Q_1^2 > \frac{R_1}{R_2} - 1$$

(b)

$$\omega L_1 = R_1/Q_1 \qquad \omega C_2 = Q_2/R_2 \qquad \omega C_3 = (1 + Q_2^2)/[R_2(Q_1 - Q_2)]$$

$$Q_2 = \sqrt{\frac{R_2}{R_1}(1 + Q_1^2) - 1} \qquad Q_1^2 > \frac{R_1}{R_2} - 1$$

(c)

$$\omega L_1 = R_1/Q_1 \qquad \omega C_2 = Q_2/R_2 \qquad \omega L_3 = R_2(Q_2 - Q_1)/(1 + Q_2^2)$$

$$Q_1 = \sqrt{\frac{R_1}{R_2}(1 + Q_2^2) - 1} \qquad Q_2^2 > \frac{R_2}{R_1} - 1$$

FIGURE 3.8
π-Transformers and relevant equations.

gate (or base) of the next-stage transistor. In this case, for a π-transformer with two shunt capacitors, the input and output capacitances of these transistors can be easily included into the matching circuit elements C_1 and C_2, respectively. Finally, a π-transformer can be directly used as the load network for a high-efficiency Class-E mode with proper calculation of its design parameters.

The π-transformer with two shunt capacitors, which is shown in Figure 3.8a, represents a face-to-face connection of the two simple low-pass L-transformers. As a result, there is no special requirement for the resistances R_1 and R_2, which means that the ratio R_1/R_2 can be greater or smaller than unity. In this case, the design equations correspond to the case of $R_1/R_2 > 1$. However, as it will be further derived, the π-transformer with a series capacitor shown in Figure 3.8b can only be used for impedance matching when $R_1/R_2 > 1$. Such a π-transformer represents a face-to-face connection of the high-pass and low-pass L-transformers, as shown in Figure 3.9a.

The design equations for a high-pass section are written using Equations 3.18 through 3.20 as

$$\omega L_1 = \frac{R_1}{Q_1} \tag{3.21}$$

$$\omega C_3' = \frac{1}{Q_1 R_0} \tag{3.22}$$

$$Q_1^2 = \frac{R_1}{R_0} - 1 \tag{3.23}$$

where R_0 is the intermediate resistance.

Similarly, for a low-pass section,

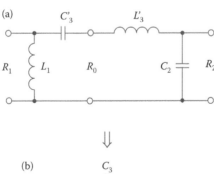

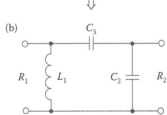

FIGURE 3.9
π-Transformer with series capacitor.

$$\omega C_2 = \frac{Q_2}{R_2} \tag{3.24}$$

$$\omega L_3' = Q_2 R_0 \tag{3.25}$$

$$Q_2^2 = \frac{R_2}{R_0} - 1 \tag{3.26}$$

Since it is assumed that $R_1 > R_2 > R_0$, from Equation 3.23, it follows that the loaded quality factor Q_1 of a high-pass L-transformer can be chosen from the condition

$$Q_2^2 > \frac{R_1}{R_2} - 1 \tag{3.27}$$

Substituting Equation 3.23 into Equation 3.26 results in

$$Q_2 = \sqrt{\frac{R_2}{R_1}(1 + Q_1^2) - 1} \tag{3.28}$$

Combining the reactances of two series elements (capacitor C_3' and inductor L_3') defined by Equations 3.22 and 3.25 yields

$$\omega L_3' - \frac{1}{\omega C_3'} = R_0(Q_2 - Q_1) = \frac{R_2(Q_2 - Q_1)}{(1 + Q_2^2)} \tag{3.29}$$

As a result, since $Q_1 > Q_2$, the total series reactance is negative, which can be provided by a series capacitance C_3 shown in Figure 3.9b with a susceptance

$$\omega C_3 = \frac{1 + Q_2^2}{R_2(Q_1 - Q_2)} \tag{3.30}$$

On the other hand, if $Q_2 > Q_1$ when $R_2 > R_1 > R_0$, the total series reactance is positive, which can be provided by a series inductance L_3, and all matching circuit parameters can be calculated according to the design equations given in Figure 3.8c. In this case, it first needs to choose the value of Q_2 for fixed resistances R_1 and R_2 to be matched, then to determine the value of Q_1, and finally to calculate the values of the shunt inductance L_1, shunt capacitance C_2, and series inductance L_3.

Some of the matching circuit configurations of the two-port T-transformers, together with the design equations, are shown in Figure 3.10 [3,4]. The T-transformers are usually used in the high-power amplifiers as the input, interstage, and output-matching circuits, especially the matching circuit with shunt and series capacitors shown in Figure 3.10b. In this case, if a high value of the inductance L_2 is chosen, then, the current waveform at the input of the transistor with a small input real part will be close to sinusoidal. By using such a T-transformer for the output matching of a power amplifier, it is easy to realize a high-efficiency Class-F operation mode, because the series inductor connected to the

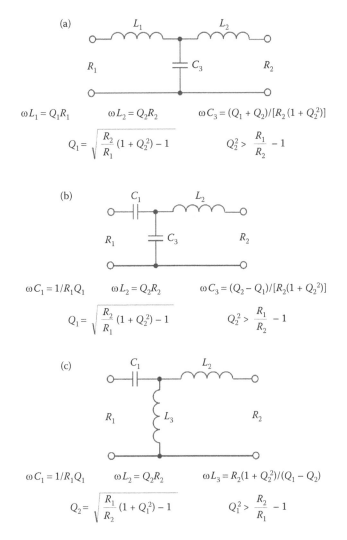

FIGURE 3.10
T-transformers and relevant equations.

drain (or collector) of the active device can create close-to-open-circuit harmonic imped-ance conditions.

The *T*-transformer with two series inductors, which is shown in Figure 3.10a, represents a back-to-back connection of the two simple low-pass *L*-transformers. In this case, the resistance ratio R_1/R_2 can be greater or smaller than unity, similar to a π-transformer with two shunt capacitors shown in Figure 3.8a. In this case, the design equations correspond to the case of $R_1/R_2 > 1$. However, the *T*-transformer with series and shunt capacitors, which is shown in Figure 3.10b, can only be used for impedance matching when $R_1/R_2 > 1$. Such a *T*-transformer represents a back-to-back connection of the high-pass and low-pass *L*-transformers, as shown in Figure 3.11a.

The design equations for a high-pass section of such a *T*-transformer are written using Equations 3.18 through 3.20 as

(a)

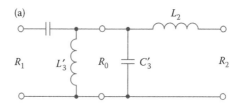

(b)

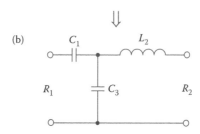

FIGURE 3.11
T-transformer with series and shunt capacitors.

$$\omega C_1 = \frac{1}{R_1 Q_1} \tag{3.31}$$

$$\omega L_3' = \frac{R_0}{Q_1} \tag{3.32}$$

$$Q_1^2 = \frac{R_0}{R_1} - 1 \tag{3.33}$$

where R_0 is the intermediate resistance.
   Similarly, for a low-pass section,

$$\omega L_2 = Q_2 R_2 \tag{3.34}$$

$$\omega C_3' = Q_2 / R_0 \tag{3.35}$$

$$Q_2^2 = \frac{R_0}{R_2} - 1 \tag{3.36}$$

Since it is assumed that $R_0 > R_1 > R_2$, from Equation 3.36, it follows that the loaded quality factor Q_2 of a low-pass *L*-transformer can be chosen from the condition

$$Q_2^2 > \frac{R_1}{R_2} - 1 \tag{3.37}$$

Substituting Equation 3.36 into Equation 3.33 results in

$$Q_1 = \sqrt{\frac{R_2}{R_1}(1 + Q_2^2) - 1}$$ (3.38)

Combining the susceptances of two shunt elements (inductor L_3' and capacitor C_3') defined by Equations 3.32 and 3.35 yields

$$\omega C_3' - \frac{1}{\omega L_3'} = \frac{Q_2 - Q_1}{R_0} = \frac{Q_2 - Q_1}{R_2(1 + Q_2^2)}$$ (3.39)

As a result, since $Q_2 > Q_1$, the total shunt susceptance is positive, which can be provided by a shunt capacitance C_3 shown in Figure 3.11b with a susceptance

$$\omega C_3 = \frac{Q_2 - Q_1}{R_2(1 + Q_2^2)}$$ (3.40)

On the other hand, if $Q_1 > Q_2$ when $R_0 > R_2 > R_1$, the total shunt susceptance is negative, which can be provided by a shunt inductance L_3, and all matching circuit parameters can be calculated according to the design equations given in Figure 3.10c. In this case, it first needs to choose the value of Q_1 for fixed resistances R_1 and R_2 to be matched, then to determine the value of Q_2, and finally to calculate the values of the series capacitance C_1, series inductance L_2, and shunt inductance L_3.

At microwave frequencies, lumped-element-matching circuits are very useful to provide a broadband operation and to reduce the size of the power amplifiers for miniaturization. Depending on the device cell size and the desired frequency range of operation, the value of the input shunt capacitance can range from 0.6 to 1.2 pF, and the value of inductances realized with bondwire inductances are of the order of 0.3–1.5 nH [5]. In this case, for the device cell with a total gate width of 6400 μm to achieve output powers up to 3 W with a 1-dB bandwidth of 2 GHz within the frequency range from 7 to 10 GHz, the input and interstage-matching circuits are implemented in the form of the *T*-transformers with two series inductors and shunt capacitors, as shown in Figure 3.10a. Figure 3.12 shows the circuit schematic of a broadband medium-power single-stage GaAs MESFET amplifier, where the matching circuits in the form of the π-transformers with two inductors and one capacitor on each side of the transistor can cover the frequency bandwidth of 6–12 GHz [6]. Here, the shunt inductors L_2 and L_3 implemented with ribbons essentially affect the low-frequency

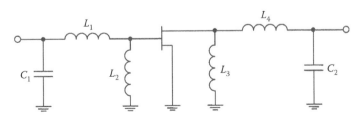

FIGURE 3.12
Schematic of a broadband lumped-element FET amplifier.

gain response, whereas the series low-pass networks (L_1, C_1, L_4, and C_2) match the high end of the frequency bandwidth and selectively mismatch the device at the low end.

Now, let us demonstrate a lumped matching circuit technique at very high frequencies to design a 150-W MOSFET power amplifier with a supply voltage of 50 V operating in a frequency bandwidth of 132–174 MHz ($\Delta f = 42$ MHz) and providing a power gain greater than 10 dB. In this case, the center bandwidth frequency is equal to $f_0 = \sqrt{132 \cdot 174} = 152$ MHz. For example, from the datasheet on the MOSFET device, it follows that the values of the input and output impedances at this frequency are $Z_{in} = (0.9 - j1.2)$ Ω and $Z_{out} = (1.8 + j2.1)$ Ω, respectively, where the input reactance $\mathrm{Im}Z_{in}$ is capacitive and the output reactance $\mathrm{Im}Z_{out}$ is inductive. To cover the required frequency bandwidth, the low-Q-matching circuits should be used that allow reduction of the in-band amplitude ripple and improvement of the input VSWR. The value of a circuit quality factor for 3-dB bandwidth level must be less than $Q = f_0/\Delta f = 152/42$ MHz = 3.6. In this case, it is very convenient to design the input- and output-matching circuits using the simple L-transformers in the form of low-pass and high-pass filter sections with a constant value of Q, which are connected in series [7].

Hence, to match the required low-value input series capacitive impedance to the standard 50-Ω source impedance in a frequency bandwidth of 27.6%, it needs to use at least three filter sections, as shown in Figure 3.13. From the negative imaginary part of the input impedance Z_{in}, it follows that the input capacitance C_{in} at the operating frequency of 152 MHz is equal to approximately 873 pF. To compensate for this capacitive reactance at the center bandwidth frequency, it is sufficient to connect an inductance of 1.3 nH in series to the device input capacitance. When the device input capacitive reactance has been compensated, the next step is to proceed with the design of the input-matching circuit. To simplify the matching design procedure, it is best to cascade the L-transformers with an equal value of Q. Although equal-Q values are not absolutely necessary, this provides a convenient guide for both analytical calculation of the matching circuit parameters and the Smith chart graphical design.

In this case, the following ratio can be written for the input-matching circuit:

$$\frac{R_1}{R_2} = \frac{R_2}{R_3} = \frac{R_3}{R_{in}} \tag{3.41}$$

resulting in $R_2 = 13$ Ω and $R_3 = 3.5$ Ω for $R_{source} = R_1 = 50$ Ω and $R_{in} = \mathrm{Re}\,Z_{in} = 0.9$ Ω. Consequently, a loaded quality factor of each L-transformer according to Equation 3.20 is equal to $Q = 1.7$. The elements of the input-matching circuit using equations given in Figure 3.6 can be calculated as $L_1 = 31$ nH, $C_1 = 47$ pF, $L_2 = 6.2$ nH, $C_2 = 137$ pF, $L_3 = 1.6$ nH, and $C_3 = 509$ pF.

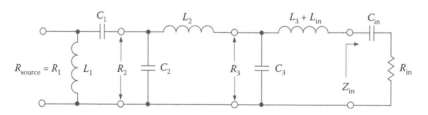

FIGURE 3.13
Complete broadband input-matching circuit.

This equal-Q approach significantly simplifies the matching circuit design using the Smith chart. In this case, it is first necessary to plot two circles of equal-Q values on the Smith chart. The circle of equal Q is plotted, taking into account that, for each point located at this circle, a ratio of X/R or B/G must be the same. Then, each element of the input-matching circuit can be readily determined, as shown in Figure 3.14. Each trace for the series inductance and capacitance must be plotted as far as the intersection point with the Q-circle, whereas each trace for the shunt capacitance and inductance should be plotted until the intersection with the horizontal real axis.

To match the output series inductive impedance to the standard 50-Ω load, it is sufficient to use only two filter sections, as shown in Figure 3.15. At the operating frequency of 152 MHz, the transistor series output inductance is equal to approximately 2.2 nH. This

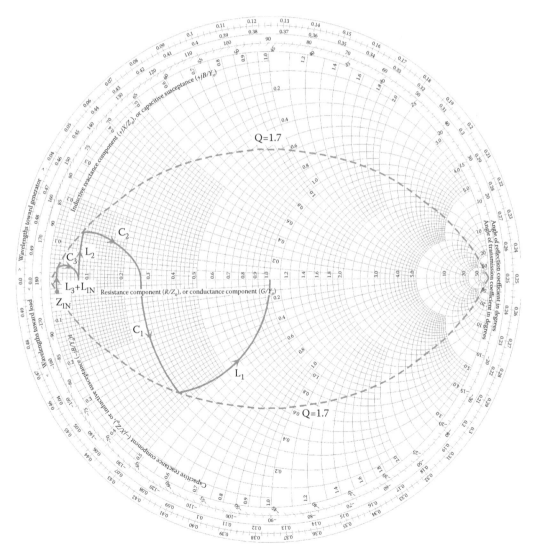

FIGURE 3.14
Smith chart with elements from Figure 3.13.

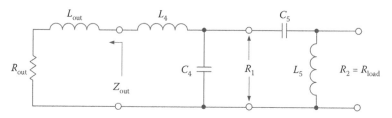

FIGURE 3.15
Complete broadband output-matching circuit.

inductance can be used as a part of the L-transformer in the form of a low-pass filter section. For an output-matching circuit, the condition of equal-Q values gives the following ratio:

$$\frac{R_2}{R_1} = \frac{R_1}{R_{out}} \tag{3.42}$$

with the value $R_1 = 9.5\,\Omega$ for $R_{load} = R_2 = 50\,\Omega$ and $R_{out} = 1.8\,\Omega$. Consequently, the loaded quality factor of each L-transformer is equal to $Q = 2.1$, which is substantially smaller than the value of Q for 3-dB bandwidth level. Then, it is necessary to check the value of a series inductance of the low-pass section, which must exceed the value of 2.2 nH for the correct matching procedure. The appropriate calculation gives the value of a total series inductance $L_4 + L_{out}$ of approximately 4 nH. As a result, the values of the output-matching circuit elements are $L_4 = 1.8$ nH, $C_4 = 231$ pF, $C_5 = 52$ pF, and $L_5 = 25$ nH.

The output-matching circuit design using the Smith chart with constant-Q circles is shown in Figure 3.16. For the final high-pass section, a trace for the series capacitance C_5 must be plotted as far as the intersection with $Q = 2.1$ circle, whereas a trace for the shunt inductance L_5 should be plotted until the intersection with the center point of the Smith chart.

3.1.3 Matching with Transmission Lines

At very high frequencies, it is very difficult to implement lumped elements with a predefined accuracy in view of a significant effect of their parasitic parameters, for example, the parasitic interturn and direct-to-ground capacitances for lumped inductors and the stray inductance for lumped capacitors. However, these parasitic parameters can represent a part of a distributed LC structure such as a transmission line. In this case, for a microstrip line, the series inductance is associated with the flow of current in the conductor and the shunt capacitance is associated with the strip separated from the ground by the dielectric substrate. If the line is wide, the inductance is reduced but the capacitance is large. However, for a narrow line, the inductance is increased but the capacitance is small.

Figure 3.17 shows an impedance-matching circuit in the form of a transmission-line transformer connected between the source impedance Z_S and load impedance Z_L. The input impedance as a function of the electrical length of the transmission line with arbitrary load impedance is defined as

$$Z_{in} = Z_0 \frac{Z_L + jZ_0 \tan\theta}{Z_0 + jZ_L \tan\theta} \tag{3.43}$$

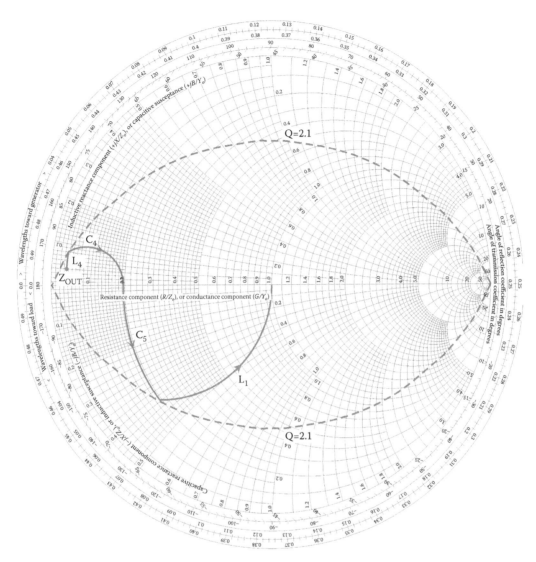

FIGURE 3.16
Smith chart with elements from Figure 3.15.

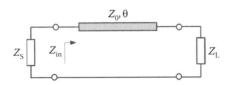

FIGURE 3.17
Transmission-line impedance transformer.

where Z_0 is the characteristic impedance, $\theta = \beta l$ is the electrical length of the transmission line, $\beta = \omega\sqrt{\mu_r \varepsilon_r}/c$ is the phase constant, c is the speed of light in free space, μ_r is the substrate permeability, ε_r is the substrate permittivity, ω is the radian frequency, and l is the geometrical length of the transmission line [8,9].

For a quarter-wavelength transmission line when $\theta = \pi/2$, the expression for Z_{in} reduces to

$$Z_{in} = Z_0^2 / Z_L \tag{3.44}$$

from which it follows that, for example, a 50-Ω load is matched to a 12.5-Ω source with the characteristic impedance of 25 Ω.

Usually, such a quarter-wavelength impedance transformer is used for impedance matching in a narrow frequency bandwidth of 10%–20%, and its length is chosen at the bandwidth center frequency. However, using a multisection quarter-wave transformer widens the bandwidth and expands the choice of the substrate to include materials with high-dielectric permittivity, which reduces the transformer size. For example, by using a transformer composed of seven quarter-wavelength transmission lines of different characteristic impedances, whose lengths are selected at the highest bandwidth frequency, the power gain flatness of ±1 dB was achieved over the frequency range of 5–10 GHz for a 15-W GaAs MESFET power amplifier [10].

To provide a complex-conjugate matching of the input transmission-line impedance Z_{in} with the source impedance $Z_S = R_S + jX_S$ when $R_S = \mathrm{Re}Z_{in}$ and $X_S = -\mathrm{Im}Z_{in}$, Equation 3.43 can be rewritten as

$$R_S - jX_S = Z_0 \frac{R_L + j(X_L + Z_0 \tan\theta)}{Z_0 - X_L \tan\theta + jR_L \tan\theta} \tag{3.45}$$

For a quarter-wavelength transformer, Equation 3.45 can be divided into two separate equations representing the real and imaginary parts of the source impedance Z_S as

$$R_S = Z_0^2 \frac{R_L}{R_L^2 + X_L^2} \tag{3.46}$$

$$X_S = -Z_0^2 \frac{X_L}{R_L^2 + X_L^2} \tag{3.47}$$

For a pure real load impedance $R_L = \mathrm{Re}Z_L$ when $X_L = 0$, a quarter-wave transmission line with the characteristic impedance Z_0 can provide impedance matching with a pure real source R_S according to

$$Z_0 = \sqrt{R_S R_L} \tag{3.48}$$

Generally, Equation 3.45 can be divided into two equations representing the real and imaginary parts as

$$R_S(Z_0 - X_L \tan\theta) - R_L(Z_0 - X_S \tan\theta) = 0 \tag{3.49}$$

$$X_S(X_L \tan\theta - Z_0) - Z_0(X_L + Z_0 \tan\theta) + R_S R_L \tan\theta = 0 \tag{3.50}$$

Solving Equations 3.49 and 3.50 for the two independent variables Z_0 and θ yields

$$Z_0 = \sqrt{\frac{R_S(R_L^2 + X_L^2) - R_L(R_S^2 + X_S^2)}{R_L - R_S}} \tag{3.51}$$

$$\theta = \tan^{-1}\left(Z_0 \frac{R_S - R_L}{R_S X_L - X_S R_L}\right) \tag{3.52}$$

As a result, the transmission line having the characteristic impedance Z_0 determined by Equation 3.51 and the electrical length θ determined by Equation 3.52 can match any source and load impedances when the impedance ratio gives a positive value inside the square root in Equation 3.51.

In practice, to simplify the power amplifier designs at microwave frequencies, the simple matching circuits are very often used, which can include an L-transformer with a series transmission line as the basic matching section. It is convenient to analyze the transforming properties of this matching circuit by substituting the equivalent transformation of the parallel RX circuit to the series circuit. For example, R_1 is the resistance and $X_1 = -1/\omega C$ is the reactance of the impedance $Z_1 = jR_1 X_1/(R_1 + jX_1)$ for the parallel RC circuit, and $R_{in} = \mathrm{Re}\, Z_{in}$ is the resistance and $X_{in} = \mathrm{Im}\, Z_{in}$ is the reactance of the input impedance $Z_{in} = R_{in} + jX_{in}$ for the loaded series transmission-line circuit shown in Figure 3.18. For a conjugate matching when $Z_1 = Z_{in}^*$, one can obtain

$$\frac{R_1 X_1^2}{R_1^2 + X_1^2} + j\frac{R_1^2 X_1}{R_1^2 + X_1^2} = R_{in} - jX_{in} \tag{3.53}$$

The solution of Equation 3.53 can be written in the form of two expressions for real and imaginary impedance parts by

$$R_1 = R_{in}(1 + Q^2) \tag{3.54}$$

$$X_1 = -X_{in}(1 + Q^{-2}) \tag{3.55}$$

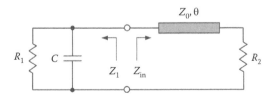

FIGURE 3.18
L-transformer with series transmission line.

where $Q = R_1/|X_1| = X_{in}/R_{in}$ is a quality factor equal for both parallel capacitive and series transmission-line circuits. By using Equation 3.43, the real and imaginary parts of the input impedance Z_{in} can be written as

$$R_{in} = Z_0^2 R_2 \frac{1 + \tan^2\theta}{Z_0^2 + (R_2\tan\theta)^2} \tag{3.56}$$

$$X_{in} = Z_0\tan\theta \frac{Z_0^2 - R_2^2}{Z_0^2 + (R_2\tan\theta)^2} \tag{3.57}$$

From Equation 3.57, it follows that an inductive input impedance (necessary to compensate for the capacitive parallel component) is provided when $Z_0 > R_2$ for $\theta < \pi/2$ and $Z_0 < R_2$ for $\pi/2 < \theta < \pi$. As a result, to transform the resistance R_1 into the other resistance R_2 at the given frequency, it is necessary to connect a two-port L-transformer with a shunt capacitor and series transmission line between them. When one parameter (usually the characteristic impedance Z_0) is known in advance, the matching circuit parameters can be calculated from the following two equations:

$$C = \frac{Q}{\omega R_1} \tag{3.58}$$

$$\sin 2\theta = \frac{2Q}{(Z_0/R_2) - (R_2/Z_0)} \tag{3.59}$$

where

$$Q = \sqrt{\frac{R_1}{R_2}\left[\cos^2\theta + \left(\frac{R_2}{Z_0}\right)^2\sin^2\theta\right] - 1} \tag{3.60}$$

is the circuit quality factor defined as a function of the resistances R_1 and R_2 and the parameters of the transmission line (characteristic impedance Z_0 and electrical length θ).

As it follows from Equations 3.59 and 3.60, the electrical length θ can be calculated as a result of the numerical solution of a transcendental equation with one unknown parameter θ. In this case, it is more convenient to rewrite them in the implicit form of

$$\frac{R_1}{R_2} = \frac{1 + ((Z_0/R_2) - (R_2/Z_0))^2\sin^2\theta\cos^2\theta}{\cos^2\theta + (R_2/Z_0)^2\sin^2\theta} \tag{3.61}$$

A π-transformer can be realized by back-to-back connection of the two L-transformers, as shown in Figure 3.19a, where the resistances R_1 and R_2 are transformed to some intermediate resistance R_0. In this case, to minimize the electrical length of the transmission line, the value of R_0 should be smaller than that of both R_1 and R_2, that is, $R_0 < (R_1, R_2)$. The same procedure for a T-transformer shown in Figure 3.19b gives the value of R_0 that is larger than that of both R_1 and R_2, that is, $R_0 > (R_1, R_2)$. Then, for a T-transformer, the two shunt-adjacent

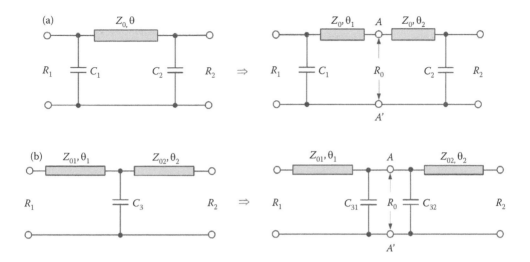

FIGURE 3.19
π- and T-transformers with transmission lines.

capacitances are combined. For a π-transformer, the two adjacent series transmission lines are combined into a single transmission line with the total electrical length.

For a π-transformer, the electrical lengths of each part of the combined transmission line can be calculated by equating the imaginary parts of the impedances from both sides at the reference plane A–A′ to zero, which means that the intermediate impedance R_0 is real. This leads to a system of the two quadratic equations to calculate the electrical lengths θ_1 and θ_2 of the combined series transmission line written as

$$\tan^2\theta_1 - \frac{R_1}{Z_0 Q_1}\left[1 - (1 + Q_1^2)\left(\frac{Z_0}{R_1}\right)^2\right]\tan\theta_1 - 1 = 0 \tag{3.62}$$

$$\tan^2\theta_2 - \frac{R_2}{Z_0 Q_2}\left[1 - (1 + Q_2^2)\left(\frac{Z_0}{R_2}\right)^2\right]\tan\theta_2 - 1 = 0 \tag{3.63}$$

where $Q_1 = \omega C_1 R_1$ and $Q_2 = \omega C_2 R_2$.

A widely used two-port π-transformer with two shunt capacitors along with its design formulas is shown in Figure 3.20 [4,11]. Such a transformer can be conveniently used as an output-matching circuit when the device collector (or drain–source capacitance) can be considered as the first shunt capacitance and the parasitic series lead inductance can easily be added to a series transmission line. Also, it is convenient to use this transformer as the matching circuits in balanced power amplifiers, where the shunt capacitors can be connected between series transmission lines due to the effect of virtual grounding. The schematic of a transmission-line two-port T-transformer with series and shunt capacitors along with the design formulas is given in Figure 3.21.

Generally, the input impedance of the transmission line at a particular frequency can be equivalently expressed as that of a lumped element. If load represents a short when $Z_L = 0$, from Equation 3.43, it follows that

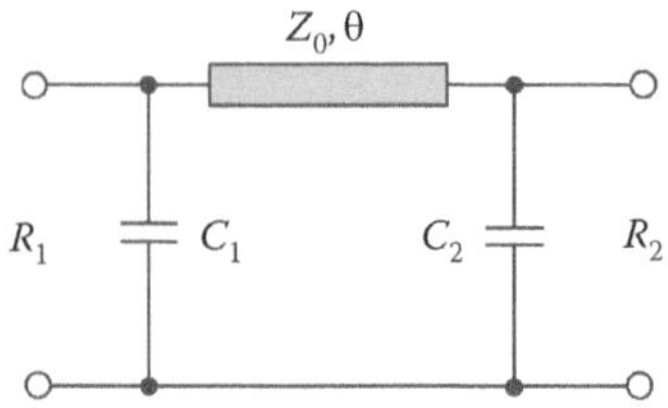

$$\omega C_1 = \frac{Q}{R_1} \qquad \omega C_2 = \frac{Q}{R_2} \qquad \theta = \theta_1 + \theta_2$$

$$Q_2 = \sqrt{\frac{R_2}{R_1}(1 + Q_1^2) - 1} \qquad Q_1 > \sqrt{\frac{R_1}{R_2} - 1}$$

FIGURE 3.20
Transmission-line π-transformer and relevant equations.

$$Z_{in} = jZ_0 \tan \theta \tag{3.64}$$

which corresponds to the inductive input impedance for $\theta < \pi/2$. The equivalent inductance at a certain radian frequency ω is calculated as

$$L = \frac{X_{in}}{\omega} = \frac{Z_0 \tan \theta}{\omega} \tag{3.65}$$

where $X_{in} = \text{Im} Z_{in}$ is the input reactance. This means that the network shunt inductor can equivalently be replaced with a short-circuited transmission line having the characteristic impedance Z_0 and electrical length θ.

Similarly, when $Z_L = \infty$,

$$Z_{in} = -jZ_0 \cot \theta \tag{3.66}$$

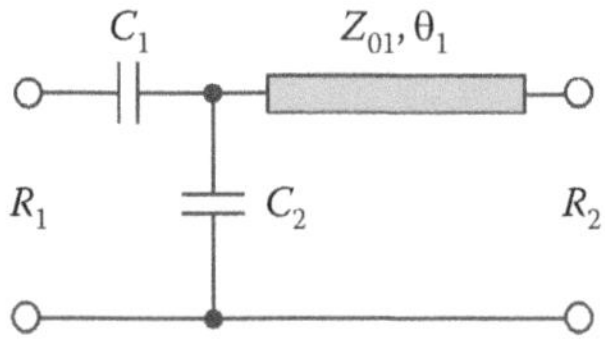

$$\omega C_1 = \frac{1}{R_1 Q_1} \qquad \sin 2\theta_1 = 2Q_2 / \left(\frac{Z_{01}}{R_2} - \frac{R_2}{Z_{01}} \right) \qquad \omega C_2 = \frac{Q_2 - Q_1}{R_1(1 + Q_1^2)}$$

$$Q_1 = \sqrt{\frac{R_2}{R_1} \frac{1 + Q_2^2}{\cos^2 \theta_1 + (R_2 / Z_{01}) \sin^2 \theta_1} - 1} \qquad Q_2 > \sqrt{\frac{R_1}{R_2} - 1}$$

FIGURE 3.21
Transmission-line T-transformer and relevant equations.

which corresponds to the capacitive input impedance for $\theta < \pi/2$. The equivalent capacitance at a certain radian frequency ω is obtained by

$$C = -\frac{1}{\omega X_{in}} = \frac{\tan \theta}{\omega Z_0} \tag{3.67}$$

Consider the design example of a broadband 150-W power amplifier for TV applications, which is required to operate over a frequency bandwidth of 470–860 MHz with a power gain of more than 10 dB at a supply voltage of 28 V. In this case, it is convenient to use a high-power-balanced LDMOS transistor as an active device, which is specially designed for UHF TV transmitters. Let us assume that the manufacturer states the input impedance for each transistor-balanced part as $Z_{in} = (1.7 + j1.3)\ \Omega$ at the center bandwidth frequency $f_0 = \sqrt{470 \times 860} = 635$ MHz. The input impedance Z_{in} represents a series combination of the input resistance and inductive reactance. To cover the required frequency bandwidth, the low-Q-matching circuits should be used to reduce an in-band amplitude ripple and to improve an input *VSWR*.

To achieve a 3-dB frequency bandwidth, the value of a circuit quality factor must be less than $Q = 635/(860 - 470) = 1.63$. On the basis of this value of Q, the next step is to define a number of matching sections. For example, for a single-stage input lumped matching circuit, the value of a quality factor Q is chosen as

$$Q > \sqrt{\frac{50}{1.7} - 1} = 5.33$$

which means that the entire frequency range can be appropriately covered using only a multistage-matching circuit.

In this case, the device input quality factor is calculated as $Q_{in} = 1.3/1.7 = 0.76$, which is smaller than the required value of 1.63 to provide the broadband performance. It is very convenient to design the input-matching circuit (as well as the output-matching circuit) by using simple low-pass L-transformers with a constant value of Q composed of a series transmission line and a shunt capacitor each for both balanced parts of the active device. Then, these two input-matching circuits are combined by inserting the shunt capacitors, the values of which are reduced twice, between the two series transmission lines.

To match the series input inductive impedance Z_{in} with the standard 50-Ω source, it is best to use three low-pass transmission-line L-transformers, as shown in Figure 3.22. In this case, the input resistance R_{in} can be assumed to be constant over the entire frequency

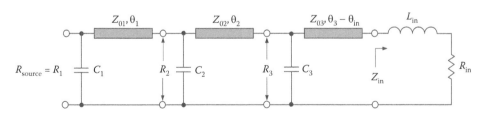

FIGURE 3.22
Complete broadband input-matching circuit.

range. At the center bandwidth frequency of 635 MHz, the input inductance is approximately equal to 0.3 nH. Taking this inductance into account, it is necessary to subtract the appropriate value of the electrical length θ_{in} from the total electrical length θ_3. Owing to the short-length size of this transmission line when $\tan \theta_{in} \approx \theta_{in}$, a value of θ_{in} can be easily calculated as

$$\theta_{in} \cong \frac{X_{in}}{Z_0} = \frac{\omega L_{in}}{Z_0} \qquad (3.68)$$

According to Equation 3.60, there are two simple possibilities to provide an input matching using a technique with equal quality factors of the L-transformers. One option is to use the same values of the characteristic impedance for all transmission lines, and the other one is to use the same electrical lengths for all transmission lines. When considering the first approach, which also allows direct use of the Smith chart, it is convenient to choose the value of the characteristic impedance as $Z_0 = Z_{01} = Z_{02} = Z_{03} = 50 \,\Omega$. In this case, the ratio of the input and output resistances can be written as

$$\frac{R_1}{R_2} = \frac{R_2}{R_3} = \frac{R_3}{R_{in}} \qquad (3.69)$$

which results in $R_2 = 16.2 \,\Omega$ and $R_3 = 5.25 \,\Omega$ for $R_{source} = R_1 = 50 \,\Omega$ and $R_{in} = 1.7 \,\Omega$. The values of the corresponding electrical lengths are determined from Equation 3.61 as $\theta_1 = 30°$, $\theta_2 = 7.5°$, and $\theta_3 = 2.4°$.

To calculate the quality factor Q (equal for each L-transformer) from Equation 3.60, it is enough to know the electrical length θ_1 of the first L-transformer. The remaining two electrical lengths θ_2 and θ_3 can be directly obtained from Equation 3.59. As a result, the quality factor of each L-transformer is equal to a value of $Q = 1.2$. The values of the shunt capacitances using Equation 3.58 are $C_1 = 6$ pF, $C_2 = 19$ pF, and $C_3 = 57$ pF.

For a constant Q, we can significantly simplify the design of the multisection-matching circuit by using the Smith chart. After calculating the value of Q, it is necessary to plot a constant Q-circle on the Smith chart. Figure 3.23 shows the input-matching circuit design using the Smith chart with a constant Q-circle, where the curves for the series transmission lines represent the arcs of the circles with the center point at the center of the Smith chart. The capacitive traces are moved along the circles with the increasing susceptances and constant conductances.

Another approach assumes the same values of the electrical lengths $\theta = \theta_1 = \theta_2 = \theta_3$ and calculates the characteristic impedances of series transmission lines from Equation 3.60 at equal ratios of the input and output resistances according to Equation 3.69. Such an approach is more convenient in practical design, because, when using the transmission lines with standard characteristic impedance $Z_0 = 50 \,\Omega$, the electrical length of the transmission line adjacent to the transistor input terminal is usually too short. In this case, it makes sense to set the characteristic impedance $Z_{01} = 50 \,\Omega$ only for the first transmission line. Then, the value of $\theta = 30°$ is determined from Equation 3.61. The subsequent calculation of Q from Equation 3.60 for fixed θ and Z_0/R_2 yields $Q = 1.2$. The characteristic impedances of the remaining two transmission lines are then calculated from Equation 3.60. Their values are $Z_{02} = 15.7 \,\Omega$ and $Z_{03} = 5.1 \,\Omega$ with the same values of the shunt capacitances.

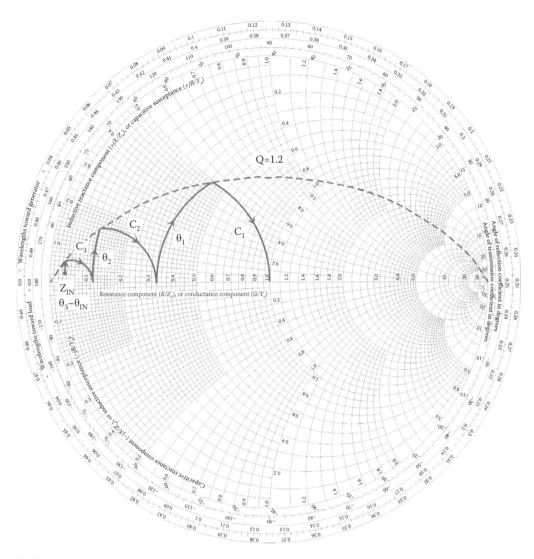

FIGURE 3.23
Smith chart with elements from Figure 5.22.

3.2 Bode–Fano Criterion

Generally, the design for a broadband-matching circuit should solve a problem with contradictory requirements when a wider-matching bandwidth is required with a minimum reflection coefficient, or how to minimize the number of the matching network sections for a given wideband specification. The necessary requirements are determined by the Bode–Fano criterion, which gives (for certain canonical types of load impedances) a theoretical limit on the maximum reflection coefficient magnitude that can be obtained with an arbitrary-matching network [12,13].

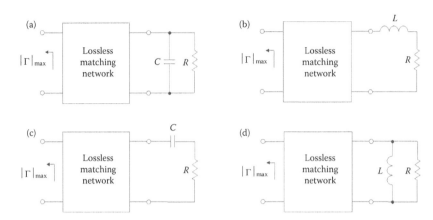

FIGURE 3.24
Loaded lossless-matching circuits.

For the lossless-matching networks with a parallel RC load shown in Figure 3.24a and with a series LR load shown in Figure 3.24b, the Bode–Fano criterion states that

$$\int_0^\infty \ln \frac{1}{|\Gamma(\omega)|}\, d\omega \le \frac{\pi}{\tau} \tag{3.70}$$

where $\Gamma(\omega)$ is the input reflection coefficient seen looking into the arbitrary lossless-matching network and $\tau = RC = L/R$.

For the lossless-matching networks with a series RC load shown in Figure 3.24c and with a parallel LR load shown in Figure 3.24d, the Bode–Fano integral is written as

$$\int_0^\infty \omega^{-2} \ln \frac{1}{|\Gamma(\omega)|}\, d\omega \le \pi\tau \tag{3.71}$$

The mathematical relationships expressed by Equations 3.70 and 3.71 reflect the flat responses of an ideal filter over the required frequency bandwidth, as shown in Figure 3.25 for two different cases. For the same load, both plots illustrate the important trade-off: the wider the matching network bandwidth, the worse the reflection coefficient magnitude.

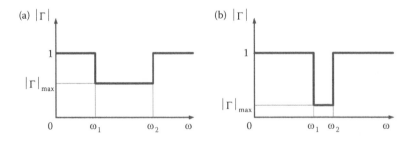

FIGURE 3.25
Ideal filter flat responses with (a) wide and (b) narrow bandwidths.

From Equation 3.70, it follows that $|\Gamma(\omega)|$ is constant and equal to $|\Gamma|_{max}$ over a frequency band of width $\Delta\omega$ and $|\Gamma(\omega)| = 1$; otherwise,

$$\int_0^\infty \ln\frac{1}{|\Gamma(\omega)|} \, d\omega = \int_{\omega_1}^{\omega_2} \ln\frac{1}{|\Gamma(\omega)|} \, d\omega = \Delta\omega \ln\frac{1}{|\Gamma|_{max}} \le \frac{\pi}{\tau} \qquad (3.72)$$

As a result,

$$|\Gamma|_{max} = \exp\left(\frac{-\pi}{\Delta\omega\tau}\right) \qquad (3.73)$$

where $\Delta\omega = \omega_2 - \omega_1$.

Similarly, for the lossless network with a series RC load and with a parallel LR load,

$$|\Gamma|_{max} = \exp\left(\frac{-\pi\omega_0^2\tau}{\Delta\omega}\right) \qquad (3.74)$$

where $\omega_0 = \sqrt{\omega_1\omega_2}$ is the center bandwidth frequency. It should be noted that the theoretical bandwidth limits can be realized only with an infinite number of matching network sections. The frequency bandwidth with a maximum reflection coefficient magnitude is determined by a loaded quality factor $Q_L = \omega_0\tau$ for the series RL or parallel RC circuit and by $Q_L = 1/(\omega_0\tau)$ for the parallel RL or series RC circuit, respectively. The Chebyshev matching transformer with a finite number of sections can be considered as a close approximation to the ideal passband network when the ripple of the Chebyshev response is made equal to $|\Gamma|_{max}$. By combining the matching theory with the closed formulas for the element values of a Chebyshev low-pass filter, explicit formulas for optimum matching networks can be obtained in certain simple but common cases [14]. For example, analytic closed-form solutions for the design of optimum matching networks up to order $n = 4$ can be derived [15].

Generally, Equations 3.73 and 3.74 can be rewritten in a simplified form

$$|\Gamma|_{max} = \exp\left(-\pi\frac{Q_0}{Q_L}\right) \qquad (3.75)$$

where $Q_0 = \omega_0/\Delta\omega$.

3.3 Broadband-Matching Networks with Lumped Elements

To correctly design the broadband-matching circuits for the transistor power amplifiers, it is necessary to transform and match the device complex impedances with the source and load impedances, which are usually resistive and equal to 50 Ω. For high-power or low-supply voltage cases, the device impedances may be small enough, and it needs to include

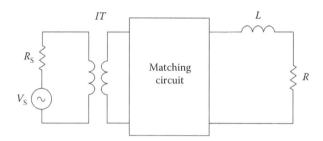

FIGURE 3.26
Matching circuit with IT.

an ideal transformer (*IT*) together with a matching circuit, as shown in Figure 3.26. In this case, such an IT provides only a required transformation between the source resistance R_S and the input impedance of the matching circuit and does not have any effect on the circuit frequency characteristics.

To implement such an IT to the impedance-transforming circuit, it is useful to operate with the Norton transform. As a result, an IT with two capacitors C_1 and C_2, which is shown in Figure 3.27a, can be equivalently replaced by three capacitors C_I, C_{II}, and C_{III} connected in the form of a π-transformer, as shown in Figure 3.27b. Their values are determined by

$$C_I = n_T(n_T - 1)C_1 \tag{3.76}$$

$$C_{II} = n_T C_1 \tag{3.77}$$

$$C_{III} = C_2 - (n_T - 1)C_1 \tag{3.78}$$

where n_T is the transformation coefficient. In this case, all the parameters of these two-port networks are assumed identical at any frequency. However, such a replacement is possible only if the capacitance C_{III} obtained by Equation 3.78 is positive and, consequently, physically realizable.

Similarly, an IT with two inductors L_1 and L_2, as shown in Figure 3.28a, can be replaced by three inductors L_I, L_{II}, and L_{III} connected in the form of a *T*-transformer, as shown in Figure 3.28b, with values determined by

$$L_I = n_T(n_T - 1)L_2 \tag{3.79}$$

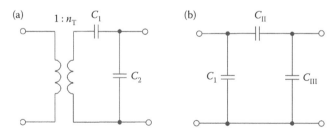

FIGURE 3.27
Capacitive impedance-transforming circuits.

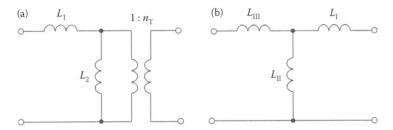

FIGURE 3.28
Inductive impedance-transforming circuits.

$$L_{II} = n_T L_2 \tag{3.80}$$

$$L_{III} = L_1 - (n_T - 1)L_2 \tag{3.81}$$

Again, this replacement is possible only if the inductance L_{III} defined by Equation 3.81 is positive and, consequently, physically realizable.

The broadband impedance-transforming circuits generally represent the transforming bandpass filters when the in-band matching requirements with a specified ripple must be satisfied. In this case, the out-of-band mismatching can be very significant. One of the design methods of such matching circuits is based on the theory of transforming the low-pass filters of a ladder configuration of series inductors alternating with shunt capacitors, whose two-section equivalent representation is shown in Figure 3.29. For a large ratio of R_0/R_5, mismatching at zero frequency is sufficiently high, and such a matching circuit can be treated as a bandpass impedance-transforming filter.

Table 3.1 gives the maximum passband ripples and coefficients g_1 and g_2 required to calculate the parameters of a two-section low-pass Chebyshev filter for different transformation ratios $r = R_0/R_5$ and frequency bandwidths $w = 2(f_2 - f_1)/(f_2 + f_1)$, where f_2 and f_1 are the high- and low-bandwidth frequencies, respectively [16]. The coefficients g_3 and g_4 are calculated as $g_3 = rg_2$ and $g_4 = g_1/r$, respectively, and the circuit elements can be obtained by

$$C_1 = \frac{g_1}{\omega_0 R_0} \quad C_3 = \frac{g_3}{\omega_0 R_0} \tag{3.82}$$

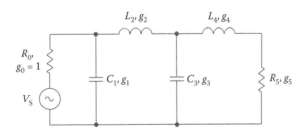

FIGURE 3.29
Two-section impedance-transforming circuit.

TABLE 3.1

Two-Section Low-Pass Chebyshev Filter
Parameters

r	w	Ripple (dB)	g_1	g_2
5	0.1	0.000087	1.26113	0.709217
	0.2	0.001389	1.27034	0.704050
	0.3	0.007023	1.28561	0.695548
	0.4	0.022109	1.30687	0.638859
10	0.1	0.000220	1.60350	0.591627
	0.2	0.003516	1.62135	0.585091
	0.3	0.017754	1.65115	0.574412
	0.4	0.055746	1.69304	0.559894
25	0.1	0.000625	2.11734	0.462747
	0.2	0.009993	2.15623	0.454380
	0.3	0.050312	2.22189	0.440863
	0.4	0.156725	2.31517	0.422868
50	0.1	0.001303	2.57580	0.384325
	0.2	0.020801	2.64380	0.374422
	0.3	0.104210	2.75961	0.358638
	0.4	0.320490	2.92539	0.338129

$$L_2 = \frac{g_2 R_0}{\omega_0} \quad L_4 = \frac{g_4 R_0}{\omega_0} \tag{3.83}$$

where $\omega_0 = \sqrt{\omega_1 \omega_2}$ is the center bandwidth frequency.

As an example, consider the design of a broadband input- matching circuit in the form of a two-section low-pass transforming filter shown in Figure 3.29, with a center bandwidth frequency $f_0 = 3$ GHz, to match the source impedance $R_S = R_0 = 50\ \Omega$ with the device input impedance $Z_{in} = R_{in} + j\omega_0 L_{in}$, where $R_{in} = R_5 = 2\ \Omega$, $L_{in} = L_4 = 0.223$ nH, and $\omega_0 = 2\pi f_0$. The value of the series input device inductance $L_{in} = L_4$ is chosen to satisfy the requirements of Table 3.1 for $r = 25$ and $w = 0.4$ with a maximum ripple of 0.156725 and $g_1 = 2.31517$. From Equation 3.83, for $g_2 = 0.422868$, it follows that

$$L_4 = \frac{g_4 R_0}{\omega_0} = \frac{g_1 R_0}{\omega_0 r} = 0.223\,\text{nH}$$

As a result, the circuit parameters shown in Figure 3.30a are calculated from Equations 3.82 and 3.83, thus resulting in the corresponding circuit frequency response shown in Figure 3.30b with the required passband from 2.6 to 3.4 GHz. The particular value of the inductance L_{in} is chosen for the design convenience. If this value differs from the required value, it means that it is necessary to change the maximum frequency bandwidth, the power ripple, or the number of ladder sections.

Another approach is based on the transformation from the low-pass impedance-transforming prototype filters, the simple L-, T-, and π-type equivalent circuits of which are

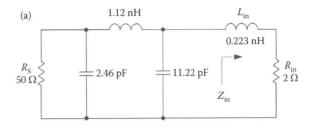

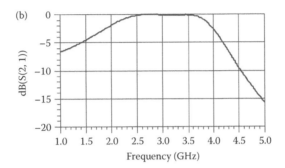

FIGURE 3.30
Two-section broadband low-pass-matching circuit and its frequency response.

shown in Figure 3.31, to the bandpass impedance-transforming filters. Table 3.2 gives the parameters of the low-pass impedance-transforming Chebyshev filters–prototypes for different maximum in-band ripples and the number of elements n [17]. This transformation can be obtained using the frequency substitution as

$$\omega \to \frac{\omega_0}{\Delta\omega}\left(\frac{\omega}{\omega_0} - \frac{\omega_0}{\omega}\right) \qquad (3.84)$$

where $\omega_0 = \sqrt{\omega_1\omega_2}$ is the center bandwidth frequency, $\Delta\omega = \omega_2 - \omega_1$ is the passband, and ω_1 and ω_2 are the low and high edges of the passband, respectively.

As a result, a series inductor L_k is transformed into a series LC-circuit according to

$$\omega L_k = \frac{\omega_0}{\Delta\omega}\left(\frac{\omega}{\omega_0} - \frac{\omega_0}{\omega}\right)L_k = \omega L'_k - \frac{1}{\omega C'_k} \qquad (3.85)$$

where

$$L'_k = \frac{L_k}{\Delta\omega} \qquad C'_k = \frac{\Delta\omega}{\omega_0^2 L_k} \qquad (3.86)$$

Similarly, a shunt capacitor C_k is transformed into a shunt LC-circuit as

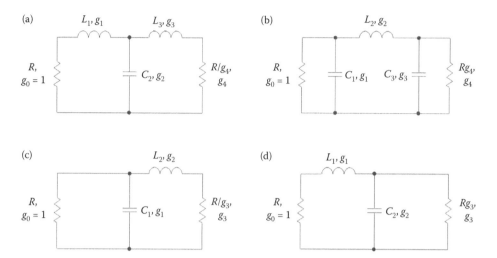

FIGURE 3.31
Lumped L-, π-, and T-type impedance-transforming circuits.

$$\omega C_k = \frac{\omega_0}{\Delta\omega}\left(\frac{\omega}{\omega_0} - \frac{\omega_0}{\omega}\right)C_k = \omega C_k' - \frac{1}{\omega L_k'} \tag{3.87}$$

where

$$C_k' = \frac{C_k}{\Delta\omega} \quad L_k' = \frac{\Delta\omega}{\omega_0^2 C_k} \tag{3.88}$$

The low-pass impedance-transforming prototype filter will be transformed to the bandpass impedance-transforming filter when all its series elements are replaced by the series-resonant circuits and all its parallel elements are replaced by the parallel resonant

TABLE 3.2
Parameters of Low-Pass Chebyshev Filters–Prototypes

Ripple (dB)	n	g_1	g_2	g_3	g_4
0.01	1	0.0960	1.0000		
	2	0.4488	0.4077	1.1007	
	3	0.6291	0.9702	0.6291	1.0000
0.1	1	0.3052	1.0000		
	2	0.8430	0.6220	1.3554	
	3	1.0315	1.1474	1.0315	1.0000
0.2	1	0.4342	1.0000		
	2	1.0378	0.6745	1.5386	
	3	1.2275	1.1525	1.2275	1.0000
0.5	1	0.6986	1.0000		
	2	1.4029	0.7071	1.9841	
	3	1.5963	1.0967	1.5963	1.0000

circuits, where each of them are tuned to the center bandwidth frequency ω_0. The bandpass filter elements can be calculated from

$$\Delta\omega C_k = \frac{g_k}{R} \tag{3.89}$$

$$\Delta\omega L_k = g_k R \tag{3.90}$$

where k is an element serial number for the low-pass prototype filter and g_k are the appropriate coefficients given by Table 3.2.

Generally, the low-pass prototype filters obtained on the basis of their bandpass filters do not perform an impedance transformation. The input and output resistances are either equal for the symmetric *T*- or π-type filters shown in Figure 3.31a and b where $g_4 = 1$ or their ratio is too small for *L*-type filters, as those shown in Figure 3.31c and d, where $g_3 < 2$. Therefore, in this case, it is necessary to use an IT concept. This approach is based on using the existing data tables, from which the parameters of such impedance-transforming networks can be easily calculated for a given quality factor of the device input or output circuit. However, they can also be very easily verified or optimized by using a CAD optimization procedure incorporating it in any comprehensive circuit simulator.

Consider the design example of the broadband interstage impedance-transforming filter with the center bandwidth frequency of 1 GHz to match the output driver-stage circuit with the input final-stage circuit of the power amplifier, as shown in Figure 3.32a [18]. In this case, it is initially convenient to convert the parallel connection of the device output resistance R_{out} and capacitance C_{out} into the corresponding series connection at the center bandwidth frequency ω_0 according to

$$R'_{out} = \frac{R_{out}}{1 + (\omega_0 R_{out} C_{out})^2} \tag{3.91}$$

$$C'_{out} = \frac{1 + (\omega_0 R_{out} C_{out})^2}{(\omega_0 R_{out})^2 C_{out}} \tag{3.92}$$

as shown in Figure 3.32b.

For the three-element low-pass impedance-transforming prototype filter shown in Figure 3.31a with the maximum in-band ripple of 0.1 dB, we can obtain $g_1 = g_3 = 1.0315$, $g_2 = 1.1474$, and $g_4 = 1$ for $n = 3$ from Table 3.2. According to Equation 3.90, the relative frequency bandwidth in this case is defined as

$$\frac{\Delta\omega}{\omega_0} = \frac{g_1 R_{in}}{\omega_0 L_{in}} = 16.5\%$$

based on a value of which the shunt capacitance C_2 can be calculated using Equation 3.89, thus resulting in a capacitive reactance equal to 0.215 Ω. The inductive reactance corresponding to a series inductance L_{in} is equal to 9.42 Ω.

To convert the low-pass filter into its bandpass prototype, it is necessary to connect the capacitor in series to the input inductor and the inductor in parallel to the shunt

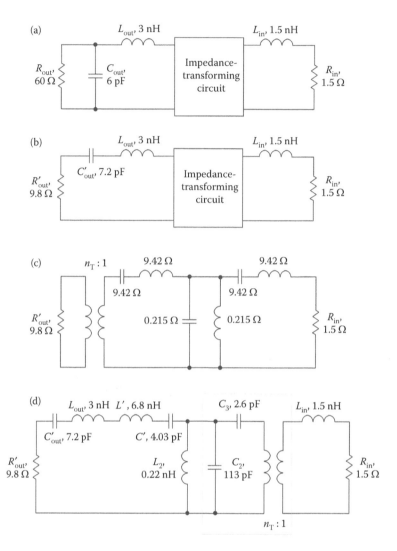

FIGURE 3.32
Impedance-transformer design procedure using a low-pass filter–prototype.

capacitor and calculate with the same reactances to resonate at the center bandwidth frequency ω_0, as shown in Figure 3.32c, where an *IT* with the transformation coefficient $n_T = \sqrt{9.8/1.5} = 2.556$ is included. Here, the reactances for each series element are equal to 9.42 Ω, whereas the reactances for each parallel element are equal to 0.215 Ω, respectively. Then, moving the corresponding elements with transformed parameters (each inductive and capacitive reactance is multiplied by n_T^2) to the left-hand side of *IT* to apply a Norton transform gives the circuit shown in Figure 3.32d, where the required series elements with reactances of $9.42 n_T^2$ Ω are realized by the inductance L_{out}, converted device output capacitance C'_{out}, and additional elements L' and C'. Finally, by using a Norton transform shown in Figure 3.27 with the *IT* and two capacitors, the resulting impedance-matching bandpass filter is obtained, as shown in Figure 3.33a. The frequency response of the filter with minimum in-band ripple and significant out-of-band suppression is shown in Figure 3.33b.

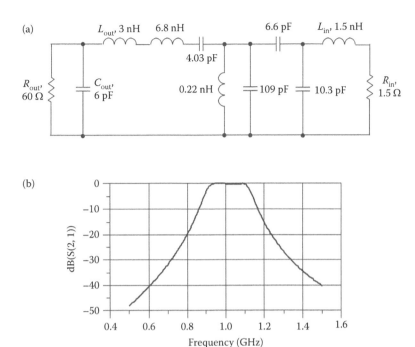

FIGURE 3.33
Impedance-transforming bandpass filter and its frequency response.

In the case of serious difficulties with practical implementation of a very small inductance of 0.22 nH or a very large capacitance of 109 pF, it is possible to design a multi-section low-pass impedance-transforming circuit.

Figure 3.34a shows the circuit schematic of a microwave broadband amplifier using a 1-μm GaAs FET-packaged transistor, where the input multisection-matching circuit is

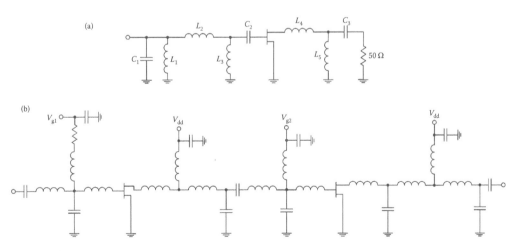

FIGURE 3.34
Schematics of broadband lumped-element microwave FET amplifiers.

designed to provide the required gain taper and both input- and output-matching circuits are optimized to provide broadband impedance transformation [19]. As a result, a nominal power gain of 8 dB with a maximum deviation of ±0.07% in a frequency range of 7–14 GHz was achieved. In the first monolithic broadband GaAs FET amplifier, the input- and output-matching circuits were based on lumped elements fabricated together with the FET device on a semi-insulating high-resistivity gallium-arsenide substrate with a total size of 1.8×1.2 mm^2, providing a power gain of 4.5 ± 0.9 dB with an output power of 11 dBm at 1-dB gain compression from 7.0 to 11.7 GHz [20]. The circuit diagram of a two-stage pHEMT MMIC power amplifier for *Ku*-band applications is shown in Figure 3.34b, where the lumped components were used in the input, interstage, and output-matching circuits to minimize the overall chip size [21]. Here, the topology of each matching network represents a double-resonant circuit to form a broadband impedance transformer, which includes a shunt inductor in series with a bypass capacitor to provide a dc path, a series-blocking capacitor, and a low-pass *L*-section transformer. In this case, for a 8.4-mm driver-stage pHEMT and a 16.8-mm power-stage pHEMT, a saturated output power of 38.1 dBm (6.5 W), a small-signal gain of 10.5 dB, and a peak *PAE* of 24.6% from 13.6 to 14.2 GHz were achieved with a chip size of MMIC as small as 3.64×2.35 mm^2. On the basis of *T*-shape-combining transformers with three individual inductors implemented in a 0.15-µm pHEMT technology, a broadband MMIC power amplifier combining two pHEMT devices with an overall 400-µm gate-width size achieved a saturated output power of 22–23.5 dBm and a power gain of more than 10 dB from 17 to 35 GHz [22]. In a 90-nm standard CMOS process, a canonical doubly terminated third-order bandpass network was converted into the output-matching topology, which provides both impedance transformation and differential-to-single-ended power combining [23]. The power amplifier achieved a 3-dB bandwidth from 5.2 to 13 GHz with a 25.2-dBm peak-saturated output power and a peak *PAE* of 21.6%.

3.4 Broadband-Matching Networks with Mixed Lumped and Distributed Elements

The matching circuits, which incorporate mixed, lumped, and transmission-line elements, are widely used both in hybrid and monolithic design techniques. Such matching circuits are very convenient when designing the push–pull power amplifiers with the effect of virtual grounding, where the shunt capacitors are connected between two series microstrip lines. According to the quasi-linear transformation technique, the basic four-step design procedure consists of an appropriate choice of the lumped-prototype schematic resulting in near-maximum gain across the required frequency bandwidth, its decomposition into subsections, their replacement by almost equivalent-distributed circuits, and then, the application of an optimization technique to minimize power variation over the operation frequency bandwidth [24].

A periodic lumped *LC* structure in the form of a low-pass ladder π-network is used as the basis for the lumped-matching prototype. Then, the lumped prototype should be split up into individual π-type sections with equal capacitances by consecutive step-by-step process and replaced by their equivalent-distributed network counterparts. Finally, the complete mixed-matching structure is optimized to improve the overall performance by employing a standard nonlinear optimization routine on the element values. Note that generally, the lumped-prototype structure can be decomposed into different subnetworks also including L-type matching sections and individual capacitors or inductors.

For a single-frequency equivalence between lumped and distributed elements, the low-pass-lumped π-type ladder section can be made equivalent to a symmetrically loaded transmission line at a certain frequency, as shown in Figure 3.35a. The transmission $ABCD$-matrices of these lumped and distributed ladder sections can be written, respectively, as

$$[ABCD]_L = \begin{bmatrix} 1 - \omega_0^2 LC & j\omega_0 L \\ j\omega_0 C(2 - \omega_0^2 LC) & 1 - \omega_0^2 LC \end{bmatrix} \tag{3.93}$$

$$[ABCD]_T = \begin{bmatrix} \cos\theta_0 - \omega_0 C_T Z_0 \sin\theta_0 & jZ_0 \sin\theta_0 \\ \dfrac{j}{Z_0}(2\omega_0 C_T Z_0 \cos\theta_0 + \sin\theta_0 - \omega_0^2 C_T^2 Z_0^2 \sin\theta_0) & \cos\theta_0 - \omega_0 C_T Z_0 \sin\theta_0 \end{bmatrix} \tag{3.94}$$

where θ_0 is the electrical length of a transmission line at the center bandwidth frequency ω_0.

Consequently, since these two circuits are equivalent, equal matrix elements $A_L = A_T$ and $B_L = B_T$ can be rewritten as

$$1 - \omega_0^2 LC = \cos\theta_0 - \omega_0 C_T Z_0 \sin\theta_0 \tag{3.95}$$

$$j\omega_0 L = jZ_0 \sin\theta_0 \tag{3.96}$$

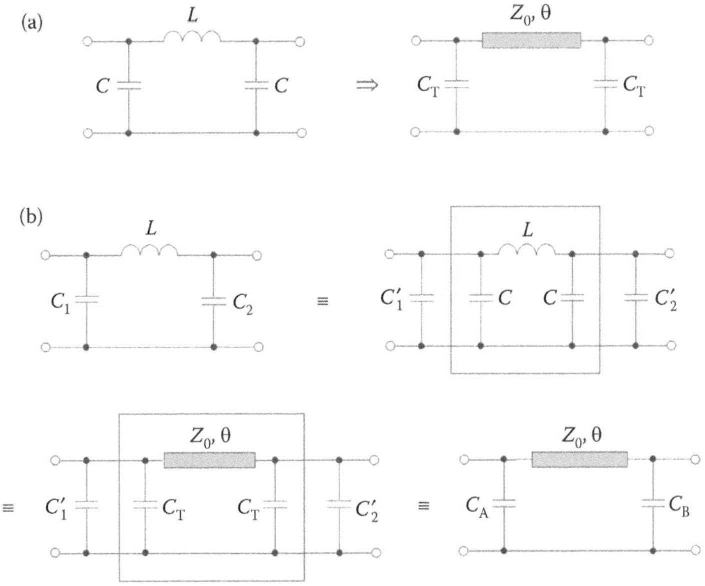

FIGURE 3.35
Transforming design procedure for lumped and distributed matching circuits.

After solving Equations 3.95 and 3.96, the characteristic impedance Z_0 and shunt capacitance C_T can be explicitly calculated by

$$Z_0 = \frac{\omega_0 L}{\sin\theta_0}$$
(3.97)

$$C_T = \frac{\cos\theta_0 + \omega_0^2 LC - 1}{\omega_0^2 L}$$
(3.98)

To provide the design method using a single-frequency-equivalent technique, the following consecutive design steps can be performed:

- Designate the lumped π-type C_1–L_1–C_2 section to be replaced.
- From a chosen π-type C_1–L_1–C_2 section, form the symmetrical C–L–C ladder section with equal capacitances C, as shown in Figure 3.35b. The choice of capacitances is arbitrary but the values cannot exceed the minimum of C_1 and C_2.
- Calculate the parameters of the symmetrical C_T–TL–C_T section using the parameters of the lumped-equivalent π-section by setting the electrical length θ_0 of the transmission line according to Equations 3.97 and 3.98. Here, it is assumed that the minimum of the capacitances C_1 and C_2 should be greater than or equal to C_T so that C_T can be readily embedded in the new C_T–TL–C_T section.
- Finally, replace the π-type C_1–L_1–C_2 ladder section by the equivalent symmetrical C_T–TL–C_T section and combine adjacent shunt capacitors, as shown in Figure 3.35b, where the loaded shunt capacitances C_A and C_B are given as $C_A = C_1' + C_T$ and $C_B = C_2' + C_T$.

Figure 3.36a shows the circuit schematic of a simulated broadband 28-V LDMOSFET power amplifier. To provide an output power of about 15 W with a power gain of more than 10 dB in a frequency range of 225–400 MHz, an LDMOSFET device with a gate geometry of 1.25 μm × 40 mm was chosen. In this case, the matching design technique is based on using multisection low-pass networks, with two π-type sections for the input-matching circuit and one π-type section for the output-matching circuit. The sections adjacent to the device input and output terminals incorporate the corresponding internal input gate–source and output drain–source device capacitances. Since a ratio between the device-equivalent output resistance at the fundamental for several tens of watts of output power and the load resistance of 50 Ω is not significant, it is sufficient to be limited to only one section for the output-matching network.

Once a matching network structure is chosen, based on the requirements for the electrical performance and frequency bandwidth, the simplest and fastest way is to apply an optimization procedure using CAD simulators to satisfy certain criteria. For such a broadband power amplifier, these criteria can be the minimum output power ripple and input return loss with maximum power gain and efficiency. In a CAD of broadband and low-noise microwave amplifier, an objective function to maximize gain while minimizing ripple and noise figure can be used [25]. To minimize the overall dimensions of the power amplifier board, the shunt microstrip line in the drain circuit can be treated as an element of the output-matching circuit and its electrical length can be considered as a variable to be optimized. Applying a nonlinear broadband CAD optimization technique

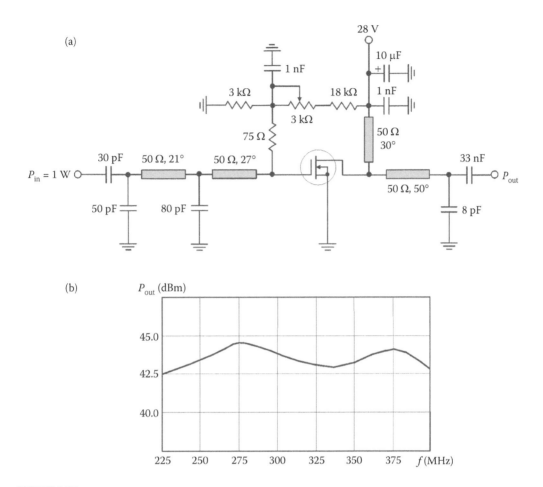

FIGURE 3.36
Circuit schematic and performance of broadband LDMOSFET power amplifier.

implemented in any high-level circuit simulator and setting the ranges of electrical length of the transmission lines between 0° and 90° and parallel capacitances from 0 to 100 pF, we can potentially obtain the parameters of the input- and output-matching circuits. The characteristic impedances of all transmission lines can be set to 50 Ω for simplicity and convenience of the circuit implementation. However, to speed up this procedure, it is best to optimize circuit parameters separately for input- and output-matching circuits with the device-equivalent input and output impedances: a series RC circuit for the device input and a parallel RC circuit for the device output. It is sufficient to use a fast linear optimization process, which will take only a few minutes to complete the matching circuit design. Then, the resulting optimized values are incorporated into the overall power amplifier circuit for each element and final optimization is performed using a large-signal-active device model. In this case, the optimization process is finalized by choosing the nominal level of input power with optimizing elements in much narrower ranges of their values of about 10%–20% for most critical elements. Figure 3.36b shows the simulated broadband power amplifier performance, with an output power of 43.5 ± 1.0 dBm and a power gain of 13.5 ± 1.0 dB in a frequency bandwidth of 225–400 MHz [4].

3.5 Matching Networks with Transmission Lines

The lumped or mixed-matching networks generally work well at sufficiently low frequencies (up to one or several gigahertz). However, the lumped elements such as inductors and capacitors are difficult to implement at microwave frequencies where they can be treated as distributed elements. In addition, the quality factors for inductors are sufficiently small so that they contribute to additional losses.

Generally, the design of a practical distributed filter circuit is based on some approximate equivalence between lumped and distributed elements, which can be established by applying Richards's transformation [26]. This implies that the distributed circuits composed of equal-length open- and short-circuited transmission lines can be treated as lumped elements under the transformation

$$s = j \tan \frac{\pi \omega}{2 \omega_0} \tag{3.99}$$

where $s = j\omega/\omega_c$ is the conventional normalized complex frequency variable and ω_0 is the radian frequency for which the transmission lines are a quarter-wavelength [27].

As a result, the one-port impedance of a short-circuited transmission line corresponds to the reactive impedance of a lumped inductor Z_L as

$$Z_L = sL = j\omega L = jL \tan \frac{\pi \omega}{2 \omega_0} \tag{3.100}$$

Similarly, the one-port admittance of an open-circuited transmission line corresponds to the reactive admittance of a lumped capacitor Y_C as

$$Y_C = sC = j\omega C = jC \tan \frac{\pi \omega}{2 \omega_0} \tag{3.101}$$

The results given in Equations 3.100 and 3.101 show that an inductor can be replaced with a short-circuited stub of the electrical length $\theta = \pi\omega/(2\omega_0)$ and characteristic impedance $Z_0 = L$, while a capacitor can be replaced with an open-circuited stub of the electrical length $\theta = \pi\omega/(2\omega_0)$ and characteristic impedance $Z_0 = 1/C$ when a unity-filter characteristic impedance is assumed.

From Equation 3.99, it follows that, for a low-pass filter prototype, the cutoff occurs when $\omega = \omega_c$, resulting in

$$\tan \frac{\pi \omega_c}{2 \omega_0} = 1 \tag{3.102}$$

which gives a stub length $\theta = 45°$ (or $\pi/4$) with $\omega_c = \omega_0/2$. Hence, the inductors and capacitors of a lumped-element filter can be replaced with the short-circuited and open-circuited stubs, as shown in Figure 3.37. Since the lengths of all stubs are the same and equal to $\lambda/8$ at the cutoff frequency ω_c, these lines are called the *commensurate lines*. At the frequency

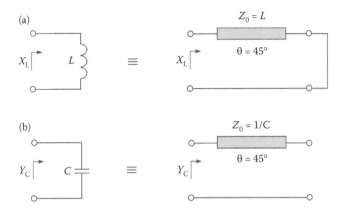

FIGURE 3.37
Equivalence between lumped elements and transmission lines.

$\omega = \omega_0$, the transmission lines will be a quarter-wavelength long, resulting in an attenuation pole. However, at any frequency away from ω_c, the impedance of each stub will no longer match the original lumped-element impedances, and the filter response will differ from the desired filter prototype response. Note that the response will be periodic in frequency, repeating every $4\omega_c$.

Since the transmission line generally represents a four-port network, it is very convenient to use a matrix technique for a filter design. In the case of cascade of several networks, the rule is that the overall matrix of the new network is simply the matrix product of the matrices for the individual networks taken in the order of connection [28]. In terms of Richards's variable, an *ABCD* matrix for a transmission line with the characteristic impedance Z_0 can be written as

$$\begin{bmatrix} A & B \\ C & D \end{bmatrix} = \frac{1}{\sqrt{1-s^2}} \begin{bmatrix} 1 & sZ_0 \\ \dfrac{s}{Z_0} & 1 \end{bmatrix} \tag{3.103}$$

representing a unit element that has a half-order transmission zero at $s = \pm 1$. The matrix of the unit element is the same as that of a transmission line of the electrical length θ and characteristic impedance Z_0. Unit elements are usually introduced to separate the circuit elements in transmission-line filters, which are otherwise located at the same physical point.

The application of Richards's transformation provides a sequence of the short-circuited and open-circuited stubs, which are then converted into a more practical circuit implementation. This can be done based on a series of equivalent circuits known as Kuroda identities, which allows these stubs to be physically separated, transforming the series stub into the shunt and changing impractical characteristic impedances into more realizable impedances [29]. The Kuroda identities use the unit elements, and these unit elements are thus commensurate with the stubs used to implement inductors and the capacitors of the prototype design. Connecting the unit element with the characteristic impedance Z_0 to the same load impedance Z_0 does not change the input impedance. The four Kuroda identities are illustrated in Figure 3.38, where the combinations of unit elements with the

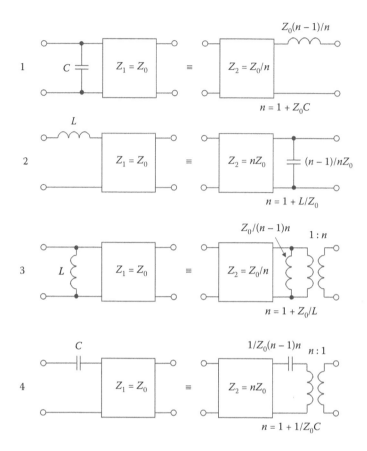

FIGURE 3.38
Four Kuroda identities.

characteristic impedance Z_0 and electrical length $\theta = 45°$, the reactive elements, and the relationships between them are given.

To prove the equivalence, consider two circuits of identity at the first row in Figure 3.38 when $ABCD$ matrix for the entire left-hand circuit can be written as

$$
\begin{bmatrix} A & B \\ C & D \end{bmatrix}_L = \frac{1}{\sqrt{1-s^2}} \begin{bmatrix} 1 & 0 \\ sC & 1 \end{bmatrix} \begin{bmatrix} 1 & sZ_1 \\ \dfrac{s}{Z_1} & 1 \end{bmatrix}
$$

$$
= \frac{1}{\sqrt{1-s^2}} \begin{bmatrix} 1 & sZ_1 \\ s\left(C + \dfrac{1}{Z_1}\right) & 1 + s^2 Z_1 C \end{bmatrix}
$$

(3.104)

where Z_1 is the characteristic impedance of the left-hand unit element.

Similarly, for the right-hand circuit,

$$\begin{bmatrix} A & B \\ C & D \end{bmatrix}_R = \frac{1}{\sqrt{1-s^2}} \begin{bmatrix} 1 & sZ_2 \\ \dfrac{s}{Z_2} & 1 \end{bmatrix} \begin{bmatrix} 1 & sL \\ 0 & 1 \end{bmatrix}$$

$$= \frac{1}{\sqrt{1-s^2}} \begin{bmatrix} 1 & s(Z_2 + L) \\ \dfrac{s}{Z_2} & 1 + \dfrac{s^2 L}{Z_2} \end{bmatrix}$$

(3.105)

where Z_2 is the characteristic impedance of the right-hand unit element.

The results in Equations 3.104 and 3.105 are identical if

$$Z_1 = Z_2 + L \qquad \frac{1}{Z_1} + C = \frac{1}{Z_2} \qquad \frac{L}{Z_2} = Z_1 C$$

or

$$Z_2 = \frac{Z_1}{n} \qquad L = \frac{n-1}{n} Z_1$$

(3.106)

where $n = 1 + Z_1 C$.

As an example, consider the design of the broadband input transmission-line-matching circuit based on a lumped two-section low-pass-transforming filter shown in Figure 3.29, with a center bandwidth frequency $f_0 = 3$ GHz to match a 50-Ω source impedance with the device input impedance $Z_{in} = R_{in} + j\omega_0 L_{in}$, where $R_{in} = 2\ \Omega$ and $L_{in} = 0.223$ nH. The value of the series input device inductance is chosen to satisfy Table 3.1 when $n = 4$, $w = 0.4$, maximum ripple of 0.156725, $r = 25$, and $g_1 = 2.31517$; from Equation 3.83, it follows that

$$L_4 = \frac{g_1 R_0}{\omega_0} = \frac{g_1 R_0}{\omega_0 r} = 0.223\,\text{nH}$$

From Table 3.1, we obtain $g_2 = 0.422868$, which gives the circuit parameters from Equations 3.82 and 3.83 shown in Figure 3.39. The inductance value is chosen for the design convenience. If this value differs from the required value, it means that it is necessary to change the maximum frequency bandwidth, the power ripple, or the number of ladder sections.

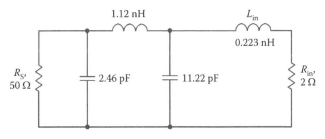

FIGURE 3.39
Two-section broadband-matching circuit.

Figure 3.40 shows the design transformation of a lumped low-pass-transforming filter to a microstrip filter using the Kuroda identities. The first step, which is shown in Figure 3.40a, is to add a 50-Ω unit element at the end of the circuit and convert a shunt capacitor into a series inductor using the second Kuroda identity, as shown in Figure 3.40b. Then, adding another unit element and applying the first Kuroda identity, as shown in Figure 3.40c, result in the circuit with two unit elements and three shunt capacitors shown in Figure 3.40d. To keep the same physical dimensions during the calculation of the circuit parameters, the inductance should be taken in nanohenries, and the capacitance is measured in

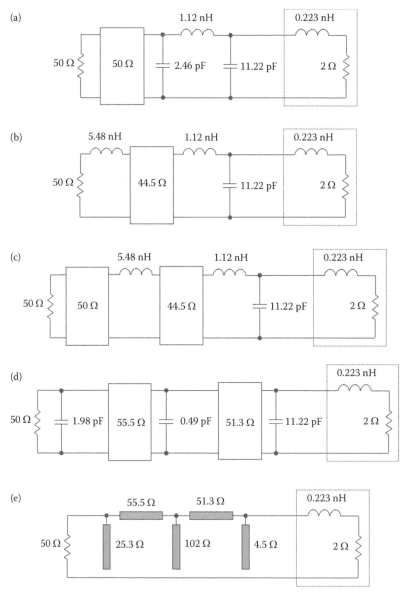

FIGURE 3.40
Design transformation from lumped low-pass to microstrip-transforming filter.

nanofarads if the operating frequency is measured in gigahertz. Finally, Richards's transformation is used to convert the shunt capacitors into the corresponding transmission-line stubs. According to Equation 3.101, the normalized characteristic impedance of a shunt stub is $1/C$, which is necessary to multiply by 50 Ω.

Figure 3.40e shows the microstrip layout of the final low-pass-transforming circuit, where the lengths of the shunt stubs are $\lambda/8$ at the cutoff frequency f_c, as well as the lengths of each unit element representing the series stubs. If the normalized frequency bandwidth and center bandwidth frequency are chosen to be $w = 0.4$ and $f_0 = 3$ GHz, respectively, the cutoff frequency becomes equal to

$$f_c = f_0 \left(1 + \frac{w}{2}\right) = 3.6\,\text{GHz}$$

In practical design of a microwave bipolar or GaAs FET amplifier, it is necessary to take into account that the intrinsic device generally exhibits a small-signal gain roll-off with increasing frequency at approximately 6 dB per octave [30,31]. Therefore, to maintain a constant gain across the design frequency band, the matching network must be designed for maximum gain at the highest frequency of interest [32]. In this case, reflective mismatching conditions are provided to compensate for the increase in the intrinsic gain of an FET when the frequency is decreased. As a result, it is necessary to selectively mismatch the input of the transistor by employing the gain-tapered input-matching circuit so that the overall gain of the amplifier will be flat [33]. Alternatively, the gain tapering could be done in the output network with input flat-matching conditions. In a simplified practical implementation when two impedance-transforming L-sections with series microstrip lines and shunt microstrip stubs in the input-matching circuit and a single impedance-transforming T-section with two series microstrip lines and one shunt microstrip stub in the output- matching network are used, a flat gain of about 6 dB was achieved across the octave band of 4–8 GHz for a single-cell GaAs FET amplifier with the device transconductance $g_m = 55$ mS [34].

Figure 3.41 shows the matching circuit design steps and the circuit schematic of a broadband microwave GaN HEMT power amplifier [35]. In this case, the first step to design the octave-band power amplifier intended to operate across the frequency bandwidth of 2–4 GHz was to find the optimum source and load impedances that maximize the performance of the device in terms of efficiency in the required bandwidth. In view of a GaN HEMT device Cree CGH60015DE, since the optimum impedances were relatively close to each other across the band and the acceptable level of degradation in *PAE* was estimated to be less than 8%, the task was simplified to provide the optimum impedances seen by the device input and output at the center bandwidth frequency across the entire bandwidth. The bandpass-matching network shown in Figure 3.41a was derived from a low-pass prototype-matching circuit, assuming that the transistor output can be approximated by an ideal current source with a parallel RC network, where R_0 is the source resistance corresponding to device-equivalent output resistance at the fundamental (or load-line resistance) and the capacitance C_0 is the total drain–source capacitance. To scale the obtained terminating resistor R_L upward to 50 Ω, a Norton transformation of an IT ($n_T = 1.173$) with two series–shunt capacitors to an arrangement of three capacitors, as shown in Figure 3.27, was used. Then, based on the transforms between lumped and distributed elements, the two resonant-parallel LC-circuits were approximated by the corresponding grounded shunt quarter-wave transmission lines TL_1 and TL_3 with the characteristic impedance of each line equal to the reactance of the inductor or capacitor multiplied by $\pi/4$, whereas the lumped π-network with a series inductor and two shunt capacitors was approximated by the series transmission line TL_2, as shown in Figure 3.41b. A similar approach can be applied to the design of the input-matching circuit, which also includes lossy elements for better input return loss and stability. The entire circuit schematic of the designed broadband GaN HEMT power amplifier is shown in Figure 3.41c, where the two series-tapered transmission lines TL_1 and TL_5 are added at the input and output of the device. As a result, an output power of 41 ± 1 dBm with a power gain of 10 ± 1 dB and a drain efficiency of 52%–72% was achieved across the frequency bandwidth of 1.9–4.3 GHz.

An alternative impedance-matching technique is based on the multisection-matching transformers consisting of the stepped transmission-line sections with different characteristic impedances and electrical lengths [36]. These transformers, in contrast to the continuously tapered transmission-line transformers, are significantly shorter and provide broader performance. Figure 3.42a shows the schematic structure of a stepped transmission-line transformer, which consists of a cascaded connection of n uniform sections of equal quarter-wave lengths $l = \lambda_0/4$, where λ_0 is the wavelength corresponding to the center bandwidth frequency. Such a stepped transmission-line transformer represents an antimetric structure, for which the ratio between the characteristic impedances of its transmission-line sections can be written in the general form as

$$Z_i Z_{n+1-i} = Z_S Z_L \tag{3.107}$$

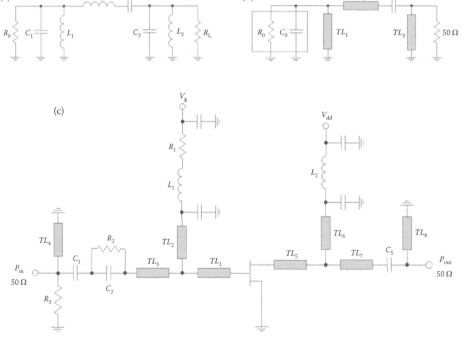

FIGURE 3.41
Schematics of broadband microwave GaN HEMT power amplifier.

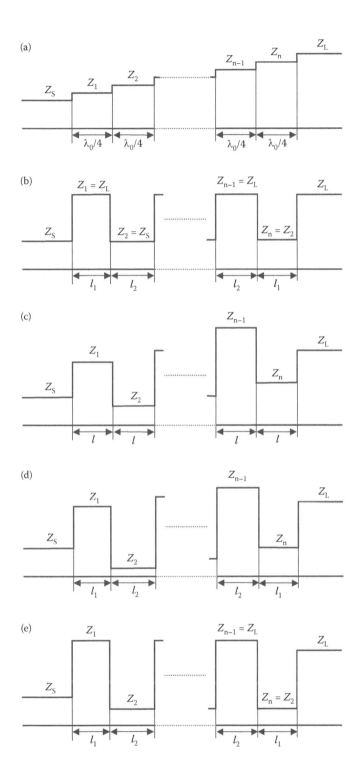

FIGURE 3.42
Schematic structures of different stepped transmission-line transformers.

where $i = 1, 2, ..., n$ and n is the number of sections, Z_S is the source impedance, and Z_L is the load impedance [37]. In Figure 3.43, as a practical example, the minimum possible *VSWR* is plotted for the five-step transmission-line impedance transformer with a total characteristic-impedance variation of 8:1, which was designed for maximum *VSWR* of 1.021 in an octave frequency bandwidth and where each section is of a quarter-wave electrical length [38].

The main drawback of the stepped quarter-wave transformers is their significant total length of $L = n\lambda_0/4$. However, it is possible to reduce the overall transformer length by applying other profiles of its structure. The stepped transformers using n-cascaded uniform transmission-line sections of various lengths with alternating impedances are shorter by 1.5–2 times. In this case, the number of sections n is always an even number and the section impedances can be equal to the source and load impedances to be matched, as shown in Figure 3.42b. For example, the input- and output-matching circuits of a microwave GaAs MESFET power amplifier, which was designed to operate in a frequency bandwidth of 4–8 GHz, were composed of the stepped microstrip lines where all the high-impedance sections were made of 50 Ω and all low-impedance sections were made of 10 Ω [34].

To define the unknown section lengths, the optimization approach to achieve the global minimum of the objective function $|\Gamma(\theta, A)|$ can be used, which is written as

$$\min_{A} \max_{\theta \in [\theta_1, \theta_2]} |\Gamma(\theta, A)| \tag{3.108}$$

where θ_1 and θ_2 are the electrical lengths at the low- and high-frequency bandwidth edges, respectively, and the vector $A = (A_1, A_2, ..., A_n)$ consists of the normalized section lengths $L_i = l_i/\lambda_0$ as components [39]. By solving Equation 3.108 numerically, the optimum Chebyshev characteristics can be provided by the stepped transmission-line structure with

$$l_i = l_{n+1-i} \tag{3.109}$$

where $i = 1, 2, ..., n/2$.

The total length of such a stepped transmission-line transformer can be further reduced by using the structure representing the cascade connection of n transmission-line sections of the same length $l < \lambda_0/4$ with

$$\begin{aligned} Z_1 &< Z_3 < ... < Z_{n-1} \\ Z_2 &< Z_4 < ... < Z_n \end{aligned} \tag{3.110}$$

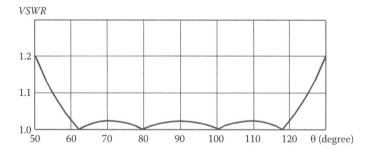

FIGURE 3.43
Theoretical frequency bandwidth of the five-step transformer.

where n is an even number and $Z_1 > Z_n$ when $Z_S < Z_L$, as shown in Figure 3.42c [40]. An example of the stepped transmission-line transformer to match the source impedance of 25 Ω with the load impedance of 50 Ω is shown in Figure 3.44a, where the electrical length of each section is equal to $\lambda_0/12$. In this case, the total transformer length is shorter by 3 times compared to the basic structure with the quarter-wave sections, and an octave passband from 2 to 4 GHz for the lossless ideal transmission-line sections is provided with an input return loss better than 25 dB, as shown in Figure 3.44b. However, it requires the use of a high-impedance ratio for its sections reaching 30–50 when the source and load impedances differ significantly.

To reduce a high-impedance ratio of the stepped transformers with a short total length, their generalized structure representing cascaded even n sections of different lengths l_i and impedances Z_i, as shown in Figure 3.42d, can be used. The optimum Chebyshev characteristics for this structure can be provided with the ratios between the lengths and characteristic impedances of its sections according to

$$l_i = l_{n+1-i}$$
$$Z_i Z_{n+1-i} = Z_S Z_L \tag{3.111}$$

where $i = 1, 2, \ldots, n/2$, and

$$Z_{n-1} > Z_{n-3} > \ldots > Z_1 > Z_n > Z_{n-2} > \ldots > Z_2 \tag{3.112}$$

where the impedances of both even and odd sections decrease in the direction from higher impedance Z_L to lower impedance Z_S and the impedance of any odd section is always larger than that of any even section [39]. The lengths of even sections decrease in the direction from the transmission line of a smaller impedance, whereas the lengths of odd sections increase in the same direction.

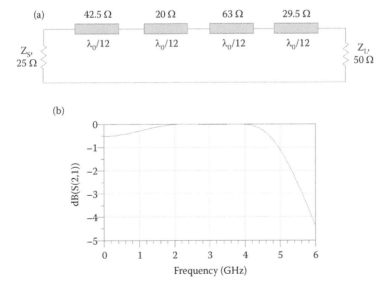

FIGURE 3.44
Stepped transmission-line transformer with equal-length sections.

TABLE 3.3

Optimum Parameters for Different Four-Section Transformers

$\vert\Gamma\vert_{min}$	*Equal-length structure of Figure 3.42c*				
	$L_{1,2,3,4}$	Z_1 (Ω)	Z_2 (Ω)	Z_3 (Ω)	Z_4 (Ω)	L
0.065	0.0833	42.38	19.80	63.13	29.49	0.3330
0.071	0.0625	55.75	13.58	92.03	22.42	0.2500
0.074	0.0418	82.58	8.45	148.01	12.14	0.1670
0.076	0.0313	109.64	6.17	202.48	11.40	0.1250
	Generalized structure of Figure 3.42d					
	$L_{1,2}$	$L_{3,4}$	Z_1 (Ω)	Z_2 (Ω)	Z_3 (Ω)	Z_4 (Ω)	L
0.068	0.0625	0.0833	51.75	18.20	68.73	24.15	0.2916
0.070	0.0525	0.0725	62.00	15.28	81.85	20.15	0.2500
0.075	0.0320	0.0510	103.00	10.13	123.39	12.14	0.1660
0.076	0.0205	0.0420	152.90	7.78	160.95	8.18	0.1250
	New structure of Figure 3.42e				
	$L_{1,2}$	$L_{3,4}$	Z_1 (Ω)	Z_2 (Ω)	L
0.064	0.0479	0.1171	52.38	23.86	0.3330
0.070	0.0405	0.0841	72.91	17.14	0.2500
0.074	0.0282	0.0553	114.55	10.91	0.1670
0.075	0.0213	0.0412	155.67	8.03	0.1250

Another structure of the stepped transmission-line transformer with the reduced total electrical length is shown in Figure 3.42e, for which Equation 3.109 can be applied and for which the same characteristic impedances for odd and even sections differ from the source and load impedances according to

$$Z_1 = Z_3 = \ldots = Z_{n-1}$$
$$Z_2 = Z_4 = \ldots = Z_n \qquad (3.113)$$

where $Z_1 Z_2 = Z_S Z_L$, $Z_n < Z_S$, and $Z_{n-1} > Z_L$ [39]. In particular situations of high-impedance-matching ratio at microwave frequencies when it is necessary to match the standard source load impedance of 50 Ω with the device input and output impedance of 1 Ω and smaller, both the length and width of the microstrip-line sections can be optimized.

Table 3.3 gives the optimum parameters for different four-section transformers ($n = 4$) designed to match the transmission lines with impedances $Z_S = 25$ Ω and $Z_L = 50$ Ω in an octave frequency range, where the section lengths L_i and total length L are normalized to λ_0 [39].

3.6 Matching Technique with Prescribed Amplitude–Frequency Response

An additional difficulty to design the broadband transistor power amplifier apart from the broadband impedance transformation with a large-impedance ratio and prescribed reactive constraints is that the matching networks must be designed with prescribed tapered magnitude characteristics to compensate for the transistor gain roll-off at higher frequencies that is inherent with the high-frequency bipolar FETs. The gain tapering is a direct consequence of the fact that the transistor maximum available gain decreases with increasing frequency. Thus, for a flat overall gain response, matching networks with tapered magnitude characteristics must be used. The exact frequency dependence of the taper or gain slope varies with specific types of transistors, as well as with power levels (linear small-signal amplifiers or large-signal power amplifiers).

For example, for a common-emitter bipolar transistor, the input-matching network must be designed to simultaneously provide broadband matching to equalize the series-reactive constraint represented by the transistor series RL-equivalent circuit (base resistance r_b and lead inductance L_b) and broadband impedance transformation of a 50-Ω source to the load impedance levels (usually r_b of the order of 1 Ω or less for a high-power Class-C-biased transistor). However, the reactive constraint for the output-matching network is a shunt collector capacitance C_c, and the equivalent load impedance depends on the supply voltage and output power. At the same time, for a common-source FET device, its input circuit can be modeled more closely to a series RC circuit (gate–source resistance R_{gs} and gate–source capacitance C_{gs}).

The input-matching circuit is responsible for achieving the corresponding gain level over the operating frequency bandwidth. The collector loading, on the other hand, can be designed independently to satisfy output power and collector efficiency considerations. A pragmatic approach to the design of an amplifier stage is to first determine a suitable output-matching network, and then design the input-matching section to satisfy gain requirements [41]. The most promising topology for the input matching consists of a two-section low-pass impedance-transforming structure, which easily absorbs the series RL nature of the bipolar transistor. The low-pass structure that employs shunt capacitors is also more readily constructed with microstrip lines than a high-pass or bandpass structure, which would require series capacitors in their realization. The design philosophy for the input-matching circuit is based on the gain response of the transistor versus frequency. Therefore, it is important to find a network that closely matches the transistor at the highest bandwidth frequency, where the transistor power gain will be the lowest. At lower frequencies, the power gain is greater but must be sacrificed by selective mismatching so that the overall gain of the amplifier will be flat. Thus, the matching networks must exhibit some prescribed tapered magnitude characteristics to compensate for the intrinsic transistor gain roll-off. Note that the best gain–bandwidth performance can be realized with a lumped-element equalizer, whereas simple transmission-line networks can usually be expected to provide near-optimum gain [31].

Figure 3.45a shows the general low-pass ladder-matching network configuration used for broadband-matching network design to approximate a prescribed tapered magnitude gain characteristic [41,42]. This ladder network must provide not only the prescribed gain slope but also provide a broadband impedance transformation. Let the gain function for an n-element low-pass ladder network be given by

$$| S_{21}(\omega) |^2 = \frac{1}{1 + [Q(\omega^2)]^2} \tag{3.114}$$

where

$$Q(\omega^2) = a_0 + a_1 \omega^2 + \cdots + a_{n/2}\, \omega^n \tag{3.115}$$

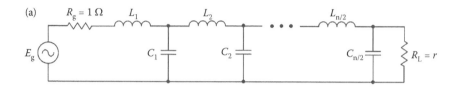

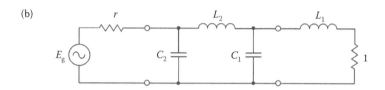

FIGURE 3.45
Low-pass tapered magnitude-matching networks.

and n must be an even number. Once an approximation is found from Equation 3.114, an optimization procedure can be used to adjust the actual element values [33].

To choose the polynomial $Q(\omega^2)$ in Equation 3.114, let $G(\omega)$ be the desired realizable gain function with a prescribed taper. Then, we need to set

$$G(\omega_k) = \frac{1}{1 + \left(\sum_{i=0}^{n/2} a_i \omega_k^{2i}\right)^2} \quad k = 0,1,\ldots,\frac{n}{2} \tag{3.116}$$

The condition given in Equation 3.116 can be rewritten as

$$\sum_{i=0}^{n/2} a_i \omega_k^{2i} = C_k \quad \text{for } k = 0,1,\ldots,\frac{n}{2} \tag{3.117}$$

where

$$C_k = \sqrt{\frac{1}{G(\omega_k)} - 1} \tag{3.118}$$

with a convenient way to choose the values ω_k to space them evenly over the band of interest using either a linear or a logarithmic frequency scale [33]. Figure 3.45b shows the low-pass tapered magnitude-matching network configuration with $n = 4$.

For a two-stage power amplifier whose block schematic is shown in Figure 3.46b, the typical specification of a good input and output matching implies a frequency response of the input- and output-matching networks that is flat over the amplifier passband and the interstage-matching network must provide a positive-sloped gain with increasing frequency, as shown in Figure 3.46c, to compensate for the gain roll-off of each transistor, as shown in Figure 3.46a, thus resulting in an overall flat transducer gain of a two-stage power amplifier [43].

The design example steps for the synthesis of the interstage-matching network of an arbitrary-specified gain versus frequency is outlined in Figure 3.47, where the output

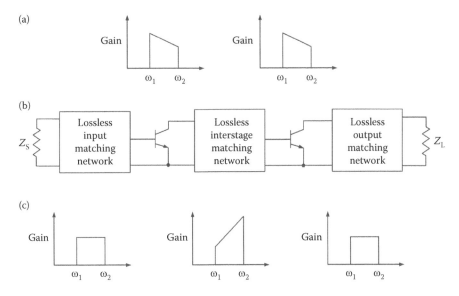

FIGURE 3.46
Block schematic and frequency responses of a two-stage power amplifier.

impedance of the first-stage device is represented by a shunt RC circuit and the input impedance of the second-stage device is approximated by a series RC circuit, as shown in Figure 3.47a [43]. The insertion loss of the lossless-matching network shown in Figure 3.47b and having all transmission zeroes at zero and infinite frequencies can be expressed as

$$IL = \frac{a_0 + a_2\omega^2 + \cdots + a_{2n}\omega^{2n}}{\omega^{2j}} \tag{3.119}$$

where $n = 3$ is the number of natural frequencies of the network, $j = 2$ is the number of transmission zeros at zero frequency, and $n - j = 1$ is the number of transmission zeroes at infinite frequency.

The overall synthesis procedure consists of two basic steps: an insertion-loss function is first obtained that approximates a flat passband response in an equiripple or maximally flat manner, and then a straightforward computational procedure yields a network as prescribed by the insertion-loss function. At the same time, the synthesis of interstage-matching networks requires that insertion-loss functions of a specified gain versus frequency slope, bandwidth, and ripple are to be found. In this case, the insertion-loss functions approximating $6k$-dB/octave gain slope, where k is an integer, are easily obtained from flat insertion-loss functions. The flat insertion-loss function is first normalized to an upper cutoff frequency of 1 rad/s and then divided by ω^{2k}. Once the sloped insertion-loss function is obtained, the computation procedure yields a network having the specified frequency response.

The practical steps in the interstage-matching network design process shown in Figure 3.47 can be listed as follows: (a) model the input and output impedances of the active devices to be used in the microwave power amplifier; (b) select a topology consistent with device parasitics; (c) adjust the gain–bandwidth performance to ensure an inclusion of parasitics; (d) select the reflection coefficient zeroes consistent with inclusion of parasitics; (e) transform impedances to desired levels; (f) transform the lumped design to a transmission-line

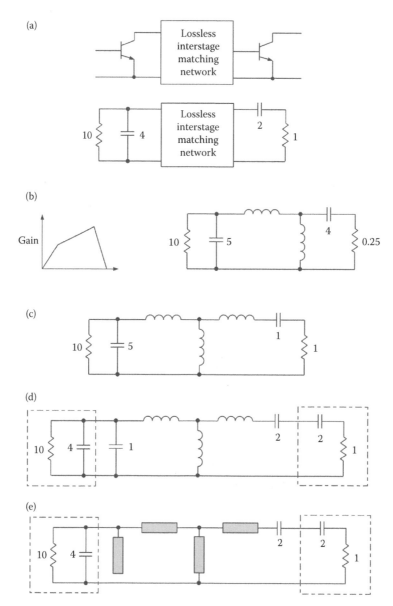

FIGURE 3.47
Design synthesis process of the interstage-matching circuit for a microwave power amplifier.

realization; (g) analyze the resultant design by itself and/or as a part of the complete amplifier design; and (h) optimize the amplifier design if needed [44]. Historically, the input and output impedances of active devices have generally been modeled by curve-fitting, optimization, or Smith chart manipulation. Besides, an algebraic method can be used to fit two- or three-element models to measured data by using the simple impedance equations and constraints at upper and lower passband frequencies. Given that many topologies of the matching networks are available, a good topology would accommodate the existent parasitics and provide a wide range of impedance transformation capabilities. The

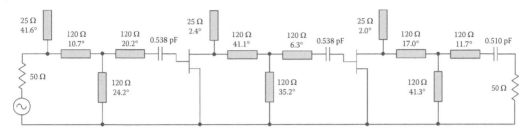

FIGURE 3.48
Design example of two-stage microstrip GaAs MESFET power amplifier.

method of gain adjustment can be based on the binary search algorithm. Figure 3.48 shows the design example of a two-stage microstrip GaAs MESFET power amplifier with three matching networks having a similar structure, where the input- and interstage-matching networks provide a gain slop of 6 dB/octave and the output-matching network has a flat frequency response, resulting in a power gain of 15 ± 2 dB over the frequency bandwidth of 6–12 GHz [44].

3.7 Practical Examples of Broadband RF and Microwave Power Amplifiers

Multisection-matching networks based on the low- and high-pass *L*-transformers for input- and output-matching circuits can provide a wide frequency bandwidth with minimum power gain ripple and significant harmonic suppression. Such a multisection matching circuit configuration using lumped elements was applied for the design of a 60-W power amplifier operating in the frequency bandwidth of 140–180 MHz. The complete circuit schematic of the power amplifier is shown in Figure 3.49 [7]. To realize such technical requirements, an internally matched bipolar transistor for VHF applications, which provides a 100-W output power level at a supply voltage of 28 V, was used. According

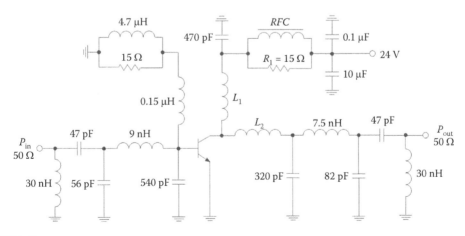

FIGURE 3.49
Circuit schematic of a broadband high-power VHF bipolar amplifier.

to the device data sheet, the input device impedance at the center bandwidth frequency $f_0 = \sqrt{140 \times 180} = 159$ MHz is equal to $Z_{in} = (0.9 + j1.8)$ Ω. Therefore, the input-matching circuit was designed as a three-section network with two low-pass sections and one high-pass section to minimize the circuit quality factor Q. In this case, the device input lead inductance of $1.8/(2\pi \times 0.159) = 1.8$ nH was considered as a series-inductive element of the second low-pass section with a shunt capacitor of 540 pF. This power amplifier is operated in Class-C due to the base bias circuit composed of the two inductors and a 15-Ω resistor, which also provides low-frequency stability.

A similar design philosophy was used to design the output- matching circuit when the three-section network maintains a value of the quality factor close to unity or within $Q = 1$ circle on a Smith chart. The output device impedance is practically resistive of 1.65 Ω because the output device capacitive reactance is compensated by the device lead inductance. The series inductance L_2 of the first matching low-pass section adjacent to the collector terminal according to the Smith chart can be realized as a section of a 50-Ω microstrip line with the electrical length of $0.011\lambda_0$, where λ_0 is the wavelength corresponding to the center bandwidth frequency f_0. The physical length of this microstrip line for a 1/16-inch Teflon fiberglass with a dielectric permittivity $\varepsilon_r = 2.55$ must be of 0.51 inch, whereas its width is equal to 0.4 inch. The collector feed is provided through the combination of an inductor L_1, a resistor $R_1 = 15$ Ω, and an RF choke (*RFC*), which behaves as a high-impedance circuit at the operating frequencies but offers a very low resistance at dc. As a result, the designed broadband power amplifier achieved a power gain of at least 8 dB with a gain ripple of less than 3 dB, a collector efficiency of more than 50%, and an input *VSWR* below 3:1 [7]. As an alternative, the broadband input- and output-matching circuits can be composed of a single low-pass-matching section followed by a 4:1 transmission-line transformer. In this case, an output power of more than 25 W with collector efficiency close to 70% was achieved across the frequency range of 118–136 MHz for the input power of 2 W using a 12.5-V bipolar device [45].

At microwave frequencies, the amplifier bandwidth performance can also be improved by using an increased number of the transmission-line transformer sections. For example, with the use of a multisection transformer with seven quarter-wave transmission lines of different characteristic impedances, a power gain of 9 ± 1 dB and a *PAE* of $37.5 \pm 7.5\%$ over 5–10 GHz were achieved for a 15-W GaAs MESFET power amplifier [10]. The simplified schematic diagram of this microwave octave-band power amplifier is shown in Figure 3.50. To achieve minimum output power flatness, the number of sections of the output-matching circuit is determined based on load-pull measurements. At the same time, the number of sections of the input-matching circuit to compensate for the frequency-dependent power gain is chosen based on the small-signal S-parameter measurements. For a 5.25-mm GaAs MESFET device, the values of the input and output impedances at the fundamental derived from its large-signal model were assumed resistive and equal to $Z_{in} = 0.075$ Ω and $Z_{out} = 1.32$ Ω, respectively. To achieve minimum gain flatness, the length of each microstrip section was initially chosen as a quarter-wavelength at the highest frequency of 10 GHz. However, because the input and output device impedances are not purely resistive in practical implementation, the final optimized length of each microstrip section was reduced to be a quarter-wavelength at around 15 GHz. The microstrip transformer sections $L_1...L_6$ and $L_9...L_{14}$ were fabricated on an alumina substrate with a dielectric permittivity $\varepsilon_r = 9.8$ and a thickness of 0.635 mm for L_1 and L_2, 0.2 mm for $L_3...L_6$ and $L_{10}...L_{12}$, and 0.38 mm for $L_{13}...L_{14}$. The microstrip section L_7 was realized on a high-dielectric substrate with $\varepsilon_r = 38$ and thickness of 0.18 mm, whereas the microstrip sections L_8 and L_9 were fabricated on a high-dielectric substrate with $\varepsilon_r = 89$ and thickness of 0.15 mm. The final power amplifier

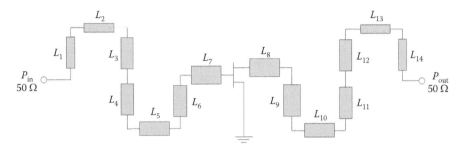

FIGURE 3.50
Microstrip broadband 15-W GaAs MESFET power amplifier.

represents a balanced configuration of the two 5.25-mm GaAs MESFETs with hybrid quadrature couplers.

The broadband power amplifier, whose circuit schematic is shown in Figure 3.51, was intended for TV transponders with complex video and audio TV signal amplification in the frequency bandwidth of 470–790 MHz. The power amplifier was implemented on a laminate substrate with $\varepsilon_r = 4.7$ of a 1.5-mm thickness. The microstrip lines are given in terms of their lengths on the high-bandwidth frequency, and both collectors *RFCs* represent the three-turn air-core inductors. The device input and output impedances measured at the base and collector terminals at 600 MHz are equal to $Z_{in} = (6 + j4)\ \Omega$ and $Z_{out} = (15 + j17.5)\ \Omega$, respectively, which allows the corresponding two-section input-matching circuit and a single-section output-matching circuit to be used. In a Class-A operation mode, such a power amplifier using a balanced TPV-595A bipolar transistor achieved a linear output power of 7 W with a power gain of about 12 dB for a quiescent collector current of 1.3 A.

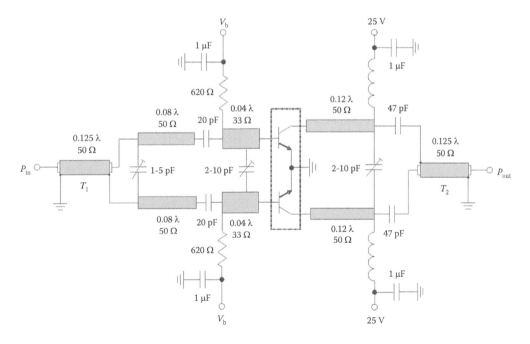

FIGURE 3.51
Circuit schematic of a bipolar UHF power amplifier for TV applications.

The circuit schematic of a high-power amplifier intended for applications in TV transmitters based on a balanced bipolar transistor BLV861 is shown in Figure 3.52 [46]. In a Class-AB operation with a quiescent current of 100 mA, it covers the frequency bandwidth of 470–860 MHz with an output power of 100 W, a power gain of about 9.5 dB with a gain ripple of ±0.5 dB, and a collector efficiency of 55%. The nominal device input and load impedances at 663 MHz are equal to $Z_{in} = (4.4 + j7.9)\ \Omega$ and $Z_L = (8.8 - j3.65)\ \Omega$, respectively. In this case, the three-section input-matching circuit and two-section output-matching circuit contain mixed microstrip-lumped elements to transform each terminal impedance level to approximately 25 Ω. The balanced-to-unbalanced transformation to 50 Ω is obtained by the transmission-line baluns, each of them represented by a 25-Ω semi-rigid coaxial cable with an electrical length of 45° at the midband and a diameter of 1.8 mm, soldered over the whole length on top of the same-length microstrip line. For low-frequency stability enhancement, the input balun stubs are connected to the bias point by means of 1-Ω series resistors. The large-value electrolytic capacitors are added at the input- and output-biasing points to improve the amplifier video response. The power amplifier is fabricated on a laminate substrate with $\varepsilon_r = 2.55$ and a thickness of 0.51 mm (20 mils).

Figure 3.53 shows the schematic diagram of a two-octave high-power transistor amplifier covering both the civil and military airbands between 100 and 450 MHz [47]. The BLF548 device is a balanced n-channel enhancement-mode VDMOS (vertical diffusion metal-oxide semiconductor) transistor designed for use in broadband amplifiers with an output power of 150 W and a power gain of more than 10 dB in a frequency range of up to 500 MHz. In a frequency bandwidth from 100 to 500 MHz, the real part of its input impedance $\mathrm{Re}Z_{in}$ is almost constant and equal to 0.43 Ω, whereas the imaginary part of the input impedance $\mathrm{Im}Z_{in}$ changes its capacitive reactance of −4.1 Ω at 100 MHz to the inductive reactance of 0.5 Ω at 500 MHz. The required load impedance seen by the device output at the fundamental is inductive and equal to $Z_L = (1.1 + j0.4)\ \Omega$ at the high-bandwidth frequency of 500 MHz. Coaxial semirigid baluns are used to transform the unbalanced 50-Ω source and load into two 180° out-of-phase 25-Ω sections, respectively, followed by coaxial 4:1 transformers with the characteristic impedance of 10 Ω for input matching and of 25 Ω for output matching. This yields the lower impedance $R_{in} = \sqrt{25 \times 10}/4 = 3.95\ \Omega$, which is then necessary to transform to the device input impedance of 0.43 Ω, and the higher impedance $R_{out} = \sqrt{25 \times 25}/4 = 6.25\ \Omega$, which is then necessary to transform to the load impedance of 2.8 Ω seen by the device output at the center bandwidth frequency of 250 MHz.

The final matching is provided by simple L-transformers with series microstrip lines and parallel variable capacitors. The microstrip lines were fabricated on a 30-mil substrate with

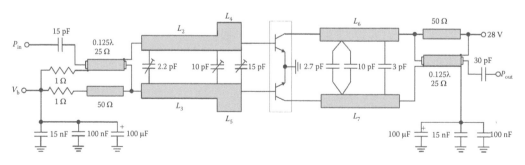

FIGURE 3.52
Bipolar high-power UHF amplifier for TV transmitters.

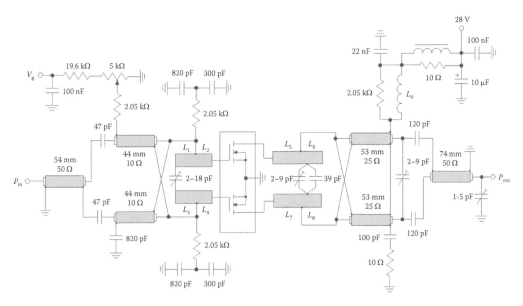

FIGURE 3.53
Circuit diagram of a broadband high-power VHF–UHF MOSFET amplifier.

a dielectric permittivity $\varepsilon_r = 2.2$. In this case, the dimensions of each microstrip line with the characteristic impedance of 20 Ω are as follows: L_1 and L_3 are 5×8 mm, L_2 and L_4 are 2.5×8 mm, L_5 and L_7 are 11.5×8 mm, and L_6 and L_8 are 4×8 mm. To compensate for the 6-dB/octave slope, conjugate matching is provided at 450 MHz, since at lower frequencies, a mismatch gives the required decrease of a power gain to provide acceptable broadband power gain flatness. As a result, the gain variation of an output power of 150 W is smaller than 1 dB with an input return loss better than 12 dB in a frequency range of 100–450 MHz.

MMICs based on GaN HEMT technology can provide wider bandwidth, higher output power density, improved reliability at high junction temperature, better thermal properties, higher breakdown voltage, and higher operating efficiency compared to MMICs based on GaAs technology. For a 0.25-μm GaN HEMT technology using an SiC substrate, the breakdown voltage of 120 V allows operation with a supply voltage up to 40 V, and the maximum output power density of 5.6 W/mm for the device gate periphery and capacitance sheet of 250 pF/mm^2 for an MIM capacitor can be provided. The three-stage reactively matched 0.25-μm GaN HEMT power amplifier MMIC with parallel matching networks can achieve an output power from 6 to 10 W over 6–18 GHz with a minimum power gain of 18 dB and a *PAE* greater than 13% at $V_{dd} = 25$ V [48]. An averaged output power of 20 W with an averaged power gain of 9.6 dB and a *PAE* of more than 15% over *C–Ku* band (6–18 GHz) was achieved for a two-stage GaN HEMT MMIC power amplifier at $V_{dd} = 35$ V [49].

Figure 3.54 shows the circuit schematic of a two-stage reactively matched GaN HEMT MMIC power amplifier, which operates as a driver amplifier of ultra-wideband high-power transmit modules for multifunctional active electronically scanned antenna radar systems [50]. The MMIC driver amplifier is based on three identical GaN HEMT cells, each with 8×100-μm gate periphery (one transistor in the first stage and two transistors in the second stage), to achieve the maximum output power of about 36 dBm with a parallel connection of two second-stage-amplifying paths. The unconditional stability of the MMIC driver amplifier from 100 MHz to 6 GHz is provided by applying parallel *RC* networks

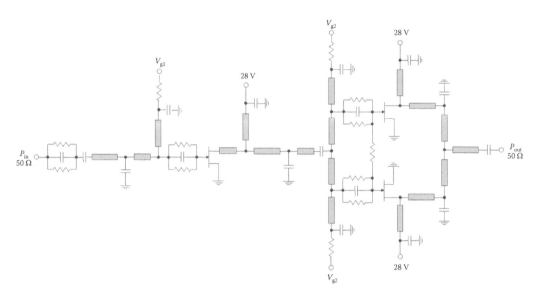

FIGURE 3.54
Circuit schematic of a broadband GaN HEMT MMIC power amplifier.

at the gates of each transistor cell. The integrated resistors are also used in the gate bias circuits of each device cell to ensure stability without sacrificing gain or efficiency. The dc-feed paths, which consist of narrow microstrip lines to provide the corresponding inductive reactances and bypass MIM capacitors to provide isolation between the dc and RF paths, are constituent parts of the input, interstage, and output-matching circuits, which are realized in the form of low-pass *L*- and *T*-transformers with the series microstrip lines and shunt MIM capacitors. The matching networks provide impedance transformation with low-*Q*-factors enabling an increased frequency bandwidth.

Figure 3.55a shows a photograph of the reactively matched MMIC driver amplifier with a chip size of 3×4 mm^2. When driven with a 20-dBm input power, the maximum-measured output power achieves 4 W with a typical output power of 1.8 W and a worst-case return loss of 7.5 dB in the frequency range from 6 to 18 GHz, as shown in Figure 3.55b. The output power of 10 W was achieved for a three-stage reactively matched GaN HEMT power amplifier, with four similar device cells in the final stage [50]. When driven with a 28-dBm input power, the maximum-measured output power achieves 15.6 W with an average output power of 10.6 W and a worst-case return loss of 7.5 dB in the frequency range from 6.4 to 18.4 GHz. Over the complete frequency range from 6 to 18 GHz, average values for a *PAE* of 18% are the same for simulations and measurements, with a typical value of 20% across the frequency range from 6 to 12 GHz.

3.8 Broadband Millimeter-Wave Power Amplifiers

In recent years, there have been increasing demands for millimeter-wave systems, such as automotive and fire-control radars, security screening and remote sensing, personal communication, and imaging systems. Millimeter-wave systems have advantages of wide

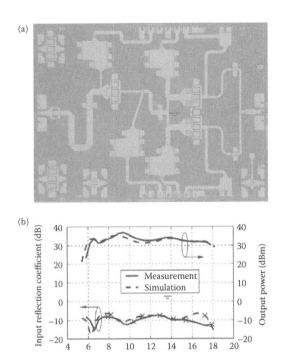

FIGURE 3.55
Circuit schematic of the reactively matched two-stage broadband GaN HEMT MMIC power amplifier. (Courtesy of Cassidian.)

frequency bandwidth, compact size, and suitability for short-range multicell communications owing to high attenuation of radiated signals. In communication systems, it is desirable to design the broadband power amplifiers that cover several frequency ranges, for example, LMDS (local multipoint distribution system) across a 28-GHz band, wireless communication system at 38 GHz, and broadband cellular communication systems at 60 GHz and beyond.

Significant progress in SiGe technology had contributed to the design and development of the fully integrated millimeter-wave broadband power amplifiers in *Ka* (26.5–40 GHz) and *W* (75–110 GHz) frequency bands with increased output power using a 0.13-μm SiGe process with the transition frequency f_T of 200 GHz and maximum frequency f_{max} of 300 GHz [51,52]. Figure 3.56a shows the simplified half-schematic of a balanced *Ka*-band power amplifier operating in a linear Class-A [51]. The input and output microstrip-line matching networks consisting of two low-pass *L*-sections with shunt short-circuited stubs and series MIM capacitors provide a positive gain slope with increasing frequency to compensate for the transistor gain roll-off, resulting in a wideband response of both power gain and output power. To stabilize the power amplifier operation at very low frequencies, where the transistor gain is extremely high, all the base and collector bias circuits and bypass capacitors present low RF impedances to all transistor terminals (<1 Ω to the collector and <50 Ω to the base). As a result, the saturated output power of greater than 17 dBm with a power gain of greater than 10 dB at a supply voltage of 1.4 V was achieved across the entire frequency range of 26–40 GHz. By using a similar circuit design approach and 0.13-μm SiGe technology, the three-stage broadband power amplifier, whose single-ended half-schematic is

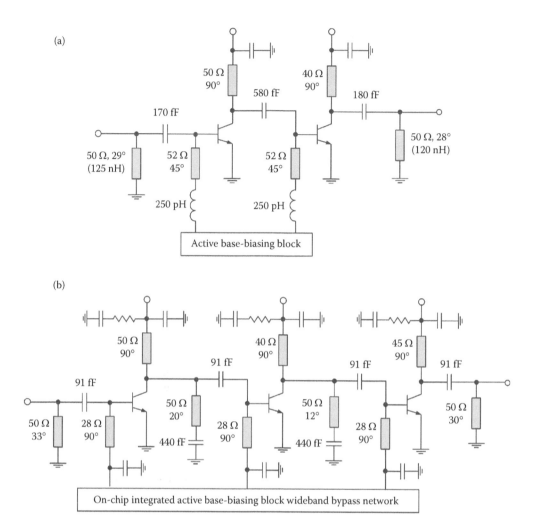

FIGURE 3.56
Circuit schematics of a broadband SiGe BiCMOS power amplifiers.

shown in Figure 3.56b, provides a small-signal gain of greater than 12 dB and a saturated power of greater than 14 dBm in a frequency bandwidth of 79.3–96.8 GHz [52].

Historically, different FET technologies have been used for millimeter-wave MMIC applications. Figure 3.57a shows the circuit schematic of a two-stage broadband power amplifier using a 0.15-μm T-shaped gate AlGaAs/InGaAs heterojunction FET process with $f_T = 70$ GHz and $f_{max} = 240$ GHz [53]. This MMIC power amplifier with equal first and second stages was designed and fabricated based on a microstrip-line circuit configuration and achieved a high gain of 16.5 ± 1.5 dB across the frequency bandwidth of 50–68 GHz. By using a 0.15-μm GaAs HEMT technology, the broadband three-stage MMIC power amplifier achieved a power gain greater than 20 dB from 41 to 63 GHz and a saturated output power of at least 16 dBm from 19 to 57 GHz, with the distributed input stage for broadband operation and the second and third single-ended stages for high gain and large output power [54]. In many high-performance low-noise applications covering the entire *D* (110–170 GHz) frequency band, InP HEMTs and metamorphic HEMTs (mHEMTs) grown on a

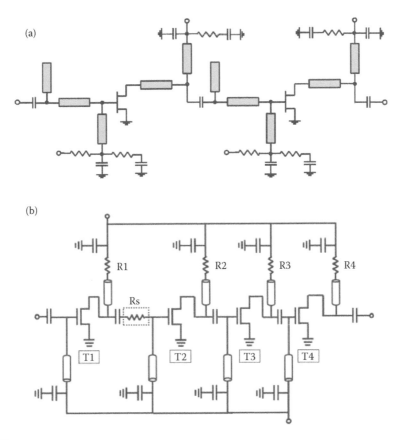

FIGURE 3.57
Circuit schematics of a broadband HEMT power amplifiers.

GaAs substrate with high indium content in the transistor channel are much demanded. The 50-nm process provides a composite channel with an indium content of 80%, resulting in a low source resistance, and a 2×15-µm device typically reaches a transition frequency of $f_T = 400$ GHz at a drain voltage of 1 V. Figure 3.57b shows the circuit schematic of a broadband four-stage lossy match MMIC mHEMT power amplifier with a small-signal gain exceeding 25 dB from 105 to 175 GHz, and an output power reaches 5 dBm at 140 GHz at a supply voltage of 1.25 V [55]. Here, the matching circuits are designed using coplanar transmission lines (CPW) and the resistors R_1 through R_4 fixed to 6.5 Ω each are used for gain equalization and stabilization.

Owing to an advanced metamorphic technology with $f_T = 500$ GHz together with a grounded CPW topology, the solid-state integrated circuit amplifiers have reached operating frequencies of greater than 300 GHz [56]. In this case, a common topology for HEMT MMIC amplifiers is to use a grounded CPW medium with a 1- or 2-mil-thick InP substrate for the circuit design. This allows for the elimination of source induction of a thru-substrate via (or a pair of vias) on the source of the transistor, which would otherwise reduce the amplifier gain at the same frequency. By using a narrow ground-to-ground spacing for the CPW ground planes that is approximately equal to the physical height of the HEMT, the source inductance is virtually eliminated. Figure 3.58 shows the circuit schematic of a cascode amplifier using a 35-nm InAlAs/InGaAs mHEMT technology,

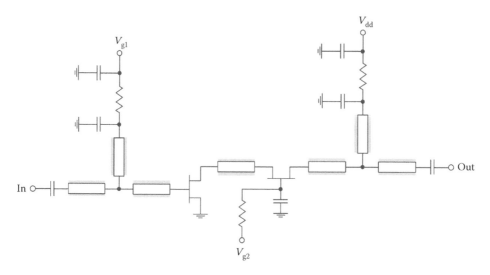

FIGURE 3.58
Circuit schematic of *H*-band cascode mHEMT amplifier stage.

achieving a small-signal gain of more than 10.5 dB across the frequency bandwidth from 220 to 320 GHz [57]. In this case, due to the very low source resistance of 0.03 Ω/mm, the transistor on-resistance at peak g_m is only 0.25 Ω/mm.

Despite the fact that HEMT technologies have superior performance compared to CMOS due to their higher electron mobility, higher breakdown voltage, and the availability of high-quality factors for passive elements, a CMOS implementation provides a higher level of integration and reduced cost. In this case, based on a 130-nm digital CMOS technology and using CPW transmission lines, the three-stage cascode amplifiers achieved a power gain of more than 15 dB in a frequency bandwidth from 34 to 45 GHz and a power gain of more than 10 dB with a saturated power in a frequency bandwidth from 53 to 64 GHz at a supply voltage of 1.5 V [58]. However, by using a high-resistivity substrate and through-silicon vias (TSVs), the performance of the integrated microwave and millimeter-wave CMOS power amplifiers can be significantly improved. Moreover, the stacking of multiple devices in a CMOS process allows increasing the supply voltage, which, in turn, allows higher output power and a broader bandwidth for the output-matching circuit. At the same time, the maximum operating frequency and quality factor of the passive elements can be improved by using a 45-nm CMOS silicon-on-insulator (SOI) process.

Figure 3.59 shows the circuit schematic of the two-, three-, and four-stack CMOS power amplifier with shunt CPW transmission lines and series inductances between the devices [59]. The circuit for a stacked power amplifier is based on a series interconnection of a common-source transistor cascaded with common-gate-like transistors. The stacked configuration differs from a cascode, in which the gate of the common-gate transistor is grounded at the frequency of operation. Here, the gate of the common-gate-like transistor is connected to a finite impedance and experiences a voltage swing. Ideally, the drain voltages of the transistors add-in phase, whereas the drain current is constant through each transistor. The gate voltage swing is controlled by introducing appropriate capacitances C_k at the gates of stacked transistors, where $k = 1, 2, \ldots, K$, and K is the number of stacked transistors. The series combination of C_k and the gate–source capacitance C_{gs} of the corresponding transistor form voltage dividers that determine the gate voltages. In contrast

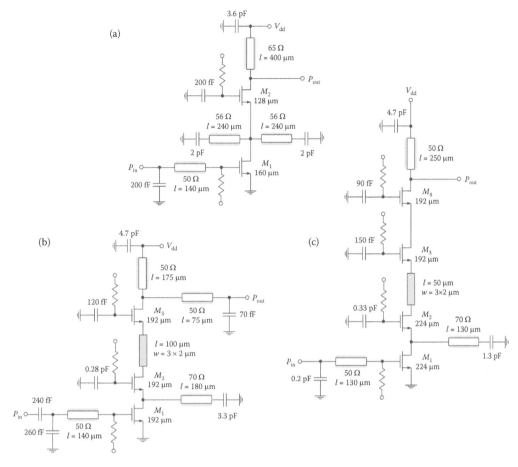

FIGURE 3.59
Circuit schematics of two-, three-, and four-stack CMOS power amplifiers.

with cascode amplifiers, this approach reduces the drain–gate and drain–source swings under large-signal conditions, allowing reliable transistor operation under large aggregate voltage swings. However, the gain of the stacked power amplifier is lower than that of the cascode power amplifier. Besides, a critical design consideration is the proper adjustment of the dc gate voltages for efficient and reliable operation. With a supply voltage much greater than the breakdown levels of the transistors, the gates of the stacked devices must be biased such that the dc and RF (V_{gs}, V_{gd}, and V_{ds}) voltages of each transistor are less than their respective breakdown voltages. Note that the on-resistance of the K-stacked power amplifier is K times larger than the on-resistance of each transistor.

Without any reactive tuning, the efficiency and output power reduction are significant at millimeter-wave frequencies. To achieve the proper complex impedance between the transistors, additional tuning elements are necessary for optimal operation. Figure 3.59a shows the circuit schematic of a two-stacked CMOS power amplifier implemented in a 45-nm CMOS SOI technology, where the shunt inductance, which is formed by two identical parallel CPW transmission lines to increase the output power, tunes out the parasitic capacitance at an intermediate node. The shunt-tuning inductance between the first two transistors is used in all amplifiers. However, due to layout constraints, a series-tuning inductance was used between the second and third devices in a three-stack power amplifier, as shown in Figure 3.59b, and in a four-stack power amplifier, as shown in Figure 3.59c. From Figure 3.60a, it follows that a higher output power of around 19 dBm and greater than 20 dBm over a wide frequency range of 40–48 GHz was obtained by stacking three and four transistors, respectively [59]. In both cases, a *PAE* of around 20% and greater was achieved, as shown in Figure 3.60b, with a supply voltage of 3.5 V for a three-stack power amplifier and of 5 V for a four-stack power amplifier.

Figure 3.61a shows the block diagram of a broadband *W*-band CMOS medium-power amplifier implemented in a 90-nm CMOS technology [60]. The power amplifier represents a three-stage design with the first single-ended stage driving a two-stage-balanced structure to achieve twice of the output power. The balanced structure cancels reflected powers and improves the overall amplifier input return loss. Each stage represents a cascode configuration with microstrip lines used for the matching circuits and all interconnections

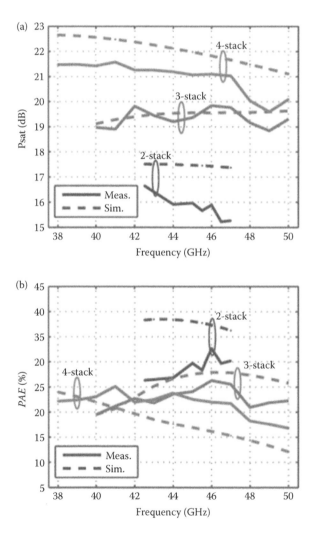

FIGURE 3.60
Measured power and efficiency for two-, three-, and four-stack CMOS power amplifiers.

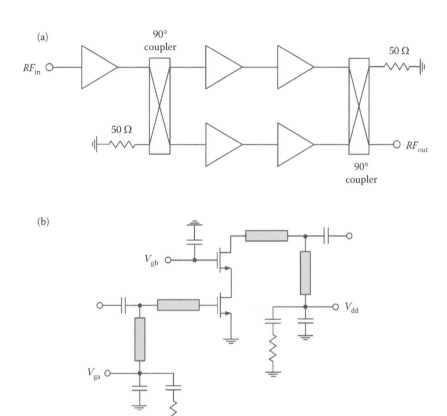

FIGURE 3.61
Schematics of a broadband *W*-band CMOS power amplifier.

to reduce chip size, as shown in Figure 3.61b. As a result, when each transistor is biased at $V_{ds} = 1.2$ V with gate–source voltage set to $V_{gs} = 0.8$ V, a measured maximum small-signal of 15.1 dB is achieved at 100.5 GHz with a 3-dB bandwidth of 18 GHz from 90 to 108 GHz. The power amplifier demonstrates an output power of 10 dBm with about 7-dB saturated power gain in a frequency bandwidth from 90 to 100 GHz. A similar balanced power amplifier with two three-stage cascode power cells achieved an output power of more than 11.8 dBm with a power gain of greater than 18.1 dB from 68 to 83 GHz using a 90-nm CMOS technology [61]. By using multiway Wilkinson power combiners with low-imped-ance transmission lines, the output power over a wide frequency range can be increased. As a result, with a 1.2-V supply voltage for a 32-way *V*-band power amplifier, for a 16-way *W*-band power amplifier, and for an eight-way *D*-band power amplifier, each of them based on a 65-nm CMOS technology, the saturation powers of 23.2, 18, and 13.2 dBm at 64, 90, and 140 GHz can be achieved with a 25.1-, 26-, and 30-GHz 3-dB bandwidth, respectively [62].

Figure 3.62a shows the circuit schematic of a three-stage single-ended *W*-band power amplifier using a 65-nm CMOS technology to provide high gain and output power under low-voltage operation [63]. Here, three nMOS cells are combined for the first stage, four nMOS cells are combined for the second stage, and six nMOS cells are used in the final stage, with a 2-μm width and 16 fingers for each nMOS cell. The interstage-matching

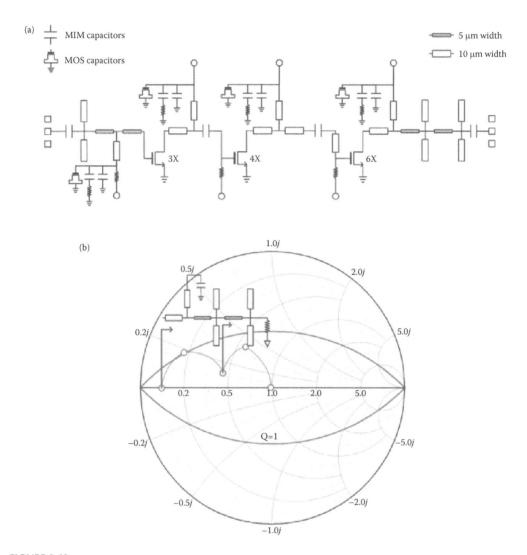

FIGURE 3.62
Circuit schematic and output matching of a broadband *W*-band CMOS power amplifier.

networks represent high-pass-matching sections with series MIM capacitors and shunt short-length stubs. The dc supply is fed through the stubs to reduce losses and the chip size. The high-pass-matching circuits can compensate the device frequency response and flatten the overall gain in the desired frequency range. The output-matching design procedure at 90 GHz is demonstrated in Figure 3.62b. To increase the bandwidth, two sections of low-pass-matching networks, each composed of the series transmission line with a shunt open-circuit stub, are used to keep the circuit quality factor inside the $Q = 1$ circle on the Smith chart. The first section transforms the impedance from a 50-Ω load to about $(25 + 5j)$ Ω and then goes to 5 Ω by the second section to optimize the output power. As a result, a small-signal gain of more than 10 dB and a saturated output power of more than 13 dBm were obtained across the frequency range of 80–105 GHz. By using four cascode stages with a final single-ended common-source stage for a broadband power amplifier based on a 65-nm CMOS process, a small-signal gain of more than 10 dB was obtained across the very wide frequency range from 110 to 180 GHz [64].

References

1. W. L. Everitt, Output networks for radio-frequency power amplifiers, *Proc. IRE*, 19, 725–737, 1931.
2. S. Roberts, Conjugate-image impedances, *Proc. IRE*, 34, 198–204, 1946.
3. V. M. Bogachev and V. V. Nikiforov, *Transistor Power Amplifiers* (in Russian), Moskva: Energiya, 1978.
4. A. Grebennikov, *RF and Microwave Power Amplifier Design*, New York: McGraw-Hill, 2015.
5. H. Q. Tserng and H. M. Macksey, Microwave GaAs power FET amplifier with lumped-element impedance matching networks, *IEEE MTT-S Int. Microw. Symp. Dig.*, Ottawa, Canada, 282–284, 1978.
6. R. L. Camisa, J. B. Klatskin, and A. Mikelsons, Broadband lumped-element GaAs FET power amplifier, *IEEE MTT-S Int. Microw. Symp. Dig.*, Los Angeles, CA, 126–128, 1981.
7. A. Tam, Network building blocks balance power Amp parameters, *Microw. RF*, 23, 81–87, 1984.
8. P. H. Smith, *Electronic Applications of the Smith Chart*, New York: Noble Publishing, 2000.
9. D. M. Pozar, *Microwave Engineering*, New York: John Wiley & Sons, 2004.
10. Y. Ito, M. Mochizuki, M. Kohno, H. Masuno, T. Takagi, and Y. Mitsui, A 5–10 GHz 15-W GaAs MESFET amplifier with flat gain and power responses, *IEEE Microw. Guid. Wave Lett.*, 5, 454–456, 1995.
11. A. V. Grebennikov, Create transmission-line matching circuits for power amplifiers, *Microw. RF*, 39, 113–122, 2000.
12. H. W. Bode, *Network Analysis and Feedback Amplifier Design*, New York: Van Nostrand, 1945.
13. R. M. Fano, Theoretical limitations on the broad-band matching of arbitrary impedances, *J. Franklin Inst.*, 249, 57–83, 139–154, 1950.
14. R. Levy, Explicit formulas for Chebyshev impedance-matching networks, filters and interstages, *Proc. IEE*, 111, 1099–1106, 1964.
15. D. E. Dawson, Closed-form solutions for the design of optimum matching networks, *IEEE Trans. Microw. Theory Tech.*, MTT-57, 121–129, 2009.
16. G. L. Matthaei, Tables of Chebyshev impedance-transforming networks of low-pass filter form, *Proc. IEEE*, 52, 939–963, 1964.
17. G. L. Matthaei, L. Young, and E. M. T. Jones, *Microwave Filters, Impedance-Matching Networks, and Coupling Structures*, New York: Artech House, 1980.
18. O. A. Chelnokov, *Radio Transmitter Devices* (in Russian), Moskva: Radio i Svyaz, 1982.
19. W. H. Ku, M. E. Mokari-Bolhassan, W. C. Petersen, A. F. Podell, and B. Kendall, Microwave octave-band GaAs FET amplifiers, *IEEE MTT-S Int. Microw. Symp. Dig.*, Palo Alto, CA, 69–72, 1975.
20. R. S. Pengelly and J. A. Turner, Monolithic broadband GaAs F.E.T. amplifiers, *Electron. Lett.*, 12, 251–252, 1976.
21. C. H. Lin, H. Z. Liu, C. K. Chu, H. K. Huang, C. C. Liu, C. H. Chang, C. L. Wu, C. S. Chang, and Y. H. Wang, A compact 6.5-W PHEMT MMIC power amplifier for Ku-band applications, *IEEE Microw. Wirel. Compon. Lett.*, 17, 154–156, 2007.
22. P. C. Huang, Z. M. Tsai, K. Y. Lin, and H. Wang, A 17–35 GHz broadband, high efficiency PHEMT power amplifier using synthesized transformer matching technique, *IEEE Trans. Microw. Theory Tech.*, MTT-60, 112–119, 2012.
23. H. Wang, C. Sideris, and A. Hajimiri, A CMOS broadband power amplifier with a transformer-based high-order output matching network, *IEEE J. Solid-State Circuits*, SC-45, 2709–2722, 2010.

24. B. S. Yarman and A. Aksen, An integrated design tool to construct lossless matching networks with mixed lumped and distributed elements, *IEEE Trans. Circuits Syst.—I: Fundam. Theory Appl.*, CAS-I-39, 713–723, 1992.

25. T. W. Houston and L. W. Read, Computer-aided design of broadband and low-noise microwave amplifiers, *IEEE Trans. Microw. Theory Tech.*, MTT-17, 612–624, 1969.

26. P. I. Richards, Resistor–transmission line circuits, *Proc. IRE*, 36, 217–220, 1948.

27. R. Saal and E. Ulbrich, On the design of filters by synthesis, *IRE Trans. Circuit Theory*, CT-5, 284–327, 1958.

28. P. I. Richards, Applications of matrix algebra to filter theory, *Proc. IRE*, 34, 145–150, 1946.

29. H. Ozaki and J. Ishii, Synthesis of transmission-line networks and the design of UHF filters, *IRE Trans. Circuit Theory*, CT-2, 325–336, 1955.

30. H. F. Cooke, Microwave transistors: Theory and design, *Proc. IEEE*, 59, 1163–1181, 1971.

31. R. S. Tucker, Gain–bandwidth limitations of microwave transistor amplifiers, *IEEE Trans. Microw. Theory Tech.*, MTT-21, 322–327, 1973.

32. C. A. Liechti and R. L. Tillman, Design and performance of microwave amplifiers with GaAs Schottky-gate field-effect transistors, *IEEE Trans. Microw. Theory Tech.*, MTT-22, 510–517, 1974.

33. W. H. Ku and W. C. Petersen, Optimum gain–bandwidth limitations of transistor amplifiers as reactively constrained active two-port networks, *IEEE Trans. Circuits Syst.*, CAS-22, 523–533, 1975.

34. R. E. Neidert and H. A. Willing, Wide-band gallium arsenide power MESFET amplifiers, *IEEE Trans. Microw. Theory Tech.*, MTT-24, 342–350, 1976.

35. P. Saad, C. Fager, H. Cao, H. Zirath, and K. Andersson, Design of a highly efficient 2–4-GHz octave bandwidth GaN-HEMT power amplifier, *IEEE Trans. Microw. Theory Tech.*, MTT-58, 1677–1685, 2010.

36. H. Q. Tserng, V. Sokolov, H. M. Macksey, and W. R. Wisseman, Microwave power GaAs FET amplifiers, *IEEE Trans. Microw. Theory Tech.*, MTT-24, 936–943, 1976.

37. H. J. Riblet, General synthesis of quarter-wave impedance transformers, *IRE Trans. Microw. Theory Tech.*, MTT-5, 36–43, 1957.

38. S. B. Cohn, Optimum design of stepped transmission-line transformers, *IRE Trans. Microw. Theory Tech.*, MTT-3, 16–21, 1955.

39. V. P. Meschanov, I. A. Rasukova, and V. D. Tupikin, Stepped transformers on TEM-transmission lines, *IEEE Trans. Microw. Theory Tech.*, MTT-44, 793–798, 1996.

40. G. L. Matthaei, Short-step Chebyshev impedance transformers, *IEEE Trans. Microw. Theory Tech.*, MTT-14, 372–383, 1966.

41. O. Pitzalis Jr. and R. A. Gilson, Broad-band microwave Class-C transistor amplifiers, *IEEE Trans. Microw. Theory Tech.*, MTT-21, 660–668, 1973.

42. O. Pitzalis Jr. and R. A. Gilson, Tables of impedance matching networks which approximate prescribed attenuation versus frequency slopes, *IEEE Trans. Microw. Theory Tech.*, MTT-19, 381–386, 1971.

43. D. J. Mellor and J. G. Linvill, Synthesis of interstage networks with prescribed gain versus frequency slopes, *IEEE Trans. Microw. Theory Tech.*, MTT-23, 1013–1020, 1975.

44. D. J. Mellor, Improved computer-aided synthesis tools for the design of matching networks for wide-band microwave amplifiers, *IEEE Trans. Microw. Theory Tech.*, MTT-34, 1276–1281, 1986.

45. B. Becciolini, *Impedance Matching Networks Applied to RF Power Transistors*, Application Note AN721, Freescale Semiconductor, Denver, CO, 2005.

46. *A Broadband 100 W Push Pull Amplifier for Band IV and V TV Transmitters Based on the BLV861*, Application Note AN98033, Philips Semiconductors, Eindhoven, The Netherlands, March 1998.

47. *MHz 250 W Power Amplifier with the BLF548 MOSFET*, Application Note AN98021, Philips Semiconductors, Eindhoven, The Netherlands, March 1998.

48. G. Mouginot, Z. Ouarch, B. Lefebvre, S. Heckmann, J. Lhortolary, D. Baglieri, D. Floriot et al., Three stage 6–18 GHz high gain and high power amplifier based on GaN technology, *IEEE MTT-S Int. Microw. Symp. Dig.*, Anaheim, CA, 1392–1395, 2010.

49. B. Kuwata, K. Yamanaka, H. Koyama, Y. Kamo, T. Kirikoshi, M. Nakayama, and Y. Hirano, C–Ku band ultra broadband GaN MMIC amplifier with 20 W output power, *Proc. 2011 Asia-Pac. Microw. Conf.*, Melbourne, Australia, pp. 1558–1561, 2011.

50. U. Schmid, H. Sledzik, P. Schuh, J. Schroth, M. Oppermann, P. Brueckner, F. van Raay, R. Quay, and M. Seelmann-Eggebert, Ultra-wideband GaN MMIC chip set and high power amplifier module for multi-function defense AESA applications, *IEEE Trans. Microw. Theory Tech.*, MTT-61, 3043–3051, 2013.

51. M. Chang and G. M. Rebeiz, A 26 to 40 GHz wideband SiGe balanced power amplifier IC, *IEEE RFIC Symp. Dig.*, Honolulu, Hawaii, 729–732, 2007.

52. M. Chang and G. M. Rebeiz, A wideband high-efficiency 79–97 GHz SiGe linear power amplifier with >90 mW output, *IEEE Bipolar/BiCMOS Circuits Technol. Meet. Dig.*, Monterey, CA, 69–72, 2008.

53. T. Inoue, K. Hosoya, and K. Ohata, High-gain wideband V-band multi-stage power MMICs, *IEEE Top. Millim. Waves Symp. Dig.*, Kanagawa, Japan, 40–43, 1997.

54. P. S. Wu, T. W. Huang, and H. Wang, An 18–71 GHz multi-band and high gain GaAs MMIC medium power amplifier for millimeter-wave applications, *IEEE MTT-S Int. Microw. Symp. Dig.*, Philadelphia, PA, 863–866, 2003.

55. T. Merkle, S. Koch, A. Leuther, M. Seelmann-Eggebert, H. Massler, and I. Kallfas, Compact 110–170 GHz amplifier in 50 nm mHEMT technology with 25 dB gain, *Proc. 8th Eur. Microw. Integr. Circuits Conf.*, Nuremberg, Germany, pp. 129–132, 2013.

56. L. A. Zamoska, An overview of solid-state integrated circuit amplifiers in the submillimeter-wave and THz regime, *IEEE Trans. Terahertz Sci. Technol.*, TST-1, 9–24, 2011.

57. A. Tessmann, A. Leuther, H. Massler, M. Kuri, and R. Loesch, A metamorphic 220–320 GHz HEMT amplifier MMIC, *IEEE Compound Semicond. Integr. Circuits Symp. Dig.*, Monterey, CA, 1–4, 2008.

58. C. H. Doan, S. Emami, A. M. Niknejad, and R. W. Brodersen, Millimeter-wave CMOS design, *IEEE J. Solid-State Circuits*, SC-40, 144–155, 2005.

59. H. T. Dabag, B. Hanafi, F. Golcuk, A. Agah, J. F. Buckwalter, and P. M. Asbeck, Analysis and design of stacked-FET millimeter-wave power amplifiers, *IEEE Trans. Microw. Theory Tech.*, MTT-61, 1543–1556, 2013.

60. Y. S. Jiang, J. H. Tsai, and H. Wang, A W-band medium power amplifier in 90 nm CMOS, *IEEE Microw. Wirel. Compon. Lett.*, 18, 818–820, 2008.

61. J. Lee, C. C. Chen, J. H. Tsai, K. Y. Lin, and H. Wang, A 68–83 GHz power amplifier in 90 nm CMOS, *IEEE MTT-S Int. Microw. Symp. Dig.*, Boston, MA, 437–440, 2009.

62. Y. H. Hsiao, Z. M. Tsai, H. C. Liao, J. C. Kao, and H. Wang, Millimeter-wave CMOS power amplifiers with high output power and wideband performances, *IEEE Trans. Microw. Theory Tech.*, MTT-61, 4520–4533, 2013.

63. J. H. Tsai, J. L. Kuo, and H. Wang, A W-band power amplifier in 65-nm CMOS with 27 GHz bandwidth and 14.8 dBm saturated output power, *IEEE RFIC Symp. Dig.*, Montreal, Canada, 69–72, 2012.

64. P. H. Chen, J. C. Kao, T. L. Yu, Y. W. Hsu, Y. M. Teng, G. W. Huang, and H. Wang, A 110–180 GHz broadband amplifier in 65-nm CMOS process, *IEEE MTT-S Int. Microw. Symp. Dig.*, Seattle, WA, 1–4, 2013.

4

Lossy Matched and Feedback
Broadband Power Amplifiers

Dissipative or lossy gain-compensation-matching circuits can provide an important trade-off between power gain, reflection coefficient, and operating frequency bandwidth. Moreover, the resistive nature of such a simple matching circuit may also improve amplifier stability and reduce its size and cost. For the first time, it was suggested to use attenuation-equalizing circuits to maintain a high-quality transmission in long-telephone circuits by H. W. Bode in the mid-1930s [1]. For example, such an attenuation equalizer can represent a four-terminal frequency-selective network together with the connected source and load impedances having the constant-resistance image impedances at its input and output terminals.

Since it was impossible to provide broadband input matching of the bipolar power transistors with low-value frequency-varying input impedances, initially, a circuit arrangement composed of a resistor in addition to pure reactances was implemented to keep the input reflection coefficient at low values across large frequency bandwidths [2]. Such an impedance network was designed in such a way as to compensate the amplification slope of approximately 6 dB as much as possible within the bandwidth of one octave. Figure 4.1 shows the circuit schematic of an octave-band single-stage microstrip bipolar amplifier covering the frequency range from 500 to 1000 MHz with a maximum output power of about 3 W, a power gain of around 7 dB, and an input *VSWR* less than 2.5, where a lossy compensation circuit consists of a series resistor R shunted by the series-resonant circuit composed of a capacitor C and an inductor L, whose parameters were properly optimized.

4.1 Amplifiers with Lossy Compensation Networks

4.1.1 Equivalent Input Device Impedance

For a broadband lossy match silicon MOSFET high-power amplifier, it is sufficient to use a simple gain-compensation network with a resistor connected in series with a lumped inductor when operating frequencies are low enough compared to the device transition frequency f_T [3]. In this case, it is very important to optimize the elements of a lossy compensation circuit to achieve minimum gain flatness over maximum frequency bandwidth. Let us consider the small-signal silicon MOSFET-equivalent circuit shown in Figure 4.2. When the load resistor R_L is connected between the drain and source terminals, an analytical expression for the input device impedance Z_{in} can be obtained as

$$Z_{in} = R_g + \frac{(R_{gs} + (1/j\omega C_{gs}))}{[1 + (C_{gd}/C_{gs})((1 + j\omega\tau_g)(1 + j\omega C_{ds}R_{L0}) + g_m R_{L0}/1 + j\omega R_{L0}(C_{ds} + C_{gd}))]} \tag{4.1}$$

where $R_{L0} = (R_L + R_d)/[1 + (R_L + R_d)/R_{ds}]$ and $\tau_g = R_{gs}C_{gs}$.

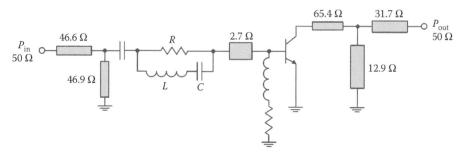

FIGURE 4.1
Schematic of octave-band microstrip lossy match bipolar power amplifier.

The modified equivalent circuit shown in Figure 4.3a adequately describes the frequency behavior of such an input impedance of Figure 4.2. In Equation 4.1, the series source resistance R_s and device transit time τ are not taken into account because of their sufficiently small values for high-power MOSFETs in a frequency range of $f \leq 0.3f_T$, where $f_T = g_m/2\pi C_{gs}$. When $\omega\tau_g \leq 0.3$ and the device output capacitive reactance is inductively compensated, the input equivalent circuit simplifies significantly and can represent a capacitor and a resistor connected in series, as shown in Figure 4.3b, where

$$R_{in} \cong R_g + R_{gs} \tag{4.2}$$

$$C_{in} \cong C_{gs} + C_{gd}\left[1 + g_m\frac{R_L + R_d}{1 + R_L/(R_{ds} + R_d)}\right] \tag{4.3}$$

To provide a constant real part of the input impedance Z_{in} in a frequency range up to $0.1f_T$, it is enough to use a simple lossy compensation circuit consisting of an inductor L_{corr} and a resistor R_{corr} connected in series, as shown in Figure 4.3c.

The total input impedance of both lossy match gain-compensation circuit and device input circuit is written as

$$Z_{in} = \frac{R_{corr} - \omega^2 C_{in}R_{in}L_{corr} + j\omega(L_{corr} + C_{in}R_{in}R_{corr})}{1 - \omega^2 L_{corr}C_{in} + j\omega C_{in}(R_{corr} + R_{in})} \tag{4.4}$$

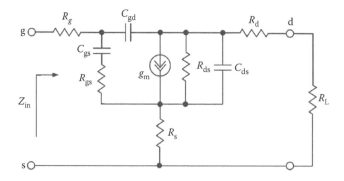

FIGURE 4.2
Small-signal silicon MOSFET-equivalent circuit.

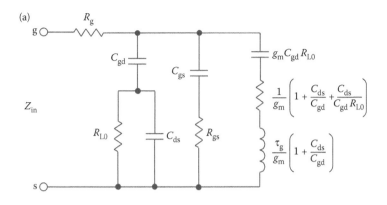

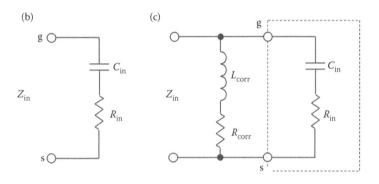

FIGURE 4.3
Equivalent circuits characterizing device input impedance.

whose real and imaginary parts can be expressed through the circuit parameters by

$$\operatorname{Re} Z_{\text{in}} = \frac{(1 - \omega^2 L_{\text{corr}} C_{\text{in}})(R_{\text{corr}} - \omega^2 C_{\text{in}} R_{\text{in}} L_{\text{corr}}) + \omega^2 C_{\text{in}}(R_{\text{corr}} + R_{\text{in}})(L_{\text{corr}} + C_{\text{in}} R_{\text{in}} R_{\text{corr}})}{(1 - \omega^2 L_{\text{corr}} C_{\text{in}})^2 + (\omega C_{\text{in}})^2 (R_{\text{corr}} + R_{\text{in}})^2} \quad (4.5)$$

$$\operatorname{Im} Z_{\text{in}} = \omega \frac{L_{\text{corr}}(1 - \omega^2 L_{\text{corr}} C_{\text{in}}) - C_{\text{in}}(R_{\text{corr}}^2 - \omega^2 L_{\text{corr}} C_{\text{in}} R_{\text{in}}^2)}{(1 - \omega^2 L_{\text{corr}} C_{\text{in}})^2 + (\omega C_{\text{in}})^2 (R_{\text{corr}} + R_{\text{in}})^2} \quad (4.6)$$

Under the condition $R = R_{\text{corr}} = R_{\text{in}}$, the equations for $\operatorname{Re} Z_{\text{in}}$ and $\operatorname{Im} Z_{\text{in}}$ can be reduced to

$$\operatorname{Re} Z_{\text{in}} = R \frac{(1 - \omega^2 L_{\text{corr}} C_{\text{in}})^2 + 2\omega^2 C_{\text{in}}(L_{\text{corr}} + C_{\text{in}} R^2)}{(1 - \omega^2 L_{\text{corr}} C_{\text{in}})^2 + (2\omega C_{\text{in}} R)^2} \quad (4.7)$$

$$\operatorname{Im} Z_{\text{in}} = \omega(1 - \omega^2 L_{\text{corr}} C_{\text{in}}) \frac{L_{\text{corr}} - C_{\text{in}} R^2}{(1 - \omega^2 L_{\text{corr}} C_{\text{in}})^2 + (2\omega C_{\text{in}} R)^2} \quad (4.8)$$

From Equation 4.8, it follows that the reactive part of the input impedance Z_{in} becomes zero, that is, $\operatorname{Im} Z_{\text{in}} = X_{\text{in}} = 0$, when

$$L_{\text{corr}} = C_{\text{in}} R^2 \quad (4.9)$$

that leads to a pure resistive input impedance Z_{in} obtained as

$$Z_{in} = R = R_{in} \qquad (4.10)$$

At microwaves, the short-circuited transmission line can be included instead of an inductor L_{corr} with the same input inductive reactance. In terms of amplifier circuit parameters, the low-frequency gain in decibels can be calculated as

$$\text{Gain} = 20 \log_{10} \left[\frac{g_m \sqrt{R_{corr} R_L}}{1 + (R_L + R_d) / R_{ds}} \right] \qquad (4.11)$$

However, when the frequency increases, the voltage amplitude applied to the input capacitance C_{in} decreases. This leads to the appropriate decrease of the operating power gain G_p at higher-bandwidth frequencies. Because the values of R_{in} for high-power MOSFETs are sufficiently small, the value of G_p may not be high enough. Therefore, it is necessary to provide an additional impedance matching with lossless-matching circuits to match with the source 50-Ω impedance or high output impedance of the active device of the previous power-amplifier stage.

4.1.2 Lossy Match Design Techniques

In many practical cases, to provide broadband matching with minimum gain flatness and input reflection coefficient, it is sufficient to use the single resistive shunt element at the transistor input. An additional matching improvement with reference to upper frequencies can be achieved by employing inductive reactive elements in series to the resistor. The resistive nature of this type of network may also improve amplifier stability and distortion. To provide a broadband performance for microwave GaAs MESFET power amplifiers, a resistively loaded shunt network, where the resistor is connected in series with a short-circuited quarter-wave microstrip line to decrease the loaded quality factor without greatly reducing the maximum available gain, was used in the load network to provide a flat gain of more than 8–12 GHz, or in the input-matching circuit to cover a frequency bandwidth of 2–6.2 GHz [4,5]. For ultra-broadband high-gain multistage amplifiers, using a simple lossy compensation shunt circuit with a resistor in series with an inductor placed at the input and output of each transistor in parallel with the second-order LC circuits allows the gain of 12±1.5 dB with a *VSWR* of less than 2.5 from 150 MHz to 16 GHz to be achieved for a three-stage GaAs MESFET amplifier [6]. A 14-dB gain was obtained over the 3-dB bandwidth from 700 kHz to 6 GHz for a two-stage microstrip GaAs MESFET power amplifier, where a flat gain performance was achieved by using a shunt lossy gain-compensation circuit having a resistor connected in series with a short-circuited microstrip line placed at the input and output of the first-stage transistor in parallel to the input and interstage LC-matching circuits [7].

A bandstop/bandpass diplexing RLC network is more useful than a simple lossy RL gain-compensation circuit because it provides an exact match at one frequency and an arbitrary amount of attenuation at any other frequency. Diplexing networks can be used in either input or output networks of the amplifier depending on noise figure, power output, and other amplifier constraints. Figure 4.4a shows the resonant-diplexing LC network for lossy gain compensation, where the series $L_s C_s$ and parallel $L_p C_p$ resonant circuits are

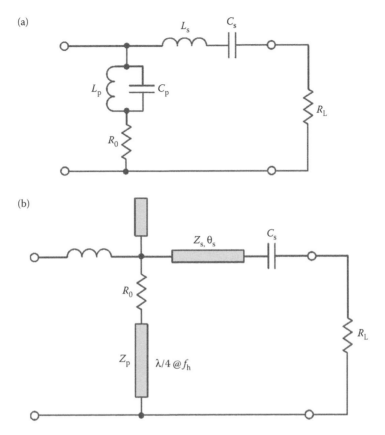

FIGURE 4.4
Circuit schematics of lossy gain-compensation circuits.

tuned to high-bandwidth frequency and $R_L = R_0$ [8]. Here, the series capacitance C_s and shunt inductance L_p are defined as $C_s = BW/\omega_h R_L$ and $L_p = BWR_L/\omega_h$, where BW is the normalized frequency bandwidth and $\omega_h = 2\pi f_h$ is the high-bandwidth radian frequency. The distributed form of a lossy gain-compensation network with additional input low-pass matching section is shown in Figure 4.4b, where $Z_p = 4\omega_h L_p/\pi$, $Z_s = \omega_h L_s/\tan\theta_s$, and θ_s is the electrical length of the series transmission line at high-bandwidth frequency.

Figure 4.5a shows the basic block of a microwave lossy match GaAs MESFET amplifier, where an input-matching circuit and an open-circuit shunt stub cascaded with a series transmission line at the device drain terminal are included to provide the amplifier-desired frequency response [9]. For frequencies up to 1 GHz, the reactive elements of the transistor- equivalent model have relatively little influence on the gain magnitude and reflection coefficients. As a result, the transistor described by S-parameters can be represented by its low-frequency model and the amplifier circuit can be significantly reduced to a simple network, where $S_{12} = 0$ and both S_{11} and S_{22} have negligible imaginary components. Then, the amplifier gain can be derived as

$$\text{Gain} = |S_{21}|^2 = \left[\frac{g_m Z_0}{2}(1 + S_{11})(1 + S_{22}) \right]^2 \tag{4.12}$$

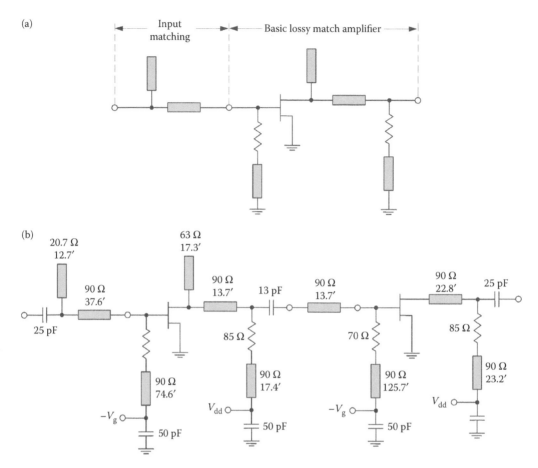

FIGURE 4.5
Circuit topologies of microstrip lossy match MESFET amplifiers.

which clearly expresses the trade-offs between the gain and the reflection coefficients, where g_m is the device transconductance and Z_0 is the characteristic impedance [9,10]. The schematic of a multistage lossy match amplifier can be divided into three basic circuit functions: input matching, amplification, and interstage matching. Figure 4.5b shows the lossy match two-stage GaAs MESFET amplifier with optimum values of the gate and drain shunt resistances to achieve flat gain performance over the frequency bandwidth from 2 to 8 GHz.

Figure 4.6a shows the circuit schematic of a broadband high-power LDMOSFET amplifier with the device geometry of 1.25 μm×40 mm. The optimized input three-element lossy-matching circuit allows a very broadband operation to be provided with minimum power gain flatness, and a 1:2 output transformer contributes to an increase in the output power level. The capacitor of 20 pF connected in parallel with the resistor of 27 Ω provides an additional increase of power gain at higher bandwidth frequencies. The simulation results are shown in Figure 4.6b, where an output power of 23.5±1.5 W with a power gain of 13.7±0.3 dB in a frequency range from 5 to 300 MHz can be achieved (curve 1). In this case, the input return loss is greater than 8 dB up to 225 MHz (curve 2). However, when a 50-Ω load is directly connected to the device drain terminal through the blocking capacitor, this results in output power levels in the range of 6.5±0.5 W.

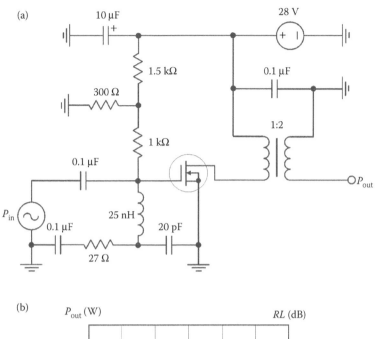

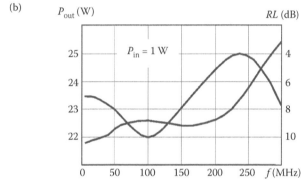

FIGURE 4.6
Schematic and performance of broadband LDMOSFET high-power amplifier.

Figure 4.7a shows an alternative *RLC*-matching network, which can be designed to provide an excellent *VSWR* performance along with the gain equalization of a microwave GaAs MESFET amplifier stage over a decade frequency bandwidth or more [11]. In this case, if $R_{gs}=0$, the input circuit can be redrawn into the bridged-*T* form shown in Figure 4.7b. This bridged-*T* circuit represents a second-order all-pass network if $L_1 = L_2 = R_0^2 C_{gs}/2$, $C_1=C_{gs}/4$, and $R_1=R_0$. For $R_{gs}>0$, the network structure becomes different from the standard all-pass form but the values of L_1, L_2, C_1, and R_1 can still be chosen such that the source generator sees a pure resistive R_0 at all frequencies. The design equations for this condition are

$$L_1 = \frac{1}{2}(1-p)R_0^2 C_{gs} \tag{4.13}$$

$$L_2 = \frac{1}{2}(1+p)R_0^2 C_{gs} \tag{4.14}$$

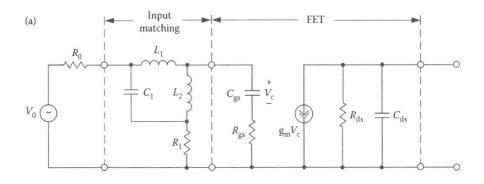

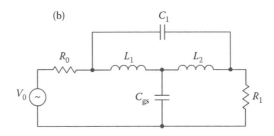

FIGURE 4.7
Second-order all-pass lossy-matching network.

$$C_1 = \frac{1}{4}(1 - p^2)C_{gs} \tag{4.15}$$

$$R_1 = R_0 \tag{4.16}$$

where

$$p = \frac{R_{gs}}{R_0} < 1 \tag{4.17}$$

which can be derived from circuit analysis [11].

Such an all-pass gain-compensation-matching circuit can also provide a flat gain over as wide frequency bandwidth as possible. In terms of the simple FET model shown in Figure 4.7a, the voltage V_c across the gate–source capacitance C_{gs} as a gain-controlled element should be constant. In this case, the voltage transfer function V_c/V_0 can be written as

$$2\frac{V_c}{V_0} = \frac{1 - j\,(\omega/2\omega_g)(1 - p)}{1 + j(\omega/2\omega_g)2p + j(\omega/2\omega_g)^3(1 - p)(1 - p^2)} \tag{4.18}$$

where $\omega_g = 1/R_0 C_{gs}$. It should be noted that, for practical cases, R_{gs} of submicron-gate FETs is very small compared to 50 Ω which is a usual value of R_0, and therefore $p \ll 1$. As a result, from Equation 4.18, it follows that V_c is approximately constant from $\omega = 0$ to

$$\omega_1 = 2\omega_g\sqrt{\frac{1 - 3p}{(1 - p)(1 - p^2)}} \approx 2\omega_g \tag{4.19}$$

at which the magnitude of the right-hand side of Equation 4.18 equals to unity, that is, the same as at $\omega=0$. As an example, for a two-stage GaAs MESFET MMIC power amplifier where the all-pass lossy match network was incorporated on the input of the first stage, a linear gain of 20 ± 0.5 dB, an input *VSWR* better than 1.7, a saturated power greater than 1 W with a flatness of only ±0.5 dB, and a *PAE* of more than 30% were achieved across the frequency bandwidth from 2 to 6 GHz [12].

4.1.3 Practical Examples

For solid-state single-sideband (SSB) and amplitude modulation (AM) communication transmitters, it is required to provide a linear amplification across the entire frequency range from 2 to 30 MHz, which was first covered by using bipolar technology based on a push–pull amplifier implementation with broadband toroidal transmission-line impedance transformers and combiners and interstage *RLC* gain-compensation networks. In this case, the driver stages are operated in a Class-A mode for increased power gain, whereas the final stages are biased in Class-AB with optimized quiescent currents for better linearity. As a result, the overall four-stage bipolar power amplifier achieved a *PAE* of more than 31% for a two-tone 60-W peak-envelope-power (PEP) signal over the entire frequency range from 2 to 30 MHz, with the third-order intermodulation products (IM_3) equal to −30 dBc or better at output powers from 5 to 60 W [13].

Figure 4.8 shows the circuit schematic of a bipolar broadband high-power amplifier designed for broadcasting VHF FM transmitters in a frequency range from 66 to 108 MHz

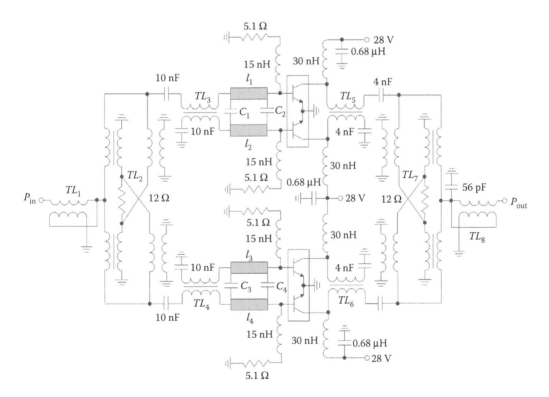

FIGURE 4.8
Bipolar broadband high-power amplifier for VHF FM transmitters.

[14]. When using the 200-W balanced VHF–UHF bipolar transistors, such as NEC 2SC3812, biased in a Class-C mode, an output power of 350 W with a power gain of 11.0±1.0 dB and a collector efficiency of about 60% can be provided across the entire frequency bandwidth by combining two transistors. An appropriate negative biasing in a Class-C mode is achieved by using a series resistor of 5.1 Ω together with a series inductor of about 15 nH in each bias circuit, which also serves as a lossy match gain-compensation circuit to provide minimum gain and power variations. The asymmetric 1:2 input TL_1 and output TL_8 transformers with the coaxial-cable characteristic impedances of 25 Ω are used to convert 12.5 Ω into the standard source and load 50-Ω impedances, respectively. The unbalanced-to-balanced stripline transformers TL_3 through TL_6 with the stripline characteristic impedances of 6 Ω are necessary to provide the 3-Ω source and load impedances for each part of the balanced bipolar transistors. Because of the small value of the device single-ended input impedance of about 1 Ω with the inductive component, the additional input two-section L-type impedance-matching circuits are used. Here, the series microstrip lines l_1 through l_4 are the inductive elements for the first section and the device lead inductances are the inductive elements for the second section. Power dividing at the input as well as power combining at the output of the high-power amplifier is realized by hybrid power splitters/combiners TL_2 and TL_7, each having 12-Ω ballast resistors and the stripline characteristic impedances of 12.5 Ω. Such a hybrid power splitter/combiner provides excellent device-to-device and device-to-load isolations and contributes to amplifier operation stability.

The circuit schematic of the input, interstage, and output networks intended to be implemented in microwave broadband power amplifiers are shown in Figure 4.9. A constant-resistance input network shown in Figure 4.9a provides the input device impedance to be pure, resistive, and equal to $Z_{in}=R_{in}$ when $L_1 = C_{gs}R_{in}^2$, $C_1=L_g /R_{in}^2$, and $R_1=R_{in}$, thus making wideband transformation of the input resistance to the source resistance much easier. In the output network shown in Figure 4.9b, a value of the drain inductance L_d is properly chosen to compensate for the capacitive device output reactance at the center bandwidth

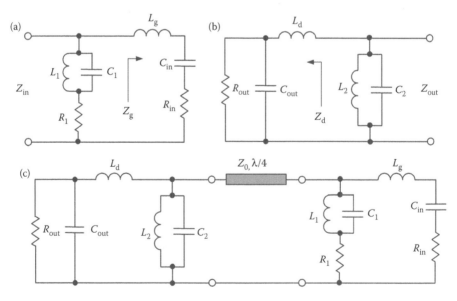

FIGURE 4.9
Schematics of input, output, and interstage broadband-matching circuits.

frequency, and a resonant frequency of the parallel L_2C_2 circuit is set to be equal to the same center bandwidth frequency. In this case, for lower frequencies where the device output impedance Z_d is capacitive, reactance of the parallel resonant circuit is inductive. On the other hand, for higher frequencies where the impedance Z_d is inductive, reactance of the parallel resonant circuit is capacitive. As a result, the wideband reactance compensation is realized when the reactive part of the overall output impedance becomes very small over a wide frequency bandwidth. For microwave applications, such a parallel resonant circuit is fabricated by using a quarter-wave short-circuit stub. The interstage network, whose circuit schematic is shown in Figure 4.9c, comprises the input and output networks described above and a quarter-wave microstrip transformer with the characteristic impedance of $Z_0 = \sqrt{R_{in}L_d/R_{out}C_{out}}$.

Figure 4.10 shows the circuit schematic diagram of a two-stage lossy match MESFET power amplifier [15]. By using a 1.05-mm device in the driver stage and two 1.35-mm devices in the final stage, a saturated output power of 27.7±2.7 dBm, a linear power gain of 8.3±2.8 dB, and a drain efficiency of 15.3%±8.3% were measured in a frequency range from 4 to 25 GHz. The input and interstage constant-resistance networks are represented by the series connection of a resistor and a high-impedance microstrip line. Two such networks connected in parallel provide pure resistive input impedance, where l_4 and l_5 are the series microstrip lines, R_1 and R_2 are the series resistors. The short-circuited microstrip lines (l_7 and l_8 in the interstage network, l_{19} and l_{21} in the output network) with quarter-wave electrical lengths at the center bandwidth frequency serve as the parallel resonant circuits connected at the device output terminals. The microstrip lines l_{10} and l_{14} in the interstage network represent the quarter-wave impedance transformers, which provide

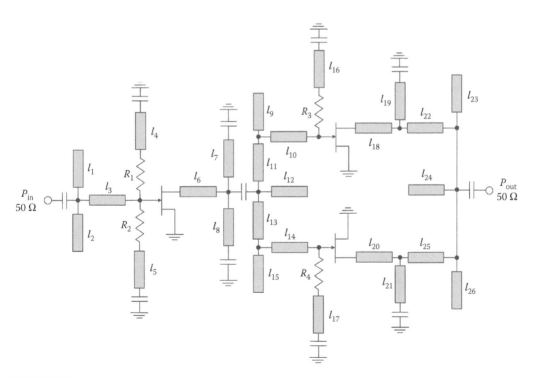

FIGURE 4.10
Microstrip two-stage lossy match MESFET power amplifier.

matching between the output impedance of the driver-stage device and the input imped-ance of the second-stage devices connected in parallel. The input- and output-matching circuits are realized in the form of T-transformers, where the series microstrip lines and parallel open-circuit microstrip stubs replace the series inductors and shunt capacitors, respectively. To further increase an output power, the number of amplifying stages with lossy input and interstage-matching circuits connected in parallel can be increased. As a result, by optimizing the output matching and combining circuits, for a three-stage MMIC 0.25-μm pHEMT power amplifier with a distributed amplifier used as a driver stage and four 1200-μm transistors in the output stage, an output power of 2.4±1.1 W with a small-signal gain of 24±3.5 dB over the frequency range from 6 to 18 GHz was measured [16].

Figure 4.11 shows the circuit schematic of a two-stage broadband microstrip lossy match 0.1-μm InP-HEMT amplifier, where the source inductor is inserted between the source terminal of each transistor and the ground to increase the input device impedance and lower the Q-factor [17]. Here, the shunt lossy match networks are connected at the outputs of both devices and at the input of the first-stage transistor. The source inductor l_s of 10-μm length increases the real part of the device input impedance from 2 to 9.5 Ω, thus reduc-ing the Q-factor from 8.0 to 1.4 at the same time. To maximize frequency bandwidth and improve an input $VSWR$, the input, interstage, and output-matching circuits incorporating the CPW transmission lines and open-circuit stubs are used. In this case, the best gain flat-ness in the gain–frequency characteristic is achieved with an optimum value $R=30$ Ω. For the transistors with the transconductance $g_m=1.4$ S/mm and transition frequency $f_T=240$ GHz, a 94-GHz bandwidth and a 10-dB gain with a power consumption of 79 mW at a sup-ply voltage of 2 V were achieved. The amplifier operates as a CR-coupled amplifier in the low-to-medium frequencies and as an LC match amplifier at high frequencies.

The output-matching network can also represent the series connection of a π-type low-pass-matching circuit and a lossy gain-compensation network, as shown in Figure 4.12 [8]. For both bipolar and MESFET broadband amplifiers, the π-type network comprises a device output capacitance and an open-circuit microstrip stub, which are connected to each side of a series-lumped inductor, respectively. The output lossy gain-compensation network is connected between the π-type matching circuit and the load. This configuration

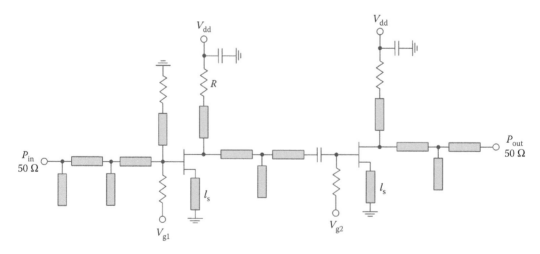

FIGURE 4.11
Circuit schematic of the two-stage microstrip lossy match InP-HEMT amplifier.

is usually used for very broadband medium-power amplifiers. For example, the two-stage cascade of an *L*-band broadband bipolar amplifier shown in Figure 4.12a was designed for a minimum input reflection coefficient with a *VSWR* of 1.78, a maximum gain variation of ±1.2 dB, and around 16.5-dB power gain over the frequency range from 1 to 2 GHz. The two-stage cascade of a microwave MESFET amplifier shown in Figure 4.12b was designed for a maximum flat gain in a frequency range from 4 to 6 GHz when the power gain varies within 15.4±0.5 dB. To provide minimum loss at high-bandwidth frequency, the short-circuited microstrip line in the output lossy gain-compensation circuit was chosen to be of a quarter-wavelength long for each amplifier. For a two-stage 1-μm MESFET amplifier with the first stage designed for minimum noise figure and the second stage designed for maximum flat gain at higher bandwidth frequencies, a power gain of 9.5±1 dB over the frequency range from 6.5 to 12 GHz was achieved [18].

Figure 4.13a shows the circuit schematic of a broadband GaN HEMT microwave power amplifier implemented in the form of a flip-chip-integrated circuit with the device geometry of 0.7 μm×1 mm, transition frequency f_T=18 GHz, and maximum frequency f_{max}=35 GHz [19]. The optimized input three-element lossy *LCR*-matching circuit provides a power gain up to 11.5 dB and a low-input reflection less than −10 dB over the frequency range from 3 to 9 GHz. As the impedance at the input of a lossy match gain-compensation circuit is only of about 10 Ω, this necessitates an additional 50-to-10-Ω broadband Tr1, which was realized using a few sections of quarter-wave coplanar transmission lines with decreasing characteristic impedances. The output network incorporates a low-pass

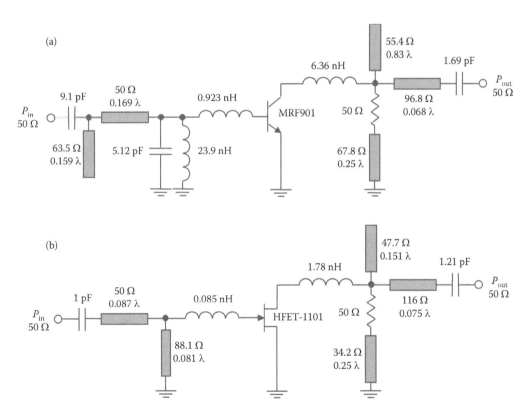

FIGURE 4.12
Broadband microstrip lossy match bipolar and MESFET power amplifiers.

LC circuit to compensate for the output device capacitance such that the intrinsic device sees approximately a real load within the entire frequency bandwidth. Since the optimum load for this 1-mm device with a supply voltage of 20 V is of about 50 Ω, no output Tr1 is needed. The output power was measured of about 1.6 W with a *PAE* from 14% to 24% across the frequency bandwidth from 4 to 8 GHz. By combining of four such GaN HEMT power amplifiers connected in parallel, the highest output power of 8 W with a *PAE* of about 20% was obtained at 9.5 GHz and the lowest output power of 4.5 W was measured at 4.5 GHz, with a small-signal gain of 7 dB across the frequency bandwidth from 3 to 10 GHz [20].

To provide multidecade bandwidth with a very good input return loss, a compact bridged-*T* all-pass input *RLC*-matching network can be used, as shown in Figure 4.13b, where the resistor R_1 was set to be of 50 Ω [21]. In this case, a GaN HEMT periphery of 2.2

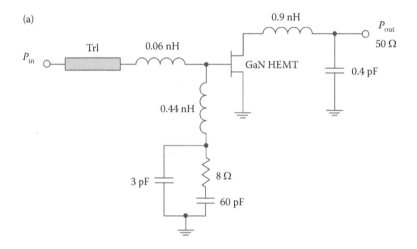

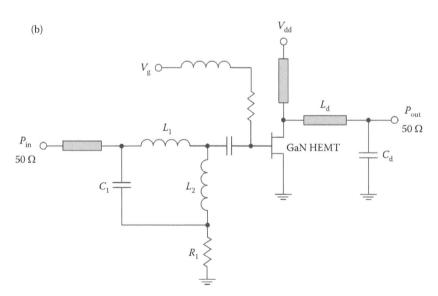

FIGURE 4.13
Schematics of microwave broadband GaN HEMT power amplifiers.

mm was chosen to obtain an output power in the range of 10 W. A simple two-element-matching circuit consisting of a series microstrip line and a shunt capacitor was used at the output to provide optimum load impedance at the upper band edge. The power amplifier was packaged in a ceramic package including GaN HEMT on an SiC device operating at 28 V and GaAs integrated passive-matching circuitry. As a result, an output power of 8 W and a power gain of 12 dB were measured over the frequency bandwidth from 50 MHz to 2 GHz with a drain efficiency from 36.7% to 65.4%.

4.2 Feedback Amplifiers

The principle of a feedback linearization and flat gain bandwidth expansion of the power amplifier was invented by H. S. Black in 1927. A year later, he filed the patent application on a vacuum-tube feedback amplifier [22]. Black recognized that using a large amount of feedback in an amplifier comprising several vacuum-tube stages in a cascade to yield a very high open-loop gain gives a glorious opportunity to make the resulting negative feedback amplifier increased in bandwidth and insensitive to nonlinearity and uncertainty in the characteristics of the vacuum tube [23]. The gain of the negative feedback amplifier decreases by an amount of the feedback or loop gain; so do the nonlinear components. In this case, the negative feedback amplifier becomes insensitive to the gain or phase variations as long as its stability conditions are satisfied. Unfortunately, the significance of this invention was not fully understood at that time, as well as the operation principle of a negative feedback amplifier. For instance, Black's director of research insisted that a negative feedback amplifier would never work; similarly, the Patent Office initially did not believe it would work and took more than 9 years to decide to issue the patent [24]. However, a three-stage vacuum-tube amplifier with a proper parallel feedback *RLC* network designed by H. W. Bode was able to demonstrate a 28-dB uniform feedback in a frequency range between 60 kHz and 2 MHz with a phase margin of stability of approximately 45° [25,26].

The design of the transistor feedback amplifiers is in some respect more involved due to the physical nature of the transistor itself. Even at low frequencies, the bipolar transistor is represented by a more complicated equivalent circuit than a vacuum tube, and its input impedance is in general low with a considerable effect by conditions in its output circuit. However, simple and useful expressions can be derived if the applied negative feedback is large enough. To achieve reasonable gain-stability performance with a single-stage amplifier, almost all available gain of the amplifier would need to be utilized for the application of negative feedback. For most practical purposes, therefore, feedback over more than one stage is necessary, and, as the common-emitter stage gives a phase reversal, the most suitable number of stages is three if a phase-inverting circuit is not used.

Figure 4.14 shows the simplified circuit schematic of a three-stage bipolar amplifier, where the resistor R_1 is responsible for series negative feedback and the resistor R_2 is responsible for parallel negative feedback. Assuming that all three transistors have the same parameter values and the source resistance R_0 is much smaller than the load resistance R_L, the gain for an amplifier with a large negative feedback can be derived as

$$\text{Gain} = \frac{R_2}{R_1} \frac{R_2}{R_2 + R_L} \tag{4.20}$$

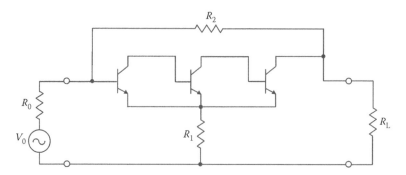

FIGURE 4.14
Schematic of the three-stage bipolar amplifier with series–parallel resistive feedback.

which shows the relationship between the gain and the feedback parameters [27]. Note that the design of feedback amplifiers using both series and parallel feedback is basically the same as when both types of feedback are applied separately.

The type of feedback can be cascaded when, for example, the first and third stages utilize the series resistive feedback type, whereas the second stage utilizes the parallel resistive feedback type [28]. To increase gain of a three-stage bipolar amplifier, the negative feedback used in the third stage can be combined with lossy-matching circuits used in the first and second stages [29]. In one of the first monolithic implementations, a two-stage balanced silicon bipolar amplifier with a series–parallel feedback in each stage achieved a flat gain of 12 dB within 0.5 dB over a frequency range from dc to 1 GHz with a noise figure of 7.5 dB at 800 MHz [30]. High stability and direct cascading of four such two-stage amplifiers resulted in a flat gain of nearly 50 dB varying within 1 dB across the same frequency range.

4.2.1 Negative Feedback Design Techniques

At microwave frequencies, the parasitic elements of a GaAs MESFET device restrict the amplifier bandwidth capability. Therefore, to extend the bandwidth of the negative feedback amplifier to higher frequencies, it is advisable to use a series inductor connected to the drain terminal and a series inductor inserted into the resistive feedback loop, as shown in Figure 4.15a [31]. In this case, the drain inductance is necessary to compensate for the device output capacitive reactance at the upper band edge, and the feedback inductance is required to improve the amplifier gain capability at higher frequencies by reducing negative feedback when frequency increases. However, it is very important to prevent any instability at higher frequencies when the feedback loop can potentially elevate the insertion gain above that of the open-loop amplifier. The degree of feedback is mainly controlled by the value of the feedback resistor R_F. To derive a simple analytic expression for the feedback resistance, the equivalent circuit of a GaAs MESFET can be reduced to the low-frequency model when the reactive elements of the transistor and matching circuits are neglected. Besides, to demonstrate the trade-off between $VSWR$ and gain, it can be assumed that $G_{ds}Z_0 \ll 1$ and $G_{ds} \ll g_m$, where G_{ds} is the drain conductance, g_m is the device transconductance, and Z_0 is the characteristic impedance equal to the source and load resistances.

As a result, by taking into account that the device series gate resistance R_g, source resistance R_s, and drain resistance R_d are very small compared to the feedback resistance R_F and

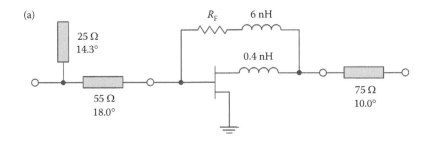

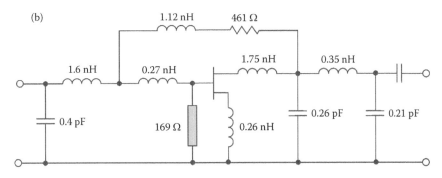

FIGURE 4.15
Circuit topologies of the microstrip negative feedback MESFET amplifiers.

load resistance $R_L = Z_0$, the low-frequency S-parameters of the negative feedback amplifier stage can be written as

$$S_{11} = S_{22} = \frac{1}{\Sigma}\left[\frac{R_F}{Z_0} - g_m Z_0\right] \tag{4.21}$$

$$S_{12} = \frac{2}{\Sigma} \tag{4.22}$$

$$S_{21} = -\frac{2}{\Sigma}[g_m R_F - 1] \tag{4.23}$$

with

$$\Sigma = 2 + g_m Z_0 + \frac{R_F}{Z_0} \tag{4.24}$$

which demonstrate that the gain, input and output *VSWR*, and reverse isolation are all fixed quantities for a particular transistor once the value of the feedback resistor R_F is chosen [31].

The ideal matching condition when $S_{11} = S_{22} = 0$ can only be satisfied for

$$R_F = g_m Z_0^2 \tag{4.25}$$

In this case, the associated gain is

$$G = 20 \log_{10}(g_m Z_0 - 1) \tag{4.26}$$

The circuit schematic of a matched feedback amplifier with simple input- and output-matching circuits consisting of an open-circuit stub and short series transmission lines is shown in Figure 4.15a. By selecting a feedback inductor of 6 nH, a drain inductor of 0.4 nH, and a feedback resistor of 160 Ω, the frequency range from 350 MHz to 14 GHz was covered with a minimum gain of 4 dB at an output power of 13 dBm using a 1-μm GaAs MESFET device with the gate width of 800 μm [31]. Since a parallel feedback resistor provides a strong negative feedback to match the transistor over a very wide frequency range, which makes the gain very low for the devices with low transconductance according to Equation 4.26, two or more transistors can be connected in a parallel configuration, which increases the overall transconductance [32].

Figure 4.15b shows the circuit schematic of a single-stage feedback GaAs FET amplifier intended for a hybrid implementation on an alumina substrate using lumped inductors, chip capacitors and resistors [33]. In this case, the transconductance of a structure with three parallel transistors (total gate width of 900 μm and gate length of 0.7 μm) was more than 72 mS, which contributed to a power gain of around 8 dB over the frequency range from 100 MHz to 6 GHz with a 1-dB gain-compressed output power of 15 dBm at 6 GHz and a worst-case 4.5-dB noise figure at 3 GHz. A monolithic negative feedback 0.7-μm GaAs FET amplifier covering the frequency range from 1 to 7 GHz was able to achieve a small-signal gain of 6.0±0.2 dB with maximum input and output *VSWR* of 2.3 and 1.7, respectively [34].

The equivalent circuit of a negative feedback MOSFET power amplifier with a parallel feedback resistor R_F and inductor L_F is shown in Figure 4.16. For medium- and high-power devices, the series source resistance R_s and feedback gate–drain capacitance C_{gd} are sufficiently small to have a substantial influence on the device RF performance, and the series drain inductor L_d is used to compensate for the device drain–source capacitance C_{ds} at high-bandwidth frequency. In this case, the device admittance matrix can be obtained in a simplified form as

$$[Y_A] = \begin{bmatrix} \dfrac{j\omega C_{gs}}{1 + j\omega C_{gs}(R_{gs} + R_g)} & 0 \\ \dfrac{g_m}{1 + j\omega C_{gs}(R_{gs} + R_g)} \dfrac{R_{ds}}{R_{ds} + R_d} & \dfrac{1}{R_{ds} + R_d} \end{bmatrix} \tag{4.27}$$

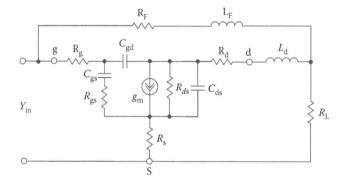

FIGURE 4.16
Equivalent circuit of the negative feedback MOSFET power amplifier.

The parallel negative feedback admittance matrix can be written as

$$[Y_F] = \begin{bmatrix} Y_F & -Y_F \\ -Y_F & Y_F \end{bmatrix} \tag{4.28}$$

where $Y_F = 1/Z_F$ and $Z_F = R_F + j\omega L_F$ is the feedback impedance.

As a result, the total admittance Y-matrix can be derived as

$$[Y] = [Y_A] + [Y_F] = \begin{bmatrix} \dfrac{j\omega C_{gs}}{1 + j\omega C_{gs}(R_{gs} + R_g)} + Y_F & -Y_F \\ \dfrac{g_m}{1 + j\omega C_{gs}(R_{gs} + R_g)} \dfrac{R_{ds}}{R_{ds} + R_d} - Y_F & \dfrac{1}{R_{ds} + R_d} + Y_F \end{bmatrix} \tag{4.29}$$

To evaluate the feedback inductance L_F, it is advisable to further simplify the calculation procedure by taking into account that usually, $R_d/R_{ds} \ll 1$ and $\omega C_{gs}(R_g + R_{gs}) \ll 1$ when $f \leq 0.1 f_T$. Then, the input two-port network admittance Y_{in} can be written as

$$Y_{in} = j\omega C_{gs} + \frac{1 + R_{ds}/R_L + g_m R_{ds}}{R_{ds} + (1 + R_{ds}/R_L)(R_F + j\omega L_F)} \tag{4.30}$$

where R_L is the load resistance.

Under the condition of $\omega L_F/[R_F + R_{ds}/(1 + R_{ds}/R_L)] \leq 0.3$ when the phase shift due to the feedback circuit is sufficiently small to cause the circuit instability, the inductance L_F obtained from the condition of $\text{Im} Y_{in} = 0$ can be calculated from

$$L_F = C_{gs} \frac{1 + R_{ds}/R_L}{1 + R_{ds}/R_L + g_m R_{ds}} \left(R_F + \frac{R_{ds}}{1 + R_{ds}/R_L} \right)^2 \tag{4.31}$$

In this case, an input impedance of the matched negative feedback amplifier circuit becomes real and equal to

$$R_{in} = \frac{R_F + R_{ds}/(1 + R_{ds}/R_L)}{1 + g_m R_{ds}/(1 + R_{ds}/R_L)} \tag{4.32}$$

By applying a condition of $R_{in} = R_{ds}/(1 + R_{ds}/R_L)$ for the input impedance, the ratio of the feedback resistance R_F and the basic parameters of the device-equivalent circuit can be written in a simple form as

$$R_F = g_m \left(\frac{R_L}{1 + R_L/R_{ds}} \right)^2 \tag{4.33}$$

The operating power gain G_P can be expressed through the amplifier-equivalent circuit parameters as

$$G_P = \left(1 - \frac{g_m R_L}{1 + R_L/R_{ds}} \right)^2 \frac{1}{1 + R_L/R_{ds}} \tag{4.34}$$

In this case, the maximum available gain $MAG = G_{Pmax}$ with unilateral amplification for an ideally matched parallel feedback power amplifier can be written as

$$G_{Pmax} = \frac{g_m^2 R_L}{(\omega C_{gs})^2 (R_{gs} + R_g)(1 + R_L/R_{ds})^2} \tag{4.35}$$

Consequently, the ratio G_{Pmax}/G_P can characterize the negative feedback depth provided that $g_m R_L \gg 1$ by

$$\frac{G_{Pmax}}{G_P} \cong \frac{1 + R_L/R_{ds}}{R_L(R_{gs} + R_g)(\omega C_{gs})^2} \tag{4.36}$$

From Equation 4.36, it follows that, at operating frequencies where $\omega C_{gs}(R_g + R_{gs}) \leq 0.1$ and under the condition of $(R_g + R_{gs})(1 + R_L/R_{ds})/R_L \geq 0.1$, the strong negative feedback level of more than 10 dB is realized, which leads to the significant improvement of the amplifier nonlinear characteristic with an appropriately reduced level of the intermodulation distortion.

Figure 4.17 shows the circuit schematic of a microstrip negative feedback power amplifier using an SiC MESFET, where an unconditional stability is provided by adding a parallel combination of R_s and C_s in series at the input [35]. In this case, the frequency range from 10 MHz to 2.4 GHz was covered with a power gain of 8±0.5 dB, an output power of 37 dBm at 1-dB gain compression point, and a *PAE* of almost 35% at a supply voltage of 30 V. In a two-stage configuration with a GaAs MESFET driver, a power gain of 22±1 dB with a noise figure of less than 4 dB was achieved across the same frequency range with a 37-dBm output power [36]. By using a GaN HEMT transistor in a single-stage negative feedback amplifier, an output power of 43±1 dBm with a power gain of 12±1 dB was achieved from 500 MHz to 2.5 GHz at a supply voltage of 28 V with a *PAE* from 40% at low frequencies below 750 MHz to 30% at higher frequencies [37]. At the same time, a linear power of more than 37.3 dBm and a power gain of more than 13.3 dB were achieved over the frequency range from 1.0 to 3.4 GHz, with an interception point IP_3 of 48.5 dBm and a *PAE* of 23.5 at 3 GHz [38].

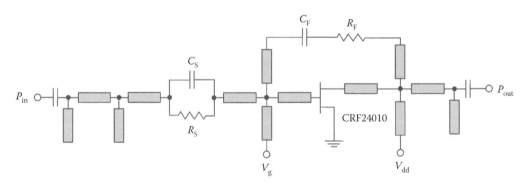

FIGURE 4.17
Circuit schematic of the microstrip negative feedback SiC MESFET power amplifier.

4.2.2 Noise Figure

The noise figure F_T for the parallel negative feedback amplifier, in which both the amplifier and feedback two-port networks have noise sources, as shown in Figure 4.18, can be written as

$$F_T = 1 + \sum_{i=1}^{2} (F_i - 1) \tag{4.37}$$

where

$$F_1 = 1 + \frac{R_S \left|(1/R_S) + g_m\right|^2}{R_F \left(g_m - (1/R_F)\right)^2} \tag{4.38}$$

is the noise figure for the circuit, in which only the feedback resistor block is noisy, g_m is the device transconductance, R_S is the input source resistance, R_F is the feedback resistance, and F_2 is the noise figure for the circuit, in which only the amplifying block has noise sources [39].

Generally, the Y-parameters for the parallel connection of two-port networks can be written as

$$Y = \begin{bmatrix} Y_{11} & Y_{12} \\ Y_{21} & Y_{22} \end{bmatrix} = \begin{bmatrix} \sum_{i=1}^{2} y_{11i} & \sum_{i=1}^{2} y_{12i} \\ \sum_{i=1}^{2} y_{21i} & \sum_{i=1}^{2} y_{22i} \end{bmatrix} \tag{4.39}$$

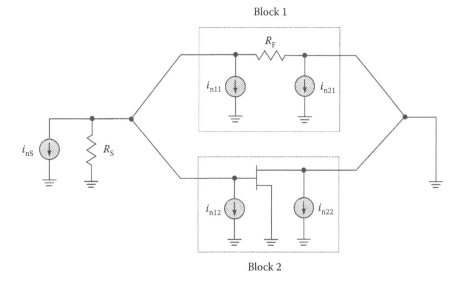

FIGURE 4.18
Noise-equivalent circuit for common-source FET with negative feedback.

where n is the number of parallel two-port networks.

Defining the noise figure F_N for the FET without the feedback resistor (Block 1), the equation expressed through the device noise currents and equivalent-circuit admittance parameters can be obtained as

$$\frac{F_2 - 1}{F_N - 1} = \frac{i_{n12u}^2 + \overline{| i_{n12} - i_{n12u} + (i_{n22} / -Y_{21} / (Y_S + Y_{11})) |^2}}{i_{n12u}^2 + \overline{| i_{n12} - i_{n12u} + (i_{n22} / -y_{21} / (Y_S + y_{11})) |^2}} \tag{4.40}$$

where the total admittance parameters Y_{11} and Y_{21} are defined from Equation 4.39, y_{11} and y_{21} are the transistor admittance parameters, Y_S is the source admittance, i_{n12} and i_{n22} are approximately treated as fully correlated noise currents, and i_{n12u} is the uncorrelated noise current component. For $i_{n12u} = i_{n12} = 0$ when it is considered that i_{n22} is more dominative than i_{n12}, Equation 4.40 can approximately be calculated for the FET with $g_m = 60$ mS and $R_F = 250 \, \Omega$ as

$$\frac{F_2 - 1}{F_N - 1} \approx 1.65 \tag{4.41}$$

Since for a 1.2-μm FET with a 400-μm gate width for a 50-Ω source impedance (without feedback), the calculated F_1 value from Equation 4.38 is equal to 1.41 and the measured F_N value was 2.8 (4.5 dB), resulting in $F_2 = 3.97$. Consequently, the noise figure for Figure 4.18 is obtained from F_1 and F_2 using Equation 4.37 as

$$F_T = 1 + \sum_{i=1}^{2} (F_i - 1) = 1 + 0.41 + 2.97 = 4.38 \; (\approx 6.4 \, \text{dB}) \tag{4.42}$$

where F_T is about 2 dB higher than $F_N = 4.5$ dB [39].

Exact formulas for the noise parameters and noise figure of the amplifiers with parallel feedback and lossy input- and output-matching circuits can be given through the equivalent noise resistance R_n, equivalent noise conductance G_n, and the corresponding correlation admittance Y_{corr} [40,41]. The value of the parallel feedback resistor represents a practical compromise between the *VSWR*, gain, and noise figure. For example, greater resistance values decrease the noise figure and increase the gain, but also increase the reflection coefficients at the input and output terminals that severely limit the capability of a broadband multistage amplification [40]. The measured noise figure for the low-noise two-stage amplifier using AlGaN/GaN HEMT on SiC technology with a source-inductive feedback in the first stage and a parallel *RLC* feedback in the second stage was less than 4 dB in an operating frequency range from 4 to 18 GHz [42]. However, for an equal amount of negative feedback, the series feedback gives a much lower noise figure than the parallel feedback [43].

4.2.3 Practical Examples

Figure 4.19 shows the circuit schematic of a linear MOSFET power amplifier designed for multioctave frequency bandwidth from 1.5 to 60 MHz with a 50-W output power and a power gain of 23±1 dB [44]. By using the negative feedback resistors of 36 Ω in the final

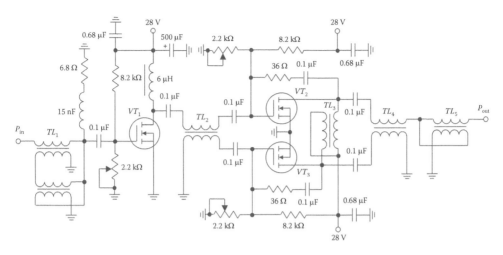

FIGURE 4.19
Circuit schematic of the multioctave linear MOSFET HF-VHF power amplifier.

stage, the power-amplifier nonlinearity can be significantly improved by reducing the level of the IM_3 down to −45 dBc. Here, the input Tr1 is provided by an asymmetrical 1:9 impedance transformer using a 17-Ω stripline and a ferrite core with permeability $\mu = 400$. An increased value of the ferrite core permeability is necessary to reduce the cable length to less than 15 cm, providing a significantly higher inductive impedance of the transformer primary winding in about 10 times than the standard 50-Ω source at low-bandwidth frequencies. Owing to the 36-Ω parallel resistive feedback, the input impedance of each device should be of 6.25 Ω. In this case, it is sufficient to use an unbalanced-to-balanced 1:1 transformer with a 12.5-Ω stripline characteristic impedance, which provides a 12.5-Ω load for the driver-stage MOSFET device. The use of the transformer TL_3 contributes to additional even-harmonic suppression in a push–pull operation mode. In terms of the circuit simplicity, it is convenient to use its common terminal for the drain voltage supply. The output transformation is achieved by using a 1:1 balun TL_4 to convert an output impedance of each 50-W MOSFET device with $f_T = 1$ GHz designed for a push–pull operation into 12.5 Ω and an asymmetrical 1:2 transformer TL_5 to transform 12.5 Ω to the standard 50-Ω load.

A linear output power of 150 W can be achieved by simultaneously using the negative feedback and lossy match techniques required for both stability and gain flatness performance in a two-stage MOSFET power amplifier based on a balanced 28-V 400-W VDMOSFET device in a final stage to cover the frequency range from 20 to 100 MHz with a *PAE* around 40%, a power gain of more than 30 dB, and a gain flatness within 1 dB [45]. The test results demonstrated a stable operation of the implemented broadband power amplifier under different *VSWR* values up to ∞ at all phases.

The circuit schematic of a bipolar push–pull power amplifier achieving an output power of 160 W (PEP) with the intermodulation distortion (IM_3 and IM_5) better than −30 dBc at a supply voltage of 28 V into a 50-Ω load is shown in Figure 4.20a [46]. For broadband linear operation, a quiescent collector current of 60–80 mA for each final-stage transistor 2N5942 (150 W, 250 MHz) should be provided. Higher quiescent current levels will reduce the fifth-order intermodulation products (IM_5), but will have a small effect on the IM_3 except at lower power levels. A biasing adjustment is provided in the amplifier circuit to compensate for variations in the transistor current gain, which allows control of the idling

(a)

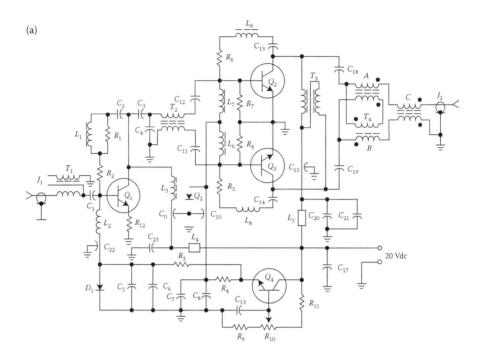

(b)

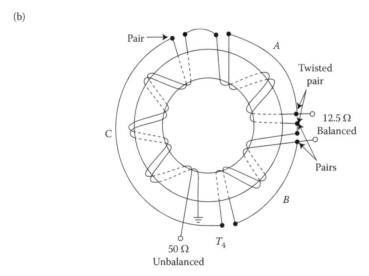

FIGURE 4.20
Circuit schematic of a broadband bipolar VHF power amplifier.

current for both the output and driver devices. A driver stage should achieve about 4.5 W (PEP) with the intermodulation products better than −40 dBc at a quiescent collector current of 10–15 mA, which is the maximum necessary to drive two devices in the final stage. Higher current levels will not improve linearity, but will degrade driver efficiency. To compensate for variations in output with changes in the operating frequency, the negative voltage feedback is employed on both the final and driver stages. At the low end of the desired frequency band, approximately 4.5 dB of feedback is inserted into the final stage and 15 dB of feedback is inserted into the driver stage, resulting in a maximum total gain variation of 0.5 dB over the frequency range from 3 to 30 MHz.

To achieve the desired broadband response, the transmission-line transformers based on twisted-pair windings and toroidal cores were used for input power dividing and output combining. The transformers T_1, T_2, and T_3 have turn ratios of 4:1, 1:1, and 1:4, respectively. The transformer T_1 consisting of six turns of two-twisted pairs wound on a toroidal core provides a Tr1 to match the 50-Ω source to the low-impedance at the base of the transistor Q_1. In this case, the two pairs (four separate wires) are twisted together and the two wires from each original pair are soldered together at each end. The transformer T_2 is a 1:1 balun consisting of six turns of two-twisted pairs of wire (four wires in total). The transformer T_3 consists of four turns of two-twisted wires, where both wires of each pair are soldered together at each end. The transformer T_4 is a 1:4 balun with three separate windings, as shown in Figure 4.20b, where the windings A and B consist of five turns of two-twisted pairs, whereas the winding C is formed from eight turns of a single-twisted pair. The ferrite core used for the transformer T_4 has a specified maximum flux density of about 100 gauss.

Figure 4.21 shows the circuit schematic of a multioctave push–pull VHF power amplifier based on a balanced MOSFET device, which achieved an output power of 300 W with a power gain of about 15 dB from 10 to 175 MHz at a supply voltage of 50 V [47]. In this circuit, the gate–bias voltage divider accommodates a thermistor–resistor combination for temperature stabilization of the MOSFET quiescent current. Without this stabilization, the drain idle current would have an approximate temperature coefficient of +15 mA/°C. The input and output impedance matching is realized with unique wideband coaxial-cable transformers, whose advantages are the dc isolation between the primary and secondary turns, automatic balanced-to-unbalanced functions, and compact size. The low-impedance side always has one turn and consists of parallel-connected segments of the coaxial outer conductor. The high-impedance side has inner-conductor segments that are connected in series. This arrangement permits only integer impedance ratios that are perfect squares, such as 1, 4, 9, and 16. The coupling coefficient between the primary and secondary turns can be controlled by varying the coaxial-cable characteristic impedance. In an optimum

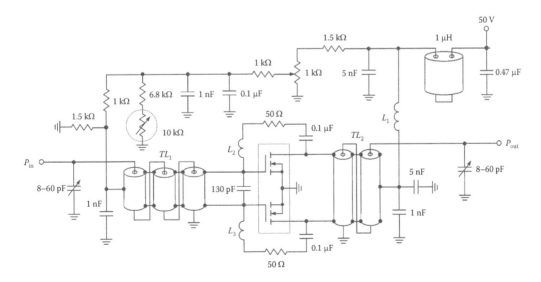

FIGURE 4.21
Circuit schematic of the multioctave MOSFET VHF power amplifier.

configuration, the low-impedance winding connection points should be brought together as close as possible, which minimizes the lengths of the uncovered inner-conductor segments. The transformer T_1 provides a 16:1 impedance ratio resulting in a closer match at lower frequencies, whereas the high end can be corrected by using an input shunt variable capacitor and a balanced capacitor connected between the device gates. The output matching is provided by a 4:1 transformer T_2, with a shunt variable capacitor at the output optimized at 175 MHz. The negative RLC feedback for each balanced transistor part improves the input return loss and flattens the power gain over the entire frequency range, where it is sufficient to use the lead lengths of the resistors as the feedback inductors L_2 and L_3.

Using LDMOSFET and GaN HEMT technologies allows the broadband high-power operation of the negative feedback amplifiers with high efficiency to be achieved across VHF and UHF frequency bands. For example, a two-stage broadband push–pull power amplifier based on two 4-W LDMOSFET devices MRF281Z in the driver stage and two 10-W LDMOSFET devices MRF282Z in the final stage with a parallel RLC feedback for each transistor and 1:1 baluns at the amplifier input and output demonstrated an output power of more than 37 dBm and a PAE of more than 43% with a power gain of 22±1.5 dB over the entire frequency bandwidth of 2–500 MHz [48]. Figure 4.22 shows the circuit

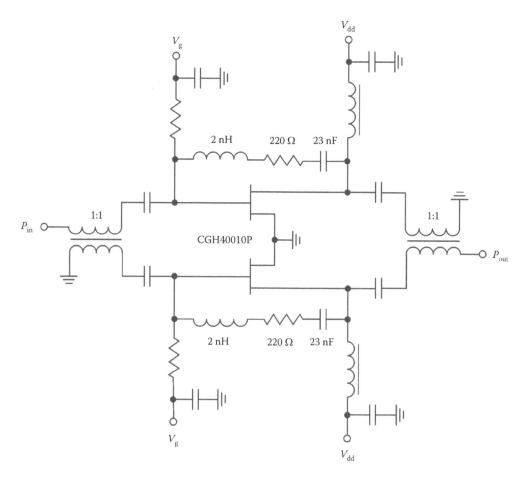

FIGURE 4.22
Circuit schematic of a broadband push–pull GaN HEMT VHF–UHF power amplifier.

schematic of a single-stage broadband push–pull power amplifier based on two 10-W GaN HEMT devices CGH40010P, which provides an operating frequency range from 8 to 800 MHz with an output power of more than 40.5 dBm, a power gain of 17.2±0.3 dB, and a *PAE* from 26.7% to 45% [49]. Here, the input and output 1:1 baluns are also used to match the required 25-Ω impedances at each device input and output, provided by optimized parallel *RLC* feedback circuits, with the standard 50-Ω source and load impedances.

A similar negative feedback approach was applied to a two-stage broadband power amplifier using four 10-W LDMOSFET devices in the driver stage and four 45-W GaN HEMT devices in the final stage, as shown in Figure 4.23, resulting in an output power of more than 100 W, a power gain of 29.2±1.8 dB, and a *PAE* of more than 43% from 10 to 500 MHz with the second-harmonic suppression of more than 29 dB [50]. Here, the four-way balun method is used for matching of the source and load 50-Ω impedances with the required input and output 12.5-Ω impedances at the input of each driver transistor and

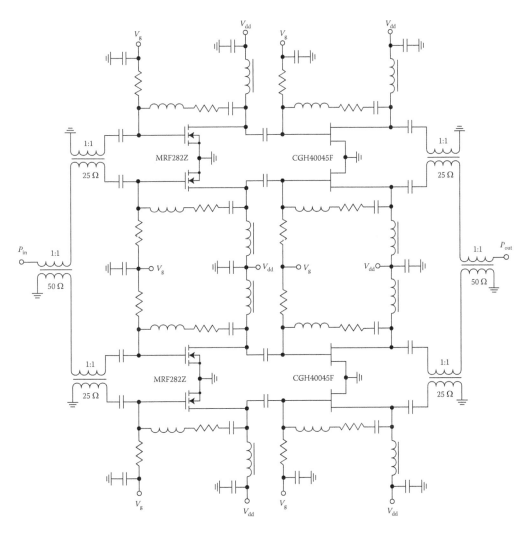

FIGURE 4.23
Circuit schematic of a broadband high-power GaN HEMT VHF–UHF amplifier.

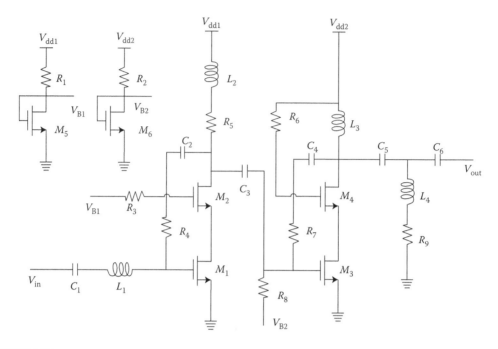

FIGURE 4.24
Circuit schematic of a broadband microwave CMOS power amplifier.

at the output of each final-stage transistor. The permeability of the ferrite core and the number of winding turns need to be carefully chosen to ensure that the transformer can provide a low insertion loss and a wide operating frequency bandwidth. With four 60-W MRF9060 LDMOSFET devices in the final stage, an output power of more than 100 W, a power gain of 40 ± 1.5 dB, and a *PAE* of more than 35% were measured in a frequency range from 30 to 500 MHz [51].

Figure 4.24 shows the broadband two-stage cascode 0.13-µm RF CMOS power amplifier where the parallel feedback in both stages and lossy output matching are used to cover a very wide frequency range with a flat power gain [52]. The parallel *RC* feedback improves the input return loss across the entire frequency bandwidth, prevents the dc current from flowing directly through the resistive feedback path, and improves the power gain at the high-bandwidth frequencies. The shunt *RL* circuit in the output-matching circuit improves the output return loss at low-bandwidth frequencies. As a result, an output power $P_{1\,dB}$ over 6 dBm with a power gain of more than 10 dB was simulated across the frequency bandwidth of 0.7–6 GHz at a supply voltage of 1.8 V.

4.3 Graphical Design of Gain-Compensating and Feedback Lossy Networks

The first step for designing feedback amplifier modules can be based on two graphical methods [32]. The first method is based on the use of a set of curves of a constant maximum available gain $G_{\max}$ (or *MAG*) and a stability factor K, plotted in polar diagrams that lead to the configuration and values of the feedback circuit in a simple way to obtain gain equalization and unconventional stability in a broad range of frequencies. The second graphical method pays more attention to obtain a flat magnitude of S_{21} over a wide band, controlling the S-parameters and making the input and output amplifier design easier. These two methods lead to design values close to the final ones, so that very little computer optimization is needed.

The transistor two-port network together with the parallel feedback circuit is shown in Figure 4.25a. Assuming, in the first approximation, that the feedback network admittances Y'_{ij} are negligible compared to the corresponding transistor admittances Y_{ij}, except Y'_{12}, then, the stability factor K and the maximum available gain G_{max} for the feedback amplifier are respectively given by

$$K = \frac{2\operatorname{Re} Y_{11} \operatorname{Re} Y_{22} - \operatorname{Re}(Y_{21} Y^{\mathsf{T}}_{12})}{|\, Y_{21} Y^{\mathsf{T}}_{12}\,|} \qquad (4.43)$$

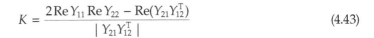

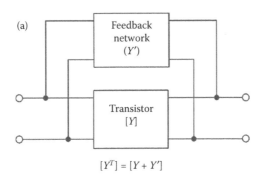

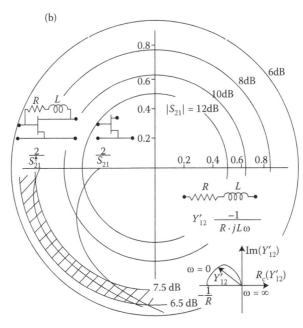

FIGURE 4.25
Feedback diagram for equalizing and matching for *RL* feedback circuit.

$$G_{\max} = \frac{|Y_{21}|}{|Y_{12}^T|}(K - \sqrt{K^2 - 1}) \quad \text{for } K > 1 \tag{4.44}$$

where $Y_{12}^T = Y_{12} + Y_{12}'$ is the feedback admittance.

From Equations 4.43 and 4.44, it follows that, for a given device, K and $G_{\max}$ are functions of the feedback admittance Y_{12}^T. By taking this admittance as a complex variable, a set of constant-K curves and a family of constant-$G_{\max}$ curves for each frequency can be plotted. In this case, the constant-K curves are ellipses and constant $G_{\max}$ curves are circles. To obtain a gain G_0 with unconditional stability when the transistor feedback admittance Y_{12} is located at the point which is conditionally stable, this point can be translated to any other point on the G_0 circle, and the resulting network represents a simple RC series circuit. This procedure can be repeated for several frequencies in the required frequency band by plotting the constant-K and constant-$G_{\max}$ curves and Y_{12} point at each bandwidth frequency and then synthesizing the entire broadband network.

A different approach can be followed if the feedback circuit to obtain an amplifier module with a constant S_{21}^T instead of an amplifier with a constant $G_{\max}$ can be used, unconditionally stable and matched in a very broad frequency range. In this case, it may be easily shown that the network S-parameters are approximately obtained by

$$\frac{2}{S_{21}^T} \cong \frac{2}{S_{21}} + \frac{Y_{12}'}{Y_0} \tag{4.45}$$

$$\frac{2S_{11}^T}{S_{21}^T} \cong \frac{2S_{11}}{S_{21}} - \frac{Y_{12}'}{Y_0} \tag{4.46}$$

$$\frac{2S_{22}}{S_{21}^T} \cong \frac{2S_{22}}{S_{21}} - \frac{Y_{12}'}{Y_0} \tag{4.47}$$

$$\frac{S_{12}^T}{S_{21}^T} \cong \frac{Y_{12} + Y_{12}'}{Y_{21}} \tag{4.48}$$

when the magnitude of S_{21} of the transistor is reasonably high, $Y_0 = 1/Z_0$, and Z_0 is the source and load characteristic impedance.

In a complex plane, constant S_{21}^T curves are a family of circles centered at the origin, whose radii are $2/|S_{21}^T|$, as shown in Figure 4.25b, where typical values $2/S_{21}$ are plotted for different bandwidth frequencies and Y_{12}' represents vector translation. Therefore, the problem is to define the type of feedback circuit that translates the curve $2/S_{21}$ into a constant $2/|S_{21}^T|$ circle. This graphical method can give the topology and initial values for the feedback circuit. In this case, the feedback circuit includes an inductor whose effect is to decrease the value of Y_{12}' for high frequencies which produces gain equalization.

Figure 4.26 shows the basic structures for interstage networks design where the two-port networks represent the active devices and the interstage networks equalize the gain roll-off with frequency [53]. If the two-port networks are characterized by their Z-parameters, as shown in Figure 4.26a, or their Y-parameters, as shown in Figure 4.26b, the corresponding S-parameters of the overall circuits are written as

$$S_{ij}^T = \frac{A + BZ_E}{C + DZ_E} \tag{4.49}$$

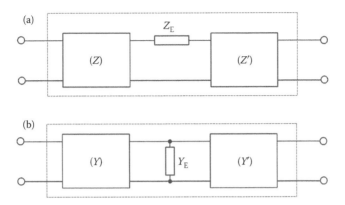

FIGURE 4.26
Basic structures for the interstage network design.

$$S_{ij}^T = \frac{A' + B'Y_E}{C' + D'Y_E} \tag{4.50}$$

where A, B, C, and D depend only on Z- and Z'-parameters and on Y- and Y'-parameters, respectively.

Equation 4.49 represents a bilinear transformation between the complex planes S_{ij}^T and Z_E. The constant $|S_{ij}^T|$ circles are transformed into other circles on the Z_E plane. Moreover, for a new bilinear transformation written as

$$\alpha = \frac{Z_E - Z_0}{Z_E - Z_0} \tag{4.51}$$

where Z_0 is the characteristic impedance, and the constant $|S_{ij}^T|$ circles are transformed into other new circles on the Smith chart. For the circuit shown in Figure 4.25b, the constant $|S_{ij}^T|$ geometrical loci on the admittance Smith chart will also be circles.

As a result, the constant $|S_{21}^T|$ circles on the Smith chart are a family of circles, whose centers and radii are

$$\alpha_0 \exp(j\theta) = \frac{2(1 + w_c^*)}{r^2 - |1 + w_c|^2} + 1 \tag{4.52}$$

$$R = \frac{2r}{|r^2 - |1 + w_c|^2|} \tag{4.53}$$

where

$$w_c = \frac{w_{12}w_{21}}{1 + w_{11}} + \frac{w'_{12}w'_{21}}{1 + w'_{22}} - w'_{11} - w_{22} \tag{4.54}$$

$$r = \frac{2}{|S_{21}^T|} \left| \frac{w_{21}w'_{21}}{(1 + w_{11})(1 + w'_{22})} \right| \tag{4.55}$$

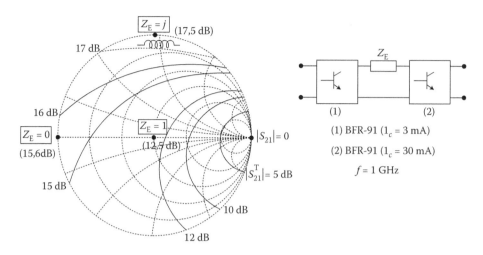

FIGURE 4.27
Constant S_{21}^T circles on impedance Smith chart ($Z_0 = 50\ \Omega$).

w_{ij} are the normalized immittance Z- or Y-parameters, depending on the configuration used, and these circles are located on the impedance or admittance Smith chart.

Figure 4.27 shows the constant S_{21}^T circles for a particular case. If both transistors are connected directly when $Z_E=0$, the circuit gain is 15.6 dB. However, for nonzero impedance Z_E, the gain is determined by the circle that passes through that impedance. For a passive impedance Z_E, the graph indicates that the maximum gain is obtained with an inductor $L=7.96$ nH when $Z_E=+j1$. In a broadband design, the circles of the desired value $|\ S_{21}^T\ |$ for several frequencies in the band are plotted, and then, the interstage networks are synthesized, whose impedance Z_E or admittance Y_E falls on the respective circle at each frequency.

The graphic design technique was applied to the design of an amplifier module operating with a 20-dB gain from 100 MHz to 1.1 GHz [53]. The amplifier module based on two bipolar transistors with $f_T=4.5$ GHz consists of two stages coupled by a lossy interstage network to equalize the gain of the devices. In this case, the first stage is biased for low noise and the second stage is biased for high gain. From Figure 4.27, it follows that it is impossible to obtain a 20-dB gain using series impedances with positive real parts. A possible matching network at 1 GHz may consist of a section of a microstrip line in series with a capacitor and include a parallel equalization network connected at an inner point of the matching network, whose parameters are optimized by trial and error on the Smith chart. The lossless input- and output-matching circuit can also be designed by graphic work on the Smith chart and computer optimization. As a result, the amplifier gain of 20±0.6 dB was achieved in a frequency band from 100 MHz to 1.1 GHz, as shown in Figure 4.28. The input *VSWR* is high because it was not taken into account in the optimization procedure, and this power reflection can be compensated by using balanced configuration, isolators, etc., or by designing more sophisticated matching networks.

4.4 Decomposition Synthesis Method

The decomposition synthesis method is based on the synthesis of the active circuits and passive impedance-correction networks from the circuit performance specifications and

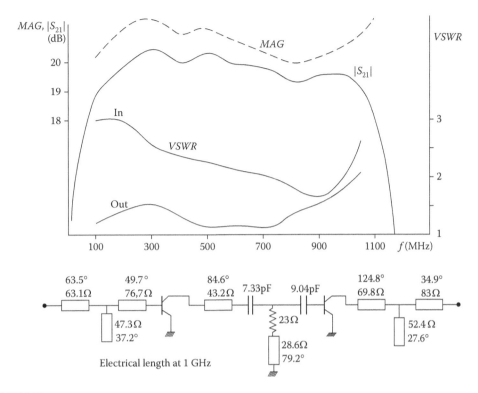

FIGURE 4.28
Broadband lossy match power amplifier: schematic and performance.

offers a general approach to the design of linear and nonlinear active circuits, which can be represented in the form of the corresponding connection of the active devices and passive networks [54]. For example, for a single-stage transistor amplifier, the input impedances of the optimum matching networks (optimum source or load impedances for active two-port networks) are determined that correspond to the extremum (maximum or minimum) value of one of the amplifier characteristics at the selected frequency. As a result, the optimum impedance points corresponding to each frequency will result in the optimum impedance locus over the selected frequency range, and this optimum impedance points should be set within the acceptable region where the impedance-correction circuit can be physically realized. The decomposition synthesis method assumes a certain sequence of steps: (1) selection of a block diagram (structural scheme) of an active circuit; (2) constructing a mathematical model of a circuit of the chosen block diagram; (3) determining the limit (extremely achievable) circuit characteristics within the frame of the chosen block diagram; (4) determining acceptable regions of the correction network parameters (in a common case, immittance or scattering parameters) at selected frequencies over a frequency band of interest; and (5) finding the correction network structures and elements from the acceptable parameter region. It should be noted that this method does not provide a full structural synthesis of active high-frequency circuits. It only allows the synthesis of the passive correction networks according to the active device requirements, with a given circuit block diagram.

To determine acceptable regions of the matching and correction network immittances incorporated into the linear active circuit, several different approaches can be used:

mapping of functions; describing circuit characteristics in the complex immittance or
S-parameter plane; analytic solution of systems of inequalities; and finding a projection of
a multidimensional acceptable region on the plane or three-dimensional subspace using
a numerical algorithm [54]. For some nonlinear active circuits with lossless-matching
networks, acceptable regions of the network input immittances can be obtained using
contours on the source/load plane that can be produced from the device load-pull measurements or computer modeling. The interactive "visual" procedure for the network
design involves two stages: (1) selection of the correction network structure in a set of
standard structures, according to the location of acceptable regions in the network immittance or S-parameter plane; and (2) computation of the network elements by solving a
corresponding system of equations or inequalities [55]. The "visual" design technique
can be applied to designing networks of moderate complexity. However, it allows the
user to select appropriate network configurations and to directly control all the network
elements.

Figure 4.29a shows the circuit schematic of a lossy- matched multioctave bipolar amplifier operated in a frequency range from 10 MHz to 4 GHz [56]. In this case, the design
requirements are a good flatness of the amplifier gain response and low input/output
VSWR, which should be achieved in a single-stage bipolar amplifier over a wide frequency
range using a low-power bipolar transistor with $f_T=7$ GHz. An amplifier block diagram
with input and output lossy two-port equalizers was used in a "visual" design procedure. On the basis of the decomposition synthesis method, configurations and elements of
equalizers have been sequentially determined. As a result, an amplifier gain of 7.0±0.8 dB
and an input/output VSWR of 1.7/1.8 were measured in a frequency range from 10 MHz to
3.6 GHz. Similarly, the decomposition synthesis method was applied to the design of a linear power amplifier based on a 3-W 28-V bipolar transistor with $f_T=1$ GHz, whose circuit
schematic is shown in Figure 4.29b. As a result, a power gain of 6.0±1.0 dB and an input/
output VSWR of 2.0/3.0 were measured in a frequency range from 10 to 750 MHz with a
linear output power $P_{1\,\text{dB}}$ of about 1 W.

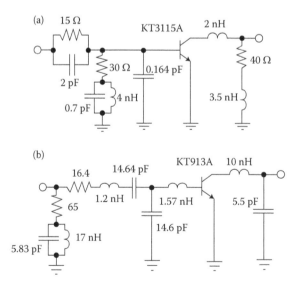

FIGURE 4.29
Schematics of broadband lossy match power amplifiers.

References

1. H. W. Bode, Attenuation equalizer, U.S. Patent 2,096,027, October 1937 (filed Jan. 1936).
2. G. Loeber, H. Overbeck, and W. Schlotterbeck, Transistorized microwave broadband power amplifiers covering the frequency range from 500 to 1000 MHz, *Proc. 1st Eur. Microw. Conf.*, London, pp. 439–442, 1969.
3. V. V. Nikiforov, T. T. Kulish, and I. V. Shevnin, Broadband HF-VHF MOSFET power amplifier design (in Russian), *Poluprovodnikovaya Elektronika v Tekhnike Svyazi*, 23, 27–36, 1983.
4. C. A. Liechti and R. L. Tillman, Design and performance of microwave amplifiers with GaAs Schottky-gate field-effect transistors, *IEEE Trans. Microw. Theory Tech.*, MTT-22, 510–517, 1974.
5. D. P. Hornbuckle and L. J. Kuhlman Jr., Broad-band medium-power amplification in the 2–12.4-GHz range with GaAs MESFET's, *IEEE Trans. Microw. Theory Tech.*, MTT-24, 338–342, 1976.
6. J. Obregon, R. Funck, and S. Barvet, A 150 MHz–16 GHz FET amplifier, *IEEE Int. Solid-State Circuits Conf. Dig.*, Philadelphia, PA, pp. 66–67, 1981.
7. K. Honjo and Y. Takayama, GaAs FET ultrabroad-band amplifiers for Gbit/s data rate systems, *IEEE Trans. Microw. Theory Tech.*, MTT-29, 629–633, 1981.
8. N. Riddle and R. J. Trew, A broad-band amplifier output network design, *IEEE Microw. Theory Tech.*, MTT-30, 192–196, 1982.
9. K. B. Niclas, On design and performance of lossy match GaAs MESFET amplifiers, *IEEE Trans. Microw. Theory Tech.*, MTT-30, 1900–1906, 1982.
10. K. B. Niclas, Multi-octave performance of single-ended microwave solid-state amplifiers, *IEEE Trans. Microw. Theory Tech.*, MTT-32, 896–908, 1984.
11. P. K. Ikalainen, An RLC matching network and application in 1–20 GHz monolithic amplifier, *IEEE MTT-S Int. Microw. Symp. Dig.*, Long Beach, CA, 3, 1115–1118, 1989.
12. T. Arell and T. Hogsmatip, A unique MMIC broadband power amplifier approach, *IEEE J. Solid-State Circuits*, SC-28, 1005–1010, 1993.
13. O. Pitzalis Jr., R. E. Horn, and R. J. Baranello, Broadband 60-W HF linear amplifier, *IEEE J. Solid-State Circuits*, SC-6, 93–103, 1971.
14. A. V. Grebennikov, V. V. Nikiforov, and A. B. Ryzhikov, The powerful transistor amplifier modules for VHF FM and TV broadcasting (in Russian), *Elektrosvyaz*, 28–31, 1996.
15. Y. Ito, M. Nii, Y. Kohno, M. Mochizuki, and T. Takagi, A 4 to 25 GHz 0.5 W monolithic lossy match amplifier, *IEEE MTT-S Int. Microw. Symp. Dig.*, San Diego, CA, 1, 257–260, 1994.
16. A. R. Barnes, M. T. Moore, and M. B. Allenson, A 6–18 GHz broadband high power MMIC for EW applications, *IEEE MTT-S Int. Microw. Symp. Dig.*, Denver, CO, 3, 1429–1432, 1997.
17. Y. Inoue, M. Sato, T. Ohki, K. Makiyama, T. Takahashi, H. Shigematsu, and T. Hirose, A 90-GHz InP-HEMT lossy match amplifier with a 20-dB gain using a broadband matching technique, *IEEE J. Solid-State Circuits*, SC-40, 2098–2103, 2005.
18. R. S. Pengelly, Broadband lumped-element X band GaAs F.E.T. amplifier, *Electron. Lett.*, 11, 58–60, 1975.
19. Y. F. Wu, R. A. York, S. Keller, B. P. Keller, and U. K. Mishra, 3–9-GHz GaN-based microwave power amplifiers with L–C–R broad-band matching, *IEEE Microw. Guid. Wave Lett.*, 9, 314–316, 1999.
20. J. J. Xu, S. Keller, G. Parish, S. Heikman, U. K. Mishra, and R. A. York, A 3–10-GHz GaN-based flip-chip integrated broad-band power amplifier, *IEEE Trans. Microw. Theory Tech.*, MTT-48, 2573–2578, 2000.
21. K. Krishnamurthy, D. Wang, B. Landberg, and J. Martin, RLC matched GaN HEMT power amplifier with 2 GHz bandwidth, *IEEE Compound Semicond. Integr. Circuits Symp. Dig.*, Monterey, CA, pp. 1–4, 2008.
22. H. S. Black, Wave translation system, U.S. Patent 2,003,282, June 1935 (filed Aug. 1923).
23. H. S. Black, Stabilized feedback amplifiers, *Bell Syst. Tech. J.*, 13, 1–18, 1934.
24. H. S. Black, Inventing the negative feedback amplifier, *IEEE Spectr.*, 14, 55–60, 1977.

25. H. W. Bode, Amplifier, U.S. Patent 2,123,178, July 1938 (filed June 1937).

26. H. W. Bode, Relations between attenuation and phase in feedback amplifier design, *Bell Syst. Tech. J.*, 19, 421–454, 1940.

27. J. Almond and A. R. Boothroyd, Broadband transistor feedback amplifiers, *Proc. IEE—Part B: Radio Electron. Eng.*, 103, 93–101, 1956.

28. E. M. Cherry and D. E. Hooper, The design of wide-band transistor feedback amplifiers, *Proc. IEE*, 110, 375–389, 1963.

29. F. H. Blecher, Design properties for single loop transistor feedback amplifiers, *IRE Trans. Circuit Theory*, CT-4, 145–156, 1957.

30. J. B. Coughlin, R. J. H. Gelsing, P. J. W. Johems, and H. J. M. van der Laak, A monolithic silicon wide-band amplifier from dc to 1 GHz, *IEEE J. Solid-State Circuits*, SC-8, 414–419, 1973.

31. K. B. Niclas, The matched feedback amplifier: Ultrawide-band microwave amplification with GaAs MESFET's, *IEEE Trans. Microw. Theory Tech.*, MTT-28, 285–294, 1980.

32. F. Perez and V. Ortega, A graphical method for the design of feedback networks for microwave transistor amplifiers: Theory and applications, *IEEE Trans. Microw. Theory Tech.*, MTT-29, 1019–1028, 1981.

33. R. S. Pengelly, Application of feedback techniques to the realisation of hybrid and monolithic broadband low-noise-and-power GaAs FET amplifiers, *Electron Lett.*, 17, 798–799, 1981.

34. P. A. Terzian, D. B. Clark, and R. W. Waugh, Broad-band GaAs monolithic amplifier using negative feedback, *IEEE Trans. Microw. Theory Tech.*, MTT-30, 2017–2020, 1982.

35. A. Sayed, S. von der Mark, and G. Boeck, An ultra wideband 5 W power amplifier using SiC MESFETs, *Proc. 34th Eur. Microw. Conf.*, Amsterdam, pp. 57–60, 2004.

36. A. Sayed and G. Boeck, Two-stage ultrawide-band 5-W power amplifier using SiC MESFET, *IEEE Trans. Microw. Theory Tech.*, MTT-53, 2441–2449, 2005.

37. J. Sim, J. Lim, M. Park, W. Kang, and B. I. Mah, Analysis and design of wide-band power amplifier using GaN, *Proc. Asia-Pac. Microw. Conf.*, Singapore, pp. 2352–2355, 2009.

38. A. Sayed and G. Boeck, 5 W highly linear GaN power amplifier with 3.4 GHz bandwidth, *Proc. 37th Eur. Microw. Conf.*, Munich, pp. 1429–1432, 2007.

39. K. Honjo, T. Sugiura, and H. Itoh, Ultra-broad-band GaAs monolithic amplifier, *IEEE Trans. Microw. Theory Tech.*, MTT-30, 1027–1033, 1982.

40. K. B. Niclas, Noise in broad-band GaAs MESFET amplifiers with parallel feedback, *IEEE Trans. Microw. Theory Tech.*, MTT-30, 63–70, 1982.

41. K. B. Niclas, The exact noise figure of amplifiers with parallel feedback and lossy matching circuits, *IEEE Trans. Microw. Theory Tech.*, MTT-30, 832–835, 1982.

42. G. A. Ellis, J. S. Moon, D. Wong, M. Micovic, A. Kurdoghlian, P. Hashimoto, and M. Hu, Wideband AlGaN/GaN HEMT MMIC low-noise amplifier, *IEEE MTT-S Int. Microw. Symp. Dig.*, Fort Worth, TX, 1, 153–156, 2004.

43. A. F. Bellomo, Gain and noise considerations in RF feedback amplifier, *IEEE J. Solid-State Circuits*, SC-3, 290–294, 1968.

44. A. V. Grebennikov, V. V. Nikiforov, and S. Y. Terentyev, Broadband HF-VHF power amplifier with improved linearity (in Russian), *Elem. Modules Mod. Receiv. Amplifier Tech.: Conf. Dig.*, Uzhgorod, USSR, 17–18, 1991.

45. N. Sahan, M. E. Inal, S. Demir, and C. Toker, High-power 20–100-MHz linear and efficient power-amplifier design, *IEEE Trans. Microw. Theory Tech.*, MTT-56, 2032–2039, 2008.

46. H. Granberg, *Broadband Linear Power Amplifiers Using Push–Pull Transistors*, Application Note AN593, Motorola Semiconductor, Phoenix, AZ, 1993.

47. H. O. Granberg, Building push–pull, multioctave, VHF power amplifiers, *Microwav. RF*, 26, 77–86, 1987.

48. S. You, K. Lim, J. Cho, M. Seo, K. Kim, J. Sim, M. Park, and Y. Yang, A 5 watt ultra-broadband power amplifier using silicon LDMOSFETs, *Proc. Asia-Pac. Microw. Conf.*, Singapore, pp. 1116–1119, 2009.

49. K. Kim, M. Seo, J. Jeon, M. Kim, H. Kim, H. Lim, C. Park, and Y. Yang, Design of a broadband power amplifier using an optimized feedback network, *Microw. Opt. Technol. Lett.*, 53, 2846–2851, 2011.

50. J. Cho, K. Lim, S. You, M. Seo, K. Kim, J. Sim, M. Park, and Y. Yang, Design of a 100 watt high-efficiency power amplifier for the 10–500 MHz band, *Proc. Asia-Pac. Microw. Conf.*, Singapore, pp. 285–288, 2009.

51. J. Sim, J. Lim, M. Park, S. W. Seo, and B. I. Mah, A 100 watt ultra-broadband power amplifier using silicon LDMOSFETs, *Proc. Asia-Pac. Microw. Conf.*, Yokohama, Japan, pp. 418–421, 2010.

52. X. Sun, F. Huang, X. Tang, and M. Shao, A 0.7–6 GHz broadband CMOS power amplifier for multi-band applications, *Proc. Int. Microw. Millim.-Wave Technol. Conf.*, Shenzhen, China, pp. 1–4, 2012.

53. J. C. Villar and F. Perez, Graphic design of matching and interstage lossy networks for microwave transistor amplifier, *IEEE Trans. Microw. Theory Tech.*, MTT-33, 210–215, 1985.

54. L. I. Babak, Decomposition synthesis approach to design of RF and microwave active circuits, *IEEE MTT-S Int. Microw. Symp. Dig.*, Phoenix, AZ, 2, 1167–1170, 2001.

55. L. I. Babak and M. V. Cherkashin, Interactive visual design of matching and compensation networks for microwave active circuits, *IEEE MTT-S Int. Microw. Symp. Dig.*, Phoenix, 3, 2095–2098, 2001.

56. L. I. Babak, M. V. Cherkashin, and M. Y. Pokrovsky, Computer-aided design of ultrawide-band transistor amplifiers using decomposition synthesis method, *Proc. 32nd Eur. Microw. Conf.*, Milan, Italy, pp. 1–4, 2002.

5

Design of Wideband RF and Microwave Amplifiers Employing Real Frequency Techniques

In Chapter 3, some practical matching networks were introduced to design narrow bandwidth amplifiers. There, the major idea was to match the input and the output of the amplifier to a given standard resistive source and load over a narrow frequency band. In this regard, first, the proper circuit topologies with two or three elements are selected for the input and the output matching networks. Then, the element values are determined by means of nonlinear optimization methods. However, when we deal with wideband matching problems, life becomes more difficult. First, we have no idea about the optimum choice for the circuit topology. Furthermore, even we choose a practical circuit topology for the matching networks, this topology would require more than two elements to achieve wideband designs. Moreover, the objective function to be optimized is highly nonlinear in terms of the component values and requires good initials. Eventually, we start the nonlinear optimization with ad-hoc component values. At the end of the optimization, most probably, we will either hit a bad-local extremum or end up with nondivergent endless iterations.

Therefore, in this chapter, we introduce novel techniques, which are called "Real Frequency Techniques (RFTs)," to design wideband amplifiers [1–4]. In essence, RFTs are wideband semi-analytic design methods to realize lossless matching networks with optimum circuit topologies.

Referring to Figure 5.1, a typical RF or single-stage microwave amplifier consists of input and output matching networks [F] and [B], respectively.* The amplifier could be driven either with small or large signals (i.e., high-power input drives).

In RFTs, instead of choosing the circuit topologies for [F] and [B], they are fully described by means of their Darlington driving point network functions, such as reflectance $\{S_F$ and $S_B\}$ or impedance $\{Z_F$ and $Z_B\}$ or, equivalently, admittance $\{Y_F = 1/Z_F, Y_B = 1/Z_B\}$, or in short immittance functions,† respectively.‡ In the impedance-based RFT, positive real functions $Z_F(p)$ and $Z_B(p)$ are determined to optimize the performance parameters of the amplifier, such as transducer power gain (TPG), noise figures (NF), voltage standing wave ratios (VSWR), etc. In this representation, "p" designates the complex Laplace domain variable such that $p = \sigma + j\omega$ and the impedance functions $Z_F(p)$ and $Z_B(p)$ are specified as rational functions of complex variable p as follows.§

---

* The letter [F] designates the "Front-End Equalizer" or equivalently "Input Matching Network." Similarly, the letter [B] stands for the "Back-End Equalizer" or equivalently "Output Matching Network," as shown in Figure 5.1.

† In circuit theory, the expression "immittance" refers to either impedance or admittance.

‡ At this point, we should emphasize that, according to Darlington's theorem, a lossless two-port such as [F] or [B] can completely be described by means of its driving point reflectance or immittance.

§ In classical circuit theory, complex Laplace variable is designated by the letter "$s = \sigma + j\omega$." This letter may be confused with those of scattering parameters of two ports $S = \{s_{ij}; i,j = 1,2\}$ or reflectance parameters S_F and S_B. Therefore, we use the notation p or the Laplace domain variable.

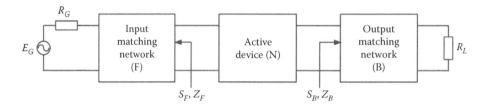

FIGURE 5.1
Schematic description of a single-stage amplifier.

$$Z_F(p) = \frac{a_F(p)}{b_F(p)} = \frac{a_{1,F}p^n + a_{2,F}p^{n-1} + \cdots + a_{n,F}p + a_{(n+1),F}}{b_{1,F}p^n + b_{2,F}p^{n-1} + \cdots + b_{n,F}p + b_{(n+1),F}} \tag{5.1}$$

$$Z_B(p) = \frac{a_B(p)}{b_B(p)} = \frac{a_{1,B}p^n + a_{2,B}p^{n-1} + \cdots + a_{n,B}p + a_{(n+1),B}}{b_{1,B}p^n + b_{2,B}p^{n-1} + \cdots + b_{n,B}p + b_{(n+1),B}} \tag{5.2}$$

Similarly, in the reflectance-based RFT, which is called "Simplified Real Frequency Technique (or in short SRFT)," bounded real reflectance functions are specified as[*]

$$S_F(p) = \frac{h_F(p)}{g_F(p)} = \frac{h_{1,F}p^n + h_{2,F}p^{n-1} + \cdots + h_{n,F}p + h_{(n+1),F}}{b_{1,F}p^n + b_{2,F}p^{n-1} + \cdots + b_{n,F}p + b_{(n+1),F}} \tag{5.3}$$

$$S_B(p) = \frac{h_B(p)}{g_B(p)} = \frac{h_{1,B}p^n + h_{2,F}p^{n-1} + \cdots + h_{n,B}p + h_{(n+1),B}}{g_{1,B}p^n + g_{2,B}p^{n-1} + \cdots + g_{n,B}p + g_{(n+1),B}} \tag{5.4}$$

In the course of matching network design, RFTs determine the coefficients of Equations 5.1 and 5.2 or Equations 5.3 and 5.4 to optimize the performance parameters of the amplifier such as *TPG* over the prescribed frequency band of operation. Then, network functions (either immittance or reflectance) are synthesized in Darlington sense [5] as lossless two-ports in resistive terminations yielding the optimum circuit topologies for [F] and [B] with element values.

Generation of realizable network functions on the computer is an art. It is fortunate that RFTs provide robust and efficient algorithms to complete these tasks [1–4]. Furthermore, our recently developed high-precision Darlington synthesis algorithms yield optimum circuit topologies for the matching networks to construct RF and microwave amplifiers [6–8].

There are many variants of RFTs as referenced at the end of this chapter [9–59]. In the following sections, we present the selected algorithms to design wideband practical power amplifiers employing real frequency techniques.

---

[*] In this context, reflectance functions are assumed to be real-normalized (unit-normalized).

5.1 Real Frequency Line Segment Technique

The real frequency line segment technique (RF-LST) was first introduced by H. J. Carlin; it generates excellent initial guess for nonlinear optimization to construct wideband matching networks by means of realizable network functions [1].

Let us consider the design of a power amplifier. Assume that we have completed the load-pull measurements for the active device under consideration.* Let the optimum source and load impedance for the active device be $Z_S(j\omega) = R_S(\omega) + jX_S(\omega)$ and $Z_L(j\omega) = R_L(\omega) + jX_L(\omega)$, respectively, which are specified over the band of operation. Under the load pull measurements, Z_S must be equal to the complex conjugate of the input impedance $Z_{in}(j\omega) = R_{in}(j\omega) + jX_{in}(\omega)$ of the active device [N], while the output port is terminated in the optimum load impedance Z_L, as shown in Figure 5.2. In other words, load pull measurement determines $Z_{in}(j\omega) = R_{in}(j\omega) + jX_{in}(\omega)$ under the large signal drives. Literally, Figure 5.2 describes a single matching problem: The measured complex impedance $Z_{in}(j\omega)$ is matched to a resistive generator R_G, which may be specified as standard 50 Ω. In this case, the real frequency matching problem is defined as the determination of the driving point impedance Z_F, which in turn yields the lossless two-port [F] in resistive termination R_G as the result of Darlington synthesis.

As mentioned above, ideally, Z_F must be equal to complex conjugate of Z_{in} in angular frequency variable $\omega = 2\pi f$ where the letter "f" designates the frequency of the operation. That is to say,

$$R_F(\omega) = R_{in}(\omega) \tag{5.5}$$

$$X_F(\omega) = -X_{in}(\omega) \tag{5.6}$$

Unfortunately, this is not possible. We can test it as follows. Let the positive real function Z_{in} be modeled as a rational function in complex variable p as

$$Z_{in}(p) = \frac{a_{in}(p)}{b_{in}(p)} \tag{5.7}$$

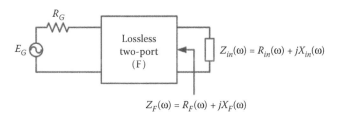

$$Z_F(\omega) = R_F(\omega) + jX_F(\omega)$$

FIGURE 5.2
Description of a single matching problem.

where the numerator polynomial $a_{in}(p)$ and the denominator polynomial $b_{in}(p)$ must be Hurwitz.[*] The frequency domain form of $Z_{in}(j\omega)$ is obtained by replacing complex variable p with $j\omega$. On the other hand, for the perfect match situation $Z_F(j\omega)$ must be equal to $Z_{in}(-j\omega) = a_{in}(-j\omega)/b_{in}(-j\omega)$ or in complex variable p the rational form of $Z_F(p)$ is obtained from $Z_{in}(p)$ by replacing p with $-p$. Thus, we have,

$$a_F(p) = a_{in}(-p) \tag{5.8}$$

$$b_F(p) = b_{in}(-p) \tag{5.9}$$

In Equations 5.8 and 5.9, neither $a_F(p)$ nor $b_F(p)$ is Hurwitz since $a_{in}(p)$ and $b_{in}(p)$ are Hurwitz polynomials. Therefore, ideal $Z_F(p)$ is not a realizable impedance. Therefore, we face the theoretical gain-bandwidth limitations of Bode, Fano, and Youla as introduced in References 13–15. However, the RF-LST is a straightforward practical method to construct lossless matching networks by means of its driving point impedance $Z_F(j\omega) = R_F(\omega) + jX_F(\omega)$.

In RF-LST, Z_F is assumed to be a minimum reactance impedance.[†] It can be proven that a realizable $Z_F(j\omega)$ can be uniquely determined from its nonnegative real part $R_F(\omega)$ over the entire angular frequency axis ω. In this case, the imaginary part $X_F(\omega)$ is determined employing the Hilbert transformation relation such that

$$X_F(\omega) = R_{F\infty} + \frac{1}{\pi} \int_{-\infty}^{+\infty} \frac{R_F(\Omega)}{\omega - \Omega} d\Omega \tag{5.10}$$

For a realizable Z_F, its real part $R_F(\omega)$ must be an even function in ω. Therefore, Equation 5.10 takes the following form:

$$X_F(\omega) = R_{F\infty} + \frac{2\omega}{\pi} \int_{0}^{+\infty} \frac{R_F(\Omega)}{\Omega^2 - \omega^2} d\Omega \tag{5.11}$$

The above integral can be expressed in logarithmic form using "integration by parts" rule such that

$$X_F(\omega) = R_{F\infty} + \frac{1}{\pi} \int_{0}^{+\infty} \left[\frac{dR_F(\Omega)}{d\Omega} \right] \ln \left| \frac{\Omega + \omega}{\Omega - \omega} \right| d\Omega \tag{5.12}$$

In the above equations, constant term $R_{F\infty}$ is the value of $R_F(\omega)$ at infinity. For many practical matching networks it is forced to be zero.

Referring to Figure 5.3, in RF-LST the real part $R_F(\omega)$ is piecewise linearized by means of line segments.

---

[*] A Hurwitz polynomial must have all its zeros in the open left half plane (LHP).
[†] A minimum reactance impedance is the one that is free of $j\omega$ poles.

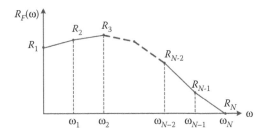

FIGURE 5.3
Piecewise linearization of the part $R_F(\omega)$.

In this case, $R_F(\omega)$ is expressed as

$$R_F(\omega) = \begin{cases} a_j\omega + b_j & \text{for } \omega_j \leq \omega \leq \omega_{j+1}; j = 1,2,\ldots,(N-1) \\ 0 & \omega \geq \omega_N \end{cases} \tag{5.13}$$

where

$$a_j = \frac{R_j - R_{j+1}}{\omega_j - \omega_{j+1}} = \frac{\Delta R_j}{\omega_{j+1} - \omega_j} \tag{5.14}$$

and

$$b_j = \frac{(R_{j+1})\omega_j - (R_j)\omega_{j+1}}{\omega_j - \omega_{j+1}} \tag{5.15}$$

$$\Delta R_j = R_{j+1} - R_j \tag{5.16}$$

Using Equation 5.13 in Equation 5.12, $X_F(\omega)$ is derived as

$$X_F(\omega) = \sum_{j=1}^{N-1} B_j(\omega)\Delta R_j \tag{5.17}$$

where

$$B_j(\omega) = \frac{1}{\pi(\omega_j - \omega_{j+1})}[F_{j+1}(\omega) - F_j(\omega)] \tag{5.18}$$

and

$$F_j(\omega) = (\omega + \omega_j)\ln(|\omega + \omega_j|) + (\omega - \omega_j)\ln(|\omega - \omega_j|) \tag{5.19}$$

In Equations 5.13 through 5.16, the points $\{R_j \geq 0; j = 1, 2, ..., N\}$ are called the break points of the real part $R_F(\omega)$ and they are all nonnegative. Sampling frequencies $\{\omega_j; j = 1, 2, ..., N\}$ are called the break frequencies and $\Delta R_j = R_{j+1} - R_j$ is called the excursions in the real part.

In RF-LST, once the break frequencies $\{\omega_j; j = 1, 2, ..., N\}$ are fixed by the designer, the break points $\{R_j \geq 0; j = 1, 2, ..., N\}$ are selected as the unknowns of the matching problems. Obviously, the pairs (ω_j, R_j) fully describe the unknown even part $R_F(\omega)$, as specified by Equations 5.13 through 5.16. Then, the imaginary part $X_F(\omega)$ is determined by means of the unknown R_js in a linear manner using Equations 5.17 through 5.19.

For the single matching problem of Figure 5.2, impedance-based TPG $T_{IF}(\omega)$ of the front-end equalizer is given by

$$T_{IF}(\omega) = \frac{4R_F(\omega)R_{in}(\omega)}{[R_F(\omega) + R_{in}(\omega)]^2 + [X_F(\omega) + X_{in}(\omega)]^2} \tag{5.20}$$

In Equation 5.20, $Z_{in}(j\omega) = R_{in}(\omega) + jX_{in}(\omega)$ is the measured large signal input impedance of the active device. $R_F(\omega)$ is the unknown real part of $Z_F(j\omega)$ and is described by means of the break points R_i. $X_F(\omega)$ is the imaginary part of $Z_F(j\omega)$ and is uniquely determined as the linear combination of the unknown break points R_i, as specified by Equations 5.17 through 5.19.

In RF-LST, once the break frequencies are selected and the unknown break points are initialized, $T(\omega)$ is maximized as flat as possible over the frequency band of operations B such that $\omega_L \leq B \leq \omega_H$; where ω_L is the lower-end and ω_H is the upper-end of the passband.

In the course of optimization of the TPG of Equation 5.20, the designer should try to hit maximum allowable flat gain level T_0, which may be determined by accessing the gain-bandwidth equations of Youla. At this point, we should note that, for specified complex termination $Z_{in}(j\omega)$, determination of T_0 is not an easy task. Depending on the complexity of the termination, it may even be impossible to determine T_0. Therefore, permissible flat gain level T_0 may be reached by "trial and error." In this regard, the optimization process may start by selecting a reasonably high-flat gain level T_0, then the objective function $\varepsilon(\omega) = T(\omega) - T_0$ is minimized over the frequency band of operation. We know that the ideal value of T_0 is unity. So, we can start sweeping T_0 from 1 and come down to a level where we can end up with a reasonable design. The sweeping step size for T_0 could be $\Delta T = 0.05$.

Employing Equation 5.20, the error function $\varepsilon(\omega) = T(\omega) - T_0$ may be evaluated as

$$\varepsilon(\omega) = \frac{4R_F(\omega)R_{in}(\omega) - T_0\{[R_F(\omega) + R_{in}(\omega)]^2 + [X_F(\omega) + X_{in}(\omega)]^2\}}{[R_F(\omega) + R_{in}(\omega)]^2 + [X_F(\omega) + X_{in}(\omega)]^2} \tag{5.21}$$

During the optimization, one should try to end with zero error level. In this case, $\varepsilon(\omega)$ can be approximated as

$$\varepsilon(\omega) = 4R_F(\omega)R_{in}(\omega) - T_0\{[R_F(\omega) + R_{in}(\omega)]^2 + [X_F(\omega) + X_{in}(\omega)]^2\} \tag{5.22}$$

In view of Equations 5.17 through 5.19, the objective function $\varepsilon(\omega)$ is quadratic in terms of the unknown break points R_js. Therefore, Equation 5.22 defines a convex objective function to be minimized. Therefore, optimization process of $\varepsilon(\omega)$ must converge to global minimum if one starts with a reasonable initial guess for R_js. A practical choice for R_js may start as follows.

To maximize *TPG*, one must try to design the lossless matching network to end up with reactance cancellation. In other words, the term in Equation 5.20 must be minimized or it should be small enough to be ignored (i.e., $X_F(\omega) + X_{in}(\omega) \cong 0$). In this case, selected flat gain level will be approximated at selected break frequency ω_j as

$$T_0 = \frac{4R_{in}(\omega_j)R_j}{[R_{in}(\omega_j) + R_j]^2} \tag{5.23}$$

Equation 5.23 results in high- and low-initial values for R_j. High value of R_j is given by

$$R_{jH} = R_{in}(\omega_j)\left[\frac{(2 - T_0) + 2\sqrt{1 - T_0}}{T_0}\right] \geq 0 \tag{5.24}$$

Similarly, low value of R_j is given by

$$R_{jL} = R_{in}(\omega_j)\left[\frac{(2 - T_0) - 2\sqrt{1 - T_0}}{T_0}\right] \geq 0 \tag{5.25}$$

Remarks: There are several numerical issues that need to be clarified in the process of implementing the RF-LST on the computer.

a. The RF-LST requires the evaluation of Hilbert transform integral. In this regard, equation set (5.13) through (5.16) is programmed carefully in the MATLAB® environment. From the numerical point of view, in Equations 5.13 through 5.16, $R(\omega)$ must be bounded to accurately evaluate imaginary part $X(\omega)$. In practice, this issue can be secured as follows:

- For low-pass matching network design problems, the last break point R_N is fixed at zero, which makes $TPG = 0$ at $\omega_N = 0$ up to infinity. In other words, $R_N = 0$ is selected. In this regard, passband must be chosen over the break frequencies ω_1 and ω_{N-1}. In this case, for $\omega_N \geq \omega$ *TPG* vanishes.
- In bandpass design problems, the objective is to make *TPG* zero from DC (i.e., $\omega = 0$) up to first break frequency ω_1 where R_1 is fixed at zero level (i.e., $R_1 = 0$). Similarly, at ω_N, the final break point is selected as $R_N = 0$. In this case, the passband is chosen between ω_2 and ω_{N-1}. Here, TPG will be zero for $0 \leq \omega \leq \omega_1$ and $\omega_N \leq \omega \leq \infty$.

b. In the course of implementation of the real frequency techniques, frequency and impedance normalizations are essential to avoid numerical errors as well as round-off and truncation errors.

- We usually select the upper edge of the frequency band as the normalization frequency f_0, which makes $\omega_{N-1} = 1$; or it may be a practical frequency unit extracted from the data sampling frequencies. In any case, it must be equal to or higher than that of the upper edge of the band (i.e., $f_0 \geq f_H$) where f_H designates the high-end or equivalently upper edge of the frequency band of operation.

- For impedance normalization, we may select a meaningful normalization resistance R_0 such as standard termination, which is 50 Ω for many communication systems. It may as well be selected as the highest level of the real parts of the terminating generator/load impedances of the matching problem under consideration.

c. The designer can select the break frequencies over the normalized angular frequency axis ω based on the nature of the matching problem. For example, break frequencies may overlap with those of sampling frequencies selected to characterize the active devices or they may be uniformly or nonuniformly distributed over the frequency band of interest.

d. If the measured immittance or reflectance data to be matched are provided over a finite frequency band of operation excluding DC (i.e., ω = 0-lower edge of the stop band) and higher edge of the stop band, then we may artificially extrapolate or augment the given data without disturbing the nature of the matching problem. On the other hand, the driving point immittance or reflectance of the equalizer must be generated over the entire frequency axis to cover the stop-bands and must also be able to complete the synthesis of the matching network as desired. Details of these issues will be elaborated on within the examples.

e. In evaluating the numerical Hilbert transform of Equations 5.17 through 5.19, one must be careful to generate the imaginary part at the break frequencies ω_j since the term $\ln(|\omega - \omega_j|)$ in Equation 5.19 becomes minus infinity when $\omega = \omega_j$. However, at $\omega = \omega_j$, the term $(\omega - \omega_j)\ln(|\omega - \omega_j|)$ approaches zero. Therefore, singularity of the logarithmic function at zero must be removed by setting the term $(\omega - \omega_j)\ln(|\omega - \omega_j|)$ to zero when $\omega = \omega_j$.

f. Real frequency techniques can be programmed either for given impedances or admittances. Equation 5.20 expresses the impedance-based TPG. The generic form of Equation 5.20 is also valid for admittance-based TPG. In other words, let $Y_{in}(j\omega) = 1/Z_{in} = G_{in}(\omega) + jX_{in}(\omega)$ and $Y_F(j\omega) = 1/Z_F = G_F(\omega) + jB_F(\omega)$ be the input admittance of the active device and the driving point input admittance of the front-end equalizer [F], respectively. Then, admittance-based *TPG* for the front-end equalizer is given by

$$T_{AF}(\omega) = \frac{4G_F(\omega)G_{in}(\omega)}{[G_F(\omega) + G_{in}(\omega)]^2 + [B_F(\omega) + B_{in}(\omega)]^2} \tag{5.26}$$

Similarly, admittance-based *TPG* for the back-end equalizer is given by

$$T_{AB}(\omega) = \frac{4G_B(\omega)G_{out}(\omega)}{[G_B(\omega) + G_{out}(\omega)]^2 + [B_B(\omega) + B_{out}(\omega)]^2} \tag{5.27}$$

where $Y_{out}(j\omega) = 1/Z_{out} = G_{out}(\omega) + jX_{out}(\omega)$ is the driving point output admittance of the active device and $Y_B(j\omega) = 1/Z_B = G_B(\omega) + jB_B(\omega)$ output admittance of the back-end equalizer [B], respectively.

In general, for single matching problems, let us designate the complex immittance termination of the lossless matching network by $C(j\omega) = A(\omega) + jB(\omega)$. Let the driving point immittance of the matching network be represented by $K(j\omega) = P(\omega) + jQ(\omega)$. Then, in terms of the immittances, generic form of the *TPG* is expressed as

$$TPG(\omega) = \frac{4A(\omega)P(\omega)}{[A(\omega) + P(\omega)]^2 + [B(\omega) + Q(\omega)]^2} \tag{5.28}$$

Now, let us run an example to exhibit the implementation of the RF-LST on MATLAB.

EXAMPLE 5.1

The large signal-load pull input and output impedances for an LDMOSFET device (RD07 MUS2B by Mitsubishi) is given over the actual frequency band of 330–530 MHz, as shown in Table 5.1.

   a. Choosing flat gain level $T_0 = 0.95$, compute the impedance-based low- and high-level initial break points for the front-end matching network [F]
   b. Choosing flat gain level $T_0 = 0.925$, compute the impedance-based low- and high-level initial break points for the back-end matching network [B]
   c. Compute the imaginary parts of the above generated real parts using numerical Hilbert transform of Equations 5.17 through 5.19
   d. Compute the initial *TPG* for the front-end of the amplifier
   e. Compute the initial *TPG* for the back-end of the amplifier

Solution

Considering remark (a) given above, for part (a) and (b), break points must cover the entire frequency axis. On the other hand, measured load-pull data are only specified over 330–530 MHz, which is assumed to be the passband of the operation. In this case, we extrapolate the given data to cover DC and stop band up to infinity. For example, the high end of the stop band frequency f_s can be selected twice as much on the upper edge of the measured frequencies. In other words, we may select $f_s = 2 \times 530 = 1060$ MHz. At DC, $R_{in}(0)$ may be selected as 20 Ω. At 1060 MHz, $R_{in}(1060)$ may be set as 9 Ω. Similarly, we have the freedom to choose $X_{in}(1060) = -16$ Ω, $R_{out}(0) = 25$ Ω, $R_{out}(1060) = 18$ Ω, $X_{out}(0) = 0$ Ω, and $X_{out}(1060) = -15$ Ω. It should be noted that the above mentioned slack values do not affect the TPG of the input matching network [F].

TABLE 5.1

Large Signal Input and Output Impedance of the LDMOSFET Device (RD07 MUS2B by Mitsubishi)

Actual Frequency (MHz)	$Z_{in}(j\omega) = R_{in}(\omega) + jX_{in}(\omega)$		$Z_{out}(j\omega) = R_{out}(\omega) + jX_{out}(\omega)$	
	$R_{in}(\omega)$	$X_{in}(\omega)$	$R_{out}(\omega)$	$X_{out}(\omega)$
330	17.61	−04.63	22.47	−07.84
350	14.50	−03.67	16.06	−12.69
370	18.70	−11.30	21.23	−12.32
390	22.00	−10.11	24.40	−13.45
410	09.16	−13.96	22.03	−12.14
430	10.23	−17.94	23.50	−14.01
450	10.40	−15.89	17.82	−09.52
470	17.18	−04.85	18.53	−01.53
490	14.33	−05.11	21.06	−04.54
510	13.36	−12.11	26.80	−11.06
530	11.04	−14.03	21.50	−13.54

For the problem under consideration, break frequencies are chosen as the sampling frequencies provided by Table 5.1 including DC and f_s. These frequencies are normalized with respect to the high end of the passband, which is $f_0 = 530$ MHz. In this case, normalized angular break frequencies are given as

$WB(1) = 0, WB(2) = 330/530 = 0.6226, WB(3) = 350/530 = 0.6604,$

$WB(4) = 370 / 530 = 0.6981, WB(5) = 390/530 = 0.7358, WB(6) = 410/530 = 0.7736,$

$WB(7) = 430/530 = 0.8113, WB(8) = 450/530 = 0.8491, WB(9) = 470/530 = 0.8868,$

$WB(10) = 490/530 = 0.9245, WB(11) = 510/530 = 0.9623, WB(12) = 530/530 = 1,$

$WB(13) = \dfrac{1060}{530} = 2$

where $WB(1) = 0$ and $WB(13) = 1060/530 = 2$ are the added slack break frequencies to be able to define passband over 330 MHz $\leq f \leq$ 530 MHz.

For impedance normalization, we use the standard termination $R_0 = 50\ \Omega$. Thus, normalized break frequencies and normalized load-pull input impedance data $Z_{in}(j\omega) = R_{in-N}(\omega) + jX_{in-N}(\omega)$ are given in Table 5.2.

With the above selections, we developed a MATLAB function called "initials" such that

$$function\ RB = initials(T0, WB, Ar, Br, sign) \tag{5.29}$$

where T_0 is the flat gain level, WB is the normalized angular break frequencies, Ar is the real part of the terminating immittance $Cr(\omega) = Ar(\omega) + jBr(\omega)$, $Br(\omega)$ is the imaginary part of the terminating immittance, and *sign* is the control flag to set low or high values of the initial break points RB.

In essence, *function initials* programs the generic form of Equations 5.24, 5.25, and 5.28. MATLAB list of function "initials" is given by Program List 5.1.

TABLE 5.2

Normalized Input Impedance Data for LD MOS RD07

Normalized Break Fr	Normalized Real Part	Normalized Imaginary Part
WB	Rin_N	Xin_N
0	0.4000	0
0.6226	0.3522	−0.0926
0.6604	0.2900	−0.0734
0.6981	0.3740	−0.2260
0.7358	0.4400	−0.2022
0.7736	0.1832	−0.2792
0.8113	0.2046	−0.3588
0.8491	0.2080	−0.3178
0.8868	0.3436	−0.0970
0.9245	0.2866	−0.1022
0.9623	0.2672	−0.2422
1.0000	0.2208	−0.2806
2.0000	0.1800	−0.3200

TABLE 5.3

Computed Initial Break Points for the Front-End Matching Network [F]

Normalized Break Frequencies	Normalized Low-Level Initial Break Points for [F]	Normalized High-Level Initial Break Points for [F]
0	0.2538	0.6304
0.6226	0.2235	0.5551
0.6604	0.1840	0.4570
0.6981	0.2373	0.5894
0.7358	0.2792	0.6934
0.7736	0.1162	0.2887
0.8113	0.1298	0.3225
0.8491	0.1320	0.3278
0.8868	0.2180	0.5415
0.9245	0.1819	0.4517
0.9623	0.1695	0.4211
1.000	0.1401	0.3480
2.0000	0	0

Thus, let us run *function initials* to complete parts (a) and (b).

Part (a): Execution of MATLAB *function initials* results in the following low- and high-initial break points for the front-end matching network [F], as shown in Table 5.3.

It should be noted that the first and last break points are fixed at zero, as shown in Figure 5.4.

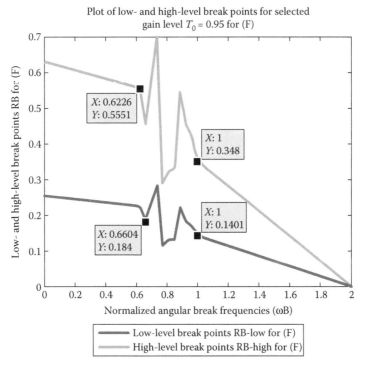

FIGURE 5.4
Low- and high-level initial break points to design front-end matching network [F] for Example 5.1.

TABLE 5.4

Computed Initial Break Points for the Back-End Matching Network [B]

Normalized Break Frequencies	Normalized Low-Level Initial Break Points for [B]	Normalized High-Level Initial Break Points for [B]
0	0.2850	0.8771
0.6226	0.2562	0.7884
0.6604	0.1831	0.5635
0.6981	0.2420	0.7449
0.7358	0.2782	0.8561
0.7736	0.2512	0.7729
0.8113	0.2679	0.8245
0.8491	0.2032	0.6252
0.8868	0.2113	0.6501
0.9245	0.2401	0.7389
0.9623	0.3055	0.9403
1.000	0.2451	0.7543
2.0000	0	0

Part (b): Execution of MATLAB *function initials* results in low- and high-initial break point values for the back-end matching network [B], as shown in Table 5.4. As in Part (a), the last break points are fixed at zero, as depicted in Figure 5.5.

Part (c): In this part of the example, we developed a MATLAB function called "num-hilbert" such that

$$function\ Q = num\_hilbert(w, WB, RB) \tag{5.30}$$

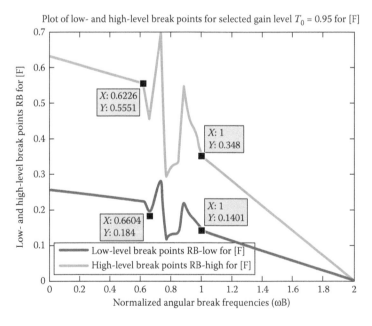

FIGURE 5.5

Low- and high-level initial break points to design back-end matching network [B] for Example 5.1.

TABLE 5.5

Hilbert Transform of the Low-Level Break Points for [F] and [B]

WB	RBF_Low	XF_Low	RBB_Low	XB_Low
0	0.2538	0.0000	0.8771	0.0000
0.6226	0.2235	−0.1243	0.7884	−0.1255
0.6604	0.1840	−0.1106	0.5635	−0.1005
0.6981	0.2373	−0.1012	0.7449	−0.0623
0.7358	0.2792	−0.1882	0.8561	−0.0942
0.7736	0.1162	−0.2041	0.7729	−0.1168
0.8113	0.1298	−0.1372	0.8245	−0.1403
0.8491	0.1320	−0.0981	0.6252	−0.1430
0.8868	0.2180	−0.1143	0.6501	−0.1073
0.9245	0.1819	−0.1600	0.7389	−0.0913
0.9623	0.1695	−0.1712	0.9403	−0.1361
1.000	0.1401	−0.1722	0.7543	−0.1800
2.0000	0	−0.1265	0	−0.1805

where w is the normalized angular frequency at which imaginary part $Q(\omega)$ of the driving point immittance is evaluated. *WB* and *RB* are the break frequencies and break points, respectively. This function takes the numerical Hilbert transform of the {break frequencies and break points} pair, as detailed by Equations 5.17 through 5.19. Results of the execution of *function num_hilbert* are listed in Tables 5.5 and 5.6, and depicted in Figures 5.6 and 5.7, for low and high values of the break points, respectively.

MATLAB list for function "num-hilbert" is given in Program List 5.2.

Part (d): Initial break points-based TPG for [F] is computed employing the impedance pairs

$$\{[R_{in}(\omega), X_{in}(\omega)]; [RBF\_Low(\omega), XF\_Low(\omega)]\}$$
$$\{[R_{in}(\omega), X_{in}(\omega)]; [RBF\_High(\omega), XF\_High(\omega)]\}$$

For this purpose, we developed a MATLAB function called *"gain_singleMatching."*

TABLE 5.6

Hilbert Transform of the High-Level Break Points for [F] and [B]

WB	RBF_High	XF_High	RBB_High	XB_High
0	0.6304	−0.0000	0.8771	−0.0000
0.6226	0.5551	−0.3087	0.7884	−0.3862
0.6604	0.4570	−0.2747	0.5635	−0.3093
0.6981	0.5894	−0.2513	0.7449	−0.1918
0.7358	0.6934	−0.4675	0.8561	−0.2900
0.7736	0.2887	−0.5069	0.7729	−0.3595
0.8113	0.3225	−0.3407	0.8245	−0.4317
0.8491	0.3278	−0.2437	0.6252	−0.4401
0.8868	0.5415	−0.2838	0.6501	−0.3303
0.9245	0.4517	−0.3973	0.7389	−0.2809
0.9623	0.4211	−0.4252	0.9403	−0.4189
1.000	0.3480	−0.4277	0.7543	−0.5541
2.0000	0	−0.3143	0	−0.5554

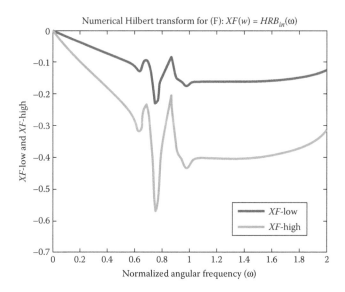

FIGURE 5.6
Hilbert transform of low- and high-level initial break points to design front-end matching network [F] for Example 5.1.

This function programs the generic form of the single matching TPG for a given complex load immittance $C(\omega) = A(\omega) + jB(\omega)$ and for a given complex matching network driving point-immittance $K(\omega) = P(\omega) + jQ(\omega)$ such that

$$TPG(\omega) = \frac{4A(\omega)P(\omega)}{[A(\omega) + P(\omega)]^2 + [B(\omega) + Q(\omega)]^2} \tag{5.31}$$

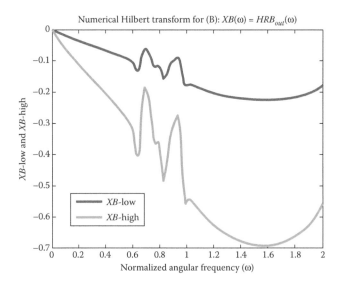

FIGURE 5.7
Hilbert transform of low- and high-level initial break points to design back-end matching network [B] for Example 5.1.

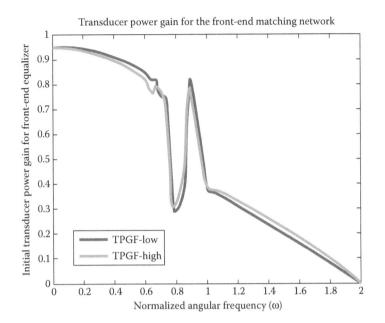

FIGURE 5.8
Initial transducer power gain plot for the front-end [F] of Example 5.1.

Execution of function *"gain_singleMatching"* results in low- and high-break points based on initial TPG for the front-end matching network, as depicted in Figure 5.8 and listed in Table 5.7.

Part (e): Similarly, we can compute the TPG for the back-end matching network using the impedance data pairs

$$\{[R_{out}(\omega), X_{out}(\omega)]; [RBB\_Low(\omega), XB\_Low(\omega)]\}$$
$$\{[R_{out}(\omega), X_{out}(\omega)]; [RBB\_High(\omega), XB\_High(\omega)]\}$$

TABLE 5.7

TPG for the Front-End [F]

WB	TPGF_Low	TPGF_High
0	0.9500	0.9500
0.6226	0.8319	0.7946
0.6604	0.8255	0.7804
0.6981	0.7385	0.7628
0.7358	0.7339	0.7044
0.7736	0.2634	0.2515
0.8113	0.2969	0.3440
0.8491	0.3804	0.4526
0.8868	0.8323	0.8018
0.9245	0.7234	0.6517
0.9623	0.5011	0.4896
1.0000	0.3690	0.3724
2.0000	0	0

TABLE 5.8

TPG for the Back-End [B]

WB	TPGB_Low	TPGB_High
0	0.9718	0.9250
0.6226	0.8641	0.7758
0.6604	0.8359	0.6581
0.6981	0.7827	0.8111
0.7358	0.8111	0.7887
0.7736	0.5327	0.7422
0.8113	0.4643	0.7103
0.8491	0.4431	0.6547
0.8868	0.8305	0.8222
0.9245	0.8742	0.8388
0.9623	0.6932	0.7787
1.0000	0.5043	0.6228
2.0000	0	0

Execution of function *"gain_singleMatching"* with the output matching network impedance data results in *TPG* for the back-end matching network, as listed in Table 5.8 and depicted in Figure 5.9.

It should be noted that imaginary parts given by Figure 5.6 are both negative (capacitive). In other words, we see that by setting *KFlag* = 1, minimum reactance impedance $ZF = RF + jXF$ is capacitive.

We can see the same capacitive behavior in the back-end driving point input impedance, as in Figure 5.7.

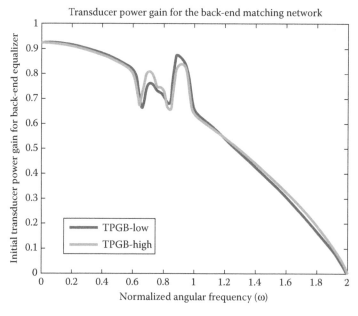

FIGURE 5.9

Initial transducer power gain plot for the back-end [B] of Example 5.1.

Close examination of Table 5.7 reveals that initial TPG of the front-end matching network drops drastically after normalized angular frequency $\omega = 0.7358$.

The plot of Table 5.7 is depicted in Figure 5.8.

Similarly, in Table 5.8, initial TPG of the back-end matching network reduces instantly after the same normalized angular frequency $\omega = 0.7358$ and increases after $\omega = 0.849$, and drops again after $\omega = 0.9245$.

The plot of Table 5.8 is shown in Figure 5.9.

All of the above computations are programed under the MATLAB main program called "Example5_1." A list of the main program is given in Program List 5.4.

It may be worth trying to generate "admittance-based initial gain." In Example 5.2, we will make an attempt to generate admittance-based break points and TPG for the front- and back-end matching networks.

Remarks: To improve computational resolution in generating numerical Hilbert transform of the break points, we can increase the total number of break points in the real part $R(\omega)$. Similarly, in computing TPG, measured load-pull data may be linearly interpolated to generate more data in between the given break frequencies. In this regard, we developed a MATLAB function called "line" such that

$$function\ ya = line\_seg(x, y, xa) \tag{5.32}$$

where x and y are the MATLAB vectors that describe the measured data on the two-dimensional x–y plane. xa is an arbitrary point on x-axis at which the value of the measured data $y = f(x)$ is interpolated connecting two adjacent points by means of a line.

In the main program Example5_1.m, TPGs for front- and back-ends are computed over 100 sampling points, as shown in Figures 5.8 and 5.9, respectively. MATLAB *functionline* is given in Program List 5.5.

5.1.1 Computation of Optimum Break Points for RF-LST

For a selected flat gain level T_0, optimum value of the unknown break points can be found employing a nonlinear optimization on the error function expressed by Equations 5.21 and 5.22. In this regard, we have experienced that MATLAB build-in function *"lsqnonlin"* successfully minimizes the error function

$$\varepsilon(\omega) = T(\omega) - T_0 \tag{5.33}$$

over the frequency band of operation.

lsqnonlin utilizes *Levenberg–Marquard* method, which minimizes the objective function expressed as the sum of squares of the error over the finite frequencies [60–65]

$$\delta = \sum_{i=1}^{M}[T(\omega_i, X) - T_0]^2 \tag{5.34}$$

where vector X includes all the unknown break points R_k, ω_i is the sampling frequency to generate TPG, M is the total sampling points selected by the user. We have found that M must be much greater than that of total number of unknowns N. A rule of thumb may be given as $M \geq 2N$.

There are several ways to call *lsqnonlin* in MATLAB. We have experienced that the following is simple.

$$X = lsqnonlin(@error\_RFLST, X0, lb, lu, options) \tag{5.35}$$

In Equation 5.35, input argument *"error_RFLST"* is the user-supplied objective function to be minimized.

Vector X_0 is the initial start for the unknown vector X. For the RF-LST, X_0 includes either low- or high-initial break points generated by our MATLAB function

$$RBA = initials(T0, WB, RLA, XLA, sign) \tag{5.36}$$

"lb" and *"lu"* stand for lower and upper bound for the unknown vector *"X."* For our case, we do not wish to impose any bounds on the initials except that they must all be positive. Therefore, we simply skip the positions of *"lb"* and *"lu"* by using empty brackets [ ], [ ].

"options" is an optimization set that includes maximum number of error function evaluations and also maximum number of iterations to reach the minimum of the error function. The options for optimization can be supplied to *"lsqnonlin"* as follows:

$$options = optimset('MaxFunEvals', 20000, 'MaxIter', 50000) \tag{5.37}$$

The way Equation 5.37 is written, maximum number of error function evaluation is bounded by 20,000. Similarly, maximum number of iteration is limited by 50,000. We have found that these numbers are sufficient to successfully run the optimizations for RF-LST.

The error function in Equation 5.35 is employed as MATLAB anonymous functions, which pass the common input parameters to *"lsqnonlin."* For example, in order to generate the error function, we have to use several input parameters such as T_0, all the real and the imaginary parts of the terminating impedances, etc. By combining *error_RFLST* function using the symbol *"@"* we automatically pass the input arguments of function *"error"* to the nonlinear least-square optimization algorithm. In other words, let the error function have the following form to be called by the optimization package.

$$function\,eps = error\_RFLST(w, T0, WB, RB, RLA, XLA, KFlag) \tag{5.38}$$

which generates variable *eps* (or $\varepsilon(\omega)$ of Equations 5.21 and 5.22) at given normalized angular frequency ω. Then, by defining *function error_RFLST* as an anonymous function with unknown input vector (X) by

$$f = @(X0)error\_RFLST(T0, WB, RB, RLA, XLA, KFlag) \tag{5.39}$$

We can simply call the optimization package as

$$X = lsqnonlin(f, X0, [\ ], [\ ], options) \tag{5.40}$$

where *options* is specified by Equation 5.37.

In Equations 5.36 and 5.39, inputs are flat gain level T_0, user selected break frequencies WB, and complex impedance termination $ZLA = RLA + jXLA$ of the lossless matching network. In function *initials*, termination impedance ZLA is defined as an array generated at the normalized angular break frequencies, WB. "*sign*" is used as a control flag to determine whether we wish to work with low- or high-level initial break points RB. In the above representation, the letter "*A*" refers to MATLAB arrays. In other words, *RBA* is a MATLAB array or vector that includes the break points. *RLA* and *XLA* are MATLAB vectors, which include real and imaginary parts of the complex impedance vectors evaluated at the break frequencies. For example, for the design of a single-stage power amplifier, *ZLA* will be the terminating impedance of the front-end matching network, which is the measured load-pull input impedance Z_{in} of the active device. Similarly, for the back-end matching network, *ZLA* is the measured load-pull output impedance Z_{out} of the active device.

Let us now run an example to exhibit the computation of the optimized break points.

EXAMPLE 5.2

    a. Choosing $T_0 = 0.975$, generate the admittance-based high-initial break points *RB* for the front-end matching network with 19 break points in the passband.

    b. Compute the imaginary part of the driving point admittance $YF = GF + jBF$ using Hilbert transform of *RB*.

    c. Optimize *TPG* over the passband of 330–530 MHz using MATLAB nonlinear least-square optimization function "*lsqnonlin*," and plot the initial and optimized gain for the front-end matching network.

Solution

Part (a): In this part, 19 distinct admittance-based break points shall be generated over the passband. On the other hand, termination impedance Z_{in} of the front-end equalizer is given over 11 points. In Example 5.1, given data were extrapolated to DC and stop band frequency 1060 MHz. Thus, we have 13 sampling points. Therefore, for this part of the example, we developed a MATLAB function called "Impedance termination" to generate the desired number of break points in the passband. This function is accessed as follows:

$$function[FC, A, B] = Impedance\_Termination(KFlag, N, FA, RA, XA, FL, FH) \quad (5.41)$$

"*Impedance_Termination*" converts the given impedance data $ZA = RA + jXA$ to an admittance $Y = A + jB$ if *KFlag* is set to zero (i.e., *KFlag* = 0) with desired number of break points employing linear interpolation. If *KFlag* = 1, then termination is preserved as an impedance $Z = A + jB$ at the output with desired number of break points "*N*" in the passband. In this regard, *FL* is the low-end, *FH* is the high-end of the passband and output argument *FC* contains "*N* + 2" sample frequencies such that "*N*" samples are placed within the passband. The other two break points are located at DC and the stop band frequency f_s. The execution result of function *Impedance_Termination* is depicted in Figure 5.10.

Function "*Impedance_Termination*" is given by Program List 5.7.

Eventually, using Equation 5.24, the high values of the initial break points are determined, as shown in Figure 5.11.

Part (b): Hilbert transform of Figure 5.11 is generated by running the MATLAB function "*num_hilbert*" over the break frequencies. The result is depicted in Figure 5.12.

Part (c): As detailed in Section 5.2, the RFLST-based gain is optimized using MATLAB optimization package "*lsqnonlin*." This function utilizes the Levenberg–Marquard technique to minimize the sum of squares of the errors generated at each sampling point in the passband. In order to access the optimization package, first the unknown break

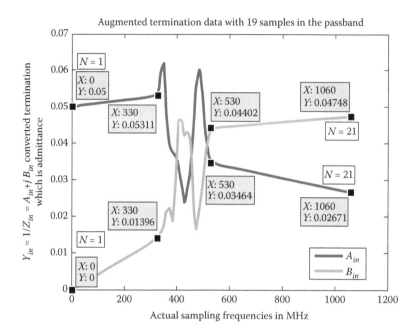

FIGURE 5.10
Conversion of the given termination data $Y_{in} = A_{in} + jB_{in}$.

points must be selected. In this regard, we have two options: We can either have the first break point $RB(1)$ as part of the unknowns or fix its value at DC. For example, for the low-pass matching problems, typical LC-ladder type of matching network topologies may be sufficient. In this case, if we do not wish to use a transformer at the far end, then the terminating resistor of the equalizer can be set to a desired level. For the problem

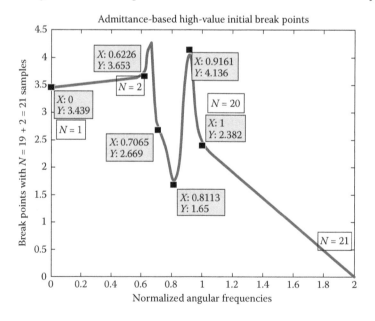

FIGURE 5.11
Admittance-based initial break points with 21 samples.

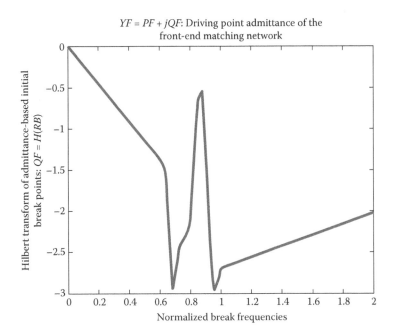

FIGURE 5.12
Numerical Hilbert transform of Figure 5.11.

under consideration, we selected flat gain level T_0 as $T_0 = 0.975$, which in turn yields the normalized admittance-based $RB(1) = 3.4390$. Thus, $RB(1)$ is fixed. $RB(N)$ is also fixed at zero. Hence, we have a total $NC = 21 - 2 = 19$ unknowns within the passband. Furthermore, total sampling points for the optimization is chosen as $M = 2 \times NC = 38$. Under these circumstances, we call the optimization within our main program as follows:

$M = 2 * NC;$

$options = optimset('MaxFunEvals', 20000, 'MaxIter', 50000);$

$eps = @(X0)error\_RFLST(X0, ktr, T0, KFlag, WB, R1, WA, Rin\_N, Xin\_N, wL, wH, M);$

$X = lsqnonlin(eps, X0, [], [], options);$ (5.42)

In Equation 5.42, the objective function is accessed by the following statement:

$"eps = error\_RFLST(X0, ktr, T0, KFlag, WB, R1, WA, Rin\_N, Xin\_N, wL, wH, M)"$

and it is minimized over the passband $wL \leq w \leq wH$ with M sampling points. Common input arguments of *"error_RFLST"* is passed to *"lsqnolin"* by the MATLAB statement

$eps = @(X0)error\_RFLST(X0, ktr, T0, KFlag, WB, R1, WA, Rin\_N, Xin\_N, wL, wH, M);$

In Equation 5.42,

- $X0$ is the initial guess and includes initial value of the unknown break points RB.
- ktr is a control flag and determines whether $RB(1)$ is part of the unknowns or not. If $ktr = 0$, $RB(1)$ is not an unknown. This choice may be preferred for low-pass designs,

which in turn result in a transformerless equalizer. If ktr = 1, then RB(1) is included among the unknown list. Then, at the far end of the matching network, we may utilize a transformer to adjust the terminating resistance value to a desired level.
- *T0 is the flat gain level selected by the user.*
- *WB is the term for the normalized break frequencies given as a vector.*
- *R1 is the first break point RB(1).*
- *WA is the normalized-angular sampling frequency array for which the complex imped-ance termination is specified.*
- *Rin_N and Xin_N are the real and imaginary parts of the complex termination speci-fied over the frequencies given by WA, respectively.*
- *wL and wH are the lower and the upper edges of the optimization frequencies.*
- *M is the total number of sampling points over which the optimization of the gain is completed by minimizing the objective function of Equation 5.33.*

MATLAB codes for the objective function are given by Program List 5.7.

Execution of MATLAB function "*lsqnonlin*" results in the optimized TPG, as depicted in Figure 5.13, which in turn determines the new break points *RBF* and its corresponding imaginary part *XF*, as shown in Figure 5.14.

Figure 5.13 reveals that, in the passband, the minimum of the TPG is $T_{min} = 0.954$.

Remarks: It should be noted that in Example 5.2, the initial value of the TPG is much bet-ter than that of Example 5.1. Therefore, the choice for the value of *KFlag* is essential. We can start the design with impedances or admittances. This choice depends on the behavior of the complex termination of the equalizer. Literally speaking, complex termination is called the load of the single matching problem. If the load is capacitive, then the equalizer driving point impedance (DPI) must start with an inductor to be able to provide reactance cancellation. Inductive DPI can be obtained by means of minimum susceptance admit-tance. In this case, we must set *KFlag* = 0. If the complex termination is inductive, then

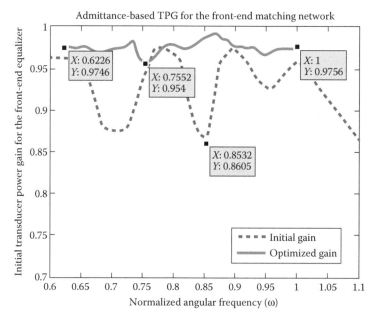

FIGURE 5.13
Initial and optimized gain of the front-end matching network.

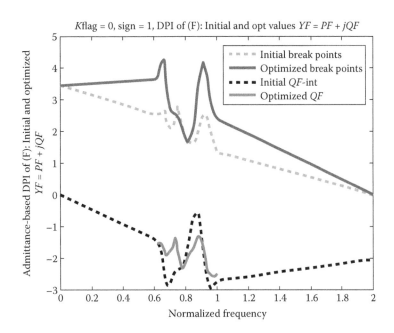

FIGURE 5.14
Initial and optimized break points.

equalizer DPI must be capacitive. This can be achieved by a minimum reactance driving point impedance. In this case, we set *KFlag* = 1. Close examination of Table 5.1 reveals that load-pull measured input and output impedance of Mitsubishi *LD–MOS RD007* is capacitive (imaginary parts of the impedances are negative) over the passband. In this case, for better optimization we must start with an inductive-minimum susceptance DPI for both front- and back-end equalizers. In this case, it might be proper to start with *KFlag* = 0. In fact, this is how we started Example 5.2. For the sake of the experiment, let us solve Example 5.2 with minimum reactance driving point impedance. In this case, we should start with *KFlag* = 1.

EXAMPLE 5.3
Repeat Example 5.2 using impedance-based break points and TPG by setting *KFlag* = 1.

Solution
Execution of MATLAB program Example5_3.m with *KFlag* = 1 results in break points, as depicted in Figure 5.15.

The corresponding Hilbert transform of the break points is depicted in Figure 5.16.

Notice that, initially, the *KFlag* = 1 choice results in capacitive imaginary part for the DPI of the front-end equalizer. However, the optimization process tries to convert it into a reactive part while creating wiggles in the break points, as shown in Figure 5.15.

Impedance-based TPG is shown in Figure 5.17.

An examination of Figure 5.17 reveals that, at the low and the high end of the passband, TPG drastically drops. Hence, selection of *KFlag* = 0 is a much better choice for the problem under consideration.

In summary, RF-LST generates an almost ideal solution for single matching problems. In this regard, for the given complex load or equivalently complex termination such as load pull input or output impedance of an active device, we can generate the driving

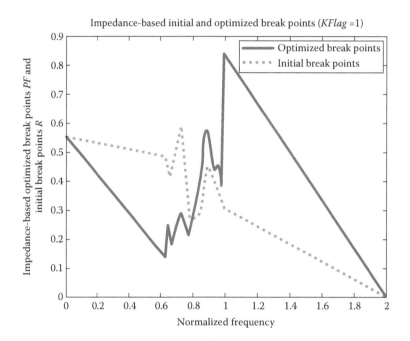

FIGURE 5.15
Initial and optimized break points.

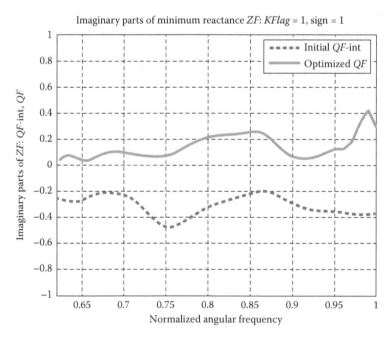

FIGURE 5.16
Initial and optimized break points.

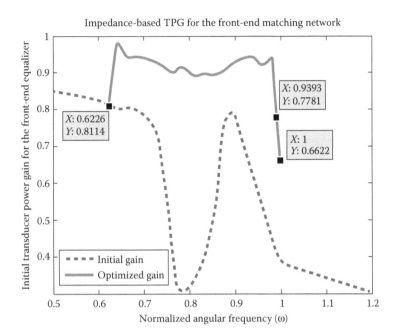

FIGURE 5.17
Initial and optimized break points.

point input immittance (DPI) for the lossless equalizer [N] as a set of data points. In order to realize [N], we must first model the computed DPI as positive real function, then synthesize it as a lossless two-port [N] in resistive termination.

Therefore, in the following subsection, we will introduce a practical approach to model the break points as an even nonnegative function $R(\omega)$ in terms of the normalized angular frequency ω. Then, using an algebraic method, we will construct complete DPI $Z(p)$ from $R(\omega)$.

5.1.2 Practical Approach to Model Break Points by Nonnegative Even Rational Function

In practice, we wish to construct matching networks as simply as possible. For example, using lumped circuit elements such as capacitors, inductors, and transformers, a simple equalizer topology can be a ladder network, as shown in Figure 5.18.

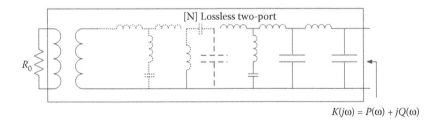

$$K(j\omega) = P(\omega) + jQ(\omega)$$

FIGURE 5.18
A typical lossless ladder described by its DPI $Z(j\omega)$.

The generic form of the real part of the DPI of a lossless ladder terminated in a resistor R_0 is given by

$$P(\omega^2) = \frac{a_0^2 \omega^{2ndc} \prod_{i=1}^{nz} (\omega_i^2 - \omega^2)^2}{B_1 \omega^{2n} + B_2 \omega^{2(n-1)} + \cdots + B_n \omega^2 + 1} = \frac{A(\omega^2)}{B(\omega^2)} \geq 0; \quad \forall \omega \qquad (5.43)$$

Obviously, $P(\omega^2)$ is a nonnegative, rational even function in ω.

Loosely speaking, zeros of $P(\omega^2)$ are called the transmission zeros of the immittance $K(j\omega)$.

In detail, Equation 5.43 has zeros at $\omega = 0$ of multiplicity $2 \times ndc$. These are called the transmission zeros at DC. Similarly, finite real frequency transmission zeros are located at $\omega = \mp\omega_i$ of multiplicity 4. Furthermore, the degree of the denominator polynomial $2n$ must be less than or equal to $2(2nz + ndc)$ to make $P(\omega)$ finite. In other words $n \leq 2nz + ndc$. If $n_\infty = n - ndc - 2nz > 0$, then we say that immittance $K(p)$ has transmission zeros at infinity of multiplicity $2n_\infty$. Some authors count the transmission zeros by dropping the multiplier 2. In Figure 5.18, the total number of reactive elements is given by the integer "n."

As far as construction of the matching network via RF-LST is concerned, the first step is to fit the optimized break points to the simple form of Equation 5.43 to end up with a ladder network. In the second step, using Hilbert transform, we should generate the analytic form of the immittance. The last step is the synthesis of the immittance $K(p)$ as a ladder network in resistive termination R_0. For single matching problems, R_0 may be considered as the internal resistance of the source network.

In the real frequency direct computational technique (RF-DCT), nonnegativity of Equation 5.43 is assured by introducing an auxiliary polynomial $c(\omega)$ to produce strictly positive denominator polynomial $B(\omega)$ such that

$$B(\omega^2) = \frac{1}{2}[c^2(\omega) + c^2(-\omega)] > 0 \text{ for all } \omega \qquad (5.44)$$

and

$$c(\omega) = c_1 \omega^n + c_2 \omega^{(n-1)} + \cdots + c_n \omega + 1 \qquad (5.45)$$

In this case, by selecting $A_0 = a_0^2 > 0$ as a strictly positive real quantity, one can run a curve-fitting algorithm to determine a_0 and the real arbitrary coefficients c_i of the auxiliary polynomial $c(\omega)$.

The curve-fitting problem can be defined as follows:

Let $\varepsilon(\omega)$ be an error function defined as

$$\varepsilon(\omega_j, c_1, c_2, \ldots, c_n, c_{n+1}) = P(\omega_j^2) - R_j \qquad (5.46)$$

where ω_j are the break frequencies and $R_j = RB(\omega_j)$ are the optimized break points via RFLST.

Thus, the curve-fitting problem is structured to minimize all $\varepsilon(\omega_j)$ to determine the unknown coefficients of $c(\omega)$. It can be shown that $\varepsilon(\omega_j, c_1, c_2, \ldots, c_n, c_{n+1})$ is quadratic or convex in terms of c_i. Therefore, the minimization algorithm is always convergent. In this case, we may as well utilize the same nonlinear least-square minimization algorithm of MATLAB, namely *"lsqnonlin,"* as in the previous subsection.

Now, let us run an example to model the break points of the front-end equalizer [F] obtained in Example 5.2.

EXAMPLE 5.4

Model the break points provided by Figure 5.14 of Example 5.2 employing Equation 5.43 with $ndc = 0$, $n = 5$ and with no finite transmission zeros.

Solution

For this example, we develop a main MATLAB program called *"Example5_4.m."* This program first initiates the unknown coefficients c_i in an ad-hoc manner by setting initials as $x0_i = c_i = \mp 1$, as in function

$$function[x0] = Initiate\_CurveFitting(ktr, n, a0)$$

where ktr is selected as $ktr = 0$ which fixes $a0$ in advance at $a0 = \sqrt{RB(1)} = 1.8545$, $n = 5$ as requested. In the second step, nonlinear minimization package *"lsqnonlin"* of MATLAB is called as follows:

```
%Curve fitting for the optimized break points:
%Initilize the unknowns :
a0 = sqrt(RB(1));
n = 5;ndc = 0;WZ = 0;nz = 0;
[x0 ]=Initiate_CurveFitting(ktr,n,a0 );
%
NB = length(WB);
wL = WB(1);wH = WB(NB);
M = 2 * NC;
options = optimset('MaxFunEvals',200000,'MaxIter',500000);
eps = @(x0)error_RFDCT(x0,ktr,WB,RB,n,ndc,WZ,a0,wL,wH,M);
x = lsqnonlin(eps,x0,[],[],options);
[c,a0,a,b,AA,BB ] = Evaluate_CurveFitting(x,ktr,n,ndc,WZ,R1);        (5.47)
```

In the above MATLAB codes, curve fitting is completed in *"lsqnonlin"* by minimizing the error function called

$$function[eps] = error\_RFDCT(x0, ktr, WB, RB, n, ndc, WZ, a0, wL, wH, M)$$

In the above function, the abbreviation *RFDCT* stands for real frequency-direct computational technique. In essence, RFDCT generates the rational function form for the break points *RB*, as in Equation 5.43. Function *"error_RFDCT"* simply programs the error expression given by Equation 5.46.

Input arguments to *"error_RFDCT"* are described as follows:

- $x0$: *This vector contains the initial values for the unknown coefficients c_i.*
- ktr: *A control flag. $ktr = 0$ fixes the value of $a0$. $ktr = 1$ puts $a0$ among the unknowns.*

- *WB: A vector that includes the break frequencies.*
- *RB: A vector that includes the break points to be modeled as in Equation 5.43.*
- *n: Degree of denominator of Equation 5.43.*
- *ndc: Number of transmission zeros at DC.*
- *WZ: A vector that includes finite frequency transmission zeros beyond DC.*
- *If there is no finite transmission zeros we set WZ = 0.*
- *a0: Numerator coefficient of Equation 5.43.*
- *wL: Low end of optimization.*
- *wH: High end of optimization.*
- *M: Total number of frequency sampling points at which the error Equation 5.46 is evaluated.*

"lsqnonlin" returns with the optimized values of the unknown coefficients c_i in vector x as in function *"Evaluate_Curvefitteng"* such that

$$function\ [c, a0, a, b, AA, BB\ ] = Evaluate\_CurveFitting(x, ktr, n, ndc, WZ, R1\ )$$

where $R1$ is the value of the first break point $RB(1)$. The output arguments are described as below.

- *c: A vector that includes optimized coefficients extracted from x.*
- *a0: Numerator coefficient of Equation 5.43.*
- *"a" and "b": "Polynomial vectors" that "describe the positive real immittance function $K(p) = a(p)/b(p)$ generated from Equation 5.43 using an algebraic Hilbert transform technique called "Parametric Approach." Thus, $K(p) = a(p)/b(p)$ is the DPI of the matching network [N] to be designed. The matching network is obtained as a result of Darlington synthesis of $K(p) = a(p)/b(p)$. Details will be presented in the following subsections.*
- *AA: Polynomial vector that includes the coefficients of the numerator of Equation 5.43.*
- *BB: Polynomial vector that includes the coefficients of the denominator of Equation 5.43.*

The results of modeling are outlined as follows:
Computed unknown coefficient vector c is given by

$$c = [0.8487\quad 1.2975\quad -1.5108\quad -1.6772\quad 1.9793]$$

Nonnegative real part $P(\omega^2)$ is found as

$$P(\omega^2) = \frac{3.4390}{0.7203\omega^{10} - 0.8810\omega^8 + 1.2901\omega^6 - 0.5729\omega^4 + 0.5635\omega^2 + 1}$$

The fit between break points and analytic form of the real part of Equation 5.43 is depicted in Figure 5.19.

As we see from Figure 5.19, the fit between break points and analytic form of Equation 5.43 may be acceptable. We have missed the fluctuations in the passband. Therefore, we expect the disturbance in the gain function at those points where the original break points deviate from the rational form of the real part.

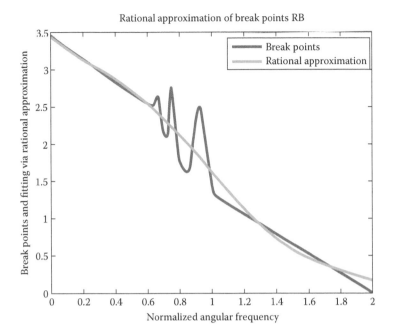

FIGURE 5.19
Approximation of break points via nonnegative rational function.

The next stage is the construction of the positive real driving point immittance function from Equation 5.43. This can easily be achieved using the parametric approach presented in the next section.

5.2 Generation of Minimum Immittance Function from Its Real Part

A minimum positive real function $K(p) = a(p)/b(p)$ can be generated uniquely from its even part using parametric approach [38].[*]
On the real frequency axis, the even part is given by

$$P(j\omega) = \frac{(a_0^2)W_z(j\omega)\omega^{2ndc}}{b(j\omega)b(j\omega)} \tag{5.48}$$

where $W_z(p)$ includes all the finite transmission zeros of $K(p)$ beyond $p = 0$.

As explained in the previous section, the denominator polynomial $C(\omega^2) = b(j\omega)b(j\omega)$ must be a strictly positive, even polynomial. It is generated by means of an auxiliary polynomial $c(\omega) = c_1\omega^n + c_2\omega^{n-1} + \cdots + c_n\omega + 1$ such that

$$b(j\omega)b(j\omega) = C(\omega^2) > 0 \tag{5.49}$$

---

[*] It should be noted that the expression "minimum positive real function" refers to either a minimum reactance impedance function or a minimum susceptance admittance function.

Replacing ω^2 by $-p^2$, a realizable ladder form of the even part is given by

$$P(p^2) = \frac{A(p^2)}{B(p^2)} \tag{5.50}$$

where

$$A(p^2) = (-1)^{ndc} a_0^2 p^{2ndc} \prod_{i=1}^{nz} (\omega_i^2 + p^2)^2 = A_0 p^{2ndc} \prod_{i=1}^{nz} (\omega_i^2 + p^2)^2 \tag{5.51}$$

with

$$A_0 = (-1)^{ndc} a_0^2 \tag{5.52}$$

and

$$
\begin{aligned}
B(p^2) &= (-1)^n C_1 p^{2n} + (-1)^{n-1} C_2 p^{n-1} + \cdots - C_1 p + 1 \\
&= B_1 p^n + B_2 p^{2(n-1)} + \cdots + B_1 p^2 + 1
\end{aligned}
\tag{5.53}
$$

where

$$B_i = (-1)^{(n-i+1)} C_i; \quad i = 1, 2, \ldots, n \tag{5.54}$$

It should be emphasized that once ndc and finite transmission zeros ω_i are selected and the coefficients $\{a_0, c_1, c_2, \ldots, c_n\}$ are initialized, coefficients of $B(p^2)$ can easily be generated on MATLAB, which, in turn yields a realizable ladder form for $P(p^2)$.

In the parametric approach, a positive real minimum immittance function is represented in terms of its poles p_i and its corresponding residues k_i as

$$K(p) = K_0 + \sum_{i=1}^n \frac{k_i}{p - p_i} \tag{5.55}$$

Using Equation 5.55, $P(p^2)$ is expressed in terms of the same residues and poles such that

$$P(p^2) = \frac{1}{2}[K(p) + K(-p)] = K_0 + \sum_{i=1}^n \frac{k_i p_i}{p^2 - p_i^2} \tag{5.56}$$

where the residues k_i are computed from the realizable form of $P(p^2)$ as specified by Equation 5.48 as follows:

$$k_i = (-1)^n \frac{A(pi)}{p_i B_1 \prod_{\substack{j=1 \\ j \neq i}}^n p_i^2 - p_j^2} \tag{5.57}$$

for the section under consideration. If $ndc < n$, then $K_0 = A_1/(-1)^n C_1 = 0$ and for $n = ndc$ $K_0 = 1/C_1$ (notice that $C_1 = c_1^2 > 0$).

Hence, positive real minimum immittance function production starts by selecting ndc and initializing the coefficients $\{a_0, c_1, c_2, \ldots, c_n\}$. Then, the even polynomial

$$C(\omega^2) = \frac{1}{2}[c^2(\omega) + c^2(-\omega)] = C_1\omega^{2n} + C_2\omega^{2(n-1)} + \cdots + C_n p^2 + 1$$

is computed. Replacing ω^2 by $-p^2$, $B(p^2)$ is obtained. Thereafter, poles p_i are computed and k_i are determined by means of Equation 5.57. Finally, $K(p)$ is generated as a minimum rational function as

$$K(p) = \frac{a(p)}{b(p)} = \frac{a_1 p^n + a_2 p^{n-1} + \cdots + a_n p + a_{n+1}}{b_1 p^n + b_2 p^{n-1} + \cdots + b_n p + b_{n+1}} \tag{5.58}$$

The above formulation is programmed as a MATLAB function

$$function[a, b] = Minimum\_Function(ndc, WZ, a0, c) \tag{5.59}$$

Now, let us run an example to generate minimum susceptance admittance function $Y(p) = a(p)/b(p)$ for Example 5.4.

EXAMPLE 5.5

 a. Generate the minimum susceptance admittance function $Y_F(p)$ from its real part $P(\omega^2)$ as determined in Example 5.4.
 b. Compute the admittance-based TPG T_{YF} for the front-end matching network of Figure 5.2 using its normalized load-pull measured input impedance as listed in Table 5.2.
 c. Compare your results with those obtained in Example 5.4.

Solution

For the example under consideration, we developed a MATLAB main program called *Example5_5.m*. This program completes the MATLAB program *Example5_4.m* by executing $[a,b] = Minimum\_Function$ to generate the minimum susceptance admittance function $K(p) = a(p)/b(p)$. Thus, we have the following results.

Part (a): Execution of $[a,b] = Minimum\_Function(ndc, WZ, a0, c)$ with

$$ndc = 0, WZ = 0, a0 = 1.8545$$

and

$$c = [0.8487 \quad 1.2975 \quad -1.5108 \quad -1.6772 \quad 1.9793]$$

yields

$$a = [0 \quad 2.0658 \quad 6.5134 \quad 10.7052 \quad 9.5211 \quad 4.0520]$$
$$b = [1.0000 \quad 3.1529 \quad 5.5819 \quad 5.8696 \quad 3.8228 \quad 1.1782]$$

or

$$Y_F(p) = \frac{2.0658p^4 + 6.5134p^3 + 10.7052p^2 + 9.5211p + 4.0520}{p + 3.1529p^4 + 5.5819p^3 + 5.8696p^2 + 3.8228p + 1.1782}$$

Part (b): Using the termination admittance values generated as in Example 5.4, we computed the TPG by executing the following loop in MATLAB:

```
ws1=0;ws2=2;Ns=100;delw=(ws2-ws1)/(Ns-1);w=ws1;j=sqrt(-1);
for i=1:Ns
    WA(i)=w;
    p=j*w;
% -- Generation Positive Real RFDCT Min. Susc.Admt. YF=PF+jQF for [F] --
    aval=polyval(a,p);
    bval=polyval(b,p);
    YFval=aval/bval;
    Pf0_RFDCT=real(YFval);        PAf0_RFDCT(i)=Pf0_RFDCT;
    Qf0_RFDCT=imag(YFval);        QAf0_RFDCT(i)=Qf0_RFDCT;
% ------ Generation of termination admittance Yin=A+jB ------    (5.60)
    A=line_seg(WB,Ain_N,w);       AA(i)=A;
    B=line_seg(WB,Bin_N,w);       BA(i)=B;
% Generation of single matching gain ----------------------------
% Part (b)
            [ TPG ] = Gain(A,B,Pf0_RFDCT,Qf0_RFDCT );
            TfA(i)=TPG;
% ------ Generation of Optimized Break Points for RFLST -------------
            Pf_RFLST=line_seg(WB,RB,w);    PAf_RFLST(i)=Pf_RFLST;
            Qf_RFLST=num_hilbert(w,WB,RB);  QAf_RFLST(i)=Qf_RFLST;
    w=w+delw;
end
```

Resulting TPG is depicted in Figure 5.20. In this figure, gain obtained using the RF-LST is also shown. It is observed that gain obtained as a result of modeling is penalized due to misfit or fitting error occurring between the optimized break points and the rational form of the even part of the driving point admittance. Of course, this is expected.

Part (c): For this part of the example, we compare the driving point input admittances that are generated based on the optimized break points and the amplifier model. Hence, the results are depicted in Figure 5.21.

Clearly, discrepancies between real and imaginary parts of the RFDCT- and RFLST-based driving point input admittances affect the TPG as explained in Part (b). However, the loss in TPG may be recovered by reoptimization of the gain function generated by the parametric approach.

5.3 Optimization of TPG Using a Parametric Approach

The aim of the immittance-based real frequency techniques is to determine the DPI of the lossless matching network to optimize the TPG as flat and as high as possible within the frequency band of operation. Dealing with lumped elements, in classical complex Laplace

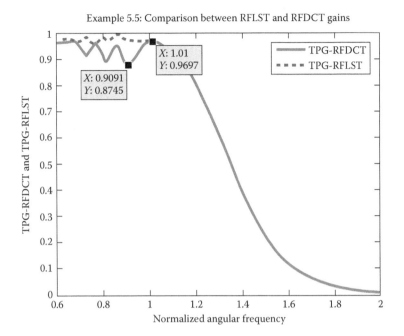

FIGURE 5.20
TPGs of [F] employing RFLST and RFDCT.

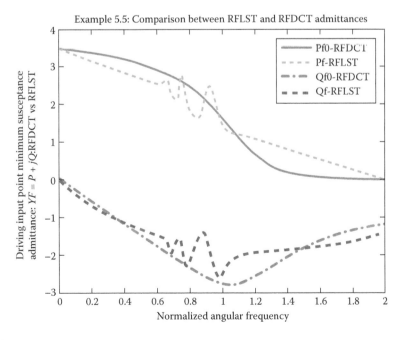

FIGURE 5.21
Comparison between RFDCT and RFLST-based DPI admittances.

variable "$p = \sigma + j\omega$," DPI is designated by a positive real function $K(p)$. Synthesis of $K(p)$ results in the desired matching network with optimum circuit topology.

In the real frequency techniques, DPI, which is described on the real frequency axis as $K(j\omega) = P(\omega) + jQ(\omega)$, is assumed to be a minimum function. In this case, imaginary part $Q(\omega)$ is uniquely generated from the real part $P(\omega)$ using Hilbert transform relation, which may be designated by an operator $H\{.\}$. In other words, $Q(\omega) = H\{P(\omega)\}$. In a similar manner, realizable analytic form of a minimum function $K(p)$ can be determined from its even part $P(p^2)$ using an algebraic Hilbert operator $HP\{.\}$ such that $K(p) = HP\{P(p^2)\}$. In this representation, $HP\{.\}$ stands for "Hilbert transform via Parametric Method."

So far, we have been trying to generate intelligent initial guesses to start the nonlinear optimization of TPG for single matching problems. These processes are completed in the following major steps.

Step 1: In this step, the real part $P(\omega)$ is piecewise linearized and expressed in terms of the unknown break points of line segments. At this point, the objective is "to generate initial guess for gain optimization." At this step, initial break points are computed by reactance cancellation. At the beginning of this step, the user decides to work with either impedance- or admittance-based gain function for the broadband matching problem under consideration. Furthermore, the designer selects the sampling frequencies, called break frequencies, and picks a flat gain level T_0 over the passband. The outcome of this step is a convergent initial guess for the break points.

Step 2: In the second step, TPG of the matching network is optimized to reach the idealized flat gain level T_0. As desired, in the course of gain optimization, one may wish to jack-up the flat gain level T_0 to high values by trial and error. The outcome of this step is the optimized break points.

All the above numerical processes are called the "real frequency line segment technique" or, in short, RFSLT. The outcome of the RFSLT is the optimum break points, which approximate the ideal flat gain level T_0.

Step 3: In this step, computed break points are modeled by means of a curve-fitting algorithm to generate the actual equalizer. The outcome of this step is the nonnegative rational form $P(\omega^2)$ for the break points. This rational form is described by means of an auxiliary polynomial

$$c(\omega) = c_1\omega^n + c_2\omega^{(n-1)} + \cdots + c_n\omega + 1$$

as described by Equations 5.43 through 5.45. Coefficients c_i provide an excellent initial guess for the actual optimization of the TPG of the matching network to be designed.

At this step, the analytic computation process of the even part $P(\omega^2)$ is called the "real frequency direct computational technique" or, in short, RFDCT.

Quality of the initial guess generated by Step 3 is checked by computing the analytic form

$$K(p) = \frac{a(p)}{b(p)} = HL\left\{P(\omega^2)\big|_{\omega^2 = -p^2}\right\}$$

and generating TPG as a function of termination as well as $K(p)$.

Construction of $K(p)$ is accomplished by employing the parametric approach.

Eventually, the best form of the matching network can be obtained by running TPG optimization on the coefficients of the auxiliary polynomial

$$c(\omega) = c_1\omega^n + c_2\omega^{(n-1)} + \cdots + c_n\omega + 1$$

To accomplish this task, we developed an objective function in MATLAB called

$$function\,[eps] = Error\_Parametric(x0, ktr, T0, n, ndc, WZ, a0, WB, Ain\_N, Bin\_N, wL, wH, M) \tag{5.61}$$

This function generates an error function when the DPI is evaluated employing the parametric approach such that

$$Error\_Paramateric = TPF(\omega_j, c_1, c_2, \ldots, c_n, a0, W_z) - T0 \tag{5.62}$$

List of function *Error_parametric* is given by Program List 5.19.
The optimization task can be completed by executing the following MATLAB codes:

$[x0] = Initiate\_Parametric(ktr, n, c, a0);$
$options = optimset('MaxFunEvals', 200000, 'MaxIter', 500000);$
$T0 = 0.97; wL = WB(2); wH = WB(N-1);$
$eps = @(x0)Error\_Parametric(x0, ktr, T0, n, ndc, WZ, a0, WB, Ain\_N, Bin\_N, wL, wH, M);$
$x = lsqnonlin(eps, x0, [], [], options);$
$\quad [a, b] = Evaluate\_Parametric(x, ktr, n, ndc, WZ, a0);$

The execution of Equation 5.62 results in an optimal solution for the matching network to be the design. In Equation 5.62, statement $[x0] = Initiate\_Parametric(ktr, n, c, a0)$ generates the initials for the final optimization based on the rational approximation of Equation 5.43, which results in the coefficients of the auxiliary polynomials. The following MATLAB codes are run to complete Step 3:

$$T0 = 0.97; wL = WB(2); wH = WB(N-1);$$
$$eps = @(x0)Error\_Parametric(x0, ktr, T0, n, ndc, WZ, a0, WB, Ain\_N, Bin\_N, wL, wH, M);$$
$$x = lsqnonlin(eps, x0, [], [], options);$$

The above statements optimize TPG in the mean least-square sense. Optimized values of the coefficients c_i are stored in vector x.

The statement $[a, b] = Evaluate\_Parametric(x, ktr, n, ndc, WZ, a0)$ generates a realizable rational form of DPI. In this regard, if $ktr = 0$, $K(p) = a(p)/b(p)$ returns as a minimum susceptance admittance. If $ktr = 1$, then $K(p) = a(p)/b(p)$ is minimum reactance impedance.

Let us now run an example to generate the best solution for Example 5.5.

EXAMPLE 5.6

Optimize TPG obtained in Example 5.5 on the initial coefficients c_i specified by

$$c0 = [0.8487 \quad 1.2975 \quad -1.5108 \quad -1.6772 \quad 1.9793]$$

Solution

For this example, we develop a MATLAB main program called *"Example5_6.m."* This program is based on *"Example5_5.m"* with the inclusion of optimization codes of Equation 5.62. The result of optimization is given by

$$c = [11.1588 \quad 5.2383 \quad -15.4989 \quad -7.0961 \quad 5.3529]$$

and

$$a = [0 \quad 1.5791 \quad 1.2501 \quad 1.9742 \quad 0.8301 \quad 0.3082]$$
$$b = [1 \quad 0.7916 \quad 1.5921 \quad 0.7964 \quad 0.5088 \quad 0.0896]$$

or driving point admittance $K(p)$ is expressed as

$$K(p) = \frac{a(p)}{b(p)} = \frac{1.5791p^4 + 1.2501p^3 + 1.9742p^2 + 0.8301p + 0.3082}{p^5 + 0.7916p^4 + 1.5921p^3 + 0.7964p^2 + 0.5088p + 0.0896}$$

Corresponding TPG is depicted in Figure 5.22.

Comparison of Figures 5.21 and 5.22 reveals that final optimization on the coefficients of $c(\omega)$ significantly improves TPG ($TPG_{min} = 0.875$ of Figure 5.21 is compared with that of $TPG_{min} = 0.93$ of Figure 5.22).

MATLAB codes of the main program *"Example5_6.m"* are listed in Program List 5.20.

Section 5.5 may be regarded as the third step of the "real frequency-immittance-based matching network design" process.

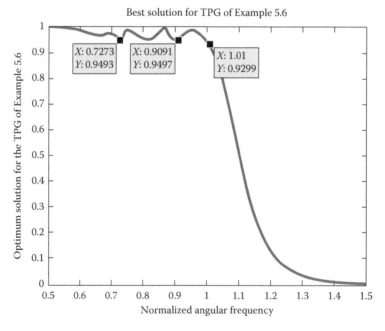

FIGURE 5.22
Reoptimized TPG on the coefficients of c_i via parametric approach.

At the final step, DPI $K(p) = a(p)/b(p)$ of the matching network must be synthesized as a lossless two port in resistive termination. Therefore, in the following section, a high-precision synthesis method of positive real functions is presented [5,6].

5.4 High-Precision Ladder Synthesis of Positive Real Functions

The last step of the real frequency technique is the synthesis of the DPI of the equalizer. Usually, we deal with immittance functions that have transmission zeros at infinity yielding low-pass LC-ladder networks. Sometimes, we may need to employ transmission zeros at DC and infinity, which results in a bandpass LC ladder. Rarely, one may wish to employ transmission zeros at finite frequencies as well as at DC and infinity. Therefore, this subsection is devoted to the synthesis of a DPI with transmission zeros at finite frequencies, at DC and at infinity.

On the computer, Darlington synthesis of positive real functions is an art. It requires special care due to the accumulation of numerical errors at the extraction of transmission zeros. Until 2013, synthesis was one of the bottlenecks limiting the real frequency techniques. We are pleased to announce that the problem affecting reliable-robust and automated synthesis has been solved, as reported in References 5–7.

First, in the new synthesis algorithms, transmission zeros at finite frequencies are extracted as Brune sections. A Brune section contains three inductors, namely, L_a, L_b, and L_c, in a T-like configuration, as shown in Figure 5.23a. One of the series inductors of the T-structure can have a negative value. However, the negative inductor can be removed employing a coupled coil, as shown in Figure 5.23b.

In the second step, we remove DC transmission zeros in a sequential manner either using a series capacitor or a shunt inductor, as shown in Figure 5.24.

In the third step, transmission zeros at infinity are removed one by one either using a shunt capacitor or a series inductor, as shown in Figure 5.25.

At the end of the synthesis, typically, one would end up with a lossless LC ladder, as depicted in Figure 5.26.

Referring to Equation 5.63, the driving point impedance of Figure 5.26 has $nz = 3$ finite transmission zeros; which are realized by means of 3-cascade connections of Brune sections, $ndc = 3$ transmission zeros at DC; realized using two-shunt inductors and one series capacitor, and $n_\infty = 5$ transmission zeros at infinity; realized by 3-shunt capacitors and

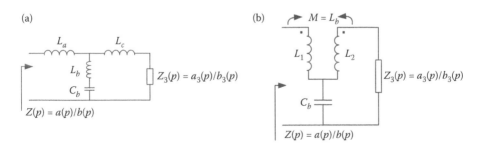

FIGURE 5.23
(a) Extraction of a finite transmission zero from a driving point impedance $Z(p)$ as a Brune section. (b) Removal of negative inductance with coupled coils.

FIGURE 5.24
Extraction of DC transmission zeros as a high-pass ladder.

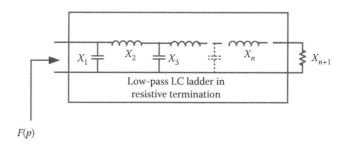

FIGURE 5.25
Extraction transmission zeros at infinity as a low-pass ladder.

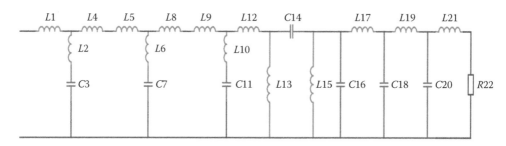

FIGURE 5.26
Typical synthesis of an immittance function $K(p) = a(p)/b(p)$ with $n = 15$, $nz = 3$, $ndc = 3$, $n_\infty = 6$, and $KFlag = 1$ (minumum reactance).

3-series inductors. Hence, half-degree n of the denominator of Equation 5.63 is $n = 2 \times nz + ndc + n_\infty = 6 + 3 + 6 = 15$.

In the following, we will briefly cover the basic principles of practical ladder synthesis with MATLAB.

5.4.1 Extraction of Transmission Zero via Zero Shifting Method

In this section, we deal with the extraction of a finite frequency transmission zero from the given impedance. In this regard, let the driving point input impedance of the matching network be

$$Z(p) = \frac{a(p)}{b(p)} = \frac{a_1 p^n + a_2 p^{n-1} + \cdots + a_n p + a_{n+1}}{b_1 p^n + b_2 p^{n-1} + \cdots + b_n p + b_{n+1}} \tag{5.63}$$

Further, let $Z(p)$ be a minimum reactance impedance with a single real frequency finite transmission zero at ω_a, DC transmission zeros of order ndc and transmission zeros at infinity of order n_∞ to be synthesized in the Darlington sense. In this case, even part of $Z(p)$ is given by

$$R(p) = A(p)/[b(p)b(-p)] \tag{5.64}$$

Finite zeros of the even polynomial $A(p)$ are called the transmission zeros of $Z(p)$, including those at DC. For the case under consideration, $A(p)$ is given by

$$F(p) = a_0^2(p^2 + \omega_a^2)^2(-1)^{ndc}p^{2ndc} = \mp f^2(p) \tag{5.65}$$

where

$$f(p) = a_0(p^2 + \omega_a^2)p^{ndc} \tag{5.66}$$

At the finite frequency transmission zeros $p = \mp j\omega_a$

$$Z(j\omega_a) = R(j\omega_a) + jX_a(j\omega_a) \tag{5.67}$$

where

$$R(j\omega_a) = 0 \tag{5.68}$$

Referring to Equations 5.68 and 5.69, we may extract an inductor L_a from $Z(p)$ such that

$$X_a = \omega_a L_a \tag{5.69}$$

In Equation 5.69, the inductor L_a could be either positive or negative without disturbing the positive real feature of $Z(p)$. Based on Equation 5.69, we can express $Z(p)$ as follows and partially synthesize it, as shown in Figure 5.27.

$$Z(p) = pL_a + Z_1(p) \tag{5.70}$$

where

$$Z_1(p) = \frac{a_1(p)}{b_1(p)} \tag{5.71}$$

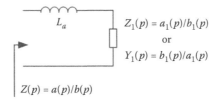

FIGURE 5.27
Extraction of an inductor L_a from $Z(p)$.

and

$$b_1(p) \equiv b(p)$$
$$a_1(p) \equiv a(p) - (pL_a)b(p) \tag{5.72}$$

As above, degree of $a_1(p)$ is degree of $(n + 1)$.

Obviously, as it is introduced above, $Z_1(p) = Z(p) - L_ap$ is zero when $= \mp j\omega_a$. In other words,

$$Z_1(j\omega_a) = Z(j\omega_a) - j\omega_a L_a = 0 \tag{5.73}$$

Therefore, the numerator polynomial $a_1(p)$ must include the term $(p^2 + \omega_a^2)$ such that

$$a_1(p) = (p^2 + \omega_a^2)a_2(p) \tag{5.74}$$

or the admittance function $Y_1(p)$ has poles at $p = \mp j\omega_a$. That is,

$$Y_1(p) = \frac{b_1(p)}{a_1(p)} = \frac{b_1(p)}{(p^2 + \omega_a^2)a_2(p)} \tag{5.75}$$

By extracting the poles at $= \mp j\omega_a$, $Y_1(p)$ can be written as

$$Y_1(p) = \frac{k_bp}{p^2 + \omega_a^2} + Y_2(p) \tag{5.76}$$

where

$$Y_2(p) = \frac{b_2(p)}{a_2(p)} \tag{5.77}$$

In the above formulation $a_2(p)$ and $b_2(p)$ are found as

$$a_2(p) = \frac{a_1(p)}{p^2 + \omega_a^2}$$
$$b_2(p) = \left[\frac{1}{p^2 + \omega_a^2}\right][b(p) - (k_bp)a_2(p)] \tag{5.78}$$

Partial synthesis of $Y_2(p)$ is depicted in Figure 5.28.

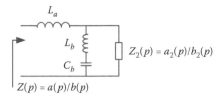

FIGURE 5.28
Extraction of the poles at $p = \pm j\omega_a$ from the admittance function $Y_2(p)$.

Finally, we set

$$Z_2(p) = \frac{a_2(p)}{b_2(p)} = L_c p + Z_3(p) \tag{5.79}$$

where

$$Z_3(p) = \frac{a_3(p)}{b_3(p)}$$

$$L_c = \frac{a_2(1)}{b_2(1)} \tag{5.80}$$

and

$$a_3(p) = a_2(p) - (L_c p)b_3(p)$$
$$b_3(p) = b_2(p) \tag{5.81}$$

It should be noted that, in the MATLAB environment, a polynomial $P(x) = p_1 x^n + p_2 x^{n-1} + \cdots + p_n x + p_{n+1}$ of degree n is described by means of a vector P that includes all the coefficients $\{p_1, p_2, \ldots, p_n, p_{n+1}\}$ such that

$$P = [p(1)p(2)\ldots p(n)p(n+1)] \tag{5.82}$$

Furthermore, the norm of a vector P is defined as

$$norm(P) = \sqrt{p^2(1) + p^2(2) + \cdots + p^2(n) + p^2(n+1)} \tag{5.83}$$

Based on the above MATLAB notation, $a_2(1)$ and $b_2(1)$ are the leading coefficients of polynomials $a_2(p)$ and $b_2(p)$, respectively. Thus, partial synthesis of Equation 5.63 is completed, as shown by Figure 5.23a.

The above realization process of the finite transmission zero ω_a is called the "Brune section extraction." In this form, one of the inductors L_a or L_c may have a negative value. However, it may be eliminated by introducing a coupled coil with mutual inductance $M > 0$, as depicted in Figure 5.29.

At first glance, it is clear that the way the inductors L_a, L_b, and L_c are derived must satisfy the following equation [66,67]:

$$L_a.L_b + L_a.L_c + L_b.L_c = 0 \tag{5.84}$$

In this regard, the negative inductor is removed employing (5.85).

$$L_1 = L_a + L_b > 0 \tag{5.85}$$

$$L_2 = L_b + L_c > 0 \tag{5.86}$$

$$M = L_b > 0 \tag{5.87}$$

Its realization is depicted in Figure 5.29.

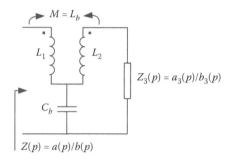

FIGURE 5.29
Realization of negative inductor.

In the following subsections, we summarize the upgraded version of our high-precision synthesis algorithms introduced in References 5, 6, and 67 to include the extraction of finite transmission zeros.

5.4.2 MATLAB® Implementation of Zero Shifting Algorithm

In cascade synthesis, transmission zeros are realized as the poles of the immittance function at each step. In this regard, DC transmission zeros are realized either as series capacitors, which are the poles of an impedance functions at $p = 0$, or as shunt inductors, which are the poles of admittance functions at $p = 0$. In a similar manner, in Brune synthesis, a finite frequency transmission zero at ω_a is realized by introducing a pole at that frequency into the admittance function in the second step of the synthesis. In this regard, synthesis algorithm can be implemented within three steps. In step 1, at a given finite frequency transmission zero ω_a, an inductance L_a is extracted from the given impedance function $Z(p) = a(p)/b(p)$, to introduce a zero into the remaining impedance function $Z_1(p) = a_1(p)/b_1(p)$, which is the pole of the admittance function $Y_1(p) = b_1(p)/a_1(p)$. This fact is described by Equations 5.67 through 5.76. Thus, in MATLAB, first we generate

$$Z(j\omega_a) = R_a + jX_a \tag{5.88}$$

In this step, R_a must be zero since the even part of the given impedance is zero at $p = \mp j\omega_a$, as specified by Equations 5.67 and 5.68. However, due to numerical computational errors, R_a will not be exactly zero, rather, it will be a small number. In this regard, we define an algorithmic zero such that $\varepsilon_{zero} = 10^{-m}$; $m > 0$. If $R_a \leq \varepsilon_{zero}$ then we can go ahead with the synthesis. Otherwise, the synthesis algorithm must be stopped; meaning that the given impedance does not include a finite transmission zero at ω_a. In this case, if $R_a \leq \varepsilon_{zero}$, then by Equation 5.69 we set

$$L_a = \frac{X_a}{\omega_a} \tag{5.89}$$

In Equation 5.89, the value of L_a may be positive or negative.

In the second step, we generate the numerator and denominator polynomials of $Z_1(p)$, as in Equation 5.72, as follows:

$$a_1(p) = a(p) - pL_a b(p)$$
$$b_1(p) = b(p) \tag{5.90}$$

At this point, we should mention that degree of $a_1(p)$ is increased by 1. Then, the numerator polynomial of $Z_2(p) = a_2(p)/b_2(p)$ is determined, as in Equation 5.78

$$a_2(p) = \frac{a_1(p)}{p^2 + \omega_a^2} \tag{5.91}$$

Employing Equation 5.76 residue k_b is found.

$$k_b = \frac{b(j\omega_a)}{(j\omega_a)a_2(j\omega_a)} \tag{5.92}$$

k_b must be a real positive number. At this point, we completed the extraction of the finite transmission zero ω_a as a series resonance circuit in shunt configuration, as shown in Figure 5.28, with element values

$$L_b = \frac{1}{k_b} > 0$$

$$C_b = \frac{1}{\omega_a^2 L_b} > 0 \tag{5.93}$$

The last computation line of this step is to determine the denominator polynomial $b_2(p)$ of $Z_2 = a_2(p)/b_2(p)$ such that

$$b_2(p) = \left[\frac{1}{p^2 + \omega_a^2}\right][b(p) - (k_b p)a_2(p)] \tag{5.94}$$

In the MATLAB environment, division by $(p^2 + \omega_a^2)$ is performed using the function

$$[q, r] = deconv(u, v)$$

deconv performs the polynomial division operation $u(p)/v(p)$, resulting in the quotient polynomial $q(p)$ and the remainder $r(p)$. Obviously, in computing $a_2(p)$ and $b_2(p)$ using Equations 5.92 through 5.94, remainders must be zero. However, due to accumulated numerical errors for both operations, remainders may be small numbers but not exactly zeros. In this case, we can compare the norm of remainders with the algorithmic zero if they are less than ε_{zero}. If so, then we can go ahead with step 3, otherwise the algorithm is stopped indicating that extraction of the given finite transmission zero is not possible. At this step, the degree of polynomial $a_2(p)$ is $n - 1$ and that of $b_2(p)$ is $n - 2$. In this case, the impedance function $Z_2(p) = a_2(p)/b_2(p)$ must include a pole at infinity.

In step 3, the remaining pole at infinity of $Z_2(p) = a_2(p)/b_2(p)$ is removed as an inductor

$$L_c = a_2(1)/b_2(1) \tag{5.95}$$

If L_a is negative, then L_c must be positive. Otherwise, L_c may take a negative value. Upon completion of this process, we end up with the remaining positive real impedance $Z_3(p) = a_3(p)/b_3(p)$. In this case,

$$a_3(p) = a_2(p) - [L_c p][b_3(p)]$$
$$b_3(p) = b_2(p) \tag{5.96}$$

In the above equation set, degree of $a_3(p)$ must be $n_{a3} = n - 3$ or $n_{a3} = n - 2$ and degree of $b_3(p)$ must be $n_{b3} = n - 2$.

It should be mentioned that the above Brune or equivalently Type-C section extraction process is also known as zero shifting and it is programmed in MATLAB under the functions called

$$[Even\_Part, La] = Zero\ shifting\_Step1(a, b, wa, eps\_zero)$$

$$[Lb, Cb, kb, a2, b2, r\_norm] = Zero\ shifting\_Step2(wa, La, a, b, eps\_zero)$$

and

$$[Lc, a3, b3, L1, L2, M] = Zero\ shifting\_Step3\ (La, Lb, a2, b2)$$

In Step 4, we plot the result of zero shifting synthesis using our general purpose MATLAB plot function

$$Plot\_Circuit4(CT, CV)$$

where CT designates the type of component to be drawn and CV is the value of that component.

On the basis of our nomenclature, $CT(i) = 1$ describes a series inductor L_i. Components of a series resonance circuit $(pL + 1/pC)$ in shunt configuration are described by $CT(i) = 10$ and $CT(i + 1) = 11$ referring to inductor L and capacitor C, respectively. Terminating resistor R_T is designated by $CT(i) = 9$ with the component value $CV(i) = R$. For the Brune section, we use the following MATLAB codes:

$$CT(1) = 1, CV(1) = L_a; \quad CT(2) = 10, CV(2) = L_b$$
$$CT(3) = 11, CV(3) = C_b; \quad CT(4) = 1, CV(4) = L_c \tag{5.97}$$

For $n = 2$, terminating resistor R_T is given by

$$CT(5) = 9, CV(5) = R_T \tag{5.98}$$

If $n > 2$, then in Step 5, the remaining impedance $Z_3(p) = a_3(p)/b_3(p)$ is synthesized using our high-precision LC ladder synthesis algorithm and the final circuit schematic is plotted. It is noted that if the driving point immittance function $K(p) = a(p)/b(p)$ is specified as admittance, then we should flip it over to make it impedance $Z(p) = b(p)/a(p)$ to be able to apply the zero shifting synthesis algorithm. In this case, $Z(p)$ may have a pole at infinity and a pole at DC. If it is so, then poles of $Z(p)$ are extracted as a Foster function as follows:

$$Z(p) = \frac{b(p)}{a(p)} = L_x p + \frac{1}{C_x p} + Z_1(p) \tag{5.99}$$

All the above steps are gathered under the major MATLAB function called

$$[CT,CV,L1,L2,M] = Synthesis\ by\ Transzeros\ (KFlag,WZ,ndc,a,b,eps\_zero)$$

This function synthesizes the immittance function $F(p) = a(p)/b(p)$ as described above. If the input variable $KFlag = 1$ is selected, $F(p)$ is an impedance; if $KFlag = 0$ is selected, then $K(p)$ is an admittance. $K(p)$ may include poles at $p = 0$ and/or at $p = \infty$. At the beginning of the synthesis, these poles are extracted, as in Equation 5.99, leaving a minimum reactance function. Then, the total number of nz finite transmission zeros are extracted in a sequential manner, as they are provided by the input vector WZ of size nz. Thereafter, total number of ndc transmission zeros at DC are removed and, finally, the remaining transmission zeros at infinity are extracted using our high-precision synthesis algorithms introduced in References 5–7.

Remark: It should be emphasized that, in the above synthesis process, after each pole extraction, the remaining impedance is corrected using the parametric approach. In this regard, we reconstruct the remaining minimum reactance impedance from its real part. This process is called "impedance correction." The impedance correction process assures the LC ladder topology for the remaining impedance. Therefore, the algorithm introduced above may be called "*Zero Shifting with Impedance Correction*," or, in short, "*ZS-with ImC*." "ZS-with ImC" is able to extract more than 10 Type-C sections with accumulated numerical error less than 10^{-3}. However, straightforward zero shifting synthesis without impedance correction fails due to over/under flows after 3 or 4 Brune section extractions.

5.4.3 Synthesis with Transmission Zeros at DC and Infinity

A minimum immittance function $K(p)$ is specified as a positive real function in rational form as

$$K(p) = \frac{a(p)}{b(p)} = \frac{a_1 p^n + a_2 p^{n-1} + a_3 p^{n-2} + \cdots + a_n p + a_{n+1}}{b_1 p^n + b_2 p^{n-1} + b_3 p^{n-2} + \cdots + b_n p + b_{n+1}} \tag{5.100}$$

where the denominator polynomial $b(p)$ is the degree of n and is free of $j\omega$ poles. For many engineering applications, the aim is to set $a_1 = 0$, which makes $K(p)$ zero as frequency approaches infinity (a transmission zero at infinity).

Transmission zeros of a minimum immittance function are specified by its zeros and poles as well as the zeros of its even part, which is defined as

$$R_{even}(p) = \frac{1}{2}[K(p) + K(-p)] = \frac{A(p)}{B(p)} = \frac{A_1 p^{2n} + A_2 p^{2(n-1)} + \cdots + A_n p^2 + A_{n+1}}{B_1 p^{2n} + B_2 p^{2(n-1)} + \cdots + B_n p^2 + B_{n+1}} \tag{5.101}$$

If $R_{even}(p)$ includes transmission zeros only at DC (meaning at $p = 0$) and infinity (meaning $p = \infty$), then $A(p)$ is simplified as

$$A(p) = A_0 p^{2(ndc)}; \quad n - ndc \geq 0; \quad (-1)^{ndc} A_0 > 0 \tag{5.102}$$

or

$$P(p^2) = \frac{A_0 p^{2(ndc)}}{B_n p^{2n} + B_2 p^{2(n-1)} + \cdots + B_n p^2 + B_{n+1}} \geq 0 \tag{5.103}$$

where $A_0 = A_{n-ndc+1}$ and all the other terms A_ks of Equation 5.103 are zero.

In the above formulation, integer "n" designates the total number of reactive elements when $K(p)$ is synthesized as a lossless two-port [N] in resistive termination. Integer "ndc" is the count of DC transmission zeros that are realized as series capacitors and shunt inductors in two-port [N]. On the other hand, integer $n_\infty = n - ndc > 0$ is the count of transmission zeros at infinity that are realized as series inductors and shunt capacitors, as shown in Figure 5.30.

It should be mentioned that the $ndc = 0$ case results in a low-pass LC ladder with all series inductors and shunt capacitors. Similarly, the $ndc = n$ case yields a high-pass ladder with all series capacitors and shunt inductors. Therefore, we refer to the series inductors and shunt capacitors as low-pass reactive elements or, in short, low-pass elements. Similarly, series capacitors and shunt inductors are referred to as high-pass reactive elements or, in short, high-pass elements.

If $ndc > 0$ then the last term a_{n+1} of $a(p)$ must be zero (i.e., $a_{n+1} = 0$), introducing zeros of transmission at zero.

Synthesis of $K(p)$ can be carried out in a sequential manner by extracting transmission zeros step by step.

In cascade synthesis, at a given pole location $p = p_i$, a transmission zero is realized by removing that pole from the given immittance function. Obviously, in this section, we are only concerned with transmission zeros at $p = 0$ and $p = \infty$. For example, assuming $ndc > 0$, first, transmission zeros at DC can be removed one by one. In this case, a_{n+1} must be zero. Therefore, at step 1, first $K(p)$ is flipped over as $H(p) = 1/K(p) = b(p)/a(p)$ to introduce a pole in $H(p)$. Then, a transmission zero at DC is extracted by removing the pole of $H(p)$ at $p = 0$. At this step,

$$H(p) = \frac{b(p)}{a(p)} = \frac{k_1}{p} + F_r(p) \tag{5.104}$$

where

$$F_r(p) = \frac{R(p)}{a_r(p)} \tag{5.105}$$

The remainder $R(p)$ and the denominator polynomial $a_r(p)$ are obtained after proper degree cancellations. At the end of this process, $F_r(p)$ may not be a minimum function. Continuing the pole extraction process with care, after ndc steps, the remaining immittance function $F_r(p)$ must include transmission zeros only at infinity. In this case, $H(p) = 1/K(p)$ is expressed as

FIGURE 5.30
Synthesis of $K(p)$ as a lossless two-port in resistive termination.

$$H(p) = \frac{k_1}{p} + \cfrac{1}{\cfrac{k_2}{p} + \cfrac{1}{\cfrac{\cdots}{\cdots} + \cfrac{\cdots}{\cdots + \cfrac{1}{\cfrac{k_{ndc}}{p} + F_r(p)}}}}$$

(5.106)

From the programming point of view, after each step, remaining positive real function $F_r(p)$ can be renamed or initialized as $F_r = a(p)/b(p)$ by setting $a(p) = R(p)$ and $b(p) = a_r(p)$, so that the same algorithm is employed to remove the poles at $p = 0$ from the inverse function $H_r(p) = b(p)/a(p)$.

As far as actual realization of $H(p)$ is concerned, if the synthesis begins with a minimum reactance impedance $K(p)$, then $H(p) = 1/K(p)$ is an admittance function. Therefore, the first element of the ladder must be a shunt inductor and it is specified by $L_1 = 1/k_1$. In this case, the second element will be a series capacitor with $C_2 = 1/k_2$, and so on. As a rule of thumb, we can say that if the initial minimum function $K(p)$ is an impedance function, then the odd index terms of $H(p)$ of Equation 5.106 will be shunt inductors and even index terms will be series capacitors. Obviously, if ndc is an even integer, the front high-pass section will end with a series capacitor; or if it is an odd integer, then the high-pass section will end in a shunt inductor, as shown in Figures 5.31a and 5.31b, respectively.

However, if $K(p)$ is a minimum susceptance function (or equivalently minimum admittance), its inverse $H(p) = 1/F(p)$ will be an impedance function.

Therefore, in this case, actual synthesis starts with a series capacitor $C_1 = 1/k_1$, and continues with a shunt inductor $L_2 = 1/k_2$, as shown in Figures 5.32a and 5.32b.

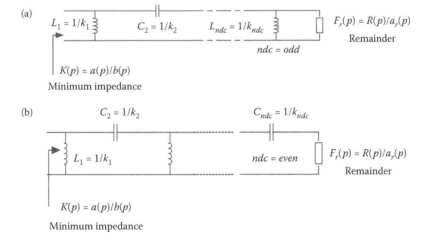

FIGURE 5.31
(a) Synthesis of a minimum impedance $K(p)$ for $ndc = odd$ case. (b) Synthesis of a minimum impedance $K(p)$ for $ndc = even$ case.

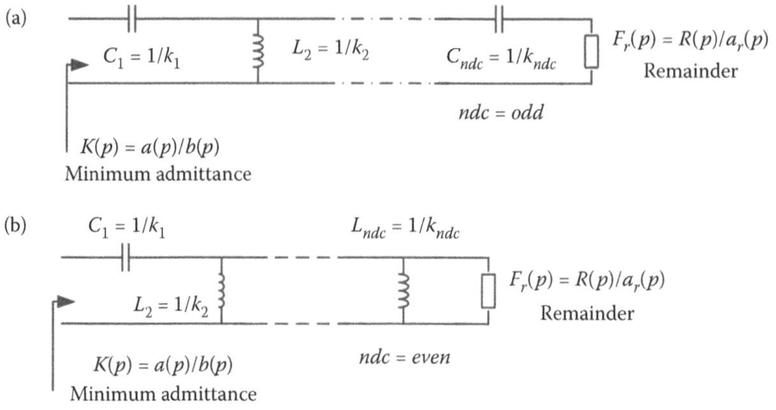

FIGURE 5.32
(a) Synthesis of a minimum admittance $K(p)$ for *ndc = odd* case. (b) Synthesis of a minimum admittance function $K(p)$ for *ndc = even* case.

The remaining low-pass section can be synthesized by extracting transmission zeros at infinity from the immittance function $F_r(p) = a(p)/b(p)$. As mentioned before, $F_r(p) = a(p)/b(p)$ may be a minimum function with $a_1 = 0$ and $b_1 \neq 0$, or it may not be minimum. In this case, $a_1 \neq 0$ and $b_1 = 0$. If $F_r(p)$ is a nonminimum function, then low-pass synthesis directly starts by expressing $F_r(p)$ as

$$\frac{k_{ndc}}{p} + F_r(p) \tag{5.107}$$

(last step of poles extractions at $p = 0$)

$$F_r(p) = \frac{a(p)}{b(p)} = q_1 p + \frac{R(p)}{b_r(p)} \tag{5.108}$$

or the last step is expressed as

$$\frac{k_{ndc}}{p} + q_1 p + \frac{R(p)}{b_r(p)} \tag{5.109}$$

In this case, at the very beginning of the synthesis process, if the starting function $F(p) = a(p)/b(p)$ is a nonminimum reactance function, and if $ndc = even$, then the low-pass section starts with a series inductor $L_{(ndc+1)} = q_1$; and it ends with a shunt capacitor $C_n = q_{n\infty}$ in parallel with a terminating conductance G if $n_\infty = n - ndc$ is an even integer, as shown in Figure 5.33a.

Contrarily, if $F_r(p)$ is a minimum function, then it must be flipped over to be able to extract a pole at infinity from the inverse function $H_r(p) = b(p)/a(p)$ such that

$$\frac{k_{ndc}}{p} + F_r(p) \tag{5.110}$$

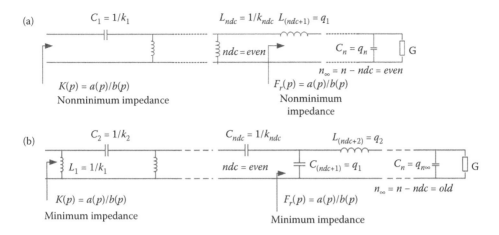

FIGURE 5.33
(a) Synthesis of a nonminimum impedance $K(p)$ for $n_\infty = n - ndc = even$ case. (b) Synthesis of a minimum imped-ance $K(p)$ for $ndc = even$ case and minimum impedance remainder $F_r(p) = R(p)/a_r(p)$ as $n - ndc = odd$ case.

(last step of pole extractions at $p = 0$)

$$\frac{k_{ndc}}{p} + \frac{1}{H_r(p)} = \frac{k_{ndc}}{p} + \frac{1}{q_1 p + (R(p) / a_r(p))} \tag{5.111}$$

In this case, series capacitor C_{ndc} is followed by a shunt capacitor $C_{(ndc+1)}$, as shown in Figure 5.33b, and so on.

Just for the sake of completion, let us mention that a sequential cascaded connection of series capacitors and shunt inductors constitute a high-pass ladder network. Therefore, these elements are called "High-pass Elements—HE" (i.e., shunt inductors and series capacitors). Similarly, a connection of series inductors and shunt capacitors form a low-pass ladder. Therefore, these elements are called "Low-pass Elements—LE."

Even though the extraction process looks straightforward throughout the numerical computations, one may end up with severe accumulation errors that destroy the ladder form of [N] by introducing nonzero terms in $A(p)$ beyond $A_0 p^{2(ndc)}$. Unfortunately, this is a classical problem of network synthesis.

In our previous work [5], for the $ndc = 0$ case, we introduced a "parametric method" to overcome severe accumulation errors by enforcing the analytic ladder form of Equation 5.103 and regenerating $F_r(p)$ from its even part at each step. In this approach, we start with a minimum function $F(p) = a(p)/b(p)$ with $a_1 = 0$. In this case, even part of $K(p)$ is given by

$$P(p^2) = \frac{A_0}{B_n p^{2n} + B_2 p^{2(n-1)} + \cdots + B_n p^2 + B_{n+1}} \tag{5.112}$$

The first transmission zero at infinity is completely removed from the inverse function $H(p) = 1/F(p) = b(p)/a(p)$ such that

$$H(p) = q_r p + F_r(p)$$
$$F_r(p) = R(p)/a_r(p) \tag{5.113}$$

The remaining function $F_r(p)$ must be minimal and possess an even part, as in Equation 5.112 of degree $(n - 1)$. However, due to accumulation errors, numerator polynomial $A(p)$ of the even part $R_{even}(p) = [1/2][F_r(p) + F_r(-p)]$ may have very small coefficients but not exactly zero beyond $A_{n+1} = A_0$. In this case, by defining an algorithmic zero such as $\varepsilon_{zero} = 10^{-m}$; $m > 1$, we can set all nonzero terms to zero if they are smaller than ε_{zero}. In this way, we force the even part $R_{even}(p) = [1/2][F_r(p) + F_r(-p)]$ to be the low-pass ladder form, then regenerate $F_r(p)$ by means of parametric approach from the corrected even part, as specified by Equation 5.112. This process continues until all the transmission zeros are extracted.

In Reference 5, it was shown that one can synthesize a low-pass ladder up to 40 elements from a given driving point function $K(p)$ with a computational resolution better than 10^{-1}, which may be acceptable for many practical problems.

5.4.4 MATLAB Implementation: Extraction of Transmission Zero at DC

In the previous sections, we generated a minimum function that has built in DC transmission zeros. These zeros can be extracted by removing the poles at $p = 0$ from the inverse function as follows.

Initially, the full form of a minimum immittance function $K(p)$ is specified as

$$K(p) = \frac{a(p)}{b(p)} = \frac{a_1 p^n + a_2 p^{n-1} + \cdots + a_n p + a_{n+1}}{b_1 p^n + b_2 p^{n-1} + \cdots + b_n p + b_{n+1}} \tag{5.114}$$

with $a_1 = 0$, which indicates that there is a transmission zero at infinity; and $a_{n+1} = 0$ indicating that there is at least one transmission zero at infinity.

The inverse of this function is given by

$$H(p) = \frac{1}{K(p)} = \frac{b_1 p^n + b_2 p^{n-1} + \cdots + b_n p + b_{n+1}}{a_2 p^{n-1} + \cdots + a_n p} \tag{5.115}$$

Thus, a DC transmission zeros is extracted by removing the pole $p = 0$ as follows:

$$H(p) = \frac{b(p)}{a(p)} = \frac{k_r}{p} + F_r(p)$$

$$F_r(p) = \frac{R}{a_r(p)} \tag{5.116}$$

where

$$k_r = \lim_{p \to 0} p H(p) = \frac{b_{n+1}}{a_n}$$

$$a_r(p) = \frac{1}{p} a(p) = a_2 p^{n-2} + a_3 p^{n-3} + \cdots + a_{n-1} p + a_n \tag{5.117}$$

$$R(p) = \frac{1}{p} [b(p) - k_r a_r(p)] = R_1 p^{n-1} + R_2 p^{n-2} + \cdots + R_{n-1} p + R_n$$

Here, it is important to note the following points:

a. If $ndc > 1$, after we remove the first transmission zero at DC, the last term R_n of $R(p)$ must be zero to introduce the leftover transmission zeros at DC.
b. Due to cancellation of the common term "p," which appears in $R(p)$ and is already built in $a_r(p)$, degree of $a_r(p)$ is $(n - 2)$.
c. Degree of $R(p)$ is $(n - 1)$, since the leading term of $R(p)$ is $(1/p)b_1 p^n = b_1 p^{n-1}$.
d. $F_r(p) = R(p)/a_r(p)$ is not a minimum function since $\lim_{p \to \infty} F_r(p) \to (R_1/a_2)p \to \infty$ indicating that $F_r(p)$ has a pole at ∞.
e. The even part of $F_r(p)$ must yield a ladder form, as in Equation 5.112, with the total of $(ndc - 1)$ transmission zeros at DC. However, due to error accumulations this may not be the case.

The above introduced essential points are programmed under the following MATLAB functions:

- Function $[A,B] = even\_part(a,b)$ generates the even part of a positive real function $K(p) = a(p)/b(p)$.
- Function $[kr, R, ar] = Highpass\_Remainder(a, b)$ extracts a pole at DC $(p = 0)$ with residue k_r, as detailed by Equation 5.117. Once extraction is completed, for the next step, the remainder $R(p)$ is set to $a(p)$ and $a_r(p)$ is set to $b(p)$ to continue for the follow-up steps of the synthesis.

In the following section, we present a fine process to remove a pole at infinity from a given proper function $F(p)$.

5.4.5 MATLAB Implementation: Extraction of a Pole at Infinity

A nonminimum function $F(p)$ with a pole at infinity is described as $F(p) = a(p)/b(p)$, such that

$$a(p) = a_1 p^n + a_2 p^{n-1} + \cdots + a_n p + a_{n+1}$$

and

$$b(p) = 0p^n + b_2 p^{n-1} + \cdots + b_n p + b_{n+1}$$

It should be noted that a pole at infinity is introduced when the coefficient b_1 is zero (i.e., $b_1 = 0$). In this case, $F(p)$ is expressed as

$$F(p) = k_\infty p + F_{min}(p)$$

where the minimum function is described as

$$F_{min}(p) = \frac{a_{min}(p)}{b_{min}(p)}$$

The residue k_∞ is given by

$$k_\infty = \lim_{p \to \infty} \frac{1}{p} \frac{a(p)}{b(p)}$$

$$= \lim_{p \to \infty} \frac{a_1 p^n + a_3 p^{n-2} + \cdots + a_n p + a_{n+1}}{b_2 p^{n-1} + \cdots + b_n p + b_{n+1}} = \frac{a_1}{b_2} \qquad (5.118)$$

Once a pole at infinity is extracted, the remaining function is expressed as

$$K(p) = \frac{a(p)}{b(p)} = k_\infty p + \frac{R(p)}{b_r(p)}$$

where the remainder $R(p)$ is specified as

$$R(p) = a(p) - [k_\infty p][b(p)] \qquad (5.119)$$

The degree of $b_r(p)$ is $(n-1)$. In order to end up with degree reduction, first two terms of $R(p)$ must be zero introducing left over transmission zeros at infinity. To execute the above computations, we developed the following MATLAB functions:

- Function [kinf, R, br] = removepole_atinfinity(anew, bnew) removes a pole at infinity as described by Equation 5.113. Then, $F_r(p) = R(p)/b_r(p)$ becomes a minimum positive real function.
- Function [a1, b1, ndc] = check_immitance(a_min, b_min).
- Checks the immittance function $F_{min}(p) = a_{min}(p)/b_{min}(p)$ if it satisfies an LC bandpass ladder form. This function first generates the even part $R(p) = A(p)/B(p)$, then compares all the coefficients of $A(p)$ with a selected small number $\varepsilon_{zero} = 10^{-m}$. It then decides whether it is a ladder or not. If it is a ladder within a precision of 10^{-m}, then from the corrected real part, $F_{min}(p)$ is regenerated as an exact ladder using the parametric approach.
- Function [a_G,b_G,ndc] = kinfFmin_ToF(kinf,a_min,b_min) generates the corrected nonminimum function from the given pole at infinity described by the residue k_∞ and corrected minimum function $F_{min}(p) = a_{min}(p)/b_{min}(p) = R(p)/b_r(p)$.

5.4.6 Algorithm: General Rules of LC Ladder Synthesis

For the sake of easy explanation, we assume here that DPI $F(p) = a(p)/b(p)$ is a minimum reactance (impedance) function without loss of generality.

1. If the synthesis process starts with the extractions of high-pass elements, the first element must be a shunt inductor, since $F(p)$ is a minimum reactance function expressed by means of its inverse with a pole at $p = 0$ as

$$H(p) = \frac{1}{F(p)} = \frac{k_1}{p} + F_r(p) = \frac{1}{L_1 p} + F_r(p).$$

2. Obviously, the second high-pass element is a series capacitor if $ndc > 1$.

3. In general, we can state that, if the synthesis of a minimum reactance function starts with the extraction of high-pass elements, the odd indexed terms are shunt inductors and the even indexed terms are series capacitors.

4. During the extraction process of odd indexed elements, the remaining function $F_r(p) = R(p)/a_r(p)$ is a nonminimum admittance function. This is useful information, helping to refine the coefficients of the numerator and denominator polynomial $F_r(p)$ to yield a high-precision LC ladder, employing the MATLAB function *Ladder_Correction*.

5. During the extraction process of even indexed elements, the remaining function $F_r(p) = R(p)/a_r(p)$ is a minimum reactance function. This is also useful information, helping to refine coefficients of the numerator and denominator polynomial $F_r(p)$ to yield a high-precision LC ladder, using the MATLAB function *check_immitance*.

6. After extracting all transmission zeros at $p = 0$ as shunt inductors and series capacitor within the "*ndc*" steps, we can start extraction of low-pass elements from the remaining function $F_r(p)$ as series inductors and shunt capacitors in $n_\infty = n - ndc$ steps.

7. If $ndc = even$, synthesis of the low-pass section starts with an impedance function.

8. If $ndc = odd$, synthesis of the low-pass section starts with an admittance function.

9. No matter what the ndc is, for both odd and even indexed components, high-pass circuit elements of the *LC* ladder network can be extracted employing our MATLAB function

$$[kr, R, ar] = Highpass\_Remainder(a, b)$$

where the immitance function to be synthesized is described as

$$K(p) = \frac{a(p)}{b(p)} = \frac{k_r}{p} + F_r(p)$$

such that

$$F_r(p) = R(p)/a_r(p).$$

10. Whether it is minimum or not, the remaining function $F_r(p) = R(p)/a_r(p)$ is redefined as $K(p) = a(p)/b(p)$ by setting $a(p) = R(p)$, and $b(p) = a_r(p)$. Then, all the high-pass elements of the LC ladder network can be extracted one by one within a loop that runs on index "i" up to "ndc."

11. Once the high-pass section is constructed, the resulting residual function $F_r(p) = R(p)/a_r(p)$ includes only transmission zeros at infinity.

12. If $F_r(p) = R(p)/a_r(p) = K(p) = a(p)/b(p)$ is a minimum function, then our MATLAB function $q = LowpassLadder\_Yarman(a, b)$ completely removes the transmission zeros at infinity as series inductors and shunt capacitors. On the other hand, if $K(p) = a(p)/b(p)$ is not a minimum function, then its inverse $H(p) = b(p)/a(p)$ must be

minimum. In this case, $q = LowpassLadder\_Yarman(b, a)$ completely removes transmission zeros at infinity.

13. All the above steps are combined under a MATLAB function

$$[k, q, Highpass\_Elements, Lowpass\_Elements] = GeneralSynthesis\_Yarman(a, b)$$

which completely removes all the transmission zeros at DC and infinity with high precision.

If the starting function is a minimum susceptance (minimum admittance) function, then all the above statements are true with the following provisions: In the high-pass section of the LC ladder, shunt inductors and series capacitors are replaced by their counterparts of series capacitors and shunt inductors, respectively. Similarly, in the low-pass section, shunt capacitors and series inductors are replaced by their counterparts of series inductors and shunt capacitors, respectively.

In immittance synthesis, resulting lossless two port is not unique. Essentially, during the synthesis process, the zeros of the even-part function $R_{even}(p)$ appear as the transmission zeros of the lossless two-port. These zeros are realized as the poles of the immittance function $F_r(p) = a(p)/b(p)$ or $H_r(p) = 1/F_r(p)$ at each step. In the course of synthesis, we are free to extract transmission zeros in any order that we wish. In other words, for the case under consideration, synthesis can start by extracting either a pole at DC (k_1/p) or a pole at infinity ($q_1 p$). This statement is true at each step of the synthesis. For the sake of simplicity, we have taken the liberty of starting with the extraction of the DC transmission zeros first; then continued with the extraction of transmission zeros at infinity. However, we could have done it the other way around or we could have extracted the transmission zeros at infinity and DC in an ad-hoc manner. Hence, we end up with different combinations of circuit topologies yielding a variety of element values. This may be desired from a practical implementation point of view. However, in this book, we first extract finite transmission zeros, then we extract transmission zeros at DC. Finally, synthesis is automatically completed by extracting the transmission zeros at infinity.

All the above synthesis procedures are combined under a MATLAB function called

$$[CTFinal, CVFinal, LL1, LL2, MM] = SynthesisbyTranszeros(KFlag, W, ndc, a, b, eps\_zero)$$

This function completely extracts the transmission zeros of a given immittance function as a ladder and plots the result of synthesis.

5.5 Automated Real Frequency Design of Lossless Two-Ports for Single Matching Problems

In this section, we develop a MATLAB code that automatically constructs a lossless 2-port in resistive termination for single matching problems. The user provides the following inputs to the program.

The measured complex impedance termination of the equalizer is described by means of its real part RA and imaginary part XA over the given actual frequencies FA. The measured data may be provided as a MATLAB text file or as a matrix such that $Ter\_Impedance = [FA\ RA\ XA]$ typed into the program. Throughout computations, we work with normalized frequencies and normalized impedances. Therefore, the user must specify frequency normalization number f_0 and impedance normalization number R_0. At this point, it is assumed that the single matching problem is defined between resistive generator R_0 and a complex impedance $ZA = RA + jXA$ for the specified frequency band $FL \leq f \leq FH$, where FL is the lower edge and FH is the upper edge of the band.

1. The first step of the design is to generate an idealized real frequency line segment solution, which is referred to as the real frequency line segment technique (RF-LST). In this regard, the user selects the following input arguments:

   - A target flat gain level T_0.
   - Number of unknown break points NC in the passband.
   - In the MATLAB codes, the last break point is fixed as zero (i.e., $RB(N) = 0$).
   - The user should decide to work with either high values of break points or low values of break points by setting control flag $sign = 1$ or -1, respectively.
   - The first break point $RB(1)$ could be either included among the unknowns or it is fixed by setting the control flag $ktr = 1$ or 0, respectively.
   - The user decides whether to work with impedance or admittance-based TPG in the course of optimizations by setting the control flag $KFlag = 1$ or 0, respectively.
   - Eventually, initial break points are computed and optimized to reach the target flat gain level T_0.
   - The outcome of the first step is the optimized break points $RB(\omega)$ over the break frequencies WB.
   - We must note that the total number of break points are chosen as $N = NC + 2$.
   - Break frequencies are uniformly distributed within the passband.
   - All the above computations are gathered under a MATLAB function called

     $$[WB, RB, TB] = RFLST\_SingleMatching(NC, ktr, KFlag, sign, T0, FL, FH,$$
     $$FA, RA, XA, f0, R0)$$

2. In the second step of design, line segments are modeled with a nonnegative rational function, as in Equation 5.43. In this regard, the user must specify the following input arguments:

   - n: Degree of the denominator polynomial.
   - ndc: Transmission zeros at DC.
   - WZ: Finite transmission zeros on the real frequency axis.
   - WB and RB: Break frequencies and break points.
   - The outcome of this step is the coefficients of the auxiliary polynomial $c(\omega)$, constant real coefficient of the numerator a_0, and open form of the rational function $P(\omega^2) = AA(\omega^2)/BB(\omega^2)$ and the minimum immittance function $K(p) = a(p)/b(p)$ generated from $P(\omega^2)$ using the parametric approach.
   - All the above computations are gathered under the MATLAB function *"CurveFitting_BreakPoints"* such that

     $$[c, a0, AA, BB, P, a, b] = CurveFitting\_BreakPoints(ktr, n, ndc, WZ, WB, RB)$$

3. The third step of the design process is the final optimization of the TPG using the analytic form of the real part $P(\omega^2)$ obtained in the second step of the design. In this step, DPI is generated using the parametric approach under a MATLAB function

$$\left[a, b, c\_final, a0, WA, PA, QA, TA\right] = FinalOptimization\_$$

Parametric $(KFlag, ktr, T0, n, ndc, WZ, c, a0, NC, FA, RA, XA, FL, FH, f0, R0)$

Input arguments of the above function were already introduced in Steps 1 and 2. The outcome of this step is given as follows:

- a and b describe the driving point impedance $K(p) = a(p)/b(p)$ to be synthesized; c_final is a vector that includes the coefficients of the auxiliary polynomial $c(\omega)$ and it is the final and best solution to single matching problems.
- $a0$ is the constant real coefficient of the numerator of the rational form of the real part $P(\omega)$.
- $K(j\omega) = PA(\omega) + jQA(\omega)$ is specified over the sampling frequencies WA by PA and QA and they are plotted for comparison purposes.
- TA is a vector that includes the final and best form of the TPG of the matched system. Eventually, it is plotted.

4. The final step of the design is the synthesis of $K(p) = a(p)/b(p)$ as a lossless two-port in resistive termination R_{n+1}. This step is programmed under the MATLAB function given below.

$[CTFinal, CVFinal, LL1, LL2, MM] = SynthesisbyTranszeros(KFlag, WZ, ndc, a, b, eps\_zero)$

- In the above function, inputs are specified by the DPI $K(p) = a(p)/b(p)$ with finite transmission zeros given by WZ, with DC transmission zeros specified by ndc. The degree of the polynomials $a(p)$ and $b(p)$ is n.
- eps_zero is the user-specified real number that determines the precision of the component values. We usually set $eps\_zero = 10^{-8}$.
- The outcome of *SynthesisbyTranszeros* is the component types (CT) and the values (CV) of lossless two port. Component types (CT) are listed as follows:

  $CT = 1 > Series\ Inductor$

  $CT = 7 > Shunt\ Inductor$

  $CT = 8 > Shunt\ Capacitor$

  $CT = 2 > Series\ Capacitor$

  $CT = 9 > Terminating\ Resistor$

  Series R//L//C: $CT = 4 > Inductor, CT = 5 > Capacitor, CT = 6 > Resistor,$

  Shunt R + L + C: $CT = 10 > Inductor, CT = 11 > Capacitor, CT = 12 > Resistor,$

If we have finite transmission zeros in the real part $P(\omega)$, during the synthesis, these transmission zeros are removed as Brune T-sections, which consist of $La - Lb - Lc$. One of the series inductors of the T is negative. As explained in Sections 5.4.1 and 5.4.2, a negative inductor is removed by a coupled coil $L1$–M–$L2$. In the output arguments, LL1, LL2, and MM include the element values of the coupled coils of each Brune section.

5.6 Computation of Actual Elements

So far, all the algebraic manipulations were completed in the normalized frequency and immittance domain. Therefore, outcome of synthesis results are in the normalized element values. At the end of the design process, we must convert normalized elements to actual elements by denormalization. Hence, the actual elements are given as follows:

An actual capacitor $C_{i-actual}$ is given by

$$C_{i-actual} = \frac{C_i}{\omega_0 R_0} = \frac{C_i}{2\pi f_0 R_0} \tag{5.120}$$

An actual inductor $L_{i-actual}$ is given by

$$L_{i-actual} = \frac{L_i R_0}{\omega_0} = \frac{L_i R_0}{2\pi f_0} \tag{5.121}$$

An actual resistor is given by

$$R_i = R_{n+1} R_0 \tag{5.122}$$

where C_i, L_i, and R_i designate the normalized values of a capacitor, an inductor, and a resistor, respectively.

Employing the outputs of our synthesis package, we develop a MATLAB function called

$$[CVA] = Actual\_Elements(CT, CV, R0, F0)$$

Inputs of the above function are given as follows:

- *CT: Component typevector*
- *CV: Normalized element values*
- *R0: Impedance normalization number*
- *F0: Normalization frequency*

Output *CA* is the actual value of the components.

Let us run a simple example to complete this section.

EXAMPLE 5.7

Develop a MATLAB program that automatically constructs the front-end matching network [F] of a single-stage power amplifier of Figure 5.1 with the load-pull impedance measurement data specified by Table 5.1 of Example 5.1.

Solution

For the problem under consideration, we develop a main program called "*Example5_7.m.*" This program executes the four major steps of immittance-based real frequency technique, namely

- *Step 1: Execution of function RFLST_SingleMatching*
- *Step 2: Execution of function CurveFitting_BreakPoints (Rational curve fitting via RFDCT)*
- *Step 3: Execution of function FinalOptimization_Parametric Final optimization via Parametric Approach*
- *Step 4: Execution of function SynthesisbyTranszeros*

The results are depicted in Figures 5.34 and 5.35.
In Figure 5.34, we compare the following quantities:

a. The optimized TPGs.

b. Optimized break points and rational form of the real part obtained via RFLST and parametric approach.

Hence, the minimum susceptance driving point input admittance $K(p) = a(b)/b(p)$ is found as

$$a = [0 \quad 1.5585 \quad 1.1704 \quad 1.9210 \quad 0.7726 \quad 0.2954]$$

$$b = [1.0000 \quad 0.7510 \quad 1.5803 \quad 0.7568 \quad 0.5030 \quad 0.0859]$$

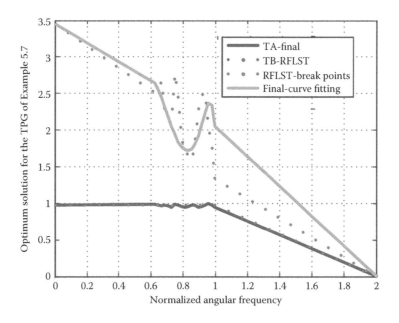

FIGURE 5.34
Optimized transducer power gains and real parts obtained via RFLTS and parametric approach for Example 5.7.

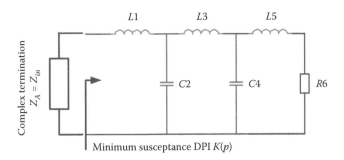

FIGURE 5.35
Synthesis of the minimum susceptance driving point admittance $K(p) = a(p)/b(p)$ for Example 5.7.

Synthesis of $K(p)$ results in the following normalized element values:

	L1	C2	L3	C4	L5	R6
Normalized Component Values	0.641	4.48	0.673	4.51	0.387	0.290

In Figure 5.35, we see the end result of the synthesis of the front-end matching network [F] with normalized element values.

The actual element values are found by denormalizing the capacitor as

$$C_{i-actual} = \frac{C_i}{\omega_0 R_0} = \frac{C_i}{2\pi f_0 R_0}$$

Similarly, actual values of the inductors are determined as

$$L_{i-actual} - \frac{L_i R_0}{\omega_0} = \frac{L_i R_0}{2\pi f_0}$$

Termination resistance is determined as

$$RT = R_{n+1} R_0$$

Thus, we have actual element values as listed below.

$$L1 = 9.63\,\text{nH}, C2 = 26.92\,\text{pF}$$
$$L3 = 10.11\,\text{nH}, C4 = 27.09\,\text{pF}$$
$$L5 = 5.8137\,\text{nH}$$
$$RT = 14.53\,\Omega$$

The above value of terminating resistor complies with the value of $RB(1)$.

The termination can be a level up to 50 Ω by using a transformer, if desired. Finally, we should mention that the front-end matching network goes to the input of the power transistor of Figure 5.1. Therefore, it must be flipped when it is physically connected to the active device, as shown in Figure 5.36.

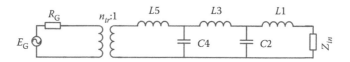

FIGURE 5.36
Filliped matching network to be placed to its physical location.

In Figure 5.36 if the source resistor is 50 Ω, then the transformer ratio n_{tr} must be

$$n_{tr} = \sqrt{\frac{R_G}{R_{n+1}}} = \sqrt{\frac{50}{14.53}} = 3.813$$

In many practical circumstances, we avoid the use of a transformer in the designs. In this case, we can rerun Example 5.7 by fixing $a_0 = 1$. Let us perform Example 5.8 to design the front-end matching network without a transformer.

EXAMPLE 5.8
Repeat Example 5.7 with $a0 = 1$.

Solution
For low-pass designs, we select $ndc = 0$. Then, we can control the far-end termination by fixing the numerator coefficient of Equation 5.43. In detail, at $\omega = 0$, the far-end termination becomes a_0^2. In this case, Equation 5.43 becomes

$$P(\omega^2)\,|_{\omega=0} = a_0^2$$

Hence, by setting $a_0 = 1$, we can fix the far-end termination at the normalized value 1. Therefore, for the actual values, the termination becomes $R_0 = 50$ Ω.
Execution of our MATLAB program Example5_8.m with fixed $a0 = 1$, reveals the following optimization results.

Coefficients of the auxiliary polynomial are given by

$$c = [5.8883 \quad 3.7782 \quad -7.9649 \quad -5.1597 \quad 2.7451]$$

The minimum susceptance driving point admittance $K(p) = a(p)/b(p)$ of the equalizer is given as

$$a = [0 \quad 1.3320 \quad 1.2853 \quad 1.6022 \quad 0.8589 \quad 0.1698]$$

$$b = [1.0000 \quad 0.9650 \quad 1.6124 \quad 1.0400 \quad 0.5224 \quad 0.1698]$$

or

$$K(p) = \frac{1.3320p^4 + 1.2853p^3 + 1.6022p^2 + 0.8589p + 0.1698}{p^5 + 0.9650p^4 + 1.6124p^3 + 1.0400p^2 + 0.5224p + 0.1698}$$

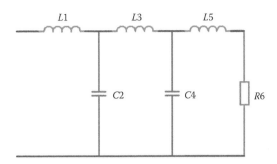

FIGURE 5.37
Synthesis of $K(p)$ for Example 5.8.

The synthesis of $K(p)$ is depicted in Figure 5.37 for Example 5.8.
Normalized and actual component values of Figure 5.37 are given by

	L1	C2	L3	C4	L5	R6
Normalized	0.7508	3.2526	1.2893	1.8048	1.0363	1.0000
Actual	11.27 nH	19.53 pF	19.36 nH	10.84 pF	15.56 nH	50 Ω

The resulting gain function is depicted in Figure 5.38.

As we see from Figure 5.38, minimum TPG is about $T_{min} = 0.94$, which corresponds to −0.25 dB in the passband (330–530 MHz), which more or less agrees with the gain level obtained in Example 5.7.

Now we are ready to fully automate the design of matching networks with lumped elements.

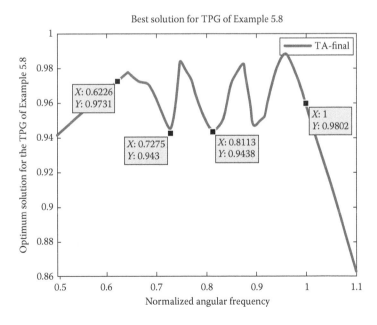

FIGURE 5.38
TPG of the front-end matching network design without transformer.

5.7 Automated Design of Matching Networks with Lumped Elements

As described in the previous sections, all the above computational processes can be combined under a MATLAB function to construct the matching networks with lumped elements for single matching problems. The design engineer only needs to input certain design preferences, such as complexity of the lossless two port (i.e., n: total number of elements, ndc: total number of DC transmission zeros, $WZ = [\omega_{z1} \omega_{z2} \dots \omega_{znz}]$: Fixed number of finite transmission zeros outside of passband, etc.). For this purpose, we developed a MATLAB function called

$$function[CT, CV, CVA, FA, TA, c, a, b, a0] = Compact\ Single\ Matching\ (data)$$

The inputs of this function are listed below.
data are an array that include the following design preferences (i.e., user selected inputs).

- $T0 = data(1)$;
- $NC = data(2)$; $ktr = data(3)$; $KFlag = data(4)$; $sign = data(5)$;
- $FL = data(6)$; $FH = data(7)$; $f0 = data(8)$; $R0 = data(9)$; $F\_unit = data(10)$;
- $n = data(11)$; $ndc = data(12)$; $WZ = data(13)$; $a0 = data(14)$.

The above 14 entries are collected under a MATLAB vector designated by

$$User\ Selected\ Inputs = [.].$$

After the 14th entry, measured complex termination impedance is placed as a vector designated by

$$Imp\_input = [FA \quad RA \quad XA].$$

In other words, array *data* are formed as

$$data = [UserSelectedInputs \quad Imp\_input]$$

where vector *FA* includes the sampling frequencies at which the real part *RA* and the imaginary part *XA* of the complex termination impedance $ZA = RA + jXA$ is measured. In this representation, vectors *FA*, *RA*, and *XA* must have the same sizes. In order to simplify data management at the input, *FA* only contains actual frequencies without units. The measured frequency unit is stored in F_unit, which is stored in $data(10)$. For example, if the frequencies are measured in a Mega Hertz range such as 330, 340, 350 MHz, etc., then vector *FA* contains only plain numbers such as $FA = [330 \quad 340 \quad 350\dots]$.

The detailed explanations are as follows:

- $R0, f0$: Impedance and frequency normalization numbers, respectively. It is noted that $f0$ is a plain number and does not have a unit as in *FA*.
- FL, FH: Lower and upper edges of the passband without units as in *FA*.

- F_{unit}: Frequency measurement unit. For example, if the frequencies are measured in MHz range, then $F\_unit = 10^6$ is selected. If we make the measurements in GHz range, then $F\_unit = 10^9$, etc.
- *T0*: Target flat gain level for the single matching problems.
- *KFlag*: Describes the type of immittance that we carry out computation for. *KFlag* = 1 is selected for impedance-based computations. *KFlag* = 0 is selected for admittance-based computation. This selection is guided by success of the optimization. For example, for capacitive load terminations, *KFlag* = 0 may result in successful optimization. Similarly, for inductive load terminations, *KFlag* = 1 may yield good results in gain optimization.
- *ktr*: This control flag may be used to control the far-end resistive termination for low-pass designs. When we select *ndc* = 0 for low-pass designs, far-end termination is either fixed to be R0, or it is left free to make the optimization more flexible. In this case, for low-pass designs we use a transformer to level up the termination resistor at the far end to R0. For bandpass design (*ndc* > 0) we should set *ktr* = 1.
- *sign*: This control flag determines the initial guess for RFLST if we start with low initials or high initials. Sign = 1 corresponds to high initials. Sign = 0 is for low initials. Choice of signs is dictated by success of gain optimization.
- *NC*: This integer is the total number of unknown break points in the passband.
- *N, ndc, WZ*: These integers determine the complexity of the matching network as described before.

ac: This slack variable is equal to the desired values of *a0*. If *a0* is fixed in advance, then *ac* must be selected as $\sqrt{R_0}$. If a0 is part of the unknowns, then it is initialized as desired.

The outputs of function *"ImmittanceBased_RealFrSingMatch"* are given as the following arrays:

- *CT*: contains the component type of the resulting matching network.
- *CV*: contains the normalized element values of the circuit topology obtained as a result of synthesis.
- *CVA*: contains the actual element values of the resulting matching network.
- *FA*: contains the sampling frequencies to plot the optimized gain TA.
- *TA*: optimized values of the TPG of the matched system in absolute values rather than in decibels.
- *c*: contains the coefficients of the auxiliary polynomial $c(\omega)$.
- *a*: contains the optimized coefficients of the numerator polynomial of immittance $K(p) = a(p)/b(p)$.
- *b*: contains the optimized coefficients of the denominator polynomial of immittance $K(p) = a(p)/b(p)$.
- *a0*: is the leading coefficient of the numerator of the real part of $K(j\omega)$, as in Equation 5.43.
- This function calls the four major steps of the real frequency technique such that-
  Step 1: Idealized solution by RF-LST employing

$$[WB, RB, TB] = RFLST\_SingleMatching(NC, ktr, KFlag, sign, T0, FL, FH, FA, RA,$$
$$XA, f0, R0)$$

Step 2: Curve fitting of the break points by means of a rational function via real frequency direct computational technique by calling

$$[c0, a0, AA, BB, P, a1, b1] = CurveFitting\_BreakPoints(ktr, n, ndc, WZ, WB, RB);$$

Step 3: Final optimization of TPG on $c(i)$ via parametric approach employing

$$[a, b, c, a0, WA, PA, QA, TA] = FinalOptimization\_Parametric(KFlag, ktr, T0, n,$$
$$ndc, WZ, c0, a0, NC, FA, RA, XA, FL, FH, f0, R0);$$

Step 4a: In this step, we complete the Darlington synthesis of the DPI $K(p) = a(p)/b(p)$ with *eps_zero* = 10^{-8} precision. In this regard, we call the synthesis function

$$[CT, CV, LL1, LL2, MM] = SynthesisbyTranszeros(KFlag, WZ, ndc, a, b, eps\_zero).$$

The above function completes the synthesis and automatically plots the network topology with element values.

Step 4b: In this step, actual element values are computed.

Hence, we complete the design in one shot.

Let us run an example to exhibit the utilization of the function *"Immittance Based_RealFrSingMatch."*

EXAMPLE 5.9

Develop a MATLAB program to complete the design of a power amplifier using the load-pull measurement data given in Table 5.1.

Solution

For this purpose we developed a MATLAB program called *"GKYExample5_9b.m,"* as detailed in the Program Listing 5.42.

In program *"Example5_9b.m,"* lines 8–10 read the termination data $Z_{in} = R_{in} + jX_{in}$ of the front-end equalizer [F]. Measurement frequencies are given in a Mega Hertz range within vector *FAin*. Target flat gain level is selected as $T0 = 0.975$ (line 25). RFLST employs $NC = 19$ unknowns break points in the passband (line 26); RFLST fixes the first break point $RB(1)$ and RF-DCT of curve-fitting fixes $a0$ ($ktr = 0$ case in line 26); we work with admittance-based gain formulas ($KFlag = 0$) and start optimization for RFLST with high value of initials (sign = +1). Passband starts at $FL = 330$ MHz and ends at $FH = 530$ MHz (line 27). Lines 27–30 provide the user-selected input arguments to construct [F]. Frequency unit *F_unit* is set as 1 MHz (i.e., *F_unit* = 1e6).

Front-end matching network [F] is constructed in line 35 as a low-pass ladder with $n = 5$ elements. In the design, $a0$ is fixed as $a0 = 1$ by choosing $ktr = 0$. This choice must result in the unity equalizer termination RF6 (i.e., RF6 = 1).

Similarly, lines 46–48 provide the input data to construct back-end equalizer [B] as a low-pass ladder with $n = 5$ elements.

Back-end matching network is constructed in line 69.

Execution of program *"Example5_9b.m"* results in both front-end [F] and back-end [B] matching networks, as shown in Figure 5.39.

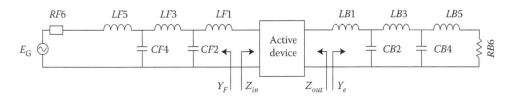

FIGURE 5.39
Complete design of a power amplifier constructed employing LD MOS device RD007.

The actual element values of the above amplifier are listed below.

Front-End	LF1	CF2	LF3	CF4	LF5	RF6
Actual values	11.27 (nH)	19.53 (pF)	19.36 (nH)	10.84 (pF)	15.56 (nH)	50 (Ω)
Back-End	LB1	CB2	LB3	CB4	LB5	RB6
Actual values	9.66 (nH)	15.87 (pF)	14.42 (nH)	7.4 (pF)	0.113 (nH)	50 (Ω)

The resulting TPG of the matching networks [F] and [B] is given in Figure 5.40.

5.8 Design of Interstage Equalizers: Double Matching Problem

In Figure 5.41, a typical two-stage power amplifier design schematic is shown.

In this figure, front-end [F] and back-end [B] matching networks are designed based on the measured input impedance Z_{in1} of Active Device-1 and the measured output impedance Z_{out2} of Active-Device-2, respectively. [F] and [B] can easily be constructed using real frequency single matching design techniques presented in previous sections. However, the design of the interstage matching network (IMN) is tricky, since the TPG for the doubly terminated equalizer is different from that of a single matching one.

Referring to Figure 5.41, the generic design problem of the interstage equalizer is denoted as "Double Matching." Literally speaking, a double matching problem is one that constructs an optimum matching network between the complex generator impedance Z_G and complex load impedance Z_L. Both Z_G and Z_L are non-Foster positive real functions (Figure 5.42).

IMN is designed to achieve TPG over a prescribed frequency band as high and as flat as possible. Maximum flat gain level T_0 may be determined by accessing the gain bandwidth theory of double matching problems developed by Yarman and Carlin [2] and Carlin and Yarman [18].

As far as the design of two-stage power amplifier design is concerned, the output of the active device −1 is modeled as a Thevenin source with internal impedance $Z_G = Z_{out1} = R_{out1} + jX_{out1}$. Similarly, input of the next stage will be the complex load termination $Z_L = Z_{in2} = R_{in2} + jX_{in2}$.

The input and the output impedances of the active devices are determined by load-pull measurements.

The interstage matching network can be described in terms of its DPI $K(p) = a(b)/b(p)$, either from Port-1 or from Port-2, as shown in Figure 5.43.

(a)

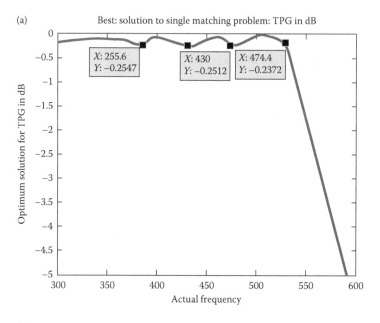

(b)

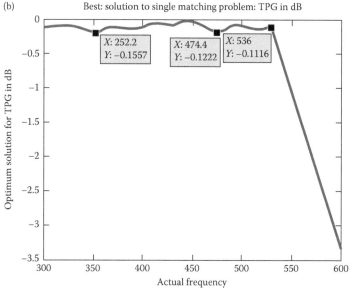

FIGURE 5.40
(a) TPG in dB for [F]. (b) TPG in dB for [B].

The even part driving point immittance $K(p) = a(p)/b(p)$ is given by

$$P(p^2) = \frac{1}{2}[K(p) + K(-p)] = \frac{1}{2}\left[\frac{a(p)b(-p) + a(-p)b(p)}{b(p)b(-p)}\right] = \frac{A(p^2)}{B(p^2)} \tag{5.123}$$

As described in Equations 5.51 and 5.52, for many practical problems, we have the free-dom to choose numerator polynomial as

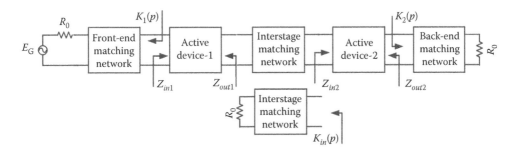

FIGURE 5.41
Design of a two-stage power amplifier.

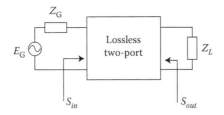

FIGURE 5.42
Definition of double matching problem.

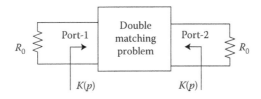

FIGURE 5.43
Description of the lossless matching network for double matching problems.

$$A(p^2) = (-1)^{ndc} a_0^2 p^{2ndc} \prod_{i=1}^{nz} (\omega_i^2 + p^2)^2 = A_0 p^{2ndc} \prod_{i=1}^{nz} (\omega_i^2 + p^2)^2 = f(p)f(-p) \quad (5.124)$$

where

$$f(p) = a_0 p^{ndc} \prod_{i=1}^{nz} (p^2 + \omega_i^2) \quad (5.125)$$

In this case, TPG of the double matched system is given by Reference 4

$$T_G(\omega) = \frac{1 - |G_{22}|^2}{|1 - G_{22}S_{in}|^2} [1 - |S_{in}|^2] \quad (5.126)$$

where

$$G_{22} = \frac{Z_G - 1}{Z_G + 1} = \frac{1 - Y_G}{1 + Y_G}; \quad Y_G = 1/Z_G \tag{5.127}$$

is the unit normalized generator reflectance.

As proven by the main theorem of Yarman and Carlin, S_{in} is given by References 2–4 and 18[*]

$$S_{in} = \eta_B(j\omega)\left[\frac{Z_L - Z(-j\omega)}{Z_L(j\omega) + Z_B(j\omega)}\right] = -\eta_B(j\omega)\left[\frac{Y_L(j\omega) - Y(-j\omega)}{Y_L(j\omega) + Y(j\omega)}\right] \tag{5.128}$$

which is the unit normalized input reflectance of the lossless two-port (*IMN*) while the output port is terminated in complex load.

It is interesting to observe that

$$1 - |S_{in}(j\omega)|^2 = \frac{4PA_L}{[P + A_L]^2 + [Q + B_L]^2} \tag{5.129}$$

which is immittance-based TPG of the conventional single matching problems.

In Equation 5.128, the all pass function $\eta_B(p)$ is given by References 2, 3, and 18

$$\eta_B(p) = \left[\frac{f(-p)}{f(p)}\right]\left[\frac{b(-p)}{b(p)}\right] = (-1)^{ndc}\frac{b(-p)}{b(p)} \tag{5.130}$$

Assuming $K(j\omega)$ is a minimum function, then $Q(\omega) = Hilbert\{P(\omega)\}$.

In the above formulation, the complex generator drives the matching network. Therefore, $T_G(\omega)$ is referred to as the generator-based TPG of the double matching problem.

However, we can turn the matching problem the other way around: feeding the matching network with a complex generator of internal impedance Z_L while terminating it in Z_G at the front-end. This type of formulation of the TPG may be referred to as the load-based TPG. $T_G(\omega)$ of Equation 5.128 is called the generator-based TPG of the double matched system.

Similarly, we can define the load-based TPG of the double matched system as

$$T_L(\omega) = \left[\frac{1 - |L_{11}|^2}{|1 - L_{11}S_{in}|^2}\right][1 - |S_{out}|^2] \tag{5.131}$$

where

$$L_{11} = \frac{Z_L - 1}{Z_L + 1} = \frac{1 - Y_L}{1 + Y_L} \tag{5.132}$$

---

[*] $S_{in}(p)$ is also known as the complex normalized–regularized reflectance in the Yarman and Carlin sense.

is the unit normalized load reflection coefficient and

$$S_{out} = \eta_B(j\omega)\left[\frac{Z_G - Z(-j\omega)}{Z_G(j\omega) + Z(j\omega)}\right] = -\eta_B(j\omega)\left[\frac{Y_G(j\omega) - Y(-j\omega)}{Y_G(j\omega) + Y(j\omega)}\right]$$ (5.133)

is the unit normalized output reflectance. Furthermore,

$$1 - |S_{out}(j\omega)|^2 = \frac{4PA_G}{[P + A_G]^2 + [Q + B_G]^2}$$ (5.134)

In this case, $K(j\omega) = P(\omega) + jQ(\omega)$ is the driving point immittance of the resistively terminated equalizer at the front-end (or at the generator $[Z_G]$ end).

Obviously, generator and load-based defined TPGs must be identical. Thus,

$$T(\omega) = T_G(\omega) \equiv T_L(\omega)$$ (5.135)

As in the single matching case, for double matching problems, the lossless matching network is described from its driving point immittance $K(p)$ either from Port-1 or Port-2. Here, we also assume that $K(p)$ is a minimum function. Therefore, it can be generated using an auxiliary polynomial $c(\omega)$, as in the real frequency direct computational and parametric approach. Coefficients of $c(\omega)$ may be initialized using RFLST and RFDCT for single matching problems. At this point, we have to make a decision to start the double matching problem either from Port-1 or Port-2. Once T_0 is selected, and the coefficients $c_j;\ j = 1, 2, ..., n$ and $A_0 = a_0^2 \geq 0$ are initialized via RFLST and RFDCT, we can generate the generator-based error function ε_G or equivalently the load-based error function ε_L as follows:

$$\varepsilon_G = \left[\frac{1 - |G_{22}|^2}{|1 - G_{22}S_{in}|^2}\right]\left\{\frac{4R_BR_L}{[R_B + R_L]^2 + [X_B + X_L]^2}\right\} - T_0$$ (5.136)

or

$$\varepsilon_L = \left[\frac{1 - |L_{11}|^2}{|1 - L_{11}S_{in}|^2}\right]\left\{\frac{4R_BR_G}{[R_B + R_G]^2 + [X_B + X_G]^2}\right\} - T_0$$ (5.137)

Then the error function is minimized, which in turn yields the realizable DPI $K(p) = a(p)/b(p)$ of the lossless equalizer. Eventually, $K(p) = a(p)/b(p)$ is synthesized yielding the desired lossless matching network in resistive termination R. Finally, resistive termination is replaced by an ideal transformer with transformer ratio $[R = n^2{:}1]$ to level R to a desired value of R_0 when necessary.

Thus, we propose the following design algorithm for the solution of double matching problems.

5.8.1 Algorithm to Construct Lossless Matching Networks under Complex Terminations at Both Ends

Inputs-1 (Design preferences):

1. *KPort*: Selection of design port to describe the lossless matching network to be designed.
   - *Kport* = 1 > Port-1.
   - *Kport* = 2 > Port-2.
2. *KFlag*: Choice of design immittance $K(p)$.
   - *KFlag* = 1 > $K(p)$ is a minimum reactance.
   - *KFlag* = 0 > $K(p)$ is a minimum susceptance.
3. *T0*: Flat gain level for the double matching problem.
4. *NC*: Total number of break points in the passband.
5. *sign*: Initiate RFLST with either high or low levels.
6. *Ktr*: Design with transformer or without transformer.
7. *n*: Total number of circuit elements in the matching network.
8. *nd*: Total number of DC transmission zeros at DC.
9. *Wz*: Location of finite frequency of transmission zeros.
10. *a0*: Leading coefficient of the numerator polynomial of $P(\omega^2)$.
11. R_0: Impedance normalization number.
12. *f0*: Normalization frequency.
13. f_L: Low end of the passband.
14. f_H: High end of the passband.
15. *f_unit*: Actual frequency unit.

Inputs-2:

- [FG RG XG]: Measured complex generator impedance $Z_G(j\omega) = R_G(\omega) + jX_G(\omega)$ as a MATLAB matrix.
- [FL RL XL]: Measured complex load impedance $Z_L(j\omega) = R_L(\omega) + jX_L(\omega)$ as a MATLAB matrix.

Computational Steps:

Step 1: For each complex termination of the double matching problem, find the real frequency line segment solution for single matching problems. Let T_{01} and T_{02} be the idealized flat gain levels of the complex generator and load impedances obtained via RFLST. Further, let $T_0 = minimum\ of\ \{T_{01}, T_{02}\}$. Then, depending on the index, generate the idealized DPI $K(j\omega) = P(\omega) + jQ(\omega)$ for the single matching problem using RFLST.

Step 2: Generate the rational form of $P(\omega)$ of Step 1 with the aid of auxiliary polynomial $c(\omega)$ by means of RFDCT.

Step 3: Optimize the double matching TPG over the passband, which in turn results in the best driving point immittance $K(p) = a(p)/b(p)$.

Step 4: Finally, synthesize $K(p) = a(p)/b(p)$ to end up with the matching network topology with actual elements.

The above steps are gathered under a MATLAB function called

$$"[aL,bL,cL,aL0,WA,TA] = CompactDoubleMatching(data,FAG,RAG,XAG)"$$

In this function, we use generator-based double matching gain. Therefore, the lossless matching network is described at Port-2. Input argument includes "data" user preferences to design a doubly matched lossless equalizer, as in single matching problems. Furthermore, it also includes the actual complex load impedance. Moreover, input arguments *FAG* contain the sampling frequencies of the measured generator impedance; *RAG* is the actual real part and *XAG* is the actual imaginary part of the generator impedance measured using load-pull over the sampling frequencies.

Function *CompactDoubleMatching* calls the following functions:

- [CTL,CVL,CVAL,FLR,TLR,cL0,aL,bL,a0L] = CompactSingleMatching(Input_Data): This function generates the initial guess for the double matching problem solving the single matching problem from the load end. In this case, *Input_Data* is defined for the single matching problems from the load end (Port-2). Thus, we have a rough estimate for the double matching problem under consideration.

- [aL,bL,cL,aL0,WA,TA] = FinalOptimization_DoubleMatching(KFlag,ktr,T0,n,ndc,WZ,cL0, a0L,WG,RG,XG,RL,XL,wL,wH): This objective function optimizes the generator-based double matching gain for which the equalizer is defined from the load end (i.e., Port-2).

Remarks: At this point, we should emphasize that the function *CompactDoubleMatching* can be utilized to optimize both generator and load-based TPG by simply changing the input arguments. More specifically, to optimize generator-based double matching gain, we use the following MATLAB codes:

a. MATLAB design of double matching equalizer using generator-based TPG for which the matching network is described from the load end (Port-2)

  UserSelectedInputs = [T0 NC ktr KFlag sign FL FH f0 R0 F_unit n ndc WZ a0];

  Data_Port2 = [UserSelectedInputs ZAL];% Load impedance data at Port - 2

  [aL,bL,cL,aL0,WA,TA] = CompactDoubleMatching(Data_Port2, FAG,RAG,XAG);

  where *ZAL* refers to actual complex load impedance such that $ZAL(j\omega) = RAL(\omega) + jXAL(\omega)$ and the triplet *FAG, RAG, XAG* describe the measured complex generator impedance as $ZAG(j\omega) = RAG(\omega) + jXAG(\omega)$ over the sampling frequencies stored in the MATLAB array *FAG*.

b. Similarly, we design a double matching equalizer using load-based TPG for which the matching network is described from the generator end (Port-1) using the following MATLAB codes:

  UserSelectedInputs = [T0 NC ktr KFlag sign FL FH f0 R0 F_unit n ndc WZ a0];

Data_Port1 = [UserSelectedInputs ZAG];% generator impedance data at Port-1

[aL,bL,cL,aL0,WA,TA] = CompactDoubleMatching(Data_Port2,FAL,RAL,XAL);

where ZAG refers to actual complex load impedance such that $ZAG(j\omega) = RAG(\omega) + jXAG(\omega)$ and the triplet FAL, RAL, XAL describes the measured complex generator impedance as $ZAL(j\omega) = RAL(\omega) + jXAL(\omega)$ over the sampling frequencies stored in the MATLAB array FAL.

Let us run an example to exhibit the implementation of the above algorithm.

EXAMPLE 5.10

Referring to Figure 5.41, load-pull measurements for the active device LD-MOS RD01 and LD-MOS RD07 by Mitsubishi is given by Table 5.9.

a. Design an interstage equalizer by solving the double matching problem employing the generator-based TPG of Equations 5.126 through 5.130, which defines the lossless two port at Port-2. The passband is selected as 330–530 MHz.
b. Design an interstage equalizer by solving the double matching problem employing the load-based TPG of Equations 5.131 through 5.134, which defines the lossless two-port at Port-1 over the same frequency band of operation as in Part (a).

Solution

For the problem under consideration, we developed a MATLAB main program called *GKY_Example5_10.m*. In this program, first, we read Tables 5.9 and 5.10 as MATLAB matrices AG and AL where AG includes complex generator impedance, which is measured as the output impedance Z_{out1}, and AL includes load impedance, which is measured as the input impedance Z_{in2}, respectively. Then, we run the function "*CompactDoubleMatching*" for part (a) and (b) as required.

Part (a): Generator-based double matching gain design at Port-2.

TABLE 5.9

Load Pull Measurement Results for the Power Transistor LDMOS 01 and LDMOS 07

LDMOS 01			LDMOS 07		
Z_{out1} for RD01			Z_{in1} for RD07		
Fr (MHz)	R_{out1}	X_{out1}	Fr (MHz)	R_{in1}	X_{in2}
330	22.47	−07.84	330	2.32	−3.97
350	16.06	−12.69	350	3.90	−1.98
370	21.23	−12.32	370	1.81	−3.21
390	24.40	−13.45	390	1.93	−2.48
410	22.03	−12.14	410	1.86	−3.87
430	23.50	−14.01	430	1.93	−1.81
450	17.82	−09.52	450	1.05	−1.29
470	18.53	−01.53	470	1.30	−1.58
490	21.06	−04.54	490	1.65	−2.98
510	26.8	−11.06	510	1.64	−2.91
530	21.5	−13.54	530	1.66	−3.33

TABLE 5.10

Load Pull Measurement Results for the Power Transistor LDMOS 07 and
LDMOS 02

	LDMOS 07			LDMOS 02	
	Z_{out1} for RD07			Z_{in2} for RD02	
Fr (MHz)	R_{out1}	X_{out1}	Fr (MHz)	R_{in2}	X_{in2}
330	22.47	−07.84	330	2.32	−3.97
350	16.06	−12.69	350	3.90	−1.98
370	21.23	−12.32	370	1.81	−3.21
390	24.40	−13.45	390	1.93	−2.48
410	22.03	−12.14	410	1.86	−3.87
430	23.50	−14.01	430	1.93	−1.81
450	17.82	−09.52	450	1.05	−1.29
470	18.53	−01.53	470	1.30	−1.58
490	21.06	−04.54	490	1.65	−2.98
510	26.8	−11.06	510	1.64	−2.91
530	21.5	−13.54	530	1.66	−3.33

For this part of the problem, passband is stretched from 330 MHz down to 310 MHz
to prevent sudden gain drops at the lower frequencies. We use the following inputs to
run our MATLAB main program GKYExample5.10:

- $FL = 310$, $FH = 530$, $R0 = 50$ $f0 = 530$, $F\_unit = 10^6$.
- Flat gain level for double matching problem is chosen as $T0 = 0.93$.
- Total number of break points in the passband for single matching problem at
  Port-2 is selected as $NC = 19$. RFLST uses load-pull measured input impedance
  $Zout2$ for the LD MOS RD-07. This impedance is taken as the complex load
  termination ZL for the lossless matching network at Port-2.
- Close examination of Table 5.9 reveals that imaginary parts of the measured
  impedances are capacitive. Therefore, it would be wise to start the RFT design
  with minimum susceptance DPI. Hence, we set $KFlag = 0$.
- RFLST is initiated using high-level break points by setting $sign = +1$.
- In the equalizer topology, we wish to work with five reactive elements as an
  LC-low-pass ladder. Therefore, we have the freedom to select $n = 5$, $ndc = 0$ (no
  DC transmission zero) and $WZ = 0$ (no finite frequency transmission zero).
- We wish to avoid a transformer in the matching network design. Therefore,
  $ktr = 1$ selected with $a0 = 1$.
- We wanted to run the optimization over passband with $M = 2 \times (NC + 2) = 42$
  sampling points.
- At this point, it should be noted that RFLST is run to generate the optimized
  break points to determine the idealized solution for the single matching problem. Then, break points are modeled by rational function using a curve fitting to generate initials for the double matching problem. Eventually, double
  matching optimization is run to end up with best solution for the driving point
  input admittance $K(p) = aL(p)/bL(p)$ at Port 2. Finally, $K(p) = aL(p)/bL(p)$ is synthesized, which yields a lossless double matching equalizer. All steps mentioned here are transparent to the user.

The execution of MATLAB program *GKYExample5_10.m* for Part (a) yields the following solutions:

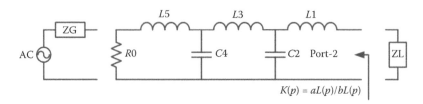

FIGURE 5.44
Synthesis of driving point input admittance $K(p) = aL(p)/bL(p)$.

$$aL0 = 1$$
$$cL = [3.5573e - 05 \quad 3.7047 \quad -0.94220 \quad -4.6560 \quad 0.5.3446]$$
$$aL = [0 \quad 6.0378e + 00 \quad 6.2880e + 05 \quad 1.5991e + 05 \quad 5.4410e + 05 \quad 2.8111e + 04]$$
$$bL = [1 \quad 1.0414e + 05 \quad 2.6486e + 04 \quad 1.3088e + 05 \quad 1.5024e + 04 \quad 2.8111e + 04]$$

Resulting matching network is depicted in Figure 5.44.
Component values of Figure 5.44 is given as follows:

	L1	C2	L3	C4	L5	R6
Actual	2.48 nH	92.6 pF	5.53 nH	23.6 pF	144.172 fH	50 Ω

The resulting TPG performance of the matched system is shown in Figure 5.45.
Part (b): For this part of the double matching problem, the passband is also stretched from both lower and upper edges of the passband such that the lower edge is shifted down to 320 MHz to avoid sudden gain drops at the edge of the bands. It is worth mentioning that in Part (b), RFLST requires low-level initial break points to end up with

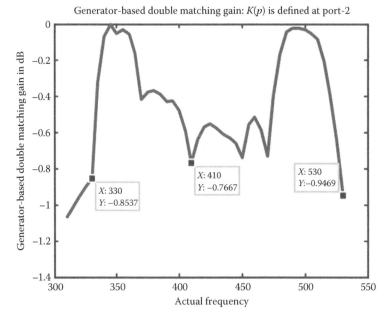

FIGURE 5.45
Generator-based TPG for the double matching problem of Example 5.10, Part (a).

acceptable solutions. Therefore, we set *sign* = –1. Thus, we have the following MATLAB codes to run Part (b):

$FL = 320; FH = 530; f0 = 530; R0 = 50; F\_unit = 1e6;$

$wL = FL/f0; wH = FH/f0;$

$T0 = 0.93;$

$NC = 19; ktr = 0; KFlag = 0; sign = -1;$

$n = 5; ndc = 0; WZ = [0]; a0 = 1; M = 2 * (NC + 2);$

$[aG, bG, cG, aG0, CTG, CVG, LL1G, LL2G, MMG, CVAG, WAL, TAL] =$
$CompactDoubleMatching(Data\_Port1, FAL, RAL, XAL).$

Hence, final results are given as

$$aG0 = 1$$
$$cG = [0.32999 \quad 2.1300 \quad -1.2482 \quad -7.2422 \quad 0.57058]$$
$$aG = [0 \quad 12.146 \quad 78.417 \quad 14.702 \quad 64.183 \quad 3.0304]$$
$$bG = [1 \quad 6.4560 \quad 3.7914 \quad 2.19471 \quad 1.7303 \quad 3.0304]$$

The resulting matching network is depicted in Figure 5.46. Component values of Figure 5.47 are given as follows:

	L1	C2	L3	C4	L5	R6
Actual	1.23 nH	28.26 pF	5.01 nH	98.94 pF	2.3257 fH	50 Ω

Resulting TPG performance of the matched system is shown in Figure 5.47.

It is very interesting to observe that resulting gain performances of Part (a) and Part (b) are very similar to each other, as expected.

Lumped elements are useful in the design of microwave filters, matching networks and amplifiers up to frequencies of a few GHz. Beyond approximately 2 or 3 GHz, the physical size of lumped components becomes comparable with that of the operating wavelength of microwave signals to be amplified or processed. In this case, use of distributed circuit elements, such as equal length transmission lines (also called commensurate transmission lines), is inevitable. In this regard, one may construct a matching network with unit elements (UE), series or shunt stubs, etc., and its DPI can be expressed in a new complex variable $\lambda = \Sigma + j\Omega$. The new complex variable λ is known as "Richards's variable."

Realizability conditions of an immittance or reflectance function are expressed in Richards's variable as same as those established in the classical Laplace variable $p = \sigma + j\omega$.

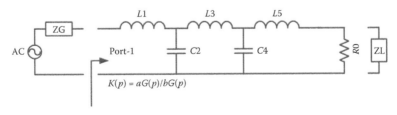

FIGURE 5.46
Load-based TPG design for the double matching problem of Example 5.10, Part (b).

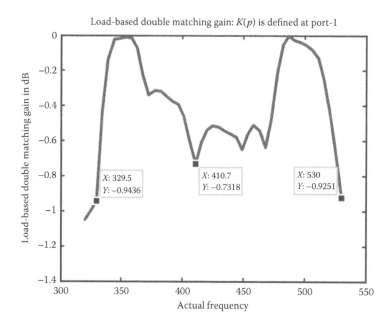

FIGURE 5.47
Load-based TPG design for the double matching problem of Example 5.10, Part (b).

In the next section, we will summarize the essential points to construct matching networks with commensurate transmission lines.

5.9 Matching Networks Constructed with Commensurate Transmission Lines

A Richards's DPI $K(\lambda) = a(\lambda)/b(\lambda)$ is a positive real rational function expressed in complex Richards's variable $\lambda = \Sigma + j\Omega$ instead of classical complex frequency or Laplace variable $p = \sigma + j\omega$. It possesses all the mathematical properties of a positive real function $K(p) = a(p)/b(p)$ described in complex p-Plane.

Complex Richards's plane $\lambda = \Sigma + j\Omega$ is a transformed domain defined on the complex surface $p = \sigma + j\omega$, which is obtained under a tangent hyperbolic mapping $\lambda = \tanh(p\tau)$.

In circuit theory, it is well established that any driving point positive real immittance function $K(p) = a(p)/b(p)$ can be synthesized as a lossless two port in resistive termination in *p-domain* using lumped circuit elements such as *p-domain* inductive impedances $Z_L(p) = pL_L$, capacitive admittances $Y_C = pC_L$, and resistor R, where L_L, C_L are the lumped inductors and capacitors, respectively. Similarly, a Richards's driving point immitance function can be synthesized as a lossless two port in resistive termination using Richards's unit elements (UE) in cascade configuration, Richards's inductive impedances $Z_L(\lambda) = \lambda L_\lambda$ and Richards's capacitive admittances $Y_C(\lambda) = \lambda C_\lambda$, as shown in Figure 5.48.

UEs in cascade configuration can be realized as equal length TEM-transmission lines with characteristic impedance Z_i and constant delay length $\tau = l_s/v_p$. In this expression, l_s is the commensurate physical length of the lines; and v_p is the velocity of the wave propagation within the transmission medium. A Richards's inductive impedance $Z_L(\lambda) = \lambda L_\lambda$ is

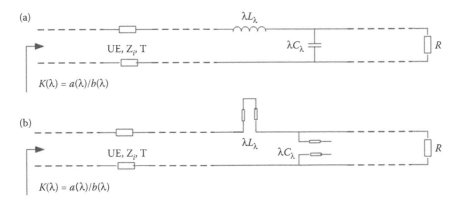

FIGURE 5.48
Lossless two-ports constructed with commensurate transmission lines. (a) Synthesis of $F(\lambda)$ in tandem connection of UEs and LC ladder sections in λ-domain. (b) Realization of inductive and capacitive immitances by means of short and open stub commensurate lines.

realized as a commensurate short stub with characteristic impedance $Z_i = L_\lambda$. Similarly, a Richards's capacitive admittance is realized as a commensurate open stub with characteristic impedance $Z_i = 1/C_\lambda$, as depicted in Figure 5.48a and b, respectively. An equal physical length l_s or corresponding delay length τ is a design parameter and is selected properly at an appropriate frequency f_s such that $\omega_s\tau$ is fixed as a fraction of π (i.e., $\omega_s\tau = 2\pi f_s\tau = \pi/m$ or $\tau = 1/2mf_s$; $m > 1, 2, 3, \ldots$).

In designing communication systems, use of lossless matching networks is inevitable. Up to GSM/UMTS frequencies, high-quality factor lumped circuit components may be preferred to design matching networks. However, if the operating frequencies go beyond a few GHz, then we are forced to utilize distributed elements such as commensurate transmission lines.

The "Real Frequency Techniques" in short RFTs are considered the best design methods to construct lossless matching networks for communication systems [1–4]. They work on network functions either in complex Laplace variable $p = \sigma + j\omega$ or in Richards's variable $\lambda = \Sigma + j\Omega$. Therefore, accurate synthesis of the network functions are in high demand.

In our previous publications [4–6], we introduced the high-precison synthesis of immitance functions generated in complex-p or λ-*Domain*. In this section, high-precision synthesis techniques introduced in p-*Domain* are expanded to synthesize network functions in *Richards's Domain*.

In the following sections, first we introduce the parametric method to generate a realizable positive immitance function $K(\lambda) = a(\lambda)/b(\lambda)$ in complex variable λ. Then, high-precision synthesis of $K(\lambda)$ is presented. Various examples are given to exhibit the utilization-proposed design algorithms to construct microwave matching networks and power amplifiers.

5.10 Generation of Realizable Positive Real Function in Richards's Domain

In classical network theory, Richards's variable is the transformed variable of $p = \sigma + j\omega$ which is expressed as

$$\lambda = \Sigma + j\Omega = \tanh(p\tau) \tag{5.138}$$

It should be noted on the real frequency axis ω (for the case where $\sigma = 0$, $p = j\omega$) Richards's frequency Ω is given by

$$\Omega = \tan(\omega\tau) \tag{5.139}$$

Many practical lossless matching networks designed with commensurate transmission lines demand minimum DPI function. We define a minimum function $K(\lambda) = a(\lambda)/b(\lambda)$ as one that is strictly analytic in the closed right half plane (RHP). If the minimum function refers to an impedance, it is called minimum reactance. Similarly, if the minimum function is an admittance, it is called minimum susceptance. For the sake of completeness, let us state the following properties of the positive real functions specified in Richards's domain.

5.10.1 Properties of Richards's Immittance Function

a. A positive real rational function $K(\lambda)$ must have all its zeros and poles in the open RHP.

b. $K(\lambda)$ may have poles and zeros on the Richards's frequency axis $\lambda = j\Omega$, but these poles and zeros must be simple (i.e., of order 1).

c. If $K(\lambda)$ is a minimum function, then it must be free of poles on the $\lambda = j\Omega$ axis. However, it may have simple zeros on the imaginary axis $\lambda = j\Omega$.

Any positive real function $F(\lambda)$ can be expressed as

$$K(\lambda) = K_F(\lambda) + K_M(\lambda) \tag{5.140}$$

where $K_F(\lambda)$ is a Foster's function that is purely imaginary for $\lambda = j\Omega$ such that

$$K_F(j\Omega) = jX_F(\Omega); \quad \frac{dX_F}{d\Omega} \geq 0 \quad \text{for all } \Omega \tag{5.141}$$

Furthermore, the Foster's function can expressed as

$$K_F(\lambda) = k_\infty \lambda + \frac{k_0}{\lambda} + \sum_{i=1}^{N_F} \frac{k_i \lambda}{\lambda^2 + \Omega_i^2} \tag{5.142}$$

d. The minimum function $K_M(\lambda)$ can be expressed in terms of its even and odd part such that

$$K_M(\lambda) = R(\lambda) + K_{M\text{-}odd}(\lambda) \tag{5.143}$$

On the Ω axis,

$$K_M(j\Omega) = R(\Omega) + jX_M(\Omega) \tag{5.144}$$

where

$$X_M(\Omega) = K_{M\text{-}odd}(j\Omega)/j \tag{5.145}$$

e. From Equations 5.140 through 5.144, we can deduce that $K(j\Omega)$ and $K_M(j\Omega)$ must possess the same real part $R(\Omega)$.

f. Let $K(j\Omega) = R(\Omega) + jX(\Omega)$. Then,

$$X(\Omega) = X_F(\Omega) + X_M(\Omega) \tag{5.146}$$

g. In designing matching networks employing the real frequency techniques, the real part $R(\Omega)$ of Equation 5.144 is specified as

$$R(\Omega^2) = \frac{[a_0^2][\Omega^{2q}(1 + \Omega^2)^k]}{\dfrac{1}{2}[c^2(\Omega) + c^2(-\Omega)]} = \frac{A(\Omega^2)}{B(\Omega^2)} \geq 0 \quad \text{for all } \Omega \tag{5.147}$$

where

$$
\begin{aligned}
A(\Omega) &= A_1\Omega^{2(q+k)} + A_2\Omega^{2(q+k-1)} + \cdots + A_{(q+k)}\Omega^2 + A_{q+k+1} \geq 0; \quad A_{q+k+1} = a_0^2 \geq 0 \\
B(\Omega^2) &= B_1\Omega^{2n} + B_2\Omega^{2(n-1)} + \cdots + B_n\Omega^2 + 1 > 0; \quad n \geq q + k
\end{aligned} \tag{5.148}
$$

and $c(\Omega)$ is an auxiliary real polynomial of degree n to generate strictly positive denominator polynomial $B(\Omega)$ such that

$$c(\Omega) = c_1\Omega^n + c_2\Omega^{n-1} + \cdots + c_n\Omega + 1 \tag{5.149}$$

Note that replacing Ω^2 by $-\lambda^2$ we can obtain Richards's domain even function $R(\lambda^2)$ as

$$R(\lambda^2) = [a_0^2]\frac{(-1)^q\lambda^{2q}(1 - \lambda^2)^k}{B(\lambda^2)} = \frac{A(\lambda^2)}{B(\lambda^2)} \tag{5.150}$$

In the above representation, zeros of $R(\lambda^2)$ are known as the zeros of transmission of the lossless two-port constructed as the result of the synthesis of the immittance function $K(\lambda)$.

h. These zeros are located at $\lambda = \mp 1$ of multiplicity k, $\lambda = 0$ of multiplicity of $2q$ and, perhaps, at infinity with multiplicity of $2n_\infty = 2(n - q - k)$ if $n > q + k$.

i. As far as the circuit components of the lossless two port are concerned, transmission zeros at $\lambda = \mp 1$ are realized as UEs in cascade configurations; the integer k refers to total number of UEs in the synthesis. The integer q refers to total number of transmission zeros at DC ($\Omega = 0$) that are realized as series Richards's capacitors and shunt Richards's inductors. The integer n_∞ refers to total number of transmission zeros at infinity that are realized as shunt Richards's capacitors and series Richards's inductors.

j. In designing broadband matching networks via real frequency techniques, the designer specifies the transmission zeros of the lossless two port under consideration by fixing the integers k, q, and n. Unknowns of the matching problem are chosen as the real coefficients $\{c_i; i = 1, 2, \ldots, n\}$ of the auxiliary polynomial $c(\Omega) = c_1\Omega^n + c_2\Omega^{n-1} + \cdots + c_n\Omega + 1$ and the real coefficient a_0 of the numerator.

k. In RFTs, if $K(\lambda)$ is assumed to be a minimum function, then it is generated from its real part $R(\Omega)$ using the parametric approach as in the following subsection.

5.10.2 Parametric Approach in Richards's Domain

A positive real minimum function $K(\lambda)$ can be generated from its nonnegative even-real part $R(\Omega^2)$ which is specified on the transformed frequency axis $\Omega = \tan(\omega\tau)$ as follows.

Once a_0, q, k, and arbitrary real coefficients $\{c_1, c_2, \ldots, c_n\}$ are initialized such that $n \geq q + k$, then minimum positive real function $K(\lambda)$ can easily be generated using the parametric approach as follows.

In the parametric approach, minimum function $K(\lambda)$ is expressed in partial fraction expansion form, as in Equation 5.55.

$$K(\lambda) = R_0 + \sum_{j=1}^{n} \frac{k_j}{\lambda - \lambda_j} = \frac{a(\lambda)}{b(\lambda)} \tag{5.151}$$

where λ_j are the closed left half plane roots of the denominator $B(\lambda) = B_1 \prod_{j=1}^{n}(\lambda^2 - \lambda_j^2)$ and the residues k_j are given by

$$k_j = \frac{1}{\lambda_j} \frac{A(\lambda_j)}{B_1 \prod_{\substack{i=1 \\ j \neq i}}^{n}(\lambda_j^2 - \lambda_i^2)} \tag{5.152}$$

and

$$R_0 = \lim_{\lambda \to \infty} R(\lambda^2) = \begin{cases} 0 & \text{for } n > (q + k) \\ \text{a positive constant} = \dfrac{a_1}{b_1} & \text{for } n = (q + k) \end{cases} \tag{5.153}$$

Now, let us run an example to generate an arbitrary positive real function in Richards's domain.

EXAMPLE 5.11

a. Let $q = 0$, $k = 5$, $a_0 = 1$, and $c = [1 - 11 - 11]$. Generate $R(\lambda^2) = A(\lambda^2)/B(\lambda^2)$, as in Equation 5.147.

b. Generate $K(\lambda) = a(\lambda)/b(\lambda)$ using Equations 5.151 through 5.153.

c. Find the roots of $a(\lambda)$ and $b(\lambda)$. Comment on the result.

d. Regenerate the even part of $K(\lambda) = a(\lambda)/b(\lambda)$ from the computed $a(\lambda)$ and $b(\lambda)$ and evaluate the numerical error in the course of computations.

Solution

a. For this purpose, we developed a MATLAB function accessed by MATLAB command

$$[A, B] = EvenPart\_Richard(k, q, a0, c).$$

This function generates the MATLAB vectors [A] and [B] to construct the rational form of $R(\lambda^2) = A(\lambda^2)/B(\lambda^2)$. Thus, the execution of this function yields $A = [-1\ 5 - 10\ 10 - 5\ 1]$ meaning that $A(\lambda^2) = -\lambda^{10} + 5\lambda^8 - 10\lambda^6 + 10\lambda^4 - 5\lambda^2 + 1 = (1 - \lambda^2)^5$ as expected and $B = [-1\ 3 - 5\ 1\ 1\ 1]$ meaning that $B(\lambda^2) = -\lambda^{10} + 3\lambda^8 - 5\lambda^6 + \lambda^4 + \lambda^2 + 1$.

b. Employing MATLAB, Equations 5.151 through 5.153 is programmed under a function

$$[a, b] = Minimum\_FRichard(k, q, a0, c).$$

Execution of the above function yields $F(\lambda) = a(\lambda)/b(\lambda)$ in Richards's domain such that

$$a = \begin{bmatrix} 1 & 6.6816 & 15.7449 & 17.3402 & 8.27681 \end{bmatrix}$$

meaning that

$$a(\lambda) = \lambda^5 + 6.6816\lambda^4 + 15.7449\lambda^3 + 17.3402\lambda^2 + 8.2768\lambda + 1$$

and

$$b = \begin{bmatrix} 1 & 4.2010 & 7.3241 & 6.6223 & 3.4992 & 1 \end{bmatrix}$$

meaning that

$$b(\lambda) = \lambda^5 + 4.2010\lambda^4 + 7.3241\lambda^3 + 6.6223\lambda^3 + 3.4992\lambda^1 + 1.$$

c. We can easily generate the roots of $a(\lambda)$ by means of MATLAB function $pa = roots(a)$. Similarly, roots of $b(\lambda)$ is given by $pb = roots(b)$. Hence, we find the following result (Table 5.11).

As we see from the above table, all the roots are located in the closed left half plane as expected.

d. Finally, we regenerate the even part from the computed $K(\lambda) = a(\lambda)/b(\lambda)$ as

$$R_1(\lambda^2) = \frac{1}{2}\left[\frac{a(\lambda)}{b(\lambda)} + \frac{a(-\lambda)}{b(-\lambda)}\right] = \frac{A_1(\lambda^2)}{B_1(\lambda^2)}$$

TABLE 5.11

Roots of $a(\lambda)$ and $b(\lambda)$ for Example 5.11

pa = roots(a)	pb = roots(b)
−3.2385	−1.3002 + 0.6248i
−1.1334 + 0.6838i	−1.3002 − 0.6248i
−1.1334 − 0.6838i	−1.0000
−1.0000	−0.3002 + 0.6248i
−0.1762	−0.3002 − 0.6248i

and compute the relative norm errors as

$$\varepsilon_{rA} = \frac{\|A_1 - A\|}{\|A\|} = 1.638 \times 10^{-15} \quad \text{and} \quad \varepsilon_{rB} = \frac{\|B_1 - B\|}{\|B\|} = 1.895 \times 10^{-15}.$$

From the above errors we can say that proposed algorithm to generate positive real functions in Richards's domain is numerically robust.

All the above computations are collected under the main program "GKYExample5_11.m," which is given by Program List 5.41.

5.10.3 Cascade Connection of *k*-Unit Elements

Referring to Figure 5.49, let us consider the driving input impedance $Z_{in}(\lambda) = a(\lambda)/b(\lambda)$ of a cascaded connection of k-UEs with

$$a(\lambda) = a_1\lambda^k + a_2\lambda^{k-1} + \cdots + a_k\lambda + a_{k+1}$$
$$b(\lambda) = b_1\lambda^k + b_2\lambda^{k-1} + \cdots + b_k\lambda + b_{k+1}$$

$Z_{in}(\lambda)$ is synthesized by extracting the UEs of characteristic impedance Z_i step by step. In order to keep track of the synthesis steps by means of proper indexing, at the first step, let us set

$$Z_{in}(\lambda) = Z_{in1}(\lambda) = \frac{a_1(\lambda)}{b_1(\lambda)}$$

In terms of the termination impedance $Z_{in2}(\lambda)$ and the characteristic impedance Z_1, $Z_{in1}(\lambda)$ is given by

$$Z_{in1}(\lambda) = Z_1 \frac{Z_{in2}(\lambda) + \lambda Z_1}{Z_1 + \lambda Z_{in2}(\lambda)} = \frac{a_1(\lambda)}{b_1(\lambda)} \tag{5.154}$$

The last termination R_0 of the *k*-cascaded UE is specified by setting $\lambda = 0$ such that

$$R_0 = \frac{a_{k+1}}{b_{k+1}}$$

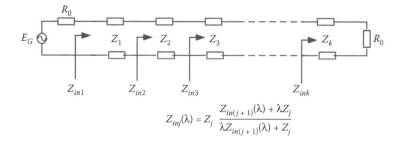

$$Z_{inj}(\lambda) = Z_j \frac{Z_{in(j+1)}(\lambda) + \lambda Z_j}{\lambda Z_{in(j+1)}(\lambda) + Z_j}$$

FIGURE 5.49
Cascade connection of *k*-UE.

Furthermore, the even part $R(\lambda)$ of $Z_{in1}(\lambda)$ can be expressed in the following form:

$$R(\lambda) = \frac{1}{2}[Z_{in1}(\lambda) + Z_{in1}(-\lambda)] = \frac{a_0^2(1 - \lambda^2)^k}{B(\lambda^2)} = \frac{A(\lambda^2)}{B(\lambda^2)} \tag{5.155}$$

where

$$A(\lambda^2) = A_1\lambda^{2k} + A_2\lambda^{2(k-1)} + \cdots + A_k\lambda + A_{k+1}$$
$$B(\lambda^2) = B_1\lambda^{2k} + B_2\lambda^{2(k-1)} + \cdots + B_k\lambda + 1$$

and

$$R_0 = A_{k+1} = [a_{k+1}]^2 = a_0^2 > 0$$

At $\lambda = 1$, Equation 5.154 yields the characteristic impedance Z_1 such that

$$Z_1 = Z_{in1}(1) = Z_1\frac{Z_{in2}(1) + Z_1}{Z_1 + Z_{in2}(1)} = \frac{a(1)}{b(1)} = \frac{\sum_{i=1}^{k} a_k}{1 + \sum_{i=1}^{k-1} b_k} \tag{5.156}$$

On the other hand, we can pull $Z_{in2}(\lambda)$ from Equation 5.154 in terms of the given polynomials $a(\lambda)$ and $b(\lambda)$ and computed characteristic impedance Z_1 such that

$$Z_{in2}(\lambda) = \frac{\lambda Z_1^2 - Z_1 Z_{in1}(\lambda)}{\lambda Z_{in1}(\lambda) - Z_1} = \frac{a_2(\lambda)}{b_2(\lambda)} \tag{5.157}$$

where

$$a_2(\lambda) = [a(1)]^2[\lambda b(\lambda)] - [a(1)b(1)][a(\lambda)] = \alpha^2[\lambda b(\lambda)] - [\alpha\beta][a(\lambda)]$$
$$b_2(\lambda) = [b(1)]^2[\lambda a(\lambda)] - [a(1)b(1)][b(\lambda)] = \beta^2[\lambda a(\lambda)] - [\alpha\beta][b(\lambda)]$$
$$\alpha = a(1)$$
$$\beta = b(1)$$

It should be noted that the degree of polynomials $a_2(\lambda)$ and $b_2(\lambda)$ is $k + 1$ due to the terms $\lambda b(\lambda)$ and $\lambda a(\lambda)$, respectively. Furthermore, at $\lambda = \pm 1$, both of them vanish. Therefore, by defining new polynomials $a_{new}(\lambda)$ and $b_{new}(\lambda)$, we can express $a_2(\lambda)$ and $b_2(\lambda)$ as

$$a_2(\lambda) = (\lambda - 1)(\lambda + 1)a_{new}(\lambda) = (\lambda^2 - 1)a_{new}$$
$$b_2(\lambda) = (\lambda - 1)(\lambda + 1)b_{new}(\lambda) = (\lambda^2 - 1)b_{new} \tag{5.158}$$

Hence, the degree of new polynomials $a_{new}(\lambda)$ and $b_{new}(\lambda)$ is reduced to $(k - 1)$. In short, we say that *a unit–element (UE)* of characteristic impedance $Z_1 = a(1)/b(1) = \alpha/\beta$ is extracted from the given input impedance $Z_{in}(\lambda) = a(\lambda)/b(\lambda)$ resulting in a one degree less positive real input impedance.

From the numerical implementation point of view, in Equation 5.158, due to multipliers α^2, β^2, and $\alpha\beta$, we may end up with exponential multipliers in the order of $[10^m; m > 1]$ in the numerator and denominator polynomials of $Z_2(\lambda) = a_{new}(\lambda)/b_{new}(\lambda)$. In this case, it may be appropriate to divide both $a_{new}(\lambda)$ and $b_{new}(\lambda)$ with the norm of the vector a_{new} such that

$$Z_{in2}(\lambda) = \frac{a_{new}(\lambda)/norm(a_{new})}{b_{new}(\lambda)/norm(a_{new})} \tag{5.159}$$

where

$$norm(a_{new}) = \sqrt{a_{new,1}^2 + a_{new,2}^2 + \cdots + a_{new,(n-1)+a_{new,n}^2}^2}$$

The above division operation improves the numerical accuracy in the course of the UE extraction process.

From the algorithmic implementation point of view, we initialize $Z_{in2}(\lambda)$ as

$$Z_{in2}(\lambda) = \frac{a(\lambda)}{b(\lambda)}$$

where $a(\lambda) = a_{new}(\lambda)/norm(a_{new})$ and $b(\lambda) = b_{new}(\lambda)/norm(a_{new})$ are the newly reset polynomials for the extraction of next UE from $Z_2(\lambda)$.

Now, let us verify the above equations by means of an example.

EXAMPLE 5.12

Let the driving point input impedance $Z_{in1}(\lambda) = a(\lambda)/b(\lambda)$ be given, as in Example 5.11.

If it can be synthesized as a connection of cascaded transmission lines, determine the characteristic impedances of each commensurate transmission line.

Solution

In Example 5.11, $Z_{in1}(\lambda) = a(\lambda)/b(\lambda)$ is given as

$$Z_{in1}(\lambda) = \frac{\lambda^5 + 6.6816\lambda^4 + 15.7449\lambda^3 + 17.3402\lambda^2 + 8.2768\lambda + 1}{\lambda^5 + 4.2010\lambda^4 + 7.3241\lambda^3 + 6.6223\lambda^2 + 3.4992\lambda + 1}$$

and its even part $R(\lambda) = A(\lambda^2)/B(\lambda^2) =$ of $Z_{in}(\lambda) = a(\lambda)/b(\lambda)$ is found as $A(\lambda^2) = (1 - \lambda^2)^5$ and

$$B(\lambda^2) = -\lambda^{10} + 3\lambda^8 - 5\lambda^6 + \lambda^4 + \lambda^2 + 1$$

Therefore, it must be synthesized as a cascaded connection of 5-UEs.

We programmed Equations 5.157 and 5.158 under a MATLAB function

$$[Z, a\_new, b\_new, ra, rb] = ImpedanceBasedRichard\_Extraction(a, b)$$

where the inputs a and b are the numerator and the denominator polynomials of $Z_{in}(\lambda)$. The output Z is the characteristic impedance of the extracted line. a_new and b_new are the reduced degree polynomials of the resulting or new input impedance; ra and rb are the remainders of the synthetic divisions $a_{new}(\lambda) = a(\lambda)/\lambda^2 - 1 + ra$ and $b_{new}(\lambda) = b(\lambda)/\lambda^2 - 1 + rb$. Obviously, remainders ra and rb are supposed to be zero.

In the first step, execution of the above function yields

$$Z1 = 2.1163; \ a2 = a\_new = [0.1410 \quad 0.5257 \quad 0.7286 \quad 0.4105 \quad 0.0666];$$
$$b2 = b\_new = [0.0315 \quad 0.1437 \quad 0.2473 \quad 0.2017 \quad 0.0666];$$
$$ra1 = 10^{-11} \times [0 \quad 0 \quad 0 \quad 0 \quad 0 \quad 0.1819 \quad 0.3638];$$
$$rb1 = 10^{-11} \times [0 \quad 0 \quad 0 \quad 0 \quad 0 \quad -0.1819 \quad -0.1592].$$

In short, at the end of the first step, a new Richards's impedance is found as

$$Z_{in2}(\lambda) = \frac{0.1410\lambda^4 + 0.5257\lambda^3 + 0.7286\lambda^2 + 0.4105\lambda + 0.0666}{0.0315\lambda^4 + 0.1437\lambda^3 + 0.2473\lambda^2 + 0.2017\lambda + 0.0666}$$

If algorithmic *zero* is less than 10^{-10}, then $ra1$ and $rb1$ are confidently set to zero. In the second step, the second line is extracted yielding,

$$Z2 = 2.7105; \ a3 = [0.2401 \quad 0.6993 \quad 0.6467 \quad 0.1874];$$
$$b3 = [0.1464 \quad 0.4571 \quad 0.4982 \quad 0.1874];$$
$$ra2 = 10^{-15} \times [0 \quad 0 \quad 0 \quad 0 \quad 0.2220 - 0.1249];$$
$$rb2 = 10^{-16} \times [0 \quad 0 \quad 0 \quad 0 \quad 0.8327 - 0.9714].$$

In short, at the end of the second step, new Richards's impedance is found as

$$Z_{in3}(\lambda) = \frac{0.2401\lambda^3 + 0.6993\lambda^2 + 0.6467\lambda + 0.1874}{0.1464\lambda^3 + 0.4571\lambda^2 + 0.4982\lambda + 0.1874}$$

In the third step,

$$Z3 = 1.3758; \ a4 = [0.4228 \quad 0.8163 \quad 0.3936]; \ b4 = [0.3664 \quad 0.7599 \quad 0.3936];$$
$$ra3 = 10^{-14} \times [0 \quad 0 \quad 0 \quad -0.1221 \quad 0.1554];$$
$$rb3 = 10^{-15} \times [0 \quad 0 \quad 0 \quad -0.9992 \quad 0.9992].$$

In short, at the end of the third step, new Richards's impedance is found as

$$Z_{in4}(\lambda) = \frac{0.4228\lambda^2 + 0.8163\lambda + 0.3936}{0.3664\lambda^2 + 0.7599\lambda + 0.3936}$$

In the fourth step,

$$Z4 = 1.0742; \ a5 = [0.7071 \quad 0.7071]; \ b5 = [0.7071 \quad 0.7071];$$
$$ra4 = 10^{-14} \times [0 \quad 0 \quad 0.2776 \quad -0.1998]; \ rb4 = 10^{-14} \times [0 \quad 0 \quad 0.2442 \quad -0.1887].$$

In short, at the end of the fourth step, Richards's impedance is found as

$$Z_{in5}(\lambda) = \frac{0.7071\lambda + 0.7071}{0.7071\lambda + 0.7071}$$

In the fifth step, we have

$$Z5 = 1.0000; \quad a6 = 1.4142; \quad b6 = 1.4142;$$

As seen from above, the terminating resistor is given by

$$Z_6 = \frac{a6}{b6} = 1$$

The above computation steps are combined under the main MATLAB program GKYExample5.12.m. Interested readers are encourged to execute this program.

In the above example, we have shown the straightforward extraction of UEs in a sequential manner, where the DC zeros q of the even part function $R(\lambda^2)$ is set to zero (i.e., $q = 0$).

In the above steps we observe that, after each extraction, the generic form of the Richards's impedance is preserved and the steps are all minimum functions of the Richards's impedance is preserved and it reflects a minimum function as we start the extraction process.

5.10.4 Correction of the Richards's Impedance after Each Extraction

In the course of the cascaded synthesis process, at each step, numerical precision is lost as we continue with multiplication and division operations. Therefore, it may be appropriate to correct the impedance using the parametric approach, as detailed in Rererences 4–6 and 67. However, in Richards's domain $\lambda = \Sigma + j\Omega$, the correction operation is a little tricky. If the total number of elements n is greater than that of the total number of cascaded UEs k (i.e., $n > k$), then, at each UE extraction, the remaining impedance function changes its character from minimum reactance to minimum susceptance or vice versa. For example, if $n > k$ and we start Richards's synthesis with a minimum reactance function $F_{in}(\lambda) = a(\lambda)/b(\lambda) = Z_{in1}(\lambda)$, then after the first UE extraction, the remaining driving point impedance $Z_{in2}(\lambda)$ becomes a minimum susceptance. In this case, we can flip over the function to correct it using our parametric method. Therefore, at odd steps of UE extraction, immittance correction is applied on the admittance function. On the other hand, at even steps, correction is directly applied on the minimum reactance impedance function. In any case, the generic form of the numerator polynomial $A_i(\lambda^2)$ of the even part $R(\lambda^2) = A_i(\lambda^2)/B_i(\lambda^2)$ is forced to be

$$A_i(\lambda^2) = (-1)^q (a_{0i}^2)\lambda^{2q}(1 - \lambda^2)^{k-i}$$
$$B_i(\lambda^2) = b(\lambda)b(-\lambda)$$

(5.160)

Ignoring the remaining terms ra and rb, the above forms can be generated from the reset polynomials $a(\lambda)$ and $b(\lambda)$ at the end of each step. In the correction process, it may be appropriate to normalize both $A_i(\lambda^2)$ and $B_i(\lambda^2)$ by (a_{0i}^2) so that $A_i(\lambda^2) = (-1)^q (a_{0i}^2)\lambda^{2q}(1 - \lambda^2)^{k-i}$ is precisely generated at each correction step. In the course of corrections, first, we determine the integers k, q, and (a_{0i}^2) from the immitance function $F_{in}(\lambda) = a(\lambda)/b(\lambda)$. This step is completed under the MATLAB function

$$[k, q, a0, nA, A1, B1] = Richard\_Numerator(a, b)$$

where the input arguments a and b are the numerator and denominator polynomials of the minimum immitance function $F_{in}(\lambda) = a(\lambda)/b(\lambda)$. Inside the above MATLAB function, first, vectors a and b are normalized with respect to norm of the original a, which is specified by *norm* (a). This process introduces numerical robustness within function *Richard_Numerator*.

The output arguments k and q are the total number of UEs and DC zeros of $R(\lambda^2) = A_i(\lambda^2)/B_i(\lambda^2)$ and are directly generated from the given a and b. $a0$ is the square root of the leading nonzero coefficient of $A_i(\lambda^2)$. $n = nA - 1$ is the total number of elements of the lossless two port consisting of commensurate transmission lines, which is obtained as the result of synthesis. $A1$ and $B1$ are the corrected normalized numerator and denominator polynomials of the even part function $R(\lambda^2) = A_1(\lambda^2)/B_1(\lambda^2)$. Let us observe the above using the following example.

EXAMPLE 5.13

Let the minimum input impedance $Z_{in1}(\lambda)$ of a tandem connection of UE and LC ladder sections be produced from the following input data:

$$k = 5, \ q = 0, \ a0 = 1 \quad and \quad c = [1 -1 1 -1 -1 1]$$

a. Determine $Z_{in1}(\lambda)$ and extract five UEs using our MATLAB function $[Z_1, a\_new, b\_new, ra, rb] = ImpedanceBasedRichard\_Extraction(a, b)$ and correct the remaining immitance function at each step by means of our MATLAB function *Richard_ImmittanceCorrection*.
b. Synthesize the remaining LC-ladder network and draw the final lossless two port in resistive termination.
c. Comment on the results.

Solution

a. Execution of $[a, b] = Richard\_MinimumFunction(k, q, a0, c)$ yields Table 5.12.

TABLE 5.12

Minimum Reactance Richards's Impedance $Z_{in}(\lambda) = a(\lambda)/b(\lambda)$ with the Inputs of $k = 5$, $q = 0$, $a0 = 1$, and $c = [1 -1 1 -1 1 -1 1]$ Using MATLAB Function $[a, b] = Richard\_MinimumFunction(k, q, a0, c)$

$a(\lambda)$	$b(\lambda)$
0	0.0197
0.0384	0.1100
0.2150	0.2784
0.5145	0.4132
0.6623	0.3912
0.4699	0.2430
0.1666	0.0957
0.0197	0.0197

The above impedance $Z_{in}(\lambda) = a(\lambda)/b(\lambda)$ will be realized using $k = 5$ cascaded UEs, $q = 0$ high-pass elements and $n_L = n - k - q = 7 - 5 - 0 = 2$ low-pass elements in Richards's domain.

First, we start extracting UE in a sequential manner. Thus, using our MATLAB function $[Z_1, a\_new, b\_new, ra, rb] = ImpedanceBasedRichard\_Extraction$ (a, b), characteristic impedances Z_i are found at each step and the remaining immittance function is corrected.

In the first step $(i = 1; iisoddcase)$, characteristic impedance of the first line is found as $Z_1 = 1.3282$, and the remaining normalized input impedance is

$$Z_{in2}(\lambda) = \frac{0.0358\lambda^6 + 0.2004\lambda^5 + 0.4902\lambda^4 + 0.65833 + 0.4971\lambda^2 + 0.1926\lambda + 0.0270}{0.0127\lambda^5 + 0.0711\lambda^4 + 0.1622\lambda^3 + 0.1882\lambda^2 + 0.1110\lambda + 0.0270}$$

The above impedance is minimum susceptance. Therefore, correction is applied on the admittance function by calling our MATLAB function $[b, a] = Richard\_ImmittanceCorrection (b, a, k, q)$.

In the second step $(i = 2; i = evencase)$, we found $Z_2 = 3.6718$ and the next input impedance is

$$Z_{in3}(\lambda) = \frac{0.0663\lambda^4 + 0.3709\lambda^3 + 0.7103\lambda^2 + 0.5714\lambda + 0.1645}{0.0595\lambda^5 + 0.3331\lambda^4 + 0.7966\lambda^3 + 0.9930\lambda^2 + 0.6327\lambda + 0.1645}$$

The above impedance is a minimum reactance. Therefore, correction is directly applied on this form by calling our function $[a, b] = Richard\_ImmittanceCorrection (a, b, k, q)$.

In the third step $(i = 3, i = oddcase)$ we have
$Z_3 = 0.6321$ and

$$Z_{in4}(\lambda) = \frac{0.0523\lambda^4 + 0.2929\lambda^3 + 0.6606\lambda^2 + 0.6501\lambda + 0.2289}{0.0631\lambda^3 + 0.3529\lambda^2 + 0.5180\lambda + 0.2289}$$

which is a minimum susceptance admittance, as expected. Hence, it is corrected as in step 1 $(i = oddcase)$.

In the fourth step, we have $Z_4 = 1.6209$ and

$$Z_{in5}(\lambda) = \frac{0.1369\lambda^2 + 0.7660\lambda + 0.6281}{0.0886\lambda^3 + 0.4959\lambda^2 + 1.0340\lambda + 0.6281}$$

which is a minimum reactance and is corrected as in step 2 $(i = evencase)$.

Finally, at step 5, we have $Z_5 = 0.6815$ and the remaining input impedance is given by

$$Z_{in6}(\lambda) = \frac{0.0844\lambda^2 + 0.4721\lambda + 0.8775}{0.1568\lambda + 0.8775}$$

b. Synthesis of the LC-ladder in Richards's domain:
The above impedance is not a minimum reactance function and it is free of UEs. This fact can easily be checked by using our MATLAB function $[k6, q6, a06, nA6, A6, B6] = Richard\_Numerator(a, b)$. Execution of this function yields $k6 = 0, q6 = 0, a06 = 1, nA6 = 3, A6 = 1, B6 = [-0.03191.0000]$. This result indicates that $Z_{in6}(\lambda)$ includes $n_L = nA6 - 1 = n - 1 = 3 - 1 = 2$ low-pass elements.

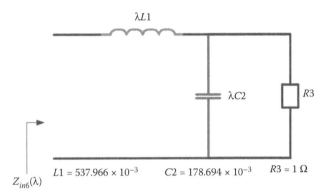

FIGURE 5.50
Synthesis of LC ladder section in Richards's domain λ.

$Z_{in}6(\lambda)$ is synthesized using our "Impedance-Based High-Precision LC-ladder Synthesis" function $[CTCV] = Synthesis\_ImpedanceBased\,(a, b, R0, f0)$. In this function $R0$ is the normalization resistance and $f0$ is the normalization frequency to calculate actual element values. They are selected as $R0 = 1$, and $f0 = 1/2\pi$ since we are dealing with normalized values in all the above examples.

Hence, execution of $[CTCV] = Synthesis\_ImpedanceBased\,(a, b, R0, f0, ndc)$ results in the following low-pass LC ladder (Figure 5.50).

In the above figure, inductor L1 is realized employing an equal length (or commensurate) short-ended transmission line with characteristic impedance $Z_6 = L1 = 0.5380$. Capacitor C2 is realized with an open-ended commensurate transmission line with characteristic impedance $Z_7 = 1/C2 = 1/0.1787 = 5.5962$. Complete synthesis is realized with seven commensurate transmission lines with different characteristic impedances, as shown in Figure 5.51.

c. Comments:

1. When the driving point Richards's impedance $Z_{in}(\lambda) = a(\lambda)/b(\lambda)$ is completely synthesized with total of n number of commensurate transmission lines, if $k = n$, then, as a result of synthesis, lossless two port must include cascaded connection of *k-commensurate* (or equal length) transmission lines.

2. If $n > k$ and $q = 0$, then synthesis must include total of $n - k$ series short stubs and open shunt stubs to realize series inductors and shunt capacitors in λ-domain.

3. If the synthesis starts with a minimum reactance impedance function and if $n > k$, then at each step, after extraction of a UE, the remaining function changes its generic form. In this example, it is shown that for $i = odd$

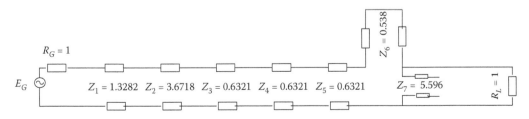

FIGURE 5.51
Lossless two-port constructed with commensurate transmission lines in resistive termination.

steps, remaining immittance function is minimum susceptance and for the $i = even$ steps, it is a minimum reactance. Therefore, immittance correction is applied on either minimum susceptance or minimum reactance functions accordingly.

4. In terms of Laplace variable $p = \sigma + j\omega$, Richards's variable is expressed as $\lambda = \Sigma + j\Omega = \tanh(p\tau)$ where τ is the constant delay length of the commensurate transmission lines and it is specified by the designer at a selected frequency f_s which may be beyond the upper end of the operating frequencies f_{c2}. On the real frequency axis $p = j\omega$, Richards's real frequency is specified as $\Omega = \tan(\omega\tau)$. At selected frequency f_s, $\omega_s\tau$ can be fixed at any real number such as $\omega_s\tau = \pi/m$ with $m = 2, 3, 4, 5, \ldots$, etc. Let us designate the velocity of propagation of the commensurate lines by v_p, then the physical length of a commensurate line is given by $l_s = v_p\tau = v_p/2mf_s$. Choice of m depends on the designer.

5. In Richards's domain λ, capacitor admittance $Y_C = C_i\lambda$ is realized as an open stub with characteristic impedance $Z_i = 1/C_i$ of a commensurate physical length l_s. On the other hand, impedance of an inductor $Z_L = L_i\lambda$ is realized as a short stub commensurate transmission line with characteristic impedance $Z_i = L_i$.

6. It must be noted that synthesis is not unique. One is free to extract UEs, and Richards's high-pass and low-pass elements, in any order as desired. Furthermore, using Kuroda identities, in the final layout, position of the series short stubs and shunt open stubs can be shifted as desired [4, Chapter 4, p. 187].

In the following Example, let us synthesize a complicated Richards's impedance function with k-UEs, q-series capacitors and shunt inductors, and $n - k - q$ series inductors and shunt capacitors in λ-domain.

EXAMPLE 5.14

Let the Richards's impedance $Z_{in}(\lambda)$ be generated with $k = 5$; $q = 5$; $a0 = 1$; $c = [1\ -1\ 1\ -1\ 1\ -1\ 1\ -1\ 1\ -1\ 1\ -1\ 1\ -1\ 1]$. That is, it includes a total of $n = 15$ elements out of which we have 5-UEs, $q = 5$ high-pass stubs (series capacitors and shunt inductors in Richards's domain) and $n - k - q = 5$ low-pass stubs (series inductors and parallel capacitors in Richards's domain)

  a. Generate the rational form of the Richards's impedance $Z_{in}(\lambda)$
  b. Synthesize it and draw the complete circuit layout
  c. Comment on the result

Solution

For the solution of this example, we developed a MATLAB program called GKYExample5.14.m.

Part (a): Program GKYExample5_14m results in the following minimum reactance input impedance $Z_{in} = (\lambda) = a(\lambda)/b(\lambda)$ as listed in Table 5.13.

TABLE 5.13

Input Impedance $Z_{in}(\lambda) = a(\lambda)/b(\lambda)$ for Example 5 $k = 5$; $q = 5$; $a0 = 1$; $c = [1\ -1\ 1\ -1\ 1\ -1\ 1\ -1\ 1\ -1\ 1\ -1\ 1\ -1\ 1]$

$a(\lambda)$	[0 0.0006 0.0060 0.0312 0.1024 0.2364 0.4036 0.5231 0.5208 0.3988 0.2331 0.1018 0.0317 0.0064 0.0006 0]
$b(\lambda)$	[0.0054 0.0589 0.3117 1.0664 2.6375 4.9951 7.4846 9.0389 8.8780 7.1023 4.5999 2.3760 0.9516 0.2806 0.0549 0.0054]

TABLE 5.14

Characteristic Impedances of the Cascaded Transmission Lines of Example 5.14

Z_1	Z_2	Z_3	Z_4	Z_5
$5.2090e - 02$	$5.8756e - 02$	$2.2217e - 03$	$2.3862e - 03$	$1.9715e - 04$

TABLE 5.15

Input Impedance $Z_{inII}(\lambda) = a_{II}(\lambda)/b_{II}(\lambda)$ for Example 5.14 $k = 5$; $q = 5$; $a0 = 1$; $c = [1 -1 1 -1 1 -1 1 -1 1 -1 1 -1 1 -1 1]$

$a_{II}(\lambda)$	[1.0000 10.8605 53.8305 157.0510 287.8958 324.3108 204.9895 77.2318 16.7636 1.6559 0.0000]
$b_{II}(\lambda)$	$10^5 \times$ [0 0.0292 0.3176 1.5096 3.8928 5.2415 2.5179 0.5884 0.0581 0.0000 0.0000]

Part (b): Synthesis of the Richards's impedance is completed with our MATLAB function called

$$[Z\_UE1, a\_new, b\_new, CT, CV] = Richard\_Complete \operatorname{Im} pedanceSynthesis(a, b, k, q, R0, f0)$$

The above function programs the impedance-based Richards's synthesis algorithm as presented in the previous sections and detailed in Reference 67.

The first part of the synthesis results in 5-cascaded connections of UEs with characteristic impedances listed in Table 5.14.

In the second part of the synthesis, the remaining Richards's impedance function $Z_{inII} = (\lambda) = a_{II}(\lambda)/b_{II}(\lambda)$ is given as Table 5.15

The synthesis is depicted in Figure 5.52 where

$$C1 = 3.51012\,\text{kF} \quad C5 = 4643.49\,\text{MF} \quad C9 = 444.671\,\text{F}$$
$$L2 = 946.012\,\mu\text{H} \quad L6 = 506.57\,\mu\text{H} \quad L10 = 57.1213\,\mu\text{H}$$
$$C3 = 4.76563\,\text{kF} \quad C7 = 1.03012\,\text{kF} \quad R11 = 620.365\,\mu\Omega$$
$$L4 = 8.70756\,\text{H} \quad L8 = 284.425\,\mu\text{H}$$

The complete synthesis with commensurate transmission lines is shown in Figure 5.53.

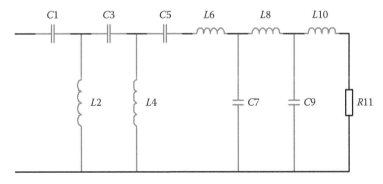

FIGURE 5.52
Synthesis of $Z_{in}\text{II}(\lambda) = a_{II}(\lambda)/b_{II}(\lambda)$ of Example 5.15.

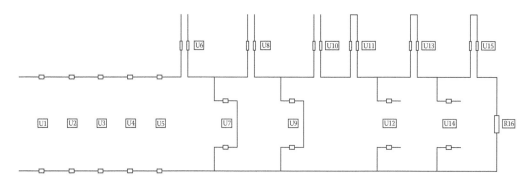

FIGURE 5.53
Complete layout of the synthesis of Example 5.14 with commensurate transmission lines.

Characteristic impedances of the above lines are given as follows:

$U1 = 52.0903\,m\Omega$ $U5 = 197.155\,\mu\Omega$ $U9 = 8.70756\,\Omega$ $U13 = 284.425\,\mu\Omega$

$U2 = 58.7563\,m\Omega$ $U6 = 284.89\,\mu\Omega$ $U10 = 215.355\,p\Omega$ $U14 = 2.24885\,m\Omega$

$U3 = 2.22171\,m\Omega$ $U7 = 946.012\,\mu\Omega$ $U11 = 506.57\,\mu\Omega$ $U15 = 57.1213\,\mu\Omega$

$U4 = 2.3862\,m\Omega$ $U8 = 209.836\,\mu\Omega$ $U12 = 970.757\,\mu\Omega$ $R16 = 620.365\,\mu\Omega$

Remarks

- The above hypothetical example is presented to exhibit implementation of the Richards's synthesis algorithm employing our MATLAB function *"Richard_CompleteImpedanceSynthesis."*
- In the course of synthesis, first cascaded UEs are extracted. Then, from the remaining impedance, Richards's high-pass sections are removed. Finally, Richards's low-pass section is extracted.
- After each transmission zeros extractions, impedance correction is used as in the lumped circuit synthesis.
- Eventually, Richards's capacitors C_i are realized as open stubs with characteristic impedance $Z_i = 1/C_i$ and Richards's inductors L_i are realized as short stubs with characteristic impedance $Z_i = L_i$.
- Chracteristic impedances are denormalized by multiplying normalized impedances by R0.
- Function *"Richard_Complete Impedance Synthesis"* results in accurate element values with relative error less than 10^{-7} up $n = 15$ commensurate line element extractions as UE and stubs. Details are omitted here. However, interested readers are encouraged to read Reference 7.

5.11 Integration of Richards's High-Precision Synthesis Module with Real Frequency Matching Algorithm

Referring to Figures 5.54 and 5.55, for the real frequency direct computation technique (RFDCT), TPG of the doubly matched system is described in terms of the driving point

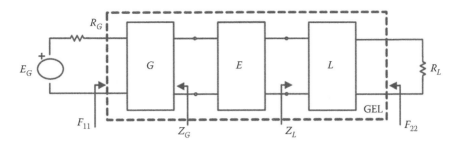

FIGURE 5.54
Double matching problem.

immittances of the generator $K_G = [Z_G$ or $Y_G]$, equalizer $K_B = [Z_B(\lambda)$ or $Y_B(\lambda)]$ and the load $K_L = [Z_L$ or $Y_L]$ networks. In Figure 5.54, both generator and load impedances are replaced by their Darlington equivalents [G] and [L] lossless two ports to compute the transducer power gain of the doubly matched system.

In this case, TPG of Figure 5.55 is given by References 3 and 4

$$T(\omega) = \frac{1 - |G_{22}|^2}{|1 - G_{22}S_{in}|^2} T_{[EL]} \tag{5.161}$$

where

$$G_{22} = \frac{Z_G - 1}{Z_G + 1} = \frac{1 - Y_G}{1 + Y_G} \tag{5.162}$$

is the unit normalized generator reflectance.

As proven by the main theorem of Yarman and Carlin [4,5], S_{in} is given by

$$S_{in}(j\omega) = \eta_B(\lambda) \left[\frac{Z_L(j\omega) - Z_B(-\lambda)}{Z_L(j\omega) + Z_B(\lambda)} \right]_{\lambda = jtan(\omega\tau)} \tag{5.163}$$

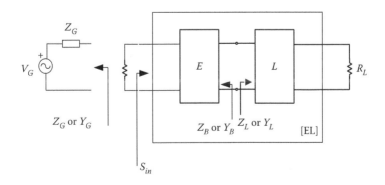

FIGURE 5.55
Cascaded connection of two lossless two-ports [G] and [EL].

$$= \eta_B(\lambda) \left[\frac{Y_B(\lambda) - Y_L(-j\omega)}{Y_B(\lambda) + Y_L(j\omega)} \right]_{\lambda = j\tan(\omega\tau)} \tag{5.164}$$

which is the unit normalized input reflectance of the lossless two-port [EL].

In Equations 5.163 and 5.164, the all pass function $\eta_B(\lambda)$ is given by

$$\eta_B(\lambda) = \left[\frac{W_B(\lambda)}{W_B(-\lambda)} \right] \tag{5.165}$$

The rational analytic function $W_B(\lambda)$ of Equation 5.165 is constructed on the explicit factorization of the even part $R_B(\lambda^2)$ such that

$$K_B(\lambda) = \frac{a(\lambda)}{b(\lambda)} \tag{5.166}$$

$$R_B(\lambda^2) = Even\{F_B(\lambda)\} = \frac{A(\lambda^2)}{B(\lambda^2)}$$

$$= \frac{(-1)^q \lambda^{2q}(1 - \lambda^2)^k}{B(\lambda^2)} \tag{5.167}$$

$$= \frac{n_B(\lambda)n_B(-\lambda)}{b(\lambda)b(-\lambda)} \tag{5.168}$$

$$= W_B(\lambda)W_B(-\lambda) \tag{5.169}$$

where the function $W_B(\lambda)$ is described by

$$W_B(\lambda) = \frac{n_B(\lambda)}{b(\lambda)}$$

The numerator polynomial $n_B(\lambda)$ must include all the proper RHP and $j\Omega$-axis zeros of $R_B(\lambda^2)$ as described by Equations 5.167 through 5.169. Thus, for a cascade connection k UE with q DC transmission zeros, $W_B(\lambda)$ and $W_B(-\lambda)$ are given as

$$W_B(\lambda) = (-\lambda)^q \frac{(1 - \lambda^2)^{k/2}}{b(\lambda)} \tag{5.170}$$

$$W_B(-\lambda) = (\lambda)^q \frac{(1 - \lambda^2)^{k/2}}{b(-\lambda)} \tag{5.171}$$

Hence, the all pass function $\eta_B(\lambda)$ is

$$\eta_B(\lambda) = (-1)^q \left[\frac{b(-\lambda)}{b(\lambda)} \right] \tag{5.172}$$

For the direct method of real frequency broadband matching, the unknown of the problem is the rational form of the even part $R_B(\lambda^2)$, as specified by Equations 5.167 through 5.172.

It is interesting to observe that in Equations 5.161 through 5.164, the TPG T_{EL} of the lossless two port [EL] is given by

$$T_{EL} = 1 - |S_{in}(j\omega)|^2 = \frac{4R_B R_L}{[R_B + R_L]^2 + [X_B + X_L]^2} \tag{5.173}$$

which is the immittance-based conventional single matching gain.

Assuming $K_B(\lambda)$ as a minimum function, TPG is expressed as a function of the real part $R_B(\Omega)$ as

$$T(\omega) = \left[\frac{1 - |G_{22}|^2}{|1 - G_{22}S_{in}|^2}\right]\left\{\frac{4R_B R_L}{[R_B + R_L]^2 + [X_B + X_L]^2}\right\} \tag{5.174}$$

where $R_B(\Omega) = Real\ Part\ \{F_B(j\Omega)\}\ X_B(\omega) = Imaginary\ Part\ \{F_B(j\Omega)\}$.

Once integers k and q are selected by the designer, $R_B(\Omega)$ is described by means of an auxiliary polynomial $c(\Omega) = c_1\Omega^n + c_2\Omega^{n-1} + \ldots + c_n\Omega + c_0$ of Equations 5.147 through 5.149. In this case, unknowns of the matching problem are the arbitrary real coefficients of $c(\Omega)$. When the coefficients $\{c_i; i = 0, 1, 2, \ldots, n\}$ are initialized, then $K_B(\lambda)$ is generated using parametric method of Section 5.11.2 and $T(\omega)$ is computed, as in Equation 5.161.

In RFDCT, TPG is optimized over the band of operation employing a nonlinear optimization algorithm. For example, in MATLAB, one may wish to employ the nonlinear optimization functions, such as *lsqnonlin, fminmax, fminsearch*, of the optimization toolbox. Optimization of TPG yields the driving point immittance $K_B(\lambda) = a(\lambda)/b(\lambda)$. Eventually, $K_B(\lambda)$ is synthesized using the newly proposed Richards's synthesis algorithm.

Let us apply the above-described process to design a wideband impedance transformer designed for a power amplifier.

EXAMPLE 5.15

Referring to Figure 5.56, an impedance transforming filter is constructed between $R_G = 12\Omega$ generator (output of an RF power amplifier that is designed employing LD-MOS

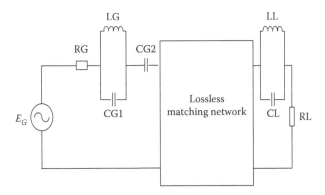

FIGURE 5.56
Double matching problem for Example 5.15.

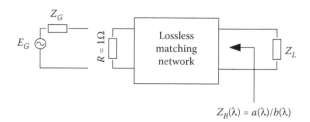

$$Z_B(\lambda) = a(\lambda)/b(\lambda)$$

FIGURE 5.57
Description of the matching network by means of its driving point input impedance $Z_B(\lambda)$.

RD07 of Mitsubishi) and a standard load of $R_L = 50\ \Omega$ using the RFDCT/Parametric algorithm. The design is completed over 850–2100 MHz. At the generator end (12 Ω), the resonance circuit $LG//CG1$ introduces a zero of transmission at 4200 GHz, which is the second harmonic at the high end of the passband. Furthermore, $CG2$ introduces a zero of transmission at DC. Similarly, at the load end, the tank circuit $LL//CL$ introduces a transmission zero at the third harmonic (6300 MHz). Thus, Figure 5.57 describes a double matching problem. Therefore, a lossless matching network is constructed between a complex generator Z_G and a complex load Z_L.

RG = 12 Ω; LG = 0.947 nH; CG1 = 1.515 pF, CG2 = 3.4 pF; CL = 1.515 pF, LL = 0.412 nH; RL = 50 Ω.

Referring to Figure 5.57, for the problem under consideration, RFDCT algorithm is implemented in MATLAB under the main program "*RichardMain_ImpTransFilter.m.*"

The matching network is described by means of its driving point impedance $Z_B(\lambda)$ employing six commensurate transmission lines ($n = 6$). In the course of design, $k = 4$ (total number of cascaded UEs) and $q = 0$ (no DC transmission zero) are selected, which in turn yield $n_\infty = k - q = 2$ transmission zeros at infinity.

Coefficients $\{c_i; i = 1, 2, ..., 6\}$ of the auxiliary polynomial $c(\Omega)$ are initialized in an ad hoc manner. Furthermore, c_0 is fixed as unity ($c_0 = 1$), so that the far-end normalized termination resistance R is set to unity.

Using the least-square nonlinear optimization tool (*lsqnonlin*) of MATLAB, driving point impedance $Z_B(\lambda)$ is obtained as below.

$$Z_B(\lambda) = \frac{a(\lambda)}{b(\lambda)}$$

$$= \frac{0\lambda^6 + 0.62\lambda^5 + 0.7259\lambda^4 + 0.2427\lambda^3 + 0.1717\lambda^2 + 0.0169\lambda + 0.0036}{2.1530\lambda^6 + 2.5204\lambda^5 + 1.1745\lambda^4 + 0.9848\lambda^3 + 0.1634\lambda^2 + 0.0754\lambda + 0.0036}$$

For interested readers, the above results may be reproduced using the accurate coefficients of the polynomials $c(\Omega)$, $a(\lambda)$, and $b(\lambda)$, as shown in Table 5.16. Coefficients of $c(\Omega)$ are obtained as a result of optimization. Coefficients of the numerator polynomial $a(\lambda)$ and the denominator polynomial $b(\lambda)$ are obtained using our MATLAB function "$[a,b] = Richard\_NewMinimumFunction(k,q,c,czero)$," where the input variables are selected, as shown at the top of the first column of Table 5.16. Coefficients of $a(\lambda)$ and $b(\lambda)$ are listed in the second and the third columns of the same table.

$Z_B(\lambda)$ is synthesized using our newly developed Richards's synthesis package called "$[Z\_UE,a\_new,b\_new,CT,CV] = Richard\_CompleteImpedanceSynthesis(a,b,k,q,R0,f0)$," with the normalization numbers R0 = 1, $f0 = 1/(2\pi)$. The output, vector Z_UE includes the normalized characteristic impedances of the cascaded transmission lines. Vectors a_new and b_new include the numerator and the denominator polynomials of the remaining impedance function $Z_{new}(\lambda) = a_{new}(\lambda)/b_{new}(\lambda)$ after cascaded line extractions. Vectors CT and CV include the synthesis result of $Z_{new}(\lambda)$. In this representation, vector CT includes

TABLE 5.16

Result of Optimization for Example 5.15 $n = 6$, $k = 4$, $q = 0$, $czero = 1$

$c(\Omega)$	$a(\lambda)$	$b(\lambda)$
1.0e + 02 *[.]		
5.925734959253541	0	2.152976016472793
6.085157899757289	0.620037880652403	2.520421163343775
−2.296435769309379	0.725858804052723	1.174454841708657
−2.380581686436587	0.242668481362066	0.984800857540397
0.157989962292003	0.171746189292335	0.163416611930619
0.175760624674885	0.016854588221304	0.075382287191566
0.010000000000000	0.003633264112008	0.003633264112008

TABLE 5.17

Result of Richards's Synthesis $CT(i) = 1 >$ *Series Richard Inductor*, $CT(i) = 8 >$ *Shunt Richard Capacitor*, $CT(i) = 9 >$ *Resistance*

Index	Z_UE	a_new	b_new	CT	CV
1	0.251700042758011	0	1.000000000000000	8	7.563713873657589
2	1.872104026360417	0.132210183608708	1.170668481236947	1	0.854212798940930
3	0.124594021679597	0.154774294849264	0.154774294850914	9	0.999999999989340
4	1.536354982668161				

the codes of the Richards components such that $CT(i) = 8$ refers to a shunt Richards's capacitor, $CT(i) = 1$ refers to a series Richards's inductor, $CT(i) = 9$ refers to termination resistor.

List of output vectors are shown in Table 5.17 and final synthesis of $Z_B(\lambda)$ is depicted in Figure 5.58.

In summary, characteristic impedances of Figure 5.58 are given by

$$Z_1 = 0.2517, \quad Z_2 = 1.8721, \quad Z_3 = 0.1246, \quad Z_4 = 1.5364$$

and Richards's components are specified as

$$C = 7.5637, \quad L = 0.8542$$

By selecting resistive normalization number $R_0 = 50\,\Omega$, actual element values are given by

$$Z\_UE = [12.5850 \quad 93.6052 \quad 6.2297 \quad 76.8177]\,\Omega$$

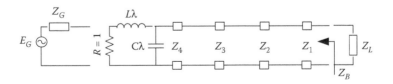

FIGURE 5.58
Synthesis of $Z_B(\lambda)$.

The Richards's capacitor C is realized as a shunt open-stub with normalized characteristic impedance $Z_{cap} = 1/C$ or with actual characteristic impedance $Z_{cap-act} = R_0/C$. Similarly, the Richards inductor L is realized as a series short-stub with normalized characteristic impedance $Z_{Ind} = L$ or with actual characteristic impedance $Z_{ind-act} = R_0 L$. Thus, it is found that $Z_{cap-act} = R_0/C = 50/7.5 = 6.6105 \; \Omega$ and $Z_{ind-act} = R_0 L = 42.7106 \; \Omega$. For the case under consideration, $Z_{3-act} = 6.2297 \; \Omega$ and $Z_{cap-act} = 6.6 \; \Omega$ could be difficult to realize.

As far as practical implementation is concerned, we may prefer to utilize microstrip technology to realize the ideal commensurate transmission lines. The shunt Richards's capacitors (i.e., open stubs in shunt configuration) can be easily realized but realization of the series Richards's inductors (i.e., short stubs in series configuration) presents serious difficulties. Nevertheless, physical implementation problems can be bypassed using the Kuroda identities [3,68]. In this regard, successive application of the Kuroda identities removes the series short stubs with those of shunt open stubs. Therefore, for interested readers, it may be appropriate to show the usage of Kuroda identities to improve the existing design.

Kuroda Identity I

Referring to Figure 5.59, a capacitive loaded transmission line can be replaced with its inductively loaded line equivalent employing the following equation set:

$$Z_B = \frac{Z_A}{C_A Z_A + 1} \tag{5.175}$$

$$L_B = \frac{C_A Z_A^2}{C_A Z_A + 1} \tag{5.176}$$

For our case, the Richards's capacitor and transmission line pair $\{C\lambda, Z_4\} = \{C\lambda = 7.5637\lambda, Z_4 = 1.5364\}$ can be replaced with that of transmission line and Richards's inductor pair $\{Z_{B1}, L_{B1}\lambda\}$ such that

$$Z_{B1} = \frac{Z_A}{C_A Z_A + 1} = 0.1217$$

$$L_{B1} = \frac{C_A Z_A^2}{C_A Z_A + 1} = 1.4146$$

Thus, at the first step, we exercised Kuroda identity I and end up with the following network topology.

FIGURE 5.59
Kuroda identity I.

Kuroda Identity II

Referring to Figure 5.60 pairs $\{L\lambda, Z_{B1}\}$ and $\{L_{B1}, Z_3\}$ can be replaced by their identical pairs using Kuroda identity II, as shown in Figure 5.61. The replacement equation set is given by

$$Z_B = Z_A + L_A \tag{5.177}$$

$$C_B = \frac{L_A}{Z_A(Z_A + L_A)} \tag{5.178}$$

In the second step, Kuroda identity II is applied twice. First, the pair $\{\lambda L_{B1}, Z_3\}$ is replaced by $\{Z_{B3}, C_{B3}\}$

$$Z_{B3} = Z_3 + L_{B1} = 1.5393$$

or actual value of Z_{B3} is $Z_{B3-act} = 76 \ \Omega$.

$$C_{B3} = \frac{L_{B1}}{Z_{B1}(L_{B1} + Z_{B1})} = 7.37$$

Similarly, the pair $\{L\lambda, Z_{B1}\}$ is replaced with the new one $\{Z_{B2}, C_{B2}\}$ such that

$$Z_{B2} = Z_{B1} + L = 0.9759$$

or actual value of Z_{B2} is found as $Z_{B2-act} = 48.7974 \ \Omega$.

$$C_{B2} = \frac{L}{Z_{B1}(L + Z_{B1})} = 7.1899$$

Hence, we end up with Figure 5.62 as the final synthesis of $Z_B(\lambda)$.

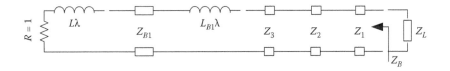

FIGURE 5.60
First step of the successive applications of Kuroda identities.

FIGURE 5.61
Kuroda identity II.

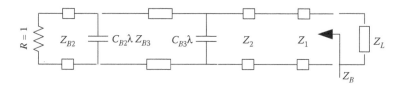

FIGURE 5.62
Synthesis of $Z_B(\lambda)$ after successive application of Kuroda identities.

The Richards's capacitor C_{B3} is realized as an open stub with actual characteristic impedance $Z_{cap3-act} = R_0/C_{B3} = 6.7787\ \Omega$. Similarly, C_{B2} is also realized as an open stub with actual characteristic impedance $Z_{cap2-act} = R_0/C_{B2} = 6.9542\ \Omega$. The final matching network is depicted in Figure 5.63.

The physical length of the commensurate transmission lines is fixed at normalized delay length $\tau = 0.5$. Then, we can compute the actual delay length as

$$\omega_a \tau_a = \omega_N \tau_N = 0.5 = 2\pi f_{c2} \tau_a$$

or

$$\tau_a = \frac{0.5}{2 \times \pi \times 2.1 \times 10^9} = 3.7894 \times 10^{-11}\ \text{s}$$

Physical length is computed upon the selection of substrate using the effective propagation velocity as

$$l = v_{eff} \tau_a$$

where v_{eff} is the effective velocity of propagation in the physical medium on which commensurate transmission lines are printed. The optimized TPG is depicted in Figure 5.65.

Alternative Design: Design with $n = 6$, $k = n$, $q = 0$ with $\tau = 0.5$

In the above design, implementation of low-characteristic impedances may create some problems. We can generate an alternative design with six cascaded elements. In this case, coefficients of the auxiliary polynomial $c(\Omega)$ is initialized in an ad hoc manner with fixed $c_0 = 1$. Then, the result of nonlinear optimization yields that

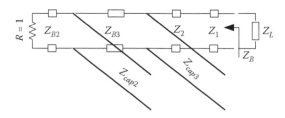

FIGURE 5.63
Synthesis of $Z_B(\lambda)$ after successive application of Kuroda identities; actual characteristic impedances: $Z_{B2} = 48.7974$, $Z_{Cap2} = 6.9542$, $Z_{B3} = 76.9633$, $Z_{Cap3} = 6.7787$, $Z_2 = 93.6050$, $Z_1 = 12.5850$.

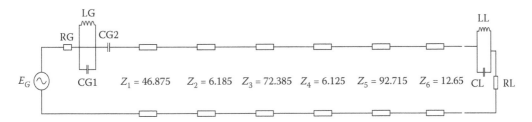

FIGURE 5.64
Alternative design with $n = 6, k = n, q = 0$ for Example 5.15.

$$Z\_UEFfipedover = [0.9375\ 0.1237\ 1.4477\ 0.1225\ 1.8543\ 0.2530]$$

$$Actual\,Z\_UEFlipedover = [46.8750\quad 6.1850\quad 72.3850\quad 6.1250\quad 92.7150\quad 12.6500]$$

$$c = [656.8151\quad 709.8196\quad -248.9498\quad -273.0388\quad 17.0399\quad 19.3983]$$

$$a = [0.0000\quad 0.6071\quad 0.7409\quad 0.2328\quad 0.1679\quad 0.0151\quad 0.0032]$$
$$b = [2.0933\quad 2.5548\quad 1.1301\quad 0.9780\quad 0.1528\quad 0.0717\quad 0.0032]$$

The resulting Richards's synthesis is shown in Figure 5.64 and the performance of the matched system is almost the same as the previous design, as depicted in Figure 5.65.

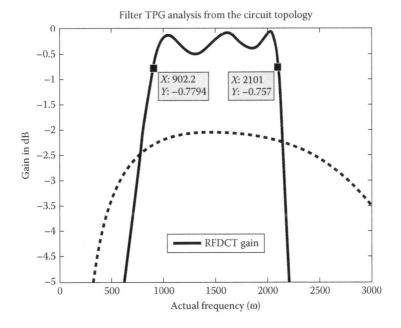

FIGURE 5.65
Performance of the matched system for Example 5.15.

5.12 SRFTs to Design RF and Microwave Amplifiers

Referring to Figures 5.54 and 5.55, in the SRFT, a lossless equalizer or matching is simply described in terms of its unit normalized scattering parameters [17,33]. Therefore, in the course of matching network design, neither the impedance nor the admittance parameters are employed. All the computations are carried out with the scattering parameters of the blocks to be matched. Therefore, the gain optimization of the matched system is well achieved numerically. The system is faster than the other existing CAD algorithms and easier to use. It is also naturally suited to design broadband microwave amplifiers.

Referring to Figure 5.66, unit normalized scattering parameters for a lossless equalizer [E] are described as

$$E = \begin{bmatrix} E_{11} & E_{12} \\ E_{21} & E_{22} \end{bmatrix} \tag{5.179}$$

Since the two port is lossless, it satisfies the following condition:

$$EE^{\dagger} = \begin{bmatrix} E_{11}(p) & E_{12}(p) \\ E_{21}(p) & E_{22}(p) \end{bmatrix} \begin{bmatrix} E_{11}(-p) & E_{21}(-p) \\ E_{12}(-p) & E_{22}(-p) \end{bmatrix} = I = \begin{pmatrix} 1 & 0 \\ 0 & 1 \end{pmatrix} \tag{5.180}$$

where the symbol "†" designates the paraconjugate transposition of a matrix.

The basis for SRFT or the scattering approach is to describe the lossless equalizer [E] in terms of its unit normalized input reflection coefficient $E_{11}(p)$ in V. Belevitch form [69] such that

$$E_{11}(p) = \frac{h(p)}{g(p)} \tag{5.181}$$

Furthermore,

$$E_{12}(p) = E_{21}(p) = \frac{f(p)}{g(p)} \tag{5.182}$$

$$E_{22}(p) = -\frac{f(p)}{f(-p)} \frac{h(-p)}{g(p)} \tag{5.183}$$

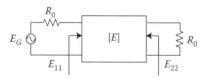

FIGURE 5.66
Description of a lossless two-port [E] by means of unit normalized scattering parameters.

where

$$h(p) = h_n p^n + h_{n-1} p^{n-1} + \cdots + h_1 p + h_0 \tag{5.184}$$

$$g(p) = g_n p^n + g_{n-1} p^{n-1} + \cdots + g_1 p + g_0 \tag{5.185}$$

Transmission zeros of the lossless equalizer [E] overlap with those zeros of the transfer scattering parameter $E_{21}(p) = f(p)/g(p)$. For many practical applications, it is practical to work with LC ladders with transmission zeros only at DC and infinity, and rarely at finite frequencies. In this case, numerator polynomial $f(p)$ is selected as

$$f(p) = p^{ndc} \prod_{i=1}^{nz} (p^2 + \omega_i^2) \tag{5.186}$$

$$f(-p) = (-1)^{ndc} p^{ndc} \prod_{i=1}^{nz} (p^2 + \omega_i^2) \tag{5.187}$$

$$\frac{f(p)}{f(-p)} = (-1)^{ndc} \tag{5.188}$$

Employing Equations 5.180 through 5.188, losslessness condition is expressed by

$$G(p^2) = g(p)g(-p) = h(p)h(-p) + f(p)f(-p) = H(p^2) + F(p^2) \tag{5.189}$$

$$G(p^2) = g(p)g(-p) = h(p)h(-p) + (1)^{ndc} p^{2ndc} \prod_{i=1}^{nz} (p^2 + \omega_i^2)^2 \tag{5.190}$$

$$G(p^2) = g(p)g(-p); H(p^2) = h(p)h(-p); F(p^2) = f(p)f(-p) \tag{5.191}$$

It should be noted that, in Equations 5.181 through 5.185, the common denominator polynomial $g(p)$ of degree "n" must be strictly Hurwitz. In other words, it must be free of closed RHP zeros. Numerator polynomial $h(p)$ of degree "n" is an arbitrary real polynomial. In other words, there is no restriction imposed on it.

On the $j\omega$ axis, the scattering parameters must be bounded by 1 such that

$$|E_{ij}(j\omega)| \le 1; \; i, j = 1, 2 \tag{5.192}$$

$$G(\omega^2) = H(\omega^2) + F(\omega^2) > 0 \tag{5.193}$$

Equation 5.193 is always true as long as even polynomials $H(\omega^2)$ and $F(\omega^2)$ are not simultaneously zero. Therefore, one can always generate a strictly Hurwitz polynomial $g(p)$ from Equation 5.191 by explicit factorization once transmission zeros are selected and the real coefficients of the polynomial $h(p)$ are initialized. This idea constitutes the crux of the SRFT. In other words, TPG of the system to be matched is expressed as an implicit function of $h(p)$.

Replacing generator and load impedances by their Darlington equivalent lossless two-ports [G] and [L], and utilizing their unit normalized scattering descriptions for [G], [E], and [L], TPG of the doubly matched system can be generated in two steps [3].

In the first step, by multiplying the transfer scattering parameters of the lossless two-ports [G] and [E], gain of the cascaded duo [GE] is generated as

$$T_1 = |G_{21}|^2 \frac{|E_{21}|^2}{|1 - E_{11}G_{22}|^2} \tag{5.194}$$

Then, in the second step, TPG of the cascaded two-ports [GE] and [L] is obtained as

$$T = T_1 \frac{|L_{21}|^2}{|1 - \hat{E}_{22}L_{11}|^2} \tag{5.195}$$

Thus, using the open form of T_1 in the above equation, the double matching gain of the trio {[G] − [E] − [L]} is expressed as follows:

$$T(\omega) = |G_{21}|^2 \frac{|E_{21}|^2}{|1 - E_{11}G_{22}|^2|1 - \hat{E}_{22}L_{11}|^2} |L_{21}|^2 \tag{5.196}$$

where the scattering parameters of the generator G_{ij} and the load L_{ij} networks are specified either by measurements or are provided as circuit models. More specifically, in terms of the generator $Z_G(j\omega)$ and the load $Z_L(j\omega)$ impedances, the generator reflectance is given by

$$G_{22} = \frac{Z_G - 1}{Z_G + 1}; \quad |G_{21}|^2 = 1 - |G_{22}|^2 \tag{5.197}$$

the load reflectance is given by

$$L_{11} = \frac{Z_L - 1}{Z_L + 1}; \quad |L_{21}|^2 = 1 - |L_{11}|^2 \tag{5.198}$$

and the back-end reflectance of the equalizer, while it is terminated in the complex generator, is given by

$$\hat{E}_{22} = E_{22} + \frac{E_{21}^2 G_{22}}{1 - E_{11}G_{22}} \tag{5.199}$$

In the "SRFT algorithm," the goal is to optimize the TPG as high and as flat as possible over the band of operation, as in the other real frequency techniques. The coefficients of the numerator polynomial $h(p)$ are selected as the unknowns of the matching problem. To construct the scattering parameters of [E], it is sufficient to generate the Hurwitz denominator polynomial $g(p)$ from $h(p)$. It can be readily shown that, for simple LC-ladder structures, once the coefficients of $h(p)$ are initialized and the complexity of the equalizer [E] specified (i.e., n and ndc are fixed in advance), then $g(p)$ is generated as a Hurwitz polynomial by explicit factorization of Equation 5.191. Thus, the bounded realness of the scattering parameters $\{E_{ij}; i, j = 1, 2\}$ is already built into the design procedure. It is noted that, in order to assure the realizability of the lossless equalizers, we cannot simultaneously allow $h(p)$ and $g(p)$ to become zero. Therefore, selection of initial coefficients of $h(p)$ must be proper. For example, if $k > 0$, we cannot let $h_0 = 0$ since it makes $g(0) = 0$.

In generating the Hurwitz denominator polynomial $g(p)$ from the initialized coefficients of $h(p)$, we first construct $G(p^2)$, as in Equation 5.190. That is,

$$G(p^2) = g(p)g(-p) = G_n p^{2n} + G_{n-1}p^{2(n-1)} + \cdots + G_1 p^2 + G_0 = h(p)h(-p) + (-1)^{ndc}p^{2ndc} \quad (5.200)$$

where the coefficients G_i are given as the convolution of $h(p)$ and $h(-p)$ with the contributing term $(-1)^{ndc}p^{2ndc}$ as follows:

$$G_0 = h_0^2$$
$$G_1 = -h_1^2 + 2h_2 h_0$$
$$\cdots$$

$$G_i = (-1)^i h_i^2 + 2\left(h_{2i}h_0 + \sum_{j=2}^{i}(-1)^{j-1}h_{j-1}h_{2j-i+1} \right) \quad (5.201)$$

$$\cdots$$

$$G_k = G_i|_{i=ndc} + (-1)^{ndc}$$

$$\cdots$$

$$G_n = (-1)^n h_n^2$$

Then, explicit factorization of Equation 5.200 follows. At the end of the factorization process, polynomial $g(p)$ is formed on the left half plane zeros of $G(p^2)$.

Hence, the scattering parameters of [E] are obtained, as in Equation 5.179, and the TPG $T(\omega)$ is generated employing Equation 5.196. Then, as in the other real frequency techniques, by selecting flat gain level T_0, the error function $error = T(\omega) - T_0$ is minimized. As the result of optimization, the unknown coefficients h_i are determined.

Details of the numerical work can be found in the reading list.

In brief, examination of Equations 5.194 through 5.196 together with Equation 5.200 indicates that TPG is almost inversely quadratic in the unknown coefficients h_i. Therefore, the SRFT algorithm is always convergent. Furthermore, the numerical stability of the algorithm is excellent, since all the scattering parameters E_{ij} and reflection coefficients G_{22} and L_{11} are bounded as real, that is, $\{|E_{ij}|, |G_{22}|, \text{and } |L_{11}|\} \leq 1$.

As is usually the case, an intelligent initial guess is important in efficiently running the program. It has been experienced that, for many practical problems, an ad hoc direct choice of the coefficients h_i (e.g., $h_i = \mp 1$) provides satisfactory initialization to start the SRFT algorithm.

As indicated previously, SRFT is naturally suited to design microwave amplifiers since it employs scattering parameters for all the units to be matched. For over 30 years, matching networks and amplifiers designed using SRFT has displayed excellent agreement with laboratory performance measurements.

5.13 SRFT to Design Microwave Amplifiers

It is well known that SRFT is well suited to design microwave amplifiers in two steps.

Referring to Figure 5.67, in the first step, front-end matching network $[E_F]$ is designed while the output port is terminated in 50 Ω (or in the normalization resistance $R_0 = 50$ Ω). At this step, TPG T_1 which is specified by

$$T_1(\omega) = \frac{|S_{21F}|^2 |S_{21}|^2}{|1 - S_{22F}S_{11}|^2} \tag{5.202}$$

is optimized to hit the ideal stable flat gain level

$$T_{flat1} = minimum \ of \ T_1(\omega) = T_{01} = \left\{\frac{|S_{21}(\omega)|^2}{1 - |S_{11}(\omega)|^2}\right\} over \ B \tag{5.203}$$

In Equations 5.202 and 5.203, $\{S_{ijF}; i, j = 1, 2\}$ and $\{S_{ij}; i, j = 1, 2\}$ are the real normalized (50 Ω) scattering parameters of the front-end matching network $[E_F]$ and the MOS transistor, respectively. B is the frequency band, which runs from 11.23 to 22.39 GHz (more than 10-GHz bandwidth).

In the second step, the back-end equalizer $[E_B]$ is designed to optimize the overall TPG $T_2(\omega)$, as shown by Figure 5.68. In this step, $T_2(\omega)$ is specified by

$$T_2(\omega) = T_1(\omega) . \frac{|S_{21B}|^2}{|1 - S_{2F}S_{11B}|^2} \tag{5.204}$$

At this step, the optimization algorithm targets the flat gain level of

$$T_{flat2} = minimum \ of \ T_2(\omega) = T_{02}(\omega) = \frac{T_1(\omega)}{1 - |S_{22}(\omega)|^2} \tag{5.205}$$

In Equation 5.205, the back-end reflection coefficient is given by

$$S_{2F} = S_{22} + \frac{S_{12}S_{21}}{1 - S_{11}S_{22F}} \tag{5.206}$$

FIGURE 5.67
Step 1: Design of the front-end matching network.

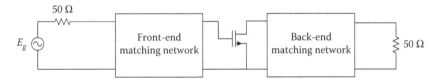

FIGURE 5.68
Step 2: Design of the back-end matching network.

In the SRFT, the front-end $[E_F]$ and the back-end $[E_B]$ matching networks are described in terms of their unit normalized input reflection coefficients in Belevitch form:

$$S_{22F} = \frac{h_F}{g_F} = \frac{h_{0F} + h_{1F}p + \cdots + h_{n_F F}p^{n_F}}{g_{0F} + g_{1F}p + \cdots + g_{n_F F}p^{n_F}} \qquad (5.207)$$

$$S_{11B} = \frac{h_{0B} + h_{1B}p + \cdots + h_{n_B F}p^{n_B}}{g_{0B} + g_{1B}p + \cdots + g_{n_B B}p^{n_B}} \qquad (5.208)$$

In Step 1, selecting the total number of equalizer elements n_F in $[E_F]$ first, initialize the coefficients $\{h_{0F}, h_{1F}, \ldots, h_{n_F}\}$. Then, the denominator polynomial $g_F(p)$ is generated as a strictly Hurwitz polynomial by the spectral factorization of

$$g_F(p)g_F(-p) = h_F(p)h_F(-p) + f_F(p)f_F(-p)$$

In this step, S_{21F} is given as

$$S_{21F} = \frac{f_F(p)}{g_F(p)}$$

and the monic polynomial $f_F(p)$ is selected by the designer to include transmission zeros of the front-end equalizer $[E_F]$. For example, if $[E_F]$ is a simple low-pass LC ladder, then $f_F(p) = 1$ selected. Eventually, coefficients $\{h_{0F}, h_{1F}, \ldots, h_{n_F}\}$ are determined by minimizing the error function

$$\varepsilon_1(\omega) = T_1(\omega) - T_{flat1}$$

over the band of interest $B = f_2 - f_1$ where f_1 and f_2 designate the lower and the upper end of the frequency band, respectively.

Similarly, in Step 2, the back-end equalizer $[E_B]$ is constructed by determining the unknown numerator coefficients $\{h_{0B}, h_{1B}, \ldots, h_{nB}\}$ of S_{11B}. In this case, the error function

$$\varepsilon_2(\omega) = T_2(\omega) - T_{flat2}$$

is minimized over the same frequency band.

Once S_{22F} and S_{11B} are determined, they are synthesized as lossless two ports, as in Darlington's procedure, yielding the circuit topologies for front-end and back-end matching networks with element values.

5.14 SRFT Single-Stage Microwave Amplifier Design Algorithm

In this section, the gain-bandwidth performance of a typical 0.18 μ NMOS-FET is assessed by designing an amplifier over 11.23–22.39 GHz.

The 50-Ω-based scattering parameters of the device are given in Table 5.18.

The biasing conditions and design parameters of the device are given as follows:

- VDD = 1.8V; ID (drain DC current) = 200 mA.
- NMOS size is W/L = 1000 μ/0.18 μm.
- Simulation temperature is 50°C.
- Expected output power is approximately 100 mW up to 23 GHz when operated as an A-Class amplifier.

For this purpose, we developed a MATLAB Main program which is called "*TSMC_amp.m*."

As explained above, this program constructs a microwave amplifier in two parts. In the first part, the front-end equalizer $[E_F]$ is constructed. Then, in the second part, the back-end matching network of the amplifier is completed.

The aim is to construct both equalizers employing LC-ladder networks without a transformer. Therefore, in the Main program, we set

$$k = 0$$

and we fix

$$h_{0F} = h_{0B} = 0$$

Thus, in the first part, the objective function

$$function\ Func\ =\ levenberg\_TR1(x,k,w,f11,f12,f21,f22,TS1)$$

minimizes $\varepsilon_1(\omega) = T_1(\omega) - T_{flat1}$, which in turn generates the numerator polynomial $h_F(p) = h_{nF}p^n + \ldots + h_{1F}p + 0$.

TABLE 5.18

Selected Samples from the Measured Scattering Parameters of 0.18 μm TSMC MOS Transistor

Fr(GH)	MS11	PS11	MS21	PS21	MS12	PS12	MS22	PS22
10.0	0.90	−2.99	1.11	1.1	0.05	−0.34	0.77	−2.96
11.2	0.90	−3.01	0.97	1.1	0.06	−0.39	0.78	−2.96
12.6	0.91	−3.01	0.84	1.1	0.05	−0.44	0.79	−2.96
14.1	0.91	−3.01	0.73	1.0	0.05	−0.49	0.80	−2.96
15.8	0.92	−3.02	0.63	0.90	0.05	−0.54	0.82	−2.96
17.8	0.92	−3.02	0.55	0.80	0.04	−0.60	0.83	−2.96
20.0	0.93	−3.03	0.46	0.81	0.04	−0.66	0.85	−2.96
22.4	0.94	−3.04	0.39	0.74	0.04	−0.72	0.86	−2.96

Note: In the following, MS*ij* are the amplitudes and PS*ij* are the phases of the S-parameters. Phases are given in radian.

In this case, the designer provides the following inputs to the main program: *TSMC_amp.m.*

5.14.1 Part I: Design of Front-End Matching Network

Inputs:
TMSC_Spar.txt: Scattering parameters of the CMOS transistor loaded from the text file.

In this file, S-parameters are read as amplitude and phase in degree in the following order:

$$MS11(i) = TMSC\_Spar(i,2); PS11(i) = TMSC\_Spar(i,3);$$
$$MS21(i) = TMSC\_Spar(i,4); PS21(i) = TMSC\_Spar(i,5);$$
$$MS12(i) = TMSC\_Spar(i,6); PS12(i) = TMSC\_Spar(i,7);$$
$$MS22(i) = TMSC\_Spar(i,8); PS22(i) = TMSC\_Spar(i,9).$$
$$(i = 1 \text{ to } 48)$$

The program converts all the phases from degree to radian, then generates the maximum stable gain of the amplifier at the first step as

$$T_{01} = \left\{ \frac{|S_{21}(\omega)|^2}{1 - |S_{11}(\omega)|^2} \right\}$$

and plots it. For the CMOS transistor under consideration, this figure is depicted in Figure 5.69.

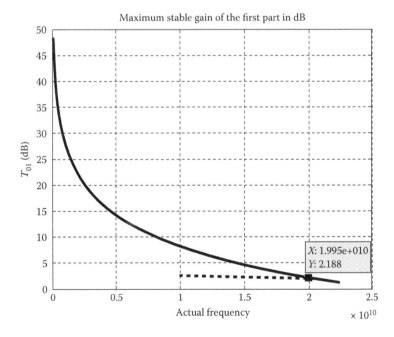

FIGURE 5.69
Maximum stable gain of part I of the single-stage amplifier design.

Close examination of the above figure reveals that one can obtain about 2.1 dB flat gain level, even beyond 20 GHz.

At this point, the designer inputs the frequency band of operation.

In this case, low (f_1) and the high (f_2) ends of the frequency band must be entered. The low end of the frequency band is denoted by

$$f_1 = FR(nd1)$$

Similarly, the high end is denoted by

$$f_2 = FR(nd2).$$

However, the user only enters the index of the low and high ends of the frequency band. The lower end of the optimization starts from the frequency $FR(nd1) = 11.23$ GHz, which is placed at the 42nd sampling point of the measured data.

Therefore, we set

$$nd1 = 42$$

The high end of the optimization frequency is $FR(nd2) = 22.39$ GHz, which is located at the 48th sampling point. Hence, we set

$$nd2 = 48$$

Now, the program asks the designer to enter the desired flat gain level T_{flat1} of Part I. Here, it may be appropriate to select a minimum or a slightly higher value of

$$T_{01} = \left\{ \frac{|S_{21}(\omega)|^2}{1- |S_{11}(\omega)|^2} \right\}$$

in the passband to assure the absolute stability of the design.

Referring to Figure 5.70, it may be proper to select

$$T_{Flat1} = 2.1 \text{ dB}$$

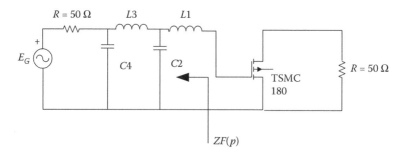

FIGURE 5.70
Completion of part I of the single-stage amplifier design employing the Si CMOS transistor TSMC 180 nm.

As the last input, we should enter the initials for the unknown coefficients such that

$$hF(n + 1) = 0, [xF(1) = hF(1), xF(2) = hF(2),\ldots, XF(n) = hF(n)]$$

Nonlinear optimization can be initialized in an ad hoc manner. For example, the following random initial coefficients yield successful optimization results.

$$[xF(1) = hF(1), xF(2) = hF(2),\ldots, XF(n) = hF(n)] = [-1 \quad -1 \quad -1 \quad 1]$$

The above input means that $nF = length(hF) = 4$. Therefore, we will employ only four elements in the LC ladder without an ideal transformer, since we set $hF(n + 1) = 0$ in advance.

5.14.1.1 Initialization of Nonlinear Optimization to Construct a Front-End Matching Network

While we construct a front-end matching network, nonlinear optimization may be automatically initialized using RF-LST introduced previously. For this purpose, we can use our MATLAB function *CompactSingleMatching* as called in the following codes:

```
1.  %  AUTOMATIC INITIALIZATION OF THE FRONT-END MATCHING NETWORK
2.  % Generation of the termination impedance for the front-end matching network
3.  NA=length(freq);j=sqrt(-1);
4.  for i=1:NA
5.  FAin(i)=freq(i)/1e9;
6.  Zin(i)=(1+S11R(i)+j*S11X(i))/(1-S11R(i)-j*S11X(i));
7.  Rin(i)=real(Zin(i));
8.  Xin(i)=imag(Zin(i));
9.  end

10. %     Initial Guess for SRFT Amplifier for front-end matching
11. %     network
12. % User selected Inputs:
13. T0=0.975;
14. NC=19;ktr=0;KFlag=1;sign=1;
15. FL=FAin(42);FH=FAin(48);f0=FAin(47);R0=50;F_unit=1e9;
16. n=5;ndc=0;WZ=0;a0=1;
17. %
18. Imp_input=[FAin Rin Xin];
19. UserSelectedInputs=[T0 NC ktr KFlag sign FL FH f0 R0 F_unit n ndc WZ a0];
20. Input_Data=[UserSelectedInputs Imp_input];
21. %
22. % --------------------------------------------------------------------
23. % Design of front-end matching network in single function:
24. %
25. [ CTF,CVF,CVAF,FAF,TAF,cF,aF,bF,a0F ] = CompactSingleMatching( Input_Data );
26. hF0=aF-bF;
27. for i=1:n
28. xF0(i)=hF0(i);
29. end
```

In the above codes, $\{S_{ij}; i, j = 1, 2\}$ designates the scattering parameters of the CMOS transistor designed using the TSMC 180 nm technology available in the Cadence library [77]. $Z_{in} = R_{in} + jX_{in}$ is the input impedance of the transistor when the output port is terminated in $R0 = 50\ \Omega$ and it is specified over the actual frequencies *FAin*. RF-LST starts with

$NC = 19$ break points. The ideal flat gain level for the single matching problem is selected as $T0 = 0.975$. We use an admittance-based design ($KFlag = 0$). The low end of the passband is selected as $= 11.22$ GHz ($FAin(42) = 11.2$ GHz). The high end of the passband is fixed at $FH = 20$ GHz (i.e., $FAin(48) = 22.38$ GHz). We use $n = 5$ elements in the low-pass ladder (with $ndc = 0$, $WZ = 0$) in the front-end matching network.

The above automatic initialization code is gathered under the MATLAB program "*GKY SRFTLumpedAmplDesignSec5_14.m.*"

5.14.2 Result of Optimization

At this part of the main program, the optimization function is described as below:

$$xF = lsqnonlin('levenberg\_TR1', xF0, [], [], OPTIONS, k, ww, f11, f12, f21, f22, T_{flat1});$$

function lsqnonlin is a least mean square minimization algorithm that employs the error function *levenber_TR1* such that

$$function\, Func = levenberg\_TR1(x, k, w, f11, f12, f21, f22, T_{flat1})$$

This function basically generates the error term in decibels.

$$\varepsilon_1 = T_1 - T_{flat1}$$

Minimization of $\varepsilon_1 = T_1 - T_{flat1}$ yields the optimum reflection coefficient

$$S_{22F}(p) = \frac{h_F(p)}{g_F(p)}$$

as shown in Table 5.19.

The above reflection coefficient leads to front-end driving point impedance

$$Z_F(p) = \frac{g_F + h_F}{g_F - h_F} = \frac{1.9293p^4 + 1.1928p^3 + 3.1183p^2 + 0.89332p + 1}{10.687p^3 + 6.6079p^2 + 10.888p + 1}$$

and is synthesized by long division as

$$Z_F(p) = 0.18052p + \cfrac{1}{9.2703p + \cfrac{1}{0.7128p + \cfrac{1}{1.6174p + 1}}}$$

TABLE 5.19

Front-End Reflection Coefficient $S_{22F} = (h_F/g_F)$

h_{oF}	0	g_{oF}	1
h_{1F}	−4.9972	g_{1F}	5.8905
h_{2F}	−1.7448	g_{2F}	4.8631
h_{3F}	−4.7473	g_{3F}	5.9401
h_{4F}	0.96463	g_{4F}	0.96463

TABLE 5.20

Normalized and Actual Element Values of Front-End Equalizer for the Microwave Amplifier Designed Employing TSMC 180-nm Si CMOS Transistor

Component-Type Inductors L Capacitors C	Normalized Element Values (CVF)	Actual Element Values (AEVF)
$L1(H)$	0.18052	6.4167e-011
$C2(F)$	9.2703	1.3181e-012
$L3(H)$	0.7128	2.5337e-010
$C4(F)$	1.6174	2.2996e-013

The normalized circuit elements of the front-end matching network are given in Table 5.20.

The actual circuit element values are generated by denormalization employing the *function AEVF = Actual(kimm,R0,fnorm,CVF)*, as shown in the same table.

Thus, the first part of the amplifier design procedure is completed, as shown in Figure 5.70.

The optimized gain performance of the first part of the design problem is depicted in Figure 5.71.

In the second part of the design process, the back-end equalizer of the amplifier is constructed by minimizing the error

$$\varepsilon_2 = T_2 - T_{flat2}$$

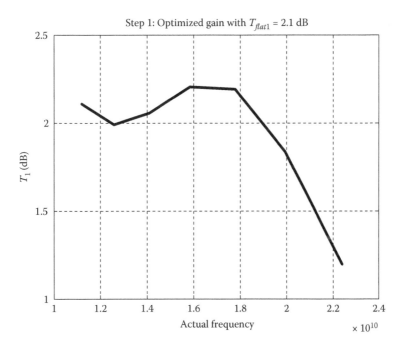

FIGURE 5.71
Optimized gain performance of part I.

At this point, the main program *TSMC_amp.m* calls the optimization routine *lsqnonlin* with error function

$$function\, Func = levenberg\_TR2(x,k,w,T1,SF2,f11,f12,f21,f22,T_{flat2})$$

The function *levenberg_TR2* generates $\varepsilon_2 = T_2 - T_{flat2}$ in decibel and is minimized by *lsqnonlin*, which in turn yields the reflectance

$$S_{11B} = \frac{h_B(p)}{g_B(p)}$$

of the back-end equalizer of the amplifier.

5.14.3 Inputs to the Main Program for Part II: Design of a Back-End Matching Network

At this point, the main program generates the maximum stable gain at the high end of the passband, which may be used as the possible target for the flat gain level for the amplifier.

For the present case, the main program returns to the designer with the following statements:

"Suggested maximum flat gain level for the second part of the design $T_{flat2} = 7.2059$

Enter the desired target gain value in dB for Step 2: $T_{flat2} =$"

In this case it may be appropriate to enter

$$T_{flat2} = 7.1\, dB$$

Then, the main program asks for the initial values for the back-end equalizer such that

"enter back-end equalizer; [xB(1)=hB(1),xB(2) = hB(2),…,XB(n) = hB(n)] ="

Here, we use an ad hoc initialization for the coefficient as

$$xB0 = [-1 \quad -1 \quad -1 \quad 1 \quad -1 \quad 1]$$

Meaning that we will have $nB = length\, (hB) = 6$ elements in the back-end matching network. We already fixed the structure to LC ladder without an ideal transformer by setting $k = 0$ and $hB0 = 0$ in advance.

Thus the optimization returns with the following results.

5.14.4 Results of Optimization

Back-end reflection coefficient is determined as in Table 5.21.

Finally, $S_{11B} = h_B/g_B$ is synthesized as in part I, as shown in Figure 5.72.

TABLE 5.21

Back-End Reflection Coefficient $S_{11B} = h_B/g_B$
(or in short S_B)

h_{oB}	0	g_{oB}	1
h_{1B}	−5.9242	g_{1B}	7.8643
h_{2B}	−1.9747	g_{2B}	13.375
h_{3B}	−15.969	g_{3B}	25.796
h_{4B}	2.8824	g_{4B}	20.767
h_{5B}	−9.2939	g_{5B}	17.764
h_{6B}	6.4072	g_{6B}	6.4072

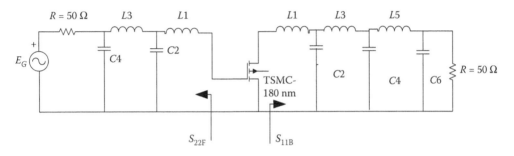

FIGURE 5.72
A single-stage amplifier designed using SRFT for a typical 0.18 μm NMOS transistor over 11.23–22.39 GHz.

TABLE 5.22

Normalized and Actual Element Values of Back-End Equalizer
for the Microwave Amplifier Designed Employing TSMC 180-nm
Si CMOS Transistor

Component Type	Normalized Element Values (CVB)	Actual Element Values (AEVB)
$L1(H)$	0.4736	1.6834e−010
$C2(F)$	6.9925	9.9422e−013
$L3(H)$	0.5019	1.7841e−010
$C4(F)$	5.2831	7.5117e−013
$L5(H)$	0.9645	3.4285e−010
$C6(F)$	1.5129	2.1511e−013

The element values of the back-end equalizer are summarized in Table 5.22.
The gain performance of the amplifier is depicted in Figure 5.73 and listed in Table 5.23.

5.14.4.1 Initialization of Nonlinear Optimization to Construct a Back-End Matching Network

A back-end matching network of the amplifier can be constructed by automatic initialization of the nonlinear optimization using the RF-LST as introduced to construct the

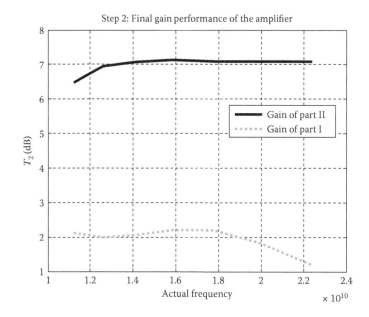

FIGURE 5.73
Ideal performance of the amplifier over the normalized frequencies from 0.55 (11.22 GHz) to 1 (22.39 GHz).

TABLE 5.23

Gain Performance of the Amplifier

Actual Frequency	Gain (dB)
1.1220e + 010	6.4730
1.2589e + 010	6.9404
1.4125e + 010	7.0587
1.5849e + 010	7.1319
1.7783e + 010	7.0882
1.9953e + 010	7.0980
2.2387e + 010	7.0987

front-end matching network. In this case, we use the following MATLAB codes under the main program *"GKYSRFTLumpedAmplDesignSec5_14.m"*

```
%   AUTOMATIC INITIALIZATION OF THE BACK-END MATCHING NETWORK
% Generation of the termination impedance for the back-end matching
network
NA=length(freq);j=sqrt(-1);
for i=1:NA
    FAout(i)=freq(i)/1e9;
    Zout(i)=(1+S22R(i)+j*S22X(i))/(1-S22R(i)-j*S22X(i));
    Rout(i)=real(Zout(i));
    Xout(i)=imag(Zout(i));
end
%              Initial Guess for SRFT Amplifier for back-end matching
%              network
% User selected Inputs:
```

```
    T0=0.975;
    NC=19;ktr=0;KFlag=1;sign=1;
    FL=FAout(41);FH=FAout(47);f0=FAout(47);R0=50;F_unit=1e9;
    n=4;ndc=0;WZ=0;a0=1;
%
Imp_input=[FAout Rout Xout];
UserSelectedInputs=[T0 NC ktr KFlag sign FL FH f0 R0 F_unit n ndc WZ a0];
Output_Data=[UserSelectedInputs Imp_input];
%
%--------------------------------------------------------------------------
% Design of back-end matching network in single function:
%
[CTB,CVB,CVAB,FAB,TAB,cB,aB,bB,a0B] = CompactSingleMatching (Output_Data);
hB0=aB-bB;
for i=1:n
    xB0(i)=hB0(i);
end
```

5.14.5 Stability of Amplifier

In order to analyze the stability of the amplifier, the input and the output impedances of the matched amplifier must be computed.

Let $Z_{in}(j\omega) = R_{in}(\omega) + jX_{in}(\omega)$ and $Z_{out}(j\omega) = R_{out}(\omega) + jX_{out}(\omega)$ be the input and the output impedances of the matched amplifier.

Then, absolute stability of the amplifier requires that

$$R_{in}(\omega) \geq 0; \forall \omega \tag{5.209}$$

and

$$R_{out}(\omega) \geq 0; \forall \omega \tag{5.210}$$

Sometimes, it may be difficult to satisfy the above equation. In this case, we may be challenged by conditional stability.

In general, let the amplifier be terminated in complex generator with internal impedance $Z_G = R_G(\omega) + jX_G(\omega)$ and a complex load $Z_L(j\omega) = R_L(\omega) + jX_L(\omega)$.

Then, conditional stability requires that

$$R_G(\omega) + R_{in}(\omega) \geq 0; \forall \omega \tag{5.211}$$

and

$$R_L(\omega) + R_{out}(\omega) \geq 0; \forall \omega \tag{5.212}$$

For the problem under consideration, $S_G = Z_G - 1/Z_G + 1$ is the input reflectance of the front-end equalizer, which is generated in the first part of the amplifier design process as $SF = hF/gF$.

Similarly, complex termination $S_L = Z_L - 1/Z_L + 1$ of the active device is generated as the input reflectance of the back-end equalizer such that $S_L = S_B = hB/gB$.

In this regard, we need to generate the input and output impedances to investigate the stability situation of the matched amplifier.

The input reflectance S_{in} is given by

$$S_{in} = f_{11} + \frac{f_{12}f_{21}}{1 - f_{22}S_B} S_B \tag{5.213}$$

where
  $\{f_{ij}; i, j = 1, 2\}$ are the S-parameters of the active device
and

$$S_B = \frac{h_B}{g_B} \tag{5.214}$$

is the input reflectance of the back-end equalizer.
  Then, the input impedance is generated form S_{in} as

$$Z_{in} = \frac{1 + S_{in}}{1 - S_{in}} \tag{5.215}$$

such that

$$Z_{in}(j\omega) = R_{in}(\omega) + jX_{in}(\omega) \tag{5.216}$$

  Similarly, output reflectance is as

$$S_{out} = f_{22} + \frac{f_{12}f_{21}}{1 - f_{11}S_{11F}} S_F \tag{5.217}$$

where

$$S_F = \frac{h_F}{g_F} \tag{5.218}$$

is the input reflectance of the front-end equalizer.
  Then, the output impedance is generated from S_{out} as

$$Z_{out} = \frac{1 + S_{out}}{1 - S_{out}} \tag{5.219}$$

such that

$$Z_{out}(j\omega) = R_{out}(\omega) + jX_{out}(\omega) \tag{5.220}$$

  All the above computations are carried out employing the following statements in the main program *TSMC_amp.m*.

5.14.5.1   Investigation on the Stability of the Amplifier

Stability check at the output port

$$SB11 = hoverg(hB, k, ww);$$

(Input reflectance of the back – end equalizertance from hB)

$$SL = SB11;$$

$$ZL = stoz(SL);$$

$$RL = real(ZL);$$

(Complex termination of the active device)

$$Sout = SF2;$$

$$Zout = stoz(Sout);$$

$$Rout = real(Zout);$$

(output reflectance of the active devices when it is terminated in

front – end equalizer i.e in SG = SF)

$$StabilityCheck\_output = RL + Rout;$$

(it must be always nonnegative)

Stability check at the input port

$$SF = hoverg(hF, k, ww);$$

$$SG = SF;$$

(Complex input termination of the active device)

$$ZG = stoz(SG);$$

$$RG = real(ZG);$$

$$[Sin] = INPUT\_REF(SB11, f11, f12, f21, f22);$$

$$Zin = stoz(Sin);$$

$$Rin = real(Zin);$$

$$StabilityCheck\_input = RG + Rin;$$

(it must be always nonnegative)

TABLE 5.24

Stability Check of the Amplifier

Frequency (GHz)	Absolute Stability Check		Conditional Stability Check	
	R_{in}	R_{out}	$RG + R_{in}$	$RL + R_{out}$
11.22	−0.0067505	0.115920	0.050926	0.22373
12.589	0.0064048	0.092387	0.059000	0.22254
14.125	0.017844	0.067221	0.067412	0.23336
15.849	0.023586	0.042898	0.071946	0.24681
17.783	0.025254	0.024308	0.073142	0.24775
19.953	0.023913	0.012860	0.068118	0.24740
22.387	0.016272	0.003359	0.048029	0.078335

Note: All the impedances given in this table are normalized with respect to $R_0 = 50\ \Omega$.

Now, let us investigate the stability of the amplifier as designed with the above front-end and back-end equalizers.

Table 5.24 indicates that the amplifier is absolutely stable from 12.6 to 22.39 GHz but is conditionally stable at 11.22 GHz.

5.14.6 Practical Design Aspects

- It has been seen that SRFT is naturally suited to design microwave amplifiers in a sequential manner. In fact, one can even design a multistage amplifier step by step, as described by Yarman and Carlin [33]. Details are omitted here. Interested readers are referred to the reading list provided at the end of this chapter. In this regard, the MATLAB main program *TSMC_Amplifier.m* can easily be modified to design multistage amplifiers.

- The stability of Equations 5.209 through 5.212 can be utilized as constraints in the course of TPG optimization.

- Upon completion of the equalizer design with ideal elements, lumped circuit components can be realized on a selected substrate.

- Physical implementation of lumped circuit elements introduces material loss, and some inductive and capacitive parasitism, as expected. These additions degrade the final electrical performance of the power transfer networks designed via other

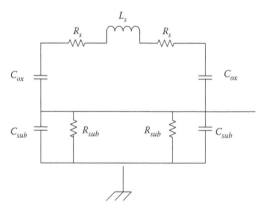

FIGURE 5.74
A 9-element model for an ideal inductor L_S.

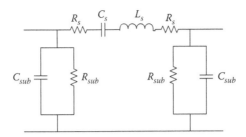

FIGURE 5.75
An 8-element model for an ideal capacitor C_S.

real frequency techniques. Nevertheless, once we have an idealized solution, we can always reoptimize it with commercially available software design packages employing the circuit models provided by the foundries. For example, an ideal inductor may be replaced by a practical model, which is shown in Figure 5.74.

• Similarly, an ideal capacitor acts as a passive device that may be described by a model, as shown in Figure 5.75.

• Parameters of the selected models are specified by geometric layouts and material properties of the components that are provided by the manufacturers.

• In order to end up with actual electrical performance of the matching networks or microwave amplifiers designed employing the real frequency techniques, we need to replace the ideal element values with those models that include loss and parasitic components; the way they are produced with the technology under consideration. On the other hand, degraded performance of the matched system can be improved by reoptimization of the gain in systems on the element values of the preset models.

• It should also be noted that, in practice, neither an ideal lumped component nor an ideal distributed element exists. To be more realistic, physical passive layout may be modeled with both lumped and distributed elements, which leads the engineer to design lossless two ports with both lumped and distributed elements. From the theoretical point of view, designs consisting of mixed elements are highly complicated.

5.15 Design of an Ultra-Wideband Microwave Amplifier Using Commensurate Transmission Lines

In the previous section, it has been emphasized that ideal transmission lines are very handy to introduce physical sizes and geometric layouts in microwave circuit design. In fact, lumped circuit components such as resistors, capacitors, and inductors are realized in terms of the geometric parameters of the layouts. A simple layout may be a TEM line. Therefore, in this section, we will exhibit the design of a wideband microwave amplifier using commensurate transmission lines.

In the present case, the active device used in the microwave amplifier is a 180 nm CMOS transistor that was designed by Dr. N. Retdian of Tokyo Institute of Technology using the VDEC foundry tools of Tokyo University [74].

The original amplifier chip consists of lumped components [74]. In the present case, however, a similar design is made with commensurate transmission lines. The scattering

TABLE 5.25

$R_0 = 50 \, \Omega$ Normalized Scattering Parameters of a 180 nm Silicon CMOS Transistor

Frequency (GHz)	S_{11R}	S_{11X}	S_{21R}	S_{21X}	S_{12R}	S_{12X}	S_{22R}	S_{22X}
0.45	0.99	−0.05	−1.71	0.10	0.000	0.010	0.95	−0.04
0.7	1.00	−0.07	−1.71	0.15	0.001	0.015	0.94	−0.07
1	0.99	−0.10	−1.70	0.22	0.002	0.022	0.94	−0.09
2	0.97	−0.20	−1.64	0.43	0.009	0.042	0.92	−0.19
6	0.76	−0.51	−1.14	1.02	0.064	0.096	0.75	−0.49
10	0.51	−0.64	−0.56	1.19	0.127	0.101	0.54	−0.65

parameters of the 180 nm silicon CMOS transistor are summarized in Table 5.25 and the information is stored on a MATLAB text file called *newcmos.txt*.

As in the previous example, here we also design the microwave amplifier in two steps. In the first step, the front-end, and in the second step, the back-end, equalizers are constructed. However, in this case, designs are made using commensurate transmission lines. Therefore, the scattering parameters of the equalizers are described in Richards's variable λ such that, on the real frequency axis $j\omega$

$$\lambda = j \tan(\omega \tau)$$

where the constant delay τ is fixed as

$$\tau = \frac{1}{4(f_{end} \times kLength)}$$

kLength is a user-defined parameter that refers to the fraction of the upper edge of the passband frequency f_{end}.

The amplifier design program developed for the previous section is revised to handle the design with commensurate transmission lines. The new program is called *Amplifier_Distributed.m*.

Let us summarize the design steps as follows.

5.15.1 The First Step of the Amplifier Design: Description of the Front-End Matching Network

The front-end equalizer is described in terms of its unit normalized scattering parameters (normalization number $R_0 = 50 \, \Omega$).

In this step, the unknown of the design problem is the numerator polynomial

$$h_F(\lambda) = h_1 \lambda^{nF} + h_2 \lambda^{(nF-1)} + \cdots + h_{nF} + h_{nF+1}$$

of the input reflection coefficient

$$S_{22F} = \frac{h_F(\lambda)}{g_F(\lambda)}$$

On the other hand, the transfer scattering parameter of the equalizer is specified by

$$S_{21}(\lambda) = \frac{\lambda^{qF}(1-\lambda^2)^{kF/2}}{g_F}$$

Once the coefficients $h_F(\lambda)$ are initialized, a strictly Hurwitz polynomial $g_F(\lambda)$ is generated by the losslessness condition, as in Equations 5.189 through 5.191, such that

$$G(\lambda^2) = g_F(\lambda)g_F(\lambda) = h_F(\lambda)h_F(-\lambda) + (-1)^{qF}(1-\lambda^2)^{kF}$$

Based on the above nomenclature, we set the following inputs to the MATLAB main program *"Amplifier_Distributed.m."*

Passband subject to optimization

$$f_{Start} = f_{C1} = 1GHz, f_{end} = f_{C2} = 10GHz$$

where
f_{start} is the beginning frequency of the optimization
f_{end} is the high end of the passband where the optimization ends
$q_F = 0$ is the count of transmission zeros at DC (It means that there is no transmission zero at DC)
$k_F = 6$ is the total number of cascaded sections
$n_F = 6$ is the total number of elements in the front-end equalizer

It should be noted that, by choosing $n_F = k_F$, we avoid using series short or open shunt stubs. This fact makes the implementation easy. Thus, the front-end equalizer will consist of six cascade connections of UEs.

$ntr = 0$ means that no ideal transformer is used in the design.

$kLength = 0.37$. This is our choice to fix the constant delay of UE.

$T_{flat1} = 6.3$ dB; idealized flat gain level subject to optimization in the first part.

$xF = [-1 -1 -1 \ 1 \ 1 -1 \ 1]$; initials for the optimization.

Here, it should be noted that the unknown of the front-end equalizer is the numerator polynomial $h_F(\lambda) = h_1\lambda^{nF} + h_2\lambda^{(nF-1)} + \cdots + h_{nF} + h_{nF+1}$. In other words, by using MATLAB polynomial convention, at the beginning of the optimization we set the numerator polynomial as

$$h_{F0} = [xF \ h_{F+1}] = [xF \ 0]$$

As you may have noticed, in the above equation we automatically set $h_{F+1} = 0$, since we do not want an ideal transformer in the front-end equalizer. Actually, in order to simplify the main program, we set $h_{nF+1} = 0.0$ in advance. In this case, a simplified version of the transfer scattering parameter is given as

$$S_{21F}(\lambda) = \frac{(1 - \lambda^2)^{kF/2}}{g_F}$$

At this step, error *error* = $T_1(\omega) - T_{flat1}$, which is generated by

$$function\, Func = Distributed\_TR1(x, kLength, qF, kF, w, f11, f12, f21, f22, Tfalt1)$$

is minimized. The list of this function together with the main program *Amplifier_ Distributed.m* is given at the end of this chapter.

5.15.2 The Second Step of the Amplifier Design: Description of the Back-End Matching Network

In this step, the back-end equalizer is described in terms of its input scattering parameters as

$$S_{11B} = \frac{h_B(\lambda)}{g_B(\lambda)}$$

where

$$h_B(\lambda) = h_1\lambda^{nB} + h_2\lambda^{(nB-1)} + \cdots + h_{nB} + h_{nB+1}$$

The transfer scattering parameter is specified by

$$S_{21B}(\lambda) = \frac{\lambda^{qB}(1 - \lambda^2)^{kB/2}}{g_B}$$

Similarly, the denominator polynomial $g_B(\lambda)$ is generated via losslessness of the back-end equalizer, as in Equations 5.189 through 5.191.

At this step, error *error* = $T_2(\omega) - T_{flat2}$, which is generated by

$$function\, Func = Distributed\_TR2(x, kLength, qB, kB, w, T1, SF2, T_{flat2})$$

is minimized. A list of this function together with the main program *Amplifier_ Distributed.m* and other related MATLAB functions is given at the end of this chapter.

Now, let us use the following inputs for the back-end equalizer design.

Passband subject to optimization

$$f_{Start} = f_{C1} = 1\,GHz, f_{end} = f_{C2} = 10\,GHz$$

$q_B = 0$: Total count of transmission zeros at DC
$k_B = 6$: Total number of cascaded sections
$n_B = 6$: Total number of elements in the back-end matching network
$ntr = 0$: No ideal transformer

TABLE 5.26

Result of Optimization of Step 1

$T_{flat1} = 6.3$ dB	$kLength = 0.37$	$nd1 = 51$ freq(51) = 1 GHz	$nd2 = 101$ freq(101) = 10 GHz
hF	gF	zF (Normalized)	ZFActual (Ω)
−15.65	15.682	1.8171	Z1F = 90.854
42.125	63.873	2.6136	Z2F = 130.68
8.5322	64.792	0.5207	Z3F = 26.035
27.25	57.638	3.1297	Z4F = 156.49
4.1329	26.387	0.52992	Z5F = 26.496
2.4754	8.0561	1.9205	Z6F = 96.026
0	1	1	R0 = 50

Actually, the main program *"Amplifier_Distributed.m"* is organized in such a way that it makes designs without a transformer automatically. Therefore, the user of the program does not need to bother with *ntr*. In this regard, h_{F+1} is also fixed at zero (i.e., $h_{F+1} = 0$)

$kLength = 0.37$

$T_{flat2} = 10.5$ dB; idealized flat gain level subject to optimization in the first part

$xB = [-1 - 1 - 1 \ 1 - 1 \ 1]$; initials for the optimization.

As explained above, initials for the polynomial for $h_B(\lambda)$ are expressed in the main program as

$$h_{B0} = [xB \quad 0].$$

5.15.3 Result of Optimization

In the first step, optimization of the TPG $T_1(\omega)$ yields the results summarized in Table 5.26.

In the second step, optimization of the TPG $T_2(\omega)$ reveals the following results (Table 5.27).

Gain performance of Step 1 is depicted in Figure 5.76.

Overall gain performance is depicted in Figure 5.77.

The schematic of the amplifier is shown in Figure 5.78.

TABLE 5.27

Result of Optimization of Step 2

$T_{flat2} = 10.5$ dB	$kLength = 0.37$	$nd1 = 51$ freq(51) = 1 GHz	$nd2 = 101$ freq(101) = 10 GHz
hB	gB	zB(Normalized)	ZBActual
5.0982	5.1954	5.7794	Z1B = 288.97
81.673	83.839	0.83948	Z2B = 41.974
23.821	57.315	4.0919	Z3B = 204.60
54.973	70.825	1.2459	Z4B = 62.293
7.3537	25.852	1.1117	Z5B = 55.587
5.8952	9.6154	2.4422	Z6B = 122.11
0	1	1	R0 = 50

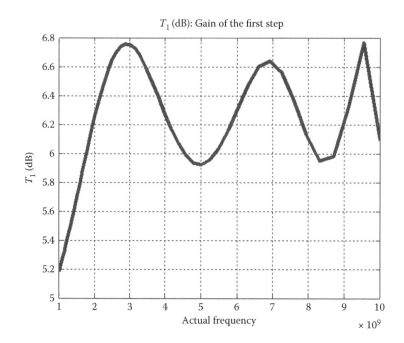

FIGURE 5.76
Gain performance of Step 1.

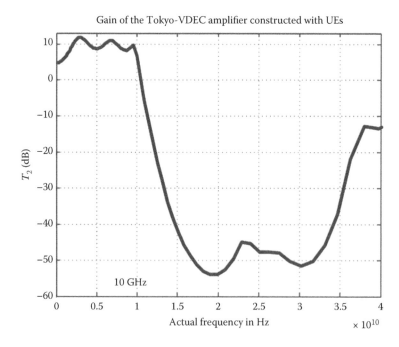

FIGURE 5.77
Gain performance of the overall amplifier constructed with commensurate transmission lines.

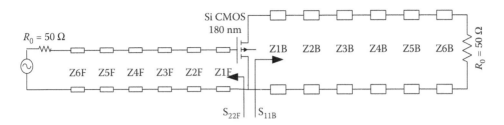

FIGURE 5.78
Design of a microwave amplifier using Si CMOS of 180 nm of VDEC of Tokyo University with TEM commensurate transmission lines.

Practical notes

- Using the same transistor and the same passband, a similar amplifier was designed employing lumped elements [74].
- When the performance of the lumped element amplifier is compared with the amplifier designed with UEs, it is observed that the lumped element design offers much better gain performance over the amplifier design with UEs. This is expected, as the lumped element design offers a smooth rolloff in the passband. In other words, power delivered to the load is pretty much confined within the band of operation. Whereas, the design made with UEs spreads the gain somewhat periodically over the entire range of frequencies by suppressing it. This is due to the periodicity of the Richards's variable $\lambda = j\tan(\omega\tau)$. This fact is clearly shown in Figure 5.80. At 10 GHz (high end of the passband), gain starts dropping drastically and it reaches its minimum value of -55 dB at 19 GHz. Then it starts increasing to -15 dB at 38 GHz. Of course, the shape is not exactly periodic as opposed to the filter design with UE. This is due to the presence of an active device. Remember that active devices are modeled with lumped elements, which suppress the gain as the frequency increases and suppresses the periodic nature of Richards's frequency $\Omega = \tan(\omega\tau)$.
- We may wish to consider that the front-end and the back-end equalizers of the amplifier designed for the above example may resemble the operation of a low-pass CLC or LCL section. On the other hand, one should also bear in mind that this is a loose statement, since we know exactly what the line impedance equations are.
- It is crucial to point out that, in the process of gain optimization, fixed delay length τ may as well be included among the unknowns of the design problem. It also affects the characteristic impedance of the UEs.

For example, with the above inputs to the main program *Amplifier_Distributed.m*, if we choose *KLength* = 0.3, we end up with similar gain performance but in different characteristic impedances such that

$$ZFActual = [102.48 \ 87.361 \ 36.289 \ 116.77 \ 34.321 \ 80.041]$$

$$ZBActual = [229.98 \ 54.205 \ 157.95 \ 65.844 \ 69.276 \ 90.29]$$

Corresponding gain performance of the amplifier is shown in Figure 5.79.

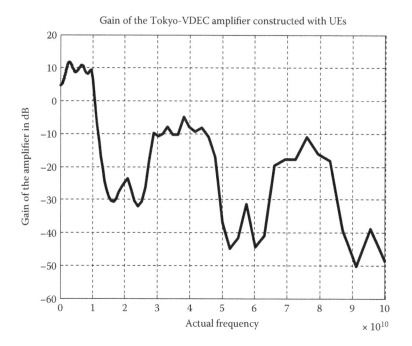

FIGURE 5.79
Gain performance of the amplifier designed for length = 0.3.

Clearly, new characteristic impedances are smaller and perhaps easier to realize as microstrip lines. The reader can run the main program *Amplifier_Distributed.m* with different values of *kLength* and investigate the effect of this parameter on the gain performance as well as on the characteristic impedances.

5.16 Physical Realization of Characteristic Impedance

Physical delay length L_{actual} of the UEs can be determined after the selection of the substrate, which is described by its thickness h, dielectric constant or permittivity $\varepsilon = \varepsilon_r \varepsilon_0$ and the permeability $\mu = \mu_r \mu_0$ [69–73].

In this case,

$$L_{actual} = v\tau_{Actual} = \frac{\tau_{Actual}}{\sqrt{\mu\varepsilon}}$$

Based on our description

$$\tau = \frac{1}{4 \times f_{e\text{-}normalized} \times kLength} = \frac{1}{4 \times kLength}$$

Then, the actual delay is given by

$$\tau_{Actual} = \frac{\tau}{f_e} \, s$$

For example, for a silicon substrate, $\varepsilon_r = 3.8$, $\mu_r = 1$. Then, for $kLength = 0.3$

$$\tau = \frac{1}{4kLength} = \frac{1}{4 \times 0.3} = 0.83333$$

and

$$\tau_{Actual} = \frac{\tau}{f_e} = \frac{0.8333}{10 \times 10^9} = 0.0833 \, ns$$

Thus, the actual physical length is

$$L_{actual} = \frac{0.0833 \times 10^{-9}}{\sqrt{3.8}} \times 3 \times 10^8 = 0.0042732 \, m = 4.2732 \, mm$$

If UEs are realized as microstrip TEM lines, the width of the lines can be determined by means of readily available formulas given for the characteristic impedance calculation as in a Bahl's book [74].

For example, the Main Program
"Microstrip_CharacteristicImpedance.m"
calculates the characteristic impedance "Z" of a given microstrip line for specified ε_r, width "w," and substrate thickness "h." A list of this program is given by Program Listing 5.90. Just to orient the reader, let us run the following example.

EXAMPLE 5.16: AN EXAMPLE FOR THE CALCULATION OF CHARACTERISTIC IMPEDANCE

Let $\varepsilon_r = 3.8$, $w = 0.02$ mm $= 20$ µm, and $h = 1$ mm, then the characteristic impedance Z is found as

$$Z = 226.87 \, \Omega$$

which is a high value of characteristic impedance; it acts like a series inductor.

On the other hand, if $w = 1$ mm then, the characteristic impedance becomes $Z = 75.3 \, \Omega$.
If $w = 4$ mm, then $Z = 32.7 \, \Omega$.
Similarly, if $w = 10$ mm, then $Z = 15.8 \, \Omega$ is shown in Figure 5.80.

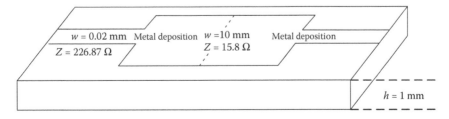

FIGURE 5.80
Realization of UE microstrip lines on a selected substrate with fixed actual length $L_{Actual} = 4273$ µm.

All the above calculations are symbolic and may not necessarily be feasible to print on a silicon chip.

We should mention that this chapter is intended to provide the reader with design capability of power transfer networks with ideal circuit elements. However, just to give some insights, we have also included an elementary design concept of circuit component stems from electromagnetic field theory. However, details and the related technologies to manufacture circuit components are out of scope of this book. Interested readers are referred to the reading list presented at the end of this chapter.

We will conclude this chapter by providing a list of the amplifier design programs and related functions developed on MATLAB. These programs and related functions have already been utilized to design the single-stage amplifiers included in this chapter.

5.17 Practical Design of Matching Networks with Mixed Lumped and Distributed Elements

In designing matching networks using lumped components, one may encounter difficulties to physically realize lumped inductors. In practice, realization of inductors is troublesome due to parasitic and coupling capacitors introduced in between winding. Therefore, it is always preferable to replace inductors with equivalent transmission lines. On the other hand, during the physical implementation of transmission lines such as microstrips, discontinuity capacitors will be introduced at both ends of the line. In this case, it may be appropriate to consider replacement of a lumped C-L-C symmetrical section with that of a symmetrical capacitive loaded transmission line section, as shown in Figure 5.81. In the following text, we present an almost-equivalent model of a C-L-C lumped-pi section with a symmetrical capacitive loaded transmission line, in short CT-TRL-CT section.

5.17.1 An Almost Equivalent Transmission-Line Model of a CLC-PI Section

A symmetrical low-pass lumped CLC-PI section can be approximated by a capacitive loaded transmission line TRL, as shown in Figure 5.81 [42].

In terms of the specified inductor L, characteristic impedance Z_0 of the line is given by

$$Z_0 = \frac{\omega_0 L}{\sin(\omega_0 \tau)} \tag{5.221}$$

If one wishes to fix the characteristic impedance Z_0 in advance, then the delay length τ of the line can be adjusted at a specified angular frequency $\omega_0 = 2\pi f_0$ yielding

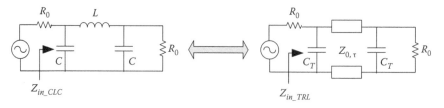

FIGURE 5.81
CLC PI equivalent is replaced by CT-TRL_CT.

$$\sin(\omega_0 \tau) = \frac{\omega_0 L}{Z_0}$$

or the corresponding delay length τ is determined as

$$\tau = \left(\frac{1}{\omega_0}\right) \sin^{-1}\left(\frac{\omega_0 L}{Z_0}\right) \tag{5.222}$$

On the other hand, in terms of the element values of the PI section, loading capacitor C_T is given by

$$C_T = \frac{\cos(\omega_0 \tau) + \omega_0^2 LC - 1}{\omega_0^2 L} \tag{5.223}$$

It is important to emphasize that, for the specified values of L, C, ω_0, and τ, the value of C_T must be nonnegative. Otherwise, an *"almost equivalent CT–TRL–CT"* counterpart of the lumped *"CLC–PI"* section does not exist. In this representation, $\omega_0 = 2\pi f_0$ is chosen as the high end of the useful passband and τ is selected in such a way that at the stopband, $\omega_{s2}\tau = 2\pi f_{s2} = (\pi/2)$.

In many practical situations, stopband frequency f_{s2} is expressed as the multiple of f_0 such that $f_{s2} = m f_0$. Thus, in terms of the high-end cutoff frequency f_0 of the passband, fixed delay τ is specified as

$$\tau = \frac{\pi}{2\omega_{s2}} = \frac{\pi}{2 \times 2\pi f_{s2}} = \frac{1}{4 f_{s2}} = \frac{1}{4 m f_0} \tag{5.224}$$

It may be useful to remember that, at the stop band frequency $\omega_{s2} = 2\pi f_{s2}$, TPG of Figure 5.82 is practically zero (perhaps less than –20 dB). For many applications, it is proper to select ω_{s2} in such a way that it satisfies the below inequality.

$$2\omega_0 \leq \omega_{s2} \leq 4\omega_0 \tag{5.225}$$

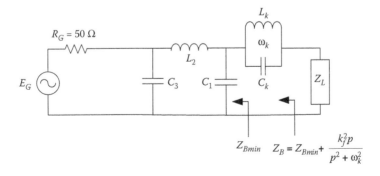

FIGURE 5.82
Input matching network of a power amplifier.

$$2 \leq m \leq 4 \tag{5.226}$$

In other words, the multiplying integer m varies between 2 and 4.

The physical length l of the transmission line can be determined by considering the actual implementation. For example, if the line is printed on the substrate with permittivity ε_r and permeability μ_r as a microstrip or coplanar line, then the velocity of propagation is given by

$$v_{sub} = \frac{v_0}{\sqrt{\mu_r \varepsilon_r}} = \frac{l}{\tau} \tag{5.227}$$

where v_0 is the speed of light in free space and is specified as $v_0 = 3 \times 10^8$ m/s. Hence, the physical length l is given by

$$l = v_{sub} \tau \tag{5.228}$$

Furthermore, employing microstrip technology, characteristic impedance of the transmission line can be approximated using the Wheeler formulas [5.70–5.72]

$$Z_0 \cong \left(\frac{120\pi}{\varepsilon_{r\text{-}eff}} \right) \frac{1}{[x + 1.98(x)^{0.172}]} \tag{5.229}$$

where $x = W/h$ the ratio of width (W) to thickness (h), the effective-relative permittivity $\varepsilon_{r\text{-}eff}$ is

$$\varepsilon_{r\text{-}eff} = 1 + \frac{\varepsilon_{r-1}}{2}\left[1 + \frac{1}{\sqrt{1 + (10/x)}} \right] \tag{5.230}$$

It must be noted that the above formulas are valid for values of $x > 0.06$. Obviously, for a fixed characteristic, impedance Z_0 and relative permittivity ε_r physical width W of the microstrip line can be found by means of a nonlinear equation solver or equivalently using optimization methods.

A simple MATLAB program is given in Program Listing 5.90.

EXAMPLE 5.17: CONSTRUCTION OF A MATCHING NETWORK WITH MIXED LUMPED AND DISTRIBUTED ELEMENTS FROM ITS LUMPED ELEMENT PROTOTYPE

In Figure 5.82, a typical input-matching network of a microwave power amplifier is shown. This network includes a 3-element low-pass PI section. It is terminated by the input of a LMOS power transistor.

The terminating impedance ZL is measured by means of Agilent load-pull equipment over 330–530 MHz. Real and imaginary parts of the measured load are depicted in Figures 5.83 and 5.84, respectively, and the measured impedance data are listed in Table 5.28.

Passband of the amplifier starts at $f_{c1} = 330$ MHz and ends at $f_{c2} = 530$ MHz. The frequency band is normalized with respect to $f_0 = 530$ MHz. The corresponding gain plot is given in Figure 5.85.

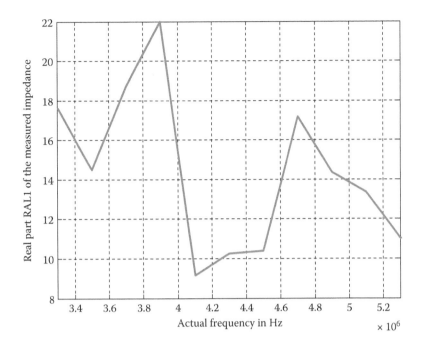

FIGURE 5.83
Real part of ZL.

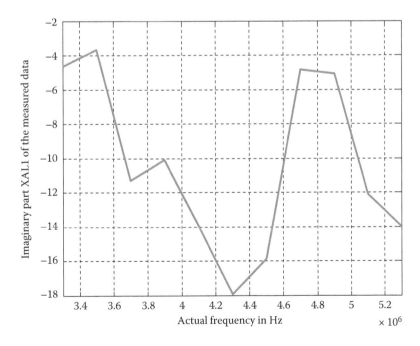

FIGURE 5.84
Imaginary part of ZL.

TABLE 5.28

Measured Input Impedance Data
$Z_{in}(j\omega) = R_{in}(\omega) + jX_{in}(\omega)$

Fr (MHz)	$R_{in}(\omega)$	$X_{in}(\omega)$
330	17.61	−04.63
350	14.50	−03.67
370	18.70	−11.30
390	22.00	−10.11
410	09.16	−13.96
430	10.23	−17.94
450	10.40	−15.89
470	17.18	−04.85
490	14.33	−05.11
510	13.36	−12.11
530	11.04	−14.03

Lumped element design yields $T_{min} = -0.9170$ dB, which is the minimum of passband gain, occurred at 410 MHz.

The matching network is designed using RF-DCT with a single Foster section (FS) extraction.

In the course of design, the driving point immittance of the equalizer was chosen as a minimum reactance function designated by Z_{Bmin} (KFlag = 1). Therefore, extracted Foster section is a parallel resonance circuit in series with Z_{Bmin} and is given by

$$Z_F(p) = \frac{k_f^2 p}{p^2 + \omega_k^2} = \frac{[1/C_k]}{p^2 + (1/L_k C_k)}$$

(5.231)

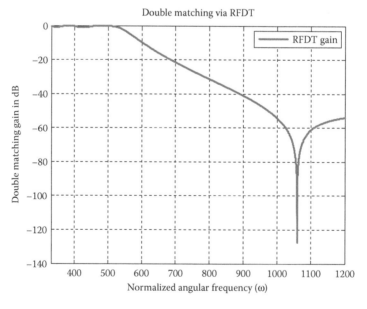

Double matching via RFDT

FIGURE 5.85
Gain performance of the matched system.

Explicit element values of the parallel resonance circuit are driven in terms of the residue $k_f^2 > 0$ and the resonance frequency ω_k.

$$C_k = \frac{1}{k_f^2}$$
(5.232)

$$L_k = \frac{k_f^2}{\omega_k^2}$$
(5.233)

The resonance frequency is selected at $f_k = 2f_{c2} = 2 \times 530 = 1060$ MHz to suppress the second harmonic of the upper edge frequency of the passband.

Results are summarized below.

$$Z_B(p) = Z_{Bmin}(p) + \frac{k_f^2 p}{p^2 + \omega_k^2} = \frac{a(p)}{b(p)} + \frac{k_f^2 p}{p^2 + \omega_k^2}$$

where

$$Z_{Bmin}(j\omega) = R_B(\omega) + jH\{R_B(\omega)\}$$

$$R_B(\omega) = \frac{1}{(1/2)[c^2(\omega) + c^2(-\omega)]}$$

$$c(\omega) = c_1\omega^n + c_2\omega^{n-1} + \cdots + c_n\omega + 1 \text{ with } n = 3$$

$$k_f = 1.5781; \quad c = [8.2573 \quad 0.6615 \quad -5.77501];$$

$$a = [0 \quad 0.2511 \quad 0.1024 \quad 0.1211]; \quad b = [1.0000 \quad 0.4077 \quad 0.7793 \quad 0.1211]$$

Normalized and actual element values of the parallel resonance circuit are computed as follows:

$$C_k = \frac{1}{k_f^2} = 0.4015$$

$$C_{Ak} = C_k/2/530/1e6/pi/50;$$

$$C_{Ak} = 2.4115e - 012 = 2.41\,\text{pF}$$

$$L_k = 1/Ck/4$$

$$L_k = 0.6226$$

$$L_{Ak} = Lk * \frac{50}{2pi530e6} = 9.3483e - 9\,H$$

$$L_{Ak} = 9.3483e - 009 = 9.35\,\text{nH}$$

Actual element values of the CLC ladder is given by

$$C_{A1} = 2.3916e - 011 \cong 23.9\,pF$$

$$C_{A2} = 1.4731e - 011 \cong 14.7\,pF$$

$$L_{A3} = 1.2694e - 008 \cong 12.7\,nH$$

Here, it is desired to find the capacitive loaded transmission line equivalent of the PI section step by step.

It is noted that the new circuit will be fabricated on a RogPers 4350B 3-layer PCB (printed circuit board) with permittivity $\varepsilon_r = 3.66$ and $h = 0.72$ mm.

Solution

Step 1: Separation of Symmetrical CLC-PI Section from a Given Arbitrary C1-L2-C3 Section

First of all, we should distinguish symmetrical CLC-PI sections from the given matching network so that they are replaced by their almost equivalent CT-TRL-CT sections.

For the example under consideration, we can only separate one PI section from the given matching network. The symmetrical capacitors C may be set to a minimum of {C1, C2}, as shown in Figure 5.86.

In this case, $C = C_2 = 14.7$ pF. Then, we have to include a residual capacitor $C_a = C_1 - C$ to the left of the symmetrical CLC-PI section to preserve the original PI.

Step 2: Construction of an Almost Equivalent CT-TRL-CT Section

Once the symmetrical PI is distinguished, it is replaced by its almost equivalent CT-TRL-CT counterpart, as shown in Figure 5.87.

In Figure 5.87, resulting load capacitors C_A and C_B are given by

$$C_A = C_a + C_T \tag{5.234}$$

$$C_B = C_T \tag{5.235}$$

Step 3: Derivation of Parameters of the CT-TRL-CT Section

Now let us calculate the element values of the equivalent circuit.

Let the normalized stop frequency be $\omega_s2 = m\omega_0 = 2$; $\omega_{s2}\tau = \pi/2$ (i.e., $m = 2$)

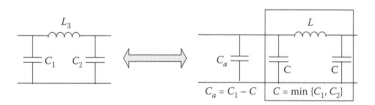

FIGURE 5.86
Extraction of symmetrical CLC-PI section.

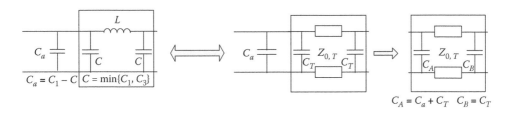

FIGURE 5.87
Replace CLC by its almost equivalent counterpart CT-TRL-CT.

Then,

$$\tau = \frac{\pi}{4\omega_0} = \frac{1}{2\pi f_0} = 2.3585e - 010 \text{ s}$$

In this case,

$$Z_0 = \frac{\omega_0 L}{\sin(\omega_0 \tau)} = \frac{2\pi \times 530 \times 10^6 \times 1.269 \times 10^{-8}}{\sin(\pi/4)}$$

$$Z_0 = 59.763 \ \Omega$$

$$C = \min(C_1, C_2) = \min(C1 = +2.392e - 011; \quad C2 = +1.473e - 011) = 1.473e - 11$$

$$C_T = \frac{\cos(\omega_0 \tau) + \omega_0^2 L C - 1}{\omega_0^2 L}$$

$$CT = 1.2649e - 011 \text{ Farad} = 12.65 \text{ pF}$$

$$C_A = C_1 - C + C_T = 2.1839e - 011 = C_A = 21.84 \text{ pF}$$

$$C_B = C_3 - C + C_T = 1.2649e - 011 =$$

$$C_B = 12.65 \text{ pF}$$

The complete matching network constructed with mixed elements is depicted in Figure 5.88.

In Figure 5.89, the gain performances of the lumped element prototype and the mixed element matching networks are depicted. It is observed that both performances are fairly close, preserving the minimum gain in the passband.

Step 4: Physical Implementation

Once the mixed element structure is obtained using ideal circuit components, distributed elements may be realized as microstrip lines. In this case, one has to pick a commercially available substrate with a proper dielectric constant to realize the characteristic impedance Z_0 and the delay length τ. For the problem under consideration, $\varepsilon_r = 3.66$ and $h = 0.72$ mm.

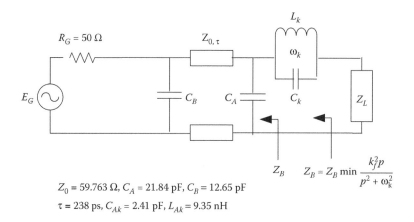

$$Z_0 = 59.763\ \Omega,\ C_A = 21.84\ \text{pF},\ C_B = 12.65\ \text{pF}$$
$$\tau = 238\ \text{ps},\ C_{Ak} = 2.41\ \text{pF},\ L_{Ak} = 9.35\ \text{nH}$$

FIGURE 5.88
Almost equivalent matching network constructed on its lumped element prototype.

Hence, using the main program called Microstrip_design.m, the physical width of the line is computed as

$$width = 0.0012\ \text{m} = 1.2\ \text{mm}$$

This result can easily be checked using Equations 5.172 and 5.173, such that

$$x = \frac{W}{h} = 1.6054$$

$$\varepsilon_{r\text{-}eff} = 1 + \frac{\varepsilon_{r-1}}{2}\left[1 + \frac{1}{\sqrt{1 + (10\ /\ x)}}\right] = 2.8247$$

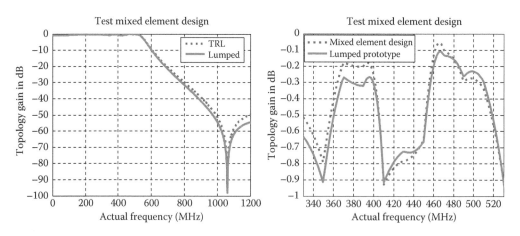

FIGURE 5.89
Gain performance of matched load using lumped prototype and mixed element two ports.

$$Z_0 \cong \left(\frac{120\pi}{\varepsilon_{r-\mathit{eff}}}\right)\frac{1}{[x + 1.98(x)^{0.172}]} = 59.78 \; \Omega$$

Eventually, physical length on the substrate is determined using Equations 5.227 and 5.228.

$$l = v_{sub}\tau = \frac{3 \times 10^8}{\sqrt{3.66}} \times 2.3585e - 010 = 0.0370 = 37 \; \mathrm{mm}$$

It should be noted that if the transmission line is realized as a microstrip rather than a parallel plate line, then one must use equivalent-permittivity $\varepsilon_{r-\mathit{eff}}$ instead of ε_r. In this case, the length l is given by

$$l = v_{sub}\tau \cong \frac{3 \times 10^8}{\sqrt{\varepsilon_{r-\mathit{eff}}}} = \frac{3 \times 10^8}{\sqrt{2.8247}} = 42 \; \mathrm{mm}$$

5.18 Physical Realization of a Single Inductor

A single inductor can be realized using an ideal parallel plate TEM line with characteristic impedance Z_0 and physical length l printed on substrate of relative permittivity ε_r. Ideal TEM line inductance is given by Reference 3 in Chapter 2 as

$$L_{TRL} = \frac{\emptyset(t,z)}{I(t,z)} = \mu\frac{h}{W}l = 4\pi \times 10^{-7}\frac{h}{W}l \qquad (5.236)$$

However, as a parallel plate TEM line, this inductor is associated with a total capacitance C_{TRL}, which is specified by

$$C_{TRL} = \frac{Q(t,z)}{V(t,z)} = \varepsilon\frac{W}{h}l \qquad (5.237)$$

As far as the physical model is concerned, this capacitor may be evenly distributed on the left and right end of the line, as shown in Figure 5.88.

Ideally, the characteristic impedance Z_0 is

$$Z_0 = \sqrt{\frac{L_{TRL}}{C_{TRL}}} = \sqrt{\frac{\mu}{\varepsilon}}\frac{h}{W} = 120\pi\sqrt{\frac{\mu_r}{\varepsilon_r}}\frac{h}{W}$$

For many practical cases, relative permeability μ_r is unity. Therefore, we can confidently state that

$$Z_0 \cong \frac{120\pi}{\sqrt{\varepsilon_r}}\frac{h}{W} \qquad (5.238)$$

In the above representations, h and W refer to the thickness and the width of the parallel plate line as printed on substrates.

It is important to note that for a specified actual inductance value L_A, once the substrate is selected among the manufacturers specified thickness h and permeability μ, width W is determined by means of Equation 5.239, such that

$$W = \mu \frac{h}{L_A} l = 4\pi \times 10^{-7} \frac{h}{L_A} l \tag{5.239}$$

where the physical length of the line is specified by

$$l = v_{sub} \tau = \frac{3 \times 10^8}{\sqrt{\varepsilon_r}} \frac{1}{4mf_0} \quad (\text{in m}) \tag{5.240}$$

Hence, the actual line width is given by

$$W = \mu \frac{h}{L_A} l = \frac{30\pi h}{\sqrt{\varepsilon_r} mf_0 L_A} \quad (\text{in m}) \tag{5.241}$$

In terms of the actual inductor L_A, the characteristic impedance Z_0 is given by

$$Z_0 = 4mf_0 L_A \tag{5.242}$$

It must be kept in mind that an inductor as a parallel plate ideal transmission line must be loaded with symmetrical capacitors C_p, as shown in Figure 5.90. In this case, C_p is determined by means of Equation 5.243, such that

$$C_p = \frac{C_{TRL}}{2} = \varepsilon \frac{W}{2h} l = \mu \varepsilon \frac{l^2}{2L_A} = \frac{1}{v_{sub}^2} \frac{v_{sub}^2 \tau^2}{2L_A} = \frac{(\omega_{s2}\tau)^2}{2\omega_{s2}^2 L_A} = \frac{(\pi/2)^2}{2(2\pi mf_0)^2 L_A}$$

$$C_p = \frac{1}{32(mf_0)^2 L_A} \tag{5.243}$$

Hence, we can confidently state that an actual inductor L_A that is associated with symmetrically loaded capacitors C_p specified by Equation 5.243, can be replaced by an ideal transmission line, as shown in Figure 5.90.

In this case, idealized characteristic impedance is computed using Equation 5.242

$$Z_0 = 4mf_0 L_A$$

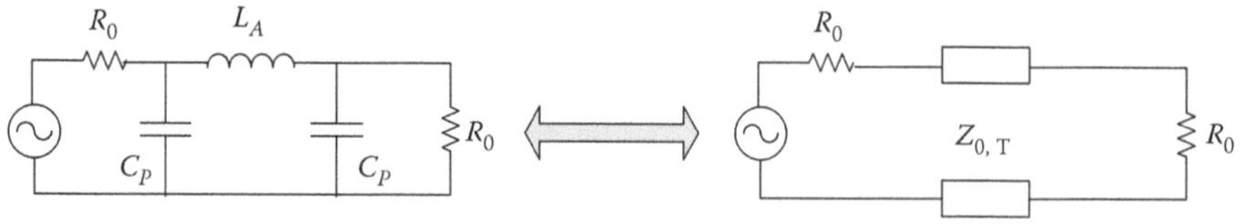

FIGURE 5.90
Approximation of an ideal inductor with a transmission line.

and the idealized delay length τ is generated employing Equation 5.243, such that

$$\tau = \frac{1}{4mf_0}$$

Hence, as in the above section, any *CLC-PI* section can be approximated by means of capacitive loaded transmission line *CA-TRL-CB* using the formulas given in Table 5.29.

EXAMPLE 5.18: REPLACEMENT OF AN INDUCTOR WITH AN IDEAL TRANSMISSION LINE

The aim here is to realize an actual inductor $L_A = 1.2694e - 008$, $H = 12.69$ nH as a microstrip transmission line with thickness $h = 0.72$ mm, permittivity $\varepsilon_r = 3.66$, stop band multiplier $m = 4$; cutoff frequency $f_0 = 530$ MHz.

Compute the following quantities:

a. The physical length l of the microstrip line
b. The characteristic impedance of the microstrip line Z_0
c. The width W
d. Accompanied capacitor C_p
e. Referring to Figure 5.91, develop a MATLAB Program to compute the driving point input impedances of lumped prototype C_p–L_{A2}–C_p-section and the physical transmission line model when they are terminated in 50 Ω

Solution

a. The physical length of the transmission line is given by Equation 5.240. Hence,

$$l = v_{sub}\tau = \frac{3 \times 10^8}{\sqrt{\varepsilon_r}} \frac{1}{4mf_0} = 0.0185\,\text{m} = 1.85\,\text{cm}$$

b. The characteristic impedance Z_0 is given by Equation 5.242 as

$$Z_0 = 4mf_0 L_A$$

Then, it is found as

$$Z_0 = 107.6451\ \Omega$$

FIGURE 5.91
Comparison of input impedances.

TABLE 5.29

Summary of the Design Equation for CLC-PI Section with an Ideal Physical Model of CA-TRL-CB Section

Given Parameters for Lumped PI Section	Computed Parameters for CA-TRL-CB Section
C1-L3-C2 PI section	CA-TRL-CB section generated by means of ideal values of lumped components
L_{A3}	
C_{A1} C_{A2}	C_A $Z_{0,\tau}$ C_B
CA1: Given actual capacitance	$C_p = \dfrac{1}{32m^2 f_0^2 L_{A2}}$
LA3: Given actual inductor	$C_A = C_{A1} - C_p > 0$
CA2: Given actual capacitor	$C_B = C_{A2} - C_p > 0$
	$Z_0 = 4mf_0 L_{A2}$
	$\tau = \dfrac{1}{4mf_0}$
	$l = v_{sub}\tau = \dfrac{3\times 10^8}{\sqrt{\varepsilon_r}}\dfrac{1}{4mf_0}$

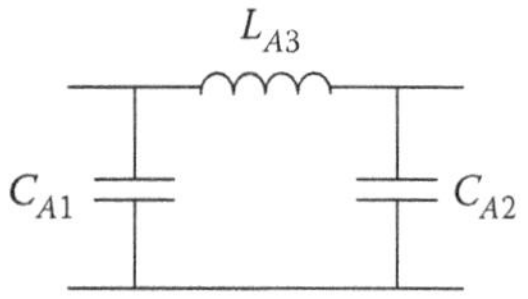

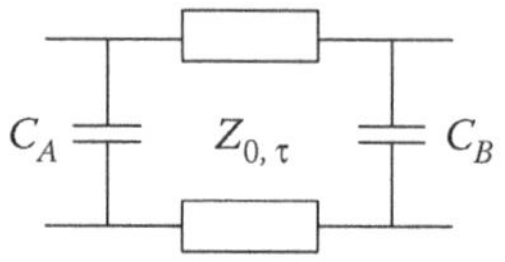

c. The line width is specified by Equation 5.239 or Equation 5.241. Then,

$$W = \mu \frac{h}{L_A} l = 4\pi \times 10^{-7} \frac{h}{L_A} l = \frac{30\pi h}{\sqrt{\varepsilon_r}\, mf_0 L_A} = 0.0037 = 3.7 \text{ mm}$$

d. The capacitor is given by Equation 5.243. Hence,

$$C_P = \frac{1}{32(mf_0)^2 L_A} = 5.4775e - 013 = 0.548\, \text{pF}$$

e. Comparison of input impedances

Input impedance of the lumped element PI section Cp-L_{A2}-Cp terminated in 50 Ω is given by

$$Z_{inLump}(p) = \frac{1}{Y_{inLump}(p)}$$

On the $j\omega$ *axis*

$$Z_{inLump}(j\omega) = R_{inLump}(\omega) + jX_{inLump}(\omega)$$

where

$$Y_{inLump} = pC_p + \frac{1}{pL_A + (1/pC_p + (1/50))}$$

Input impedance of the transmission line terminated in 50 Ω is given by

$$Z_{inTRL}(j\omega) = R_{inTRL}(\omega) + jX_{inTRL}(\omega) = Z_0 \frac{50 + jZ_0 \tan(\omega\tau)}{Z_0 + j50 \tan(\omega\tau)}$$

The above equations are programmed on MATLAB. Comparative real and imaginary parts are depicted in Figures 5.92 and 5.93, respectively.

Close examination of Figures 5.93 and 5.94 reveals maximum relative error between the curves as

$$Error_{Rin} = \frac{\max(R_{inLump} - R_{inTRL})}{\max(R_{inLump})} = 0.028 < 3\%$$

$$Error_{Xin} = \frac{\max(X_{inLump} - X_{inTRL})}{\max(X_{inLump})} = 0.0362 < 4\%$$

As observed, the fit between real and imaginary parts can be acceptable for many practical implementations. On the other hand, if we plot the above figures over a wide frequency band, we should be able to see the periodic behavior of the transmission line impedance, as shown in Figures 5.94 and 5.95.

The above results are produced using a MATLAB program called INDviaTRL.m

A list of the programs is in Program Listing 5.92.

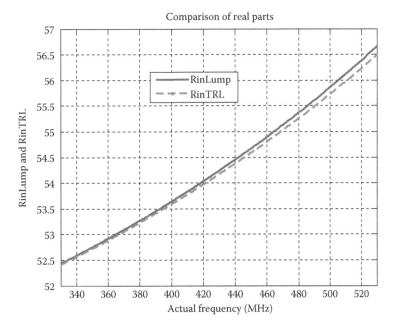

FIGURE 5.92
Comparison of real parts of the input impedances.

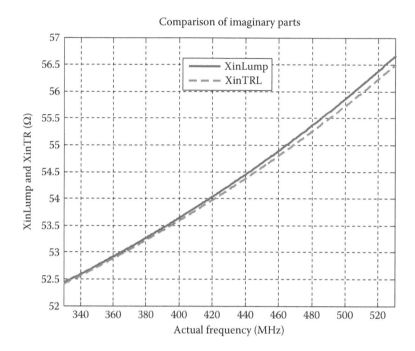

FIGURE 5.93
Comparison of imaginary parts of the input impedances.

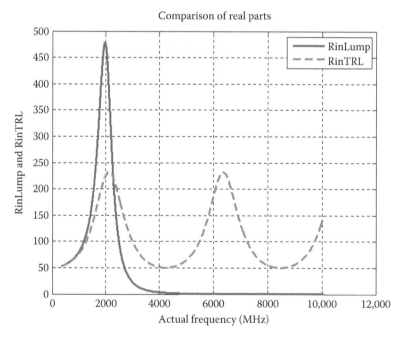

FIGURE 5.94
Periodic behavior of RinTRL.

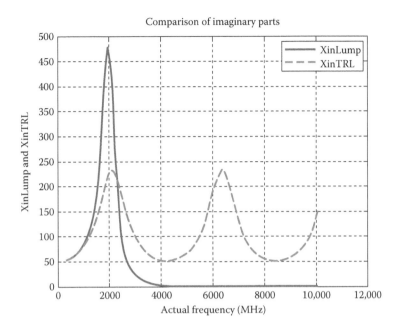

FIGURE 5.95
Periodic behavior of XinTRL.

EXAMPLE 5.19: REPLACEMENT OF AN INDUCTOR WITH AN IDEAL TRANSMISSION LINE

Using the single matching lumped element prototype circuit of Example 5.18, generate a mixed element design using the physical model method introduced in Table 5.29.

The lumped element CLC PI section consists of $C_{A1} = 2.3916e - 011 \cong 23.9$ pF, $C_{A2} = 1.4731e - 011 \cong 14.7$ pF, $L_{A3} = 1.2694e - 008 \cong 12.7$ nH. The parallel resonance circuit in series configuration has $C_{Ak} = 2.41$ pF and $L_{Ak} = 9.35$ nH and $m = 2, f_0 = 530$ MHz, $h = 0.72$ mm, epsr = 3.66.

Solution

Step 1: Computation of Line Parameters

First of all, inductor L_{A2} must be associated with its counterpart capacitor C_p, which is due to the physical width of the microstrip line. Then, the characteristic impedance, length and width of the microstrip line is determined. For this purpose, we developed a MATLAB program called TRL_Model, which yields

$$Cp = 2.1910e - 012\,\text{F}; \ width = 0.0073 \ \text{m}; \ length = 0.0370 \ \text{m};$$

$$Z0 = 53.8226 \ \Omega; \ tau = 2.3585e - 010\,\text{s}.$$

As a result of approximation, the relative errors between real and imaginary parts are given as

$$Rin_{error} = 0.0724; Xin_{error} = 0.5578$$

Step 2: Computation of Loading Capacitors C_A and C_B as in Table 5.29

In the second step, one must examine the situation to see if the inductor-related capacitor C_p is less than C_1 and C_2. If the answer is no, then the solution does not exist. If the answer is yes, then

$$C_A = C_{A1} - C_p$$

and

$$C_B = C_{A2} - C_p$$

For our case, indeed, $C_p < (C_1$ and $C_3)$.
Hence,

$$CA = 2.1725e - 011; \; CB = 1.2540e - 011$$

Step 3: Computation of the Gain from the Physical Model

At this step, gain of the single matching problem is computed using the physical model circuit elements. For comparison purposes, performances of the lumped prototype and the mixed element model are depicted in Figure 5.96.

It is clearly seen that the physical model follows the gain performance of the lumped prototype at the lower end of the passband. However, it deviates from the original as the frequency increases.

MATLAB programs written for the gain generation are given in Program Listing 5.93–5.94.

It should be noted that once the mixed element matching network prototype is obtained, it may be reoptimized to improve the TPG by means of a commercially available S/W tool such as AWR [75] or ADS [76], etc.

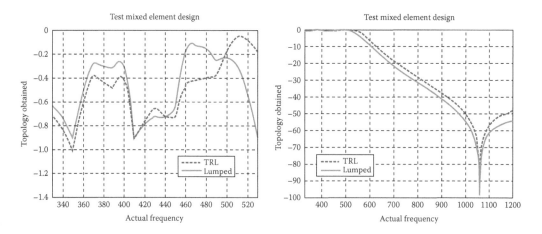

FIGURE 5.96
Comparison of the gain performances obtained from the circuit topologies using lumped element prototype and physical model.

5A.1 Appendix

In this appendix, chain parameters of a UE, CLC lumped, and C-UE-C mixed sections are derived and it is exhibited that CLC and C-UE-C sections possess almost equivalent electrical performances within acceptable practical limits.

5A.1.1 Chain Parameters of a UE

Referring to Figure 5A.1, the output voltage and current pair of a UE can be expressed as

$$V_2 = A + B$$
$$Z_0 I_2 = A - B$$

(5A.1)

where A and B are the "voltage based" incident and the reflected waves of the output port of the UE (see reference [3]; Chapter 3, Equation 3.17, p. 127).

Similarly, the input voltage and current pair is given as follows:

$$V_1 = Ae^{+j(\omega\tau)} + Be^{-j(\omega\tau)}$$
$$Z_0 I_1 = Ae^{+j(\omega\tau)} - Be^{-j(\omega\tau)}$$

(5A.2)

In the above equations, delay length τ is expressed in terms of the physical lenghth l and the propagation velocity v_p as

$$\tau = l/v_p$$

(5A.3)

The delay length τ may be associated with a specified frequency $f_s = 1/\tau$ such that

$$l = \frac{v_p}{f_s}$$

(5A.4)

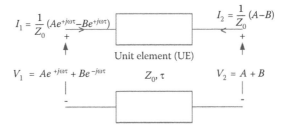

$$I_1 = \frac{1}{Z_0}(Ae^{+j\omega\tau} - Be^{+j\omega\tau})$$

$$I_2 = \frac{1}{Z_0}(A - B)$$

$$V_1 = Ae^{+j\omega\tau} + Be^{-j\omega\tau}$$

Unit element (UE)

Z_0, τ

$$V_2 = A + B$$

FIGURE 5A.1
An ideal transmission line with characteristic impedance Z_0 and delay length τ as a UE.

If f_0 is the cutoff frequency of the passband of operation, then, we have the freedom to choose f_s as

$$f_s = mf_0 \; ; \; m > 1 \tag{5A.5}$$

It should be noted that when the output port is open, then $I_2 = 0$. In this case, $B = A$. On the other hand, when output port is shortened, then $V_2 = 0$. In this case $B = -A$.

By means of actual port voltages and currents, the chain parameters A, B, C, and D are defined as follows:

$$\begin{bmatrix} V_1 \\ I_1 \end{bmatrix} = \begin{bmatrix} A & B \\ C & D \end{bmatrix} \begin{bmatrix} V_2 \\ -I_2 \end{bmatrix} = T_{UE} \begin{bmatrix} V_2 \\ -I_2 \end{bmatrix} \tag{5A.6}$$

where

$$T_{UE} = \begin{bmatrix} A & B \\ C & D \end{bmatrix}$$

is called chain or transmission matrix of the UE and its entries A, B, C, and D are given by

$$A = \frac{V_1}{V_2} \Big|_{I_2=0} = \frac{A(e^{+j(\omega\tau)} + e^{-j(\omega\tau)})}{2A} = \cosh(j\omega\tau) = \cos(\omega\tau)$$

$$B = \frac{V_1}{-I_2} \Big|_{V_2=0} = Z_0 \frac{A(e^{+j(\omega\tau)} - e^{-j(\omega\tau)})}{2A} = Z_0 \sinh(j\omega\tau) = jZ_0 \sin(\omega\tau)$$

$$C = \frac{I_1}{V_2} \Big|_{I_2=0} = \frac{A(e^{+j(\omega\tau)} - e^{-j(\omega\tau)})}{2Z_0 A} = Y_0 \sinh(j\omega\tau) = jY_0 \sin(\omega\tau) \tag{5A.7}$$

$$D = \frac{I_1}{-I_2} \Big|_{V_2=0} = \frac{Z_0 A(e^{+j(\omega\tau)} + e^{-j(\omega\tau)})}{2Z_0 A} = \cosh(j\omega\tau) = \cos(\omega\tau)$$

It should be noted that

$$\cosh(j\omega\tau) = \frac{e^{+j(\omega\tau)} + e^{-j(\omega\tau)}}{2} = \cos(\omega\tau)$$

$$\sinh(j\omega\tau) = \frac{e^{+j(\omega\tau)} + e^{-j(\omega\tau)}}{2} = j\sin(\omega\tau)$$

It is noted that "voltage based" incident and reflected wave notations A and B should not be confused with the classical notation of the chain parameters A, B, C, and D.

5A.1.2 Chain Parameters for CLC and CT-UE-CT Sections

Referring to Figure 5A.2, let T_A designate the chain matrix of CLC-π circuit. Let T_C and T_L be the chain matrices of individual circuit elements, namely, shunt capacitor C and series inductor L, respectively. Then,

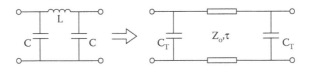

FIGURE 5A.2
Almost equivalent circuit of a symmetrical low-pass C-L-C π-section by means of a CT-UE-CT section.

$$T_A = [T_C][T_L][T_C] \tag{5A.8}$$

where

$$T_C = \begin{bmatrix} 1 & 0 \\ pC & 1 \end{bmatrix}$$

$$T_L = \begin{bmatrix} 1 & pL \\ 0 & 1 \end{bmatrix} \tag{5A.9}$$

Therefore,

$$T_A = [T_C][T_L][T_C] = \begin{bmatrix} 1 + LCp^2 & pL \\ LC^2p^3 + 2Cp & 1 + LCp^2 \end{bmatrix} \tag{5A.10}$$

on $j\omega$ axis

$$T_A = \begin{bmatrix} 1 - LC\omega^2 & j\omega L \\ j\omega C(2 - LC\omega^2) & 1 - LC\omega^2 \end{bmatrix} \tag{5A.11}$$

Similarly, let T_B be the chain matrix of CT-UE-CT section, then

$$T_B = [T_{CT}][T_{UE}][T_{CT}] \tag{5A.12}$$

where

$$T_{CT} = \begin{bmatrix} 1 & 0 \\ j\omega C_T & 1 \end{bmatrix}$$

$$T_{UE} = \begin{bmatrix} \cos(\omega\tau) & jZ_0 \sin(\omega\tau) \\ jY_0 \sin(\omega\tau) & \cos(\omega\tau) \end{bmatrix} \tag{5A.13}$$

Then,

$$T_B = [T_{CT}][T_{UE}][T_{CT}]$$

$$= \begin{bmatrix} \cos(\omega\tau) - \omega Z_0 C_T \sin(\omega\tau) & jZ_0 \sin(\omega\tau) \\ \dfrac{j}{Z_0}(2Z_0 C_T \cos(\omega\tau)) + (1 - \omega^2 Z_0^2 C_T^2)\sin(\omega\tau) & \cos(\omega\tau) - \omega Z_0 C_T \sin(\omega\tau) \end{bmatrix} \tag{5A.14}$$

Obviously, at $\omega = 0$, chain matrices $T_A(0)$ and $T_B(0)$ are equal to each other and they are given by

$$T_A(0) = T_B(0) = \begin{bmatrix} 1 & 0 \\ 0 & 1 \end{bmatrix} \tag{5A.15}$$

On the other hand, we can equate chain matrices T_A and T_B at a given frequency ω_0 such that

$$\begin{aligned} T_{A11}(j\omega_0) &= T_{B11}(j\omega_0) \; yields \quad 1 - \omega_0^2 LC = \cos\omega\tau - \omega Z_0 C_T \sin(\omega\tau) \\ T_{A12}(j\omega_0) &= T_{B12}(j\omega_0) \; yields \quad j\omega L = jZ_0 \sin(\omega\tau) \end{aligned} \tag{5A.16}$$

Solving the above equations for Z_0 and C_T we have

$$\begin{aligned} Z_0 &= \frac{\omega_0 L}{\sin(\omega_0\tau)} \\ C_T &= \frac{\cos(\omega\tau) + \omega_0^2 LC - 1}{\omega_0^2 L} \end{aligned} \tag{5A.17}$$

Using Equation 5A.17, it can be shown that $T_{A21}(j\omega_0) = T_{B21}(j\omega_0)$ and $T_{A22}(j\omega_0) = T_{B22}(j\omega_0)$.

Thus, it is expected that between $\omega = 0$ and ω_0, CLC and CT-UE-CT sections exhibit similar electrical performances due to their smooth low-pass behavior over the entire frequency axis.

Now, let us answer the following question.

5A.1.3 Do CLC and CT-UE-CT Sections Have Equal Chain Matrices at ω_0?

Indeed, the answer is yes, and it is verified as follows.

By Equation 5A.14, $T_B(2, 2)$ is given by

$$T_B(2,2) = \cos(\omega\tau) - \omega C_T Z_0 \sin(\omega\tau)$$

At $\omega = \omega_0$, employing the values of C_T and Z_0 as derived by Equation 5A.17, it is straightforward to show that

$$T_B(2,2) = 1 - \omega_0^2 LC \tag{5A.18}$$

which is equal to $T_A(2, 2)$.

Similarly, by Equation 5A.14 $T_B(2, 1)$ is given by

$$T_B(2,1) = \frac{j}{Z_0} \{ 2\omega C_T Z_0 \cos(\omega\tau) + (1 - Z_0^2 \omega^2 C_T^2)\sin(\omega\tau) \}$$

At $\omega = \omega_0$, using the values of C_T and Z_0 as derived by Equation 5A.17 in the above equation and carrying out the algebraic manipulations it is found that

$$T_B(2,1) = j\omega C(2 - \omega_0^2 LC) \tag{5A.19}$$

which is equal to $T_A(2, 1)$. Thus, the question has ended.

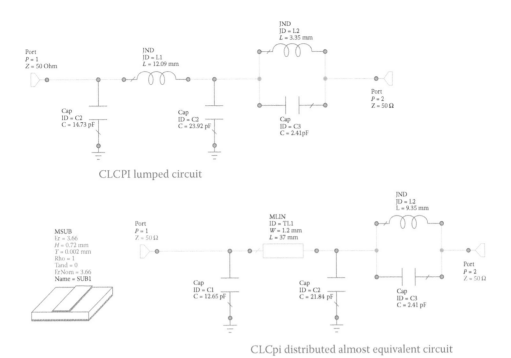

FIGURE 5A.3
The lumped (CLC) and its almost equivalent mixed element (CT-UE-CT) circuits.

Referring to Figure 5A.3, let us work on the electrical performance simulation of *CLC* and C_T-*UE*-C_T sections under practical circumstances.

The above circuit is simulated in the AWR environment [75]. Notice that a notch filter composed of inductor L2 = 9.35 nH and a C3 = 2.41 pF inserted before the output port to suppress the harmonic occurred at ~1.1 GHz.

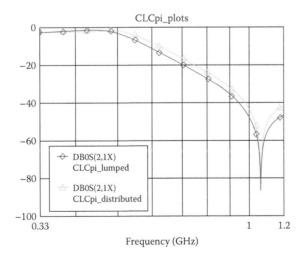

FIGURE 5A.4
Performance analysis of Figures 5A.2a and 5A.2b.

In Figure 5A.4, TPG of the circuits given by Figure 5A.3 is depicted. It is seen that gain performances of the CLC section and its almost equivalent counterpart CT-UE-CT are very close to each other, within 1% accuracy over the frequency band of $0 \leq \omega \leq \omega_0$, which may be considered acceptable for many practical applications.

5A.2 Appendix: A List of MATLAB Programs

Program List 5.1: MatLab function initials

```
function RBA=initials(T0,WB,RLA,XLA,sign)
% Inputs:
%      T0: Ideal flat gain level
%      WB: Angular break frequencies of terminating immittance
%      RLA: Real Part of the immittance data as an array vector
%      XLA: Imaginary part of the immittance data as an array vector
%      sign: It is a control integer.
%           If sign=0, low values of the break points are generated.
%           If sign=1, high values of the break points are generated.
% -------------------------------------------------------------------
% Output:
%      RBA: Break points in array form
% -------------------------------------------------------------------
% It should be noted that size N of WB, RLA, and XLA must be same
%
N=length(WB);
for j=1:N
    w=WB(j);
    RL=RLA(j);XL=XLA(j);
        RB=RL*((2-T0)+2*sign*sqrt(1-T0))/T0;%See Eqs.(5.11a)and (5.11b)
    RBA(j)=RB;
end
```

Program List 5.2: MatLab function "num-hilbert"

```
function XA=num_hilbert(w,W,R)
%Numerical Hilbert Transformation of (5.7)
% Inputs:
%      w: Given single angular frequency to evaluate numerical Hilbert
%      transform.
%      W: Break Frequencies
%      R: Break points of the real part
% Output:
%      XA: Value of the imaginary part at a given single angular
% frequency w.
% -------------------------------------------------------------------
N=length(W);
%
for k=2:N
    DW(k)=W(k)-W(k-1);
    DR(k)=R(k)-R(k-1);
```

```
        M(k-1)=DR(k)/DW(k);
end
        for k=1:N
        F(k)=(w+W(k))*log(abs(w+W(k)))+(w-W(k))*log(abs(w-W(k)));
        end
    for k=2:N
        DF=(1/pi)*(F(k)-F(k-1));
        B(k)=(1/pi)*(F(k)-F(k-1))/DW(k);
        X(k)=B(k)*DR(k);
    end
    %
            XA=0;
            for k=2:N
                XA=XA+X(k);
            end
```

Program List 5.3: MatLab function "Single Matching Transducer Power Gain"

```
function Gain_SM=gain_singleMatching(WBR,RLA,XLA,RBA,XBA)
% Computation of Single Matching Gain as an array.
% Inputs:
%       WBR,RLA,XLA: Load immittance data with break frequencies WBR
%       WBR,RBA,XBA: Driving point immittance of the equalizer
% Note that all the above arrays are equal length of N
%
%   Output:
%       Gain_SM: Single Matching Gain as a MatLab array.
% ----------------------------------------------------------------------
N=length(WBR);
%
for j=1:N
    Num(j)=4*RLA(j)*RBA(j);
    Denom(j)=(RLA(j)+RBA(j))^2+(XLA(j)+XBA(j))^2;
    Gain_SM(j)=Num(j)/Denom(j);
End
```

Program List 5.4: MatLab Main Program "Example5_1.m"

```
% Main Program Example5_1.m
% November 16 2014 by BS Yarman, Vanikoy, Istanbul, Turkey
% This program provides the solution for Example 5 given in the new
%  book of
% Andrei Grebennikov-Narendra Kumar B. Siddik Yarman Book
%
% Example 5.1
%
% Inputs:
% Load-Pull Measurement for LD MOS RD07 by Mitsubishi
% Sampling Frequencies of the actual measurements in MHz
% ----------------------------------------------------------------------
clc
close all
% Here, we add two more break points to cover DC and to make R(N)=0 at
% the
```

```
% stop band frequency, say at F(N)=2*530=1060 MHz.
% Under these assumptions we set F(1)=0 and F(N)=1060 MHz
% which are the first and the last break frequencies.
% Augmented actual frequencies:
F=[0 330 350 370 390 410 430 450 470 490 510 530 1060];
f0=530;% Selected normalization frequency.
WB=F/f0; % Frequency normalization.
% Actual large signal input impedance of the LD MOS Device RD07 MUS2B
% by Mitsubishi Zin_A=Rin_A+jXin_A
% Augmented real part:
Rin_A=[20 17.61 14.50 18.70 22.00 9.16 10.23 10.40 17.18 14.33 13.36
11.04 9];
% Augmented imaginary part:
Xin_A=[0 -04.63 -03.67 -11.30 -10.11 -13.96 -17.94 -15.89 -04.85 -05.11
-12.11 -14.03 -16 ];
%
% Note that, in the above arrays Rin_A and Xin_A are augmented at DC
% and
% stop-band frequencies Rin(1)=20 ohm and Rin(N)=9 ohm selected.
% Similarly, augmented Xin(1)=0 ohm and Xin(N)= -16 ohm is selected.
% The above choices do not affect the optimization of transducer power
% gain.
%
N=length(F);
figure
plot(F,Rin_A,F,Xin_A)
title('Plot of augmented input impedance at DC and Stop-Band Frequency
2*530 MHz, Zin=Rin+jXin')
legend('Rin','Xin')
xlabel('Actual Frequency in MHz')
ylabel('Actual input impedance Zin=Rin+jXin: Rin and Xin')
% Impedance normalization number R0=50 ohm.
R0=50;
% Normalized input impedance: Rin_N + jXin_N
Rin_N=Rin_A/R0;
Xin_N=Xin_A/R0;
%
% Frequency Normalization with respect to f0=530 MHz:
% Note that normalization is carried out on the actual angular
% frequencies:
% 2*pi*F/(2*pi*f0)=F/f0
% Part (a):
% Compute the low level break points for the input matching network [F]
sign=-1; T0=0.95;
RBF_Low=initials(T0,WB,Rin_N,Xin_N,sign);
RBF_Low(N)=0;
% Compute the high level break points for the input matching network
% [F]
sign=+1;
RBF_High=initials(T0,WB,Rin_N,Xin_N,sign);
RBF_High(N)=0;
%
figure
plot(WB,RBF_Low,WB,RBF_High)
```

```
title('Plot of Low and High level Break Points for selected gain level
T0=0.95 for [F]')
xlabel('Normalized Angular Break Frequencies WB')
ylabel('Low and High Level Break Points RB for [F]')
legend('Low Level Break Points RB-Low for [F]','High Level Break Points
RB-High for [F]')
% End of part (a)
% ---------------------------------------------------------------------
% Example 5.1 Part (b):
% Augmented output impedance of the active device
Rout_A=[25 22.47 16.06 21.23 24.40 22.03 23.50 17.82 18.53 21.06 26.80
21.50 18];
Xout_A=[0 -07.84 -12.69 -12.32 -13.45 -12.14 -14.01 -09.52 -01.53 -04.54
-11.06 -13.54 -15 ];
% Impedance normalization with respect to R0=50 ohms.
Rout_N=Rout_A/R0;
Xout_N=Xout_A/R0;
% Computations of the low and high break points for the back-end
% matching network [B]
T0=0.925;
% Low-Level initial break points for [B]:
sign=-1;
RBB_Low=initials(T0,WB,Rout_N,Xout_N,sign);
RBB_Low(N)=0;
%
% High-Level initial break points for [B]:
sign=+1;
RBB_High=initials(T0,WB,Rout_N,Xout_N,sign);
RBB_High(N)=0;
%
figure
plot(WB,RBB_Low,WB,RBB_High)
title('Plot of Low and High level Break Points for selected gain level
T0=0.925 for [B]')
xlabel('Normalized Angular Break Frequencies WB')
ylabel('Low and High Level Break Points RB for [B]')
legend('Low Level Break Points RB-Low for [B]','High Level Break Points
RB-High for [B]')
%
% Example 5.1 Part (c): Computation of imaginary parts by numerical
% Hilbert transformation
% For the fron-end matching network:
wL=WB(1);wH=WB(N);
Ns=100; wdel=(wH-wL)/(Ns-1);
w=wL;
for j=1:Ns
    W(j)=w;
% Computation of numerical hilbert transformations for [F] and [B]
XF_Low(j)=num_hilbert(w,WB,RBF_Low);
XF_High(j)=num_hilbert(w,WB,RBF_High);
%
XB_Low(j)=num_hilbert(w,WB,RBB_Low);
XB_High(j)=num_hilbert(w,WB,RBB_High);
% ---------------------------------------------------------------------
```

```
% Computation of RF_Low,RF_High, RB_Low, RB_High
%               and
% Rin and Rout at w for line interpolation
RF_Low(j)=line_seg(WB,RBF_Low,w);
RF_High(j)=line_seg(WB,RBF_High,w);
%
RB_Low(j)=line_seg(WB,RBB_Low,w);
RB_High(j)=line_seg(WB,RBB_High,w);
%
Rout_Low(j)=line_seg(WB,RBB_Low,w);
Rout_High(j)=line_seg(WB,RBB_High,w);
%
Rin(j)=line_seg(WB,Rin_N,w);
Xin(j)=line_seg(WB,Xin_N,w);
Rout(j)=line_seg(WB,Rout_N,w);
Xout(j)=line_seg(WB,Xout_N,w);
%
w=w+wdel;
end
%
figure
plot(W,XF_Low,W,XF_High)
xlabel('Normalized Angular Frequency w')
ylabel('XF-Low and XF-High')
legend('XF-Low','XF-High')
title('Numerical Hilbert Transform for [F]: XF(w)=H{RBin(w)}')
%
figure
plot(W,XB_Low,W,XB_High)
xlabel('Normalized Angular Frequency w')
ylabel('XB-Low and XB-High')
legend('XB-Low','XB-High')
title('Numerical Hilbert Transform for [B]: XB(w)=H{RBout(w)}')
% -----------------------------------------------------------------------
% Part (d): Computation of TPG for the front end:
TPGF_Low=gain_singleMatching(W,RF_Low,XF_Low,Rin,Xin);
TPGF_High=gain_singleMatching(W,RF_High,XF_High,Rin,Xin);
%
% Plot of Transducer Power Gain for the front-end:
figure
plot(W,TPGF_Low,W,TPGF_High)
xlabel('Normalized Angular Frequency w')
ylabel('Initial Transducer Power Gain for Front-End Equalizer')
legend('TPGF-Low','TPGF-High')
title('Transducer Power Gain for the Front-End Matching Network')
% -----------------------------------------------------------------------
% Part (e): Computation of TPG for the Back-End:
%
TPGB_Low=gain_singleMatching(W,RB_Low,XB_Low,Rout,Xout);
TPGB_High=gain_singleMatching(W,RB_High,XB_High,Rout,Xout);
%
% Plot of Transducer Power Gain for the back-end:
figure
plot(W,TPGB_Low,W,TPGB_High)
```

```
xlabel('Normalized Angular Frequency w')
ylabel('Initial Transducer Power Gain for Back-End Equalizer')
legend('TPGB-Low','TPGB-High')
title('Transducer Power Gain for the Back-End Matching Network')
% ----------------------------------------------------------------------
% Special loop to evaluate hilbert transform at break frequencies:
wL=WB(1);wH=WB(N);
N=length(WB);
clear XF_Low;clear XF_High;clear XB_Low;clear XB_High
%
for j=1:N
    w=WB(j);
% Computation of numerical Hilbert transformations for [F] and [B]at
% the
% break frequencies:
XF_Low(j)= num_hilbert(w,WB,RBF_Low); % Front-End Low
XF_High(j)=num_hilbert(w,WB,RBF_High);% Front-End High
%
XB_Low(j)= num_hilbert(w,WB,RBB_Low); % Back-End Low
XB_High(j)=num_hilbert(w,WB,RBB_High);% Back-End High
end
% ----------------------------------------------------------------------
% Computation of Gain at Break frequencies:
% Part (d): Power Gain at Front-End:
GainF_Low=gain_singleMatching( WB,RBF_Low, XF_Low, Rin_N,Xin_N);
GainF_High=gain_singleMatching(WB,RBF_High,XF_High,Rin_N,Xin_N);
% ----------------------------------------------------------------------
% Part (e): Power Gain at Back-End
GainB_Low= gain_singleMatching(WB,RBB_Low, XB_Low, Rin_N, Xin_N);
GainB_High=gain_singleMatching(WB,RBB_High,XB_High,Rout_N,Xout_N);
% ----------------------------------------------------------------------
% Plot of Gain at Front-End & Back-End
figure
plot(WB,GainF_Low,WB,GainF_High)
xlabel('Normalized Angular Frequency w')
ylabel('Initial Transducer Power Gain for Front-End Equalizer')
legend('TPGF-Low','TPGF-High')
title('Transducer Power Gain for the Front-End Matching Network')
% ----------------------------------------------------------------------
figure
plot(WB,GainB_Low,WB,GainB_High)
xlabel('Normalized Angular Frequency w')
ylabel('Initial Transducer Power Gain for Back-End Equalizer')
legend('TPGB-Low','TPGB-High')
title('Transducer Power Gain for the Back-End Matching Network')
```

Program List 5.5: function Line

```
function ya=line_seg(x,y,xa)
% Given piecewise linear approximation of any function y=f(x)with
% Inputs:
%       n: Total number of break points.
%       Break points pairs:
%       y(1),y(2),..,y(n-1),y(n)
%       x(1),x(2),..,x(n-1),x(n)
```

```
%        xa: a point on the x-axis
%   Output:
%        ya: corresponding point to xa on the y-axis
% It should be noted that
%   the algorithm first determines the interval of x(j)<x(j+1)
%   then, generates ya on the line specified by
%   {y(j),y(j+1) and x(j),x(j+1)}
% In other words, generates ya=f(xa)=(a)xa+b
% where real coefficients a and b are specified as in Eq. (11.1)
% See the Wiley Book written by Yarman.
% Note that this function is useful to generate data for RFLT.
%-------------------------------------------------------------------------
%
n=length(x);
for i=1:n-1
    if xa==x(i)
        ya=y(i);
    end
    if xa>x(i)
        if xa<x(i+1)
            A1=y(i)-y(i+1);
            A2=x(i)-x(i+1);
            A=A1/A2;
            B1=x(i)*y(i+1)-x(i+1)*y(i);
            B=B1/A2;
            ya=A*xa+B;
        end
    end
        if xa==x(i+1)
            ya=y(i+1);
        end
                if xa>x(i+1)
                    A1=y(n-1)-y(n);
                    A2=x(n-1)-x(n);
                    A=A1/A2;
                        B1=x(n-1)*y(n)-x(n)*y(n-1);
                        B=B1/A2;
                            ya=A*xa+B;
                end
end
```

Program List 5.6: MatLab function Impedance_Termination

```
function [ FC, A, B ] = Impedance_Termination(KFlag, N, FA,RA,XA,FL,FH )
% This function converts the given impedance termination to an impedance or
% admittance termination with total number of N+2-Sampling points
% using linear interpolation technique.
% N samples are placed within the passband (FL-FH).
% First point is FA(1). The last point is FA(NA).
%-------------------------------------------------------------------------
% Inputs:
%        KFlag: Control Flag. If KFlag=1, output is impedance termination
%        N: Number of new sampling points subject to linear interpolation
%        technique.
```

```
%         FA: Original actual frequency sampling points specified over
%         NA sampling points.
%         RA: Real part of the impedance termination with NA samples
%         XA: Imaginary part of the impedance termination with NA samples
%         FL: Low-end of the passband
%         FH: High-end of the passband
%   Outputs:
%         A: Real part of the augmented immittance with N samples
%         B: Imaginary part of the augmented immittance with N samples
%   Note:
%         Given impedance termination is expressed as ZA=RA+jXA
%         At the output augmented termination C is given by C=A+jB
%         If KFlag=1> C is impedance with N samples
%         If KFlag=0> C is and admittance with N samples
% -------------------------------------------------------------------------
NA=length(FA);
    j=sqrt(-1);
% KFlag loop for the given data:
for i=1:NA
    ZA(i)=RA(i)+j*XA(i);% Original impedance as it is measured
    if KFlag==0; % Change the impedance to admittance
        C(i)=1/ZA(i);
        Ar(i)=real(C(i));
        Br(i)=imag(C(i));
    end
    %
    if KFlag==1; % Preserve the impedance as it is
        C(i)=ZA(i);
        Ar(i)=real(C(i));
        Br(i)=imag(C(i));
    end
end
 % Augmentation loop:
 F=FL;
 delF=(FH-FL)/(N-1);
 for i=1:N
     Fk(i)=F;
  Ak(i)=line_seg(FA,Ar,F);
  Bk(i)=line_seg(FA,Br,F);
   F=F+delF;
 end
FC=[FA(1) Fk FA(NA)];
A=[Ar(1) Ak Ar(NA)];
B=[Br(1) Bk Br(NA)];
end
```

Program List 5.7: Objective function error_RFLST subject to minimization

```
function [ eps ] = error_RFLST(X0,ktr,T0,KFlag,WB,R1,WA,RA,XA,wL,wH,M )
% This objective function generates error eps(i) computed at
% sampling points wi for a given T0 for the real frequency line
% segment technique.
% -------------------------------------------------------------------------
```

```
%     Inputs:
%               X: The point for which the error function eps(i) is generated.
%                 Note that for RF-LST X0 must include the break points RB.
%               ktr: Control Flag. ktr=0>Transformerless design where RB(1) is
%               fixed. ktr=1> Design with transformer where R(1) is part of the
%               unknowns.
%               T0: Flat gain level subject to optimization
%               KFlag: Control flag to define the complex terminations for [N]
%               either impedance or admittance based optimization
%               WB: Normalized angular break frequencies (Array)
%               R1: Firt Break Point for Real Frequency-Line Segment Technique.
%               WA: Normalized angular frequency sampling points for the
%               measured complex impedance data.
%               RA: Real Part of the measured complex impedance termination.
%               XA: Imaginary part of the measured impedance termination.
%               wL: Lower frequency bound for the optimization.
%               wH: Upper frequency bound for the optimization
%               M: Total number of sampling points for the optimization
%-------------------------------------------------------------------------
% Output:
%               eps: value of the error computed at wi. This is a vector for
%               lsqnonlin.
%-------------------------------------------------------------------------
% Note that if ktr=0, then X0 consist of N-2 break points namely
% R(2),R(3),..,R(N-1). If ktr=1, then X0 contains N-1 break points.
%-------------------------------------------------------------------------
N=length(WB);
if ktr==0;% Design with no transformer. R1=R(1) is fixed.
    RB=[R1 X0 0];% The last break point R(N) is fixed at zero
end
        if ktr==1; % Design with transformer: R(1) is among the unknowns
            RB=[X0 0];% The last break point R(N) is fixed at zero
        end
% Generation of error=eps
delw=(wH-wL)/(M-1);
    w=wL;

    for i=1:M
    [ A,B ] = Measured_Impedance( w,WA,RA,XA,KFlag );
    P=line_seg(WB,RB,w);
    Q=num_hilbert(w,WB,RB);
    [ TPG ] = Gain(A,B,P,Q );
    eps(i)=TPG-T0;
    w=w+delw;
    end
end
```

Program List 5.8: Main Program Example 5.2

```
% Main Program Example5_2.m
% November 25 2014 by BS Yarman, Vanikoy, Istanbul, Turkey
% This program provides the solution for Example 5.2 given in the new book of
% Andrei Grebennikov-Narendra Kumar B. Siddik Yarman Book
%
```

```
% Example 5.2
%
% Inputs:
% Load-Pull Measurement for LD MOS RD07 by Mitsubishi
% Sampling Frequencies of the actual measurements in MHz
%-----------------------------------------------------------------------
clear
clc
close all
%-----------------------------------------------------------------------
% ------ Inputs I: -----------------------------------------------------
% Termination data for the front-end matching network: Zin(jw)=Rin+jXin
% Augmented actual frequencies:
FA=[0 330 350 370 390 410 430 450 470 490 510 530 1060];
NA=length(FA);
f0=530;% Selected normalization frequency.
R0=50; % Selected impedance normalization number.
% Frequency normalization on the sampling frequencies of the measurement
WA=FA/f0;
% Actual large signal input impedance of the LD MOS Device RD07 MUS2B by
% Mitsubishi Zin_A=Rin_A+jXin_A
% Augmented real part:
Rin_A1=[20 17.61 14.50 18.70 22.00 9.16 10.23 ];
Rin_A2=[10.40 17.18 14.33 13.36 11.04 9];
Rin_A=[Rin_A1 Rin_A2];
%
Rin_N=Rin_A/R0;% Impedance based real part normalization.
% Augmented imaginary part:
Xin_A1=[0 -04.63 -03.67 -11.30 -10.11 -13.96 -17.94 ];
Xin_A2=[-15.89 -04.85 -05.11 -12.11 -14.03 -16];
Xin_A=[Xin_A1 Xin_A2];
Xin_N=Xin_A/50; % Impedance based imaginary part normalization.
% ----------- Example 5.2 Par(a) ---------------------------------------
%
% Generation of unknown initial break points using linear interpolation
% Choice for total number of unknowns in the passband is NC
% Break points must be selected depending on the choice of immittance
% If KFlag1=1>Impedance based break points will be generated.
% If KFlag1=0>Admittance based break points will be generated.
% C=A+jB is the generated immittance for the termination.
% K=P+jQ is the driving point input immittance of the matching network.
% ---------- Inputs: II ------------------------------------------------
KFlag1=0;% Design with admittance functions
NC=19;% Total number of unknowns within the passband.
% Generation of actual termination immittance for the single matching problem.
FL=330;% Actual lower end of the passband frequency in MHz.
FH=530;% Actual higher end of the passband frequency in MHz
% Computation of the termination impedance for selected number of unknowns
%-----------------------------------------------------------------------
[ FC, Ain, Bin ] = Impedance_Termination(KFlag1, NC, FA, Rin_A, Xin_A,FL,FH );
%-----------------------------------------------------------------------
figure
plot(FC,Ain, FC, Bin)
xlabel('Actual sampling frequencies in MHz')
```

```
ylabel('Yin=1/Zin=Ain+jBin converted termination which is admittance')
legend('Ain','Bin')
title('Augmented termination data with 19 samples in the passband')
%-------------------------------------------------------------------
% Normalized angular break frequency array for the initial break points.
WB=FC/f0;
N=length(FC);% Total number of break points
% Impedance normalization number R0=50 ohm.
% R0=50;
% Normalized input immittance: Cin= Ain_N + jBin_N
if KFlag1==1;
Ain_N=Ain/R0;
Bin_N=Bin/R0;
end
if KFlag1==0
    Ain_N=Ain*R0;
    Bin_N=Bin*R0;
end
%-------------------------------------------------------------------
% Computation of initial break points:
T0=0.975;% Initial flat gain level
sign=1; % High initial values
RB=initials(T0,WB,Ain_N,Bin_N,sign);
RB(N)=0;
figure
plot(WB,RB)
xlabel('Normalized angular frequencies')
ylabel('RB: Admittance based break points computed for T0=0.975')
title('Admittance based high value initial break points RB with N=19+2=21
samples')
% Computation of initial transducer power gain
% ------------- Part (b) and Part (c) ---------------------------------
for i=1:N
    w=WB(i);
    A=line_seg(WB,Ain_N,w); % Real part of the termination
    B=line_seg(WB,Bin_N,w); % Imaginary part of the termination
    P=line_seg(WB,RB,w);    % Real part of the driving pint immittance
    Q=num_hilbert(w,WB,RB); % Imaginary part of driving point immittance
    QF(i)=Q;
    [ TPG ] = Gain(A,B,P,Q );% Transducer power gain
    T(i)=TPG;
end
plot(WB,QF)
xlabel('normalized break frequencies')
ylabel(' Hilbert transform of admittance based initial break points:
QF=H(RB)')
title ('YF=PF+jQF: Driving point admittance of the front-end matching
network')
% ---------- OPTIMIZATION LOOP ----------------------------------------
% optimization loop:
wL=WB(2)/WB(N-1);% Normalized lower end of the frequency band
wH=WB(N-1);% Normalized upper end of the frequency band
KFlag=KFlag1;% Admittance based design (Input)
```

```
ktr=0; %Design without transformer (Input)
R1=RB(1);% First break point
% Preparation for the initial guess for the unknowns X:
if ktr==0; % Design without transformer
    for i=1:N-2
    X0(i)=RB(i+1);
    end
end
if ktr==1; % Design with transformer
    for i=1:N-1
        X0=RB(i+1);
    end
end
M=2*NC;
options=optimset('MaxFunEvals',20000,'MaxIter',50000);
eps=@(X0)error_RFLST(X0,ktr,T0,KFlag,WB,R1,WA,Rin_N,Xin_N,wL,wH,M );
X=lsqnonlin(eps,X0,[],[],options);
%X=lsqnonlin('error_RFLST',X0,[],[],options,ktr,T0,KFlag,WB,R1,WA,Rin_N,X
in_N,wL,wH,M);
%
% ------ END OF OPTIMIZATION -------------------------------------------
if ktr==0
    for i=1:N-2
        RB(i+1)=X(i);
    end
    RB(1)=R1; RB(N)=0;
end
if ktr==1
    for i=1:N-1
        RB(i)=X(i);
    end
    RB(N)=0;
end
%----------------------------------------------------------------------
delw=(wH-wL)/(M-1);
    w=wL;
for i=1:M
WM(i)=w;
            P=line_seg(WB,RB,w);
            Q=num_hilbert(w,WB,RB);
            A=line_seg(WB,Ain_N,w);
            B=line_seg(WB,Bin_N,w);
            [ TPG ] = Gain(A,B,P,Q );
            TA(i)=TPG;
    w=w+delw;
end
figure
plot(WB,T,WM,TA)
xlabel('Normalized Angular Frequency w')
ylabel('Initial Transducer Power Gain for Front-End Equalizer')
legend('Inital Gain', 'Optimized Gain')
title('Admittance Based TPG the Front-End Matching Network:')
```

Program List 5.9 Main Program Example5_3.m

```
% Main Program Example5_3.m
% November 29 2014 by BS Yarman, Vanikoy, Istanbul, Turkey
% This program provides the solution for Example 5.3 given in the new
  book of
% Andrei Grebennikov-Narendra Kumar and B. Siddik Yarman Book
% This program optimizes the break points using the impedance based break
% points and the impedance based transducer power gain
%
% Example 5.3
%
% Inputs:
% Load-Pull Measurement for LD MOS RD07 by Mitsubishi
% Sampling Frequencies of the actual measurements in MHz
%-------------------------------------------------------------------------
clear
clc
close all
%-------------------------------------------------------------------------
% ------ Inputs I: ----------------------------------------------------
% Termination data for the front-end matching network: Zin(jw)=Rin+jXin
% Augmented actual frequencies:
FA=[0 330 350 370 390 410 430 450 470 490 510 530 1060];
NA=length(FA);
f0=530;% Selected normalization frequency.
R0=50; % Selected impedance normalization number.
% Frequency normalization on the sampling frequencies of the measurement
  WA=FA/f0;
% Actual large signal input impedance of the LD MOS Device RD07 MUS2B by
% Mitsubishi Zin_A=Rin_A+jXin_A
% Augmented real part:
Rin_A1=[20 17.61 14.50 18.70 22.00 9.16 10.23 ];
Rin_A2=[10.40 17.18 14.33 13.36 11.04 9];
Rin_A=[Rin_A1 Rin_A2];
%
Rin_N=Rin_A/R0;% Impedance based real part normalization.
% Augmented imaginary part:
Xin_A1=[0 -04.63 -03.67 -11.30 -10.11 -13.96 -17.94 ];
Xin_A2=[-15.89 -04.85 -05.11 -12.11 -14.03 -16];
Xin_A=[Xin_A1 Xin_A2];
Xin_N=Xin_A/50; % Impedance based imaginary part normalization.
%
% ----------- Example 5.3 Part(a) -----------------------------------------
%
% Generation of unknown initial break points using linear interpolation
% Choice for total number of unknowns in the passband is NC
% Break points must be selected depending on the choice of immittance
% If KFlag1=1>Impedance based break points will be generated.
% If KFlag1=0>Admittance based break points will be generated.
% C=A+jB is the generated immittance for the termination.
% K=P+jQ is the driving point input immittance of the matching network.
% ---------- Inputs: II ----------------------------------------------
```

```
KFlag1=1;% Design with impedance functions
NC=19;% Total number of unknowns within the passband.
% Generation of actual termination immittance for the single matching
problem.
FL=330;% Actual lower end of the passband frequency in MHz.
FH=530;% Actual higher end of the passband frequency in MHz
% Computation of the termination impedance for selected number of unknowns
% break points in the passband
%--------------------------------------------------------------------------
[FC,Ain,Bin] = Impedance_Termination(KFlag1, NC, FA, Rin_A, Xin_A,FL,FH );
%--------------------------------------------------------------------------
figure
plot(FC,Ain, FC, Bin)
xlabel('Actual sampling frequencies in MHz')
ylabel('Zin=Ain+jBin complex impedance termination with 19 points in
passband')
legend('Ain','Bin')
title('Augmented impedance termination data with 19 samples in the
passband')
%--------------------------------------------------------------------------
% Normalized angular break frequency array for the initial break points.
WB=FC/f0;
N=length(FC);% Total number of break points including DC and stop band
frequency fs
% Impedance normalization number R0=50 ohm.
% R0=50;
% Normalized input immittance: Cin= Ain_N + jBin_N
if KFlag1==1;
Ain_N=Ain/R0;
Bin_N=Bin/R0;
end
if KFlag1==0
    Ain_N=Ain*R0;
    Bin_N=Bin*R0;
end
%--------------------------------------------------------------------------
% Computation of initial break points:
T0=0.975;% Initial flat gain level
sign=1; % High initial values
RB=initials(T0,WB,Ain_N,Bin_N,sign);
RB(N)=0;
figure
plot(WB,RB)
xlabel('Normalized angular frequencies')
ylabel('RB: Admittance based break points computed for T0=0.975')
title('Admittance based high value initial break points RB with N=19+2=21
samples')
% Computation of initial transducer power gain
% ------------- Part (b) and Part (c) ----------------------------------
for i=1:N
    w=WB(i);
    A=line_seg(WB,Ain_N,w);% Real part of the termination
    B=line_seg(WB,Bin_N,w);% Imaginary part of the termination
    P=line_seg(WB,RB,w);% Real part of the driving pint immittance (DPI)
```

```
     P_int(i)=P;% Real part of driving point immittance of [F]
     Q=num_hilbert(w,WB,RB); % Imaginary part of (DPI)
     QF_int(i)=Q;% Imaginary part of DPI of [F]
     [ TPG ] = Gain(A,B,P,Q );% Transducer power gain
     T(i)=TPG;
end
plot(WB,QF_int)
xlabel('normalized break frequencies')
ylabel(' Hilbert transform of impedance based initial break points:
QF=H(RB)')
title ('ZF=PF+jQF: High value initial DPI of [F]: KFlag=1, sign=1 case')
% ---------- OPTIMIZATION LOOP --------------------------------------------
% optimization loop:
wL=WB(2)/WB(N-1);% Normalized lower end of the frequency band
wH=WB(N-1);% Normalized upper end of the frequency band
KFlag=KFlag1;% Admittance based design (Input)
ktr=0; %Design without transformer (Input)
R1=RB(1);% First break point
% Preparation for the initial guess for the unknowns X:
if ktr==0; % Design without transformer
    for i=1:N-2
    X0(i)=RB(i+1);
    end
end
if ktr==1; % Design with transformer
    for i=1:N-1
        X0=RB(i+1);
    end
end
M=2*NC;
options=optimset('MaxFunEvals',20000,'MaxIter',50000);
eps=@(X0)error_RFLST(X0,ktr,T0,KFlag,WB,R1,WA,Rin_N,Xin_N,wL,wH,M );
X=lsqnonlin(eps,X0,[],[],options);
%X=lsqnonlin('error_RFLST',X0,[],[],options,ktr,T0,KFlag,WB,R1,WA,Rin_N,X
in_N,wL,wH,M);
%
% ------ END OF OPTIMIZATION ----------------------------------------------
if ktr==0
    for i=1:N-2
        RB(i+1)=X(i);
    end
    RB(1)=R1; RB(N)=0;
end
if ktr==1
    for i=1:N-1
        RB(i)=X(i);
    end
    RB(N)=0;
end
% ------------------------------------------------------------------------
delw=(wH-wL)/(M-1);
    w=wL;
for i=1:M
WM(i)=w;
```

```
                P=line_seg(WB,RB,w);
                Q=num_hilbert(w,WB,RB);
                A=line_seg(WB,Ain_N,w);
                PF(i)=P;QF(i)=Q;
                B=line_seg(WB,Bin_N,w);
                [ TPG ] = Gain(A,B,P,Q );
                TA(i)=TPG;
        w=w+delw;
end
%
figure
plot(WB,RB,WB,P_int)
legend('optimized break points','Initial Break Points')
xlabel('normalized frequency')
ylabel('Impedance based optimized break points PF and Initial Break
Points RB')
title('Impedance Based initial and optimized break points (KFlag=1)')
%
figure
plot(WB,QF_int, WM,QF)
legend('Initial QF-int','optimized QF')
xlabel('normalized angular frequency')
ylabel('imaginary parts of ZF: QF-int, QF')
title('Imaginary parts of Minimum reactance ZF: KFlag=1, sign=1')
%
figure
plot(WB,T,WM,TA)
xlabel('Normalized Angular Frequency w')
ylabel('Initial Transducer Power Gain for the Front-End Equalizer')
legend('Initial Gain','Optimized Gain')
title('Impedance Based TPG for the Front-End Matching Network')
%
```

___

Program List 5.10: Main Program Example5_4.m

___

```
% Main Program Example5_4.m
% November 30 2014 by BS Yarman, Vanikoy, Istanbul, Turkey
% This program provides the solution for Example 5.2 given in the new
  book of
% Andrei Grebennikov-Narendra Kumar B. Siddik Yarman Book
%
% Example 5.4: Model for the break points
%
% Inputs:
% Load-Pull Measurement for LD MOS RD07 by Mitsubishi
% Sampling Frequencies of the actual measurements in MHz
%------------------------------------------------------------------------
clear
clc
close all
%
% ------ Inputs I: --------------------------------------------------------
% Termination data for the front-end matching network: Zin(jw)=Rin+jXin
% Augmented actual frequencies:
```

```
FA=[0 330 350 370 390 410 430 450 470 490 510 530 1060];
NA=length(FA);
f0=530;% Selected normalization frequency.
R0=50; % Selected impedance normalization number.
% Frequency normalization on the sampling frequencies of the measurement
  WA=FA/f0;
% Actual large signal input impedance of the LD MOS Device RD07 MUS2B by
% Mitsubishi Zin_A=Rin_A+jXin_A
% Augmented real part:
Rin_A1=[20 17.61 14.50 18.70 22.00 9.16 10.23 ];
Rin_A2=[10.40 17.18 14.33 13.36 11.04 9];
Rin_A=[Rin_A1 Rin_A2];
%
Rin_N=Rin_A/R0;% Impedance based real part normalization.
% Augmented imaginary part:
Xin_A1=[0 -04.63 -03.67 -11.30 -10.11 -13.96 -17.94 ];
Xin_A2=[-15.89 -04.85 -05.11 -12.11 -14.03 -16];
Xin_A=[Xin_A1 Xin_A2];
Xin_N=Xin_A/50; % Impedance based imaginary part normalization.
%
% ----------- Generation of initial guess --------------------------------
%
% Generation of unknown initial break points using linear interpolation
% Choice for total number of unknowns in the passband is NC
% Break points must be selected depending on the choice of immittance
% If KFlag1=1>Impedance based break points will be generated.
% If KFlag1=0>Admittance based break points will be generated.
% C=A+jB is the generated immittance for the termination.
% K=P+jQ is the driving point input immittance of the matching network.
% ---------- Inputs: II --------------------------------------------------
KFlag1=0;% Design with admittance functions
NC=19;% Total number of unknowns within the passband.
% Generation of actual termination immittance for the single matching
problem.
FL=330;% Actual lower end of the passband frequency in MHz.
FH=530;% Actual higher end of the passband frequency in MHz
% Computation of the termination impedance for selected number of
unknowns
%-----------------------------------------------------------------------
[FC,Ain,Bin] = Impedance_Termination(KFlag1, NC, FA, Rin_A, Xin_A,FL,FH
);
%-----------------------------------------------------------------------
figure
plot(FC,Ain, FC, Bin)
xlabel('Actual sampling frequencies in MHz')
ylabel('Yin=1/Zin=Ain+jBin converted termination which is admittance')
legend('Ain','Bin')
title('Augmented termination data with 19 samples in the passband')
%-----------------------------------------------------------------------
% Normalized angular break frequency array for the initial break points.
WB=FC/f0;
N=length(FC);% Total number of break points
% Impedance normalization number R0=50 ohm.
% R0=50;
```

```
% Normalized input immittance: Cin= Ain_N + jBin_N
if KFlag1==1;
Ain_N=Ain/R0;
Bin_N=Bin/R0;
end
if KFlag1==0
    Ain_N=Ain*R0;
    Bin_N=Bin*R0;
end
%------------------------------------------------------------------------
% Computation of initial break points:
T0=0.975;% Initial flat gain level
sign=1; % High initial values
RB=initials(T0,WB,Ain_N,Bin_N,sign);
RB(N)=0;
%------------------------------------------------------------------------
figure
plot(WB,RB)
xlabel('Normalized angular frequencies')
ylabel('RB: Admittance based break points computed for T0=0.975')
title('Admittance based high value initial break points RB with N=19+2=21
samples')
% Computation of initial transducer power gain
% ------------- Generation of initial TPG ---------------------------------
for i=1:N
    w=WB(i);
    A=line_seg(WB,Ain_N,w);% Real part of the termination
    B=line_seg(WB,Bin_N,w);% Imaginary part of the termination
    P=line_seg(WB,RB,w);% Real part of the driving pint immittance
    P_int(i)=P;

    Q=num_hilbert(w,WB,RB); % Imaginary part of driving point immittance
    QF_int(i)=Q;
    [ TPG ] = Gain(A,B,P,Q );% Transducer power gain
    T(i)=TPG;
end
%------------------------------------------------------------------------
figure
plot(WB,QF_int)
xlabel('normalized break frequencies')
ylabel(' Hilbert transform of admittance based initial break points:
QF=H(RB)')
title ('YF=PF+jQF: Driving point admittance of the front-end matching
network')
% ---------- OPTIMIZATION LOOP ---------------------------------------------
% optimization loop:
wL=WB(2)/WB(N-1);% Normalized lower end of the frequency band
wH=WB(N-1);% Normalized upper end of the frequency band
KFlag=KFlag1;% Admittance based design (Input)
ktr=0; %Design without transformer (Input)
R1=RB(1);% First break point
% Preparation for the initial guess for the unknowns X:
[ X0 ] = InitiateRFLST_lsqnonlin( ktr,RB );
M=2*NC;
```

```
options=optimset('MaxFunEvals',20000,'MaxIter',50000);
eps=@(X0)error_RFLST(X0,ktr,T0,KFlag,WB,R1,WA,Rin_N,Xin_N,wL,wH,M );
X=lsqnonlin(eps,X0,[],[],options);
% ------ END OF OPTIMIZATION -----------------------------------------
[ RB ] = EvaluateRFLST_lsqnonlin( ktr,X,R1 );
%--------------------------------------------------------------------
delw=(wH-wL)/(M-1);
    w=wL;
for i=1:M
WM(i)=w;
            P=line_seg(WB,RB,w);
            Q=num_hilbert(w,WB,RB);
            A=line_seg(WB,Ain_N,w);
            PF(i)=P;QF(i)=Q;
            B=line_seg(WB,Bin_N,w);
            [ TPG ] = Gain(A,B,P,Q );
            TA(i)=TPG;
    w=w+delw;
end
%--------------------------------------------------------------------
figure
plot(WB,T,WM,TA)
xlabel('Normalized Angular Frequency w')
ylabel('Initial Transducer Power Gain for the Front-End Equalizer')
legend('Initial Gain','Optimized Gain')
title('Admittance Based TPG for the Front-End Matching Network: Kfalg=0,
sign=1')
%--------------------------------------------------------------------
figure
plot(WB,RB,WB,P_int,WB,QF_int,WM,QF)
legend('Initial Break Points','optimized break points','initial
QF-int','optimized QF')
xlabel('normalized frequency')
ylabel('Admittance-Based DPI of [F]:Initial and optimized YF=PF+jQF')
title('Kfalg=0, sign=1, DPI of [F]: Initial & opt. values YF=PF+jQF')
%--------------------------------------------------------------------
% Curve fitting for the optimized break points:
% Initilize the unknowns:
a0=sqrt(RB(1));
n=5;ndc=0; WZ=0;nz=0;
[ x0 ] = Initiate_CurveFitting( ktr,n,a0 );
%
NB=length(WB);
wL=WB(1); wH=WB(NB);
M=2*NC;
options=optimset('MaxFunEvals',200000,'MaxIter',500000);
eps=@(x0)error_RFDCT( x0,ktr,WB,RB,n,ndc,WZ,a0,wL,wH,M );
x=lsqnonlin(eps,x0,[],[],options);
%--------------------------------------------------------------------
[ c,a0,a,b,AA,BB ] = Evaluate_CurveFitting(x,ktr,n,ndc,WZ,R1 );
%--------------------------------------------------------------------
%
NB=length(WB);w1=WB(1);w2=WB(NB);m=100;
w=w1;
```

```
delw=(w2-w1)/(m-1);
for j=1:m
    Wr(j)=w;
    Pr(j)=line_seg(WB,RB,w);
    AAVAL=polyval(AA,w);
    BBVAL=polyval(BB,w);
    P(j)=AAVAL/BBVAL;
    w=w+delw;
end
%------------------------------------------------------------------------
figure
plot(Wr,Pr,Wr,P)
legend('Break points','Rational approximation')
xlabel('normalized angular frequency')
ylabel('Brk. Points & Fitting via Rat. Apprx')
title('Rational Approximation of Break Points RB')
```

Program List 5.11 Main Program Example5_5.m

```
% Main Program Example5_5.m
% December 3 2014 by BS Yarman, Vanikoy, Istanbul, Turkey
% This program provides the solution for Example 5.2 given in the new
  book of
% Andrei Grebennikov-Narendra Kumar B. Siddik Yarman Book
%
% Example 5.5: Model for the break points
%
% Inputs:
% Load-Pull Measurement for LD MOS RD07 by Mitsubishi
% Sampling Frequencies of the actual measurements in MHz
%------------------------------------------------------------------------
clear
clc
close all
%
% ------ Inputs I: ------------------------------------------------------
% Termination data for the front-end matching network: Zin(jw)=Rin+jXin
% Augmented actual frequencies:
FA=[0 330 350 370 390 410 430 450 470 490 510 530 1060];
NA=length(FA);
f0=530;% Selected normalization frequency.
R0=50; % Selected impedance normalization number.
% Frequency normalization on the sampling frequencies of the measurement
WA=FA/f0;
% Actual large signal input impedance of the LD MOS Device RD07 MUS2B by
% Mitsubishi Zin_A=Rin_A+jXin_A
% Augmented real part:
Rin_A1=[20 17.61 14.50 18.70 22.00 9.16 10.23 ];
Rin_A2=[10.40 17.18 14.33 13.36 11.04 9];
Rin_A=[Rin_A1 Rin_A2];
%
Rin_N=Rin_A/R0;% Impedance based real part normalization.
% Augmented imaginary part:
```

```
Xin_A1=[0 -04.63 -03.67 -11.30 -10.11 -13.96 -17.94 ];
Xin_A2=[-15.89 -04.85 -05.11 -12.11 -14.03 -16];
Xin_A=[Xin_A1 Xin_A2];
Xin_N=Xin_A/50; % Impedance based imaginary part normalization.
%
% ----------- Generation of initial guess --------------------------------
%
% Generation of unknown initial break points using linear interpolation
% Choice for total number of unknowns in the passband is NC
% Break points must be selected depending on the choice of immittance
% If KFlag1=1>Impedance based break points will be generated.
% If KFlag1=0>Admittance based break points will be generated.
% C=A+jB is the generated immittance for the termination.
% K=P+jQ is the driving point input immittance of the matching network.
% ---------- Inputs: II --------------------------------------------------
KFlag1=0;% Design with admittance functions
NC=19;% Total number of unknowns within the passband.
% Generation of actual termination immittance for the single matching
problem.
FL=330;% Actual lower end of the passband frequency in MHz.
FH=530;% Actual higher end of the passband frequency in MHz
% Computation of the termination impedance for selected number of unknowns
% -----------------------------------------------------------------------
[FC,Ain,Bin] = Impedance_Termination(KFlag1, NC, FA, Rin_A, Xin_A,FL,FH);
% -----------------------------------------------------------------------
% Normalized angular break frequency array for the initial break points.
WB=FC/f0;
N=length(FC);% Total number of break points
% Impedance normalization number R0=50 ohm.
% R0=50;
% Normalized input immittance: Cin= Ain_N + jBin_N
if KFlag1==1;
Ain_N=Ain/R0;
Bin_N=Bin/R0;
end
if KFlag1==0
    Ain_N=Ain*R0;
    Bin_N=Bin*R0;
end
% -----------------------------------------------------------------------
% Computation of initial break points:
T0=0.975;% Initial flat gain level
sign=1; % High initial values
RB=initials(T0,WB,Ain_N,Bin_N,sign);
RB(N)=0;
% ------------- Generation of initial TPG ----------------------------
for i=1:N
    w=WB(i);
    A=line_seg(WB,Ain_N,w);% Real part of the termination
    B=line_seg(WB,Bin_N,w);% Imaginary part of the termination
    P=line_seg(WB,RB,w);% Real part of the driving pint immittance
    P_int(i)=P;
```

```
      Q=num_hilbert(w,WB,RB); % Imaginary part of driving point immittance
      QF_int(i)=Q;
      [ TPG ] = Gain(A,B,P,Q );% Transducer power gain
      T(i)=TPG;
end
% ---------- OPTIMIZATION LOOP ----------------------------------------------
% optimization loop:
wL=WB(2)/WB(N-1);% Normalized lower end of the frequency band
wH=WB(N-1);% Normalized upper end of the frequency band
KFlag=KFlag1;% Admittance based design (Input)
ktr=0; %Design without transformer (Input)
R1=RB(1);% First break point
% Preparation for the initial guess for the unknowns X:
[ X0 ] = InitiateRFLST_lsqnonlin( ktr,RB );
M=2*NC;
options=optimset('MaxFunEvals',20000,'MaxIter',50000);
eps=@(X0)error_RFLST(X0,ktr,T0,KFlag,WB,R1,WA,Rin_N,Xin_N,wL,wH,M );
X=lsqnonlin(eps,X0,[],[],options);
% ------ END OF OPTIMIZATION -------------------------------------------------
[ RB ] = EvaluateRFLST_lsqnonlin( ktr,X,R1 );
%---------------------------------------------------------------------------
delw=(wH-wL)/(M-1);
    w=wL;
for i=1:M
WM(i)=w;
            P=line_seg(WB,RB,w);
            Q=num_hilbert(w,WB,RB);
            A=line_seg(WB,Ain_N,w);
            PF(i)=P;QF(i)=Q;
            B=line_seg(WB,Bin_N,w);
            [ TPG ] = Gain(A,B,P,Q );
            T_RFLST(i)=TPG;
    w=w+delw;
end
%---------------------------------------------------------------------------
% Curve fitting for the optimized break points:
% Initilize the unknowns:
a0=sqrt(RB(1));
n=5;ndc=0; WZ=0;nz=0;
[ x0 ] = Initiate_CurveFitting( ktr,n,a0 );
%
NB=length(WB);
wL=WB(1); wH=WB(NB);
M=2*NC;
options=optimset('MaxFunEvals',200000,'MaxIter',500000);
eps=@(x0)error_RFDCT( x0,ktr,WB,RB,n,ndc,WZ,a0,wL,wH,M );
x=lsqnonlin(eps,x0,[],[],options);
%---------------------------------------------------------------------------
[ c,a0,a,b,AA,BB ] = Evaluate_CurveFitting(x,ktr,n,ndc,WZ,R1 );
%---------------------------------------------------------------------------
%
NB=length(WB);w1=WB(1);w2=WB(NB);m=100;
w=w1;
```

```
delw=(w2-w1)/(m-1);
for j=1:m
    Wr(j)=w;
    Pr(j)=line_seg(WB,RB,w);
    AAVAL=polyval(AA,w);
    BBVAL=polyval(BB,w);
    P(j)=AAVAL/BBVAL;
    w=w+delw;
end
%-------------------------------------------------------------------------
% Example 5.5
% Part (a)
% Generation of minimum susceptance driving point admittance for [F]
% YF=a(p)/b(p) from the given a0, ndc WZ and c(i)
[a,b]=Minimum_Function(ndc,WZ,a0,c);
%-------------------------------------------------------------------------
ws1=0; ws2=2;Ns=100;delw=(ws2-ws1)/(Ns-1);w=ws1;j=sqrt(-1);
for i=1:Ns
    WA(i)=w;
    p=j*w;
% ------- Generation Positive Real Minimum Susceptance YF=PF+jQF for [F] --
    aval=polyval(a,p);
    bval=polyval(b,p);
    YFval=aval/bval;
    Pf0_RFDCT=real(YFval);        PAf0_RFDCT(i)=Pf0_RFDCT;
    Qf0_RFDCT=imag(YFval);        QAf0_RFDCT(i)=Qf0_RFDCT;
% ------ Generation of termination admittance Yin=A+jB --------------------
    A=line_seg(WB,Ain_N,w);       AA(i)=A;
    B=line_seg(WB,Bin_N,w);       BA(i)=B;
% Generation of single matching gain -------------------------------------
% Part (b)
            [ TPG ] = Gain(A,B,Pf0_RFDCT,Qf0_RFDCT );
            TfA(i)=TPG;
% ------ Generation of Optimized Break Points for RFLST -------------------
            Pf_RFLST=line_seg(WB,RB,w);       PAf_RFLST(i)=Pf_RFLST;
            Qf_RFLST=num_hilbert(w,WB,RB);    QAf_RFLST(i)=Qf_RFLST;
    w=w+delw;
end
% Part (c)
figure
plot(WA,TfA,WM,T_RFLST)
title('Example 5.5: Comparision between RFLST and RFDCT Gains')
xlabel('normalized angular frequency')
ylabel('TPG-RFDCT & TPG-RFDCT')
legend('TPG-RFDCT','TPG-RFLST')
%
figure
plot(WA,PAf0_RFDCT,WA,PAf_RFLST,WA,QAf0_RFDCT,WA,QAf_RFLST)
title('Example 5.5: Comparision between RFLST and RFDCT admittances')
xlabel('normalized angular frequency')
ylabel('Driv. Input Pt. Min. Susc. Admt: YF=P+jQ:RFDCT vs RFLST')
legend('Pf0-RFDCT','Pf-RFLST','Qf0-RFDCT','Qf-RFLST')
%
```

Program List 5.12: function error_RFDCT

```
function [ eps ] = error_RFDCT( x0,ktr,WB,RB,n,ndc,WZ,a0,wL,wH,M )
% This function generates an error for the nonlinear data fitting for
% non-negative rational even function.
% Problem is stated as: Given data points WB,RB. Find the best fit to
% equation (5.28).
%   Inputs:
%       X0: Initials for the unknowns.
%       ktr: Control flag about the use of transformers at the far end.
%       ktr=0> Design without transformer: a0 is fixed. It is not
%       unknown.
%       ktr=1> Design with transformer: a0 is an unknown variable
%       WB: Break frequencies
%       RB: Break points
%       n: Degree of the denominator
%       ndc=Total number of transmission zeros at DC
%       nz: Total number of finite transmission zeros at w=wi
%       a0: Coefficient of the numerator is A0=a0*a0
%       wL: Low-end of the passband frequency
%       wH: High-end of the passband frequency
%       M: Number of sampling points for the optimization
%   Output:
%       error eps(j) over M sampling points
%--------------------------------------------------------------------
% Design by data fitting without transformer
if ktr==0;
    for j=1:n
        c(j)=x0(j);
    end
end
% Design by data fitting with transformer
if ktr==1
    for j=1:n
        c(j)=x0(j);
    end
        a0=x0(n+1);
end
% Generate the denominator polynomial B(w^2) of (5.28)in w-domain:
        C=[c 1];
        BB=Poly_Positive(C);% This positive polynomial is in w-domain
        B=polarity(BB);% Now, it is transferred to p-domain
        nB=length(BB);
% Generate the numerator polynomial in w domain:
        A=(a0*a0)*R_Num(ndc,WZ);% A is specified in p-domain
        AA=polarity(A);% This polynomial is in w domain
% Note that function RtoZ requires same length vectors A and B
% Convert A to same length of B
nA=length(A);
if (abs(nB-nA)>0)
    A=fullvector(nB,A);% This polynomial is in p-domain
end
```

```
% Generation of analytic form of Driving Point Input Immittance
                  [a,b]=RtoZ(A,B);
%--------------------------------------------------------------------
NB=length(WB);
w=wL;
delw=(wH-wL)/(M-1);
for j=1:M
    Pr=line_seg(WB,RB,w);
    AAVAL=polyval(AA,w);
    BBVAL=polyval(BB,w);
    P=AAVAL/BBVAL;
    eps(j)=P-Pr;
    w=w+delw;
end
end
```

Program List 5.13: Function Minimum_Funtion

```
function [a,b]=Minimum_Function(ndc,W,a0,c)
% Generate analytic form of Fmin(p)=a(p)/b(p)
% Generate B(-p^2)
        C=[c 1];
        BB=Poly_Positive(C);% This positive polynomial is in w-domain
        B=polarity(BB);% Now, it is transferred to p-domain
% Generate A(-p^2) of R(-p^2)=A(-p^2)/B(-p^2)
nB=length(B);
        A=(a0*a0)*R_Num(ndc,W);% A is specified in p-domain
nA=length(A);
if (abs(nB-nA)>0)
  A=fullvector(nB,A);% work with equal length vectors
end
% Generation of minimum immitance function using Bode or Parametric
method
        [a,b]=RtoZ(A,B);% Here A and B are specified in p-domain
        na=length(a);
        if ndc>0;a(na)=0;end;
end
```

Program List 5.14: function [a,b]=RtoZ(A,B)

```
function [a,b]=RtoZ(A,B)
% This MatLab function generates a minimum function Z(p)=a(p)/b(p)
%       from its even part specified as R(p^2)=A(p^2)/B(p^2)
%       via Bode (or Parametric)approach
% Inputs: In p-domain, enter A(p) and B(p)
% A=[A(1) A(2) A(3)...A(n+1)]; for many practical cases we set A(1)=0.
% B=[B(1) B(2) B(3)...B(n+1)]
% Output:
%         Z(p)=a(p)/b(p) such that
%         a(p)=a(1)p^n+a(2)p^(n-1)+...+a(n)p+a(n+1)
%         b(p)=b(1)p^n+b(2)p^(n-1)+...+b(n)p+a(n+1)
% Generation of an immitance Function Z by means of Parametric Approach
% In parametric approach Z(p)=Z0+k(1)/[p-p(1)]+...+k(n)/[p-p(n)]
```

```
% R(p^2)=Even{Z(p)}=A(-p^2)/B(-p^2) where Z0=A(n+1)/B(n+1).
%
% Given A(-p^2)>0
% Given B(-p^2)>0
%
BP=polarity(B);%BP is in w-domain
AP=polarity(A);%AP is in w-domain
% Computational Steps
% Given A and B vectors. A(p) and B(p) vectors are in p-domain
% Compute poles p(1),p(2),...,p(n)and the residues k(i) at poles
p(1),p(2),...,p(n)
[p,k]=residue_Z0(AP,BP);
%
% Compute numerator and denominator polynomials
Z0=abs(A(1)/B(1));
[num,errorn]=num_Z0(p,Z0,k);
[denom,errord]=denominator(p);
%
a=num;
b=denom;
```

Program List 5.15: function B=polarity(A)

```
function B=polarity(A)
% MatLab Programs for Parameteric Approach: Program 1: sign
% This program changes the sign of the coefficients for a given
polynomial
% Polynomial A is given as an even polynomial in w domain.
% By setting p=-w^2 we compute the coefficients of the new even
polynomial
% in p domain.
% Polynomial A is an even polynomial in w domain
% P(w^2)=A(1)[w^(2n)]+A(2)[w^2(n-1)]+A(3)[w^2(n-2)+...+A(n)[w]+A(n+1)
% Inputs
%------------- MatlLab Vector A=[A(1) A(2) A(3)...A(n) A(n+1)] :
%Coefficients of the even polynomial in w domain
%
% Outputs
%------------- MatLab Vector B=[B(1) B(2) B(3) ...B(n) B(n+1)
%
sigma=-1;
NN=length(A);
n=NN-1;
for i=1:NN
    j=NN-i+1;
    B(j)=-sigma*A(j);
    sigma=-sigma;
end
```

Program List 5.16: function [p,k]=residue_Z0(A,B)

```
function [p,k]=residue_Z0(A,B)
% This function computes the residues k of a given real part R(w^2)
% R(w^2)=A(w^2)/B(w^2)
```

```
% It should be noted the R(w^2) is given in w domain as an even function of
% w rather than complex variable p=sigma+jw.
% Inputs:
% ------- A(w^2); full coefficient numerator polynomial
% ------- B(w^2); full coefficient denominator polynomial
%Step 1: Find p domain versions of A and B
% That means we set w^2=-p^2
AA=polarity(A);
BB=polarity(B);
% Step 2: Find the LHP poles of R(p^2)
x=roots(BB);
p=-sqrt(x);
% Step 3: Compute the product terms
prd=product(p);
n=length(p);
% Step4: Generate residues
for j=1:n
    y=p(j)*p(j);
    Aval=polyval(AA,y);
    k(j)=(-1)^n*Aval/p(j)/B(1)/prd(j);
end
end
```

Program List 5.17: function [num,errorn]=num_Z0(p,Z0,k)

```
function [num,errorn]=num_Z0(p,Z0,k)
% This function computes the numerator polynomial of an
% immittance function: Z(p)=Z0+sum{k(1)/[p-p(i)]
% where we assume that Z0=A(1)/B(1)which is provided as input.
%
% Input:
%-------- poles p(i) of the immittance function Z(p)
%         as a MatLab row vector p
%-------- Residues k(i) of poles p(i)
% Output:
%-------- num; MatLab Row-Vector
% which includes coefficients of numerator polynomial of an immittance
function.
% num=Sum{k(j)*product[p-p(i)]} which skips the term when j=i
%
%----- Step 1: Determine total number of poles n
%
n=length(p);
nn=n-1;
%
%----- Step 2: Generation of numerator polynomials:
% numerator polynomial=sum of
% sum of
% {Z0*[p-p(1)].[p-p(2)]...(p-p(n)]; n the degree-full product
% +k(1)*[p=p(2)].[p-p(3)]..[p-p(n)];degree of (n-1); the term with p(1)is
skipped.
% +k(2)*[p-p(1)].[p-p(3)]..[p-p(j-1)].[p-p(j+1)]..[p-p(n)];degree of(n-
1)-the term with p(2)is skipped
```

```
% +.......................................
% +k(j)*[p-p(1)].[p-p(2)]..[p-p(j-1)].[p-p(j+1)]..[p-p(n)];degree of
(n-1)-the term with p(j)is skipped.
% +.......................................
% +k(n)[p-p(1)].[p-p(2)]...[p-p(n-1)];degree of (n-1)-the term with p(n)is
% skipped.
%
% Note that we generate the numerator polynomial within 4 steps.
% In Step 2a, product polynomial pra of k(1)is evaluated.
% In Step 2b, product polynomial prb of k(j)is evaluated by skipping the
term when i=j.
% In Step 2c, product polynomial prc of k(n)is evaluated.
% In Step 2d, denominator of Z0 is generated.
%------------------------------------------------------------------------
%
% Step 2a: Generate the polynomial for the residue k(1)
pra=[1];
for i=2:n
  simpA=[1 -p(i)];
% pra is a polynomial vector of degree n-1; total number of entrees are n.
  pra=conv(pra,simpA);% This is an (n-1)th degree polynomial.
end
na=length(pra);
% store first polynomial onto firs row of A i.e. A(1,:)
for r=1:na
  A(1,r)=pra(r);
end
% Step 2a: Compute the product for 2<j<(n-1)

for j=2:nn
    prb1=[1];
        for i=1:j-1
        simpB=[1 -p(i)];
        prb1=conv(prb1,simpB);
        end
    % Skip j th term
    prb2=[1];
        for i=(j+1):n
            simpB1=[1 -p(i)];
            prb2=conv(prb2,simpB1);
        end
        prb=conv(prb1,prb2);
        nb=length(prb);
%
% Store j polynomials on to j th row of A; i.e. A(j,:)
for r=1:nb
A(j,r)=prb(r);
end
%
    end
% Step 2c: Compute the product term for j=n
prc=[1];
```

```
for i=1:nn
    simpC=[1 -p(i)];
    prc=conv(prc,simpC);
end
nc=length(prc);
% store n the polynomial onto n the row of A(n,:)
for r=1:nc
A(n,r)=prc(r);
end
%-------------------------------------------------------------------------
for i=1:n
    for j=1:n
        C(i,j)=k(i)*A(i,j);
    end
end
%
%-------- Step 4: Generate the numerator as a MatLab row vector.
for i=1:n
D(i)=0;%Perform the sum operation to compute numerator polynomial
end
for j=1:n
    for r=1:n
    D(j)=D(j)+C(r,j);
    end;% Here is the numerator polynomial of length n.
end
[denom,errord]=denominator(p);
prd_n=Z0*denom; % this is n the degree polynomial vector with length n+1
a(1)=prd_n(1);
for i=2:(n+1)
    a(i)=D(i-1)+prd_n(i);
end
%
num=real(a);
errorn=imag(a);
```

Program List 5.18: `function [denom,errord]=denominator(p)`

```
function [denom,errord]=denominator(p)
% This function computes the denominator polynomial of an
% immittance function from the given poles.
% It should be noted that this form of the denominator is normalized with
% the leading coefficient b(1)=1: denom=(p-p(1))(p-p(2))....(p-p(n))
% Input:
%-------- poles p(i) as a MatLab row vector p
%-------- Residues k(i) of poles p(i)
% Output:
%-------- denom; MatLab Row-Vector
% ---which includes coefficients of the denominator polynomial of an
immittance function.
% denom=product[p-p(i)]
%
% --------- Step 1: Determine n
n=length(p);
% -------- Step 2: Form the product term.
```

```
    pr=[1]; %Define a simple polynomial pr=1.
for j=1:n
     simple=[1 -p(j)];%this is the simple polynomial [p-p(j)]
% Generate multiplication of polynomials starting with 1.
     pr=conv(pr,simple);
end
denom=real(pr);
errord=imag(pr);
end
```

Program List 5.19: function Apoly=R_allzero(ndc,nz,W)

```
function Apoly=R_allzero(ndc,nz,W)
% This function computes the value of the RA
% RA(-p2)=(-1)^(ndc)*(p*p)^(ndc)*{[(p^2+W(1)^2]^2}...{[p^2+W(n)^2]^2}
% Inputs
%         ndc=transmission zeros at DC
%         nz=Transmission zeros at W(i)
%         W(i)=Transmission zeros at W(i)
%         p=jw a frequency point at which RA is computed.
% Output
%         A: Even polynomial coefficients of the numerator polynomial
%         R(-p^2)
%
% Computation Steps
%
% Initialization
        Apoly=[1];
if (nz>=1)
        for i=1:nz
        Wi2=W(i)*W(i);
        Cpoly=[1 0 Wi2];
        Dpoly=conv(Cpoly,Cpoly);
        Apoly=conv(Apoly,Dpoly);
     end
end
if(ndc>0)
    D(2*ndc+1)=0.0;%in MatLab we shift all the terms by one.
    for i=1:2*ndc
        D(i)=0.0;
    end
    D(1)=(-1)^ndc;
end
        if ndc==0
            D=[1];
        end
        Apoly=conv(Apoly,D);
end
```

Program List 5.20: function AT=clear_oddpower(AA)

```
function AT=clear_oddpower(AA)
% This function clears the odd power terms in a given MatLab Polynomial AA
na=length(AA);
```

```
r=fix(na/2);
        for j=1:r
        AT(j)=AA(na-2*j+2);
        end
        for i=1:r
            ATT(r-i+1)=AT(i);
        end
        for i=1:r
            AT(i+1)=ATT(i);
        end
        AT(1)=AA(1);
end
```

Program List 5.21: Main Program Example5_6B.m

```
% Main Program Example5_6B.m
% December 5 2014 by BS Yarman, Vanikoy, Istanbul, Turkey
% This program provides solution for Example 5.6 using compact functions
% Andrei Grebennikov-Narendra Kumar B. Siddik Yarman Book
%-------------------------------------------------------------------------
% Compact optimization functions:
%          Step 1: Execute function RFLST_SingleMatching
%          Step 2: Execute function CurveFitting_BreakPoints
%                          (Rational curve fitting via RFDCT)
%          Step 3: Execute function FinalOptimization_Parametric
%                          (Final optimization via Pramteric Approach)
%          Step 4: Execute function SynthesisbyTranszeros
% Inputs:
% Load-Pull Measurement for LD MOS RD07 by Mitsubishi
% Sampling Frequencies of the actual measurements in MHz
%-------------------------------------------------------------------------
clear
clc
close all
%
% ------ Inputs I: ------------------------------------------------------
% Termination data for the front-end matching network: Zin(jw)=Rin+jXin
% Augmented actual frequencies:
FA=[0 330 350 370 390 410 430 450 470 490 510 530 1060];
NA=length(FA);
f0=530;% Selected normalization frequency.
R0=50; % Selected impedance normalization number.
% Frequency normalization on the sampling frequencies of the measurement
WA=FA/f0;
% Actual large signal input impedance of the LD MOS Device RD07 MUS2B by
% Mitsubishi Zin_A=Rin_A+jXin_A
% Augmented real part:
Rin_A1=[20 17.61 14.50 18.70 22.00 9.16 10.23 ];
Rin_A2=[10.40 17.18 14.33 13.36 11.04 9];
Rin_A=[Rin_A1 Rin_A2];
Rin_N=Rin_A/R0;% Impedance based real part normalization.
% Augmented imaginary part:
Xin_A1=[0 -04.63 -03.67 -11.30 -10.11 -13.96 -17.94 ];
Xin_A2=[-15.89 -04.85 -05.11 -12.11 -14.03 -16];
```

```
Xin_A=[Xin_A1 Xin_A2];
Xin_N=Xin_A/50; % Impedance based imaginary part normalization.
%------------------------------------------------------------------------
% INPUTS FOR RFLST:
%
        T0=0.975;
        NC=19;ktr=0;KFlag=0;sign=1;
        FL=330;FH=530;f0=530;R0=50;
%------------------------------------------------------------------------
% Step 1: Idealized solution by RFLST optimization
[ WB,RB,TB ] = RFLST_SingleMatching(NC,ktr,KFlag,sign,T0,FL,FH,FA,Rin_A,X
in_A,f0,R0 );
%------------------------------------------------------------------------
% Step 2: Rational function Curve fitting of the break points by RFDCT
% INPUTS FOR RFDCT Curve fitting:
    n=5;ndc=0;WZ=0;
[ c0,a0,AA,BB,P,a1,b1 ] = CurveFitting_BreakPoints (ktr,n,ndc,WZ,WB,RB );
%------------------------------------------------------------------------
% Step 3: Final optimization of TPG on c(i) via Parametric Approach
[ a,b,c,a0,WA,PA,QA,TA ] = FinalOptimization_Parametric( KFlag,ktr,T0,
n,ndc,WZ,c0,a0, NC, FA, Rin_A, Xin_A,FL,FH,f0,R0 );
figure
plot(WA,TA,WB,TB,WB,RB,WA,PA)
legend('TA-Final','TB-RFLST','RB','Final PA=Rational')
xlabel('normalized angular frequency')
ylabel('Optimum solution for the TPG of Example 5.6')
title('Best solution for TPG of Example 5.6')
%------------------------------------------------------------------------
% Step 4:Darlington Synthesis of the driving point input immittance K(p)
%       K(p)=a(p)/b(p)
%
eps_zero=1e-8;
[ CT, CV,LL1,LL2,MM ] = SynthesisbyTranszeros(KFlag,WZ,ndc,a,b,eps_zero);
F0=f0*1e6 %(in Hz)
[ CVA ] = Actual_LumpedElements( F0,R0,CT,CV )
```

Program List 5.22: Main Program Example5_7.m

```
% Main Program Example5_7.m
% This main program solves the single matching problem to construct the
% front-end matching network for the power LD MOS RD07 by Mitsubishi
% In this example far end termination of [F] is fixed as a0=1.8569 which
% yields T0=0.975
clc;clear all;close all;
% Load-pull measured input impedance of the LD MOS RD07.
FAin=[0 330 350 370 390 410 430 450 470 490 510 530 1060];
Rin=[20 17.61 14.50 18.70 22.00 9.16 10.23 10.40 17.18 14.33 13.36 11.04 9];
Xin=[0 -04.63 -03.67 -11.30 -10.11 -13.96 -17.94 -15.89 -04.85 -05.11
-12.11 -14.03 -16];
% Inputs for the passband and impedance normalization
FL=330;FH=530;f0=530;R0=50;F_unit=1e6;
%------------------------------------------------------------------------
% User selected inputs to design single matching network via RFT
```

```
%
   T0=0.975;
   NC=19;ktr=0;KFlag=0;sign=1;
   n=5;ndc=0;WZ=[0]; a0=1.8569;
%
% ------------------------------------------------------------------------
% Design of front-end matching network in single function:
%
[FF,TFdB,aF,bF,cF,CTF,CVF,CVAF]=ImmittanceBased_RealFrSingMatch(FAin,Rin,
Xin,R0,f0,FL,FH,F_unit,T0,KFlag,ktr,sign,NC,n,ndc,WZ,a0);
%
% ------------------------------------------------------------------------
figure
plot(FF,TFdB)
xlabel('Actual frequencies')
ylabel('TPG in dB')
title('TPG for the front-end matching network of power amplifier designed
with LD MOS 07')
```

Program List 5.23: Function Error_Parametric

```
function [ eps ] = Error_Parametric(x0,ktr,T0,n,ndc,WZ,a0,WB,Ain_N,Bin_N,
wL,wH,M )
[ a,b ] = Evaluate_Parametric(x0,ktr,n,ndc,WZ,a0 );
%
Ns=M;delw=(wH-wL)/(Ns-1);w=wL;j=sqrt(-1);
for i=1:Ns
    p=j*w;
% ------- Generation Positive Real Minimum Susceptance YF=PF+jQF for [F] --
    aval=polyval(a,p);
    bval=polyval(b,p);
    YFval=aval/bval;
    Pf0_RFDCT=real(YFval);
    Qf0_RFDCT=imag(YFval);
% ------ Generation of termination admittance Yin=A+jB --------------------
    A=line_seg(WB,Ain_N,w);
    B=line_seg(WB,Bin_N,w);
% Generation of single matching gain -------------------------------------
% Part (b)
            [ TPG ] = Gain(A,B,Pf0_RFDCT,Qf0_RFDCT );
            eps(i)=TPG-T0;
   w=w+delw;
end
end
```

Program List 5.24: Main program Example5_6.m

```
% Main Program Example5_6.m
% December 3 2014 by BS Yarman, Vanikoy, Istanbul, Turkey
% This program provides the solution for Example 5.6 given in the new book of
% Andrei Grebennikov-Narendra Kumar B. Siddik Yarman Book
%
% Example 5.5: Model for the break points
```

```
%
% Inputs:
% Load-Pull Measurement for LD MOS RD07 by Mitsubishi
% Sampling Frequencies of the actual measurements in MHz
% -----------------------------------------------------------------------
clear
clc
close all
%
% ------ Inputs I: -------------------------------------------------------
% Termination data for the front-end matching network: Zin(jw)=Rin+jXin
% Augmented actual frequencies:
FA=[0 330 350 370 390 410 430 450 470 490 510 530 1060];
NA=length(FA);
f0=530;% Selected normalization frequency.
R0=50; % Selected impedance normalization number.
% Frequency normalization on the sampling frequencies of the measurement
WA=FA/f0;
% Actual large signal input impedance of the LD MOS Device RD07 MUS2B by
% Mitsubishi Zin_A=Rin_A+jXin_A
% Augmented real part:
Rin_A1=[20 17.61 14.50 18.70 22.00 9.16 10.23 ];
Rin_A2=[10.40 17.18 14.33 13.36 11.04 9];
Rin_A=[Rin_A1 Rin_A2];
%
Rin_N=Rin_A/R0;% Impedance based real part normalization.
% Augmented imaginary part:
Xin_A1=[0 -04.63 -03.67 -11.30 -10.11 -13.96 -17.94 ];
Xin_A2=[-15.89 -04.85 -05.11 -12.11 -14.03 -16];
Xin_A=[Xin_A1 Xin_A2];
Xin_N=Xin_A/50; % Impedance based imaginary part normalization.
%
% ----------- Generation of initial guess ---------------------------------
%
% Generation of unknown initial break points using linear interpolation
% Choice for total number of unknowns in the passband is NC
% Break points must be selected depending on the choice of immittance
% If KFlag1=1>Impedance based break points will be generated.
% If KFlag1=0>Admittance based break points will be generated.
% C=A+jB is the generated immittance for the termination.
% K=P+jQ is the driving point input immittance of the matching network.
% ---------- Inputs: II --------------------------------------------------
KFlag1=0;% Design with admittance functions
NC=19;% Total number of unknowns within the passband.
% Generation of actual termination immittance for the single matching
problem.
FL=330;% Actual lower end of the passband frequency in MHz.
FH=530;% Actual higher end of the passband frequency in MHz
% Computation of the termination impedance for selected number of unknowns
% -----------------------------------------------------------------------
[FC,Ain,Bin] = Impedance_Termination(KFlag1, NC, FA, Rin_A, Xin_A,FL,FH );
% -----------------------------------------------------------------------
% Normalized angular break frequency array for the initial break points.
WB=FC/f0;
```

```
N=length(FC);% Total number of break points
% Impedance normalization number R0=50 ohm.
% R0=50;
% Normalized input immittance: Cin= Ain_N + jBin_N
if KFlag1==1;
Ain_N=Ain/R0;
Bin_N=Bin/R0;
end
if KFlag1==0
    Ain_N=Ain*R0;
    Bin_N=Bin*R0;
end
%-------------------------------------------------------------------------
% Step 1: Computation of initial break points:
T0=0.975;% Initial flat gain level
sign=1; % High initial values
RB=initials(T0,WB,Ain_N,Bin_N,sign);
RB(N)=0;
% Step 2:  Optimization of TPG on the break points RB for RFLST
%
wL=WB(2)/WB(N-1);% Normalized lower end of the frequency band
wH=WB(N-1);% Normalized upper end of the frequency band
KFlag=KFlag1;% Admittance based design (Input)
ktr=0; %Design without transformer (Input)
R1=RB(1);% First break point
% Preparation for the initial guess for the unknowns X:
[ X0 ] = InitiateRFLST_lsqnonlin( ktr,RB );
M=2*NC;
options=optimset('MaxFunEvals',20000,'MaxIter',50000);
eps=@(X0)error_RFLST(X0,ktr,T0,KFlag,WB,R1,WA,Rin_N,Xin_N,wL,wH,M );
X=lsqnonlin(eps,X0,[],[],options);
% ------ END OF OPTIMIZATION -----------------------------------------
[ RB ] = EvaluateRFLST_lsqnonlin( ktr,X,R1 );
%-------------------------------------------------------------------------
% Step 3: Model for the real part P(w) as in (5.28).
% Curve fitting for the optimized break points:
% Initilize the unknowns:
a0=sqrt(RB(1));
n=5;ndc=0; WZ=0;nz=0;
[ x0 ] = Initiate_CurveFitting( ktr,n,a0 );
%
NB=length(WB);
wL=WB(1); wH=WB(NB);
M=2*NC;
options=optimset('MaxFunEvals',200000,'MaxIter',500000);
eps=@(x0)error_RFDCT( x0,ktr,WB,RB,n,ndc,WZ,a0,wL,wH,M );
x=lsqnonlin(eps,x0,[],[],options);
%-------------------------------------------------------------------------
[ c,a0,a,b,AA,BB ] = Evaluate_CurveFitting(x,ktr,n,ndc,WZ,R1 );
c0=c;
% Step 4: Final optimization of TPG on c(i) via Parametric Approach
[ x0 ] = Initiate_Parametric( ktr,n,c,a0 );
%-------------------------------------------------------------------------
options=optimset('MaxFunEvals',200000,'MaxIter',500000);
```

```
T0=0.97;wL=WB(2);wH=WB(N-1);
eps=@(x0)Error_Parametric(x0,ktr,T0,n,ndc,WZ,a0,WB,Ain_N,Bin_N,wL,wH,M );
x=lsqnonlin(eps,x0,[],[],options);
%
[ a,b,c,a0 ] = Evaluate_Parametric(x,ktr,n,ndc,WZ,a0 );
%
ws1=0; ws2=2;Ns=100;delw=(ws2-ws1)/(Ns-1);w=ws1;j=sqrt(-1);
for i=1:Ns
    WA(i)=w;
    p=j*w;
% ------- Generation Positive Real Minimum Susceptance YF=PF+jQF for [F] --
    aval=polyval(a,p);
    bval=polyval(b,p);
    YFval=aval/bval;
    Pf0_RFDCT=real(YFval);
    Qf0_RFDCT=imag(YFval);
% ------ Generation of termination admittance Yin=A+jB --------------------
    A=line_seg(WB,Ain_N,w);     AA(i)=A;
    B=line_seg(WB,Bin_N,w);     BA(i)=B;
% --- Result of final optimization ------------------------------------------
            [ TPG ] = Gain(A,B,Pf0_RFDCT,Qf0_RFDCT );
            TfA(i)=TPG;
 w=w+delw;
end
figure
plot(WA,TfA)
xlabel('normalized angular frequency')
ylabel('Optimum solution for the TPG of Example 5.6')
title('Best solution for TPG of Example 5.6')
```

Program List 5.25: RFLST_SingleMatching

```
function [ WB,RB,TB ] = RFLST_SingleMatching(NC,ktr, KFlag1, sign, T0,
FL, FH, FA, Rin_A, Xin_A,f0,R0 )
%    Inputs:
%           FL: Lower edge of the passband. FL=FA(1) or FL(2)
%           FH: Upper edge of the passband. FH=FA(N-1).
%           ktr: Control Flag.
%               ktr=1> RB(1) is among the unknowns
%               ktr=0> RB(1) is fixed
%           NC: total number of uknowns in the passband
%           KFlag1: Immittance type of the optimization;
%               KFlag1=1> impedance based optimization
%               KFlag1=0> admittance based optimization
%           sign: Control Flag;
%               sign=1> Start with high values of initial break points
%               sign=-1> Start with low values of initial brak points
%           T0: Flat gain leven over passband subject to optimization
%           Single matching with termination impedance Zin=Rin+jXin
%           WA: Vector Sampling frequencies of the complex terminal
%           measurement. Note that terminal is specified as a complex
%           impedance with real and imaginary parts (ZL=Zin=Rin+jXin)
%           Rin: Vector, real par of Zin
%           Xin: Vector, imaginary part of Zin
```

```
%            f0: Normalization frequency
%            R0: Normalization resistance (Internal resistor of the
%            generator for single matching problem.
% Note: Single Matching problem is defined as: Construction of [N]
between
% resitive generator [R0] and complex impedance termination [Zin]
%
%   Output:
%            WB: Normalized break points.
%            RB: Vector. Normalized and optimized break points
%-----------------------------------------------------------------------
% Frequency normalization on the sampling frequencies of the measurement
WA=FA/f0;
% Impedance Normalization
Rin_N=Rin_A/R0; % Impedance based real part normalization.
Xin_N=Xin_A/50; % Impedance based imaginary part normalization.
%-----------------------------------------------------------------------
[FC,Ain,Bin] = Impedance_Termination(KFlag1, NC, FA, Rin_A, Xin_A,FL,FH);
%-----------------------------------------------------------------------
% Normalized angular break frequency array for the initial break points.
WB=FC/f0;
N=length(FC);% Total number of break points
% Impedance normalization number R0=50 ohm.
% R0=50;
% Normalized input immittance: Cin= Ain_N + jBin_N
if KFlag1==1;
Ain_N=Ain/R0;
Bin_N=Bin/R0;
end
if KFlag1==0
    Ain_N=Ain*R0;
    Bin_N=Bin*R0;
end
%-----------------------------------------------------------------------
% Step 1: Computation of initial break points:
%
RB=initials(T0,WB,Ain_N,Bin_N,sign);
RB(N)=0;
% Step 2:  Optimization of TPG on the break points RB for RFLST
% ktr=0 Case:
wL=WB(2)/WB(N-1);% Normalized lower end of the frequency band
if ktr==1
    wL=WB(1)/WB(N-1);
end
wH=WB(N-1);% Normalized upper end of the frequency band
KFlag=KFlag1;% Admittance based design (Input)
R1=RB(1);% First break point
% Preparation for the initial guess for the unknowns X:
[ X0 ] = InitiateRFLST_lsqnonlin( ktr,RB );
M=2*NC;
options=optimset('MaxFunEvals',20000,'MaxIter',50000);
eps=@(X0)error_RFLST(X0,ktr,T0,KFlag,WB,R1,WA,Rin_N,Xin_N,wL,wH,M );
X=lsqnonlin(eps,X0,[],[],options);
% ------ END OF OPTIMIZATION ----------------------------------------------
```

```
[ RB ] = EvaluateRFLST_lsqnonlin( ktr,X,R1 );
%--------------------------------------------------------------------------
%--------------------------------------------------------------------------
for i=1:N
    w=WB(i);
            P=line_seg(WB,RB,w);
            Q=num_hilbert(w,WB,RB);
            A=line_seg(WB,Ain_N,w);
            PF(i)=P;QF(i)=Q;
            B=line_seg(WB,Bin_N,w);
            [ TPG ] = Gain(A,B,P,Q );
            TB(i)=TPG;
end
end
```

Program List 5.26: function CurveFitting

```
function [ c,a0,AA,BB,P,a,b ] = CurveFitting_BreakPoints
(ktr,n,ndc,WZ,WB,RB )
% Curve fitting of the optimized break points RB by Equation (5.28) via
% Real Frequency Direct Computational Technique (RFDCT)
% Inputs:
%       ktr: Control Flag
%            ktr=1> a0 is unknown
%            ktr=0> a0 is fixed
%       n: Degree of denominator polynomial of (5.28)
%       ndc: Transmizzion zeros at DC
%       WZ: Vector. Finite transmission zeros on the frequency axis
%       WB: Vector. Normalized Angular Break Frequencies
%       RB: Normalized Break Points
% Output:
%       c: optimized coefficients of the auxiliary polynomial
%       c(w)=c(1)w^n+c(1)w^(n-1)++...+c(n)w+1
%       AA(p): Numerator polynomial of R(w^2)=AA((w^2)/BB(W^2)
%       BB(p): Denominator polynomial of R(w^2)
%       P: Vector. Values of R(W^2) evaluated at break frequencies.
%       K(p)=H(R(w^2))=a(p)/b(p); Driving Point Input Immittance of the
%       equalizer
%       a(p): Numerator polynomial of K(p)
%       b(p): Denominator polynomial of K(p)
% Curve fitting for the optimized break points:
% Initilize the unknowns:
a0=sqrt(RB(1));R1=RB(1);
[ x0 ] = Initiate_CurveFitting( ktr,n,a0 );
%
NB=length(WB);
wL=WB(1); wH=WB(NB);
M=3*NB;
options=optimset('MaxFunEvals',200000,'MaxIter',500000);
eps=@(x0)error_RFDCT( x0,ktr,WB,RB,n,ndc,WZ,a0,wL,wH,M );
x=lsqnonlin(eps,x0,[],[],options);
%--------------------------------------------------------------------------
[ c,a0,a,b,AA,BB ] = Evaluate_CurveFitting(x,ktr,n,ndc,WZ,R1 );
%--------------------------------------------------------------------------
```

```
% Check the quality of the curve fitting at the break frequencies
NB=length(WB);
for j=1:NB
    w=WB(j);
    Pr(j)=line_seg(WB,RB,w);
    AAVAL=polyval(AA,w);
    BBVAL=polyval(BB,w);
    P(j)=AAVAL/BBVAL;
end
end
```

Program List 5.27: FinalOptimization_Parametric

```
function [ a,b,c,a0,WA,PA,QA,TA ] = FinalOptimization_Parametric(
KFlag1,ktr,T0,n,ndc,WZ,c0,a0, NC, FA, Rin_A, Xin_A,FL,FH,f0,R0 )
% Final optimization of TPG on c(i) via Parametric Approach
[ x0 ] = Initiate_Parametric( ktr,n,c0,a0 );
N=length(FA);M=3*N;
options=optimset('MaxFunEvals',200000,'MaxIter',500000);
wL=FL/f0;wH=FH/f0;
% T0=0.97;wL=WB(2);wH=WB(N-1);
[ WA,Ain_N,Bin_N ] = NormalizedImmittance_Termination(KFlag1, NC, FA,
Rin_A, Xin_A,FL,FH,f0,R0 );
eps=@(x0)Error_Parametric(x0,ktr,T0,n,ndc,WZ,a0,WA,Ain_N,Bin_N,wL,wH,M );
x=lsqnonlin(eps,x0,[],[],options);
%
[ a,b,c,a0 ] = Evaluate_Parametric(x,ktr,n,ndc,WZ,a0 );
%
NA=length(WA);j=sqrt(-1);
for i=1:NA
    w=WA(i);
    p=j*w;
% ------- Generation Positive Real Minimum Susceptance YF=PF+jQF for [F] --
    aval=polyval(a,p);
    bval=polyval(b,p);
    YFval=aval/bval;
    Pf0_RFDCT=real(YFval);   PA(i)=Pf0_RFDCT;
    Qf0_RFDCT=imag(YFval);   QA(i)=Qf0_RFDCT;
% ------ Generation of termination admittance Yin=A+jB --------------------
    A=line_seg(WA,Ain_N,w);
    B=line_seg(WA,Bin_N,w);
% --- Result of final optimization -----------------------------------------
            [ TPG ] = Gain(A,B,Pf0_RFDCT,Qf0_RFDCT );
            TA(i)=TPG;
end
end
```

Program List 5.28: function Actual_LumpedElements

```
function [ CVA ] = Actual_LumpedElements( f0,R0,CT,CV )
% December 9, 2014
% -------------------------------------------------------------------------
% CT=1> Series Inductor
% CT=7> Shunt Inductor
```

```
% CT=8> Shunt Capacitor
% CT=2> Series Capacitor
% CT=9> Terminating Resistor
% Series R//L//C: CT=4> Inductor, CT=5> Capacitor,CT=6> Resistor,
% Shunt R+L+C: CT=10>Inductor,CT=11> Capacitor,CT=12>Resistor,
%-----------------------------------------------------------------
% Actual element values of inductors
n=length(CT);
for i=1:n
% Denormalization of inductors
if CT(i)==1
    CVA(i)=CV(i)*R0/f0;
end
if CT(i)==7
    CVA(i)=CV(i)*R0/f0;
end
% Denormalization of capacitors
for i=1:n
% Denormalization of inductors
if CT(i)==8
    CVA(i)=CV(i)*R0/f0;
end
if CT(i)==2
    CVA(i)=CV(i)*R0/f0;
end
% Denormalization of Resistors
if CT(i)==9
    CVA(i)=CV(i)/R0/f0;
end
if CT(i)==6
    CVA(i)=CV(i)/R0/f0;
end
if CT(i)==12
    CVA(i)=CV(i)*R0/f0;
end
end
end
```

Program List 5.29: Function FinalOptimization

```
function [ a,b,c,a0,WA,PA,QA,TA ] = FinalOptimization_Parametric( KFlag1,
ktr, T0, n, ndc, WZ, c0,a0, NC, FA, Rin_A, Xin_A,FL,FH,f0,R0 )
% Final optimization of TPG on c(i) via Parametric Approach
[ x0 ] = Initiate_Parametric( ktr,n,c0,a0 );
N=length(FA);M=3*N;
options=optimset('MaxFunEvals',200000,'MaxIter',500000);
wL=FL/f0;wH=FH/f0;
% T0=0.97;wL=WB(2);wH=WB(N-1);
[ WA,Ain_N,Bin_N ] = NormalizedImmittance_Termination(KFlag1, NC, FA,
Rin_A, Xin_A,FL,FH,f0,R0 );
eps=@(x0)Error_Parametric(x0,ktr,T0,n,ndc,WZ,a0,WA,Ain_N,Bin_N,wL,wH,M );
x=lsqnonlin(eps,x0,[],[],options);
%
[ a,b,c,a0 ] = Evaluate_Parametric(x,ktr,n,ndc,WZ,a0 );
```

```
%
NA=length(WA);j=sqrt(-1);
for i=1:NA
    w=WA(i);
    p=j*w;
% ------- Generation Positive Real Minimum Susceptance YF=PF+jQF for [F] --
    aval=polyval(a,p);
    bval=polyval(b,p);
    YFval=aval/bval;
    Pf0_RFDCT=real(YFval);   PA(i)=Pf0_RFDCT;
    Qf0_RFDCT=imag(YFval);   QA(i)=Qf0_RFDCT;
% ------ Generation of termination admittance Yin=A+jB --------------------
    A=line_seg(WA,Ain_N,w);
    B=line_seg(WA,Bin_N,w);
% --- Result of final optimization -----------------------------------------
            [ TPG ] = Gain(A,B,Pf0_RFDCT,Qf0_RFDCT );
            TA(i)=TPG;
end
end
```

Program List 5.30: SynthesisbyTranszeros

```
function [CTFinal, CVFinal,LL1,LL2,MM] = SynthesisbyTranszeros(KFlag,W,nd
c,a,b,eps_zero)
% This function synthesizes general form of an input impedance
% Z(p)=a(p)/b(p).
%
% -------------------------------------------------------------------------
aa=a;bb=b;
if norm(W)>0
if KFlag==1;
[ CTFinal, CVFinal,LL1,LL2,MM ] = ZeroShifting_AccurateImpedanceSynthesis
(1,W,ndc,a,b,eps_zero );
end
if KFlag==0,
    a=bb;b=aa; %Y(p)=aa/bb; Z(p)=bb(p)/aa(p)
    [ index ] = CheckIfab_samedegreewithnonzeroterms( a,b );
    if index==1;% Flip over the admittance function Y(p)=a(p)/b(p) to
synthesize impedance Z(p)=b(p)/a(p)
        [ CTFinal, CVFinal,LL1,LL2,MM ] = ZeroShifting_
AccurateImpedanceSynthesis(1, W,ndc,bb,aa,eps_zero );
    end
    if index==0; % Start with admittance function to extract poles of
Y(p)=a(p)/b(p)at infinity and DC
        [ CTFinal, CVFinal,LL1,LL2,MM ] = ZeroShifting_
AccurateImpedanceSynthesis(0, W,ndc,aa,bb,eps_zero );
    end
end
% -------------------------------------------------------------------------
end
%
if W==0
    LL1=' There is no finite transmission zero'
    LL2='There is no finite transmission zero'
```

```
        MM='There is no finite transmission zero'
        R0=1;f0=1/2/pi;
    [CTFinal,CVFinal] = CircuitPlot_Yarman(KFlag,R0,f0,a,b,ndc);
end
n=length(a)-1;
if n>2;%Plot_Circuitv1(CTFinal,CVFinal);end
end
```

Program List 5.31: ZeroShifting_AccurateImpedanceSynthesis

```
function [ CTFinal, CVFinal,LL1,LL2,MM ] = ZeroShifting_AccurateImpedance
Synthesis(KFlag, W,ndc,a,b,eps_zero )
% This function synthesizes the complete minimum reactance function
% Z(p)=a(p)/b(p) by extracting finite transmission zeros using zero
% shifting technique, by extracting DC transmission zeros using our high
% precision bandpass synthesis algorithm and by extracting transmission
% zeros at infinity employing our High Precision Lowpass Ladder Synthesis
% algorithm.
%-------------------------------------------------------------------------
% Inputs:
%       KFlag=1> F(p)=a(p)/b(p) is a minimum reactance
%       KFlag=0> F(p)=a(p)/b(p) ia minimum susceptance
%       W: Finite transmission zeros on the jw axis. This is a Matlab
%       vector. if there is no finite transmission zeros then we set W=0.
%       ndc: Count for the total transmission zeros at DC.
%       a(p): Numerator polynomial of Z(p)
%       b(p): Denominator polynomial of Z(p)
%       eps_zero: Algorithmic zero of the synthesis.
%-------------------------------------------------------------------------
% Outputs:
%       CTFinal: Codes for the complete circuit elements
%       CVFinal: Element values of CTFinal
%       LL1: Inductors for the primary coils of the Brune Section
%       LL2: Inductors of the secondary coils of the Brube Sections
%       MM: Mutal inductance between primary and secondary coils.
%-------------------------------------------------------------------------
% Part (c): Synthesis using zero shifting within 5 steps
% C1: Correct the impedance to yield the specified transmission zeros
nb=length(b)-1;
if KFlag==0;
    [ a,b,CTx,CVx,ndc ] = NonminimumToMinimum( b,a,ndc );
% Here it is assumed that at the input of the above function
%                         Z(p)=b(p)/b(p) is a nonminimum impedance.
% At the output the resulting Zr(p)=a(p)/b(p) is a minimum function
end
%[ a,b,eps_A,eps_a,eps_b ] = ImpedanceCheck_General(W,ndc,a,b,eps_zero );
[a,b]=General_immitCheck(ndc,W,a,b);
%[a,b,LA,LB,LC,LL1,LL2,MM,CT,CV]=
SynthesiswithC_Section(KFlag,W,ndc,a,b,eps_zero)
nW=length(W);
CT=[]; CV=[];
for i=1:nW
    wa=W(i);
[Even_Part, La ]=Zeroshifting_Step1( a,b,wa,eps_zero );
```

```
[ Lb,Cb,kb,a2,b2, r_norm ] = Zeroshifting_Step2(wa,La,a,b,eps_zero );
[ Lc, a3,b3, L1, L2,M ] =Zeroshifting_Step3(La, Lb, a2,b2 );
%-----------------------------------------------------------------------
% Store the element values of Brune Sections
        LA(i)=La;LB(i)=Lb;LC(i)=Lc;
        LL1(i)=L1;LL2(i)=L2;MM(i)=M;
%-----------------------------------------------------------------------
n=length(b3)-1;
%
[ CT1, CV1 ] =Zeroshifting_Step4( La, Lb, Lc, Cb,nb, a3, b3 );
CT=[CT CT1]; CV=[CV CV1];
%
%
    [ WW ] = Sweep_FiniteZeros(W,i );
    if WW==0;
        [a,b,ndc]=Check_immitance(a3,b3,ndc);
    end
%
if norm(WW)>0;[a,b]=General_immitCheck(ndc,WW,a3,b3);end
end
% Plot the circuit
if nb>2
%w0=1;
f0=1/2/pi;R0=1;
[CTA,CVA] = CircuitPlot_Yarman(1,R0,f0,a3,b3,ndc);
%-----------------------------------------------------------------------
if KFlag==1;CTFinal=[CT CTA]; CVFinal=[CV CVA];end;
end
% if KFlag=1, there is no pole at infinity extraction  at the begining of
the synthesis.
if KFlag==0;CTFinal=[CTx CT CTA]; CVFinal=[CVx CV CVA];end;
% if Kflag=0, firstly, we extract the pole at infinity to end up with
% a minimum reactance function then, go ahead with the zero shifting
algorithm
%-----------------------------------------------------------------------
%
%Plot_Circuitv1(CTFinal,CVFinal);
if nb==2
    CTFinal=CT;
    CVFinal=CV;
end
end
```

Program List 5.32: InitiateRFLST_lsqnonlin

```
function [ X0 ] = InitiateRFLST_lsqnonlin( ktr,RB )
%
N=length(RB);
if ktr==0; % Design without transformer
    for i=1:N-2
    X0(i)=RB(i+1);
    end
end
```

```
if ktr==1; % Design with transformer
    for i=1:N-1
        X0(i)=RB(i+1);
    end
end

end
```

Program List 5.33: Initiate_Parametric

```
function [ x0 ] = Initiate_Parametric( ktr,n,c,a0 )
%
c0=c;
if ktr==0; % Design without transformer
    for i=1:n
    x0(i)=c0(i);% the only unknowns
    end
end
if ktr==1; % Design with transformer
    for i=1:n
        x0(i)=c0(i);
    end
    x0(n+1)=a0;
end
end
```

Program List 5.34: Initiate_CurveFitting

```
function [ x0 ] = Initiate_CurveFitting( ktr,n,a0 )
%
%ktr=0;
mu=-1;c0(1)=1;
for i=2:n
    c0(i)=mu*c0(i-1);
end
%c1=[ -0.0648   -1.3689   -0.0007   2.1483   -0.0050    0.0008    0.1666
0.0036   -0.0127   -1.5000 ];
%c2=[-0.0695   -0.0002    0.0013   1.1124   -0.0400   -0.3829    0.0161
-0.0154    0.0095];
%c0=[c1 c2];
if ktr==0; % Design without transformer
    for i=1:n
    x0(i)=c0(i);% the only unknowns
    end
end
if ktr==1; % Design with transformer
    for i=1:n
        x0(i)=c0(i);
    end
    x0(n+1)=a0;
end
end
```

Program List 5.35: EvaluateRFLST_lsqnonlin

```
function [ RB ] = EvaluateRFLST_lsqnonlin( ktr,X,R1 )
%
if ktr==0
    N=length(X)+2;
end
if ktr==1
    N=length(X)+1;
end
if ktr==0
    for i=1:N-2
        RB(i+1)=X(i);
    end
    RB(1)=R1; RB(N)=0;
end
if ktr==1
    for i=1:N-1
        RB(i)=X(i);
    end
    RB(N)=0;
end
end
```

Program List 5.36: Evaluate_Parametric

```
function [ a,b,c,a0 ] = Evaluate_Parametric(x,ktr,n,ndc,WZ,a0 )
if ktr==0
    for i=1:n
        c(i)=x(i);
    end
end
if ktr==1
    for i=1:n
    c(i)=x(i);
    end
    a0=x(n+1);
end
% ------- Result of Parametric Optimization ----------------------------
% Generate the denominator polynomial B(w^2) of (5.28) in w-domain:
        C=[c 1];
        BB=Poly_Positive(C);% This positive polynomial is in w-domain
        B=polarity(BB);% Now, it is transferred to p-domain
        nB=length(BB);
% Generate the numerator polynomial in w domain:
        A=(a0*a0)*R_Num(ndc,WZ);% A is specified in p-domain
        AA=polarity(A);% This polynomial is in w domain
% Note that function RtoZ requires same length vectors A and B
% Convert A to same length of B
nA=length(A);
if (abs(nB-nA)>0)
    A=fullvector(nB,A);% This polynomail is in p-domain
end
```

```
% Generation of analytic form of Driving Point Input Immittance
            [a,b]=RtoZ(A,B);
end
```

Program List 5.37: Evaluate_CurveFitting

```
function [ c,a0,a,b,AA,BB ] = Evaluate_CurveFitting(x,ktr,n,ndc,WZ,R1 )
if ktr==0
    for i=1:n
        c(i)=x(i);
    end
end
if ktr==1
    for i=1:n
    c(i)=x(i);
    end
    a0=x(n+1);
end
if ktr==0
    a0=sqrt(R1);
end
% ------- Result of Curve Fitting -----------------------------------------
% Generate the denominator polynomial B(w^2) of (5.28)in w-domain:
        C=[c 1];
        BB=Poly_Positive(C);% This positive polynomial is in w-domain
        B=polarity(BB);% Now, it is transferred to p-domain
        nB=length(BB);
% Generate the numerator polynomial in w domain:
        A=(a0*a0)*R_Num(ndc,WZ);% A is specified in p-domain
        AA=polarity(A);% This polynomial is in w domain
% Note that function RtoZ requires same length vectors A and B
% Convert A to same length of B
nA=length(A);
if (abs(nB-nA)>0)
    A=fullvector(nB,A);% This polynomail is in p-domain
end
% Generation of analytic form of Driving Point Input Immittance
            [a,b]=RtoZ(A,B);
end
```

Program List 5.38: ImmittanceBased_RealFrSingMatch

```
function
[FA,TdB,a,b,c,CT,CV,CVA]=ImmittanceBased_RealFrSingMatch(FAin,RinA,XinA,R
0,f0,FL,FH,F_unit,T0,KFlag,ktr,sign,NC,n,ndc,WZ,ac)
% This function completely solves the single matching problem from the
% given measured data.
%    Inputs:
%    FA,RA,XA: Vectors of measured complex impedance termination data such
that ZA=RA+jXA over measurement frequencies FA. In regard, FA is shorten
by skipping its measurement scale. For example, if the frequencies are
measured in Mega Hertz range such as 330 MHz, 340 MHz,  350 MHz etc.
Then, vector FA contains only plain number as FA=[330 340 350 … ].
```

%   R0,f0: Impedance and frequency normalization numbers respectively. It is noted that f0 is plain number does not have a unit as in FA.

%   FL,FH: Lower and upper edge of the passband without unit as in FA.

%   F_unit: Frequency measurement unit. For example, if the frequencies are measure in MHz range, then, F_unit=10^6 selected. If we make the measurements in GHz range, then, ?F_unit=10?^9 etc.

%   T0: Target flat gain level for the single matching problems.

%   KFlag: Describes the type of immittance that we carry out computation. KFlag=1 selected for impedance based computations. KFlag=0 selected for admittance based computation. This selection is guided by success of the optimization. For example, for capacitive load terminations, KFlag=0 may result in successful optimization. Similarly, for inductive load termination KFlag=1 may yield food results in gain optimization.

%   ktr: This control flag may be used to control the far end resistive termination for low pass designs. When we select ndc=0 for low pass designs, far end termination is either fixed to be R0 or it is left free to make the optimization more flexible. In this case, for low pass designs we use a transformer to level up the termination resistor at the far end to R0. For bandpass design (ndc>0) we should set ktr=1.

%   sign: This control flag determines the initial guess for RFLST if we start with low initials or high initial. Sign=1 corresponds to high initials. Sign=0 is for the low initials. Choice of signs is dictated with success of gain optimization.

%   NC: This integer is the total number of unknown break points in the passband.

%   n,ndc,WZ: These integers determine the complexity of the matching network as described before.

% ac: This slack variable is the equal to the desired values of a0. If a0 is fixed in advance ac must be selected as ?(R_0). If a0 is part of the unknowns then it is initialized as we wish.

%

%   Outputs of function "ImmittanceBased_RealFrSingMatch" is given as

%   F_actual: Frequency sampling vector. Actual frequencies in F_unit range linked to the gain computation.

%   TdB: TPF for the equalizer in dB. It is generated at the frequencies specified by the vector F_actual.

%   a,b: Driving point input immittance of the matching network such that K(p)=a(p)/b(p)

%   c: Coefficients of the auxiliary polynomial which generated K(p)=a(p)/b(p) using parametric approach.

%   CT,CV,CVA: These MatLab vector describes the matching network topology with element values at the end of the synthesis of K(p)=a(p)/b(p). CT describes the component types like series capacitor, shunt inductor etc...

% CV stores the normalized element values. Eventually, in CVA we have the actual element values.

%

%

%   Step 1: Execute function RFLST_SingleMatching
[WB,RB,TB] = RFLST_SingleMatching(NC,ktr, KFlag,sign,T0,FL,FH,FAin,RinA ,XinA,f0,R0);

%   Step 2: Execute function CurveFitting_BreakPoints

%   (Rational curve fitting via RFDCT)

```
[ c0,a0,AA,BB,P,a,b ] = CurveFitting_BreakPoints (ktr,n,ndc,WZ,WB,RB );
%         Step 3: Execute function FinalOptimization_Parametric
%                       (Final optimization via Parametric Approach)
a0=ac;
[ a,b,c,a0,WA,PA,QA,TA ] = FinalOptimization_Parametric(
KFlag,ktr,T0,n,ndc,WZ,c0,a0, NC, FAin, RinA, XinA,FL,FH,f0,R0 );
FA=WA*f0;
%         Step 4: Execute function SynthesisbyTranszeros
TdB=10*log10(TA);
eps_zero=1e-8;
[ CT, CV,LL1,LL2,MM ] = SynthesisbyTranszeros(KFlag,WZ,ndc,a,b,eps_zero);
F0=f0*F_unit;
[ CVA ] = Actual_LumpedElements( F0,R0,CT,CV);
end
```

Program List 5.39: Actual_LumpedElements

```
function [ CVA ] = Actual_LumpedElements( f0,R0,CT,CV )
% Th?s funct?on generates the actual element values of lumped elements
%    Inputs:
%        f0: Normalization frequency in Hz (actual value)
%        R0: Normalization resistor. It is usually 50 ohms.
%        CT: Component type comes from the synthesis package
%        CV: Normalized component values
%    Output:
%        CVA: Actual element values
%-------------------------------------------------------------------------
%
% December 9, 2014
%-------------------------------------------------------------------------
% CT(i)=1> Series Inductor
% CT(i)=7> Shunt Inductor
% CT(i)=8> Shunt Capacitor
% CT(i)=2> Series Capacitor
% CT(i)=9> Terminating Resistor
% Series R//L//C: CT(i)=4> Inductor, CT(i)=5> Capacitor,CT(i)=6>
Resistor,
% Shunt R+L+C: CT(i)=10>Inductor,CT(i)=11> Capacitor,CT(i)=12>Resistor,
%-------------------------------------------------------------------------
% Computat?on of actual element values
n=length(CT);
for i=1:n
% Denormalization of inductors
if CT(i)==1
    CVA(i)=CV(i)*R0/f0/2/pi;
end
if CT(i)==7
    CVA(i)=CV(i)*R0/f0/2/pi;
end
% Denormalization of capacitors
if CT(i)==8
    CVA(i)=CV(i)/R0/f0/2/pi;
end
if CT(i)==2
```

```
    CVA(i)=CV(i)/R0/f0/2/pi;
end
% Denormalization of Resistors
if CT(i)==9
    CVA(i)=CV(i)*R0;
end
if CT(i)==6
    CVA(i)=CV(i)*R0;
end
if CT(i)==12
    CVA(i)=CV(i)*R0;
end
end
```

Program List 5.40: Main Program GKYExample5_8.m

```
% Main Program GKYExample5_8.m
% This main program solves the single matching problem to construct the
% front-end matching network for the power LD MOS RD07 by Mistsibushi
% In this example far end termination of [F] is fixed as a0=1 which
% yields T0=0.975
clc;clear all;close all;
% Load-pull measured input impedance of the LD MOS RD07.
FAin=[0 330 350 370 390 410 430 450 470 490 510 530 1060];
Rin=[20 17.61 14.50 18.70 22.00 9.16 10.23 10.40 17.18 14.33 13.36 11.04 9];
Xin=[0 -04.63 -03.67 -11.30 -10.11 -13.96 -17.94 -15.89 -04.85 -05.11
-12.11 -14.03 -16];
% Inputs for the passband and impedance normalization
FL=330;FH=530;f0=530;R0=50;F_unit=1e6;
%------------------------------------------------------------------------
% User selected inputs to design single matching network via RFT
%
    T0=0.975;
    NC=19;ktr=0;KFlag=0;sign=1;
    n=5;ndc=0;WZ=[0]; a0=1;
%
%------------------------------------------------------------------------
% Design of front-end matching network in single function:
%
[FF,TFdB,aF,bF,cF,CTF,CVF,CVAF]=ImmittanceBased_RealFrSingMatch(FAin,Rin,
Xin,R0,f0,FL,FH,F_unit,T0,KFlag,ktr,sign,NC,n,ndc,WZ,a0);
%
%------------------------------------------------------------------------
figure
plot(FF,TFdB)
xlabel('Actual frequencies')
ylabel('TPG in dB')
title('TPG for the front-end matching network of power amplifier designed
with LD MOS 07')
```

Program List 5.41: Main Program GKYExample5_9.m

```
% Main Program GKYExample5_9.m
% This program designs the single matching equalizers within
```

```
% one MatLab function called
%               "ImmittanceBased_RealFrSingMatch"
%
% December 7 2014 by BS Yarman,Vanikoy,Istanbul,Turkey
% This program provides solution for Example 5.9 using a compact
  functions
% Andrei Grebennikov-Narendra Kumar B.Siddik Yarman Book
%------------------------------------------------------------------------
% function "ImmittanceBased_RealFrSingMatch" executes the major functions
  developed
% specifically for the book under consideration.
%
% Compact optimization functions are given as follows:
%          Step 1: Execute function RFLST_SingleMatching
%          Step 2: Execute function CurveFitting_BreakPoints
%                        (Rational curve fitting via RFDCT)
%          Step 3: Execute function FinalOptimization_Parametric
%                        (Final optimization via Parametric Approach)
%          Step 4: Execute function SynthesisbyTranszeros
% Inputs:
% Load-Pull Measurement for LD MOS RD07 by Mitsubishi
% Sampling Frequencies of the actual measurements in MHz
%------------------------------------------------------------------------
    clear
    clc
    close all
% ------ Inputs I: --------------------------------------------------------
% Termination data for the front-end matching network:
%               Zin_A(jw)=Rin_A+jXin_A
%
    FAin=[0 330 350 370 390 410 430 450 470 490 510 530 1060];
    Rin=[20 17.61 14.50 18.70 22.00 9.16 10.23 10.40 17.18 14.33 13.36
11.04 9];
    Xin=[0 -04.63 -03.67 -11.30 -10.11 -13.96 -17.94 -15.89 -04.85 -05.11
-12.11 -14.03 -16];
%
% User selected Inputs:
    T0=0.975;
    NC=19;ktr=0;KFlag=0;sign=1;
    FL=330;FH=530;f0=530;R0=50;F_unit=1e6;
    n=5;ndc=0;WZ=0;a0=1;
%
%------------------------------------------------------------------------
% Automated design of [F]:
[FF,TFdB,aF,bF,cF,CTF,CVF,CVAF]=ImmittanceBased_RealFrSingMatch(FAin,Rin,
Xin,R0,f0,FL,FH,F_unit,T0,KFlag,ktr,sign,NC,n,ndc,WZ,a0);
%------------------------------------------------------------------------
% Design of back-end matching network [B]
% Inputs for the termination Zout=Rout+jXout of [B]
    FAout=[0 330 350 370 390 410 430 450 470 490 510 530 1060];
    Rout=[30 22.47 16.06 21.23 24.40 22.03 23.50 17.82 18.53 21.06 26.80
21.50 13];
    Xout=[ 0 -7.84 -12.69 -12.32 -13.45 -12.14 -14.01 -9.52 -1.53 -4.54
-11.06 -13.54 -10];
```

```
% User selected inputs to design back-end MN [B]:
    T0=0.975;
    NC=19;ktr=0;KFlag=0;sign=1;
    FL=330;FH=530;f0=530;R0=50;F_unit=1e6;
    n=5;ndc=0;WZ=0;a0=1;
  % Automated design of [B]
    [FB,TBdB,aB,bB,cB,CTB,CVB,CVAB]=ImmittanceBased_RealFrSingMatch(FAout,
Rout,Xout,R0,f0,FL,FH,F_unit,T0,KFlag,ktr,sign,NC,n,ndc,WZ,a0);
%-------------------------------------------------------------------------
    % TPG plot for [F]
figure
plot(FF,TFdB)
title('Design of front-end matching network')
ylabel('TPG for front-end matching network [F]')
xlabel('Actual frequency in MHz')
%
figure
plot(FB,TBdB)
title('Design of back-end matching network')
ylabel('TPG for back-end matching network [B]')
xlabel('Actual frequency in MHz')
%
plot(FAin, Rin, FAin, Xin, FAout,Rout, FAout, Xout)
legend('Rin','Xin','Rout','Xout')
xlabel('frequency')
```

Program List 5.42: Main Program GKYExample5_9b.m

```
% Main Program GKY_Example5_9b.m
% This main program solves the single matching problem to construct the
% front-end matching network for the power LD MOS RD07 by Mistsibushi
% In this example far end termination of [F] is fixed as a0=1 which
% yields T0=0.975
clc;clear all;close all;
% Load-pull measured input impedance of the LD MOS RD07.
FAin=[0 330 350 370 390 410 430 450 470 490 510 530 1060];
Rin=[20 17.61 14.50 18.70 22.00 9.16 10.23 10.40 17.18 14.33 13.36 11.04 9];
Xin=[0 -04.63 -03.67 -11.30 -10.11 -13.96 -17.94 -15.89 -04.85 -05.11
-12.11 -14.03 -16];
%
figure
plot(FAin,Rin,FAin,Xin)
legend('Rin','Xin')
xlabel('Actual frequencies')
ylabel('Zin=Rin+jXin')
title('Load-Pull Measured input impedances for LD-MOS RD07')
%
%-------------------------------------------------------------------------
% Inputs for the passband and impedance normalization
FL=330;FH=530;f0=530;R0=50;F_unit=1e6;
%-------------------------------------------------------------------------
% User selected inputs to design single matching network via RFT
%
    T0=0.975;
```

```
         NC=19;ktr=0;KFlag=0;sign=1;
         n=5;ndc=0;WZ=[0]; a0=1;
%
Imp_input=[FAin Rin Xin];
UserSelectedInputs=[T0 NC ktr KFlag sign FL FH f0 R0 F_unit n ndc WZ a0];
Input_Data=[UserSelectedInputs Imp_input];
%
%--------------------------------------------------------------------------
% Design of front-end matching network in single function:
%
[ CTF,CVF,CVAF,FAF,TAF,cF,aF,bF,a0F ] = CompactSingleMatching( Input_Data );
%
TFrontdB=10*log10(TAF);
%--------------------------------------------------------------------------
figure
plot(FAF,TFrontdB)
xlabel('Actual frequencies')
ylabel('TPG in dB')
title('TPG in dB of [F] for the power amplifier designed using
LDMOS-RD07')
%
% % Inputs for the termination Zout=Rout+jXout of [B]
    FAout=[0 330 350 370 390 410 430 450 470 490 510 530 1060];
    Rout=[30 22.47 16.06 21.23 24.40 22.03 23.50 17.82 18.53 21.06 26.80
21.50 13];
    Xout=[ 0 -7.84 -12.69 -12.32 -13.45 -12.14 -14.01 -9.52 -1.53 -4.54
-11.06 -13.54 -10];
% %
% % User selected inputs to design back-end matching network via RFT
figure
plot(FAout,Rout,FAout,Xout)
legend('Rout','Xout')
xlabel('Actual frequencies')
ylabel('Zout=Rout+jXout')
title('Load-Pull Measured output impedances for LD-MOS RD07')
% % RFT Inputs for the back-end matching network
    T0=0.975;
    NC=19;ktr=0;KFlag=0;sign=1;
    n=5;ndc=0;WZ=[0]; a0=1;
%
 Imp_output=[FAout Rout Xout];
 UserSelectedInputs=[T0 NC ktr KFlag sign FL FH f0 R0 F_unit n ndc WZ
a0];
 Output_Data=[UserSelectedInputs Imp_output];
%--------------------------------------------------------------------------
% Design of back-end matching network in single function:
%
 [ CTB,CVB,CVAB,FAB,TAB,cB,aB,bB,a0B ] = CompactSingleMatching( Output_
Data );
%
 TBackdB=10*log10(TAB);
figure
plot(FAB,TBackdB)
xlabel('actual frequencies')
```

```
ylabel('TPG in dB for the back-end matching network')
title('TPG in dB of [B] for the power amplifier designed using LDMOS-RD07')
```

Program List 5.43: GKY_Example5_10.m

```
% Main Program GKY_Example5_10.m
% This main program solves the double matching problem to construct the
% Interstage Equalizer for a two-stage amplifier over 330-530 MHz.
% Design of Interstage equalizer
% Generator side is the output impedance of LD-MOS RD-01 by Mitsubishi
%-----------------------------------------------------------------------
clc;clear all;close all;
% Port-2 based design:
KPort=2;
% Load-pull measured Zout1 LD MOS RD01: ZG=RG+jXG
%
AG=[330 22.47    -07.84
350      16.06   -12.69
370      21.23   -12.32
390      24.40   -13.45
410      22.03   -12.14
430      23.50   -14.01
450      17.82   -09.52
470      18.53   -01.53
490      21.06   -04.54
510      26.8    -11.06
530      21.5    -13.54];
%
FAG=AG(:,1)';RAG=AG(:,2)';XAG=AG(:,3)';
FAG=[0 FAG 1060]; RAG=[25 RAG 0]; XAG=[0 XAG -15];% Extrapolated
generator
%-----------------------------------------------------------------------
% % Load-pull measured Zinput2 LD MOS RD07: ZL=RL+jXL
AL=[330 2.32    -3.97
350      3.90   -1.98
370      1.81   -3.21
390      1.93   -2.48
410      1.86   -3.87
430      1.93   -1.81
450      1.05   -1.29
470      1.30   -1.58
490      1.65   -2.98
510      1.64   -2.91
530      1.66   -3.33];
%
FAL=AL(:,1)';RAL=AL(:,2)';XAL=AL(:,3)';
FAL=[0 FAL 1060]; RAL=[4 RAL 0]; XAL=[0 XAL -5];% Extrapolated load
%-----------------------------------------------------------------------
figure
plot(FAG,RAG,FAG,XAG)
legend('RAG','XAG')
xlabel('Actual frequencies')
ylabel('ZAG=RAG+jXAG')
title('Load-Pull Measured output impedances for LD-MOS RD07')
```

```
%------------------------------------------------------------------
figure
plot(FAL,RAL,FAL,XAL)
legend('RAL','XAL')
xlabel('Actual frequencies')
ylabel('ZAL=RAL+jXAL')
title('Load-Pull Measured input impedances for LD-MOS RD01')
%------------------------------------------------------------------
% Inputs for the passband and impedance normalization
FL=310;FH=530;f0=530;R0=50;F_unit=1e6;% with extended bandwidth(310 vs
330)
wL=FL/f0;wH=FH/f0;
%------------------------------------------------------------------
% Example 5.10 Part (a): Equalizer is described from Port-2
% User selected inputs to design double matching network via RFT
% Design at Port-2: Generator based double matching gain
FL=310;FH=530;f0=530;R0=50;F_unit=1e6;% with extended bandwidth(310 vs
330)
wL=FL/f0;wH=FH/f0;
    T0=0.93;
    NC=19;ktr=0;KFlag=0;sign=1;% with high line segment initials
    n=5;ndc=0;WZ=[0]; a0=1;M=2*(NC+2);
%------------------------------------------------------------------
% Impedance normalization: Generator side
FAG=[0 FAG 1060]; RAG=[25 RAG 0]; XAG=[0 XAG -15];
WG=FAG/f0; RG=RAG/R0; XG=XAG/R0;
% Impedance normalization: Load side
FAL=[0 FAL 1060]; RAL=[4 RAL 0]; XAL=[0 XAL -5];
WL=FAL/f0; RL=RAL/R0; XL=XAL/R0;
%------------------------------------------------------------------
%
ZAG=[FAG RAG XAG]; ZG=[RG XG];
ZAL=[FAL RAL XAL];ZL=[RL XL];
%
% Example 5.10 Part (a): Equalizer is described from Port-2
UserSelectedInputs=[T0 NC ktr KFlag sign FL FH f0 R0 F_unit n ndc WZ a0];
Data_Port2=[UserSelectedInputs ZAL];% Actual load impedance data set at
Port-2
%
[ aL,bL,cL,aL0,CTL, CVL,LL1L,LL2L,MML,CVAL,WAG,TAG ] =
CompactDoubleMatching( Data_Port2, FAG,RAG,XAG );
%------------------------------------------------------------------
TG_dB=10*log10(TAG);F_actual=WAG*f0;
figure
plot(F_actual, TG_dB)
xlabel('Actual Frequency')
ylabel('Generator Based Double Matching Gain in dB')
title('Generator Based Double Matching Gain: K(p) is defined at Port-2')
%------------------------------------------------------------------
% Example 5.10 Part (b): Equalizer is described from Port-1
% User selected inputs to design double matching network via RFT
% Design at Port-1: Load based double matching gain
FL=320;FH=530;f0=530;R0=50;F_unit=1e6;% with extended bandwidth(320 vs
330)
```

```
wL=FL/f0;wH=FH/f0;
    T0=0.93;
    NC=19;ktr=0;KFlag=0;sign=-1;% with low line segment initials
    n=5;ndc=0;WZ=[0]; a0=1;M=2*(NC+2);
%
    UserSelectedInputs=[T0 NC ktr KFlag sign FL FH f0 R0 F_unit n
ndc WZ a0];
Data_Port1=[UserSelectedInputs ZAG];% Actual load impedance data set at
Port-2
%
[ aG,bG,cG,aG0,CTG, CVG,LL1G,LL2G,MMG,CVAG,WAL,TAL ] =
CompactDoubleMatching( Data_Port1, FAL,RAL,XAL );
%------------------------------------------------------------------------
TL_dB=10*log10(TAL);F_actual=WAL*f0;
figure
plot(F_actual, TL_dB)
xlabel('Actual Frequency')
ylabel('Load Based Double Matching Gain in dB')
title('Load Based Double Matching Gain: K(p) is defined at Port-1')
%------------------------------------------------------------------------
```

Program List 5.44: Gain_DoubleMatching

```
function [T_Double] = Gain_DoubleMatching(w,KFlag,W,RGA,XGA,RLA,XLA,a,b,ndc)
% This function generates the double matching gain.
% This function generates the input reflectance of the lossless two-port
% [EL] when it is terminated in complex impedance ZL=RL+jXL
% In this function lossless equalizer is defined from the back-end
  (Port-2)
% immitance K(p)=a(p)/b(p). K(p)=ZB if KFlag=1, K(p)=YB if KFalg=0
% Inputs:
%       w: normalized angular frequency
%       KFlag: Control flag for immittance-based defined reflectance
%           KFlag=1> Impedance based reflectance definition
%           KFlag=0> Admittance based reflectance definition
%       a(p): Numerator polynomial of K(p)=a(p)/b(p)
%       b(p): Denominator polynomial of K(p)=a(p)/b(p)
%       ndc: # of transmission zeros at DC
%       W: Array. Normalized sampling frequencies refers to measurement
%       of generator ZG and load ZL impedances
%       RAL: Array. Normalized real part of the impedance ZL
%       XAL: Array. Normalized imaginary part of the impedance ZL
% Output:
%       T_Double: Double matching gain evaluated at the angular frequency w
%------------------------------------------------------------------------
% Generation of complex generator impedance ZG at Port-1:
 RG=line_seg(W,RGA,w);
 XG=line_seg(W,XGA,w);
 ZG=complex(RG,XG);
 if KFlag==0
     ZG=1/ZG;
 end
% Generate real normalized reflectance Sin:
  Sin=inputref_EL(w,KFlag,a,b,ndc,W,RLA,XLA);
```

```
% Generate double matching gain:
        TEL=1-abs(Sin)*abs(Sin);
        G22=(ZG-1)/(ZG+1);
        if KFlag==0
            G22=-G22;
        end
        Weight=(1-abs(G22)*abs(G22))/abs((1-G22*Sin))^2;
        T_Double=Weight*TEL;
end
```

Program List 5.45: FinalOptimization_DoubleMatching

```
function [ a,b,c,a0,WA,TA ] = FinalOptimization_DoubleMatching( KFlag,ktr,
T0,n,ndc,WZ,c0,a0,WG,RG,XG,RL,XL,wL,wH )
% Final optimization of TPG on c(i) via Parametric Approach
[ x0 ] = Initiate_Parametric( ktr,n,c0,a0 );
N=length(WG);M=3*N;
options=optimset('MaxFunEvals',200000,'MaxIter',500000);
%
eps=@(x0)Error_DoubleMatching(x0,KFlag,ktr,T0,n,ndc,WZ,a0,WG,RG,XG,RL,XL,
wL,wH,M );
x=lsqnonlin(eps,x0,[],[],options);
%
[ a,b,c,a0 ] = Evaluate_Parametric(x,ktr,n,ndc,WZ,a0 );
%
dw=(wH-wL)/(M-1);w=wL;
for i=1:M
    WA(i)=w;
[ T_Double ] = Gain_DoubleMatching(w,KFlag,WG,RG,XG,RL,XL,a,b,ndc  );
    TA(i)=T_Double;
    w=w+dw;
end
end
```

Program List 5.46: CompactSingleMatching

```
function [ CT,CV,CVA,FA,TA,c,a,b,a0 ] = CompactSingleMatching( data )
%
% This function completely solves the single matching problem from the
% given measured data.
%    Inputs:
%    FA,RA,XA: Vectors of measured complex impedance termination data such
that ZA=RA+jXA over measurement frequencies FA. In regard, FA is shorten
by skipping its measurement scale. For example, if the frequencies are
measured in Mega Hertz range such as 330 MHz, 340 MHz,  350 MHz etc.
Then, vector FA contains only plain number as FA=[330 340 350 … ].
%    R0,f0: Impedance and frequency normalization numbers respectively. It
is noted that f0 is plain number does not have a unit as in FA.
%    FL,FH: Lower and upper edge of the passband without unit as in FA.
%    F_unit: Frequency measurement unit. For example, if the frequencies
are measure in MHz range, then, F_unit=10^6 selected.  If we make the
measurements in GHz range, then, ?F_unit=10?^9 etc.
%    T0: Target flat gain level for the single matching problems.
```

% KFlag: Describes the type of immittance that we carry out computation. KFlag=1 selected for impedance based computations. KFlag=0 selected for admittance based computation. This selection is guided by success of the optimization. For example, for capacitive load terminations, KFlag=0 may result in successful optimization. Similarly, for inductive load termination KFlag=1 may yield food results in gain optimization.

% ktr: This control flag may be used to control the far end resistive termination for low pass designs. When we select ndc=0 for low pass designs, far end termination is either fixed to be R0 or it is left free to make the optimization more flexible. In this case, for low pass designs we use a transformer to level up the termination resistor at the far end to R0. For bandpass design (ndc>0) we should set ktr=1.

% sign: This control flag determines the initial guess for RFLST if we start with low initials or high initial. Sign=1 corresponds to high initials. Sign=0 is for the low initials. Choice of signs is dictated with success of gain optimization.

% NC: This integer is the total number of unknown break points in the passband.

% n,ndc,WZ: These integers determine the complexity of the matching network as described before.

% ac: This slack variable is the equal to the desired values of a0. If a0 is fixed in advance ac must be selected as ?(R_0). If a0 is part of the unknowns then it is initialized as we wish.

%

% Outputs of function "ImmittanceBased_RealFrSingMatch" is given as

% F_actual: Frequency sampling vector. Actual frequencies in F_unit range linked to the gain computation.

% TdB: TPF for the equalizer in dB. It is generated at the frequencies specified by the vector F_actual.

% a,b: Driving point input immittance of the matching network such that K(p)=a(p)/b(p)

% c: Coefficients of the auxiliary polynomial which generated K(p)=a(p)/b(p) using parametric approach.

% CT,CV,CVA: These MatLab vector describes the matching network topology with element values at the end of the synthesis of K(p)=a(p)/b(p). CT describes the component types like series capacitor, shunt inductor etc...

% CV stores the normalized element values. Eventually, in CVA we have the actual element values.

%

```
T0=data(1);
NC=data(2);ktr=data(3);KFlag=data(4);sign=data(5);
FL=data(6);FH=data(7);f0=data(8);R0=data(9);F_unit=data(10);
n=data(11);ndc=data(12);WZ=data(13);a0=data(14);
ac=a0;
NI=length(data);
ND=NI-14;
Nin=ND/3;
for j=1:Nin
    FAin(j)=data(14+j);
end
for j=1:Nin
```

```
        RinA(j)=data(14+Nin+j);
end
for j=1:Nin
        XinA(j)=data(14+Nin*2+j);
end
%
%           Step 1: Execute function RFLST_SingleMatching
%
[ WB,RB,TB ] = RFLST_SingleMatching(NC,ktr,KFlag,sign,T0,FL,FH,FAin,RinA,
XinA,f0,R0 );
%           Step 2: Execute function CurveFitting_BreakPoints
%                        (Rational curve fitting via RFDCT)
[ c0,a0,AA,BB,P,a,b ] = CurveFitting_BreakPoints (ktr,n,ndc,WZ,WB,RB );
%           Step 3: Execute function FinalOptimization_Parametric
%                        (Final optimization via Parametric Approach)
a0=ac;
[ a,b,c,a0,WA,PA,QA,TA ] = FinalOptimization_Parametric(
KFlag,ktr,T0,n,ndc,WZ,c0,a0, NC, FAin, RinA, XinA,FL,FH,f0,R0 );
FA=WA*f0;
%           Step 4: Execute function SynthesisbyTranszeros
TdB=10*log10(TA);
eps_zero=1e-8;
[ CT, CV,LL1,LL2,MM ] = SynthesisbyTranszeros(KFlag,WZ,ndc,a,b,eps_zero);
F0=f0*F_unit;
[ CVA ] = Actual_LumpedElements( F0,R0,CT,CV);
%
Plot_Circuit4(CT,CVA)
% figure
% plot(FA,TdB)
% xlabel('Actual frequencies')
% ylabel('TPG in dB')
% title('TPG for the single matching problem under consideration')
end
```

Program List 5.47: Sin=inputref_EL(w,KFlag,a,b,ndc,W,RLA,XLA)

```
function Sin=inputref_EL(w,KFlag,a,b,ndc,W,RLA,XLA)
% This function generates the input reflectance of the lossless two-port
% [EL] when it is terminated in complex impedance ZL=RL+jXL
% In this function lossless equalizer is defined from the back-end
  (Port-2)
% immittance KB(p)=a(p)/b(p). K(p)=ZB if KFlag=1, K(p)=YB if KFalg=0
% Inputs:
%        w: normalized angular frequency
%        KFlag: Control flag for the immittance based defined reflectance
%           KFlag=1> Impedance based reflectance definition
%           KFlag=0> Admittance based reflectance definition
%        a(p): Numerator polynomial of K(p)=a(p)/b(p)
%        b(p): Denominator polynomial of K(p)=a(p)/b(p)
%        ndc: # of transmission zeros at DC
%        W: Array. Normalized sampling frequencies to refer load ZL
%        RAL: Array. Normalized real part of the impedance ZL
%        XAL: Array. Normalized imaginary part of the impedance ZL
```

```
% Output:
%        Sin: Real normalized-driving point input reflectance of [EL]
%--------------------------------------------------------------------
% Define complex variable p=jw.
p=sqrt(-1)*w;
 % Generation of complex termination (load) at Port-2.
 RL=line_seg(W,RLA,w);
 XL=line_seg(W,XLA,w);
 ZL=complex(RL,XL);
 if KFlag==0
     ZL=1/ZL;
 end
 % Generation of the back-end driving point input impedance ZB(p)=K(p).
    aval=polyval(a,p);
    bval=polyval(b,p);
    bval_conj=conj(bval);
% Definition of all pass function eta:
         eta=(-1)^ndc*bval_conj/bval;
%
         ZB=aval/bval;
         ZBC=conj(ZB);
% Definition of complex input reflectance in Youla sense
       S=(ZL-ZBC)/(ZL+ZB);
% Definiation of real normalized input reflectance of [EL] in Yarman
sense
       Sin=eta*S;
if KFlag==0
    Sin=-Sin;
end
end
```

Program List 5.48: CompactDoubleMatching

```
function [ aL,bL,cL,aL0,WA,TA ] = CompactDoubleMatching( data,
FAG,RAG,XAG )
% This function solves the double matching problem in a compact way
% directly from the measured impedance data from Port-2.
% Inputs:
%        data: MatLab array. Defined from Port-2 using complex load
%        termination ad also T0, KFlag, ktr,..etc
%        FAG: Sampling frequency array to measure actual complex generator
impedance
%        RAG: Real part of the actual generator impedance
%        XAG: imaginary part of the actual generator impedance
%    Outputs:
%        aL: Numerator polynomial of K(p)=aL(p)/bL(p) defined at Port-2
%        bL: Denominator polynomial of K(p)=aL(p)/bL(p) defined at Port-2
%        cL(w^2): Coefficients of the auxiliary polynomial
%        aL0: Leading coefficients of the numerator polynomial of P(w^2)
%        WA: Normalized frequency array to generate Double Matching Gain TA
%        TA: MatLab array of TPG for the double matching problem
%--------------------------------------------------------------------
T0=data(1);
NC=data(2);ktr=data(3);KFlag=data(4);sign=data(5);
```

```
FL=data(6);FH=data(7);f0=data(8);R0=data(9);F_unit=data(10);
n=data(11);ndc=data(12);WZ=data(13);a0=data(14);
ac=a0;
NI=length(data);
ND=NI-14;
Nin=ND/3;
for j=1:Nin
    FAL(j)=data(14+j);
end
for j=1:Nin
    RAL(j)=data(14+Nin+j);
end
for j=1:Nin
    XAL(j)=data(14+Nin*2+j);
end
% Frequency normalization
wL=FL/f0; wH=FH/f0;
% Impedance normalization on the generator side
WG=FAG/f0; RG=RAG/R0; XG=XAG/R0;
% Impedance normalization: Load side
WL=FAL/f0; RL=RAL/R0; XL=XAL/R0;
%---------------------------------------------------------------------------
%
ZAG=[FAG RAG XAG]; ZG=[RG XG];
ZAL=[FAL RAL XAL]; ZL=[RL XL];
%
UserSelectedInputs=[T0 NC ktr KFlag sign FL FH f0 R0 F_unit n ndc WZ a0];
Input_Data=[UserSelectedInputs ZAL];% Actual load impedance data set at
Port-2
%
%---------------------------------------------------------------------------
% Generation of initials for Port-2 based double matching problem:
% Port-2 based design: Generator based double matching via RFLST
%
% Here, we assume that design is carried out at Port-2.
% Initials for the double matching problem via RFLST at port-2:
% In the function CompactSingleMatching, input_Data includes actual load
% impedance as measured. Result is the inital K(p)=aL(p)/bL(p) at
  port-2.
[ CTL,CVL,CVAL,FLR,TLR,cL0,aL,bL,a0L ] = CompactSingleMatching( Input_
Data );% With actual measured data
%---------------------------------------------------------------------------
% in the function "FinalOptimization_DoubleMatching" WG,RG and XG are
% normalized values of the generator impedance.
[aL,bL,cL,aL0,WA,TA] = FinalOptimization_DoubleMatching( KFlag,ktr,T0,n,
ndc,WZ,cL0,a0L,WG,RG,XG,RL,XL,wL,wH );
end
```

Program List 5.49: GKYExample5_11.m

```
% Main Program GKYExample5_11.m
% This program produces the solutions for Example 5.11
%    (a) Let q=0, k=5, a_0 and c=[1 -1  1 -1  1].
```

```
%          Generate R(λ ^2 )=(A(λ ^2))/(B(λ ^2)) as in (5.121)
%    (b)  Generate K(λ)=(a(λ))/(b(λ)) using (5.114).
%    (c)  Find the roots of a(λ) and b(λ). Comment on the result.
%    (d)  Regenerate the even part of K(λ)=(a(λ))/(b(λ)) from the computed
%         a(λ) and b(λ) and evaluate the numerical error in the course
%         of computations.
%-------------------------------------------------------------------------
clc;clear;close all;
% Solutions:
% Part (a):
q=0; k=5; a0=1; c=[1 -1  1 -1  1];
[A,B]=EvenPart_Richard(k,q,a0,c);
% Part (b):
[a,b]=Minimum_FRichard(k,q,a0,c);
% Part (c)
pa=roots(a);
pb=roots(b);
% Part (d):
[A1,B1]=even_part(a,b)
error_a=norm(A1-A)/norm(A)
error_b=norm(B1-B)/norm(B)
```

Program List 5.50: EvenPart_Richard(k,q,a0,c)

```
function [A,B]=EvenPart_Richard(k,q,a0,c)
% Generate even rational function R(lambda)=A(lambda)/B(lambda)as the
%    even part of F(lambda)analytic form of
Fmin(lambda)=a(lamda)/b(lambda)
% Generate A(lambda) and B(-lambda^2)
%-------------------------------------------------------------------------
% Inputs:
%          k: Total number of Cascaded UEs
%          q: Total number of Transmission zeros at DC (q=ndc)
%          a0:Leading constant term of the numerator: A0=a0*a0
%          c(Omega): Auxiliary polynomial of the denominator B(Omega^2)
% Outputs:
%          A(lambda): Numerator polynomial in lambda domain
%          B(lambda): Denominator polynomial in lambda domain
%-------------------------------------------------------------------------
% It is important to note the following points:
% (a) Degree n=length(c) must be greater or equal to k+q
% (b) q=0 yields lowpass structure
% (c) if q=0 and if k=n then we only have n-cascaded UE
% (d) if q=0 and n>k then, we must have series short and shunt open stubs
%-------------------------------------------------------------------------
n=length(c);
if n<(q+k)
    Attention='n<(q+k) therefore you have unrealizable demand'
end
if n>=(q+k); % then start computations
        C=[c 1];
        BB=Poly_Positive(C);% This positive polynomial is in Omega-domain
        B=polarity(BB);% Now, it is transferred to lamda-domain
```

```
% Generate A(-lamda^2) of R(-lamda^2)=A(-lamda^2)/B(-lamda^2)
nB=length(B);
A=a0*a0*lambda2q_UE2k(k,q); % A is specified in lamda-domain
nA=length(A);
if (abs(nB-nA)>0)
  A=fullvector(nB,A);% work with equal length vectors
end
end
```

Program List 5.51: Minimum_FRichard(k,q,a0,c)

```
function [a,b]=Minimum_FRichard(k,q,a0,c)
% Generate analytic form of Fmin(lambda)=a(lamda)/b(lambda)
% Generate B(-lambda^2)
%-----------------------------------------------------------------
% Inputs:
%         k: Total number of Cascaded UEs
%         q: Total number of Transmission zeros at DC (q=ndc)
%         a0:Leading constant term of the numerator: A0=a0*a0
%         c(Omega): Auxiliary polynomial of the denominator B(Omega^2)
% Outputs:
%         a(lambda): Numerator polynomial in lambda domain
%         b(lambda): Denominator polynomial in lambda domain
%-----------------------------------------------------------------
% It is important to note the following points:
% (a) Degree n=length(c) must be greater or equal to k+q
% (b) q=0 yields lowpass structure
% (c) if q=0 and if k=n then we only have n-cascaded UE
% (d) if q=0 and n>k then, we must have series short and shunt open stubs
%-----------------------------------------------------------------
n=length(c);
if n<(q+k)
    Attention='n<(q+k) therefore you have unrealizable demand'
end
if n>=(q+k); % then start computations
        C=[c 1];
        BB=Poly_Positive(C);% This positive polynomial is in Omega-domain
        B=polarity(BB);% Now, it is transferred to lamda-domain
% Generate A(-lamda^2) of R(-lamda^2)=A(-lamda^2)/B(-lamda^2)
nB=length(B);
A=a0*a0*lambda2q_UE2k(k,q); % A is specified in lamda-domain
nA=length(A);
if (abs(nB-nA)>0)
  A=fullvector(nB,A);% work with equal length vectors
end
% Generation of minimum immitance function using Bode or Parametric
method
ndc=q;
        [a,b]=RtoZ(A,B);% Here A and B are specified in p-domain
        na=length(a);
        if ndc>0;a(na)=0;end;
end
%
end
```

Program List 5.52: function [A,B]=even_part(a,b)

```
function [A,B]=even_part(a,b)
% Generate even part R(p^2)=A(p^2)/B(p^2) of a given immitance function
F(p)=a(p)/b(p)
% Notice that A(p^2) and B(p^2) are specified as an even polynomials
% Computation of Numerator of Even Part
na=length(a);
nb=length(b);
    sign=-1;
    for i=1:na
        sign=-sign;
        a_(na-i+1)=sign*a(na-i+1);
    end
    sign=-1;
    for i=1:nb
        sign=-sign;
        b_(na-i+1)=sign*b(nb-i+1);
    end
    Num_Even=(conv(a,b_)+conv(a_,b))/2;
      A=clear_oddpower(Num_Even);
    n_Even=length(Num_Even);
        BB=conv(b,b_);
    B=clear_oddpower(BB);
end
```

Program List 5.53: NewMinimumFunction

```
function [a,b]=Richard_NewMinimumFunction(k,q,c,c0)
% Generate analytic form of Fmin(lambda)=a(lamda)/b(lambda)
% Generate B(-lambda^2)
%-------------------------------------------------------------------------
% In this function first A(lambda^2)=(-1)^q*(1-lambda^2)^k is generated
% with unit leading coefficient (a0=1).
% Then, B(Omega^2)=(1/2)[C(Omega)^2+C(-Omega)^2] is computed.
% where C(omega) is a full polynomial: C=[c c0].
% At the last step, a(lambda) and b(lambda) are normalized with respect
% to norm of a(lambda).
%-------------------------------------------------------------------------
% Inputs:
%          k: Total number of Cascaded UEs
%          q: Total number of Transmission zeros at DC (q=ndc)
%          c(Omega): Auxiliary polynomial of the denominator B(Omega^2)
%          c0=last coefficient (n+1) of c(Omega)
% Outputs:
%          a(lambda): Numerator polynomial in lambda domain
%          b(lambda): Denominator polynomial in lambda domain
%-------------------------------------------------------------------------
% It is important to note the following points:
% (a) Degree n=length(c) must be greater or equal to k+q
% (b) q=0 yields lowpass structure
% (c) if q=0 and if k=n then we only have n-cascaded UE
% (d) if q=0 and n>k then, we must have series short and shunt open stubs
%-------------------------------------------------------------------------
```

```
n=length(c);
if n<(q+k)
    Attention='n<(q+k) therefore you have unrealizable demand'
end
if n>=(q+k); % then start computations
        C=[c c0];
        BB=Poly_Positive(C);% This positive polynomial is in Omega-domain
        B=polarity(BB);% Now, it is transferred to lamda-domain
% Generate A(-lamda^2) of R(-lamda^2)=A(-lamda^2)/B(-lamda^2)
nB=length(B);
A=Richard_kq(k,q); % A is specified in lambda-domain
nA=length(A);
if (abs(nB-nA)>0)
  A=fullvector(nB,A);% work with equal length vectors
end

% Generation of minimum immitance function using Bode or Parametric
method
ndc=q;
        [a,b]=RtoZ(A,B);% Here A and B are specified in p-domain
        na=length(a);
        if ndc>0;a(na)=0;end;
end
%
Norm=norm(a);a=a/Norm;b=b/Norm;
end
```

Program List 5.54: Main Program: GKYExample5_12.m

```
% Main Program: GKYExample5_12.m
%---------------------------------------------------------------------------
% High Precision Synthesis of a Richards Immittance
% via Parametric Approach by B.S. Yarman
% December 19, 2014
%
%
%---------------------------------------------------------------------------
clc;clear;close all
q=0; k=5; a0=1; c=[1 -1  1 -1  1];
n=length(c);
nc=length(c);c0=1;
nL=n-k-q;
R0=1;f0=1/2/pi;% Normalization numbers to end up with actual elements.
%---------------------------------------------------------------------------
[a1,b1]=Richard_NewMinimumFunction(k,q,c,c0);
a=a1;b=b1;
[Z_UE,a_new,b_new,CT,CV]=Richard_CompleteImpedanceSynthesis(a,b,k,q,R0,f0);
BPrint=(' Characteristic impedance of the lines:Z_UE')
Z_UE'
%
% Computation of Input Impedance over the normalized frequencies
if n>k
    w=0.1;% Initialize normalized frequency
    dw=0.01; % Frequency step size
```

```
        j=sqrt(-1);
        tau=pi/2/1.5;% tau=pi/2/we; we=Stopband frequency
        for i=1:101
            FR(i)=w;
            Omega=tan(w*tau);
            lambda=j*Omega;
        %
        [ ZL,YL ] = InputimmitanceLadder_viacodes( CT,CV,q,Omega );% Input
impedance of the Richards Ladder
        if k==0; Zin=ZL;end
        if k>0
        Zin= Richard_InputImpedancekUE(lambda,Z_UE,ZL,k );%k-Cascaded UE
terminated in Richards Ladder
        end
        Rin(i)=real(Zin);
        Xin(i)=imag(Zin);
      % Compute Original Input Impedance from the Given
      % Z(lambda)=a(lambda)/b(lambda)
        aval=polyval(a,lambda);
        bval=polyval(b,lambda);
        Z_org=aval/bval;
        R_org(i)=real(Z_org);
        X_org(i)=imag(Z_org);
        w=w+dw;
        end
end
if k==n
    ZL=a_new/b_new;
     w=-1;% Initialize normalized frequency
    dw=0.01+1e-15; % Frequency step size
    j=sqrt(-1);
    tau=pi/2/1.5;% tau=pi/2/we; we=Stopband frequency
    for i=1:501
        FR(i)=w;
        Omega=tan(w*tau);
        lambda=j*Omega;
        p=j*w;
    %
    Zin= Richard_InputImpedancekUE(lambda,Z_UE,ZL,k );%k-Cascaded UE
terminated in Richard Ladder
    Rin(i)=real(Zin);
    Xin(i)=imag(Zin);
     % Compute Original Input Impedance from the Given
   % Z(lambda)=a(lambda)/b(lambda)
    aval=polyval(a,lambda);
    bval=polyval(b,lambda);
    Z_org=aval/bval;
    R_org(i)=real(Z_org);
    X_org(i)=imag(Z_org);
    bval=polyval(b,p);
    %
    w=w+dw;
    end
end
```

```
figure
plot(FR,Rin,FR,R_org)
title(' Real-Part of the input impedance')
legend('Rin','R org')
ylabel(' Rin(w)')
xlabel('Normalized angular freuency w')
%
figure
plot(FR,Xin,FR,X_org)
title(' Imaginary-Part of the input impedance')
legend('Xin','X org')
ylabel(' Xin(w)')
xlabel('Normalized angular freuency w')
%
format short e
error_R=norm(Rin-R_org);
error_X=norm(Xin-X_org);
%[FR' Rin' R_org' Xin' X_org']
[FR' (Rin-R_org)' (Xin-X_org)']
nkqnL=[n k q nL]
Error_RX=[error_R error_X]
```

Program List 5.55: Richard_Numerator

```
function [ k,q,a0,nA,A,B ] = Richard_Numerator( a,b )
% This function generates k, q and a0 of the even part F=a/b specified in
% lambda domain. Note that R=A/B in lambda domain-(Richard Domain)
% A(lambda^2)=(a0)^2*(-1)^q*(lambda)^2q*[1-(lambda)^2]^k
% Here, in this function we count non zero terms both from bottom-up
  with
% q=x; k=nA-y-q;
% Inputs:
%       a(lambda)
%       b(lambda
% Outputs:
%       k: Total number of cascaded UEs.
%       q: Total number of transmission zeros
%       (a0)^2:Leading coefficients of A(lambda)
%-------------------------------------------------------------------------
zero=1e-4;
% Norma=norm(a);a=a/Norma;b=b/Norma;
nb=length(b);B0=b(nb)*b(nb);
[A,B]=even_part(a,b);
Norm=norm(A);A=A/B0;B=B/B0;
%
nA=length(A);
% Determine q & k: start counting the unity roots (i.e. find non zero
A(i))
% Count from bottom-up
  [ q,nA ] = Richard_DCzeros( A );
% ----- end of bottom up counting ---------------------------------------
j=1;s=0;
% Count from top-down
```

```
while abs(A(j))<zero;
    j=j+1;
    s=s+1;
end
    y=j;
% ----- end of top down  counting ----------------------------------------
%
        k=nA-y-q;
%-------------------------------------------------------------------------
a0=sqrt(abs(A(j)));
% Determine q
n=length(A)-1;
%
%[ q ] = Richard_DCzeros( A );
%
A=A/a0/a0;B=B/a0/a0;% Normalization of vectors A and B with respect to
a0*a*
% This means that non-zero leading coefficient of A(lambda^2) is 1.
end
```

Program List 5.56: Richard_NumeratorNew

```
function [ a0,A,B ] = Richard_NumeratorNew( a,b,k,q )
% This function is revised on May 30,2013.
% This function generates k, q and a0 of the even part F=a/b specified in
% lambda domain. Note that R=A/B in lambda domain-(Richard Domain)
% A(lambda^2)=(a0)^2*(-1)^q*(lambda)^2q*[1-(lambda)^2]^k
% Inputs:
%       k: Total # of Cascaded Unit Elements
%       q: Total # of DC Transmission zeros
%       a(lambda)
%       b(lambda)
% Outputs:
%       k: Total number of cascaded UEs.
%       q: Total number of transmission zeros
%       (a0)^2:Leading coefficients of A(lambda)
%-------------------------------------------------------------------------
nb=length(b);nA=length(a);
a=a/b(nb);b=b/b(nb);%Normalize Z(lambda)=a/b with respect to b(nb).
%a0=sqrt(a(nb)*b(nb));
a0=1;
A=a0*a0*Richard_kq(k,q); % A is specified in lambda-domain
[ B ] = EvenpartDenom_B( b );
end
```

Program List 5.57: GKYExample5_13.m

```
% Main Program GKYExample5_13.m
% This main MatLab program solves Example 5.13
clc;clear;close all
%
% Inputs:
k=5; q=0;  a0=1; c=[1 -1   1 -1    1    -1 1 ];
```

```
% Part (a): Generation of the minimum function in Richards domain
    [a,b]=Richard_MinimumFunction(k,q,a0,c);
n=length(c);
nL=n-k;
for i=1:k
    k1=k-i;
[ Z1,a_new,b_new,ra,rb ] = ImpedanceBasedRichard_Extraction( a,b );
j=even_odd(i);
clear a; clear b;
a=a_new;b=b_new;
Z_UE(i)=Z1;
if j==1
    [b,a]=Richard_ImmittanceCorrection(b,a,k1,q);
end
if j==0
    [a,b]=Richard_ImmittanceCorrection(a,b,k1,q);
end
CT1(i)=20; CV1(i)=Z1;
end
% Part (b): Synthesis of the remaning LC ladder
R0=1;f0=1/2/pi;
if j==1
    [CT2,CV2] = SynthesisMinimumSuseptance_Yarman(R0,f0,b,a,q);
end
if j==0
    [CT2,CV2] = SynthesisMinimumReactance_Yarman(R0,f0,a,b,ndc);
end
CT=[CT1 CT2]; CV=[CV1 CV2];
% Final result of the synthesized circuit:
[ CTD,CVD ] = RichardsLadderTo_TrLineTopology( CT2,CV2 );
CTF=[CT1 CTD];CVF=[CV1 CVD];
Plot_Circuit4(CT,CV)
Plot_Circuit4(CTF,CVF)
```

Program List 5.58: GKYExample5_14.m

```
% Main Program GKYExample5_14.m
% This main MatLab program solves Example 5.14
clc;clear;close all
%
% Inputs:
k=5; q=5;  a0=1; c=[1 -1 1 -1 1  -1 1 -1  1 -1 1 -1 1 -1 1]; c0=1;
% Part (a): Generation of the minimum function in Richards domain
    [a,b]=Richard_MinimumFunction(k,q,a0,c);
%
% Part (b): Complete synthesis of the Richards driving point input
% impedance
R0=1;f0=1/2/pi;
[Z_UE1,a_new,b_new,CT,CV ] = Richard_CompleteImpedanceSynthesis(a,b,k,q,R
0,f0 );
%-------------------------------------------------------------------
```

Program List 5.59: Main Program GKY_Example5_15.m

```
% Main Program: GKY_Example5_15.m
% This MatLab program designs an impedance transforming filter in Richards
% domain lambda=SIGMA+jomega between complex generator RG=12 ohm+LG//CG
  and
% RL=50 ohm+LL//CL where RG=12 ohm; LG=0.947nH; CG1=1.515 pF, CG2=3.4 pF;
% CL=1.515 pF, LL=0.412 nH; RL=50 ohm.
% Frequency band is 850 MHz - 2100 MHz
% By Siddik Yarman
% December 25, 2014
%-------------------------------------------------------------------------
%      Double Matching in Richard Domain without Transformer
% ------------------------------------------------------------------------
% This program designs a Real Frequency Impedance Transforming Filter for
% given resistive terminations in Richards Domain
% Design is accomplished using Real Frequency Direct Computational
  Technique
% where Parametric Approach is employed to generate ZB(p) from RB(p)
% Optimization via lsqnonlin Algorithm
% error: Error function - Output
% Inputs:
%           x0: Initialized unknowns
%           T0: Flat gain level
%           f0(MHz) : Frequency Normalization
%           fs1(MHz): Left Stop band frequency (for print purpose)
%           fs2(MHz): Right Stop band frequency (for print purpose)
%           fc1(MHz): Left Cut-off frequency (begining of optimization)
%           fc2(Mhz): Right Cut-off frequency (end of optimization)
%           wc1=2*pi*fc1/f0 Beginning of optimization
%           wc2=2*pi*fc2/f0 End of optimization
%           KFlag: =1>Work with impedance functions
%           KFlag: =0>Work with admittance functions
%           q=0 Lowpass Ladder structure
%           W=0 No finite transmission zero.
%           zero.
%           c0=1 No transformer in the Circuit,
%               (Initial for R(omega)=omega^(2q)(1+omega^2)^k/B(omega^2)
%           WNG,RNG1,XNG1: Normalized Generator Data Freq, Real Part RNG,
%           Imaginary Part XNG
%           WNL,RNL1,XNL1: Normalized Generator Data Freq, Real Part RNL,
%           Imaginary Part XNL
%
clc
clear
close all
%
%-------------------------------------------------------------------------
% Direct Computational Technique for Double Matching Problems
% Design in Richards domain lambda=SIGMA+j*OMEGA
%-------------------------------------------------------------------------
% Optimize the back-end Immitance F(lambda)=a(lambda)/b(lambda)
% Real Part of K(p)is expressed as in direct computational
% techniques R(OMEGA^2)=AA(OMEGA^2)/BB(OMEGA^2)
%-------------------------------------------------------------------------
```

```
% Inputs:
fs1=600;fs2=3000;NSample=50;tau=0.5;
fc1=850; fc2=2100; RN=50;
T0=1; KFlag=1;czero=1;
c0=[592 608 -229 -238 15.798 17.57 ];
k=4;q=0;ntr=0;
%-------------------------------------------------------------------------
[FA,RG,XG]=Generator_Impedance(fs1,fs2,NSample);
[FA,RL,XL]=Load_Impedance(fs1,fs2,NSample);
GEN =[FA',RG',XG'];% 50 ohm generator up to 50,000 MHz=50 GHz.
Load =[FA',RL',XL'];% 50 ohm load up to 50,000 MHz=50 GHz.
%-------------------------------------------------------------------------
% Generator Data Matrices
A1=GEN; % Data Matrix For Generator Termination
A2=Load; % Data Matrix For Load Termination
%
% Dimensions of the Collected Data
NA1=length(A1(:,1));% This defines the first column length of A1.
NA2=length(A2(:,1));% This defines the first column length of A2.
%
% Generator
for i=1:NA1
    FAG(i)= A1(i,1);
    RAG1(i)=A1(i,2);
    XAG1(i)=A1(i,3);
end
% Load
R0=50;
for i=1:NA2
    FAL(i)=A2(i,1);
    RAL1(i)=A2(i,2);
    XAL1(i)=A2(i,3);
    ZL=complex(RAL1(i),XAL1(i));
    SL=(ZL-R0)/(ZL+R0);
    ROL=abs(SL);
    L21(i)=10*log10(1-ROL*ROL);% Unmatched Gain
end
%
figure
plot(FAG,L21)
title('Gain without Matching')
xlabel('Actual Frequency in MHz')
ylabel('Gain in dB')
%
% Frequency Normalization with f0=fc2.
ws1=fs1/fc2;
ws2=fs2/fc2;
wc1=fc1/fc2;
wc2=fc2/fc2;
% Generator and Load impedance Normalization
FNG=fc2;
FNL=fc2;
% Generator Normalization
RNG1=RAG1/RN;
```

```
XNG1=XAG1/RN;
WNG=FAG/FNG;
% Load Normalization
RNL1=RAL1/RN;
XNL1=XAL1/RN;
WNL=FAL/FNL;
%
figure
plot(FAL,RAL1)
title('Actual Real Part Measurement for the Load Data');
xlabel('Actual Frequency in MHz for the Load Data');
ylabel('Actual Real Part of the Load Impedance');
%
figure
plot(FAL,XAL1)
title('Actual Imaginary Part Measurement for the Load Data');
xlabel('Actual Frequency in MHz for the Load Data');
ylabel('Actual Imaginary Part of the Load Impedance');
%
%----------------------------------------------------------------------------
% Step 2: Generic form of RB(omega^2):
%
% Input Set #2:
%      czero: Constant DC multiplier of the denominator
%      c0; Initial coefficients of BB(omega^2)
%      Design without transformer
% Output:
%      Optimized values of coefficients of c(lambda)
%      Direct form of Even Part R(lambda^2)=A(-lambda^2)/B(-lambda^2)
%      a(lambda) and b(lambda) of Driving Point Impedance
ZB(lambda)=a(lambda)/b(lambda)
%
%----------------------------------------------------------------------------
%
Nc=length(c0);
% Step 4: Enter extraction parameters:
x0=c0;% Transformerless Design. czero=1 is fixed at 50 ohm.
if q>0;x0=[c0 czero];end
%----------------------------------------------------------------------------
OPTIONS=optimset('MaxFunEvals',50000,'MaxIter',50000);
% x0,[],[],OPTIONS,
% Minimax Optimization
for i=1:5
%                       (x0,T0,wc1,wc2,KFlag,czero,WNG,RNG, XNG, WNL,
RNL,XNL, k,q,tau)
f = @(x)Richard_OBJECTIVE(x,OPTIONS,T0,wc1,wc2,KFlag,czero,WNG,RNG1,XNG1,
WNL,RNL1,XNL1,k,q,tau);
%
        %[x,fval] = fminimax(f,x0);
        % [x]=fminsearch(f,x0);

        x=lsqnonlin(f,x0);
        x0=x;
%----------------------------------------------------------------------------
```

```
%
end
% After optimization separate x into coefficients
%
Nx=length(x);
    c=x;
%
% Step 5: Generate analytic form of Fmin(p)=a(p)/b(p)
        [a,b]=Richard_NewMinimumFunction(k,q,c,czero);
% Generate analytic form of Fmin(p)=a(p)/b(p)
        cmplx=sqrt(-1);
% Step 6: Print and Plot results
Nprint=10001;
DW=(ws2-ws1)/(Nprint-1);
w=ws1;
Tmax=-10000;
Tmin=0;
for j=1:Nprint
            WA(j)=w;
%
            [ TPG ] = Richard_DoubleMatchingGain( w,tau,q,WNG,RNG1,XNG1,WN
L,RNL1,XNL1,KFlag,a,b);
            TA(j)=10*log10(TPG);
 % Computation of reference gain in Tchebyshev form:
 eps_sq=.1;n=7;
            Tcheby=Tchebyshev_Gain(w,n,T0,eps_sq,wc1,wc2);
            dB_Tcheby(j)=10*log10(Tcheby);
%
% Compute the performance parameters:Tmax,Tmin,Tave and detT
    if max(TA(j))>Tmax
        wmax=WA(j);Tmax=TA(j);
    end
if w>=wc1
    if w<=wc2
    if min(TA(j))<Tmin
        wmin=WA(j);Tmin=TA(j);
    end
    end
end
    w=w+DW;
end
%
FA=WA*fc2;
        figure
        plot(FA,TA,FA,dB_Tcheby)
        title('Impedance Transforming Filter via RFDCT Parametric
Approach')
        xlabel('Actual frequency f')
        ylabel('Gain in dB')
        legend('RFDCT Gain','Tchebyshev Gain')
        %
        figure
        plot(FA,TA)
```

```
          title('Impedance Transforming Filter via RFDCT Parametric
Approach')
          xlabel('Actual frequency f')
          ylabel('Gain in dB')
          legend('RFDCT Gain')
%------- New Synthesis Package ------------------------------------------------
f0=1/2/pi;% Work with normalized impedance/admittance
R0=1;% Normalized resistance
%
if KFlag==1; [Z_UE,a_new,b_new,CT,CV ] = Richard_CompleteImpedanceSynthes
is(a,b,k,q,R0,f0 );end
if KFlag==0; [Z_UE,a_new,b_new,CT,CV ] = Richard_CompleteImpedanceSynthes
is(b,a,k,q,R0,f0 );end
   %
   % Print Performance parameters:
[wmax*fc2, Tmax,wmin*fc2,Tmin],
T_Average=(Tmax+Tmin)/2,delT=T_Average-Tmin
```

Program List 5.60: Richard_OBJECTIVE

```
function
eps=Richard_OBJECTIVE(x0,OPTIONS,T0,wc1,wc2,KFlag,czero,WNG,RNG,XNG,WNL,R
NL,XNL,k,q,tau)
% In this function impedance is generated using Rihard_NewMinimumFunction
% --------------------------------------------------------------------
% DESIGN IN RICHARD DOMAIN
% --------------------------------------------------------------------
% Optimization via Direct - Parametric approach
% error: Error function - Output
% Inputs:
%          x0: Initialized unknowns
%          T0: Flat gain level
%          wc1: Beginning of optimization
%          wc2: End of optimization
%          KFlag: =1>Work with impedance functions
%          KFlag: =0>Work with admittance functions
%          czero: zero degree coefficient of vector c0[c1 c2 .. cn] and
%          czero.
%          R(omega)=nB(j*omega)nB(-j*omega)/[b(j*omega)b(-j*omega)]
%          WNG,RNG1,XNG1: Normalized Generator Data Freq, Real Part RNG,
%          Imaginary Part XNG
%          WNL,RNL1,XNL1: Normalized Generator Data Freq, Real Part RNL,
%          Imaginary Part XNL
%
%--------------------------------------------------------------------
          c=x0;
%
% Step 5: Generate analytic form of K(lambda)=a(lambda)/b(lambda)in
Richard Domain
%
          [a,b]=Richard_NewMinimumFunction(k,q,c,czero);
%
          N_Opt1=150;
          %
```

```
       w=wc1;
       del=(wc2-wc1)/(N_Opt1-1);
%
       for j=1:N_Opt1
% Compute Generator and Load immittance:
% Compute Double Matching Gain (Transducer Power Gain : TPG)
          [ TPG ] = Richard_DoubleMatchingGain( w,tau,q,WNG,RNG,XNG,WNL,
RNL,XNL,KFlag,a,b);
             eps(j)=atan(abs(TPG))-atan(T0);
             w=w+del;
       end
%        sum=0;
%        for i=1:N_Opt1
%        error=sum+eps(i)*eps(i);
%        end
    end
```

___

Program List 5.61: RichardsInputRef_EL

___

```
function Sin=RichardsInputRef_EL(w,tau,KFlag,eta,WLA,RLA,XLA,a,b)
% December 25, 2014
% This function generates the input reflectance of the lossless two-port
% [EL] when it is terminated in complex impedance ZL=RL+jXL
% In this function lossless equalizer is defined from the back-end (Port-2)
% immitance KB(lambda)=a(lambda)/b(lambda). K(lambda)=ZB if KFlag=1,
  K(lambda)=YB if KFalg=0
% Inputs:
%       w: normalized angular frequency
%       KFlag: Control flag for the immitance based defined reflectance
%           KFlag=1> Impedance based reflectance definition
%           KFlag=0> Admittance based reflectance definition
%       WLA: Array. Normalized sampling frequencies to refer load ZL
%       RAL: Array. Normalized real part of the impedance ZL
%       XAL: Array. Normalized imaginary part of the impedance ZL
% Output:
%       Sin: Real normalized-driving point input reflectance of [EL]
%-------------------------------------------------------------------------
 % Generation of complex termination (load) at Port-2.
 omega=tan(w*tau);
 lambda=sqrt(-1)*omega;
 aval=polyval(a,lambda);
 bval=polyval(b,lambda);
 ZB=aval/bval;
 ZBC=conj(ZB);
 RL=line_seg(WLA,RLA,w);
 XL=line_seg(WLA,XLA,w);
 ZL=complex(RL,XL);
 if KFlag==0
     ZL=1/ZL;
 end
         S=(ZL-ZBC)/(ZL+ZB);
% Definiation of real normalized input reflectance of [EL] in Yarman
sense
         Sin=eta*S;
```

```
if KFlag==0
    Sin=-Sin;
end
end
```

Program List 5.62: function Richard_DoubleMatchingGain

```
function [ TPG ] = Richard_DoubleMatchingGain(
w,tau,q,WGA,RGA,XGA,WLA,RLA, XLA, KFlag,a,b)
% This function computes the Double Matching Gain in Richard Domain
% --------------------------------------------------------------------
% Inputs:
%       w: Normalized angular real frequency w=2*pi*f
%       tau: Delay length of the commensurate lines
%       Note-1: omega: Angular frequency in Richard Domain
%       Note-2: lambda=j*tan(w*tau)
%       q: Total number of DC transmission zeros in lambda domain
%       WGA: Break Frequencies for generator immittance
%       RGA: Real part of the generator immitance
%       XGA: Imaginary part of the generator immittance
%       RLA: Real Part of the load immittance
%       XLA: Imaginary part of the load immittance
%       KFlag: KFlag>1 impedance, KFlag>0 addimittance
%       a(lambda): Numerator Polynomial of the Back-End immitance
%       K(lambda)=a(lambda)/b(lambda)
%       b(lambda): Denominator Polynomial of K(lambda)=a/b
% Output:
%       TPG(omega): Double Matching Gain
% --------------------------------------------------------------------
%   Definition of lambda:
            cmplx=sqrt(-1);j=cmplx;
            omega=tan(w*tau);
            lambda=j*omega;
% F(lambda)=a(lambda)/b(lambda)
            bval=polyval(b,lambda);
% Generation of allpass function eta in lambda domain:
%
        bval_conj=conj(bval);
        eta=(-1)^q*bval_conj/bval;
%
% --------------------------------------------------------------------
%
            Sin=RichardsInputRef_EL(w,tau,KFlag,eta,WLA,RLA,XLA,a,b);
%
[RG,XG]=Line_Impedance(w,WGA, RGA, XGA,KFlag);
            TEL=1-abs(Sin)*abs(Sin);
            ZG=complex(RG,XG);
            G22=(ZG-1)/(ZG+1);
            if KFlag==0
                G22=-G22;
            end
            Weight=(1-abs(G22)*abs(G22))/abs((1-G22*Sin))^2;
            TPG=Weight*TEL;
end
```

Program List 5.63: *function* Line_Impedance

```
function [R,X]=Line_Impedance(w,WN, RN, XN,KFlag)
% This function generates the impedance data for a given impedance
  triplet
% Angular Frequency Array WN,
% Real Part Array RN
% Imaginary Part Array XN
% Inputs:
% w: Normalized Angular Frequency data at a single point
% WN: Angular Frequency Array
% RG Real Part Array
% XG: Imaginary Part Array
% KFlag: KFlag=1 for impedance, KFlag=0 for admittance computations
% Generate the load impedance
j=sqrt(-1);
NG=length(WN);
R=line_seg(WN,RN,w);
X=line_seg(WN,XN,w);
Z=R+j*X;
Y=1/Z;
        if KFlag==1;% Work with impedances
          R=real(Z);
          X=imag(Z);
        end
          if KFlag==0;%Work with admittances
             R=real(Y);
             X=imag(Y);
          end
```

Program List 5.64: *function* Richard_CompleteImpedanceSynthesis

```
function [Z_UE,a_new,b_new,CTF,CVF ] = Richard_CompleteImpedanceSynthesis
(a,b,k,q,R0,f0 )
% This function carries out complete synthesis in lambda domain
% Inputs:
%       k; Total number of cascaded UEs
%       q: Total number of high-pass elements as series open-stubs,
parallel shorted-stubs
%       Zin=a/b
%       R0: Normalization resistance
%       f0: Normalization frequency but it must be fixed at f0=1/2/pi
% Outputs:
%       Z_UE: Chararcteristic impedances of the Unit Elements
%       a_new,b_new: Z_new=a_new/b_new is the remaning impedance after
%       extraction of k-UE.
%       CT,CV: Circuit codes and values of the ladder synthesis
%
ndc=q;
%
if k==0;a_new=a;b_new=b;
    Z_UE='k=0. There is no UE in the synthesis'
end
```

```
%
% ------ Definition of Algorithmic zero ----------------------------------
zero=1e-10;
zeroA=1e-10;
aux=[a b];
Kmax=max(aux);
a1=a/Kmax;b1=b/Kmax;na=length(a);
    if q>0;
    zeroA=a1(na);
    end
%
if zero<zeroA;zero=zeroA;end
% ------ End of Definition of Algorithmic zero ------------------------
if abs(a(1))<zero;[a,b]=Richard_ImmittanceCorrection(a,b,k,q);end
if abs(b(1))<zero;[b,a]=Richard_ImmittanceCorrection(b,a,k,q);end
%----------------------------------------------------------------------
n=length(a)-1;
if k<=n
    if k>0
%High precision-impedance based-Richard Extraction:
% [Z,a_new,b_new] = Richard_HighPrecisionUEExtraction( k,q,a,b );
% [Z,a_new,b_new] = Richard_RoughUEExtraction( k,q,a,b );
 [ Z,a_new,b_new ] = Richards_CameronExtractions( k,q,a,b );
   Z_UE=Z;
%----------------------------------------------------------------------
    end
%----------------------------------------------------------------------
a1=a_new;b1=b_new;
na=length(a_new);nb=length(b_new);
% a_new(na) might be zero and b_new(na) might be zero. In this case
degree
% reduction in "q" occours.
if abs(a_new(na))<zero;
if abs(b_new(nb))<zero;
    clear a_new;clear b_new;
    for i=1:na-1
        a_new(i)=a1(i);
        b_new(i)=b1(i);

    end
    q=q-1;% Degree reduction in q
        Attention=('Degree reduction occours in q=q-1')
        q1=q
end
end
% End of degree reduction loops
%----------------------------------------------------------------------
end
%
nc=n;
if nc>k
%
if abs(a_new(1))<zero;[a_new, b_new, q]=Check_immitance(a_new, b_new, q);end
```

```
if abs(b_new(1))<zero;[b_new, a_new, q]=Check_immitance(b_new, a_new, q);end
    if abs(a_new(1))>zero;
        if abs(b_new(1))>zero;
            [a_new, b_new, q]=Check_immitance(a_new, b_new, q);
        end
    end
    %
[ CT CV ] = Synthesis_ImpedanceBased( a_new, b_new, q,R0,f0 );
end
if k==n
    CT=('k=n; Therefore, synthesis is completed only with cascade
connection of n-unit elemensts')
    CV=CT;
end
% Plot the remainin synthesized Circuit in lambda domain
% if n>k; Plot_Circuitv1( CT, CV );end
%-------------------------------------------------------------------------
% Plot of the complete commensurate transmission line circuit
nue=length(Z_UE);
for i=1:nue
    CT1(i)=20; CV1(i)=Z_UE(i);
end
if n>k
[ CTD, CVD ] = RichardsLadderTo_TrLineTopology( CT,CV );
CTF=[CT1 CTD];CVF=[CV1 CVD];
end
if n==k
  CTF=[CT1 9]; CVF=[CV1 a_new/b_new];
end
Plot_Circuit4(CTF, CVF)
end
```

Program List 5.65: function Synthesis_ImpedanceBased

```
function [ CT CV ] = Synthesis_ImpedanceBased( a,b,ndc,R0,f0 )
% This function completes the impedance based synthesis
%    F(p)=a(p)/b(p) is an impedance.
n1=length(a);zero=1e-5;
norma=norm(a);
normb=norm(b);
norma=1;normb=1;
%norma=1;normb=1;
%-------------------------------------------------------------------------
% Case 1:a(1)=0,b(1)>0. Then, F(p) is a minimum
% reactance function. Therefore we set KFlag=1.
%[ ndc ] = DCZeros_Evenpart( a,b )
if abs(a(1))/norma<=zero
        if abs(b(1))/normb>zero
            if abs(b(n1))/normb>zero
                KFlag=1;

                [CT,CV] = CircuitPlot_Yarman(KFlag,R0,f0,a,b,ndc);
```

```
Case='Case 1> a(1)=0,b(1)>0,b(n1)>0: Minimum Reactance Input Impedance F(p)'
            end
        end
end
%End of Case 1
%-------------------------------------------------------------------------
% Case 2:a(1)>0, or b(1)=0, b(n1)=0. Then, F(p) is a minimum
% suseptance function. Therefore we set KFlag=0 and flip over
% F(p) as H(p)=b(p)/a(p)=1/F(p).
if abs(a(1))/norma>zero
        if abs(b(1))/normb<=zero
            if abs(b(n1))/normb<=zero
                KFlag=0;
                %[ ndc] = DCZeros_Evenpart( b,a )
                [CT,CV] = CircuitPlot_Yarman(KFlag,R0,f0,b,a,ndc);
Case='Case 2> a(1)>0,b(1)=0,b(n1)=0: Minimum Suseptance Input Impedance F(p)'
            end
        end
end
%End of Case 2
%-------------------------------------------------------------------------
% Case 3:a(1)>0, b(1)=0, b(n1)>0. Then, F(p) is not a minimum
%  reactance. It has a pole only at infinity. Therefore, we extract the
%  pole.
if abs(a(1))/norma>zero
        if abs(b(1))/normb<=zero
            if abs(b(n1))/normb>zero
                [L0,a1,b1]=removepole_atinfinity(a,b);
                L_Act=L0*R0/2/pi/f0;
                KFlag=1;
                %[ ndc] = DCZeros_Evenpart( a1,b1 )
                [CT,CV] = CircuitPlot_Yarman(KFlag,R0,f0,a1,b1,ndc);
                CT=[1 CT]; CV=[L_Act CV];
Case='Case 3> a(1)>0,b(1)=0,b(n1)>0: F(p) has a pole at infinity'
            end
        end
end
%End of Case 3
%-------------------------------------------------------------------------
% Case 4:a(1)=0, or b(1)>0, b(n1)=0. Then, F(p) is not a minimum
%  reactance function. It has pole at p=0. Therefore, we extract the
%  pole at p=0.
if abs(a(1))/norma<=zero
        if abs(b(1))/normb>zero
            if abs(b(n1))/normb<=zero
                [C0,a1,b1]=removepole_atzero(a,b);
                C_Act=C0/R0/2/pi/f0;
                KFlag=1;
                ndc1=ndc-1;
                %[ ndc1] = DCZeros_Evenpart( a1,b1 )
                [CT,CV] = CircuitPlot_Yarman(KFlag,R0,f0,a1,b1,ndc1);
                CT=[2 CT]; CV=[C_Act CV];
Case='Case 4> a(1)=0,b(1)>0,b(n1)=0: F(p) has a pole at p=0'
            end
```

```
            end
end
%End of Case 4
%-------------------------------------------------------------------------
% Case 5:F(p)>Highpass Circuit Structure as a minimum suseptance function
if abs(a(1))>zero
    if abs(b(1))>zero
        if abs(b(n1))/normb<=zero
            KFlag=0;
            %[ ndc] = DCZeros_Evenpart( b,a )
            [CT,CV] = CircuitPlot_Yarman(KFlag,R0,f0,b,a,ndc);
Case='Case 5> a(1)>0,b(1)>0,b(n1)=0: Highpass F(p) as Minimum Suseptance'
        end
    end
end
% End of Case 5
%-------------------------------------------------------------------------
% Case 6:F(p)> Highpass Circuit Structure as a minimum reactance function
if abs(a(1))>zero
    if abs(b(1))>zero
        if abs(a(n1))/norma<=zero
            KFlag=1;
            %[ ndc] = DCZeros_Evenpart( a,b )
            [CT,CV] = CircuitPlot_Yarman(KFlag,R0,f0,a,b,ndc);
Case='Case 6> a(1)=0,b(1)>0,a(n1)=0: Highpass F(p) as Minimum Reactance'
        end
    end
end
% End of Case 6
%-------------------------------------------------------------------------
Plot_Circuitv1(CT,CV);
end
%End of function
```

Program List 5.66: Main Program TSMC_amp.m

```
% Program TSMC_amp.m
clc
clear
close all
%   DESIGN OF A SINGLE STAGE AMPLIFIER
%   USING A TSMC 0.18 micron NMOS SILICON TRANSISTOR
%   via SIMPLIFIED REAL FREQUENCY TECHNIQUE (SRFT)
%-------------------------------------------------------------------------
% General Information about the device NMOS Transistor, 0.18 micron gate
%               Frequency range : 100MHz - 22.4 GHz;
%               VDD = 1.8V; ID (drain DC current) = 200mA,
%               NMOS size W/L = 1000u/0.18u;
%               Simulation temperature = 50C.
%               Output power= 200mW up to 20 GHz
%               Operaton Class: Class A amplifier.
%-------------------------------------------------------------------------
```

```
% INPUTS:
%                 FR(i): ACTUAL FREQUENCIES
%                 SIJ: SCATTERING PARAMETERS OF THE TSMC-CMOS FET.
%                 THESE PARAMETERS ARE READ FROM A FILE CALLED ersad_1.txt
%                 Tflat1: FLAT GAIN LEVEL OF STEP 1 in dB.
%                 Tflat2: FLAT GAIN LEVEL OF STEP 2 in dB.
%                 FR(nd1); nd1: BEGINING OF THE OPTIMIZATION FREQUENCY
%                 (nd1=42)
%                 FR(nd2); nd2: END OF OPTIMIZATION FREQENCY: nd2=48.
%
% OUTPUT: FRONT & BACK END EQUALIZERS ARE GENERATED EMPLOYING SRFT
%                 Step 1:
%                         hF: Front End Equalizer;
%                         T1: Gain of the First Step
%                         CVF:Normalized Element Values of the Front-End
%                         equalizer
%                         AEVF: Actual Element Values of the Front-End
%                 Step 2:
%                         hB: Back End Equalizer
%                         T2: Gain of the Second Step
%                         CVB:Normalized Element Values of the Back-End
%                         Equalizer
%                         AEVB: Actual Element Values of the Back-End
%
%
%                         DESIGN OF A MICROWAVE AMPLIFIER
% READ THE TRANSISTOR DATA ersad_1.txt
%
%-------------------------------------------------------------------------
% Special Notes: From the given data file one should obtain that
% nd=48 (Total Number of sampling points)
% For front-end Equalizer:
%     freq(48)=22.38 GHz;   freq(41)=10 GHz;       freq(35)=5.02 GHz
%       T01(48)=1.317         T01(41)=6.54           T01(35)=26.1
%       (MSG for front-end)
%       T02(48)=5.255         T02(41)=16.53          T02(35)=58.25
%       MSG(48)=7.2dB;      MSGdB(41)=12.1dB;      MSGdB(35)=17.65dB
%     (MSG for the back-end)
%-------------------------------------------------------------------------
load TMSC_Spar.txt; % Load cmos data to the program
nd=length(TMSC_Spar)
pi=4*atan(1);
rad=pi/180;
%
%   EXTRACTION OF SCATTERING PARAMETERS FROM THE GIVEN FILE ersad_1.txt
for i=1:nd
   % Read the Actual Frequencies from the file ersad_1.txt
                 freq(i)=TMSC_Spar(i,1);
   % Read the magnitude and phase of the scattering parameters
   % from the data file ersad_1.txt
MS11(i)=TMSC_Spar(i,2);PS11(i)=rad*TMSC_Spar(i,3);
MS21(i)=TMSC_Spar(i,4);PS21(i)=rad*TMSC_Spar(i,5);
MS12(i)=TMSC_Spar(i,6);PS12(i)=rad*TMSC_Spar(i,7);
MS22(i)=TMSC_Spar(i,8);PS22(i)=rad*TMSC_Spar(i,9);
```

```
% Convert magnitude and phase to real and imaginary parts.
S11R(i)=MS11(i)*cos(PS11(i));S11X(i)=MS11(i)*sin(PS11(i));
S12R(i)=MS12(i)*cos(PS12(i));S12X(i)=MS12(i)*sin(PS12(i));
S21R(i)=MS21(i)*cos(PS21(i));S21X(i)=MS21(i)*sin(PS21(i));
S22R(i)=MS22(i)*cos(PS22(i));S22X(i)=MS22(i)*sin(PS22(i));
 % Get rid off the numerical errors due to extraction of the scattering
 % paramters from cadnace...(Vdec extarction
  if S11R(i)>=1;S11R(i)=0.99999999;S11X(i)=0.0;end
 end
%
%    CONSTRUCT THE APMLITUDE SQUARES OF THE SACTTERING PARAMETERS
  for i =1:nd
 SQS21(i)=S21R(i)*S21R(i)+S21X(i)*S21X(i);
 SQS12(i)=S12R(i)*S12R(i)+S12X(i)*S12X(i);
 SQS11(i)=S11R(i)*S11R(i)+S11X(i)*S11X(i);
 SQS22(i)=S22R(i)*S22R(i)+S22X(i)*S22X(i);
 S1(i)=1.0/(1-SQS11(i));S2(i)=1.0/(1-SQS22(i));
   end
     % COMPUTATION OF MAXIMUM STABLE GAIN "MSG'
%
  for i=1:nd;if SQS11(i)>=1;SQS11(i)=0.999999999;end
if   SQS22(i)>=1;SQS22(i)=0.999999999;end
%Ideal Flat Gain Levels
     % Step 1:
                    T01(i)=SQS21(i)/(1-SQS11(i));
                    T01dB(i)=10*log10(T01(i));
     % Step 2:
                    T02(i)=T01(i)/(1-SQS22(i));
                    T02dB(i)=10*log10(T02(i));
       % Maximim Stable Gain T02=MSG
                     MSG(i)=SQS21(i)/(1-SQS11(i))/(1-SQS22(i));
                     DB(i)=10*log10(MSG(i));
   end
%
%-------------------------------------------------------------------------
% Part I: DESIGN OF THE FRONT-END EQUALIZER
%-------------------------------------------------------------------------
 figure
 plot(freq,T01dB);
 title('Maximum Stable Gain of the First Part in dB')
 ylabel('T01 (dB)');xlabel('Actual Frequency')
% Selection of data points from the given data
% between the points nd1 and nd2 out of nd points.
%First Frequency given given by data file
FR1=freq(1)
%Last Frequency given by data file
FR2=freq(nd)
nd1=input('10GHz is at nd1=41,11.2GHz is at nd1=42 Enter start point
nd1=')
nd2=input('20GHz is at nd2=47;22.4GHz is at nd2=48;Enter end point nd2=')
%nd1=42; %This corresponds to 11.23 GHz
%nd2=48;% This corresponds to 22.39 GHz
Fstart=freq(nd1)
Fend=freq(nd2)
```

```
%Define complex number j
j=sqrt(-1);
%
for k=1:(nd2-nd1+1)
            i=nd1-1+k;
% ww(i) is the normalized angular frequency indexed from k=1 to (nd2-nd1)
            ww(k)=freq(i)/Fend;
% FR(k) is the actual frequency
            FR(k)=freq(i);
%
% Scattering Parameters of CMOS Field effect transistor
f11r(k)=S11R(i); f11x(k)=S11X(i); f11(k)=f11r(k)+j*f11x(k);
f12r(k)=S12R(i); f12x(k)=S12X(i); f12(k)=f12r(k)+j*f12x(k);
f21r(k)=S21R(i); f21x(k)=S21X(i); f21(k)=f21r(k)+j*f21x(k);
f22r(k)=S22R(i); f22x(k)=S22X(i); f22(k)=f22r(k)+j*f22x(k);
 end
%
% Step1:
Tflat1=10*log10(T01(nd2))
Tflat1=input('Enter the desired targeted gain  value in dB for Step 1:
Tflat1=')
% For low pass case k=0
ndc=0;h0=0;ntr=0;
xF0=input('hF(0)=0, enter front end equalizer;
[xF(1)=hF(1),xF(2)=hF(2),...,XF(n)=hF(n)]=')
OPTIONS=optimset('MaxFunEvals',20000,'MaxIter',50000);
xF= lsqnonlin('levenberg_TR1',xF0,[],[],OPTIONS,ndc,ww,f11,f12,f21,f22,Tf
lat1,ntr,h0);
 %
 % Here it assummed that there is no transformer in the circuit. That is
 % h(0)=1. This fact is used in the function xtoh(xF)
 if ntr==0
     n=length(xF);
     for i=1:n
         hF(i)=xF(i);
     end
     hF=[hF h0];
 end
 if ntr==1
     n=length(xF);
     for i=1:n
         hF(i)=xF(i);
     end
 end
[ gF,f ] = SRFT_htogLump( hF,ndc );
T1=Gain1(hF,ndc,ww,f11,f12,f21,f22);
T1dB=10*log10(T1);
%
        figure
        plot(ww*Fend,10*log10(T1))
        title('Step 1: Optimized Gain with Tflat1=2.1 dB')
        xlabel('Actual Frequency')
        ylabel(' T1(dB)')
%
```

```
% Computation of the actual element values of the front-end
            aF=gF+hF;bF=gF-hF;
f0=Fend; R0=50;
[ CTF, CVF ] = Synthesis_ImpedanceBased( aF,bF,ndc,R0,f0 );
Plot_Circuit4(CTF,CVF)
%--------------------------------------------------------------------------
% Part II: DESIGN OF THE BACK-END EQUALIZER
%--------------------------------------------------------------------------
    SF2=reflection(hF,ndc,ww,f11,f12,f21,f22);
% output reflectance of the active device when it is terminated in SF=hF/
gF
%
% Step1:
figure
plot(freq,T02dB);
 title('Maximum Stable Gain of the Second Part in dB')
 ylabel('T02 (dB)');xlabel('Actual Frequency')
 display('Suggested maximum flat gain level for the second part of the
design')
Tflat2=10*log10(T02(nd2))
Tflat2=input('Enter the desired targeted gain  value in dB for Step 2:
Tflat2=')
    k=0;
    xB0=input('hB(0)=0, enter back end equalizer;
[xB(1)=hB(1),xB(2)=hB(2),...,XB(n)=hB(n)]=')
    OPTIONS=optimset('MaxFunEvals',20000,'MaxIter',50000);
    xB= lsqnonlin('levenberg_TR2',xB0,[],[],OPTIONS,k,ww,T1,SF2,f11,f12,
f21,f22,Tflat2,ntr,h0);
%
 if ntr==0
    hB=xB;
    hB=[hB h0];
 end
 if ntr==1
    hB=xB;
 end
%
 [ gB,fB ] = SRFT_htogLump( hB,ndc );
T2=Gain2(hB,ndc,ww,T1,SF2,f11,f12,f21,f22);
T2dB=10*log10(T2);
%
   figure (3);
   plot(ww*Fend,T2dB,ww*Fend,T1dB);
   title('Step 2: Final Gain Performance of the Amplifier')
   ylabel('T2(dB)')
   xlabel('Actual Frequency')
   legend('Gain of Part II','Gain of Part I')
%
   aB=gB+hB;bB=gB-hB;
f0=Fend;R0=50;
[ CTB, CVB ] = Synthesis_ImpedanceBased( aB,bB,ndc,R0,f0 );
figure
Plot_Circuit4(CTB,CVB)
%
```

```
% Investigation on the stability of the amplifier:
%
% Stability Check at the output port:
        SB11=hoverg(hB,k,ww);%input ref. Cof. Back-end from hB(p)
        SL=SB11;
        ZL=stoz(SL);
        RL=real(ZL);
        Sout=SF2;
        Zout=stoz(Sout);
        Rout=real(Zout);
        StabilityCheck_output=RL+Rout;
%
%Stability Check at the input port:
        SF=hoverg(hF,k,ww);
        SG=SF;
        ZG=stoz(SG);
        RG=real(ZG);
[Sin]=INPUT_REF(SB11,f11,f12,f21,f22);
            Zin=stoz(Sin);
            Rin=real(Zin);
StabilityCheck_input=RG+Rin;
```

***Program List 5.67: function** levenberg_TR1*

```
function Func=levenberg_TR1(x,ndc,w,f11,f12,f21,f22,TS1,ntr,h0)
%Design with no transformer. h(0)=0 case.
% This function is the error function to minimize error Func(i)
%    Inputs:
%    x: the unknown. It includes coefficients of numerator polynomial h(i)
%    in standard MatLab form.
%    ndc: total number of transmission zeroes at DC.
%    w(i): MatLab array which includes all the measurement frequencies for
%    the FET fij
%    fij: Scatering parameters for the FET under conideration
%    TS1: Target flat gain level for the front end stage [F]
%    ntr: Control flag to make design with or without transformer
%    h0: last coefficient of h(p). h0=0 corresponds transformerless design
Na=length(w);
%
        if ntr==0
            n=length(x);
    % For Lowpass case with no transformer
        h(1)=h0;
        for j=1:n
            h(j)=x(j);
        end
        h=[h h0];
        end
   if ntr==1
       n=length(x);
       h(1)=x(n);
       for j=1:n
           h(j)=x(j);
       end
```

```
        end
        T1=Gain1(h,ndc,w,f11,f12,f21,f22);
                        for i=1:Na
                                Func(i)=10*log10(T1(i))-TS1;
                        end
end
```

Program List 5.68: function `levenberg_TR2`

```
function Func=levenberg_TR2(x,ndc,w,T1,SF2,f11,f12,f21,f22,TS2,ntr,h0)
%Design with no transformer. h(0)=0 case.
% Optimization function for the second step
if ntr==0
        n=length(x);
    % For Lowpass case with no transformer
        for j=1:n
            h(j)=x(j);
        end
        h=[h h0];
end
  if ntr==1
      for j=1:n
          h(j)=x(j);
      end
  end
%-----------------------------------------------------------------------
%
T2=Gain2(h,ndc,w,T1,SF2,f11,f12,f21,f22);
Na=length(w);
                    for i=1:Na
                        Func(i)=10*log10(T2(i))-TS2;
                    end
end
```

Program List 5.69: function `Gain1`

```
function T1=Gain1(h,ndc,w,f11,f12,f21,f22)
%Gain1 of step 1. With front end equalizer
%        INPUTS:
                        % h: Coefficients of the h(p) polynomial for
                        % the iput equilezer
                        %ndc=0 for lowpass transformerless design
                        % w(i): Array, Normalized angular frequency
                        %fij: Scattering paramaeters of the active
                        %device. In the present case 0.18 Micron Si
                        %CMOSFET...
  %                      %
  %        OUTPUT: T1; gain of the frist satge......
Na=length(w);
[ g,f ] = SRFT_htogLump( h,ndc );
%Compute the complex variable p=jw
j=sqrt(-1);
for i=1:Na
    p=j*w(i);
```

```
      hval=polyval(h,p);
      gval=polyval(g,p);
      S22=hval/gval;
      fval=p^ndc;
      S21=fval/gval;
      MS21=abs(S21);
      MS21SQ=MS21*MS21;
      Mf21(i)=abs(f21(i));
      Mf21SQ(i)=Mf21(i)*Mf21(i);
      MD(i)=abs(1-S22*f11(i));
      MDSQ(i)=MD(i)*MD(i);
      T1(i)=MS21SQ*Mf21SQ(i)/MDSQ(i);
end
end
```

Program List 5.70: *function* Gain2

```
function T2=Gain2(h,ndc,w,T1,SF2,f11,f12,f21,f22)
%Gain2 of step 2.
%With back-end equalizer
Na=length(w);
[ g,f ] = SRFT_htogLump( h,ndc );
%
%Compute the complex variable p=jw
j=sqrt(-1);
for i=1:Na
    p=j*w(i);
    hval=polyval(h,p);
    gval=polyval(g,p);
    S11=hval/gval;
    fval=p^ndc;
    S21=fval/gval;
    MS21=abs(S21);
    MS21SQ=MS21*MS21;
    MD(i)=abs(1-S11*SF2(i));
    MDSQ(i)=MD(i)*MD(i);
    T2(i)=T1(i)*MS21SQ/MDSQ(i);
end
end
```

Program List 5.71: *function* SRFT_htogLump

```
function [ g,f ] = SRFT_htogLump( h,ndc )
% This function generates g(p) from given h(p) and transmission zeroes at
% DC of order ndc. Hereö all the polynomials are in MatLab format.
%    Inputs:
%            h(p)=h(1)p^n+h(2)p^(n-1)+...+h(n)p+h(n+1)
%            ndc: Total number of transmission zeros at DC
% Note that S21(p)=f(p)/g(p)=p^(ndc)/g(p)
%    Output:
%            G(p^2)=g(p)g(-p)
%            g(p) of S21(p)=f(p)/g(p), S11(p)=h(p)/g(p),
%            S22(p)=[f(p)/f(-p)]*h(-p)/g(p)
% --------------------------------------------------------------------------
```

```
n1=length(h);
n=n1-1;
h_=paraconj(h);
H1=conv(h,h_);
H=clear_oddpower(H1);
[ f,F ] = SRFT_fFLump( n,ndc );
G=vectorsum(H,F);
r=roots(G);
z=sqrt(r);
g1=sqrt(abs(G(1)));
g=g1*real(poly(-z));
end
```

Program List 5.72: *function* SRFT_fFLump

```
function [ f,F ] = SRFT_fFLump( n,ndc )
% This function generates numerator polynomial f(p) of S21(p)0f(p)/g(p)
  ın
% complex p=sigma +jw plane (Lumped element design)
%    Inputs:
%            n: Degree of f(p)
%            ndc: Total number trnasmission zeros at DC
%    Outputs:
%            f(p): Numnerator polynomial of S21(p)=f(p)/g(p) in MatLab
            form.
%            F(p^2)=f(p)f(-p) in MatLab form.
f0=1;
for i=1:ndc
    f0=conv(f0,[1 0]);
end
for i=1:n+1
    f1(i)=0;
end
f=vector_sum(f1,f0);
f_=paraconj(f);
FO=conv(f_,f);
F=clear_oddpower(FO);
end
```

Program List 5.73: *Main Program* Amplifier_Distributed.m

```
% Program Amplifier_Distributed.m
clc
clear
% DESIGN OF A SINGLE STAGE AMPLIFIER USING A 0.18 micron CMOS SILICON
TRANSISTOR
% with Didstributed elements via SIMPLIFIED REAL FREQUENCY TECHNIQUE
% INPUTS:
%              FR(i): ACTUAL FREQUENCIES
%              SIJ: SCATTERING PARAMETERS OF THE CMOS FET.
%              THESE PARAMETERS ARE READ FROM A FILE CALLED nico_1.txt
%%
%
%    for x = 1:10
```

```
%          disp(x)
%    end
%
%                 Tflat1: FLAT GAIN LEVEL OF STEP 1 in dB.
%                 Tflat2: FLAT GAIN LEVEL OF STEP 2 in dB.
%                  nd1: BEGINING OF THE OPTIMIZATION FREQUENCY
%   (Note that f(nd1) is the begining of the passband;
%%
%
%    for x = 1:10
%          disp(x)
%    end
%
%    nd1 is the index of the frequency f(nd1)
%                   nd2: END OF OPTIMIZATION FREQENCY
%  (Note that f(nd2) is the end of the passband;
%    nd2 is the index of the frequency f(nd2)
%                  q: number transmission zeros at DC for
%                  k: total number of cascaded section
% Notes: f(lmbda)=(lmbda)^q(1-lmbda^2)^k/2
% In the optimization, all the frequencies are normalized with respect to
% upper   edge of the passband.In other words, wc2=1 is fixed.
% All the polynomials are in MatLab format.
% No Transformer is employed in the equlizers.
% Therefore, hF(nF+1)=0
% Front end equlizer is described by
%          hF=[hF(1) hF(2)...hF(nF) 0]; Standard MatLab form
% Back-end equalizer is described by
%          hB=[hB(1) hB(2)...hB(nB) 0]
%
% OUTPUT: FRONT & BACK END EQUALIZERS ARE GENERATED EMPLOYING SRFT
%     Step 1:
%     hF: Optimized Front End Equalizer;
%     T1: Gain of the First Step in dB
%     [ZF]:Normalized Cascaded Characteristic impedances of the Front-End
%     [ZFActual]:Actual Characteristic impedances
%     Step 2:
%     hB: Optimized Back End Equalizer
%     T2: Gain of the Second Step in dB
% [ZB]:Normalized Cascaded Characteristic impedances of the Back-End
% [ZFActual]:Actual Characteristic impedances
%
%DESIGN OF A MICROWAVE AMPLIFIER
%    READ THE TRANSISTOR DATA from newcmos.txt
load newcmos.txt
display('nd=Total number of sampling points for the S-Par measurements')
nd=length(newcmos)
%
%    EXTRACTION OF SCATTERING PARAMETERS FROM THE GIVEN FILE newcmos.txt
for i=1:nd
   % Read the Actual Frequencies from the file newcmos.txt
                    freq(i)=newcmos(i,1);
                       % Read the real and the imaginary parts of the
scattering parameters
```

```
    % from the data file nico_1.txt
                        S11R(i)=newcmos(i,2);
                        S11X(i)=newcmos(i,3);
                        %
                        S12R(i)=newcmos(i,4);
                        S12X(i)=newcmos(i,5);
                        %
                        S21R(i)=newcmos(i,6);
                        S21X(i)=newcmos(i,7);
                        %
                        S22R(i)=newcmos(i,8);
                        S22X(i)=newcmos(i,9);
%
% Get rid off the numerical errors due to extraction of the scattering
% paramters from cadnace...(Vdec extarction
                                        if S11R(i)>=1;
                                        S11R(i)=0.99999999;
                                        S11X(i)=0.0;
                                        end
  end
%
display('Begining of the frequency measurements freqn(1) for
S-Parameters')
freq(1),
display('End of frequency measurements freq(nd)')
freq(nd),
%    CONSTRUCT THE APMLITUDE SQUARES OF THE SACTTERING PARAMETERS
            for i =1:nd
                        SQS21(i)=S21R(i)*S21R(i)+S21X(i)*S21X(i);
                        %
                        SQS12(i)=S12R(i)*S12R(i)+S12X(i)*S12X(i);
                        %
                        SQS11(i)=S11R(i)*S11R(i)+S11X(i)*S11X(i);
                        %
                        SQS22(i)=S22R(i)*S22R(i)+S22X(i)*S22X(i);
                %
                        S1(i)=1.0/(1-SQS11(i));
                        S2(i)=1.0/(1-SQS22(i));
            end
    % COMPUTATION OF MAXIMUM STABLE GAIN "MSG'
    %
                for i=1:nd
                                if SQS11(i)>=1
                                    SQS11(i)=0.999999999;
                                end
                                %
                                        if  SQS22(i)>=1;
                                        SQS22(i)=0.999999999;
                                        end
%Ideal Flat Gain Levels
    % Step 1:
                T01(i)=SQS21(i)/(1-SQS11(i));
    % Step 2:
                T02(i)=T01(i)/(1-SQS22(i));
```

```
        % Maximim Stable Gain T02=MSG
                    MSG(i)=SQS21(i)/(1-SQS11(i))/(1-SQS22(i));
                    DB(i)=10*log10(MSG(i));
                end
            %
            figure
            plot(freq,DB)
            title('Maximum Stable Gain of the 180 nm CMOS of VDEC of
Tokyo')
            xlabel('Actual Frequency')
            ylabel('MSG(dB)')
            display('i=51;      freq(51)= 1GHz')
            display('i=101;     freq(101)= 10GHz')
                % COMPUTATION OF NORMALIZED INPUT (zin) OUTPUT (zout)
IMPEDANCES
                            j=sqrt(-1);
                            for i=1:nd
                                    if S11R(i)>=1;
                                    S11R(i)=0.999999;
                                    end
                                zin(i)=(1+S11R(i)+j*S11X(i))/
(1-S11R(i)-j*S11X(i));
                                zout(i)=(1+S22R(i)+j*S22X(i))/
(1-S22R(i)-j*S22X(i));
                                rin(i)=real(zin(i));
                                xin(i)=imag(zin(i));
                                rout(i)=real(zout(i));
                                xout(i)=imag(zout(i));
                            end
        % Selection of data points from the given data between the points
nd1and nd2 out of nd points.
                    %First Frequency given given by data file
                    FR1=freq(1)
                    %Last Frequency given by data file
                    FR2=freq(nd)
  nd1=input('Enter start point nd1=')
  nd2=input('Enter end point nd2=')
  display('Your optimization starts at ')
  Fstart=freq(nd1)
  display(' Your optimization ends at')
  Fend=freq(nd2)
                    for ka=1:(nd2-nd1+1)
                        i=nd1-1+ka;
% ww(i) is the normalized angular frequency indexed from ka=1 to
(nd2-nd1)
                        ww(ka)=freq(i)/Fend;
                    %
                    % FR(ka) is the actual frequency
                        FR(ka)=freq(i);
                        %
                        j=sqrt(-1);
                        % Scattering Parameters of CMOS
                                f11r(ka)=S11R(i);
                                f11x(ka)=S11X(i);
```

```
                                    f11(ka)=f11r(ka)+j*f11x(ka);
                                    %
                                    f12r(ka)=S12R(i);
                                    f12x(ka)=S12X(i);
                                    f12(ka)=f12r(ka)+j*f12x(ka);
                                    %
                                    f21r(ka)=S21R(i);
                                    f21x(ka)=S21X(i);
                                    f21(ka)=f21r(ka)+j*f21x(ka);
                                    %
                                    f22r(ka)=S22R(i);
                                    f22x(ka)=S22X(i);
                                    f22(ka)=f22r(ka)+j*f22x(ka);
                    end
%
display('UE has a delay tou=1/4/Fend/kLength')
kLength=input('Enter kLength=')
% Step1:
display('Ideal flat gain leven in dB')
Tflat1=10*log10(T01(nd2))
display(' It is suggested that Tfalt1=6dB')
Tflat1=input('Enter the desired targeted gain  value in dB for Step 1:
Tflat1=')
% For low pass case q=0; n=k total number of cascaded elements
qF=0;%Transmission zeros at DC
    xF0=input('hF(0)=0, enter front end equalizer;
[xF(1)=hF(1),xF(2)=hF(2),...,XF(n)=hF(n)]=')
    OPTIONS=optimset('MaxFunEvals',20000,'MaxIter',50000);
nF=length(xF0);%Total number of UE or cascaded elements
kF=nF;
% Here we assume that n=k;
% in other words, total number of cascaded elements are equal to total
% elements
% number of
    xF= lsqnonlin('Distributed_TR1',xF0,[],[],OPTIONS,kLength,qF,kF,ww,f1
1,f12,f21,f22,Tflat1);
 %
 %
 %          Here it assummed that there is no transformer in the circuit.
 % That is q=0 and h(n+1)=0. Hence,
                        hF=[xF 0];
            qF=0.0;
            T1=Richards_Gain1(hF,kLength,qF,kF,ww,f11,f12,f21,f22);
figure
plot(Fend*ww,10*log10(T1))
title(' T1(db): Gain of the first step')
xlabel(' Actual Frequency')
ylabel('T1 (dB) ')
%Syntehsis By Richards extarctions
[GF,HF,FF,gF]=RichardsSRFT_htoG(hF,qF,kF);
%-----------------------------------------------------------------------
R0=50;
aF=gF+hF;bF=gF-hF;
```

```
[Z_UE,a_new,b_new,CTF,CVF ] = Richard_CompleteImpedanceSynthesis(aF,bF,kF,
qF,1,1/2/pi );
CVFActual=R0*CVF;
Plot_Circuit4(CTF,CVFActual)
%-------------------------------------------------------------------------
 SF2=Richards_reflection(hF,kLength,qF,kF,ww,f11,f12,f21,f22);
% DESIGN OF THE BACK-END EQUALIZER
%
% Step2:
qB=0;
Tflat2=10*log10(T02(nd2))
display(' It is suggested that Tflat2=10.5 dB')
Tflat2=input('Enter the desired targeted gain  value in dB for Step 1:
Tflat2=')
 xB0=input('hB(0)=0, enter back end equalizer;
[xB(1)=hB(1),xB(2)=hB(2),...,XB(n)=hB(n)]=')
 nB=length(xB0);
 kB=nB;%Total number of cascaded sections
 OPTIONS=optimset('MaxFunEvals',20000,'MaxIter',50000);
 xB= lsqnonlin('Distributed_TR2',xB0,[],[],OPTIONS,kLength,qB,kB,ww,T1,SF
2,Tflat2);
 %
 %
            hB=[xB 0];
            T2=Richards_Gain2(hB,kLength,qB,kB,ww,T1,SF2);
%
figure
plot(Fend*ww,10*log10(T2))
title('Gain of the amplifier designed with distributed elements')
xlabel('Actual Frequency')
ylabel('Gain of the amplifier designed with UE s in dB')
[GB,HB,FB,gB]=RichardsSRFT_htoG(hB,qB,kB);
%-------------------------------------------------------------------------
% Synthesis of the back-end equalizer
aB=gB+hB;bB=gB-hB;
[Z_UEB,a_new,b_new,CTB,CVB ] = Richard_CompleteImpedanceSynthesis(aB,bB,k
B,qB,1,1/2/pi );
CVBActual=R0*CVB;
Plot_Circuit4(CTB,CVBActual)
%-------------------------------------------------------------------------
% Print the result over the measured frequencies of the S-Parameters
[wprint,f11 f12 f21 f22]=Scattering_Parameters(newcmos,Fend);
 SF_Print=Richards_reflection(hF,kLength,qF,kF,wprint,f11,f12,f21,f22);
 T1_Print=Richards_Gain1(hF,kLength,qF,kF,wprint,f11,f12,f21,f22);
 T2_Print=Richards_Gain2(hB,kLength,qB,kB,wprint,T1_Print,SF_Print);
 Gain_dB=10*log10(T2_Print);
%
 figure
 plot(Fend*wprint,Gain_dB)
 title('Gain of the Tokyo-VDECamplifier constructed with UEs')
 xlabel('Actual Frequency')
 ylabel('Gain of the amplifier in dB')
```

Program List 5.74: *function* Distributed_TR1

```
function Func=Distributed_TR1(x,kLength,q,k,w,f11,f12,f21,f22,TS1)
%Design with no transformer. h(0)=0 case.
Na=length(w);

%Evaluate the polynomial values for h and g.
% Here use standard MatLab function Polyval. In this case change the
order
% of the coefficients h and g
n=length(x);
      n1=n+1;
    % For Lowpass case with no transformer
      h(n1)=0;
       for j=1:n
           h(j)=x(j);
       end
       T1=Richards_Gain1(h,kLength,q,k,w,f11,f12,f21,f22);
                        for i=1:Na
                               Func(i)=10*log10(T1(i))-TS1;
                        end
```

Program List 5.75: *function* Richards_Gain1

```
function T1=Richards_Gain1(h,kLength,q,k,w,f11,f12,f21,f22)
%Gain1 of step 1. With front end equalizer in Richard domain
% This gain is computed in lmbda domain.
% lmbda=j*OMEGA
% OMEGA=tan(w*tau)
% tau=1/4/w(na)
%        INPUTS:
                          % h: Coefficients of the h(p) polynomial for
                          % the iput equilezer
                          %k=0 for lowpass transformerless design
                          % w(i): Array, Normalized angular frequency
                          %fij: Scattering paramaeters of the active
                          %device. In the present case 0.18 Micron Si
                          %CMOSFET...
%                      %
%        OUTPUT: T1; gain of the frist satge......
Na=length(w);
[G,H,F,g]=RichardsSRFT_htoG(h,q,k);%
%Compute the complex variable lmbda=j*tan(w*tau)
% Note that fe=stap band frequency=1.5*fc2; wc2=ww(nd2)
tau=1/4/1/kLength;
%
j=sqrt(-1);
for i=1:Na
    teta=w(i)*tau;
    omega=tan(teta);
    lmbda=j*omega;
    %
    fval=(-1)^q*(lmbda)^q*(1-lmbda^2)^(k/2);
  %
```

```
    hval=polyval(h,lmbda);
    gval=polyval(g,lmbda);
    S22=hval/gval;
    S21=fval/gval;
    MS21=abs(S21);
    MS21SQ=MS21*MS21;
    Mf21(i)=abs(f21(i));
    Mf21SQ(i)=Mf21(i)*Mf21(i);
    MD(i)=abs(1-S22*f11(i));
    MDSQ(i)=MD(i)*MD(i);
    T1(i)=MS21SQ*Mf21SQ(i)/MDSQ(i);
end
%
end
```

Program List 5.76: function RichardsSRFT_htoG

```
function [G,H,F,g]=RichardsSRFT_htoG(h,q,k)
%This function generates the complete scattering parameters
% from te given h and F in lambda domain
% all the functionas are given in lmbda square.
%    Inputs:
%            given h and q,k
%            where F=(lmbd)^2q*(1-lmbd^2)^k
%    Outputs:
%            G,H,F and g
% Note that F(lmbda)=(-1)^q*(lmbda^2q)*(1-lmbda^2)^k
F=lambda2q_UE2k(k,q);% Generation of F=f(lambda)*f(-lambda)
h_=paraconj(h);
He=conv(h,h_);%Generation of h(lmbda)*h(-lambda)
H=poly_eventerms(He);%select the even terms of He to set H
G=vector_sum(H,F);% Generate G=H+F with different sizes
% Construction of g(lmbda) from h(lmbda)
r=roots(G);
z=sqrt(r);
g1=sqrt(abs(G(1)));
g=g1*real(poly(-z));
end
```

Program List 5.77: function lambda2q_UE2k(k,q)

```
function F=lambda2q_UE2k(k,q)
% This function computes F=(-1)^q(lambda)^2q(1-lambda^2)^k
Fa=cascade(k);
nf=length(Fa);
nq=nf+q;
    for i=1:nq
    F(i)=0.0;
    end
    if q==0;
        F=Fa;
    end
if(q>0)
    for i=1:nf
```

```
            F(i)=((-1)^q)*Fa(i);
    end
    for i=1:q
        F(nf+i)=0.0;
    end
end
end
```

Program List 5.78: function cascade

```
function F=cascade(k)
% This function generates the F=f*f_ polynomial of
% cascade connections of k UEs in Richard Domain.
% F=(1-1ambda^2)^k
% F=F(1)(Lambda)^2n+...F(k)(lambda)^2+F(k+1)
F=[1]; %Unity Polynomial;
UE=[-1 1];% Single Unit-Element; [1-(lambda)^2]
for i=1:k
    F=conv(F,UE);
end
end
```

Program List 5.79: function paraconj

```
function h_=paraconj(h)
% This function generates the para-conjugate of a polynomial h=[]
na=length(h);
n=na-1;
sign=-1;
h_(na)=h(na);
for i=1:n
    h_(n-i+1)=sign*h(n-i+1);%Para_conjugate coefficients
    sign=-sign;
end
end
```

Program List 5.80: function Richard_CompleteImpedanceSynthesis

---

```
function [Z_UE,a_new,b_new,CTF,CVF ] = Richard_CompleteImpedanceSynthesis
(a,b,k,q,R0,f0 )
% This function carries out complete synthesis in lambda domain
% Inputs:
%       k; Total number of cascaded UEs
%       q: Total number of high-pass elements as series open-stubs,
parallel shorted-stubs
%       Zin=a/b
%       R0: Normalization resistance
%       f0: Normalization frequency but it must be fixed at f0=1/2/pi
% Outputs:
%       Z_UE: Chararcteristic impedances of the Unit Elements
%       a_new,b_new: Z_new=a_new/b_new is the remaning impedance after
%       extraction of k-UE.
%       CT,CV: Circuit codes and values of the ladder synthesis
%
```

```
ndc=q;
%
if k==0;a_new=a;b_new=b;
    Z_UE='k=0. There is no UE in the synthesis'
end
%
% ------ Definition of Algorithmic zero ----------------------------------
zero=1e-10;
zeroA=1e-10;
aux=[a b];
Kmax=max(aux);
a1=a/Kmax;b1=b/Kmax;na=length(a);
    if q>0;
    zeroA=a1(na);
    end
%
if zero<zeroA;zero=zeroA;end
% ------ End of Definition of Algorithmic zero -------------------------
if abs(a(1))<zero;[a,b]=Richard_ImmittanceCorrection(a,b,k,q);end
if abs(b(1))<zero;[b,a]=Richard_ImmittanceCorrection(b,a,k,q);end
%-----------------------------------------------------------------------
n=length(a)-1;
if k<=n
    if k>0
%High precision-impedance based-Richard Extraction:
% [Z,a_new,b_new] = Richard_HighPrecisionUEExtraction( k,q,a,b );
% [Z,a_new,b_new] = Richard_RoughUEExtraction( k,q,a,b );
  [Z,a_new,b_new] = Richards_CameronExtractions( k,q,a,b );
   Z_UE=Z;
%-----------------------------------------------------------------------
    end
%-----------------------------------------------------------------------
a1=a_new;b1=b_new;
na=length(a_new);nb=length(b_new);
% a_new(na) might be zero and b_new(na) might be zero. In this case
degree
% reduction in "q" occours.
if abs(a_new(na))<zero;
if abs(b_new(nb))<zero;
    clear a_new;clear b_new;
    for i=1:na-1
        a_new(i)=a1(i);
        b_new(i)=b1(i);

    end
    q=q-1;% Degree reduction in q
        Attention=('Degree reduction occours in q=q-1')
        q1=q
end
end
% End of degree reduction loops
%-----------------------------------------------------------------------
end
%
```

```
nc=n;
if nc>k
%
if abs(a_new(1))<zero;[a_new,b_new,q]=Check_immitance(a_new,b_new,q);end
if abs(b_new(1))<zero;[b_new,a_new,q]=Check_immitance(b_new,a_new,q);end
    if abs(a_new(1))>zero;
        if abs(b_new(1))>zero;
            [a_new,b_new,q]=Check_immitance(a_new,b_new,q);
        end
    end
    %
[ CT CV ] = Synthesis_ImpedanceBased( a_new,b_new,q,R0,f0 );
end
if k==n
    CT=('k=n; Therefore, synthesis is completed only with cascade
connection of n-unit elemensts')
    CV=CT;
end
% Plot the remainin synthesized Circuit in lambda domain
% if n>k; Plot_Circuitv4(CT,CV);end
%---------------------------------------------------------------------
% Plot of the complete commensurate transmission line circuit
nue=length(Z_UE);
for i=1:nue
    CT1(i)=20; CV1(i)=Z_UE(i);
end
if n>k
[ CTD,CVD ] = RichardsLadderTo_TrLineTopology( CT,CV );
CTF=[CT1 CTD];CVF=[CV1 CVD];
end
if n==k
  CTF-[CT1 9]; CVF=[CV1 a_new/b_new];
end
%Plot_Circuit4(CTF,CVF)
end
```

Program List 5.81: function Richard_ImmittanceCorrection

```
function [a1,b1]=Richard_ImmittanceCorrection(a,b,k,q)
% In this function original forms of a(lambda) and b(lambda) are used as
% input. Here, the last coefficients of a(lambda) and b(lambda) are not one.
% Therefore, we compute a0:
%---------------------------------------------------------------------
% Revision on January 20, 2013
 % [a0,A,B,k1,q1] = Richard_a0ofAandB( a,b,k,q );

%---------------------------------------------------------------------
% June 3, 2013 Revision
% Revision July 27, 2013
% Generate the Richard Numerator of the even part R(lambda^2)=A/B with high
% precision:
if k>0
F=Richard_kq(k,q);
% Compute the leading coefficient A0:
```

```
[A1,B1]=even_part(a,b);
nA=length(A1);
nB=length(B1);B0=B1(nB);
B=B1/B0;A2=A1/B0;
A0=abs(A2(nA-k-q));
A=A0*F;A=fullvector(nB,A);
[a1,b1]=RtoZ(A,B);
%
b0=b1(nB);b1=b1/b0;a1=a1/b0;
end
if k==0
[a1,b1]=General_immitCheck(q,0,a,b)
    Note='k=0. Therefore, we use immittance correction algorithm called
General_immitCheck'
    na=length(a1);if a1(na)<1e-12;a1(na)=0;end

end
end
```

Program List 5.82: function `Richard_kq(k,q)`

```
function F=Richard_kq(k,q)
% This function computes F=(-1)^q(lambda)^2q(1-lambda^2)^k
Fa=cascade(k);
nf=length(Fa);
nq=nf+q;
    for i=1:nq
    F(i)=0.0;
    end
    if q==0;
        F=Fa;
    end
if(q>0)
    for i=1:nf
        F(i)=((-1)^q)*Fa(i);
    end
    for i=1:q
        F(nf+i)=0.0;
    end
end
end
```

Program List 5.83: function `even_part`

```
function [A,B]=even_part(a,b)
% Generate even part R(p^2)=A(p^2)/B(p^2) of a given immitance function
F(p)=a(p)/b(p)
% Notice that A(p^2) and B(p^2) are specified as an even polynomials
% Computation of Numerator of Even Part
na=length(a);
nb=length(b);
    sign=-1;
    for i=1:na
```

```
        sign=-sign;
        a_(na-i+1)=sign*a(na-i+1);
    end
    sign=-1;
    for i=1:nb
        sign=-sign;
        b_(na-i+1)=sign*b(nb-i+1);
    end
    Num_Even=(conv(a,b_)+conv(a_,b))/2;
      A=clear_oddpower(Num_Even);
    n_Even=length(Num_Even);
        BB=conv(b,b_);
    B=clear_oddpower(BB);
end
```

Program List 5.84: function RtoZ

```
function [a,b]=RtoZ(A,B)
% This MatLab function generates a minimum function Z(p)=a(p)/b(p)
%       from its even part specified as R(p^2)=A(p^2)/B(p^2)
%       via Bode (or Parametric)approach
% Inputs: In p-domain, enter A(p) and B(p)
% A=[A(1) A(2) A(3)...A(n+1)]; for many practical cases we set A(1)=0.
% B=[B(1) B(2) B(3)...B(n+1)]
% Output:
%       Z(p)=a(p)/b(p) such that
%       a(p)=a(1)p^n+a(2)p^(n-1)+...+a(n)p+a(n+1)
%       b(p)=b(1)p^n+b(2)p^(n-1)+...+b(n)p+a(n+1)
% Generation of an immitance Function Z by means of Parametric Approach
% In parametric approach Z(p)=Z0+k(1)/[p-p(1)]+...+k(n)/[p-p(n)]
% R(p^2)=Even{Z(p)}=A(-p^2)/B(-p^2) where Z0=A(n+1)/B(n+1).
%
% Given A(-p^2)>0
% Given B(-p^2)>0
%
BP=polarity(B);%BP is in w-domain
AP=polarity(A);%AP is in w-domain
% Computational Steps
% Given A and B vectors. A(p) and B(p) vectors are in p-domain
% Compute poles p(1),p(2),...,p(n)and the residues k(i) at poles
p(1),p(2),...,p(n)
[p,k]=residue_Z0(AP,BP);
%
% Compute numerator and denominator polynomials
Z0=abs(A(1)/B(1));
[num,errorn]=num_Z0(p,Z0,k);
[denom,errord]=denominator(p);
%
a=num;
b=denom;
end
```

Program List 5.85: *function* General_immitCheck

```
function [a1,b1]=General_immitCheck(ndc, W,a,b)
% Given F(p)=a(p)/b(p)as a minimum function
%
% Find zeros of transmissions at dc
% Re-compute a(p) and b(p) to make F(p) minimum reactance
%[ndc] = DCZeros_Evenpart( a,b );
eps_zero=1e-8;
[A1,B1]=even_part(a,b);% Outcome is even polynomials in p^2
na=length(a);
if norm(W)>0;nz=length(W);end
if norm(W)==0;nz=0;end
%
%for i=1:na; zero(i)=0;end
A0=A1(na-ndc-2*nz);
A=1;p=[1 0];% p^2
if ndc>0
for i=1:ndc
    A=conv(A,p);%p^2*ndc
end
end
A=A0*A;
%--------------------------------------------------------------------
% Generate polynomial Apoly constructed by means of
% finite transmission zeros W(i)
%
%
        Apoly=[1];
if (nz>=1)
        for i=1:nz
        Wi2=W(i)*W(i);
        Cpoly=[1 Wi2];% (p^2+wa^2) in p^2
        Dpoly=conv(Cpoly,Cpoly);%(p^2+wa^2)^2 in p^2
        Apoly=conv(Apoly,Dpoly);
        end
end
%
%--------------------------------------------------------------------
A2=conv(A,Apoly);
nb=length(B1);for j=1:nb;zero(j)=0;end;A3=vector_sum(zero,A2);
[a1,b1]=RtoZ(A3,B1);
if abs(a1(na))<eps_zero;a1(na)=0;end
if abs(a1(1))<eps_zero;a1(1)=0;end
%--------------------------------------------------------------------
%
end
```

Program List 5.86: *function* Check_immitance

```
function [a1,b1,ndc]=Check_immitance(a,b,ndc)
% December 12, 2012
% Upgradded version of the old Check_immittance function with fixed ndc
% Given F(p)=a(p)/b(p)as a minimum function
```

```
% F(p) is free of finite jW zeros but it may have zeros at dc
% Find zeros of transmissions at DC (Actually, DC Tr. zeros is given at the
% input in the upgraded version of the function)
% Re-compute a(p) and b(p) to make F(p) minimum reactance
na=length(a);
if na>1
%[ndc] = DCZeros_Evenpart( a,b );
[A1,B1]=even_part(a,b);
for i=na
    A(i)=0.0;
end
A(na-ndc)=A1(na-ndc);
[a1,b1]=RtoZ(A,B1);
end
if na==1
a1=a;
b1=b;
ndc=0;
end
end
```

Program List 5.87: function Richards_reflection

```
function SF2=Richards_reflection(h,kLength,q,k,w,f11,f12,f21,f22)
% Computation of the back-end reflectance of the active in Ricahrd domain
% when it is terminated in S22F=hF/gF
%Design with no transformer. h(0)=1 case.
% INPUTS
%                           h(lmbda): Numerator polynomial (Coefficients h
                            array)
%                           w: Normalized angular frequencies; array of size Na
%                           fij: Scattering paramaters of the active device
% OUTPUT:
%                           SF2: Back-end reflection Coefficients looking at
%                           the back-end of the active device.
Na=length(w);
%
n1=length(h);
%Compute the complex variable lmbda=j*tn(w*tau)
j=sqrt(-1);
Na=length(w);
[G,H,F,g]=RichardsSRFT_htoG(h,q,k);%
%Compute the complex variable lmbda=j*tan(w*tau)
% Note that fe=stap band frequency=1.5*fc2; wc2=ww(nd2)
tau=1/4/1/kLength;
%
j=sqrt(-1);
for i=1:Na
    teta=w(i)*tau;
    omega=tan(teta);
    lmbda=j*omega;
    %
    hval=polyval(h,lmbda);
    gval=polyval(g,lmbda);
```

```
        S22=hval/gval;
            SF2(i)=f22(i)+f12(i)*f21(i)*S22/(1-f11(i)*S22);
end
```

Program List 5.88: *function* Distributed_TR2

```
function Func=Distributed_TR2(x,kLength,q,k,w,T1,SF2,TS2)
%Design with no transformer. h(0)=0 case.
% Inputs:
%        x: The unknown vector which includes h(i) of the back-end
%        matching
%        network.
%        q: Count of the DC transmision zeros in lambda domain.
%        k: Count of the cascade connected tUEs.
%        w: Normalized angular frequency array vector subject to
%        optimization.
%        T1: Gain of the input match (Front-end) as an array over w
%        SF2:Unit normalized reflection coefficient of the FET when
%        from-end
%        matching is present.
%        TS2: Traget flat gain level of the overall amplifier.
% Output:
%        Func: Error function
%-----------------------------------------------------------------------
% Optimization function for the second step
n=length(x);
        n1=n+1;
        h(n1)=0;
        for j=1:n
            h(j)=x(j);
        end
%
T2=Richards_Gain2(h,kLength,q,k,w,T1,SF2);
Na=length(w);
                        for i=1:Na
                            Func(i)=10*log10(T2(i))-TS2;
                        end
end
```

Program List 5.89: *function* Richards_Gain2

```
function T2=Richards_Gain2(h,kLength,q,k,w,T1,SF2)
%Gain2 of step 2 in Richard domain.
%With back-end equalizer
Na=length(w);
%Na=length(w);
[G,H,F,g]=RichardsSRFT_htoG(h,q,k);%
%Compute the complex variable lmbda=i*tan(w*tau)
tau=1/4/1/kLength;
j=sqrt(-1);
for i=1:Na
    teta=w(i)*tau;
```

```
        omega=tan(teta);
        lmbda=j*omega;
        %
        fval=(-1)^q*(lmbda)^q*(1-lmbda^2)^(k/2);
    %
        hval=polyval(h,lmbda);
        gval=polyval(g,lmbda);
        S11=hval/gval;
        S21=fval/gval;
        MS21=abs(S21);
        MS21SQ=MS21*MS21;
        MD(i)=abs(1-S11*SF2(i));
        MDSQ(i)=MD(i)*MD(i);
        T2(i)=T1(i)*MS21SQ/MDSQ(i);
end
end
```

Program List 5.90: Microstrip_CharateristicImpedance.m

```
% Main Program Microstrip_CharateristicImpedance.m
clear
% Inputs:
% er : relative dielectric constant
% W : width of track
% h : thickness of the substrate
%
er=input('Enter relative permittivity: ');
W=input('Enter width of the substrate (mm): ');
h=input('Enter thickness of the substrate (mm): ');
if (W/h<1)
e_eff=(er+1)/2+((er-1)/2)*((1/(sqrt(1+(12*h/W))))+0.04*(1-W/h)^2);
else
    e_eff=((er+1)/2)+((er-1)/2)*(1/sqrt(1+12*h/W));
end
%
if (W/h<1)
    Z_0=60/sqrt(e_eff)*log(8*h/W+W/4*h);
else
  Z_0=(120*pi/sqrt(e_eff))*(1/(W/h+1.393+0.677*log((W/h)+1.444)));
end
%
[s,errmsg]=sprintf('Z_0 = %d',Z_0);
disp(s);
```

Program List 5.91: **Design of a Microstrip Line for a specified Characteristic Impedance** Z_0**, substrate** *thickness h and relative di-electric constant (or permittivity)* ε_r

---

```
% Main program Microstrip_design.m
% Inputs:
%        Z0: Fixed Characteristic Impedance of the Microstrip Line
%        epsr: Relative permittivity of the substrate
%        h:Thickness of the substrate
% For a Given Microstrip characteristic impedance Z0, find w/h
```

```
close all
clc
clear
%
Z0=input('Characteristic Impedance of the Microstrip Line in ohm  Z0=');
epsr=input('Relative permittivity of the substrate epsr=');
hmm=input('Enter substrate thickness in millimeter h=');
h=hmm/1000;
% Initialize x=W/h
x0=1.0;
% Optimization via Minimax Algorithm
 f=@(x)Microstrip(x,epsr,Z0)
[x,fval] = fminimax(f,x0);
% Computation of a microstrip Characteristic impedance for given width
  and epsr
% See the book Fields and Waves in Communication Electronics by Simon Ramo,
% John R. Whinnery, Theodore Van Duzer, Third Edition, John Wiley,1994,
% pp.412
width=x*h
woverh=x;
if woverh>0.06
Z00=120*pi/(x+1.98*(x)^0.172);
epseff=1+((epsr-1)/2)*(1+1/sqrt(1+10/x))
Z_Microstrip=Z00/sqrt(epseff)
end
if woverh<0.06
    'Microstrip impedance can not to be realized'
End
```

Program List 5.92: **Error function to compute the actual x=(W/h) ratio of a microstrip line for a specified characteristic impedance Z_0 and permittivity ((epsilon) r in subscript)see symbol for author original file**

```
function error=Microstrip(x,epsr,Z0)
% Inputs
%       epsr: relative permittivity of the substrate
%       Z0: Desired characteristic impedance to be realized
%       x:the unknown the ratio of x=w/h
%          where (w) is the width and (h) is the thickness
%------------------------------------------------------------------------
Z00=120*pi/(x+1.98*(x)^0.172);
epseff=1+((epsr-1)/2)*(1+1/sqrt(1+10/x));
Z_Microstrip=Z00/sqrt(epseff);
error=abs(Z0-Z_Microstrip);
end
```

Program List 5.93: Main Program INDviaTRL

```
% Main Program: INDviaTRL
% Inputs:
%       LA: actual impedance to be realized by a single transmission line
%       f0: actual operating frequency
%       h: substrate thickness
%       epsr: Relative permittivity
```

```
%          m: pass band multiplier
% Outputs:
%          l: find physical length (meter)
%          Z0: Characteristic impedance (ohm)
%          W: Actual line width (meter)
%          Cp: Parasitic capacitance of the given actual inductor LA
%
close all
clc
clear
%
% Inputs:
LA= 1.2694e-008;f0=530e6;epsr=3.66;h=2e-3;m=4
%
% Outputs
tau=1/4/m/f0;
c=3*1e8;
mu0=4*pi*1e-7;
eps0=1e-9/36/pi;
v0=1/sqrt(mu0*eps0)
vsub=v0/sqrt(epsr);
length=vsub/4/m/f0;
width=mu0*(h/LA)*length;
W=30*pi*h/sqrt(epsr)/m/f0/LA;
Z0=4*m*f0*LA;
Chr_imp=120*pi*h/W/sqrt(epsr)
Cp=epsr*eps0*W*length/2/h
Cap_Parasitic=LA/2/Z0/Z0
Cpt=1/32/m/f0/m/f0/LA
% Generate input impedance of the transmission line as it is terminated
in
% 50 ohms:
fs1=330e6;fs2=10060e6;
N=10000;
df=(fs2-fs1)/N;DW=2*pi*df;
f=fs1;w=2*pi*f;
RN=50;
j=sqrt(-1);
for i=1:N
    FA(i)=f/1e6;
    p=j*w;
    YN=1/RN;Y1=p*Cp+YN;
    Z2=p*LA+1/Y1;Y3=p*Cp+1/Z2;
    ZinLump=1/Y3;
    RinLump(i)=real(ZinLump);
    XinLump(i)=imag(ZinLump);
    ZinTRL=Z0*(RN+Z0*j*tan(w*tau))/(Z0+RN*j*tan(w*tau));
    RinTRL(i)=real(ZinTRL);
    XinTRL(i)=imag(ZinTRL);
    w=w+DW;
    f=f+df;
end
figure (1)
plot(FA,RinLump,FA,RinTRL)
```

```
title('Comparison of real parts');xlabel('Actual Frequency (MHz)')
ylabel('RinLump & RinTRL');legend('RinLump','RinTRL')
%
figure (2)
plot(FA,RinLump,FA,RinTRL)
title('Comparison of imaginary parts');xlabel('Actual Frequency (MHz)')
ylabel('XinLump & XinTRL');legend('XinLump','XinTRL')
```

Program List 5.94: Based on the lumped proto-type, matching network design with mixed elements

---

```
% Main Program Mixed_Design.m
% Inputs:
% Element Values:
close all
clc
clear
%
RGEN = [
     0        50       0
    330       50       0
    350       50       0
    370       50       0
    390       50       0
    410       50       0
    430       50       0
    450       50       0
    470       50       0
    490       50       0
    510       50       0
    530       50       0
   2000       50       0];
%
Load = [
000      19.00     -0.000
330      17.61     -04.63
350      14.50     -03.67
370      18.70     -11.30
390      22.00     -10.11
410      09.16     -13.96
430      10.23     -17.94
450      10.40     -15.89
470      17.18     -04.85
490      14.33     -05.11
510      13.36     -12.11
530      11.04     -14.03
2000      7.00     -17.00];
%
A1=RGEN; % Data Matrix For Gen_Data1
A2=Load; % Data Matrix Load_Dat1
%
NA1=length(A1);
NA2=length(A2);
%
```

```
% Resistive Generator Data for Single Matching Problem
for i=1:NA1
    FAG(i)=A1(i,1)*1e6;WAG(i)=2*pi*FAG(i);
    RAG1(i)=A1(i,2);
    XAG1(i)=A1(i,3);
end
% Actual Measurement for the Input Load Data 3; Single Matching
for i=1:NA2
    FAL(i)=A2(i,1)*1e6;WAL(i)=2*pi*FAL(i);
    RAL1(i)=A2(i,2);
    XAL1(i)=A2(i,3);
end
%
figure
plot(FAL,RAL1)
figure
plot(FAL,XAL1)
%
%
m=2;
f0=530e6;fs1=06;fs2=3*f0;epsr=3.66
ws1=2*pi*fs1;ws2=2*pi*fs2;KFlag=1;
%------------------------------------------------------------------------
CAk=2.4115e-012;LAk=9.3483e-009;%Lumped Element Prototype Resonance
Circuit
CA1=2.3916e-011;CA3= 1.4731e-011;LA2=1.2694e-008;% Lumped element CLC
CA=2.1725e-011;CB=1.2540e-011;Z0=53.8226;tau= 2.3585e-010%Physical TRL
Model
%CA=2.1839e-011;CB=1.2649e-011;Z0=59.763;tau= 2.3585e-010;%Parameters of
%almost equivalent circuit
%------------------------------------------------------------------------
%
%Z_fixed=0.0;
%[Z0,tau,length,C,Ca,CT,CA,CB]=CLCPItoTRL(f0,m,CA1,LA2,CA3,epsr,Z_fixed);
Nprint=1001;
cmplx=sqrt(-1);
DW=(ws2-ws1)/(Nprint-1);
w=ws1;
for j=1:Nprint
            WA(j)=w;
            p=cmplx*w;
            [RG,XG]=Line_Impedance(w,WAG, RAG1, XAG1,KFlag);
            [RL,XL]=Line_Impedance(w,WAL, RAL1, XAL1,KFlag);
            ZL=complex(RL,XL);
            ZG=complex(RG,XG);
% Impedance of the resonance circuit:
            YRes=p*CAk+1/p/LAk;
            ZRes=1/YRes;
% Transmission Line input impedance
        %   Load impedance of the Line: CB//ZL
            ZLoad=ZRes+ZL;
            YLA=p*CA+1/ZLoad;% Line load: CB//ZL
            ZLA=1/YLA;% TRL Load
            Num=ZLA+cmplx*Z0*tan(w*tau);
```

```
            Den=Z0+cmplx*ZLA*tan(w*tau);
            ZTRL=Z0*Num/Den;% Input impedance of the loaded line in ZLB
            Yin=1/ZTRL+p*CB;
            Zin=1/Yin;
            SinTrl=(Zin-conj(ZG))/(Zin+ZG);
            GainTrl=1-abs(SinTrl)*abs(SinTrl);
             T_Double=GainTrl;
             TATrl(j)=10*log10(T_Double);
             % Computation of gain with lumped elements
             YC1=p*CA1+1/(ZLoad);
             ZL2=p*LA2+1/YC1;
             YC3=p*CA3+1/ZL2;
             ZinLmp=1/YC3;
             SinLmp=(ZinLmp-conj(ZG))/(ZinLmp+ZG);
             GainLmp=1-abs(SinLmp)*abs(SinLmp);
             TALmp(j)=10*log10(GainLmp);
     w=w+DW;
end
   %
   FA=WA/2/pi/1e6;

   %
         figure
         plot(FA,TATrl,FA,TALmp)
         title('Test Mixed Element Design')
         xlabel('Actual Frequency')
         ylabel('Topology Gain in dB')
         legend('TRL','Lumped')
```

Program List 5.95: function CLCPItoTRL

```
function [ Z0,tau,length,C,Ca,CT,CA,CB] = CLCPItoTRL(f0,m,CA1,LA2,CA3,eps
ilonr,Z_fixed)
%Generate almost equivalent CT-TRL-CT from the given C1-L2-C3 PI section
%-------------------------------------------------------------------------
%    Inputs: Element values of the lumped PI Section
%            f0: Actual normalization frequency
%            m: Integer which specifies the stop-band frequency as
%            fs2=m*f0
%               Note m cannot be less 2. Perhaps it is chosen as 2, 3 or 4
%            RN: Normalization Resistance. It may be 50 ohm.
%            CA1: Input Capacitor
%            LA2: Mid Inductor
%            CA3: Output Capacitor
%            epsilon_r: Di-electric constant of the substrate
%            Z_fixed: fixed characteristic impedance of the line.
% Note: if Z_fixed==0, characteristic impedance Z0 is computed as a
% function of tau.
% If Z_fixed>0 then, tau is adjusted to yield the desired characteristic
% impedance Z_fixed.
%    Output:
%            Z0: Characteristic Impedance of the transmission Line
%            tau: Delay length of the transmission line
%            length: Physical length of the transmission line
```

```
%               C-min(C1,C3)
%               Ca: Residual Capacitor for the PI section to make it
%               symmetrical
%               CT: Symmetrical loading capacitor of the transmission line
%               (CT_TRL_CT)
%               CA: Equivalent left loading capacitance of the line
%               CB: Equivalent left loading capacitance
% ----------------------------------------------------------------------
% Normalization:
fs2=m*f0;
if Z_fixed==0
    C=min(CA1,CA3);% find symmetrical Capacitors C
    Ca=max(CA1,CA3)-C;% Generate residual capacitor Ca
    tau=1/4/fs2;% Compute delay length tau
    Z0=2*pi*f0*LA2/sin(2*pi*f0*tau);
    CT=(cos(2*pi*f0*tau)+(2*pi*f0)*(2*pi*f0)*LA2*C-1)/(2*pi*f0)/
    (2*pi*f0)/LA2;
    if CA1>CA3;CA=Ca+CT;CB=CT;end
    if CA3>CA1; CA=CT;CB=Ca+CT;end
end
if Z_fixed>0
    Z0=Z_fixed;
    C=min(CA1,CA3);% find symmetrical Capacitors C
    Ca=max(CA1,CA3)-C;% Generate residual capacitor Ca
    Q=2*pi*f0*LA2/Z0;
    teta=asin(2*pi*f0*LA2/Z0);
    tau=(1/2/pi/f0)*teta;
    CT=(cos(2*pi*f0*tau)+(2*pi*f0)*(2*pi*f0)*LA2*C-1)/(2*pi*f0)/
    (2*pi*f0)/LA2;
    if CA1>CA3;CA=Ca+CT;CB=CT;end
    if CA3>CA1; CA=CT;CB=Ca+CT;end
end
v0=3*1e8; vr=v0/sqrt(epsilonr);
length=vr*tau;
end; % end for the function
```

Program List 5.96: Main Program "GKYSRFTLumpedAmplDesignSec5_14.m

---

```
% Main Program GKYSRFTLumpedAmplDesignSec5_14.m
clc
clear
close all
%   DESIGN OF A SINGLE STAGE AMPLIFIER
%   USING A TSMC 0.18 micron NMOS SILICON TRANSISTOR
%   via SIMPLIFIED REAL FREQUENCY TECHNIQUE (SRFT)
% ----------------------------------------------------------------------
% General Information about the device NMOS Transistor, 0.18 micron gate
%               Frequency range: 100MHz - 22.4 GHz;
%               VDD = 1.8V; ID (drain DC current) = 200mA,
%               NMOS size W/L = 1000u/0.18u;
%               Simulation temperature = 50C.
%               Output power= 200mW up to 20 GHz
%               Operaton Class: Class A amplifier.
% ----------------------------------------------------------------------
```

```
% INPUTS:
%                  FR(i): ACTUAL FREQUENCIES
%                  SIJ: SCATTERING PARAMETERS OF THE TSMC-CMOS FET.
%                  THESE PARAMETERS ARE READ FROM A FILE CALLED ersad_1.txt
%                  Tflat1: FLAT GAIN LEVEL OF STEP 1 in dB.
%                  Tflat2: FLAT GAIN LEVEL OF STEP 2 in dB.
%                  FR(nd1); nd1: BEGINING OF THE OPTIMIZATION FREQUENCY
%                  (nd1=42)
%                  FR(nd2); nd2: END OF OPTIMIZATION FREQENCY: nd2=48.
%
% OUTPUT: FRONT & BACK END EQUALIZERS ARE GENERATED EMPLOYING SRFT
%                  Step 1:
%                          hF: Front End Equalizer;
%                          T1: Gain of the First Step
%                          CVF:Normalized Element Values of the
%                          Front-End
%                          equalizer
%                          AEVF: Actual Element Values of the Front-End
%                  Step 2:
%                          hB: Back End Equalizer
%                          T2: Gain of the Second Step
%                          CVB:Normalized Element Values of the Back-End
%                          Equalizer
%                          AEVB: Actual Element Values of the Back-End
%
%
%                          DESIGN OF A MICROWAVE AMPLIFIER
% READ THE TRANSISTOR DATA ersad_1.txt
%
%-----------------------------------------------------------------------
% Special Notes: From the given data file one should obtain that
% nd=48 (Total Number of sampling points)
% For front-end Equalizer:
%     freq(48)=22.38 GHz;    freq(41)=10 GHz;       freq(35)=5.02 GHz
%       T01(48)=1.317          T01(41)=6.54           T01(35)=26.1
%       (MSG for front-end)
%       T02(48)=5.255          T02(41)=16.53          T02(35)=58.25
%       MSG(48)=7.2dB;       MSGdB(41)=12.1dB;      MSGdB(35)=17.65dB
%     (MSG for the back-end)
%-----------------------------------------------------------------------
load TMSC_Spar.txt; % Load cmos data to the program
nd=length(TMSC_Spar)
pi=4*atan(1);
rad=pi/180;
%
%   EXTRACTION OF SCATTERING PARAMETERS FROM THE GIVEN FILE ersad_1.txt
for i=1:nd
   % Read the Actual Frequencies from the file ersad_1.txt
                   freq(i)=TMSC_Spar(i,1);
   % Read the magnitude and phase of the scattering parameters
   % from the data file ersad_1.txt
MS11(i)=TMSC_Spar(i,2);PS11(i)=rad*TMSC_Spar(i,3);
MS21(i)=TMSC_Spar(i,4);PS21(i)=rad*TMSC_Spar(i,5);
MS12(i)=TMSC_Spar(i,6);PS12(i)=rad*TMSC_Spar(i,7);
```

```
MS22(i)=TMSC_Spar(i,8);PS22(i)=rad*TMSC_Spar(i,9);
% Convert magnitude and phase to real and imaginary parts.
S11R(i)=MS11(i)*cos(PS11(i));S11X(i)=MS11(i)*sin(PS11(i));
S12R(i)=MS12(i)*cos(PS12(i));S12X(i)=MS12(i)*sin(PS12(i));
S21R(i)=MS21(i)*cos(PS21(i));S21X(i)=MS21(i)*sin(PS21(i));
S22R(i)=MS22(i)*cos(PS22(i));S22X(i)=MS22(i)*sin(PS22(i));
%
 % Get rid off the numerical errors due to extraction of the scattering
 % paramters from cadnace...(Vdec extarction
  if S11R(i)>=1;S11R(i)=0.99999999;S11X(i)=0.0;end
 end
%
%    CONSTRUCT THE APMLITUDE SQUARES OF THE SACTTERING PARAMETERS
  for i =1:nd
 SQS21(i)=S21R(i)*S21R(i)+S21X(i)*S21X(i);
 SQS12(i)=S12R(i)*S12R(i)+S12X(i)*S12X(i);
 SQS11(i)=S11R(i)*S11R(i)+S11X(i)*S11X(i);
 SQS22(i)=S22R(i)*S22R(i)+S22X(i)*S22X(i);
 S1(i)=1.0/(1-SQS11(i));S2(i)=1.0/(1-SQS22(i));
   end
      % COMPUTATION OF MAXIMUM STABLE GAIN "MSG'
%
  for i=1:nd;if SQS11(i)>=1;SQS11(i)=0.999999999;end
if   SQS22(i)>=1;SQS22(i)=0.999999999;end
%Ideal Flat Gain Levels
      % Step 1:
                    T01(i)=SQS21(i)/(1-SQS11(i));
                    T01dB(i)=10*log10(T01(i));
      % Step 2:
                    T02(i)=T01(i)/(1-SQS22(i));
                    T02dB(i)=10*log10(T02(i));
       % Maximim Stable Gain T02=MSG
                     MSG(i)=SQS21(i)/(1-SQS11(i))/(1-SQS22(i));
                     DB(i)=10*log10(MSG(i));
  end
%
%-----------------------------------------------------------------------
% Part I: DESIGN OF THE FRONT-END EQUALIZER
%-----------------------------------------------------------------------
 figure
 plot(freq,T01dB);
 title('Maximum Stable Gain of the First Part in dB')
 ylabel('T01 (dB)');xlabel('Actual Frequency')
% Selection of data points from the given data
% between the points nd1 and nd2 out of nd points.
%First Frequency given given by data file
FR1=freq(1)
%Last Frequency given by data file
FR2=freq(nd)
%nd1=input('10GHz is at nd1=41,11.2GHz is at nd1=42 Enter start point
nd1=')
%nd2=input('20GHz is at nd2=47;22.4GHz is at nd2=48;Enter end point
nd2=')
nd1=41; %This corresponds to 11.23 GHz
```

```
nd2=47;% This corresponds to 22.39 GHz
Fstart=freq(nd1)
Fend=freq(nd2)
%Define complex number j
j=sqrt(-1);
%
for k=1:(nd2-nd1+1)
            i=nd1-1+k;
% ww(i) is the normalized angular frequency indexed from k=1 to (nd2-nd1)
            ww(k)=freq(i)/Fend;
% FR(k) is the actual frequency
            FR(k)=freq(i);
%
% Scattering Parameters of CMOS Field effect transistor
f11r(k)=S11R(i); f11x(k)=S11X(i); f11(k)=f11r(k)+j*f11x(k);
f12r(k)=S12R(i); f12x(k)=S12X(i); f12(k)=f12r(k)+j*f12x(k);
f21r(k)=S21R(i); f21x(k)=S21X(i); f21(k)=f21r(k)+j*f21x(k);
f22r(k)=S22R(i); f22x(k)=S22X(i); f22(k)=f22r(k)+j*f22x(k);
 end
%
% Step1:
Tflat1=10*log10(T01(nd2))
Tflat1=input('Enter the desired targeted gain  value in dB for Step 1:
Tflat1=')
% For low pass case k=0
%-----------------------------------------------------------------------
%   AUTOMATIC INITIALIZATION OF THE FRONT-END MATCHING NETWORK
% Generation of the termination impedance for the front-end matching
network
NA=length(freq);j=sqrt(-1);
for i=1:NA
    FAin(i)=freq(i)/1e9;
    Zin(i)=(1+S11R(i)+j*S11X(i))/(1-S11R(i)-j*S11X(i));
    Rin(i)=real(Zin(i));
    Xin(i)=imag(Zin(i));
end
%               Initial Guess for SRFT Amplifier for front-end matching
%               network
% User selected Inputs:
    T0=0.975;
    NC=19;ktr=0;KFlag=0;sign=1;
    FL=FAin(42);FH=FAin(48);f0=FAin(47);R0=50;F_unit=1e9;
    n=4;ndc=0;WZ=0;a0=1;
%
Imp_input=[FAin Rin Xin];
UserSelectedInputs=[T0 NC ktr KFlag sign FL FH f0 R0 F_unit n ndc WZ a0];
Input_Data=[UserSelectedInputs Imp_input];
%
%-----------------------------------------------------------------------
% Design of front-end matching network in single function:
%
[ CTF,CVF,CVAF,FAF,TAF,cF,aF,bF,a0F ] = CompactSingleMatching( Input_Data );
hF0=aF-bF;
for i=1:n
```

```
    xF0(i)=hF0(i);
end
%
%-------------------------------------------------------------------------------
ndc=0;h0=0;ntr=0;
% xF0=input('hF(0)=0, enter front end equalizer;
[xF(1)=hF(1),xF(2)=hF(2),...,XF(n)=hF(n)]=')
OPTIONS=optimset('MaxFunEvals',20000,'MaxIter',50000);
xF= lsqnonlin('levenberg_TR1',xF0,[],[],OPTIONS,ndc,ww,f11,f12,f21,f22,Tf
lat1,ntr,h0);
 %
 % Here it assummed that there is no transformer in the circuit. That is
 % h(0)=1. This fact is used in the function xtoh(xF)
 if ntr==0
     n=length(xF);
     for i=1:n
         hF(i)=xF(i);
     end
     hF=[hF h0];
 end
 if ntr==1
     n=length(xF);
     for i=1:n
         hF(i)=xF(i);
     end
 end
[ gF,f_s21 ] = SRFT_htogLump( hF,ndc );
T1=Gain1(hF,ndc,ww,f11,f12,f21,f22);
T1dB=10*log10(T1);
%
        figure
        plot(ww*Fend,10*log10(T1))
        title('Step 1: Optimized Gain with Tflat1=2.1 dB')
        xlabel('Actual Frequency')
        ylabel(' T1(dB)')
%
% Computation of the actual element values of the front-end
            aF=gF+hF;bF=gF-hF;
f0=Fend;  R0=50;
[ CTF, CVF ] = Synthesis_ImpedanceBased( aF,bF,ndc,R0,f0 );
Plot_Circuit4(CTF,CVF)
%-------------------------------------------------------------------------------
% Part II: DESIGN OF THE BACK-END EQUALIZER
%-------------------------------------------------------------------------------
   SF2=reflection(hF,ndc,ww,f11,f12,f21,f22);
% output reflectance of the active device when it is terminated in SF=hF/gF
%
% Step2:
figure
plot(freq,T02dB);
 title('Maximum Stable Gain of the Second Part in dB')
 ylabel('T02 (dB)');xlabel('Actual Frequency')
 display('Suggested maximum flat gain level for the second part of the
design')
```

```
Tflat2=10*log10(T02(nd2))
Tflat2=input('Enter the desired targeted gain  value in dB for Step 2:
Tflat2=')
%-------------------------------------------------------------------------
%  AUTOMATIC INITIALIZATION OF THE BACK-END MATCHING NETWORK
% Generation of the termination impedance for the back-end matching
network
NA=length(freq);j=sqrt(-1);
for i=1:NA
    FAout(i)=freq(i)/1e9;
    Zout(i)=(1+S22R(i)+j*S22X(i))/(1-S22R(i)-j*S22X(i));
    Rout(i)=real(Zout(i));
    Xout(i)=imag(Zout(i));
end

%               Initial Guess for SRFT Amplifier for back-end matching
%               network
% User selected Inputs:
    T0=0.975;
    NC=19;ktr=0;KFlag=0;sign=1;
    FL=FAout(42);FH=FAout(48);f0=FAout(47);R0=50;F_unit=1e9;
    n=6;ndc=0;WZ=0;a0=1;
%
Imp_input=[FAout Rout Xout];
UserSelectedInputs=[T0 NC ktr KFlag sign FL FH f0 R0 F_unit n ndc WZ a0];
Output_Data=[UserSelectedInputs Imp_input];
%
%-------------------------------------------------------------------------
% Design of back-end matching network in single function:
%
[CTB,CVB,CVAB,FAB,TAB,cB,aB,bB,a0B] = CompactSingleMatching(Output_Data);
hB0=aB-bB;
for i=1:n
    xB0(i)=hB0(i);
end
%
%-------------------------------------------------------------------------
    k=0;
  % xB0=input('hB(0)=0, enter back end equalizer;
[xB(1)=hB(1),xB(2)=hB(2),...,XB(n)=hB(n)]=')
    OPTIONS=optimset('MaxFunEvals',20000,'MaxIter',50000);
    xB= lsqnonlin('levenberg_TR2',xB0,[],[],OPTIONS,k,ww,T1,SF2,f11,f12,f
21,f22,Tflat2,ntr,h0);
 %
 if ntr==0
    hB=xB;
    hB=[hB h0];
 end
 if ntr==1
    hB=xB;
 end
 %
 [ gB,fB ] = SRFT_htogLump( hB,ndc );
T2=Gain2(hB,ndc,ww,T1,SF2,f11,f12,f21,f22);
```

```
T2dB=10*log10(T2);
%
    figure;
    plot(ww*Fend,T2dB,ww*Fend,T1dB);
    title('Step 2: Final Gain Performance of the Amplifier')
    ylabel('T2(dB)')
    xlabel('Actual Frequency')
    legend('Gain of Part II','Gain of Part I')
%
    aB=gB+hB;bB=gB-hB;
f0=Fend;R0=50;
[ CTB, CVB ] = Synthesis_ImpedanceBased( aB,bB,ndc,R0,f0 );

Plot_Circuit4(CTB,CVB)
%
% Investigation on the stability of the amplifier:
%
% Stability Check at the output port:
        SB11=hoverg(hB,k,ww);%input ref. Cof. Back-end from hB(p)
        SL=SB11;
        ZL=stoz(SL);
        RL=real(ZL);
        Sout=SF2;
        Zout=stoz(Sout);
        Rout=real(Zout);
        StabilityCheck_output=RL+Rout;
%
%Stability Check at the input port:
        SF=hoverg(hF,k,ww);
        SG=SF;
        ZG=stoz(SG);
        RG=real(ZG);
[Sin]=INPUT_REF(SB11,f11,f12,f21,f22);
            Zin=stoz(Sin);
            Rin=real(Zin);
StabilityCheck_input=RG+Rin;
```

References

1. H. J. Carlin, A new approach to gain-bandwidth problems, *IEEE Trans. Circuits Syst.*, 23, 170–175, 1977.
2. B. S. Yarman, *Broadband Matching a Complex Generator to a Complex Load*, PhD thesis, Cornell University, 1982.
3. B. S. Yarman, *Design of Ultra Wideband Power Transfer Networks*, Wiley, West Sussex, United Kingdom, 2010.
4. B. S. Yarman, *Design of Ultra Wideband Antenna Matching Networks*, Springer, Berlin, Germany; New York, USA; Singapore, 2008.
5. A. Kilinc and B. S. Yarman, High precision LC ladder synthesis part I: Lowpass ladder synthesis via parametric approach, *IEEE Trans. CAS Part I*, 60(8), 2074–2083, 2013.
6. B. S. Yarman and A. Kilinc, High precision LC ladder synthesis part II: Immittance synthesis with transmission zeros at DC and infinity, *IEEE Trans. CAS Part I*, 60(10), 2719–2729, 2013.

7. B. S. Yarman, R. Kopru, N. Kumar, and P. Chacko, High precision synthesis of a Richards immittance via parametric approach, *IEEE Trans. CAS Part I*, 61(4), 1055–1067, 2014.

8. D. C. Atilla, C. Aydin, R. Kopru, T. Nesimoglu, and B. S. Yarman, A tunable inductance topology to realize frequency tunable matching networks and amplifiers, *IEEE International Symposium on Circuits and Systems, ISCAS*, Beijing, China, pp. 77–80, 2013.

9. R. Kopru, H. Kuntman, and B. S. Yarman, A novel method to design wideband power amplifier for wireless communication, *IEEE International Symposium on Circuits and Systems, ISCAS*, Beijing, China, pp. 1942–1945, 2013.

10. R. Kopru, H. Kuntman, and B. S. Yarman, 2 W wideband microwave PA design for 824–2170 MHz band using normalized gain function method, *8th International Conference on Electrical and Electronics Engineering (ELECO)*, Bursa, Turkey, pp. 344–348, 2013.

11. N. Kumar, E. Limiti, C. Paoloni, J. M. Collantes, R. Jansen, and B. S. Yarman, Vectorially combined distributed power amplifiers for software-defined radio applications, *IEEE Trans. MTT*, 60(10), 3189–3200, 2012.

12. H. W. Bode, *Network Analysis and Feedback Amplifier Design*, Princeton, NJ: Van Nostrand, 1945.

13. R. M. Fano, Theoretical limitations on the broadband matching of arbitrary impedances, *J. Franklin Inst.*, 249, 57–83, 1950.

14. D. C. Youla, A new theory of broadband matching, *IEEE Trans. Circuit Theory*, 11, 30–50, 1964.

15. S. Darlington, Synthesis of reactance 4-poles, *J. Math. Phys.*, 18, 257–353, 1939.

16. D. C. Youla, H. J. Carlin, and B. S. Yarman, Double broadband matching and the problem of reciprocal reactance 2n-port cascade decomposition, *Int. J. Circuit Theory Appl.*, 12, 269–281, 1984.

17. B. S. Yarman, A simplified real frequency technique for broadband matching complex generator to complex loads, *RCA Rev.*, 43, 529–541, 1982.

18. H. J. Carlin and B. S. Yarman, The double matching problem: Analytic and real frequency solutions, *IEEE Trans. Circuits Syst.*, 30, 15–28, 1983.

19. A. Fettweis, Parametric representation of Brune functions, *Int. J. Circuit Theory Appl.*, 7, 113–119, 1979.

20. J. Pandel and A. Fettweis, Broadband matching using parametric representations, *IEEE Int. Symp. Circuits Syst.*, 41, 143–149, 1985.

21. B. S. Yarman and A. Fettweis, Computer-aided double matching via parametric representation of Brune functions, *IEEE Trans. Circuits Syst.*, 37, 212–222, 1990.

22. B. S. Yarman and A. K. Sharma, Extension of the simplified real frequency technique and a dynamic design procedure for designing microwave amplifiers, *IEEE Int. Symp. Circuits Syst.*, 3, 1227–1230, 1984.

23. H. J. Carlin and P. Amstutz, On optimum broadband matching, *IEEE Trans. Circuits Syst.*, 28, 401–405, 1981.

24. H. J. Carlin and P. P. Civalleri, On flat gain with frequency-dependent terminations, *IEEE Trans. Circuits Syst.*, 32, 827–839, 1985.

25. B. S. Yarman, Modern approaches to broadband matching problems, *Proc. IEE*, 132, 87–92, 1985.

26. A. Aksen, *Design of Lossless Two-Ports with Lumped and Distributed Elements for Broadband Matching*, PhD thesis, Ruhr University at Bochum, 1994.

27. M. Gewertz, Synthesis of a finite, four-terminal network from its prescribed driving point functions and transfer functions, *J. Math. Phys.*, 12, 1–257, 1933.

28. J. Pandel and A. Fettweis, Numerical solution to broadband matching based on parametric representations, *Archiv. Elektr. Übertragung*, 41, 202–209, 1987.

29. S. Kuh and J. D. Patterson, Design theory of optimum negative-resistance amplifiers, *Proc. IRE*, 49(6), 1043–1050, 1961.

30. W. H. Ku, M. E. Mokari-Bolhassan, W. C. Petersen, A. F. Podell, and B. R. Kendall, Microwave octave-band GaAs–FET amplifiers, *Proc. IEEE MTT-S Int. Microw. Symp.*, Palo Alto, pp. 69–72, 1975.

31. B. S. Yarman, Real frequency broadband matching using linear programming, *RCA Rev.*, 43, 626–654, 1982.

32. H. J. Carlin and J. J. Komiak, A new method of broadband equalization applied to microwave amplifiers, *IEEE Trans. Microw. Theory Tech.*, 27, 93–99, 1979.

33. B. S. Yarman and H. J. Carlin, A simplified real frequency technique applied to broadband multi-stage microwave amplifiers, *IEEE Trans. Microw. Theory Tech.*, 30, 2216–2222, 1982.

34. L. Zhu, B. Wu, and C. Cheng, Real frequency technique applied to synthesis of broad-band matching networks with arbitrary nonuniform losses for MMIC's, *IEEE Trans. Microw. Theory Tech.*, 36, 1614–1620, 1988.

35. P. Jarry and A. Perennec, Optimization of gain and VSWR in multistage microwave amplifier using real frequency method, *Eur. Conf. Circuit Theory Des.*, Paris, France, vol. 23, pp. 203–208, September 1987.

36. A. Perennec, R. Soares, P. Jarry, P. Legaud, and M. Goloubkof, Computer-aided design of hybrid and monolithic broad-band amplifiers for optoelectronic receivers, *IEEE Trans. Microw. Theory Tech.*, 37, 1475–1478, 1989.

37. L. Zhu, C. Sheng, and B. Wu, Lumped lossy circuit synthesis and its application in broad-band FET amplifier design in MMIC's, *IEEE Trans. Microw. Theory Tech.*, 37, 1488–1491, 1989.

38. B. S. Yarman, A. Aksen, and A. Fettweis, An integrated design tool to construct lossless matching networks with lumped and distributed elements, *Eur. Conf. Circuit Theory Des.*, Copenhagen, Denmark, vol. 3, pp. 1280–1290, September 1991.

39. C. Beccari, Broadband matching using the real frequency technique, *IEEE Int. Symp. Circuits Syst.*, Montreal, Canada, vol. 3, pp. 1231–1234, May 1984.

40. G. Pesch, *Breitbandanpassung mit Leitungstransformatoren (Wideband Matching with Line Transformers)*, PhD thesis, Technische Hochschule, Aachen, 1978.

41. R. Pauli, *Breitbandanpassung Reeller und Komplexer Impedanzen mit Leitungsschaltungen (Broadband Matching of Real and Complex Impedances with Transmission Lines)*, PhD thesis, Technische Universitat, München, 1983.

42. B. S. Yarman and A. Aksen, An integrated design tool to construct lossless matching networks with mixed lumped and distributed elements, *IEEE Trans. Circuits Syst.*, 39, 713–723, 1992.

43. B. S. Yarman, A. Aksen, and A. Fettweis, An integrated design tool to construct lossless two-ports with mixed lumped and distributed elements for matching problems, *Int. Symp. Recent Adv. Microw. Tech.*, Reno, USA, vol. 2, pp. 570–573, August 1991.

44. A. Aksen and B. S. Yarman, Construction of low-pass ladder networks with mixed lumped and distributed elements, *Eur. Conf. Circuit Theory Des.*, Davos, Switzerland, vol. 1, pp. 1389–1393, August–September 1993.

45. A. Aksen and B. S. Yarman, A semi-analytical procedure to describe lossless two-ports with mixed lumped and distributed elements, *ISCAS'94*, 5–6, 205–208, 1994.

46. A. Aksen and B. S. Yarman, Cascade synthesis of two-variable lossless two-port networks of mixed lumped elements and transmission lines: A semi-analytic procedure, NDS-98, *First Int. Workshop Multidimens. Syst.*, Poland, pp. 12–14, July 1998.

47. S. Yarman, A dynamic CAD technique for designing broad-band microwave amplifiers, *RCA Rev.*, 44, 551–565, December 1983.

48. M. R. Wohlers, Complex normalization of scattering matrices and the problem of compatible impedances, *IEEE Trans. Circuit Theory*, 12, 528–535, 1965.

49. W. K. Chien, A theory of broadband matching of a frequency dependent generator and load, *J. Franklin Inst.*, 298, 181–221, 1974.

50. W. K. Chen and T. Chaisrakeo, Explicit formulas for the synthesis of optimum bandpass Butterworth and Chebyshev impedance-matching networks, *IEEE Trans. Circuits Syst.*, CAS-27(10), 928–942, 1980.

51. W. K. Chen and K. G. Kourounis, Explicit formulas for the synthesis of optimum broadband impedance matching networks II, *IEEE Trans. Circuits Syst.*, 25, 609–620, 1978.

52. Y. S. Zhu and W. K. Chen, Unified theory of compatibility impedances, *IEEE Trans. Circuits Syst.*, 35(6), 667–674, 1988.

53. C. Satyanaryana and W. K. Chen, Theory of broadband matching and the problem of compatible impedances, *J. Franklin Inst.*, 309, 267–280, 1980.

54. B. S. Yarman, A. Aksen, and A. Kılınç, An immittance based tool for modelling passive one-port devices by means of Darlington equivalents, *Int. J. Electron. Commun. (AEÜ)*, 55(6), 443–451, 2001.
55. B. S. Yarman, A. Kılınc, and A. Aksen, Immittance data modeling via linear interpolation techniques, , *IEEE International Symposium on Circuits and Systems, ISCAS*, Scottsdale, Arizona ABD, May 2002.
56. A. Kılınç, H. Pınarbaşı, M. Sengul, and B. S. Yarman, A broadband microwave amplifier design by means of immittance based data modelling tool, *IEEE Africon 02, 6th Africon Conf. Afr.*, George, South Africa, vol. 2, pp. 535–540, October 2–4, 2002.
57. B. S. Yarman, A. Kılınç, and A. Aksen, Immittance data modelling via linear interpolation techniques: A classical circuit theory approach, *Int. J. Circuit Theory Appl.*, 32, 537–563, 2004.
58. B. S. Yarman, M. Sengul, and A. Kilinc, Design of practical matching networks with lumped elements via modeling, *IEEE Trans. CAS-I*, 54(8), 1829–1837, 2007.
59. M. Sengul and B. S. Yarman, Broadband equalizer design with commensurate transmission lines via reflectance modelling, *IEICE Trans. Fundam.*, E91-A(12), 3763–3771, 2008.
60. T. F. Coleman and Y. Li, An interior, trust region approach for nonlinear minimization subject to bounds, *SIAM J. Optim.*, 6, 418–445, 1996.
61. T. F. Coleman and Y. Li, On the convergence of reflective Newton methods for large-scale nonlinear minimization subject to bounds, *Math. Program.*, 67(2), 189–224, 1994.
62. J. E. Dennis Jr., Nonlinear least-squares, *State of the Art in Numerical Analysis*, ed. D. Jacobs, Academic Press, Waltham, Massachusetts, USA, pp. 269–312, 1977.
63. K. Levenberg, A method for the solution of certain problems in least-squares, *Q. Appl. Math.*, 2, 164–168, 1944.
64. D. Marquardt, An algorithm for least-squares estimation of nonlinear parameters, *SIAM J. Appl. Math.*, 11, 431–441, 1963.
65. J. J. Moré, The Levenberg–Marquardt algorithm: Implementation and theory, *Numerical Analysis*, ed. G. A. Watson, *Lecture Notes in Mathematics*, Berlin, Germany, vol. 630, Springer-Verlag, pp. 105–116, 1977.
66. N. Balabanian, *Network Synthesis*, Englewood Cliffs, NJ: Prentice-Hall, Inc., 1958.
67. B. S. Yarman et al., Computer aided high precision Darlington synthesis for real frequency matching, *IEEE Benjamin Franklin Symp. Microw. Antenna Sub-Syst.*, Philadelphia, September 27, 2014.
68. K. Kuroda, A method to derive distributed constant filters from constant filters, *Joint Conv. Elec. Inst. Japan*, Kansai, Chapters 9–10, October 1952.
69. I. Bahl, *Lumped Elements for RF and Microwave Circuits*, Artech House, ISBN 1-58053-309-4, 2003; I. Bahl and P. Bhartia, *Microwave Solid State Circuit Design*, Wiley-InterScience, Hoboken, NJ, USA, ISBN 0-471-20755-1, 2003.
70. V. Belevitch, *Classical Network Theory*, San Francisco: Holden Day, 1968.
71. H. A. Wheeler, Transmission-line properties of parallel wide strips by a conformal-mapping approximation, *IEEE Trans. Microw. Theory Tech.*, MTT-12, 280–289, 1964.
72. H. A. Wheeler, Transmission-line properties of parallel strips separated by a dielectric sheet, *IEEE Trans. Microw. Theory Tech.*, MTT-13, 172–185, 1965.
73. H. A. Wheeler, Transmission-line properties of a strip on a dielectric sheet on a plane, *IEEE Trans. Microw. Theory Tech.*, MTT-25, 631–647, 1977.
74. B. S. Yarman, N. Retdian, S. Takagi, and N. Fujii, Gain–bandwidth limitations of 0.18 μ Si-CMOS RF technology, *Proc. ECCTD*, Sevilla, Spain, August 26–30, 2007.
75. AWR is RF/Microwave Circuit Design Software of National Instruments, http://www.awrcorp.com.
76. ADS (Advance Design System) by Keysight Technologies, http://www.keysight.com/.
77. http://www.cadence.com/en/default.aspx.

6

High-Efficiency Broadband Class-E Power Amplifiers

In modern wireless communication systems, it is required that the power amplifier operates with high efficiency over a wide frequency range to simultaneously provide multiband and multistandard signal transmission. The conventional design of a high-efficiency switchmode Class-E power amplifier requires a high value of the loaded quality factor Q_L to satisfy the necessary harmonic impedance conditions at the output device terminal. However, if a sufficiently small value of Q_L is selected, a high-efficiency broadband operation of the Class-E power amplifier can be realized by applying the reactance compensation technique. Usually, the bandwidth limitation in power amplifiers comes from the device's low-transition frequency and large output capacitance; therefore, silicon LDMOSFET technology has been the preferred choice up to 2.2 GHz. As an alternative, GaN HEMT technology enables high efficiency, large breakdown voltage, high-power density, and significantly higher broadband performance due to higher transition frequency and smaller periphery, resulting in the smaller input and output capacitances and less parasitics.

6.1 Reactance Compensation Technique

The high-efficiency broadband operation of a switchmode Class-E power amplifier using the reactance compensation technique can be realized if a simple network consisting of a series resonant LC circuit tuned to the fundamental frequency and a parallel inductor provides a constant load phase angle of 50° in a frequency range of about 50% [1]. From theoretical considerations, it was found in the mid-1960s that the bandwidth response of a parametric amplifier can be improved using multiple-resonant bandpass filters for the signal and idling circuits rather than simple resonant circuits [2,3]. At the same time, it was analytically calculated that the added resonant circuits should have an appropriate Q_L-factor to optimally reduce the rate of change of reactance of both the signal and idling circuits [4]. Adding more resonators can increase the potential amplifier bandwidth even further, but the amount of improvement per additional resonator will decrease rapidly as the number of resonators is increased. Such a reactance compensation technique using a single-resonant circuit had also been applied to the varactor-tuned Gunn oscillator [5,6]. Moreover, it became possible to increase the tuning range of an oscillator by adding more stages of reactance compensation. For instance, for a resonant circuit having a 50-Ω load, an improvement of 4% in the tuning range can theoretically be achieved as a result of applying a double-resonant circuit reactance compensation, whereas an increase in the tuning range can reach 17% for a resonant circuit operating into a 100-Ω load [7].

6.1.1 Load Networks with Lumped Elements

To describe reactance compensation circuit technique, let us consider the simplified equivalent load networks, one with a shunt resonant L_pC_p circuit followed by a series resonant L_sC_s circuit shown in Figure 6.1a and the other with a series resonant L_sC_s circuit followed by a shunt resonant L_pC_p circuit shown in Figure 6.1b. In this case, all resonant circuits are tuned to the fundamental frequency and R is the load resistance. The reactances of the series and shunt resonant circuits vary with frequency, increasing in the case of a series resonant circuit and reducing in the case of a loaded parallel resonant circuit near the resonant frequency. As a result, near the resonant frequency of the series circuit with positive slope of its reactance, the slope of a shunt circuit reactance is negative and that reduces the overall reactance slope of the load network. By correct choice of the components in the shunt circuit, the rate of change of reactance with frequency can be made exactly opposite to that of the series circuit, thus producing a zero total variation over a wide frequency bandwidth.

Consider the load-network admittance Y_{net} corresponding to a single-reactance compensation circuit shown in Figure 6.1a, which can be written as

$$Y_{net}(\omega) = \left(j\omega C_p + \frac{1}{j\omega L_p} + \frac{1}{R + j\omega' L_s} \right) \tag{6.1}$$

where

$$\omega' = \omega\left(1 - \frac{\omega_0^2}{\omega^2} \right) \tag{6.2}$$

and $\omega_0 = 1/\sqrt{L_sC_s} = 1/\sqrt{L_pC_p}$ is the radian resonant frequency.

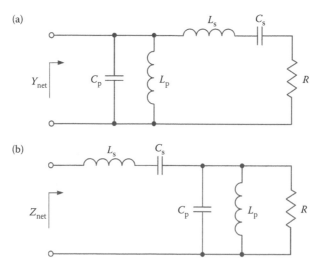

FIGURE 6.1
Single-susceptance (a) and single-reactance (b) compensation circuits.

At the resonant frequency when $\omega' = 0$, the load-network admittance $Y_{net}(\omega)$ reduces to

$$Y_{net}(\omega) = \left(j\omega C_p + \frac{1}{j\omega L_p} + G \right) \tag{6.3}$$

where $G = 1/R$ is the load conductance.

The frequency bandwidth with zero susceptance will be maximized if, at a resonant radian frequency ω_0,

$$\frac{dB_{net}(\omega)}{d\omega}\bigg|_{\omega=\omega_0} = 0 \tag{6.4}$$

where

$$B_{net}(\omega) = \text{Im}\, Y_{in}(\omega) = \omega C_p - \frac{1}{\omega L_p} - \frac{\omega' L_s}{R^2 + (\omega' L_s)^2} \tag{6.5}$$

is the load-network susceptance.

As a result, an additional equation can be written as

$$C_p + \frac{1}{\omega_0^2 L_p} - \frac{2L_s}{R^2} = 0 \tag{6.6}$$

based on which the values of the series components L_s and C_s can respectively be obtained through the values of the shunt components L_p and C_p by

$$L_s = C_p R^2 \tag{6.7}$$

$$C_s = \frac{L_p}{R^2} \tag{6.8}$$

In a similar manner, it may be shown that, for the load network with a series resonant $L_s C_s$ circuit followed by a shunt resonant $L_p C_p$ circuit shown in Figure 6.1b, the maximum bandwidth with zero reactance can be achieved if

$$\frac{dX_{net}(\omega)}{d\omega}\bigg|_{\omega=\omega_0} = 0 \tag{6.9}$$

where

$$X_{net}(\omega) = \text{Im}\, Z_{net}(\omega) = \omega L_s - \frac{1}{\omega C_s} - \frac{\omega' C_p}{G^2 + (\omega' C_p)^2} \tag{6.10}$$

is the load-network reactance, resulting in Equations 6.7 and 6.8. From Equation 6.7, it follows that the loaded quality factor of the shunt circuit $Q_L = \omega C_p R$ is equal to the loaded quality factor of the series compensating circuit $Q_L = \omega L_s/R$.

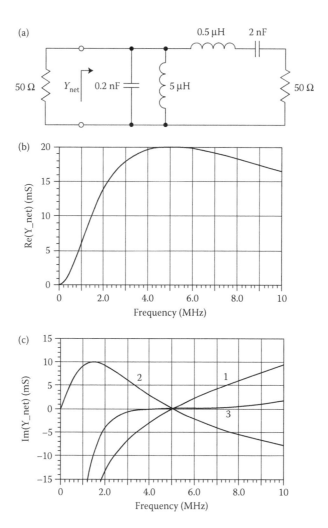

FIGURE 6.2
Single-susceptance compensation circuit (a) with its conductance (b) and susceptance (c) vs. frequency.

Figure 6.2a shows the example of a susceptance compensation load network, whose conductance $\text{Re}Y_{net}$ is almost constant across the frequency range of 40% (from 4 to 6 MHz), as shown in Figure 6.2b. The susceptance $\text{Im}Y_{net}$ of a shunt circuit varies with frequency, as shown in Figure 6.2c by curve 1, with the gradient at ω_0 being equal to $2C_p$. The addition of a series circuit with the same resonant frequency of 5 MHz between the shunt circuit and the load of the shunt circuit gives an additional susceptance term with a negative slope, as shown in Figure 6.2c by curve 2. Proper selection of the components of the series circuit enables the magnitude of the two slopes to be made identical so that the total susceptance slope around resonance is zero in an octave frequency range from 3.5 to 7 MHz, as shown in Figure 6.2c by curve 3.

The load network that provides reactance compensation is shown in Figure 6.3a, where the shunt resonant circuit is connected between the series resonant circuit and the load. In this case, the resistance and reactance curves, the frequency behavior of which is similar to that for the conductance and susceptance curves characterizing the behavior of a

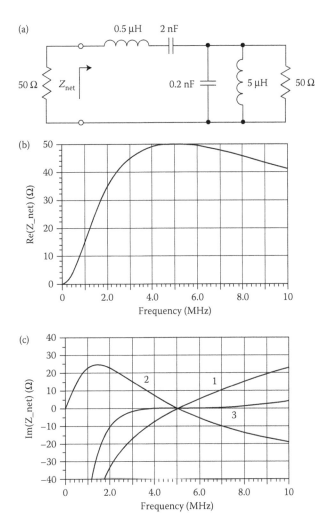

FIGURE 6.3
Single-reactance compensation circuit (a) with its resistance (b) and reactance (c) vs. frequency.

susceptance compensation load network, are shown in Figure 6.3b and c, respectively. Here, the reactance of a series resonant circuit with a positive slope is shown by curve 1, the reactance of a shunt resonant circuit with a negative slope is shown by curve 2, and the total reactance slope shown by curve 3 is zero from 3.5 to 7 MHz.

Wider frequency bandwidth can be achieved using a double-susceptance compensation circuit shown in Figure 6.4a, where $L_s C_s$ and $L_1 C_1$ are the series and parallel compensating circuits, respectively. In this case, a system of two additional equations to maximize the frequency bandwidth can be used, where the first and the third derivatives are set to zero according to

$$\frac{dB_{net}(\omega)}{d\omega}\bigg|_{\omega=\omega_0} = \frac{d^3 B_{net}(\omega)}{d\omega^3}\bigg|_{\omega=\omega_0} = 0 \qquad (6.11)$$

as the second derivative cannot provide an appropriate analytical expression.

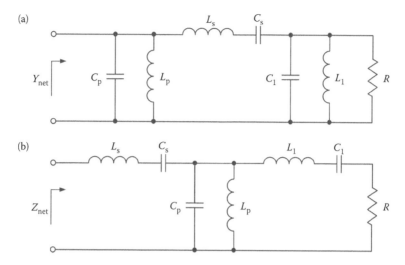

FIGURE 6.4
Double-susceptance (a) and double-reactance (b) compensation circuits.

To determine the load-network parameters for a double-susceptance compensation circuit with the load-network susceptance

$$B_{net}(\omega) = \omega C_p - \frac{1}{\omega L_p} + \omega' \frac{C_1 R^2 [1 - (\omega')^2 L_s C_1] - L_s}{R^2 [1 - (\omega')^2 L_s C_1]^2 + (\omega' L_s)^2} \tag{6.12}$$

where $B_{net} = \mathrm{Im} Y_{net}$, it is necessary to solve simultaneously the two following equations at the resonant frequency ω_0:

$$C_{p1} + \frac{1}{\omega_0^2 L_p} - 2\frac{C_1 R^2 - L_s}{R^2} = 0 \tag{6.13}$$

$$\frac{1}{\omega_0^2 L_p} + \frac{C_1 R^2 - L_s}{R^2} - 8\omega_0^2 L_s \left[C_1^2 + \frac{(C_1 R^2 - L_s)(L_s - 2C_1 R^2)}{R^4} \right] = 0 \tag{6.14}$$

As a result, the parameters of the series and shunt compensating resonant circuits with the corresponding loaded quality factors $Q_s = \omega_0 L_s / R$ and $Q_1 = \omega_0 C_1 R$, which are close to unity and greater, can be calculated as a starting point for circuit optimization from

$$L_s = \frac{R}{\omega_0} \frac{2}{\sqrt{5} - 1} \qquad C_s = \frac{1}{\omega_0^2 L_s} \tag{6.15}$$

$$C_1 = \frac{L_s}{R^2} \frac{3 - \sqrt{5}}{2} \qquad L_1 = \frac{1}{\omega_0^2 C_1} \tag{6.16}$$

Similarly, the elements for the double-reactance compensation load network shown in Figure 6.4b can be calculated from

$$L_1 = L_s \frac{\sqrt{5}-1}{2} \qquad C_1 = C_s \frac{2}{\sqrt{5}-1} \tag{6.17}$$

$$L_p = C_s \frac{2R^2}{\sqrt{5}+1} \qquad C_p = L_s \frac{\sqrt{5}+1}{2R^2} \tag{6.18}$$

where an inductance L_s and a capacitance C_s are known in advance [7]. An example of the load network that provides double-reactance compensation is shown in Figure 6.5a,

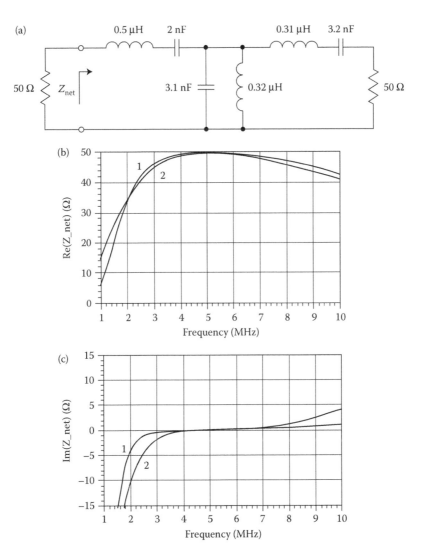

FIGURE 6.5
Double-reactance compensation circuit (a) with its resistance (b) and reactance (c) vs. frequency.

whose resistance $\text{Re}Z_\text{net}$ shown in Figure 6.5b by curve 1 provides less deviation from 50 Ω in a slightly wider frequency bandwidth compared to the single-resonance load network (curve 2) with $L_\text{s} = 0.5\ \mu\text{H}$, $C_\text{s} = 2\ \text{nH}$, $L_\text{p} = 5\ \mu\text{H}$, and $C_\text{p} = 0.2\ \text{nF}$. The reactance $\text{Im}Z_\text{net}$ of a double-reactance compensation circuit shown in Figure 6.5c by curve 1 is close to zero near resonance across the frequency range from 3 to 8 MHz, which is wider than that for a single-resonance compensation circuit (curve 2).

6.1.2 Load Networks with Transmission Lines

The reactance compensation circuit technique can also be used for bandwidth improvement of microwave transistor amplifiers because the input and output transistor impedances generally can be represented by series or shunt RLC circuits. For compensating the reactive part and transforming the real part of the equivalent output transistor impedance to the conventional load impedance at the fundamental frequency, the quarter- and half-wavelength transmission lines can be used. For the first time, a quarter-wavelength transmission-line transformer was used for active reactance compensation when, by connecting two identical active devices together with a quarter-wavelength transformer, the inverted impedance of one device compensates the impedance of the other by reducing the total circuit reactance [8].

Let us consider the characteristics of the transmission line as an element of a susceptance compensation circuit shown in Figure 6.6. For a parallel equivalent circuit, which represents the device output, the load-network input susceptance $B_\text{net} = \text{Im}Y_\text{net}$ can be defined as

$$B_\text{net}(\omega) = \omega L_p C_p \left(1 - \frac{\omega_0^2}{\omega^2}\right) + \frac{\tan\theta}{Z_0}\frac{R_\text{L}^2 - Z_0^2}{R_\text{L}^2 + Z_0^2\tan^2\theta} \tag{6.19}$$

where

$$\theta = \frac{\pi}{2}\frac{f}{f_0}k \tag{6.20}$$

is the transmission-line electrical length, Z_0 is the transmission-line characteristic impedance, $f_0 = \omega_0/2\pi$ is the transmission-line resonant frequency, and $k = 1, 2, \dots, \infty$.

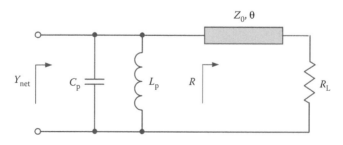

FIGURE 6.6
Transmission-line susceptance compensation circuit.

Applying the zero susceptance-derivative condition given by Equation 6.4 allows us to obtain the susceptance-compensation circuit parameters for different electrical lengths of a transmission line in accordance with

$$2C_p + \frac{\pi}{2Z_0\omega_0} \frac{R_L^2 - Z_0^2}{\cos^2\theta} \frac{R_L^2 - Z_0^2\tan^2\theta}{(R_L^2 + Z_0^2\tan^2\theta)^2} = 0 \qquad (6.21)$$

For a quarter-wavelength transmission line when $k = 1$ and $\theta = \pi/2$, the susceptance compensation will be performed under the condition $Z_0 < R_L$ with the characteristic impedance Z_0 defined from a quadratic equation

$$Z_0^2 + 4\frac{QR_L}{\pi}Z_0 - R_L^2 = 0 \qquad (6.22)$$

where $Q = \omega_0 C_p R$ and $R = Z_0^2/R_L$.

As a result, the required value of the characteristic impedance Z_0 is obtained by

$$Z_0 = R_L\left(-\frac{2Q}{\pi} + \sqrt{\left(\frac{2Q}{\pi}\right)^2 + 1}\right) \qquad (6.23)$$

or

$$Z_0 = \frac{R}{\left(-(2Q/\pi) + \sqrt{(2Q/\pi)^2 + 1}\right)} \qquad (6.24)$$

By using the quarter- and half-wavelength transformers, the reactance- or susceptance-compensation load network generally can be realized differently for shunt and series equivalent output transistor circuits, as shown in Table 6.1 along with respective design equations [9,10]. The two most important device parameters in the equations are the loaded quality factor Q and the real part R of the equivalent device output impedance. Depending on the values of the transmission-line characteristic impedances Z_1 and Z_2, each circuit provides either positive or negative parallel-resonant slope-reactance compensation.

Figure 6.7a shows the example of a single-susceptance compensation load network with a series quarterwave transmission line having a characteristic impedance of 61.2 Ω to match a 50-Ω real part of the device equivalent output admittance to a 75-Ω load and an electrical length of 90° at 50 MHz. The combination of the resistances of a shunt LC circuit (curve 1) and a series quarterwave transmission line (curve 2) provides minimum variations of the total resistance $\mathrm{Re}Z_{net}$ shown in Figure 6.7b by curve 3 around 50 Ω in a very wide frequency range. The susceptance $\mathrm{Im}Y_{net}$ of a shunt circuit having a resonant frequency of 50 MHz varies with frequency with a positive slope, as shown in Figure 6.7c by curve 1. The addition of a series quarter-wavelength transmission-line transformer between the shunt circuit and the load results in a negative slope provided by an additional susceptance, as shown in Figure 6.7c by curve 2. Selection of the proper characteristic impedance of the series quarterwave transmission line and the load resistance enables the magnitude of two slopes to be made identical, so that the total susceptance slope around resonance is zero in a frequency range from 45 to 65 MHz, as shown in Figure 6.7c by curve 3.

TABLE 6.1

Transmission-Line Reactance Compensation Circuits and Design Equations

Output Circuit Type	Matching Network	Design Equation
$Q = \omega_0 RC$	Z_1 (90°), Z_2 (180°)	$Z_1 = \sqrt{RR_L}$ $T = \left(-\dfrac{2Q}{\pi} - \dfrac{1}{2}\right)\left(\dfrac{R_L}{Z_1} - \dfrac{Z_1}{R_L}\right)$ $Z_2 = -\dfrac{TR_L}{2} + \sqrt{\left(\dfrac{TR_L}{2}\right)^2 + R_L^2}$
	Z_1 (90°), Z_2 (90°)	$Z_1 = \sqrt{RR_L}$ $Z_2 = \dfrac{\pi Z_1 R_L^2}{\pi Z_1^2 - \pi Z_L^2 - 4Q Z_1 R_L}$
$Q = 1/\omega_0 RC$	Z_1 (90°), Z_2 (90°)	$Z_1 = A Z_2$ $Z_2 = B R_L$ $A = \sqrt{R/R_L}$ $B = \dfrac{A}{1+A}\left[-\dfrac{2Q}{\pi} + \sqrt{\left(\dfrac{2Q}{\pi}\right)^2 + \dfrac{(1+A)^2}{A}}\right]$
	θ_1, Z_1 (90°), Z_2 θ_2	$Z_2 = \sqrt{RR_L}$ $Z_1 = \left(\dfrac{TZ_2}{R_L}\right)^2 \dfrac{1}{((R_L/Z_2) - (Z_2/R_L)) + 4Q/\pi}$ $T = \dfrac{2}{\pi\cos\theta_2}(\theta_1 + \theta_2)$ $\theta_1 + \theta_2 = \dfrac{n\pi}{2} \to n = 1 \text{ or } 3$

From Equation 6.23, it follows that the maximum value of the characteristic impedance Z_0 is limited by the load resistance R_L, and its value in some cases, especially for high value of Q, can be substantially smaller than 50 Ω, which causes a problem in the practical implementation of a transmission line. In this case, it is best to apply a single-frequency equivalence technique when a quarterwave transmission line can be replaced by a symmetrical π-type low-pass transmission-line section with two equal shunt capacitances at a frequency ω_0, as shown in Figure 6.8.

The transmission A-matrix (or $ABCD$-matrix) for a quarterwave transmission line can be written as

$$A_{90°} = \begin{bmatrix} \cos 90° & jZ_0 \sin 90° \\ j\dfrac{\sin 90°}{Z_0} & \cos 90° \end{bmatrix} = \begin{bmatrix} 0 & jZ_0 \\ j\dfrac{1}{Z_0} & 0 \end{bmatrix} \tag{6.25}$$

whereas, for a π-type low-pass transmission-line section, we can write

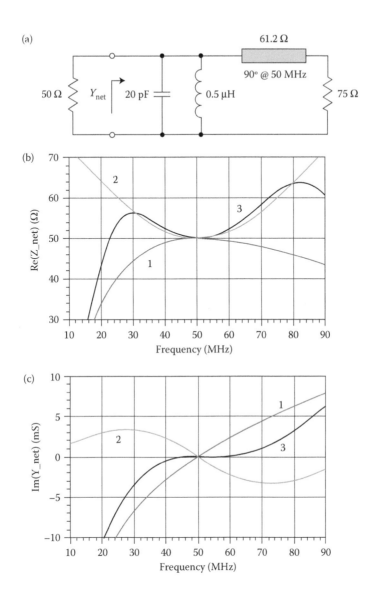

FIGURE 6.7
Susceptance compensation circuit with quarter-wavelength transmission line (a) with its resistance (b) and susceptance (c) vs. frequency.

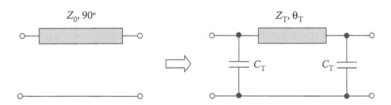

FIGURE 6.8
Transmission-line single frequency equivalence technique.

$$A_\pi = \begin{bmatrix} 1 & 0 \\ j\omega C_T & 1 \end{bmatrix} \begin{bmatrix} \cos\theta_T & jZ_T\sin\theta_T \\ j\dfrac{\sin\theta_T}{Z_T} & \cos\theta_T \end{bmatrix} \begin{bmatrix} 1 & 0 \\ j\omega C_T & 1 \end{bmatrix}$$

$$= \begin{bmatrix} \cos\theta_T - \omega C_T Z_T \sin\theta_T & jZ_T\sin\theta_T \\ \dfrac{j}{Z_T}(2Z_T\omega C_T\cos\theta_T + \sin\theta_T - Z_T^2\omega^2 C_T^2\sin\theta_T) & \cos\theta_T - \omega C_T Z_T \sin\theta_T \end{bmatrix} \quad (6.26)$$

Hence, equating A and B elements from each matrix yields

$$Z_T = \frac{Z_0}{\sin\theta_T} \tag{6.27}$$

$$C_T = \frac{\cos\theta_T}{\omega Z_0} \tag{6.28}$$

As a result, the electrical length of the transmission line can be reduced significantly with the increase of its characteristic impedance. Also, such a transformation is very important when the value of the device output capacitance exceeds the required optimum value for the optimum Class-E operation. In this case, the excess capacitance can be used as a part or entire shunt capacitance in the π-type low-pass section, and the optimum switching Class-E conditions will be completely satisfied at the fundamental frequency.

6.2 High-Efficiency Switching Class-E Modes

The single-ended switchmode power amplifier with a shunt capacitance as a Class-E power amplifier was introduced by Sokals in 1975 and has found widespread application due to its design simplicity and high-operation efficiency [11,12]. This type of high-efficiency power amplifier is widely used in different frequency ranges and output power levels ranging from several kilowatts at low-RF frequencies up to about 1 W at microwaves [13]. The reasons for the high efficiencies is that, due to a proper choice of transistor and circuit parameters, the transistor operates in a switching mode, and the voltage across the transistor and the current flowing through it can both be made equal to zero during the switching transient interval. To satisfy this condition, the current and voltage must be zero at the time just prior to the conduction interval when the transistor goes into the saturation mode and the series-tuned circuit must appear inductive at the operating frequency. In this case, a loaded quality factor of the series-tuned circuit of about 10 will give a good sinusoidal shape to the load current. The characteristics of a switchmode Class-E power amplifier can be obtained by determining its steady-state collector voltage and current waveforms.

6.2.1 Class E with Shunt Capacitance

The basic circuit of a Class-E power amplifier with shunt capacitance is shown in Figure 6.9a, where the load network consists of a capacitance C shunting the transistor, a series inductance L, a series fundamentally tuned $L_0 C_0$ circuit, and a load resistance R. In a

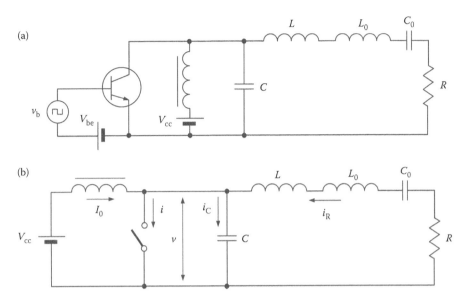

FIGURE 6.9
Basic circuits of Class-E power amplifier with shunt capacitance.

common case, a shunt capacitance C can represent the intrinsic device output capacitance and external circuit capacitance added by the load network. The collector of the transistor is connected to the supply voltage by an RF choke with high reactance at the fundamental frequency. The active device is considered an ideal switch that is driven in such a way as to provide the device switching between its on-state and off-state operation conditions. As a result, the collector voltage waveform is determined by the switch when it is turned on and by the transient response of the load network when the switch is turned off.

To simplify an analysis of the Class-E power amplifier, a simple equivalent circuit of which is shown in Figure 6.9b, the following several assumptions are introduced:

- The transistor has zero saturation voltage, zero saturation resistance, and infinite off-resistance, and its switching action is instantaneous and lossless.
- The total shunt capacitance is independent of the collector and is assumed linear.
- The RF choke allows only a constant DC current and has no resistance.
- The loaded quality factor $Q_L = \omega L_0/R = 1/\omega C_0 R$ of the series resonant $L_0 C_0$ circuit tuned to the fundamental frequency is high enough for the output current to be sinusoidal at the switching frequency.
- There are no losses in the circuit except only in the load R.
- For an optimum operation mode, a 50% duty cycle is used.

For a lossless operation mode, it is necessary to provide the following idealized optimum (or nominal) conditions for voltage across the switch (just prior to the start of switch on) at the moment $\omega t = 2\pi$, when the transistor is saturated:

$$v(\omega t)\big|_{\omega t=2\pi} = 0 \tag{6.29}$$

$$\frac{dv(\omega t)}{d\omega t}\bigg|_{\omega t=2\pi} = 0 \tag{6.30}$$

where $v(\omega t)$ is the voltage across the switch.

The detailed theoretical analysis of a Class-E power amplifier with shunt capacitance for any duty cycle is given in Reference 14, where the load current is assumed to be sinusoidal,

$$i_R(\omega t) = I_R \sin(\omega t + \varphi) \tag{6.31}$$

where φ is the initial phase shift.

When the switch is turned on for $0 \le \omega t < \pi$, the current through the capacitance

$$i_C(\omega t) = \omega C \frac{dv(\omega t)}{d\omega t} = 0 \tag{6.32}$$

and, consequently,

$$i(\omega t) = I_0 + I_R \sin(\omega t + \varphi) \tag{6.33}$$

under the initial on-state condition $i(0) = 0$. Hence, the DC current can be defined as

$$I_0 = -I_R \sin \varphi \tag{6.34}$$

and the current through the switch can be rewritten by

$$i(\omega t) = I_R[\sin(\omega t + \varphi) - \sin \varphi] \tag{6.35}$$

When the switch is turned off for $\pi \le \omega t < 2\pi$, the current through the switch $i(\omega t) = 0$, and the current flowing through the capacitor C can be written as

$$i_C(\omega t) = I_0 + I_R \sin(\omega t + \varphi) \tag{6.36}$$

producing the voltage across the switch by the charging of this capacitor according to

$$\begin{aligned} v(\omega t) &= \frac{1}{\omega C} \int_{\pi}^{\omega t} i_C(\omega t) d\omega t \\ &= -\frac{I_R}{\omega C}[\cos(\omega t + \varphi) + \cos \varphi + (\omega t - \pi)\sin \varphi] \end{aligned} \tag{6.37}$$

Applying the first idealized optimum (or nominal) condition given by Equation 6.29 enables the phase angle φ to be determined as

$$\varphi = \tan^{-1}\left(-\frac{2}{\pi}\right) = -32.482° \tag{6.38}$$

Consideration of trigonometric relationships shows that

$$\sin\varphi = \frac{-2}{\sqrt{\pi^2 + 4}} \quad \cos\varphi = \frac{\pi}{\sqrt{\pi^2 + 4}} \tag{6.39}$$

By using Fourier series expansion and Equations 6.34 and 6.39, the expression to determine the supply voltage V_{cc} can be written as

$$V_{cc} = \frac{1}{2\pi} \int_0^{2\pi} v(\omega t) d\omega t = \frac{I_0}{\pi\omega C} \tag{6.40}$$

As a result, the normalized steady-state collector voltage waveform for $\pi \leq \omega t < 2\pi$ and current waveform for period of $0 \leq \omega t < \pi$ are

$$\frac{v(\omega t)}{V_{cc}} = \pi\left(\omega t - \frac{3\pi}{2} - \frac{\pi}{2}\cos\omega t - \sin\omega t\right) \tag{6.41}$$

$$\frac{i(\omega t)}{I_0} = \frac{\pi}{2}\sin\omega t - \cos\omega t + 1 \tag{6.42}$$

Figure 6.10 shows the normalized (a) load current, (b) collector voltage waveform, and (c) collector current waveforms for an idealized optimum Class E with shunt capacitance. From collector voltage and current waveforms, it follows that, when the transistor is turned on, there is no voltage across the switch and the current $i(\omega t)$ consisting of the load sinusoidal current and DC current flows through the device. However, when the transistor is turned off, this current flows through the shunt capacitance C. The jump in the collector current waveform at the instant of switching off is necessary to obtain nonzero output power at the fundamental frequency delivered to the load, which can be defined as an integration of the product of the collector voltage and current derivatives over the entire period [15].

As a result, there is no nonzero voltage and current simultaneously, which means a lack of the power losses and gives an idealized collector efficiency of 100%. This implies that the DC power and fundamental-frequency output power delivered to the load are equal:

$$I_0 V_{cc} = \frac{I_R^2}{2} R \tag{6.43}$$

Consequently, the value of DC supply current I_0 can be determined using Equations 6.34 and 6.39 by

$$I_0 = \frac{V_{cc}}{R} \frac{8}{\pi^2 + 4} = 0.577\frac{V_{cc}}{R} \tag{6.44}$$

Then, the amplitude of the output voltage $V_R = I_R R$ can be obtained from

$$V_R = \frac{4V_{cc}}{\sqrt{\pi^2 + 4}} = 1.074V_{cc} \tag{6.45}$$

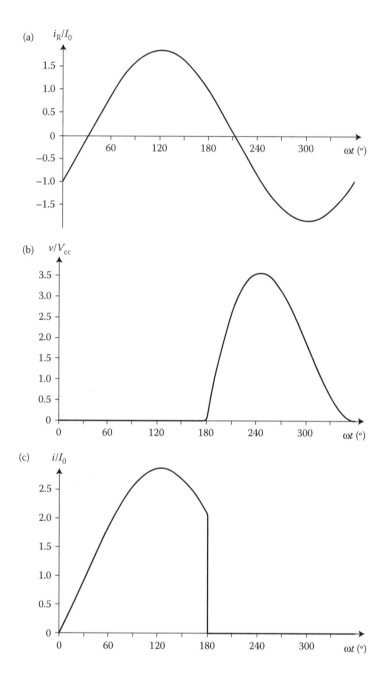

FIGURE 6.10
Normalized (a) load current and collector, (b) voltage, and (c) current waveforms for idealized optimum Class E with shunt capacitance.

The peak collector voltage V_{max} and current I_{max} can be determined by differentiating the appropriate waveforms given by Equations 6.41 and 6.42, respectively, and setting the results equal to zero, which gives

$$V_{max} = -2\pi\varphi V_{cc} = 3.562 V_{cc} \tag{6.46}$$

and

$$I_{max} = \left(\frac{\sqrt{\pi^2 + 4}}{2} + 1\right) I_0 = 2.8621 I_0 \tag{6.47}$$

The fundamental-frequency voltage $v_1(\omega t)$ across the switch consists of the two quadrature components, as shown in Figure 6.11, whose amplitudes can be found using Fourier formulas and Equation 6.41 as

$$V_R = \frac{1}{\pi}\int_0^{2\pi} v(\omega t)\sin(\omega t + \varphi)d\omega t = \frac{I_R}{\pi\omega C}\left(\frac{\pi}{2}\sin 2\varphi + 2\cos 2\varphi\right) \tag{6.48}$$

$$V_L = \frac{1}{\pi}\int_0^{2\pi} v(\omega t)\cos(\omega t + \varphi)d\omega t = -\frac{I_R}{\pi\omega C}\left(\frac{\pi}{2} + \pi\sin^2\varphi + 2\sin 2\varphi\right) \tag{6.49}$$

As a result, the idealized optimum series inductance L and shunt capacitance C can be calculated from

$$\frac{\omega L}{R} = \frac{V_L}{V_R} = 1.1525 \tag{6.50}$$

$$\omega CR = \frac{\omega C}{I_R}V_R = 0.1836 \tag{6.51}$$

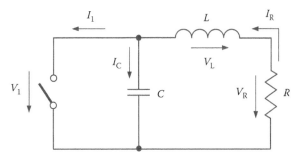

FIGURE 6.11
Equivalent Class-E load network at fundamental frequency.

The idealized optimum load resistance R can be obtained using Equations 6.43 and 6.45 for the supply voltage V_{cc} and fundamental-frequency output power P_{out} delivered to the load as

$$R = \frac{8}{\pi^2 + 4} \frac{V_{cc}^2}{P_{out}} = 0.5768 \frac{V_{cc}^2}{P_{out}} \qquad (6.52)$$

Finally, the phase angle of the load network seen by the switch at the fundamental frequency and required for an idealized optimum Class E with shunt capacitance can be determined through the load network parameters using Equations 6.50 and 6.51 by

$$\phi = \tan^{-1}\left(\frac{\omega L}{R}\right) - \tan^{-1}\left(\frac{\omega C R}{1 - (\omega L/R)\omega C R}\right) = 35.945° \qquad (6.53)$$

When realizing the optimum Class-E operation mode, it is very important to know up to which maximum frequency such an idealized efficient operation mode can be extended. In this case, it is possible to establish a relationship between the maximum frequency f_{max}, shunt capacitance C, and supply voltage V_{cc}. As a result, substituting Equation 6.51 into Equation 6.44 results in

$$I_0 = \pi \omega C V_{cc} \qquad (6.54)$$

Then, by taking into account the relationship between I_0 and I_{max} given in Equation 6.47, the maximum frequency of a nominal Class-E power amplifier with shunt capacitance can be evaluated from

$$f_{max} = \frac{1}{\pi^2} \frac{1}{\sqrt{\pi^2 + 4} + 2} \frac{I_{max}}{C_{out} V_{cc}} = \frac{I_{max}}{56.5 C_{out} V_{cc}} \qquad (6.55)$$

where $C = C_{out}$ is the device output capacitance limiting the maximum operation frequency of an ideal Class-E circuit [16].

The high-Q_L assumption for the series resonant $L_0 C_0$ circuit can lead to considerable errors if its value is substantially small in real circuits [17]. For example, for a 50% duty cycle, the values of the circuit parameters for the loaded quality factor less than unity can differ by several tens of percents. At the same time, for $Q_L \geq 7$, the errors are found to be less than 10% and become less than 5% for $Q_L \geq 10$. To match the optimum Class-E load network resistance R with standard load impedance $R_L = 50\ \Omega$, the series resonant $L_0 C_0$ circuit should be followed or fully replaced by the matching circuit, in which the first element must represent the series inductor to provide high impedance at harmonics [18].

6.2.2 Class E with Finite DC-Feed Inductance

In real practice, it is impossible to realize an RF choke with infinite impedance at the fundamental frequency and its harmonic components. Moreover, using a finite DC-feed inductance has an advantage of minimizing size, cost, and complexity of the overall circuit. The detailed approach to analyze the effect of a finite DC-feed inductance on the idealized Class-E mode with shunt capacitance and series filter was firstly described in Reference 19. It was based on Laplace-transform technique to solve a second-order differential equation describing the behavior of a Class-E load network with finite DC-feed inductance. However, since the results of numerical calculations are given only for a few particular cases, it is difficult to figure out the basic behavior of the load network elements and define simple equations for their parameters. Analytically, it was shown for a duty cycle of 50% based on the idealized optimum Class-E conditions that the series excessive reactance can be either inductive or capacitive depending on the values of the DC-feed inductance and shunt capacitance [20]. Based on the certain numbers of cases, a Lagrange polynomial interpolation was used to obtain explicit and directly usable design equations for an idealized Class E with finite DC-feed inductance and series inductive reactance [21].

The generalized second-order load network of a switchmode Class-E power amplifier with finite DC-feed inductance is shown in Figure 6.12a [18,22]. The load network consists of a shunt capacitance C, a parallel inductance L, a series reactance X, a series resonant L_0C_0 circuit tuned to the fundamental frequency, and a load resistance R. In a common case, a shunt capacitance C can represent the intrinsic device output capacitance and external circuit capacitance added by the load network, a parallel inductance L represents the finite DC-feed inductance, and a series reactance X can be positive (inductance), negative (capacitance), or zero depending on the Class-E mode. The active device is considered an ideal switch that is driven to provide the device switching between its on-state and off-state operation conditions. To simplify an analysis of the general-circuit Class-E power amplifier, a simplified equivalent circuit of which is shown in Figure 6.12b, it makes sense to introduce the preliminary assumptions similar to those for the Class-E power amplifier with shunt capacitance. The moments of switch-on is $\omega t = 0$ and switch-off is $\omega t = \pi$ with period of repeatability of input driving signal $\omega T = 2\pi$ determined by the input drive to the power amplifier. Assume the losses in the reactive circuit elements are negligible and the loaded quality factor of the series L_0C_0 circuit is sufficiently high. For lossless operation, it is necessary to provide the optimum zero-voltage and zero voltage-derivative conditions

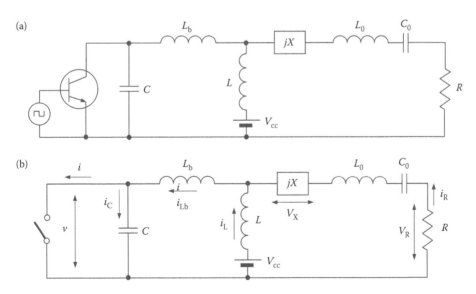

FIGURE 6.12
Equivalent circuits of the Class-E power amplifiers with generalized load network.

for voltage $v(\omega t)$ across the switch (just prior to the start of switch-on at the moment $\omega t = 2\pi$ when the transistor is saturated) given by Equations 6.29 and 6.30.

The output current flowing through the load is written as sinusoidal by

$$i_R(\omega t) = I_R \sin(\omega t + \varphi) \tag{6.56}$$

where I_R is the load current amplitude and φ is the initial phase shift.

When the switch is turned on for $0 \le \omega t < \pi$, the voltage on the switch $v(\omega t) = V_{cc} - v_L(\omega t) = 0$, the current flowing through the capacitance $i_C(\omega t) = \omega C(di_L/d\omega t) = 0$, and

$$i(\omega t) = i_L(\omega t) + i_R(\omega t) = \frac{1}{\omega L}\int_0^{\omega t} V_{cc}d\omega t + i_L(0) + I_R \sin(\omega t + \varphi)$$

$$= \frac{V_{cc}}{\omega L}\omega t + I_R[\sin(\omega t + \varphi) - \sin\varphi] \tag{6.57}$$

where the initial value for the current $i_L(\omega t)$ flowing through the DC-feed inductance L at $\omega t = 0$ can be found using Equation 6.56 for $i(0) = 0$ as $i_L(0) = -I_R \sin \varphi$.

When the switch is turned off for $\pi \le \omega t < 2\pi$, the switch current $i(\omega t) = 0$, and the current $i_C(\omega t) = i_L(\omega t) + i_R(\omega t)$ flowing through the capacitance C can be rewritten as

$$\omega C \frac{dv(\omega t)}{d\omega t} = \frac{1}{\omega L}\int_\pi^{\omega t} [V_{cc} - v(\omega t)]d\omega t + i_L(\pi) + I_R \sin(\omega t + \varphi) \tag{6.58}$$

under the initial off-state conditions $v(\pi) = 0$ and

$$i_L(\pi) = i(\pi) - i_R(\pi) = \frac{V_{cc}\,\pi}{\omega L} - \omega L I_R \sin\varphi \tag{6.59}$$

Equation 6.58 can be represented in the form of the linear nonhomogeneous second-order differential equation as

$$\omega^2 L C \frac{d^2v(\omega t)}{d(\omega t)^2} + v(\omega t) - V_{cc} - \omega L I_R \cos(\omega t + \varphi) = 0 \tag{6.60}$$

the general solution of which can be obtained in the normalized form of

$$\frac{v(\omega t)}{V_{cc}} = C_1 \cos(q\omega t) + C_2 \sin(q\omega t) + 1 - \frac{q^2 p}{1 - q^2}\cos(\omega t + \varphi) \tag{6.61}$$

where

$$q = \frac{1}{\omega\sqrt{LC}} \tag{6.62}$$

$$p = \frac{\omega L I_R}{V_{cc}} \tag{6.63}$$

and the coefficients C_1 and C_2 are determined from the initial off-state conditions by

$$
\begin{aligned}
C_1 &= -(\cos q\pi + q\pi \sin q\pi) \\
&\quad - \frac{qp}{1-q^2}[q\cos\varphi\cos q\pi - (1-2q^2)\sin\varphi\sin q\pi]
\end{aligned}
\tag{6.64}
$$

$$
\begin{aligned}
C_2 &= q\pi\cos q\pi - \sin q\pi \\
&\quad - \frac{qp}{1-q^2}[q\cos\varphi\sin q\pi + (1-2q^2)\sin\varphi\cos q\pi]
\end{aligned}
\tag{6.65}
$$

The DC supply current I_0 can be found using Fourier formula and Equation 6.57 by

$$I_0 = \frac{1}{2\pi}\int_0^{2\pi} i(\omega t)d\omega t = \frac{I_R}{2\pi}\left(\frac{\pi^2}{2p} + 2\cos\varphi - \pi\sin\varphi\right) \tag{6.66}$$

In an idealized Class-E operation mode, there is no nonzero voltage and current simultaneously that means a lack of the power losses and gives an idealized collector efficiency of 100%. This implies that the DC power P_0 and fundamental output power P_{out} are equal, which can be written similarly to that given in Equation 6.43. As a result, by using Equations 6.43 and 6.66 and taking into account that $R = V_R^2/2P_{out}$, the idealized optimum load resistance R for the specified values of a supply voltage V_{cc} and fundamental output power P_{out} can be obtained by

$$R = \frac{1}{2}\left(\frac{V_R}{V_{cc}}\right)^2 \frac{V_{cc}^2}{P_{out}} \tag{6.67}$$

where

$$\frac{V_R}{V_{cc}} = \frac{1}{\pi}\left(\frac{\pi^2}{2p} + 2\cos\varphi - \pi\sin\varphi\right) \tag{6.68}$$

The normalized load-network inductance L and capacitance C can be appropriately defined using Equations 6.62, 6.63, and 6.66 by

$$\frac{\omega L}{R} = \frac{p}{((\pi/2p) + (2/\pi)\cos\varphi - \sin\varphi)} \tag{6.69}$$

$$\omega CR = \frac{1}{(q^2(\omega L/R))} \tag{6.70}$$

The series reactance X, which may generally have an inductive, capacitive, or zero reactance, depending on the load network parameters, can be calculated using the two quadrature fundamental-frequency voltage Fourier components

$$V_R = -\frac{1}{\pi} \int_0^{2\pi} v(\omega t) \sin(\omega t + \varphi) d\omega t \qquad (6.71)$$

$$V_X = -\frac{1}{\pi} \int_0^{2\pi} v(\omega t) \cos(\omega t + \varphi) d\omega t \qquad (6.72)$$

The fundamental-frequency current flowing through the switch consists of the two quadrature components, the amplitudes of which can be found using Fourier formulas and Equation 6.57 by

$$\begin{aligned} I_R &= \frac{1}{\pi} \int_0^{2\pi} i(\omega t) \sin(\omega t + \varphi) d\omega t \\ &= \frac{I_R}{\pi} \left[\frac{\pi \cos\varphi - 2\sin\varphi}{p} + \frac{\pi}{2} - \sin 2\varphi \right] \end{aligned} \qquad (6.73)$$

$$\begin{aligned} I_X &= -\frac{1}{\pi} \int_0^{2\pi} i(\omega t) \cos(\omega t + \varphi) d\omega t \\ &= \frac{I_R}{\pi} \left[\frac{\pi \sin\varphi + 2\cos\varphi}{p} - 2\sin^2\varphi \right] \end{aligned} \qquad (6.74)$$

Generally, Equation 6.61 for a normalized collector voltage contains three unknown parameters q, p, and φ, which must be analytically or numerically determined. In a common case, the parameter q can be considered a variable, and the other two parameters p and φ are calculated from a system of the two equations by applying the two optimum zero-voltage and zero-voltage-derivative conditions given by Equations 6.29 and 6.30 into Equation 6.61. Figure 6.13 shows the dependences of the optimum parameters p and φ versus q for a Class E with finite DC-feed inductance.

Based on the calculated optimum parameters p and φ as the functions of q, the idealized optimum parameters of the Class-E load network with finite DC-feed inductance can be determined using Equations 6.67 through 6.70. The series reactance X can be calculated through the ratio of the two quadrature fundamental-frequency voltage Fourier components given in Equations 6.71 and 6.72 as

$$\frac{X}{R} = \frac{V_X}{V_R} \qquad (6.75)$$

The dependences of the normalized optimum DC-feed inductance $\omega L/R$ and series reactance X/R are shown in Figure 6.14a, while the dependences of the normalized optimum

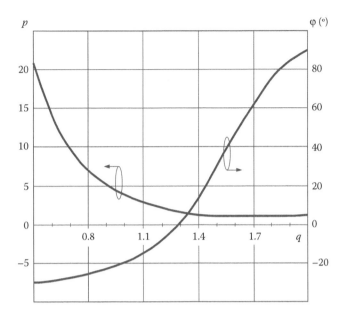

FIGURE 6.13
Optimum Class-E parameters p and φ vs. q.

shunt capacitance ωCR and load resistance RP_{out}/V_{cc}^2 are plotted in Figure 6.14b. Here, we can see that the subharmonic case of $q = 0.5$ is very close to a Class-E mode with shunt capacitance, since the value of the normalized inductance $\omega L/R$ is sufficiently high and the variations of normalized values of ωCR and RP_{out}/V_{cc}^2 are insignificant. The value of the series reactance X changes its sign from positive to negative, which means that the inductive reactance is followed by the capacitive reactance. As a result, there is a special case of the load network with a parallel circuit and a load resistance only when $X = 0$ at $q = 1.412$. In this case, the maximum value of the optimum load resistance R is provided for the same supply voltage and output power, thus simplifying the matching with the standard load of 50 Ω. Also, the values of a DC-feed inductance L become sufficiently small, thus making a parallel-circuit Class E very attractive for monolithic applications. The maximum operation frequency f_{max} is realized at $q = 1.468$, where the normalized optimum shunt capacitance ωCR reaches its maximum.

The graphical solutions for the optimum load-network parameters can be replaced by the analytical design equations represented in terms of the simple second-order and third-order polynomial functions given by Tables 6.2 and 6.3 for different ranges of the parameter q [23]. The maximum difference between the polynomial approximations and exact numerical solutions given in the graphic form is of about 2%.

6.2.3 Parallel-Circuit Class E

The theoretical analysis of a switchmode parallel-circuit Class-E power amplifier using a series filter, whose basic circuit is shown in Figure 6.15a, was first done by Kozyrev with the calculation of the voltage and current waveforms and some graphical results [24,25]. The load network consists of a finite DC-feed inductance L, a shunt capacitance C, a series L_0C_0 resonant circuit tuned to the fundamental frequency, and a load resistor R. In this

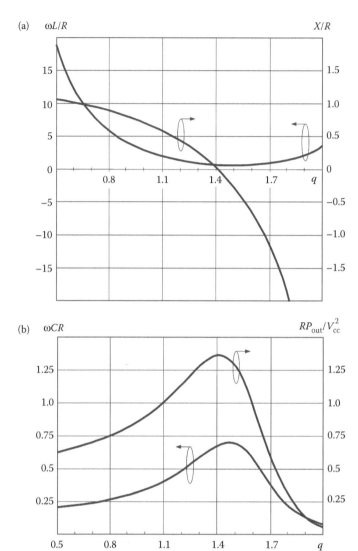

FIGURE 6.14
Normalized optimum Class-E load network parameters.

TABLE 6.2

Load Network Parameters for $0.6 < q < 1.0$

Parameter	Design Equation
$\omega L/R$	$44.93q^2 - 94.32q + 52.46$
ωCR	$0.426q^2 - 0.379q + 0.3$
X/R	$-0.73q^2 + 0.411q + 1.03$
$P_{out}R/V_{cc}^2$	$0.74q^2 - 0.6q + 0.76$

TABLE 6.3

Load Network Parameters for $1.0 < q < 1.65$

Parameter	Design Equation
$\omega L/R$	$8.085q^2 - 24.53q + 19.23$
ωCR	$-6.97q^3 + 25.93q^2 - 31.071q + 12.48$
X/R	$-2.9q^3 + 8.8q^2 - 10.2q + 5.02$
$P_{out}R/V_{cc}^2$	$-11.9q^3 + 42.753q^2 - 49.63q + 19.7$

case, the switch sees a parallel connection of the load resistor R and parallel LC circuit at the fundamental frequency, as shown in Figure 6.15b, where the real and imaginary collector fundamental-frequency current components I_X and I_R and real collector fundamental-frequency voltage component V_R are also indicated.

In the case of a parallel-circuit Class-E load network without series phase-shifting reactance, since the parameter q is unknown *a priori*, generally it is necessary to solve a system of three equations to define the three unknown parameters q, p, and φ. The two equations are the result of applying two optimum zero voltage and zero voltage-derivative conditions given by Equations 6.29 and 6.30 into Equation 6.61. Since the fundamental-frequency collector voltage is fully applied to the load, this means that its reactive part must have zero value, resulting in an additional equation

$$V_X = -\frac{1}{\pi}\int_0^{2\pi} v(\omega t)\cos(\omega t + \varphi)d\omega t = 0 \tag{6.76}$$

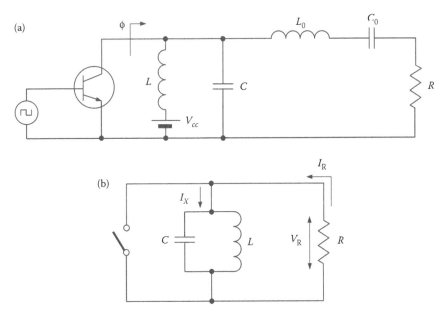

FIGURE 6.15
Equivalent circuits of parallel-circuit Class-E power amplifier.

Solving the system of three equations with three unknown parameters numerically gives the following values [18,26,27]:

$$q = 1.412 \tag{6.77}$$

$$p = 1.210 \tag{6.78}$$

$$\varphi = 15.155° \tag{6.79}$$

Figure 6.16 shows the normalized (a) load current and collector, (b) voltage, and (c) current waveforms for an idealized optimum parallel-circuit Class-E operation. From collector voltage and current waveforms, it follows that there is no nonzero voltage and current simultaneously. When this happens, no power loss occurs and an ideal collector efficiency of 100% is achieved.

By using Equations 6.67 through 6.70, the optimum load resistance R, parallel inductance L, and parallel capacitance C can be appropriately obtained by

$$R = 1.365 \frac{V_{cc}^2}{P_{out}} \tag{6.80}$$

$$L = 0.732 \frac{R}{\omega} \tag{6.81}$$

$$C = \frac{0.685}{\omega R} \tag{6.82}$$

The DC supply current I_0 can be calculated from Equation 6.66 as

$$I_0 = 0.826 I_R \tag{6.83}$$

The phase angle ϕ seen from the device collector at the fundamental frequency can be represented either through the two quadrature fundamental-frequency current Fourier components I_X and I_R or as a function of load network elements by

$$\phi = \tan^{-1}\left(\frac{R}{\omega L} - \omega RC \right) = 34.244° \tag{6.84}$$

If the calculated value of the optimum Class-E resistance R is too small or differs significantly from the required load impedance, it is necessary to use an additional matching circuit to deliver maximum output power to the load. It should be noted that, among a family of the Class-E load networks, a parallel-circuit Class-E load network offers the largest value of R, thus simplifying the final matching design procedure. In this case, the first series element of such matching circuits should be the inductor to provide high-impedance conditions for harmonics, as shown in Figure 6.17.

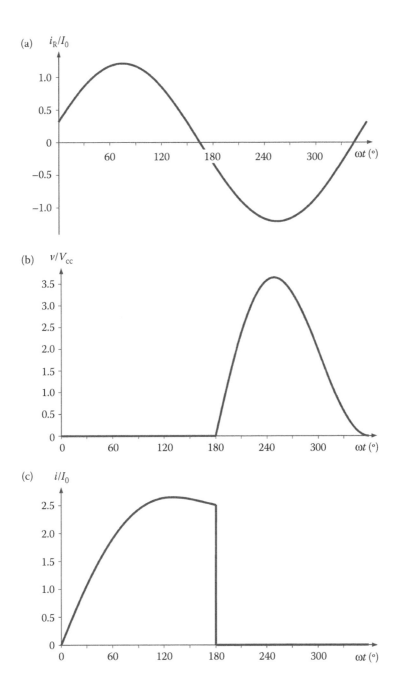

FIGURE 6.16
Normalized (a) load current and collector, (b) voltage, and (c) current waveforms for idealized optimum parallel-circuit Class E.

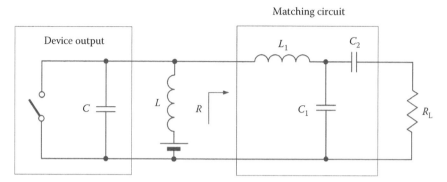

FIGURE 6.17
Parallel-circuit Class-E power amplifier with lumped matching circuit.

The peak collector current I_{max} and peak collector voltage V_{max} can be determined from Equations 6.57, 6.61, and 6.83 as

$$I_{max} = 2.647I_0 \tag{6.85}$$

$$V_{max} = 3.647V_{cc} \tag{6.86}$$

The maximum frequency f_{max} can be calculated using Equations 6.80 and 6.82 when $C = C_{out}$, where C_{out} is the device output capacitance, as

$$f_{max} = 0.0798\frac{P_{out}}{C_{out}V_{cc}^2} \tag{6.87}$$

which is 1.4 times higher than maximum operation frequency for an idealized optimum Class-E power amplifier with shunt capacitance [28].

At microwave frequencies, the parallel inductance L can be replaced by a short-length short-circuited transmission line TL according to

$$Z_0 \tan\theta = \omega L \tag{6.88}$$

where Z_0 and θ are the characteristic impedance and electrical length of such a transmission line, respectively [29]. By using Equation 6.81 determining the optimum parallel inductance L for a parallel-circuit Class-E mode, Equation 6.88 can be rewritten as

$$\tan\theta = 0.732\frac{R}{Z_0} \tag{6.89}$$

6.3 Broadband Class E with Shunt Capacitance

In the basic circuit of a Class-E power amplifier with shunt capacitance shown in Figure 6.18a, the harmonic impedance of the series fundamentally tuned $L_0 C_0$ circuit is assumed to be high due to its high loaded quality factor. The value of the shunt capacitor C must also be correct to produce the correct voltage when the switch is turned off to satisfy the steady-state switching conditions. In this case, the load phase angle of the series-tuned circuit composed of the total inductor $(L + L_0)$ and capacitor C_0, which determines the optimum angle for producing the correct voltage waveform, can be obtained according to Equation 6.50 at the resonant radian frequency $\omega_0 = 1/\sqrt{L_0 C_0}$ as

$$\theta = \tan^{-1}\left(\frac{\omega_0 L}{R}\right) = \tan^{-1} 1.1525 = 49.052° \tag{6.90}$$

If the load network is designed without incorporating the shunt capacitance, a simple broadband network with an optimum load angle $\theta = 49.052°$ given in Equation 6.90 can be designed. Then, this phase angle reduces to the required angle $\phi = 35.945°$ given by Equation 6.53 when a shunt capacitance is added. The circuit schematic of a simple load network capable of presenting a constant load angle over a very large bandwidth is shown in Figure 6.18b [1]. The load network consists of a low-Q series $L_0 C_0$ circuit connected in parallel with an inductance L that allows a constant susceptance to be maintained over a wide bandwidth. The frequency behavior of the conductance $\text{Re}Y_{\text{net}}$ and susceptance $\text{Im}Y_{\text{net}}$

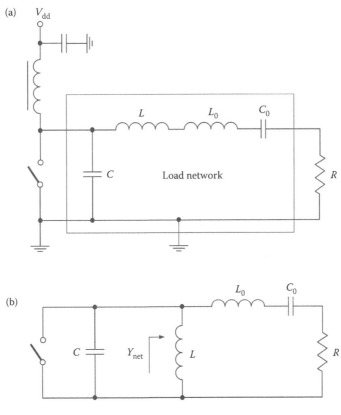

FIGURE 6.18
Load networks of Class E with shunt capacitance.

of this load network with parameters $L = 42$ nH, $L_0 = 30$ nH, and $C_0 = 35$ pF are shown in Figure 6.19a and b, respectively, where combining of the susceptance of the series resonant circuit with negative slope (curve 1) and the susceptance of the shunt inductance with positive slope (curve 2) provides a constant total susceptance over a very wide frequency range (curve 3).

In order to maintain the load angle constant in a wide frequency range, the slope of the susceptance provided by the inductance L should be canceled by the slope provided by the resonant L_0C_0 circuit. The load-network admittance of Figure 6.18b can be written as

$$Y_{net} = -\frac{j}{\omega L} + \frac{1}{R + j(\omega L_0 - (1/\omega C_0))} \tag{6.91}$$

which reduces at the resonant frequency to

$$Y_{net} = \frac{1}{R} - \frac{j}{\omega L} \tag{6.92}$$

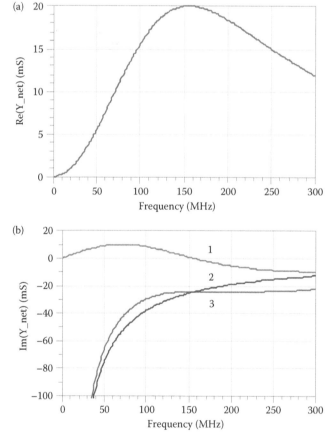

FIGURE 6.19
Conductance and susceptance of broadband Class-E circuit.

For slope cancellation, it is necessary to apply a zero-derivative condition of Equation 6.4 to Equation 6.91 for the load-network susceptance $B_{net} = \mathrm{Im}Y_{net}$ at the radian resonant frequency ω_0. As a result

$$\frac{1}{\omega_0^2 L} = \frac{2}{\omega_0^2 C_0 R^2} \tag{6.93}$$

Thus, the design equations to calculate the parameters of a broadband Class-E load network providing maximum flatness can be calculated from

$$L = \frac{R}{\omega_0 \tan\theta} \tag{6.94}$$

$$C_0 = \frac{2L}{R^2} \tag{6.95}$$

$$L_0 = \frac{1}{\omega_0^2 C_0} \tag{6.96}$$

To reduce the output power at the harmonics, such a simple load network can be combined with a broadband matching network and a bandpass filter. As an example, a complete circuit based on a low-pass *L*-type matching section and a third-order Chebyshev bandpass filter, as shown in Figure 6.20a, was designed to deliver 12 W into a 50-Ω load across the frequency bandwidth from 130 to 180 MHz using a 12-V power supply [1]. From Figure 6.20b, it follows that this load network presents a constant magnitude of input impedance of 12 Ω (curve 1) and a load phase angle of around 36° (curve 2) over the required wide frequency range. As a result, the broadband MOSFET Class-E power amplifier was capable of providing a fairly constant efficiency at approximately 60% with suppression of the second, third, and fourth harmonics better than 45 dB below fundamental. The drain efficiency of a GaN HEMT power amplifier with a Butterworth bandpass filter in the load network can be increased to more than 80% in a frequency bandwidth from 600 to 800 MHz with an output power more than 45 dBm [30,31]. To provide the frequency bandwidth of 30% around the center bandwidth frequency of 1 GHz, the load network can be composed of a series transmission line and a shunt open-circuit stub [32].

Figure 6.21a shows the example of a reactance compensation load network for a Class-E power amplifier with shunt capacitance including a series transmission line and a parallel resonant circuit. In this case, the reactance of a Class-E load network with shunt capacitance and series inductance varies similarly to that of the series resonant circuit with positive slope, whereas the required negative slope is provided by the parallel resonant circuit. Selection of the proper characteristic impedance and electrical length of the series transmission line enables the magnitude of two slopes to be made identical, so as to achieve a constant total reactance and phase of the load network impedance Z_{net} over a wide frequency range. The simulation results at the fundamental frequency show that the resistance $\mathrm{Re}Z_{net}$ varies from 35 Ω at 30 MHz to 68 Ω at 70 MHz, as shown in Figure 6.21b by curve 1, whereas the load-network phase varies between 27° and 40° in more than octave bandwidth from 33 to 80 MHz (curve 2).

Generally, the design of a practical multisection *LC* filter is based on some approximate equivalence between lumped and distributed elements, which can be established

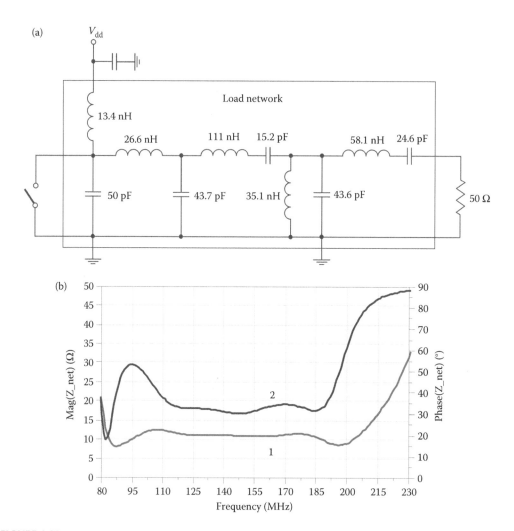

FIGURE 6.20
Broadband Class-E load network with bandpass filter (a) and its magnitude and phase angle (b) vs. frequency.

by applying a Richards's transformation [33]. This implies that the distributed circuits composed of equal-length open- and short-circuited transmission lines can be treated as lumped elements under the transformation

$$s = j \tan \frac{\pi \omega}{2\omega_0} \qquad (6.97)$$

where $s = j\omega/\omega_c$ is the conventional normalized complex frequency variable, ω_c is the cutoff radian frequency, and ω_0 is the radian frequency, for which the transmission lines are a quarter wavelength [34].

As a result, for a unity characteristic impedance and cutoff frequency, the one-port impedance of a short-circuited transmission line corresponds to the reactive impedance of a lumped inductor Z_L as

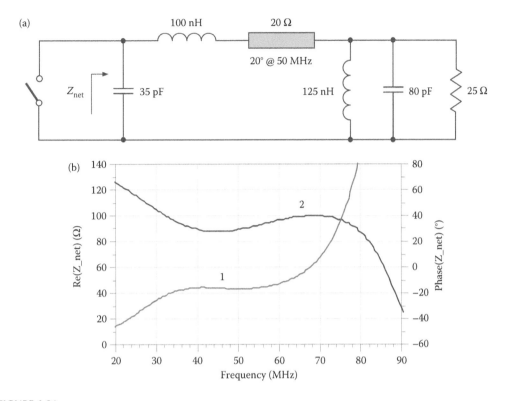

FIGURE 6.21
Class-E reactance compensation circuit with lumped elements and transmission line (a) and its resistance and phase angle (b) vs. frequency.

$$Z_L = sL = jL \tan \frac{\pi \omega}{2\omega_0} \tag{6.98}$$

Similarly, the one-port admittance of an open-circuited transmission line corresponds to the reactive admittance of a lumped capacitor Y_C as

$$Y_C = sC = jC \tan \frac{\pi \omega}{2\omega_0} \tag{6.99}$$

The results obtained by Equations 6.98 and 6.99 show that an inductor L can be replaced with a short-circuit stub of electrical length $\theta = \pi\omega/2\omega_0$ and characteristic impedance $Z_0 = L$, while a capacitor C can be replaced with an open-circuit stub of electrical length $\theta = \pi\omega/2\omega_0$ and characteristic impedance $Z_0 = 1/C$.

From Equation 6.97, it follows that the cutoff occurs when $\omega = \omega_c$, resulting in

$$\tan \frac{\pi\omega_c}{2\omega_0} = 1 \tag{6.100}$$

which gives a stub length $\theta = 45°$ (or $\pi/4$) with $\omega_c = \omega_0/2$. Hence, the inductors and capacitors of a lumped-element filter can be replaced with short-circuit and open-circuit stubs,

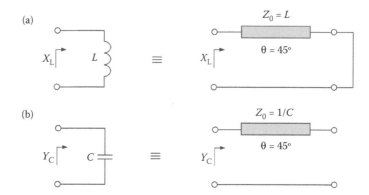

FIGURE 6.22
Equivalence between lumped elements and transmission lines.

as shown in Figure 6.22. Since the lengths of all stubs are the same and equal to $\lambda/8$ at the cutoff frequency ω_c, these lines are called the commensurate lines. At the frequency $\omega = \omega_0$, the transmission lines will be a quarter-wavelength long, resulting in an attenuation pole. However, at any frequency away from ω_c, the impedance of each stub will no longer match the original lumped-element impedances, and the filter response will differ from the desired filter prototype response. Note that the response will be periodic in frequency, repeating every $4\omega_c$.

Figure 6.23a shows the idealized simulation setup of a 10-W 28-V broadband Class-E power-amplifier circuit designed to operate over a frequency bandwidth from 1.7 to 2.7 GHz and based on a GaN HEMT CGH40010 device, where both the input matching circuit and load network are composed of ideal transmission lines. To provide an input broadband matching, it is possible to use a multisection matching transformer consisting of the stepped transmission-line sections with different characteristic impedances and electrical lengths [35,36]. Such an input matching structure is convenient in practical implementation since there is no need to use any tuning capacitors. The nominal Class-E load resistance can be calculated for $P_{out} = 15$ W, $V_{dd} = 28$ V, and $V_{sat} = 2.5$ V according to Equation 6.52 as

$$R = 0.5768 \frac{(V_{dd} - V_{sat})^2}{P_{out}} = 25\ \Omega \tag{6.101}$$

where P_{out} is the output power at the fundamental frequency, V_{dd} is the drain supply voltage, and V_{sat} is the saturation voltage defined from the device output current–voltage characteristics. In this case, the parallel resonant circuit in the broadband Class-E load network connected in parallel to a 25-Ω load is represented by the open- and short-circuit stubs, each having a characteristic impedance of 50 Ω and electrical length of 45° at 2 GHz. Simulation results show that drain efficiencies of 75% and greater can be achieved over whole required frequency bandwidth with a power gain of about 11 dB and an output power more than 42 dBm.

Figure 6.23b shows the implementation of an idealized circuit of a broadband GaN HEMT Class-E power amplifier shown in Figure 6.23a into a RO4360 substrate, where an additional series transmission line with low-characteristic impedance is used to match an idealized 25-Ω load with a standard 50-Ω load. As a result, an output power around

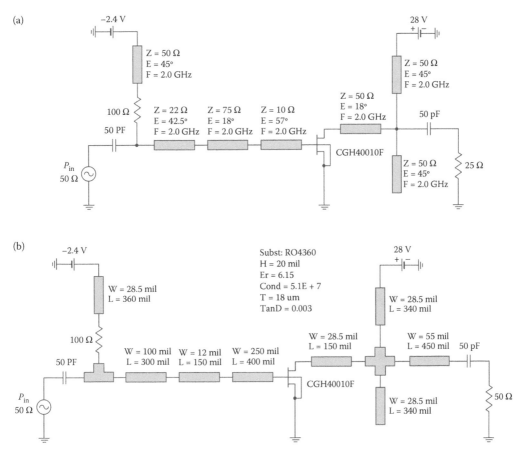

FIGURE 6.23
Idealized (a) and simulated (b) circuit schematics of broadband GaN HEMT Class-E power amplifier.

42 dBm with a power gain of more than 10 dB was simulated for an input power of 31 dBm, as shown in Figure 6.24a. In this case, the drain efficiency over 72% was achieved across the required frequency range from 1.7 to 2.7 GHz, as shown in Figure 6.24b. Previously, a *PAE* above 60% was achieved between 1.87 and 2.11 GHz with an output power varying from 20 to 23 dBm for a medium-power broadband pHEMT Class-E power amplifier using a transmission-line parallel resonant circuit with short- and open-circuit stubs [37]. For a harmonically tuned GaN HEMT broadband power amplifier using open-circuit stubs in the load network incorporating a three-section bandpass filter, an output power of around 100 W with a drain efficiency of more than 65% in a frequency bandwidth from 1.55 to 2.25 GHz was achieved [38].

Ideally, the requirements to the output matching network for a broadband Class-E power amplifier should include not only achieving the inductive fundamental-frequency impedance across the desired bandwidth, but it is also necessary to provide high reactance at harmonics. In this case, to provide an impedance matching with high-transformation ratio and satisfy Class-E requirements over octave bandwidth with minimum in-band ripple, at least three stages for the low-pass ladder-type matching network are needed. Here, the series inductor as a first matching element can provide high-impedance condition at harmonics and the device output capacitance should provide the required capacitive

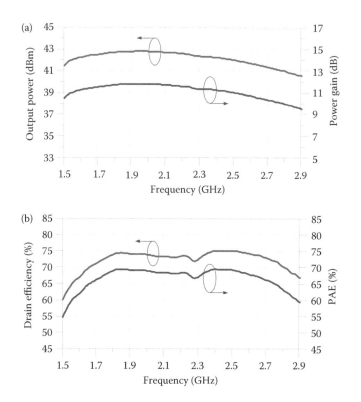

FIGURE 6.24
Output power, power gain, and efficiency vs. frequency.

harmonic reactances. By using a three-stage six-order low-pass filter-matching load net-work in a GaN HEMT Class-E power amplifier where the series inductors are replaced by the short-length high-impedance transmission-line sections and the shunt capacitors are replaced by the open-circuit low-impedance stubs, a drain efficiency of 63%–89% with an output power of 10–20 W and a power gain of 10–13 dB was measured in a frequency bandwidth from 0.9 to 2.2 GHz at a supply voltage of 26 V [39].

6.4 Broadband Parallel-Circuit Class E

The susceptance compensation technique can be directly applied to the switchmode paral-lel-circuit Class-E power amplifier because its load-network configuration has exactly the same structure with shunt and series resonant circuits, as shown in Figure 6.25a [40,41]. In this case, the nominal load resistance R and phase angle ϕ of the parallel-circuit Class-E load network can be obtained from Equations 6.80 and 6.84, respectively. The parallel inductance L and shunt capacitance C required for an idealized optimum parallel-circuit Class-E operation are calculated as functions of the load resistance R at the operating fre-quency from Equations 6.81 and 6.82, respectively. The parameters of the series resonant L_0C_0 circuit must be chosen to provide a constant phase angle of the load network over a required wide frequency bandwidth.

As a result, by substituting Equations 6.81 and 6.82 into Equation 6.6, the series capacitance C_0 and inductance L_0 can be calculated at the center bandwidth frequency ω_0 by

$$L_0 = 1.026 \frac{R}{\omega_0} \qquad (6.102)$$

$$C_0 = \frac{1}{\omega_0^2 L_0} \qquad (6.103)$$

Wider frequency bandwidth with high-efficiency performance can be achieved using a double-susceptance compensation circuit shown in Figure 6.25b, where $L_0 C_0$ and $L_1 C_1$ are the series and parallel resonant circuits, respectively [42]. In this case, similarly to the broadband design in a Class-E mode with shunt capacitance using a double-susceptance compensation, the parameters of the series and shunt resonant circuits for the broadband design in a parallel-circuit Class-E mode with the corresponding loaded quality factors $Q_0 = \omega_0 L_0 / R$ and $Q_1 = \omega_0 C_1 R$, which are close to unity and greater, can approximately

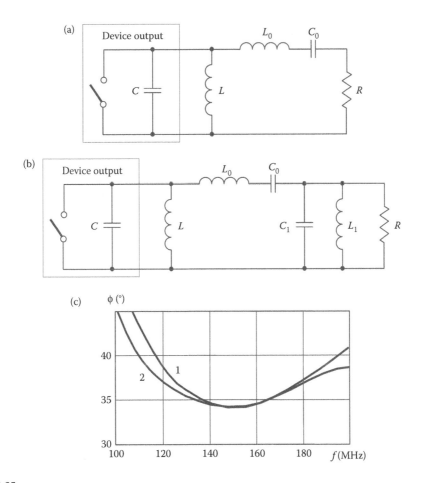

FIGURE 6.25
Single- and double-susceptance compensation circuits and their phase angles vs. frequency.

be calculated from Equations 6.15 and 6.16, where the load angle $\theta = \tan^{-1}(R/\omega_0 L)$ using Equation 6.81 is taken into account. Such a load network can be considered as a broadband matching-forming circuit, which provides simultaneously the Class-E switching conditions and matching with a standard 50-Ω load over wide frequency bandwidth [43].

The circuit simulations for these two types of susceptance compensation load networks were performed at a center bandwidth frequency $f_0 = 150$ MHz for a standard load resistance $R = 50\ \Omega$. Figure 6.25c shows the frequency dependences of the load-network phase angle ϕ for the single-susceptance (curve 1) and double-susceptance (curve 2) compensation circuits, demonstrating their very broadband operation capability. Using just a single-susceptance load network yields a significant widening of the operating frequency bandwidth with a minimum deviation of the magnitude and phase of the load-network impedance. A double-susceptance compensation load network obtains a maximum deviation from the optimum value of about 34° by only 3° in a frequency range from 120 to 180 MHz.

To achieve the high-efficiency broadband operation mode with a high-power gain in the VHF frequency band, it is best to design the power amplifier based on silicon LDMOSFET devices. It is easy to provide a very good broadband input matching using lossy-matching circuit, especially at operating frequencies about 10 times lower than the device transition frequency f_T. Figure 6.26 shows the circuit schematic of an LDMOSFET power amplifier designed for operation in a 2:1 frequency bandwidth from 100 to 200 MHz using a double-susceptance compensation load network with broadband matching properties at the fundamental frequency [36]. The input lossy-matching circuit includes a simple L-transformer connected in parallel with a series circuit consisting of an inductor of 20 nH and a resistor of 50 Ω. This provides a minimum input return loss at 200 MHz of about 15 dB and an input $VSWR$ less than 1.4 over the entire frequency bandwidth from 100 to 200 MHz.

From Figure 6.27a, it follows that, for such an octave-band VHF Class-E power amplifier with an input power of 1 W using a 1.25-μm LDMOSFET device with a total gate width

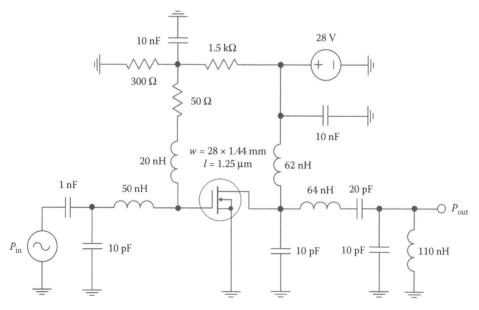

FIGURE 6.26
Simulated broadband Class-E LDMOSFET power amplifier.

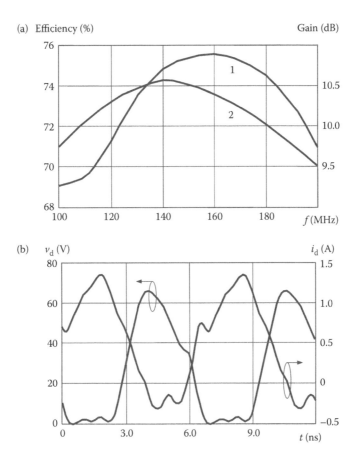

FIGURE 6.27
Broadband performance of Class-E LDMOSFET power amplifier.

of 28×1.44 mm, a power gain of 10 dB with deviation of only ±0.5 dB (curve 2) can be achieved with a drain efficiency of about 70% and higher (curve 1). An analysis of the simulated drain voltage and current waveforms at the center bandwidth frequency of 150 MHz shown in Figure 6.27b demonstrates that the broadband operating mode is very close to a nominal parallel-circuit Class-E operation mode, although the impedance conditions at higher harmonics are not controlled properly. As seen from the plots when the transistor is turned on, high values of drain current (up to 1.3 A) are achieved with small saturation voltages of 0 to 4 V. On the other hand, when the transistor is turned off, the drain current continues to flow, but now through the device gate–drain capacitance C_{gd} and drain–source capacitance C_{ds} but not through the active channel. A drain efficiency of 74% with an output power of 8 W across the frequency range from 136 to 174 MHz with a power flatness of 0.7 dB was measured for a parallel-circuit Class-E LDMOSFET power amplifier with a low-supply voltage of 7.2 V [44]. A power-added efficiency can be increased to 80% and more in a frequency range of 140–180 MHz with an output power of 34.4 ± 1.5 dBm using a GaN HEMT device [45].

Similarly, the transmission-line susceptance compensation technique can also be applied to a parallel-circuit Class-E power amplifier where the series transmission line of a quarter wavelength at the center bandwidth frequency can be used instead of a series

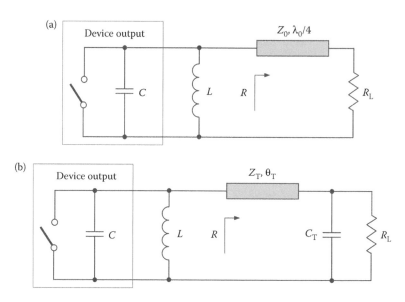

FIGURE 6.28
Transmission-line susceptance compensation circuits.

$L_0 C_0$ resonant circuit, as shown in Figure 6.28a. In some practical cases, the series quarterwave line can be replaced by an equivalent low-pass π-type circuit consisting of a series transmission line with higher characteristic impedance and electrical length much less than 90°, and two shunt capacitors when the capacitance adjacent to the device output can be counted within the total shunt capacitance required for a nominal parallel-circuit Class-E mode. If it is necessary to additionally provide an output matching between the nominal Class-E resistance R and standard load $R_L = 50\ \Omega$, a series quarterwave line can be replaced by a low-pass L-type matching circuit with a series transmission line and a shunt capacitor, as shown in Figure 6.28b.

Figure 6.29a shows the example of a transmission-line broadband Class-E load network, where the parallel inductor is replaced by a short-length short-circuited transmission line, which can be easily implemented on a printed-circuit board to minimize insertion losses. The electrical lengths of the transmission lines are given at the center bandwidth frequency of 300 MHz. In this case, the input load-network resistance varies from 17 Ω at 225 MHz to 47.5 Ω at 400 MHz, as shown in Figure 6.29b by curve 1, with much less variation from 18.5 to 27 Ω in a frequency range from 250 to 350 Ω. The phase stays almost constant around 33° in a frequency range from 250 to 350 Ω and varies from 22.5° at 225 MHz to 39.5° at 400 MHz (curve 2).

Figure 6.30a shows the circuit schematic of a broadband high-efficiency microstrip LDMOSFET power amplifier with an output power of around 20 W and a power gain of more than 12 dB in a frequency range from 225 to 400 MHz at a DC supply voltage of 28 V. Here, to approximate the parallel-circuit Class-E mode in a wide frequency range, the load network was designed to realize a single-susceptance compensation technique using a parallel short-length transmission line in conjunction with a single L-type transmission-line transformer, since a ratio between the device equivalent output resistance required for an optimum Class-E operation and the standard load of 50 Ω is not significant. The input matching circuit includes the two low-pass L-type matching sections to compensate for the device input capacitance over the entire frequency range. A lossy parallel resistance of 75 Ω

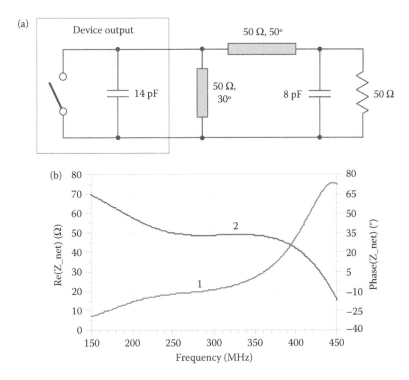

FIGURE 6.29
Transmission-line Class-E load network with susceptance compensation (a) and its resistance and phase angle (b) vs. frequency.

is necessary to simplify the matching procedure and improve the input return loss. As a first step, each matching network structure is calculated at the center-band frequency based on the technical requirements and device equivalent circuit parameters. Then, to optimize the power amplifier performance over the entire frequency band, the simplest and fastest way is to apply an optimization procedure using computer simulators to satisfy certain criteria. For such a broadband power amplifier, the minimum output power ripple and input return loss with maximum power gain and efficiency can be chosen as the criteria. Generally, by applying a nonlinear broadband optimization technique and setting the ranges of electrical length of the transmission lines between 0° and 90° and parallel capacitances from 0 to 100 pF, we can obtain the parameters of the input matching circuit and output load network.

However, to speed up this procedure, it is best to optimize circuit parameters separately for the input and output circuits. In this case, the input matching circuit is loaded by the device equivalent input series *RC* circuit, consisting of its gate resistance and gate–source capacitance. The load network must include at its input the device equivalent output shunt *RC* circuit consisting of an optimum Class-E load resistance required for a specified output power and supply voltage and drain–source capacitance. In this case, it is sufficient to use a fast linear optimization process, which will take only a few minutes to complete the circuit design procedure. Finally, the resulting optimized values are incorporated into the overall power amplifier circuit for each element and final optimization is performed using a nonlinear active device model. The optimization process is finalized by choosing the nominal level of input power with optimizing elements in narrower ranges of their values of about 10%–20% for most critical elements. For practical convenience, it is advisable to

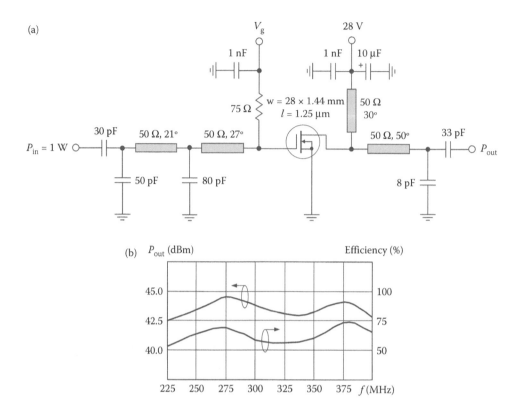

FIGURE 6.30
Broadband high-efficiency microstrip LDMOSFET power amplifier.

choose the characteristic impedances of all transmission lines of 50 Ω. Figure 6.30b shows the simulated broadband high-efficiency power amplifier performance achieving an output power of 42.5–44.5 dBm, a power gain of 13.5 ± 1 dB, and a drain efficiency of 64 ± 10% in a frequency bandwidth from 225 to 400 MHz.

The circuit schematic of a broadband two-stage InGaP/GaAs HBT power amplifier intended to operate in the WCDMA handset transmitters is shown in Figure 6.31 [40,41]. The MMIC part of this power amplifier contains the transistors with emitter areas of the first and second stage as large as 540 and 3600 μm², input matching circuit, interstage matching circuit, and bias circuits on a die with dimensions of less than 1 mm². The MMIC packaged in a 3 × 3 mm² package was mounted on an FR4 substrate, which contains the output matching circuit and microstrip lines. Standard ceramic chip capacitors were used in the output matching circuit, and no further additional tuning was necessary. In this case, a very short microstrip line operating as a DC-feed inductance is required to approximate parallel-circuit Class-E switching conditions.

Figure 6.32a shows that the small-signal gain varies within 22.5 ± 0.5 dB and the input return loss is greater than 13 dB in a frequency range from 1.6 GHz to more than 2 GHz, thus confirming the broadband operation of the amplifier. Without any tuning of the output matching circuit, a saturated output power greater than 30 dBm and a *PAE* greater than 50% were obtained. These single-tone measurements shown in Figure 6.32b were performed at the respective center-band frequencies 1.75 and 1.88 GHz. Using high-*Q* capacitors in output matching circuit can improve the power-added efficiency by about 8%,

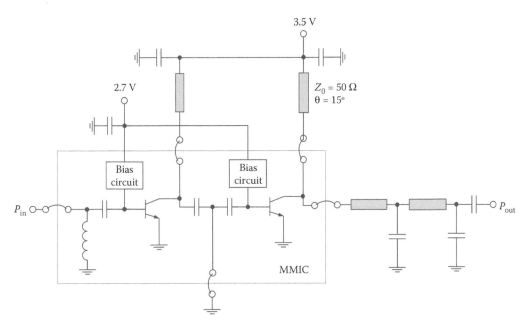

FIGURE 6.31
Circuit schematic of parallel-circuit GaAs HBT Class-E MMIC power amplifier.

resulting in close to 60% being obtained. At the same time, the power amplifier provides high-linearity performance in a handset WCDMA band (1920–1980 MHz) at a 3.5-dB back-off output power of 27 dBm with a power gain of 22.6 dB and a sufficiently high efficiency. The measured *PAE* reached value of 38.3% at center bandwidth frequency of 1.95 GHz with an *ACLR* of −37 dBc at a 5-MHz offset and an *ACLR* of −56 dBc at a 10-MHz offset.

6.5 High-Power RF Class-E Power Amplifiers

The load network of a high-power broadband VHF Class-E power amplifier can be based on lumped low-pass *LC* matching sections to match the nominal Class-E load with the standard 50-Ω load. When the frequency bandwidth is not very wide, of about 50%, it is possible to use the simplest single-section low-pass π-transformer to provide a drain efficiency greater than 60% over a frequency bandwidth from 80 to 135 MHz with an average output power of 20 W using a 28-V 60-W SiC MESFET device [46]. However, variation of both the output power (more than 3 dB) and drain efficiency (more than 15%) across this frequency bandwidth is significant, since such a simple load network cannot maintain the required real and imaginary part of the Class-E load network constant enough over the entire frequency range. In this case, the imaginary part of the load network increases from a low-band frequency to a high-band frequency, characterized by the positive slope.

Figure 6.33a shows the circuit schematic of a push–pull Class-E power amplifier using a 50-V balanced MOSFET device, producing an output power up to 200 W and operating in a frequency range from 1.8 to 128 MHz [47]. The circuit incorporates a broadband input matching circuit so that the frequency changes require only retuning or switching

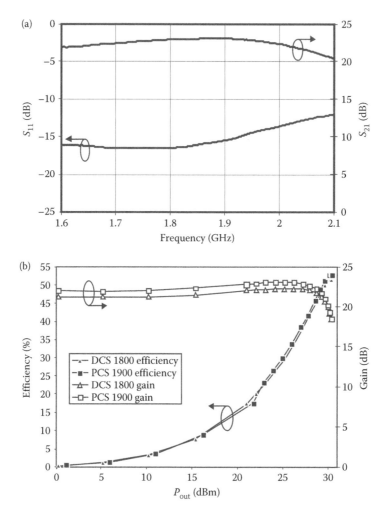

FIGURE 6.32
Input return loss, gain, and efficiency vs. frequency.

of the output filter. In this case, the drain efficiency varies from 90% at 1.8 MHz to 70% at 128 MHz. The input signal is transformed and split by a broadband 4:1 Guanella transformer, whose outputs represent two 10-W, 6.25-Ω resistors shunting the MOSFET gates and providing the required resistive loads for the input transformer. Such an input circuit is operated in a broadband mode until the reactances of the MOSFET gates become less than 6.25 Ω. Operation at higher frequencies requires the addition of inductors to cancel the MOSFET gate capacitances. The drain loads of 3.125 Ω in a Class-E mode with shunt capacitance allow about 100 W to be produced in each side of the MOSFET with a peak drain voltage below the breakdown voltage. The total drain capacitance is approximately correct for optimum Class-E operation at about 85 MHz. At lower frequencies, the drain-shunt capacitors are therefore added to achieve optimum Class-E operation. The broadband output transformer represents a balun that splits its 6.25-Ω load into two 3.125-Ω drain loads. The output impedance-transforming filter is a low-pass two-section ladder-type network where the shunt variable capacitors are responsible for load control and

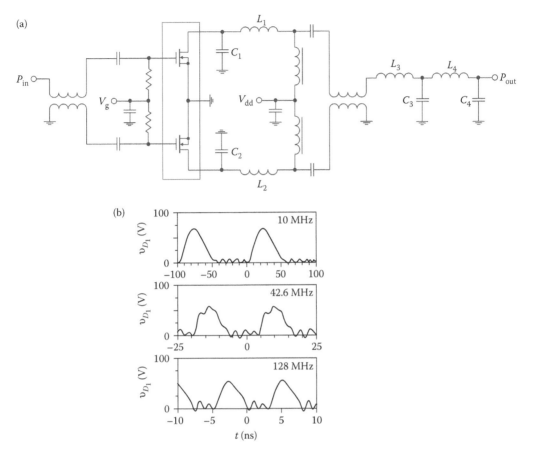

FIGURE 6.33
Push–pull MOSFET Class-E power amplifier and drain-voltage waveforms.

efficiency adjustment and the first series inductor provides sufficient reactance to ensure adequate suppression of the third harmonic. The drain-voltage waveforms for operation at 10, 42.6, and 128 MHz are shown in Figure 6.33b, where the ideal optimum Class-E waveform is satisfied at 10 MHz and close at 42.6 MHz. At 128 MHz, the power amplifier is operating in a suboptimum Class-E mode.

The third-order bandpass broadband Class-E load network can be optimized not only to provide required optimum inductive impedance at the fundamental frequency, but also purely capacitive optimized reactances at the second and third harmonics. Figure 6.34a shows such a third-order parallel-circuit Class-E load network, where the impedance-transforming network is used to transform the standard resistive 50-Ω load to the load required for proper termination of the Class-E load network [48]. As a result, the drain efficiency near 70% and better can be achieved over the frequency bandwidth from 95 to 135 MHz for a 12-W, 12-V MOSFET power amplifier with a 4:1 Ruthroff transmission-line transformer in an impedance-transforming network providing 12.5 Ω to the Class-E load network at the fundamental frequency and several harmonics. Figure 6.34b shows the output circuit of a high-power broadband 28-V LDMOSFET power amplifier with a broadband low-loss impedance transformer located right at the output port of the power transistor [49]. The transformer is made of three rings of low impedance 15-Ω semirigid

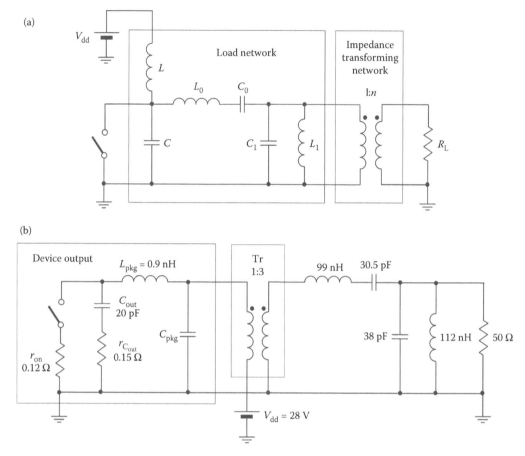

FIGURE 6.34
Output circuits of broadband VHF Class-E power amplifiers.

coaxial cable, where the cable outer jackets are connected in parallel to form the primary of the transformer and the inner conductors are connected series to form the secondary of the transformer. In order to reduce losses to a minimum, no magnetic core is used. A lumped network is located after the transformer to provide both proper loads at the harmonics and at the fundamental frequency for a nominal Class-E with shunt capacitance. In this case, high drain efficiencies up to 86% with output powers of greater than 120 W were measured in a frequency bandwidth of VHF broadcasting from 88 to 114 MHz.

6.6 Microwave Monolithic Class-E Power Amplifiers

Generally, by providing an open-circuit termination for the second- and third-harmonic components, the collector efficiency of a microwave Class-E power amplifier can be increased by 10% [50]. In this case, the second-harmonic termination has the most impact on the collector efficiency, while effect of an open-circuit termination for the fourth harmonic is negligible. Moreover, the variation of the second-harmonic load reflection coefficient by 10% in magnitude from 1 to 0.9 and ±20° in phase angle results in an insignificant efficiency variation within 1% only. Figure 6.35a shows the circuit schematic of a monolithic broadband Class-E power amplifier with a chip size of 2 mm × 2.2 mm. The load network is a compromise solution between having a low-insertion loss and meeting the necessary requirements for the optimum Class-E operation with nonzero voltage and voltage-derivative switching conditions [51]. This is accomplished by using two open-circuit stubs in conjunction with a shunt capacitor, where the first open-circuit stub in combination with the series transmission line presents broadband high-impedance terminations for the second harmonics within 18–22 GHz, while the combination of the second open-circuit stub and the shunt capacitor presents broadband low impedances at the third harmonics within 27–33 GHz and also transforms the optimum load impedances at the fundamental frequencies. The simulated loading conditions presented to the output of the device at the fundamental (inductive impedance), second (high impedance), and third harmonic (low impedance) frequencies are shown in Figure 6.35b, which are proved to be adequate for broadband Class-E power amplifiers. As a result, using an indium phosphide (InP) double HBT (DHBT) technology, a *PAE* of 49%–65% with an output power of 18–22 dBm

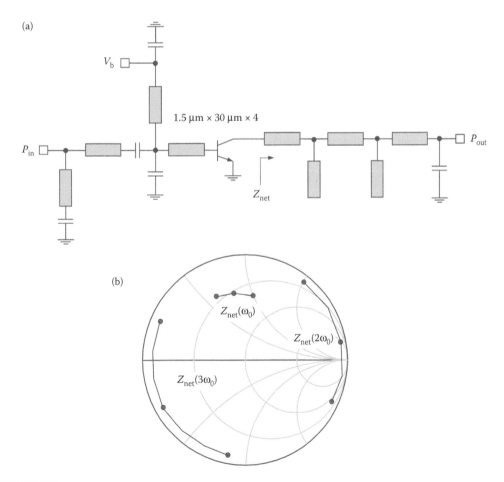

FIGURE 6.35
Broadband *X*-band In DHBT Class-E power amplifier and impedance conditions.

was achieved over the frequency bandwidth from 9 to 11 GHz [50]. Based on an InP DHBT technology, a single-stage broadband *X*-band Class-E power amplifier can also achieve a *PAE* of 45%–60% with an output power of 19–21.5 dBm and a power gain of 9–11.5 dB over a 34% bandwidth, from 8.2 to 11.6 GHz [52].

To increase the overall efficiency of a two-stage power amplifier, it may be assumed that it is worthwhile to optimize both amplifying stages to operate in a Class-E mode. For example, for a hybrid microwave GaAs MESFET Class-E power amplifier using the same devices in both stages, the maximum two-stage power-added efficiency was achieved as high as 52% (including connector loss) with a corresponding power gain of 16 dB and an output power of 20 dBm at a carrier frequency of 10 GHz and a supply voltage of 4.2 V [53]. However, owing to Class-E operation mode of the driver stage, the overall power gain is sufficiently small, thus affecting the overall efficiency. Therefore, by using a Class-AB driver stage, similar efficiency can be achieved with substantially higher power gain. As a result, for a monolithic microwave two-stage high-efficiency InP DHBT power amplifier shown in Figure 6.36a where the driver stage is operated in a Class-AB mode and the output stage is operated in a Class-E mode, a *PAE* of 52% with an output power of 24.6 dBm and a power gain of 24.6 dB was achieved at a carrier frequency of 8 GHz and a supply voltage of 4 V. The total emitter area of the driver-stage device was chosen to be 90 μm^2,

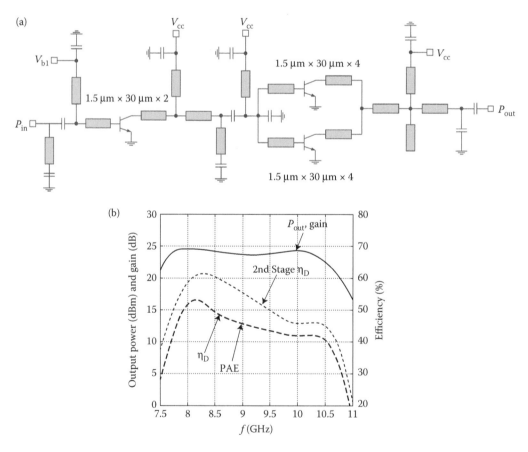

FIGURE 6.36
Broadband two-stage *X*-band In DHBT Class-E power amplifier and its performance.

providing a *PAE* of the driver stage above 40% and an adequate power to push the output stage deep into compression, as required for a switchmode Class-E operation. The output stage consists of two active devices with a total emitter area of 360 μm^2 combined in parallel reactively, taking care to provide odd-mode instability suppression resistors between the base and collector of each transistor. The power-added efficiency is maintained greater than 40% over a frequency bandwidth from 7.7 to 10.5 GHz, as shown in Figure 6.36b [53].

Figure 6.37 shows the circuit schematic of a two-stage broadband Class-E power amplifier implemented in a 0.5-μm enhancement/depletion pHEMT process with a chip size of 2×2 mm^2, which is intended to operate in a frequency range from 1.5 to 3.8 GHz with a *PAE* better than 62% and an output power of more than 27 dBm at $V_{dd} = 6$ V [54]. In this case, to provide high operation efficiency in a wide frequency range, the Class-E load network with reactance compensation technique followed by the low-pass matching network is used. The driver stage is designed to operate in a Class-AB mode with a small quiescent current for high gain and high efficiency when both input and interstage matching circuits are conjugately matched. For a 0.5-μm pHEMT two-stage broadband Class-E power amplifier with a chip size of 5.25×2.8 mm^2, a *PAE* above 50% with an output power over 36 dBm at a drain supply voltage of 6 V was obtained in a frequency range of 3.0–3.75 GHz [55].

Figure 6.38 shows the circuit schematic of a compact single-stage broadband Class-E GaN HEMT power amplifier, where the load network is based on reactance compensation technique with a parallel circuit followed by the low-Q series resonant circuit [56]. The use of a finite DC-feed inductance has advantages in terms of the output power and maximum frequency of operation and results in a higher load resistance than in the classical Class-E configuration with infinite RF choke. To shape the gate voltage waveform, the second-harmonic signal is short-circuited at the gate by the series resonant circuit representing a second-harmonic trap. The high- and low-pass matching sections form a bandpass input matching circuit, where a 300-Ω parallel resistor with a 37-Ω bias resistor provide an unconditional stability of the power amplifier both at low and high frequencies. The actual size of a broadband Class-E power amplifier with the input matching circuit and

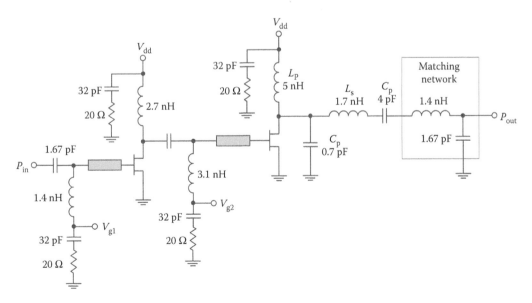

FIGURE 6.37
Circuit schematic of broadband Class-E pHEMT power amplifier.

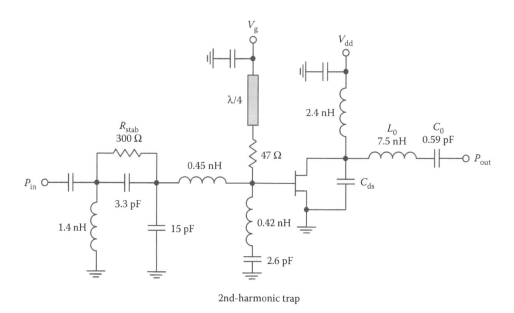

FIGURE 6.38
Circuit schematic of broadband Class-E GaN HEMT power amplifier.

load network implemented in a two-layer Rogers laminate with a dielectric permittivity of 3.5 is only 1.1 × 1.6 cm². As a result, a drain efficiency above 74% and an output power more than 7 W with an input power of 600 mW at $V_{dd} = 40$ V can be achieved across the bandwidth of 2.0–2.5 GHz. Wider frequency bandwidth of 2.1–2.7 GHz with a drain efficiency exceeding 63% and an output power above 9.3 W can be provided without input second-harmonic trap and retuning the finite DC-feed inductance. By using a two-section *LC* ladder output matching circuit in a GaN HEMT Class-E power amplifier, a drain efficiency over 68% with an output power of 42–65 W in a frequency bandwidth from 1.7 to 2.3 GHz at a supply voltage of 35 V was achieved for a compact hybrid implementation of the power amplifier with an effective area of 2 × 2 cm², where the bondwire inductors and MIM capacitors are used [57].

6.7 CMOS Class-E Power Amplifiers

The use of cascode topologies is extremely attractive for CMOS power amplifiers, especially at high-output powers and DC supply voltages. Optimizing the cascode topology requires setting the bias voltage V_g of the common-gate transistor shown in Figure 6.39a to minimize the voltage drop across the oxide of each transistor M_1 and M_2 when these voltage drops become equal allowing the use of approximately twice the supply voltage [58,59]. However, there is an additional power loss mechanism as a specific property of a cascode configuration in a switching Class-E mode when the common-source device M_1 is turned off, which is associated with the charging and discharging processes of the shunt parasitic capacitor C_p consisting of the drain-bulk capacitance of the device M_1 and gate–source and source–bulk capacitances of the device M_2. This results in a finite switching

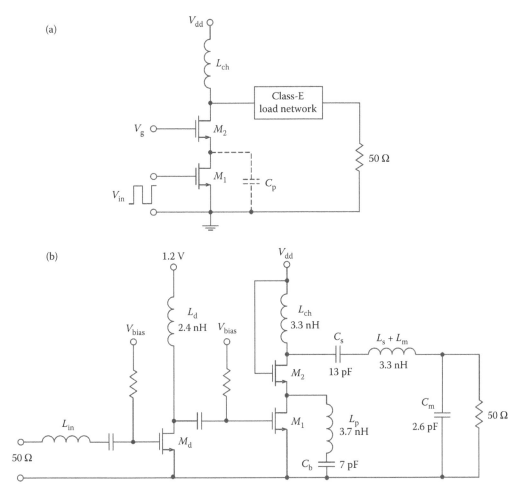

FIGURE 6.39
Broadband cascode Class-E power amplifier with compensating inductor.

time of a common gate device M_2 when it cannot be instantly switched from the saturation mode to the pinch-off mode and operates in the active region when simultaneously output current and output voltage are positive with the output power dissipation within the device. The parasitic capacitance C_p can be 3–4 times larger than the drain-bulk capacitance of the device M_2, resulting in a power loss as large as 20% of the output power. A simple and effective way to minimize this power loss contribution is to use a parallel inductor L_p resonating the parasitic capacitor C_p at the operating frequency, as shown in Figure 6.39b, where C_b is the blocking capacitor. The series resonant circuit required to provide a sinusoidal current flowing to the load is replaced by the series inductor L_m and shunt capacitor C_m forming an *L*-type lumped transformer to match the optimum Class-E load resistance with a standard resistance of 50 Ω. As a result, the two-stage cascode Class-E power amplifier with a compensating inductor implemented in a 0.13-μm CMOS process achieved a drain efficiency of 71% and a *PAE* of 67% when delivering an output power of 23 dBm at an operating frequency of 1.7 GHz with a supply voltage of 2.5 V. The driving stage with a supply voltage of 1.2 V is biased in a Class C. The value of an inductor L_d is

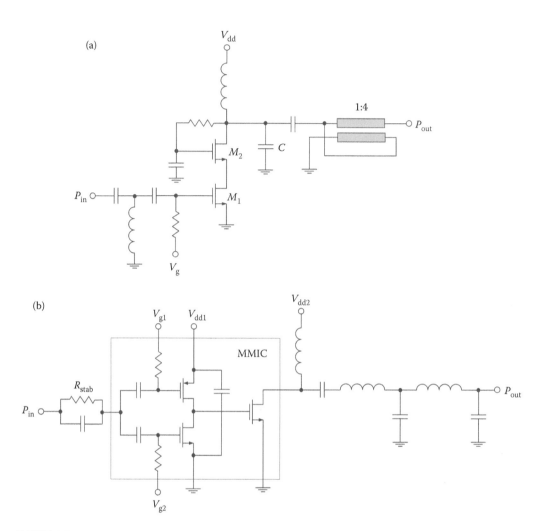

FIGURE 6.40
Circuit schematics of broadband Class-E CMOS power amplifiers.

chosen to compensate for the gate–source capacitance of the device M_1. The power-added efficiency higher than 60% was measured over the frequency bandwidth of 1.4–2.0 GHz.

To realize a broadband high-efficiency operation of the fully integrated CMOS Class-E power amplifier, a broadband and low-loss 1:4 Ruthroff-type transmission-line transformer based on the broadside-coupled transmission lines can provide an impedance transformation from 12.2 ± 0.1 to 50 Ω [60]. In a six-layer 0.18-µm CMOS process, the thickest top metal 6 is used as the primary winding, the identical thick metal stacked from metal 1 to metal 4 is used as the secondary winding to improve insertion loss, and both windings are wound in loops keeping the 1:1 turns ratio to reduce the transformer size. Figure 6.40a shows the circuit schematic of a 0.18-µm CMOS Class-E power amplifier composed of the two nMOS transistors in a cascode configuration and one shunt capacitor in the load network required for optimum Class-E operation. Here, the series LC resonant circuit at the fundamental frequency of the Class-E power amplifier is replaced by the 1:4 transmission-line transformer operating as a broadband bandpass filter. To enhance the reliability of the transistors, the thick-oxide transistor M_2 is used for the common-gate stage, and the thin-oxide transistor M_1 is used for the common-source stage. The fully integrated CMOS Class-E power amplifier with a 1:4 transmission-line transformer exhibits a broadband output power level of 24 ± 0.2 dBm from 2.4 to 3.5 GHz at a supply voltage of 3.6 V, with a maximum *PAE* of 33.2% at 2.6 GHz.

Figure 6.40b shows the circuit schematic of a two-stage broadband Class-E CMOS power amplifier, where the power output stage is formed by a high-voltage, extended-drain, thick-oxide nMOS device implemented in a standard 65-nm CMOS technology [61]. The total gate width of the transistor is 3.84 mm and the channel length is 0.28 μm, realizing an on-resistance of 0.7 Ω, an off-resistance of 10 kΩ, and a drain–source capacitance of approximately 4.14 pF. To drive the output stage as a switch, a square-wave signal is generated by an inverter-based driver implemented using standard thick-oxide MOS devices with a gate length of 0.28 μm. To reduce the peak drain voltage and improve reliable operation, a suboptimum Class-E operation is applied. The broadband load network represents an off-chip two-section *LC* ladder circuit. As a result, a measured output power of 30.5 ± 0.5 dBm, a power gain of 16.5 ± 0.5 dB, a drain efficiency above 67%, and a *PAE* above 52% are achieved across the frequency bandwidth from 550 to 1050 MHz.

References

1. J. K. A. Everard and A. J. King, Broadband power efficient Class E amplifiers with a non-linear CAD model of the active MOS device, *J. IERE*, 57, 52–58, 1987.
2. G. L. Matthaei, A study of the optimum design of wide-band parametric amplifiers and up-converters, *IRE Trans. Microwave Theory Tech.*, MTT-9, 23–38, 1961.
3. J. T. DeJaeger, Maximum bandwidth performance of a nondegenerate parametric amplifier with single-tuned idler circuit, *IEEE Trans. Microwave Theory Tech.*, MTT-12, 459–467, 1964.
4. B. L. Humphreys, Characteristics of broadband parametric amplifiers using filter networks, *Proc. IEE*, 111, 264–274, 1964.
5. C. S. Aitchison and R. V. Gelsthorpe, A circuit technique for broadbanding the electronic tuning range of Gunn oscillators, *IEEE J. Solid-State Circuits*, SC-12, 21–28, 1977.
6. C. S. Aitchison, Method of improving tuning range obtained from a varactor-tuned Gunn oscillator, *Electron. Lett.*, 10, 94–95, 1974.
7. R. V. Gelsthorpe and C. S. Aitchison, Analytical evaluation of the components necessary for double reactance compensation of an oscillator, *Electron. Lett.*, 12, 485–486, 1976.
8. G. Chapman and C. S. Aitchison, Circuit technique for broadband impedance matching of passive loads, *IEE J. Microwaves Optics Acoustics*, 3, 43–50, 1979.
9. E. Camargo and D. Consoni, Reactance compensation matches FET circuits, *Microwaves*, 24, 93–95, 1985.
10. R. Soares, *GaAs MESFET Circuit Design*, Boston: Artech House, 1988.
11. N. O. Sokal and A. D. Sokal, Class E—A new class of high-efficiency tuned single-ended switching power amplifiers, *IEEE J. Solid-State Circuits*, SC-10, 168–176, 1975.
12. N. O. Sokal and A. D. Sokal, High-efficiency tuned switching power amplifier, US Patent 3,919,656, November 1975.
13. N. O. Sokal, Class E high-efficiency power amplifiers, from HF to microwave, *1998 IEEE MTT-S Int. Microwave Sym. Dig.*, Baltimore, MD, vol. 2, pp. 1109–1112, 1998.
14. F. H. Raab, Idealized operation of the Class E tuned power amplifier, *IEEE Trans. Circuits Systems*, CAS-24, 725–735, 1977.
15. B. Molnar, Basic limitations on waveforms achievable in single-ended switching-mode tuned (Class E) power amplifiers, *IEEE J. Solid-State Circuits*, SC-19, 144–146, 1984.

16. T. B. Mader, E. W. Bryerton, M. Marcovic, M. Forman, and Z. Popovic, Switched-mode high-efficiency microwave power amplifiers in a free-space power-combiner array, *IEEE Trans. Microwave Theory Tech.*, MTT-46, 1391–1398, 1998.

17. M. Kazimierczuk and K. Puczko, Exact analysis of Class E tuned power amplifier at any Q and switch duty cycle, *IEEE Trans. Circuits Systems*, CAS-34, 149–158, 1987.

18. A. Grebennikov, N. O. Sokal, and M. J. Franco, *Switchmode RF and Microwave Power Amplifiers*, Oxford, UK: Academic Press, 2012.

19. R. E. Zulinski and J. W. Steadman, Class E power amplifiers and frequency multipliers with finite DC-feed inductance, *IEEE Trans. Circuits Systems*, CAS-34, 1074–1087, 1987.

20. C.-H. Li and Y.-O. Yam, Maximum frequency and optimum performance of Class E power amplifiers, *IEE Proc. Circuits Devices Syst.*, 141, 174–184, 1994.

21. D. Milosevic, J. van der Tang, and A. van Roermund, Explicit design equations for Class-E power amplifiers with small DC-feed inductance, *Proc. 2005 Eur. Conf. Circuit Theory Design*, Cork, Ireland, vol. 3, pp. 101–104, 2005.

22. A. Grebennikov, Load network design techniques for Class E RF and microwave amplifiers, *High Frequency Electron.*, 3, 18–32, 2004.

23. M. Acar, A. J. Annema, and B. Nauta, Analytical design equations for Class-E power amplifiers, *IEEE Trans. Circuits Systems I: Regular Papers*, CAS-I-54, 2706–2717, 2007.

24. V. B. Kozyrev, Single-ended switching-mode tuned power amplifier with filtering resonant circuit (in Russian), *Poluprovodnikovye Pribory v Tekhnike Svyazi*, 6, 152–166, 1971.

25. A. Grebennikov, Class E high-efficiency power amplifiers: Historical aspect and future prospect, *Appl. Microwave Wireless*, 14, 64–71, 2002, 64–72, 2002.

26. A. Grebennikov and H. Jaeger, Class E with parallel circuit—A new challenge for high-efficiency RF and microwave power amplifiers, *2002 IEEE MTT-S Int. Microwave Sym. Dig.*, Seattle, WA, vol. 3, pp. 1627–1630, 2002.

27. A. Grebennikov, Switched-mode RF and microwave parallel-circuit Class E power amplifiers, *Int. J. RF Microwave Computer-Aided Eng.*, 14, 21–35, 2004.

28. M. K. Kazimierczuk and W. A. Tabisz, Class C-E high-efficiency tuned power amplifier, *IEEE Trans. Circuits Systems*, CAS-36, 421–428, 1989.

29. A. Grebennikov and H. Jaeger, High efficiency transmission line tuned power amplifier, U.S. Patent 6,552,610, April 2003.

30. Al Tanany, A. Sayed, and G. Boeck, Broadband GaN switch mode Class E power amplifier for UHF applications, *2009 IEEE MTT-S Int. Microwave Symp. Dig.*, Boston, MA, pp. 761–764, 2009.

31. Al Tanany, A. Sayed, O. Bengtsson, and G. Boeck, Time domain analysis of broadband GaN switch mode Class-E power amplifier, *Proc. 5th German Microwave Conf.*, Berlin, Germany, pp. 254–257, 2010.

32. V. S. Rao Gudimetla and A. Z. Kain, Design and validation of the load networks for broadband Class E amplifiers using nonlinear device models, *1999 IEEE MTT-S Int. Microwave Symp. Dig.*, Anaheim, CA, pp. 823–826, 1999.

33. P. I. Richards, Resistor-transmission-line circuits, *Proc. IRE*, 36, 217–220, 1948.

34. R. Saal and E. Ulbrich, On the design of filters by synthesis, *IRE Trans. Circuit Theory*, CT-5, 284–327, 1958.

35. V. P. Meschanov, I. A. Rasukova, and V. D. Tupikin, Stepped transformers on TEM-transmission lines, *IEEE Trans. Microwave Theory Tech.*, MTT-44, 793–798, 1996.

36. A. Grebennikov, *RF and Microwave Power Amplifier Design*, New York: McGraw-Hill, 2004.

37. Y. Qin, S. Gao, P. Butterworth, E. Korolkiewicz, and A. Sambell, Improved design technique of a broadband Class-E power amplifier at 2 GHz, *Proc. 35th Eur. Microwave Conf.*, Paris, 1, 1–4, 2005.

38. Al Tanany, D. Gruner, and G. Boeck, Harmonically tuned 100 W broadband GaN HEMT power amplifier with more than 60% PAE, *Proc. 41st Eur. Microwave Conf.*, Manchester, UK, 159–162, 2011.

39. K. Chen and D. Peroulis, Design of highly efficient broadband Class-E power amplifier using synthesized low-pass matching networks, *IEEE Trans. Microwave Theory Tech.*, MTT-59, 3162–3173, 2011.

40. H. Jaeger, A. Grebennikov, E. P. Heaney, and R. Weigel, Broadband high-efficiency monolithic InGaP/GaAs HBT power amplifiers for 3G handset applications, *2002 IEEE MTT-S Int. Microwave Symp. Dig.*, Seattle, WA, vol. 2, pp. 1035–1038, 2002.

41. H. Jaeger, A. Grebennikov, E. P. Heaney, and R. Weigel, Broadband high-efficiency monolithic InGaP/GaAs HBT power amplifiers for wireless applications, *Int. J. RF Microwave Computer-Aided Eng.*, 13, 496–519, 2003.

42. A. Grebennikov, Simple design equations for broadband Class E power amplifiers with reactance compensation, *2001 IEEE MTT-S Int. Microwave Symp. Dig.*, Phoenix, AZ, vol. 3, pp. 2143–2146, 2001.

43. V. I. Degtev and V. B. Kozyrev, Transistor single-ended switching-mode power amplifier with forming circuit (in Russian), *Poluprovodnikovaya Elektronika v Tekhnike Svyazi*, 26, 178–188, 1986.

44. N. Kumar, C. Prakash, A. Grebennikov, and A. Mediano, High-efficiency broadband parallel-circuit Class E power amplifier with reactance-compensation technique, *IEEE Trans. Microwave Theory Tech.*, MTT-56, 604–612, 2008.

45. E. Khansalee, N. Puangngernmak, and S. Chalermwisutkul, Design of 140–170 MHz Class E power amplifier with parallel circuit on GaN HEMT, *Proc. Int. Electrical Eng./Electron. Computer Telecom. Inform. Technol. Conf.*, Chiang Mai, Thailand, pp. 570–574, 2010.

46. W. Chen, X. Li, L. Wang, Z. Feng, and X. Xue, A novel broadband VHF MESFET Class-E high power amplifier, *Microwave Optical Technol. Lett.*, 52, 272–276, 2010.

47. F.H. Raab, Broadband Class-E power amplifier for HF and VHF, *2006 IEEE MTT-S Int. Microwave Symp. Dig.*, San Francisco, CA, pp. 902–905, 2006.

48. F. J. Ortega-Gonzalez, Load-pull wideband Class-E amplifier, *IEEE Microwave Wireless Components Lett.*, 17, 235–237, 2007.

49. F. J. Ortega-Gonzalez, High power wideband Class-E power amplifier, *IEEE Microwave Wireless Components Lett.*, 20, 569–571, 2010.

50. T. K. Quach, P. M. Watson, W. Okamura, E. N. Kaneshiro, A. Gutierrez-Aitken, T. R. Block, J. W. Eldredge et al., Ultra-high efficiency power amplifier for space radar applications, *IEEE J. Solid-State Circuits*, SC-37, 1126–1134, 2002.

51. P. Watson, R. Neidhard, L. Kehias, R. Welch, T. Quach, R. Worley, M. Pacer, R. Pappaterra, R. Schweller, and T. Jenkins, Ultra-high efficiency operation based on an alternative Class-E mode, *2000 IEEE GaAs IC Symp. Dig.*, Seattle, WA, pp. 53–56, 2000.

52. P. Watson, T. Quach, H. Axtel, A. Gutierrez-Aitken, E. Kaneshiro, W. Lee, A. Mattamana, et al. An indium phosphide X-band Class-E power MMIC with 40% bandwidth, *2005 IEEE CSIC Symp. Dig.*, Palm Springs, CA, pp. 220–223, 2005.

53. S. Pajic, N. Wang, P. M. Watson, T. K. Quach, and Z. Popovic, X-band two-stage high-efficiency switched-mode power amplifiers, *IEEE Trans. Microwave Theory Tech.*, MTT-53, 2899–2907, 2005.

54. C.-H. Lin and H.-Y. Chang, A high efficiency broadband Class-E power amplifier using a reactance compensation technique, *IEEE Microwave Wireless Components Lett.*, 20, 507–509, 2010.

55. M. van Wanum, R. van Dijk, P. de Hek, and F. E. van Vliet, Broadband S-band Class E HPA, *Proc. 4th Eur. Microwave Integrated Circuits Conf.*, Roma, Italy, pp. 29–32, 2009.

56. M. P. van der Heijden, M. Acar, and J. S. Vromans, A compact 12-W high-efficiency 2.1–2.7 GHz Class-E GaN HEMT power amplifier for base stations, *2009 IEEE MTT-S Int. Microwave Symp. Dig.*, Boston, MA, pp. 657–660, 2009.

57. K. Shi, D. A. Calvillo-Cortes, L. C. N. de Vreede, and F. van Rijs, A compact 65 W 1.7–2.3 GHz Class-E GaN power amplifier for base stations, *Proc. 6th Eur. Microwave Integrated Circuits Conf.*, Manchester, UK, pp. 542–545, 2011.

58. A. Mazzanti, L. Larcher, R. Brama, and F. Svelto, A 1.4 GHz–2 GHz wideband CMOS Class-E power amplifier delivering 23 dBm peak with 67% PAE, *2005 IEEE RFIC Symp. Dig.*, Long Beach, CA, pp. 425–428, 2005.

59. A. Mazzanti, L. Larcher, R. Brama, and F. Svelto, Analysis of reliability and power efficiency in cascode Class-E PAs, *IEEE J. Solid-State Circuits*, SC-41, 1222–1229, 2006.
60. H.-Y. Liao, M.-W. Pan, and H.-K. Chiou, Fully-integrated CMOS Class-E power amplifier using broadband and low-loss 1:4 transmission-line transformer, *Electron. Lett.*, 46, 1490–1491, 2010.
61. R. Zhang, M. Acar, M. P. van der Heijden, M. Apostolidou, L. C. N. de Vreede, and M. W. Leenaerts, A 550–1050 MHz +30 dBm Class-E power amplifier in 65 nm CMOS, *2011 IEEE RFIC Symp. Dig.*, Baltimore, MD, pp. 289–292, 2011.

7

Broadband and Multiband Doherty Amplifiers

This chapter describes the historical aspect of the Doherty approach to power amplifier design and modern trends in Doherty amplifier design techniques using broadband and multiband Doherty architectures. To increase efficiency over the power-backoff range, the switchmode broadband Class-E mode can be used in the load network.

7.1 Historical Aspect and Conventional Doherty Architectures

A new power amplifier technique for amplitude-modulated (AM) radio-frequency signals was introduced by William H. Doherty in broadcasting in the mid-1930s as a more efficient alternative to both conventional amplitude-modulation techniques and Chireix outphasing [1,2]. This new technique achieves plate circuit efficiencies of up to 60%–65% independent of modulation by means of a combined action of the variation of load distribution of the vacuum tubes, and the variation of the circuit impedance over the modulation cycle. When Doherty joined the radio development department of Bell Telephone Laboratories in June 1929, he was engaged in the development of high-power radio transmitters for transoceanic radiotelephony and broadcasting. As a result, in 1936, he invented a means to greatly improve the efficiency of radio-frequency power amplifiers, quickly termed the "Doherty amplifier." It was first used in a 50-kW transmitter with audio-frequency feedback providing a resulting distortion level from less than 1% at lower frequencies to a few percent at high-audio frequencies. In this case, the power amplifier was operated at an efficiency of 60% representing a reduction of nearly one half in the all-day power consumption as compared with the power required in the conventional type of linear power amplifier operating at 33% efficiency [3]. The IRE Morris Liebmann Memorial Award was voted to Doherty in May 1937 for his improvement in the efficiency of radio-frequency power amplifiers [4]. By 1940, Doherty amplifiers were incorporated in 35 commercial radio stations worldwide, at powers up to 50 kW.

In subsequent years, Doherty amplifiers continued to be used in a number of medium- and high-power low-frequency (LF) and medium-frequency (MF) vacuum-tube AM transmitters [5,6]. In August 1953, a 1-MW vacuum-tube transmitter where the outputs of two 500-kW Doherty amplifiers were joined in a bridge-type combiner began regular operation in the long-wave band in Europe. Doherty amplifiers had also been considered for use in solid-state MF and high-frequency (HF) systems, as well as in high-power ultra-high-frequency (UHF) transmitters [7–9]. The practical implementation of a classical triode-based Doherty scheme was restricted by its substantial nonlinearity for both linear amplification of AM signals and grid-type signal modulation that required complicated envelope correction and feedback linearization circuits. At the same time, Doherty amplifiers employing tetrode transmitting tubes could improve their overall performance when the modulation was applied to the screen grids of both the carrier and peaking tubes, while the control grids of both tubes are fed by an essentially constant level of RF excitation [10]. This resulted in the peaking tube being modulated upward during the positive half of the modulating cycle and the carrier tube being modulated downward during the negative half of the modulating cycle.

The Doherty amplifier still remains in use in very high-power AM transmitters, but for lower-power AM transmitters, vacuum-tube amplifiers in general were eclipsed in the 1980s by solid-state power amplifiers due to their advantages of smaller size and cost, lower operating voltages, higher reliability and greater physical ruggedness, insensitivity to mechanical shock and vibration, and the possibility to use fully automated manufacturing processes and a high level of integration. However, the transistor (bipolar or field-effect) as a three-electrode solid-state device is characterized by the significant nonlinearity of its transfer characteristic and intrinsic capacitances that require complicated linearization schemes that were difficult to realize in the analog domain at that time. Interest in the Doherty configuration has been later revived due to significant progress in radio communication systems based on complex digital modulation schemes such as WiMAX in OFDM (orthogonal frequency division multiplexing) systems, CDMA2000, WCDMA, or LTE (long-term evolution) enhancement to the UMTS wireless standard, where the sum of several constant-envelope signals creates an aggregate AM signal with high peak-to-average ratio (*PAR*) and digital linearization techniques can be applied to reduce the Doherty amplifier distortion. Recently, Doherty amplifiers of different architectures have found widespread application in cellular base station transmitters operating at gigahertz frequencies.

7.1.1 Basic Structures

Generally, a Doherty amplifier system combines the outputs of two (or more) linear RF power amplifiers (PAs) through an impedance-inverting network composed of lumped elements or represented by a quarterwave transmission line. The two fundamental forms of the Doherty amplifier are shown in Figure 7.1, with a shunt-connected load in Figure 7.1a and with a series-connected load in Figure 7.1b [2]. In the former case, the load impedance used is $R_L = R/2$, which is the same as would be employed if the tubes were to be connected in parallel in the conventional type of power amplifier, where R is the load impedance seen by each tube at maximum output power. In the latter case, as long as the right-hand or *peaking* tube does not conduct, the impedance-inverting network provides zero impedance being terminated as an open circuit, and the left-hand or *carrier* tube operates into a load impedance $R_L = 2R$. However, when the peaking tube is permitted to conduct, each tube is operating into the impedance R at the peak of modulation and delivering twice the carrier power (or power of the transmitted unmodulated signal), so that the total instantaneous output is the required value of four times the carrier power. The shunt connection appears to be more advantageous for most practical applications because the load circuit is grounded, while the load is neither grounded nor balanced to ground in the series arrangement.

The ideal anode voltage and current behavior in the carrier and peaking tubes as the amplitude of the grid excitation is varied is shown in Figure 7.1c. Here, for the classical Doherty amplifier with equal-power tubes, the transition voltage is half the peak-envelope point (PEP), and the total output power of the amplifier comes from the carrier tube for input amplitudes less than or equal to the transition point. The region between the transition-point and PEP values represents the load modulation region and the voltage on the carrier tube remains constant at the PEP level. At the same time, the voltage across the

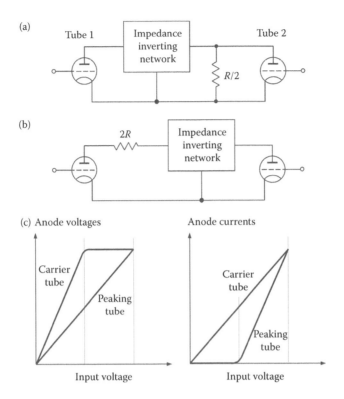

FIGURE 7.1
Doherty fundamental load-network structures and their ideal voltage and current behavior.

peaking tube continues to rise linearly, with its current commencing and rising twice as fast as the current in the carrier tube in order to reach its PEP value at maximum output power. Thus, at low-output power levels, the carrier amplifier operates linearly, reaching saturation that corresponds to maximum efficiency at some transition voltage below the system peak-output voltage. However, at higher output power levels, the carrier amplifier remains saturated, whereas the peaking amplifier operates linearly.

The corresponding shapes of the envelopes of anode current and voltages during complete load variation are shown in Figure 7.2 [2]. In a simple case of a sinusoidal modulating signal $v_m(t) = V_m \cos \Omega t$, where V_m is the modulating amplitude and Ω is the modulating frequency, the average output power P_{avr} for the modulated signal with a time-varying amplitude $V = V_0 + v_m(t) = V_0(1 + m \cos \Omega t)$, where V_0 is the carrier amplitude and $m = V_m/V_0$ is the modulation index, can be found as

$$P_{avr} = \frac{V_0^2}{4\pi R_L} \int_0^{2\pi} (1 + m \cos \Omega t)^2 d\Omega t = P_0 \left(1 + \frac{m^2}{2} \right) \tag{7.1}$$

where $P_0 = V_0^2/2R_L$ is the carrier power without modulation and m denotes the ratio of the variation of the modulated carrier amplitude to the unmodulated carrier amplitude. Since the average anode current in the carrier amplifier remains constant for all cycles of load variation but the duration of anode current in the peaking amplifier increases

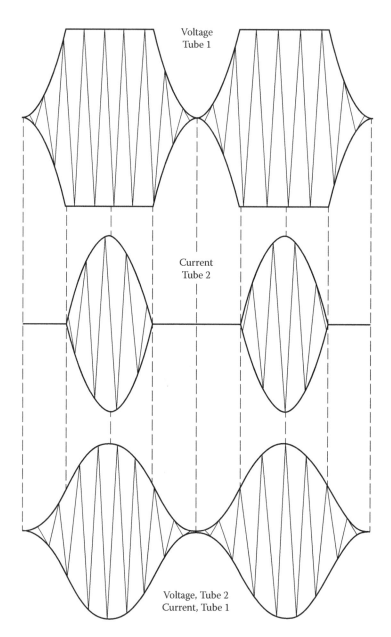

FIGURE 7.2
Envelopes of anode currents and voltages during load variation.

gradually with the amplitude, the average efficiency through the modulation cycle is then found to be

$$\eta_{\mathrm{avr}} = \eta_0 \frac{1 + (m^2/2)}{1 + (2mq/\pi)} \tag{7.2}$$

where η_0 is the efficiency for zero modulation and q is the variable factor that ranges from about 0.7 for zero modulation to 0.93 for full modulation [1].

Figure 7.3a shows the Doherty amplifier schematic, where the anodes of the carrier and peaking tubes are connected together by a π-type 90° lumped network, which introduces a lagging phase shift of 90° from the anode of the carrier tube to the anode of the peaking tube [2,10]. Such a 90° lumped network is used due to its impedance-inverting characteristics. This means that, if the terminating impedance at the point in the network where the peaking tube is located is reduced, the impedance seen by the carrier tube will increase. In this case, since the load network of the high-power transmitter was approximately of 35 Ω, the 90° lumped network was set to provide an impedance at the carrier tube of about 140 Ω. As a result, the series inductance and two shunt capacitors are each selected to have a reactance equal to $\sqrt{35 \times 140} = 70\,\Omega$. To compensate for the output phase shift of 90°, the input to the grid of the peaking tube is delayed by 90° by similar means. If the input excitation is applied to the grid of the peaking tube, a π-type lumped network consisting of the series capacitor and two shunt inductances is added between the grids of the carrier and peaking tubes to compensate for the output phase shift of 90°, as shown in Figure 7.3b.

The simplified two-stage transmission-line Doherty power-amplifier architecture shown in Figure 7.4a incorporates the carrier and peaking power amplifiers, separated by a quarterwave transmission line in the carrier amplifier path [11,12]. Such a section of line, known as a quarterwave transformer, has the ability to invert impedances according to

$$Z_0 = \sqrt{Z_{in}\,Z_{out}} \tag{7.3}$$

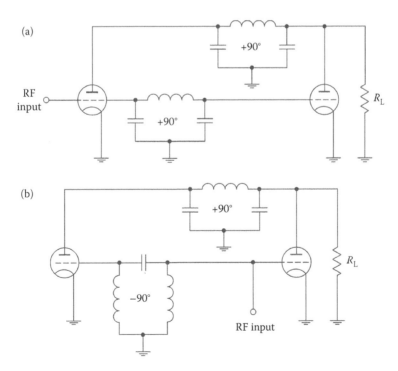

FIGURE 7.3
Doherty amplifier basic schematics with lumped elements.

(a)

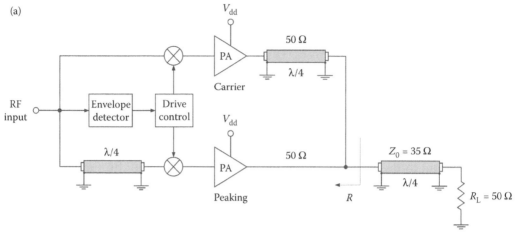

(b)

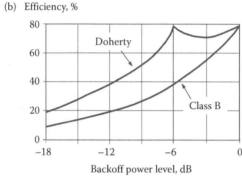

Backoff power level, dB

FIGURE 7.4
Doherty amplifier architecture with quarterwave lines and collector efficiencies.

where Z_0 is the characteristic impedance of the transmission line. This property can be seen more clearly by rewriting Equation 7.3 as

$$Z_{out} = \frac{Z_0^2}{Z_{in}} \tag{7.4}$$

from which it follows that the output impedance Z_{out} increases inversely with the input impedance Z_{in} for constant Z_0. The quarterwave transmission line at the input of the peaking amplifier is required to compensate for the 90° phase shift caused by the quarterwave transmission line at the output of the carrier amplifier. The output quarterwave line with $Z_0 = \sqrt{25 \times 50} = 35\,\Omega$ is required to match the standard load impedance of 50 Ω when both carrier and peaking amplifiers deliver maximum power, each of which designed in a 50-Ω environment.

An input drive controller is used to turn on the peaking amplifier (bias control) when the carrier amplifier starts to saturate since it is assumed that the carrier and peaking amplifiers are biased in Class B for idealized system analysis. However, in practice, the carrier amplifier is biased in a Class-B mode, whereas the peaking amplifier is biased in a Class-C mode. At a backoff power level of −6 dB, the saturated output power of the carrier amplifier is four times lower than the peak output power P_{PEP}. This indicates that its collector (or drain) efficiency when operated in an ideal Class-B mode is twice than that of a conventional Class-B power amplifier, achieving a maximum efficiency of 78.5%, as shown in Figure 7.4b.

7.1.2 Operation Principle

The basic operation principle of a conventional Doherty power-amplifier architecture shown in Figure 7.5a can be analyzed for low, medium, and peak output power regions separately [12]. Figure 7.5b shows the current and voltage behavior for ideal transistors and lossless matching circuits, where V_L is the load voltage and I_L is the load current. The condition of power conservation for a lossless output transmission line results in

$$I_3 = I_1 \sqrt{\frac{R_1}{R_3}} \qquad (7.5)$$

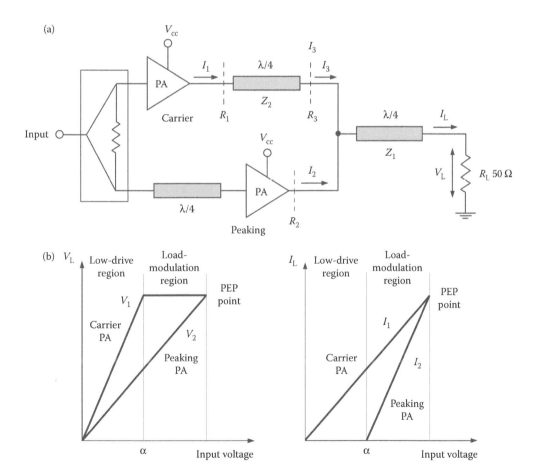

FIGURE 7.5
Basic two-stage Doherty amplifier architecture and its operation principle.

whereas the current division ratio β is defined by

$$\beta = \frac{I_3}{I_2 + I_3} \tag{7.6}$$

As a result, the overall output power P_{out} is the sum of the carrier (main) amplifier output power $P_1 = \beta P_{\text{out}}$ and peaking (auxiliary) amplifier output power $P_2 = (1 - \beta)P_{\text{out}}$. The impedance seen at the output of the transmission line in the carrier amplifier path is

$$R_3 = \frac{I_2 + I_3}{I_3} \frac{Z_1^2}{R_L} = \frac{Z_1^2}{\beta R_L} \tag{7.7}$$

and the impedance seen by the peaking power amplifier is

$$R_2 = \frac{I_2 + I_3}{I_2} \frac{Z_1^2}{R_L} = \frac{Z_1^2}{(1 - \beta)R_L} \tag{7.8}$$

At peak output power P_{PEP}, when both carrier and peaking amplifiers are saturated, the resultant collector efficiency is equal to the maximum achievable efficiency $\eta = \pi/4 \approx 78.5\%$ for an ideal Class-B operation. For the conventional Doherty power-amplifier architecture shown in Figure 7.5a with the current and power division ratios $\beta = \alpha = 0.5$, respectively, when both carrier and peaking amplifiers produce equal output powers, their load impedances are equal to $R_1 = R_3 = R_2 = Z_2 = 2Z_1^2/R_L$. If the characteristic impedance of the output transmission line is chosen to be $Z_1 = 35 \ \Omega$, then $R_1 = R_3 = R_2 = Z_2 = R_L = 50 \ \Omega$.

At lower power levels in a low-drive region, the peaking amplifier is turned off because the instantaneous amplitude of the input signal is insufficient to overcome the negative Class-C bias and appears as an open circuit, whereas the carrier amplifier operates in the active region. In this case, the load impedance seen by the carrier amplifier is

$$R_1 = \left(\frac{Z_2}{Z_1}\right)^2 R_L \tag{7.9}$$

resulting in $R_1 = 2R_L = 100 \ \Omega$ when $Z_1 = 35 \ \Omega$ and $Z_2 = R_L = 50 \ \Omega$. Because the output power of the carrier amplifier in saturation is four times less than the peak output power P_{PEP}, the collector efficiency of the carrier amplifier in an ideal Class-B mode will be twice than that of a conventional Class-B power amplifier, achieving maximum of 78.5% at backoff power level of −6 dB, as shown in Figure 7.4b.

At medium power levels in a load-modulation region, the carrier amplifier is saturated, whereas the peaking amplifier is turned on and operates in the active region. Since the output voltage of the carrier amplifier $V_1 = I_1 R_1$ is constant under saturation conditions, from Equation 7.5 it follows that the current I_3 is constant in the medium power region as well. The collector efficiency of the carrier amplifier remains at its maximum value, whereas the collector efficiency of the peaking amplifier increases up to its maximum value for Class-B operation at peak output power P_{PEP}. As a result, the Doherty amplifier architecture achieves maximum efficiency at both the transition −6 dB backoff point and the peak output power, and remains relatively high in between, as seen from Figure 7.4b.

For high-quality transmission in vacuum-tube power amplifiers, it is important to obtain a linear relation between grid exciting voltage and output anode voltage. The high impedance used for the carrier tube over the lower half of the modulation envelope causes the dynamic characteristic to be quite straight in this region. To obtain linearity from the point where curvature begins on the carrier tube, up to a point representing the peaks of modulation, is a matter involving both the point, at which the peaking tubes comes into operation, and the rate, at which its contribution increases with drive. For securing a satisfactory adjustment, there are two variables: the bias on the carrier tube and the amplitude of the excitation on this tube, whose careful selection may allow the power amplifier to operate with low distortion [2]. Besides, in a low-power region, linearity of the Doherty amplifier is entirely determined by the carrier amplifier, which should be highly linear even though the load impedance is high. In a high-power region, linearity can be improved by the harmonic cancellation from the carrier and peaking amplifiers using appropriate gate bias voltages. For example, in terms of gain characteristics of each amplifier using LDMOSFET devices, a late gain expansion of the Class-C biased peaking amplifier compensates the gain compression of the Class-AB biased carrier amplifier, thus improving the third-order intermodulation (IM_3) level [13].

In the mid-1990s, it was found that, by using existing microwave design techniques and by implementing a few modifications, a microwave version of the Doherty amplifier could be realized [14]. In this case, an efficiency of 61% was achieved at 1-dB gain compression point, and this level of efficiency was maintained through a 5.5-dB reduction in output power at an operating frequency of 1.37 GHz. A few years later, a possibility to achieve a high-efficiency performance of the microwave monolithic Doherty amplifier was demonstrated in the *Ku*- and *K*-band frequencies using pHEMT and InP DHBT technologies, respectively [15,16]. The first fully integrated Doherty amplifier monolithic microwave integrated circuit (MMIC) with a chip size of 2 mm^2 operating in a frequency range of 38–46 GHz was developed using a 0.15-μm GaAs HEMT process [17]. By using modern high-voltage HBT and GaN HEMT technologies, a high-average efficiency of more than 50% can be achieved for multicarrier WCDMA signals using a high-power two-way symmetrical Doherty amplifier [18,19].

7.1.3 Offset Lines

At high frequencies, it is necessary to take into account that the transistor input impedance is varied with bias voltage, and the transistor output reactance should be compensated to provide the required open-circuit condition by using so-called *offset lines* with optimized electrical lengths [13,20]. Figure 7.6 shows the basic schematic diagram of a fully matched microwave Doherty amplifier with offset lines in the output circuit and phase-compensating circuits in the input circuit required to reduce amplitude modulation/phase modulation (AM/PM) variations because the output power level can change significantly with phase variations between carrier and peaking paths [20,21]. At high frequencies, it is enough to use an input phase-compensating transmission line (offset line) at the input of the peaking amplifier. Because of different biasing of the carrier (Class AB) and peaking (Class C) amplifiers, an offset line at the input of the peaking amplifier can be used to optimize linearity and efficiency for multicarrier applications [22]. To closer approximate ideal Doherty amplifier performance, an adaptive power-dependent input power distribution between the carrier and peaking amplifier can be provided so as to deliver more power to the carrier amplifier in the low-power region and to the peaking amplifier in the high-power region, which can result in a linearity improvement of 5–7 dB

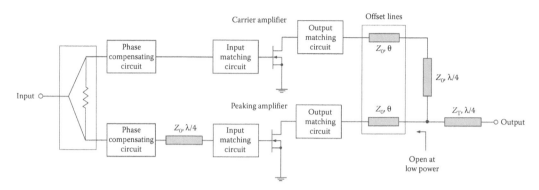

FIGURE 7.6
Block diagram of microstrip LDMOSFET Doherty amplifier.

over a wide range of output powers and an increased efficiency up to 5% for WCDMA signals at 2.4 GHz [23].

The additional offset lines with the characteristic impedance of 50 Ω and fixed electrical lengths θ are connected after the matching circuits of the carrier and peaking amplifiers, as shown in Figure 7.7a. In a low-power region, the phase adjustments of the offset lines cause

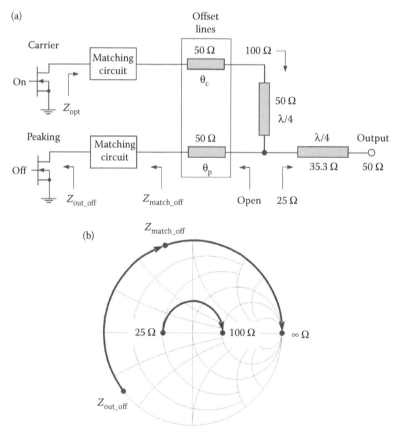

FIGURE 7.7
Load-network schematic and impedances.

the peaking amplifier to be open-circuited, and the load impedance seen by the carrier amplifier is doubled to 100 Ω due to an impedance-transforming property of a 50-Ω quarterwave transmission line, assuming that the matching circuit in conjunction with offset line provides the required impedance transformation to the optimum high-impedance Z_{opt} seen by the device output at the 6-dB power backoff, as shown in Figure 7.7b. At the same time, the offset line of the peaking amplifier is adjusted to provide high impedance (ideally infinite) from the matching-circuit impedance Z_{match_off} so that it prevents power leakage to the peaking path when the peaking transistor is turned off. In real device, the effect of the knee voltage should be considered due to nonzero value of the device on-resistance, which may be smaller or larger depending on the transistor implementation technology. To maximize efficiency at the 6-dB backoff power, the carrier amplifier with nonzero knee voltage should have a load impedance larger than 100 Ω, or larger than $2Z_0$ if $Z_0 \neq 50$ Ω, otherwise the carrier amplifier does not reach the saturation region where maximum efficiency can be achieved [24]. In this case, the offset-line length θ_c of the carrier amplifier is optimized to increase the load impedance, whereas the offset-line length θ_p of the peaking amplifier is adjusted to block the output-power leakage, and the phase-compensating line with electrical length of $90° + (\theta_c - \theta_p)$ is used at the input of the peaking amplifier.

7.1.4 Linearity

Generally, the nonideal power gain and phase performance in a high-power nonlinear region can cause a significant linearity problem for transmitting signals with nonconstant envelope in wireless communication transmitters. To solve this problem, an improved Doherty amplifier architecture can be used by using an envelope tracking technique to control the gate bias voltage of the peaking amplifier in accordance with the input signal envelope. Such an approach can also provide a higher efficiency with a lower bias voltage for higher output powers. This ensures both high efficiency and good linearity requirements over a wide range of output powers. Figure 7.8a shows the block diagram of a 2.14-GHz LDMOSFET microstrip Doherty power amplifier for WCDMA applications with adaptive gate bias control [25]. For the same average output powers of 32.7 dBm, such a two-stage Doherty power amplifier demonstrates an improvement in a *PAE* of 15.2% at an *ACLR* of −30 dBc compared to its Class-AB counterpart. This is because the quiescent current in a Doherty architecture is maintained constant only for the carrier amplifier, whereas the bias point of the peaking amplifier is varied according to the input signal envelope. However, it should be noted that the wider the modulation bandwidth of the transmitting signal, the more problematic it is to implement such a technique in practice to achieve significant linearity improvement.

The linearity of the power amplifier can be improved by using digital signal processing (DSP) to provide more accurate gate bias control alongside of digital predistortion (DPD) needed for the simultaneous correction of the gain and phase characteristics in a high-power region. Figure 7.8b shows the block schematic of an 840-MHz MESFET Doherty power amplifier for CDMA applications with the DSP implemented externally on a board controlled by a personal computer [26]. In this case, the DSP generates both the baseband in-phase (*I*) and quadrature (*Q*) signals, which are upconverted to form an RF signal using the quadrature modulator. The DSP unit also generates the voltage signal V_{g2}, which is applied to the peaking amplifier as the gate bias. This results in an efficiency improvement by the dynamic gate biasing of the peaking amplifier according to the instantaneous envelope of the input signal. At the same time, the phase performance is corrected by the phase predistortion at baseband level based on the dynamic gate bias-voltage values from

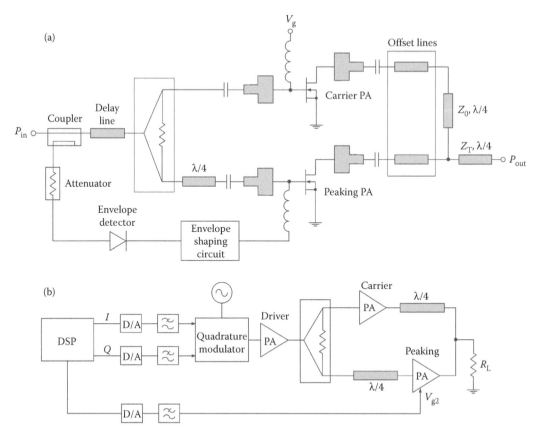

FIGURE 7.8
Block diagrams of Doherty amplifier architectures with adaptive control.

the gain correction, thus resulting in a linearity improvement. An overall improvement of *PAE* from 3% to 5% and an *ACPR* of about 10 dB at an average output power of 23 dBm can be achieved by utilizing such a DSP technique. Besides, a simple bias-switching technique can be used in Doherty-type amplifiers, so that they can satisfy the linearity requirements of the power amplifiers for CDMA handset applications over the entire dynamic power range [27]. In addition to the bias-switching technique, a dual-mode matching approach can be used to optimally design a dual-mode Doherty amplifier operated simultaneously in HPSK (hybrid phase shift keying) and OFDM 64-QAM (quadrature amplitude modulation) modes for mobile terminals [28].

7.1.5 Series-Connected Load

In comparison with the shunt-connected load type, the series-connected load combines the output powers of the carrier and peaking amplifiers in a manner similar to that of the push–pull amplifiers, having the same capability to suppress even-harmonic components in the output signal spectrum. Figure 7.9 shows the block diagram of a Doherty amplifier with a series-connected load, where the input and output baluns are implemented with the lumped inductances and capacitors [29]. The lumped-element balun can be designed for an arbitrary unbalanced-load value by proper selection of its element values. In this case, an unbalanced load of 50 Ω and a balanced port load of 100 Ω were chosen. At low-power

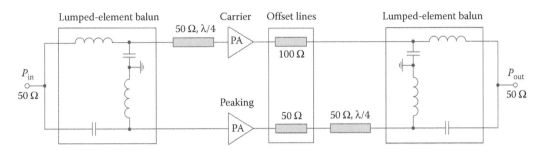

FIGURE 7.9
Block diagram of Doherty amplifier with series-connected load.

levels when the peaking amplifier is turned off, one balanced port of the balun connected to the open-circuited quarterwave transmission line is shorted, which results in a load of 100 Ω for carrier amplifier. At high-power level when the peaking amplifier becomes active, both carrier and peaking amplifiers are operated in a near push–pull mode. For such a GaN HEMT Doherty amplifier operating at 1.8 GHz, high efficiencies of 31% and 56% were achieved at 24- and 31-dBm saturated output powers, respectively.

In some cases, by inserting a short-length transmission line, the output impedance of the peaking amplifier can be transformed from capacitive to near-short impedance. As a result, the quarterwave impedance-inverting line can be removed. Figure 7.10a shows the circuit schematic of a 1.9-GHz GaN HEMT Doherty amplifier without an impedance-transforming network, which was designed in the form of a series-connected-load configuration using an output-combining balun [30]. Here, a microstrip Wilkinson power divider is used for input power splitting, and the matching circuits and balun with a transformation ratio 2:1 are implemented using lumped-element components. Therefore, the load impedance presented to the carrier amplifier varies from 100 Ω at low-power level to 50 Ω at saturation level. The circuit schematics of the carrier and peaking amplifiers are shown in Figure 7.10b and c, respectively. A two-section output matching circuit for the carrier amplifier is required to provide the two impedance conditions, matching at high-power level and impedance-transforming at low-power level. At the same time, the single-section output matching circuit for the peaking amplifier provides a near-zero condition. As a result, a *PAE* above 48% at a saturated output power of more than 31 dBm at a supply voltage of 20 V was obtained over a wide frequency range from 1.67 to 1.97 GHz.

7.2 Inverted Doherty Amplifiers

Figure 7.11 shows the schematic diagram of an inverted Doherty amplifier configuration with an impedance transformer based on a quarterwave line connected to the output of the peaking amplifier. Such architecture can be very helpful if it is easier to provide a short circuit in a low-power region rather than an open circuit at the output of the peaking amplifier, which depends on the characteristic of the transistor. The quarterwave line can be implemented in a compact form suitable for use in mobile applications [31]. In this case, at low-power levels, a quarterwave line is used to transform very low-output impedance after the offset line to high impedance seen from the load junction. In particular, by taking into account the device package parasitic elements of the peaking

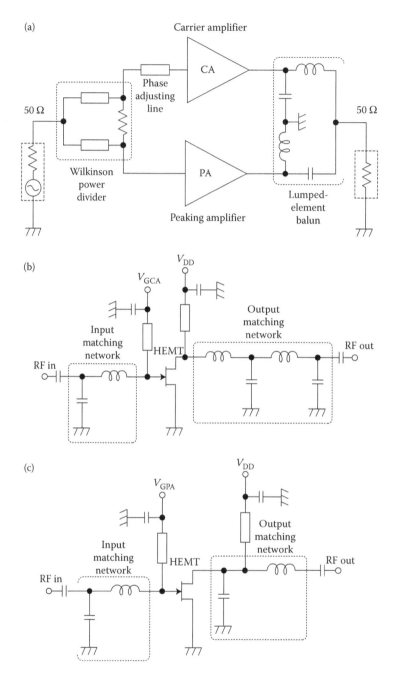

FIGURE 7.10
Schematics of Doherty amplifier without quarterwave impedance inverter.

amplifier, an optimized output matching circuit and a proper offset line are designed to provide the maximum output power from the carrier device [32]. At a high-power level, for the matched phase difference between identical carrier and peaking amplifiers, the load impedance seen from each amplifier after the offset lines is equal to the standard 50-Ω load impedance.

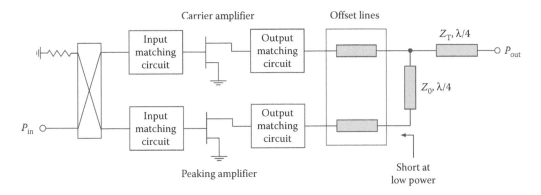

FIGURE 7.11
Schematic diagram of inverted Doherty amplifier.

Applying a four-carrier WCDMA signal, a *PAE* of 32% with an *ACLR* of −30 dBc at an average output power level as high as 46.3 dBm was achieved for an inverted 2.14-GHz LDMOSFET Doherty amplifier. This provides a 9.5% improvement in efficiency and 1-dB improvement in the output power under the same *ACLR* conditions as for the balanced Class-AB operation using the same devices [33]. For a 64-QAM modulated signal with a 24-MHz channel bandwidth, a *PAE* higher than 31% with a 0.5-dB output-power flatness at 27 dBm was achieved across a frequency bandwidth from 2.4 to 2.5 GHz with an *ACPR* better than −40 dBc [34].

To better understand the operation principle of an inverted Doherty amplifier, consider separately the load network shown in Figure 7.12a, where the peaking amplifier is turned off. In a low-power region, the phase adjustment of the offset line with electrical length θ causes the peaking amplifier to be short-circuited (ideally equal to 0 Ω), and the matching circuit in conjunction with offset line provides the required impedance transformation from 25 Ω to the optimum high-impedance Z_{opt} seen by the carrier device output at the 6-dB power backoff, as shown in Figure 7.12b. In this case, the short circuit at the end of the quarterwave line transforms to the open circuit at its input so that it prevents power leakage to the peaking path when the peaking transistor is turned off. In a high-power region, both carrier and peaking amplifier are operated in a 50-Ω environment in parallel, and the output quarterwave line with the characteristic impedance of 35.3 Ω transforms the obtained 25 Ω to the required 50-Ω load.

In a Doherty configuration, both the Class-AB carrier amplifier and Class-C peaking amplifier are not fully isolated from each other. This can result in a serious problem to robustly design the optimum load-impedance shift presented to both transistors for high efficiency and low distortion [35]. From the load-pull measurements for a unit-cell 28-V GaAs HJFET device, it was observed that, in order to obtain high efficiency and low distortion, the carrier amplifier load impedance should change from the maximum efficiency point to the maximum output power point at Class AB, whereas the peaking amplifier load impedance should vary from the small-signal gain point to the maximum output power point in Class C [36]. In this case, the load impedance corresponding to the maximum efficiency point is lower than that corresponding to the maximum output power point. An inverted Doherty architecture can be suitable to realize the carrier amplifier load impedance variation from lower impedance to higher impedance in accordance with the increase in the input power level. The external input and output matching circuits are necessary to optimize the load impedance shift presented to both carrier and peaking amplifiers as a

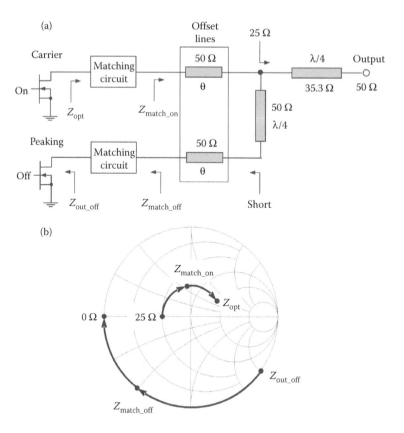

FIGURE 7.12
Load-network schematic and impedances.

function of the input power level. As a result, a drain efficiency of 42% at an output power of 49 dBm around 6-dB backoff level was achieved for a two-carrier WCDMA signal of 2.135 and 2.145 GHz with an IM_3 of −37 dBc.

Figure 7.13 shows the three-stage inverted Doherty amplifier configuration, where the quarterwave transmission lines are added in the outputs of the carrier and peaking amplifiers to provide a proper load modulation ratio [37]. The half-wave transmission line in the input path of the carrier amplifier is used to compensate for the delay provided by the output load network. The characteristic impedances of the quarterwave transmission lines are optimized to provide a high efficiency over wide output power backoff range. If the device size ratio of the carrier, first peaking, and second peaking amplifiers is 1:m_1:m_2, respectively, the characteristic impedances of the quarterwave transmission lines at the full power loading condition can be obtained by

$$Z_T = Z_1 \sqrt{\frac{1}{1 + m_1 + m_2}} \tag{7.10}$$

$$Z_4 = Z_1 \frac{Z_3}{Z_2} \sqrt{\frac{1}{m_1}} \tag{7.11}$$

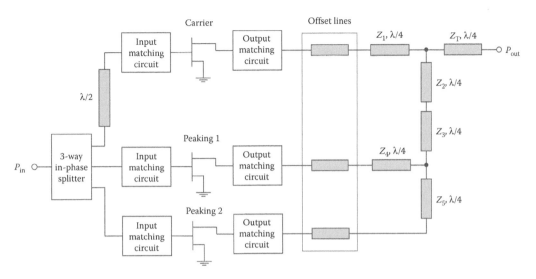

FIGURE 7.13
Schematic diagram of three-stage inverted Doherty amplifier.

$$Z_5 = Z_1 \frac{Z_3}{Z_2} \sqrt{\frac{1}{m_2}} \tag{7.12}$$

assuming the same 50-Ω loading conditions for the standard load and the carrier and peaking amplifiers at full loading conditions. As a result, for the same device sizes for the carrier and peaking amplifiers when $m_1 = m_2 = 1$, $Z_2 = Z_3 = 50\ \Omega$, and $Z_1 = 70\ \Omega$, from Equations 7.10 through 7.12, it follows that $Z_T = 40.4\ \Omega$ and $Z_4 = Z_5 = 50\ \Omega$, respectively. In this case, the drain efficiency for a single-carrier 2.14-GHz WCDMA signal with a *PAR* of 10.5 dB can be improved by 5% over a wide range of output powers.

7.3 Integration

The transmission-line two-stage Doherty amplifier can easily be implemented into the MMIC by using a pHEMT or CMOS process. For example, a fully integrated *Ku*-band MMIC Doherty amplifier using a 0.25-µm pHEMT technology achieved a two-tone *PAE* of 40% with a corresponding IM_3 of −24 dBc at 17 GHz, whereas a single-tone *PAE* of 38.5% at 1-dB compression point was measured for a 20-GHz MMIC Doherty amplifier implemented in a 0.15-µm pHEMT process for use in digital satellite communication (DSC) systems [15,38]. Furthermore, by using a 0.13-µm RF CMOS technology, a transmission-line MMIC Doherty amplifier based on cascode configuration of the carrier and peaking amplifiers achieved a saturation output power of 7.8 dBm from a supply voltage of 1.6 V at an operating frequency of 60 GHz for use in wireless personal area network (WPAN) transceivers [39]. On the other hand, the efforts to directly apply the Doherty technique to the design of power-amplifier integrated circuits with a high level of integration at lower frequencies face difficulties, since the physical size of the quarterwave transmission lines in this case is too large. For example, for an FR4 substrate with effective dielectric permittivity of $\varepsilon_r = 3.48$, the geometrical lengths of the quarterwave transmission lines are 48, 19, and 8.7 mm at the operating frequencies of 900 MHz, 2.4 GHz, and 5.2 GHz, respectively. Therefore, one of the acceptable solutions for the fabrication of small-size Doherty amplifier MMICs intended to operate in WLAN or WiMAX transmitter systems is to replace each quarterwave line in the input combining circuit and output impedance transformer by its low-pass π-type lumped-distributed equivalent with a short-length series transmission line and two shunt capacitors connected to its both ends [40,41]. Additionally, simple and small-size second-harmonic termination circuits can be realized with integrated MIM capacitors and bondwires at the end of the carrier and peaking amplifier collectors [40].

In order to minimize the inherently high-substrate loss and increase the level of integration to implement the Doherty amplifier in a CMOS process, the branch-line coupler and quarterwave transformer in the amplifier input and output circuits are fully substituted by their lumped equivalents [42]. By considering the transmission $ABCD$-matrices for a quarterwave transmission line shown in Figure 7.14a and a π-type low-pass lumped circuit consisting of a series inductance and two shunt capacitors shown in Figure 7.14b and equating the corresponding elements of both matrices, the ratio between the circuit elements can be written as

$$Z_0 \omega C = \frac{Z_0}{\omega L} = 1 \tag{7.13}$$

where Z_0 is the characteristic impedance of the quarterwave transmission line. A high-power Doherty amplifier MMIC can be integrated with lumped elements in a standard discrete package, where the compensation series circuits (each consisting of an inductance and a capacitor) are connected to the drain terminals of the carrier and peaking transistors to compensate for their output capacitances [43]. For example, an integrated solution based on four 10-W MMIC Doherty amplifier cells combined in parallel achieves a drain efficiency of 39.8% at an average output power of 7.5 W with an $ACLR$ of -50 dBc using a DPD technique for a two-carrier 2.14-GHz WCDMA signal with a PAR of 7.6 dB [44].

Similarly, the input in-phase transmission-line two-way Wilkinson divider can be replaced by its lumped equivalent, where a π-type low-pass LC circuit is used in its each branch. Figure 7.15a shows an example of the simplified schematic of a two-stage lumped Doherty amplifier, where the output quarterwave transmission line connected to the carrier amplifier output and the input phase-shifting quarterwave transmission line connected to the peaking amplifier input are replaced by equivalent π-type low-pass lumped circuits. In addition, the output quarterwave transformer is replaced by an L-type high-pass matching circuit, whereas two L-type low-pass matching circuits are used to provide the input matching of the carrier and peaking amplifiers. At the peaking amplifier input path, the right-hand shunt capacitor as a part of the equivalent quarterwave phase shifter

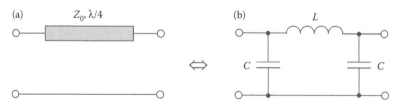

FIGURE 7.14
Quarterwave transmission line and its single-frequency lumped equivalent.

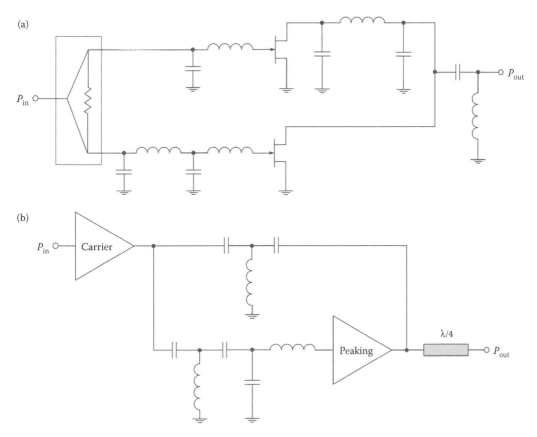

FIGURE 7.15
Circuit schematics of lumped Doherty amplifiers for handset applications.

and the shunt capacitor as a part of the input L-type low-pass matching circuit can be combined into a single shunt capacitor.

To remove a 3-dB hybrid input divider, the conventional Doherty amplifier can be rearranged to be more suitable for handset applications. Figure 7.15b shows the circuit schematic of a "series-type" Doherty architecture, where the subcircuit of the peaking amplifier and impedance transformer is connected to the output of the carrier amplifier in series rather than in parallel configuration [45,46]. The impedance transformers in both paths are composed of a high-pass lumped-element T-network each. This is because the high-value inductances or relatively long series microstrip lines are required for the low-pass T-networks, making chip-level integration impractical. The shunt inductance in the high-pass T-network can be implemented using a high-impedance microstrip line instead of a lumped-element inductor. Besides, a high-pass T-network helps to prevent occasional low-frequency oscillations. The output quarterwave transformer fabricated externally can be replaced by the equivalent low-pass π-type matching circuit with a series short-length microstrip line and two shunt chip capacitors. In a low-power region, the input impedance from the peaking amplifier path, which is turned off due to deep Class-C bias mode is sufficiently high. However, in a high-power region, it reduces significantly when the peaking amplifier is turned on, so the load seen by the carrier amplifier reduces significantly, by about three times for a practical case of a series-type MMIC Doherty amplifier with optimum device sizes based on a 2-μm InGaP HBT technology [47]. As a result, for a

1.9-GHz IS-95A CDMA signal, a *PAE* of 18% and 42.8% were achieved at 16 and 28 dBm, respectively.

As the Doherty amplifier for handset applications should be compact, a direct input-dividing circuit considering the impedance variations of the carrier and peaking amplifiers can be used instead of the Wilkinson power combiner. In this case, since the input impedance of the carrier amplifier remains almost constant, while that of the peaking amplifier changes significantly because of the Class-C bias, this effect can be utilized for the uneven input dividing [47]. As a result, large power is delivered to the carrier amplifier at the low-power region, and the power gain at the low-power region becomes much higher than that at the high-power region, deteriorating the gain flatness and linearity of the Doherty amplifier. Figure 7.16 shows the full circuit schematic of a lumped Doherty amplifier with two-stage carrier and peaking amplifiers [47]. Here, the output matching circuit takes a role of a quarterwave transformer, including parasitics, with the phase compensation network employed at the input of the carrier path, and the offset line employed at the output of the peaking path. The second and third harmonics are properly controlled to enhance the efficiency of both carrier and peaking amplifiers. Moreover, the second- and third-harmonic control circuits are also utilized for the quarterwave transformer by connecting the capacitor C_q, forming a π-network where the device output capacitance C_p and second-harmonic control circuit are considered as one capacitor and the parallel resonant LC-circuit is inductive at the fundamental. The capacitors C_c of the offset line can be combined with capacitors C_q to reduce the number of components. As a result, the Doherty amplifier MMIC implemented in a 2-μm InGaP/GaAs HBT process presents a *PAE* of 40.2% at an output power of 26 dBm with an error vector magnitude (EVM) of 3% for a 16-QAM m-WiMAX signal having a 9.54-dBc crest factor and 8.75-MHz bandwidth.

Since a frequency-dependent quarterwave transformer and output matching circuits generally provide a narrowband operation of a conventional Doherty amplifier, a quarterwave transmission line as an additional matching element can be added at the output of the peaking amplifier in series with the offset line to minimize the loaded quality factor

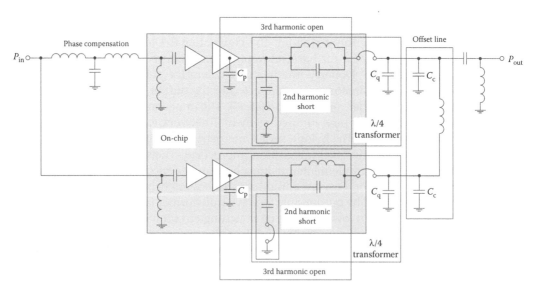

FIGURE 7.16
Schematic of Doherty amplifier with harmonic control for handset applications.

for broader operation by increasing the impedance at the output junction of the carrier and peaking amplifiers [48]. In a handset monolithic application, the transmission-line quarterwave impedance transformer and offset line are implemented with lumped elements, representing the equivalent π-type low-pass and π-type high-pass LC networks, respectively, as shown in Figure 7.17a. In this case, the network parameters are optimized to provide an open-circuit condition at the output of the peaking branch over broadband frequency range when the peaking amplifier is turned off. To simplify the load-network

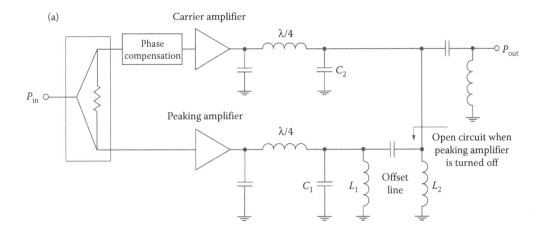

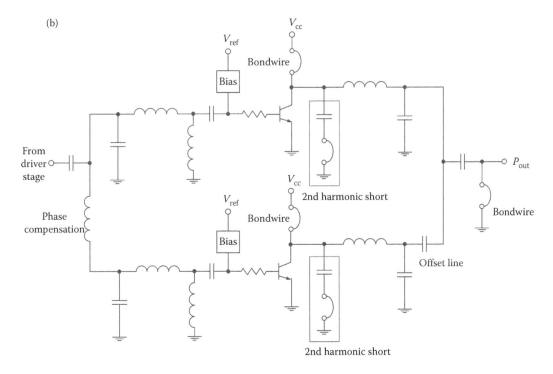

FIGURE 7.17
Schematics of Doherty amplifiers for handset applications with bandwidth enhancement.

structure, the values of the inductances L_1 and L_2 are chosen so as to merge them with the corresponding capacitances C_1 and C_2. Figure 7.17b shows the circuit schematic of a 2-μm GaAs HBT Doherty amplifier, where all of the components are fully integrated on a chip. In this case, the inductors are implemented using bondwires and slab inductors, the input-dividing circuits are broadband based on low-Q matching networks, and the second- and third-harmonic impedances are controlled for high efficiency across the bandwidth. The open-circuit conditions are achieved by optimizing all load-network elements, including drain bondwires and device output capacitances. For a mobile 8.75-MHz 16-QAM m-WiMAX application with a 9.6-dB crest factor, such a lumped Doherty amplifier exhibits a *PAE* of over 27% and an output power of over 23.6 dBm across 2.2–2.8 GHz using a DPD technique. A similar Doherty amplifier with broadband lumped networks can provide a *PAE* over 30% and an output power of over 28 dBm across 1.6–2.1 GHz for a 10-MHz LTE signal with a *PAR* of 7.5 dB [49].

Figure 7.18 shows the test chip of a wideband monolithic GaN HEMT asymmetric Doherty power amplifier, where the input network consists of a lumped-element Wilkinson power divider and phase shifter, whereas the load network represents a *T*-line impedance inverter with optimized characteristic impedances and electrical lengths of microstrip lines [50]. The input matching and stability of operation is provided with a series resistor at the input of the carrier device and with two shunt *LR* circuits connected at the inputs of the carries and peaking devices, respectively. The Doherty power amplifier was implemented in a 0.25-μm GaN HEMT process, having a 100-μm-thick SiC substrate, a relative permittivity of 9.7, a maximum drain current density of 900 mA/mm, and a maximum power density of 5–7 W/mm, with a total chip size of 2.1 mm × 1.5 mm. To obtain the highest possible output power with model verified device sizes, the total gate widths of the carrier and peaking devices were chosen as 4 × 100 μm and 10 × 100 μm, respectively. As a result, a *PAE* of greater than 30% at 9 dB power backoff within the frequency range of 6.7–7.8 GHz and a maximum output power of 35 ± 0.5 dBm from 6.6 to 8.5 GHz were achieved. For a 10-GHz MMIC Doherty power amplifier using Class-E approximation in both carrier and peaking amplifiers, which are based on 140-nm GaN HEMTs of a 1-mm gate width each, the simulated two-tone results demonstrated a *PAE* of 40.4% at peak output power of 25.6 dBm and a *PAE* of 24% at 6-dB backoff [51].

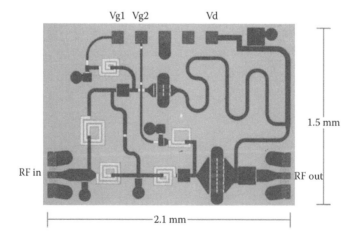

FIGURE 7.18
Test chip of wideband monolithic GaN HEMT Doherty power amplifier. (Courtesy of Chalmers Institute of Technology.)

7.4 Digitally-Driven Doherty Amplifier

In a digitally-driven dual-input Doherty amplifier architecture, the input signal of each branch is digitally preprocessed and supplied separately to each branch of the Doherty amplifier to optimize its overall performance. In this case, DSP is applied to reduce the performance degradation due to phase impairment in the Doherty amplifier branches achieved by adaptively aligning the phases of the carrier and peaking paths for all power levels after the peaking amplifier is turned on. Generally, a DSP includes a DPD system to improve linearity, which can be configured to provide a carrier signal component along a carrier amplifier path and a peaking signal component along a peaking amplifier path from a digital input signal [52]. In this case, the carrier and peaking amplifiers can amplify the signal components according to the programmable proportions of the split input signal, and not based on a saturation condition of the carrier amplifier, thus resulting in a higher efficiency. Because the signals are isolated prior to being input to the Doherty amplifier, the Doherty amplifier need not include an asymmetric splitting with input phase-matching delay, and input impedance-matching circuitry is simplified. The DPD system also performs phase and gain adjustments to each of the signal components. To further improve efficiency performance of a two-stage Doherty amplifier, the separated amplitude- and phase-modulated signals, produced by the DSP, drive through the corresponding quadrature upconverters both the carrier and peaking amplifiers, each operated in a Class-E mode [53].

Figure 7.19a shows the block diagram of a dual-input digitally-driven Doherty amplifier with DSP and dual-channel upconverter [54]. In this case, direct access and software control of the individual inputs can bring an improvement in efficiency of a Doherty amplifier between the two efficiency peaking points at maximum and 6-dB backoff powers, as shown in Figure 7.19b. The Doherty amplifier design is performed by deriving the offset line with electrical length θ_p to be inserted at the output of the peaking branch to ensure a quasi-open circuit condition and prevent leakage from the carrier amplifier to the output of the peaking amplifier at the low-power region. The offset line with electrical length θ_c was optimized to maximize efficiency around the turn-on point of the Doherty amplifier. Because of the different bias conditions for the carrier amplifier (Class AB) and the peaking amplifier (Class C), the degradation in output power due to phase imbalance condition can be as high as 40% after the peaking amplifier is fully turned on, which directly translates into significant deterioration in a drain efficiency of the Doherty amplifier. However, the dual-input digitally-driven Doherty architecture, allowing for the adoption and implementation of a power adaptive phase-alignment mechanism, can minimize the adverse effects of phase imbalance between the carrier and peaking branches. The power-dependent phase offset is adjusted using a power-indexed lookup table (LUT) to correct for the phase disparity at all power levels, where both the carrier and peaking amplifiers contribute to the total output power of the Doherty amplifier. As a result, the phase difference between the carrier and peaking branches is reduced to 0° over the input power range spanning from the turn on of the peaking amplifier until the saturation of the Doherty amplifier. The phase-aligned Doherty amplifier based on two 10-W GaN HEMT transistors demonstrates a *PAE* higher than 50% over an 8-dB output-power backoff range and a *PAE* of 57% at an average output power of 37 dBm for a single-carrier WiMAX signal with a *PAR* of 7 dB, thus resulting in an improvement of 7% in *PAE* and 1 dB in average output power with similar linearity performance corresponding to an *ACPR* of –22 dBc compared to the fully analog Doherty amplifier [54].

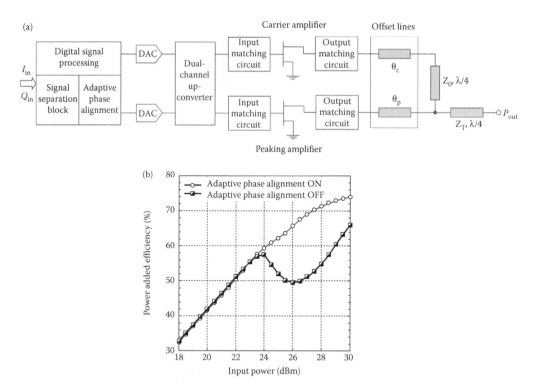

FIGURE 7.19
Block diagram and simulated performance of dual-input digital Doherty amplifier.

Efficiency enhancement in a digital Doherty amplifier over a wide power range can also be achieved by using a digitally controlled dynamic input power distribution scheme to minimize the drive power waste into the peaking branch at backoff power levels [55]. In this case, the carrier amplifier should get significantly more input power in comparison to the peaking amplifier at low-power drive, whereas the carrier amplifier should get slightly less input power in comparison to the peaking amplifier after turn-on point. As a result, the efficiency can be improved by 7% compared to the conventional fully analog symmetrical Doherty amplifier based on two 10-W GaN HEMT devices and operating at 2.14 GHz for a single-carrier WiMAX signal with a 9-dB *PAR* and 10-MHz bandwidth.

7.5 Multiband and Broadband Capability

7.5.1 Multiband Doherty Configurations

A multiband capability of the conventional two-stage Doherty amplifier can be achieved when all of its components are designed to provide their corresponding characteristics over the required bands of operation, as shown in Figure 7.20a [56]. In this case, the carrier and peaking amplifiers should provide broadband performance when, for example, their input and interstage matching circuits are designed as broadband, and the load network generally can represent a low-pass structure with two or three sections tuned to the required

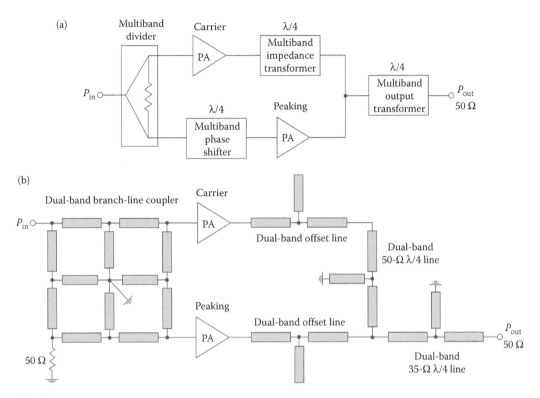

FIGURE 7.20
Block diagrams of multiband Doherty amplifiers.

frequencies. Some bandwidth extension can be achieved by simply optimizing the characteristic impedances of the quarterwave impedance transformer and quarterwave output combiner in a combining load network [57,58]. For a multiband operation with the center frequency ratio at each of the frequency bands of two or greater, the input divider can be configured by a multisection Wilkinson power divider or coupled-line directional coupler. In a dual-band operation mode, a dual-frequency Wilkinson power divider can represent a structure, where each quarterwave branch of a conventional Wilkinson power divider is substituted by the two transmission-line sections with different characteristic impedances and electrical lengths [59,60]. In practical applications, especially if the operating frequencies of one of the frequency band are sufficiently low, the miniaturized version of a dual-band Wilkinson power divider can be designed based on the concept of slow wave periodic structure [61].

A dual-band input power splitter can also represent a π- or *T*-shape stub-tapped branch-line coupler, as well as an impedance transformer network, which introduces a 90° phase shift. Similarly, the offset lines and an output quarterwave transformer can be based on a π-type or *T*-type transmission-line impedance-inverting section with proper selected transmission-line characteristic impedances and electrical lengths, where the shunt element is realized by an open- or short-circuit stub, as shown in Figure 7.20b [62,63]. However, it should be noted that it is not easy to design a multiband impedance transformer, which should adequately provide two separate matching options simultaneously: first, to operate in a 50-Ω environment without affecting the power amplifier performance in a high-power region, and second, to provide an impedance matching from 25 to 100 Ω in a low-power region. In this case, as an alternative, it is also possible to switch between two quarterwave transmission lines in a dual-band operation when each of the quarterwave transmission line is tuned to the corresponding center bandwidth frequency. However, it may not be so simple in practical implementation because of the load-network complexity and additional power losses.

A concurrent tri-band and quad-band operation of the two-stage Doherty amplifier can be achieved by using a transmission-line multiband impedance inverter, which is realized based on the idea that the transmission-line section having an electrical length equal to a quarter-wavelength at a given low frequency should be capable of acting as a multiple quarterwave line at higher frequencies [64,65]. For a quad-band application, the input matching circuit can represent a modified two-way two-stage Wilkinson splitter, whereas a four-section Chebyshev matching transformer can cover the required broadband frequency range. As a result, the peak drain efficiencies of 60.5%, 58.1%, 52.7%, and 43.3% were achieved at midband frequencies of 960 MHz, 1.5 GHz, 2.14 GHz, and 2.65 GHz, respectively, with the output powers varying between 41.7 and 44.2 dBm. In this design, 10- and 25-W Cree GaN HEMT devices are used for carrier and peaking amplifiers, respectively.

Generally, the multiband impedance transformer can represent a configuration with N ($N \geq 2$) cascade-connected transmission lines with different characteristic impedances. In this case, a simple two-stepped transmission-line impedance transformer can provide a two-pole response with different characteristic impedance ratio and different electrical lengths of the transmission-line sections [66]. It can be used as a dual-band output transformer since it is necessary to provide an impedance transformation from the output impedance of 25 Ω to the standard 50-Ω load in both low- and high-power regions [63].

As an example, the dual-band output transformer can be realized using a two-section transmission line, where the characteristic impedance of the first quarterwave transmission-line section is equal to 30 Ω and the characteristic impedance of the second quarterwave transmission-line section is set to 42 Ω, as shown in Figure 7.21a. In this case, the amplitude variations of ±0.5 Ω shown in Figure 7.21b by curve 1 and phase variations of ±1° shown in Figure 7.21c by curve 1 can be achieved across the frequency range from 2.0 to 2.8 GHz covering simultaneously 2.1-GHz (2.11–2.17 GHz) and 2.6-GHz (2.62–2.69 GHz) WCDMA/LTE bands. For comparison, the narrowband amplitude and phase responses of a quarterwave single-line impedance transformer are shown in Figure 7.21b and c by curves 2, respectively. From Figure 7.21b and c, it follows that the amplitude variations of ±1.0 Ω and phase variations of ±2° can be achieved with a 1-GHz bandwidth from 1.9 to 2.9 GHz, which means that reducing the midband frequency to 2.3 GHz can result in a simultaneous tri-band operation with inclusion of an additional 1.8-GHz (1805–1880 MHz) WCDMA/LTE band.

Figure 7.22 shows the block schematic of a broadband Doherty amplifier where the impedance transformer is based on a three-section microstrip line with two shunt reactances and the output transformer represents a three-section microstrip line with different characteristic impedances to provide the trade-off for optimum impedances seen by the carrier and peaking amplifier at backoff and saturation conditions [67]. In this case, the parasitic output capacitances of the identical carrier and peaking devices are respectively absorbed by the impedance transformer. The input broadband network provides a 2:1 asymmetric split and has been optimized to ensure that the currents at the summing node point are amplitude and phase aligned at the maximum drive level. The measured results demonstrate an average output power of 43.5 dBm with an average drain efficiency of 33.5% over the frequency bandwidth of 450–750 MHz for an 8-MHz OFDM (64QAM modulation) signal with a *PAR* of 10.5 dB. The Doherty amplifier based on two 50-V LDMOSFET devices

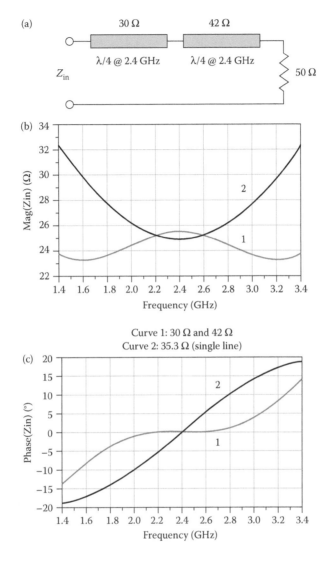

(a)

30 Ω   42 Ω

Z_{in}

λ/4 @ 2.4 GHz   λ/4 @ 2.4 GHz

50 Ω

(b)

Mag(Zin) (Ω)

Frequency (GHz)

Curve 1: 30 Ω and 42 Ω
Curve 2: 35.3 Ω (single line)

(c)

Phase(Zin) (°)

Frequency (GHz)

FIGURE 7.21
Stepped transmission-line transformer and its input impedances.

achieves the target *ACLR* of −50 dBc at the upper limit of the bandwidth if DPD is applied. Compared to a standard Class-AB push–pull power amplifier developed with the same devices, which demonstrates a 22% drain efficiency under the same biasing conditions, the Doherty amplifier provides more than 50% efficiency improvement.

7.5.2 Bandwidth Extension Using Reactance Compensation Technique

An *LC* tank at the output of the peaking amplifier, as shown in Figure 7.23a, can provide bandwidth extension of the Doherty amplifier due to reactance compensation effect when it provides an inductive reactance at low frequencies and capacitive reactance at high frequencies, which is opposite to the frequency behavior of the series loaded quarterwave transmission line with the characteristic impedance Z_T near the resonant frequency f_0 when

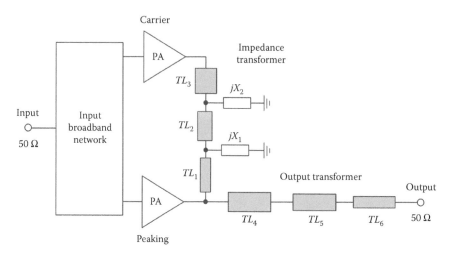

FIGURE 7.22
Block schematic of broadband Doherty amplifier.

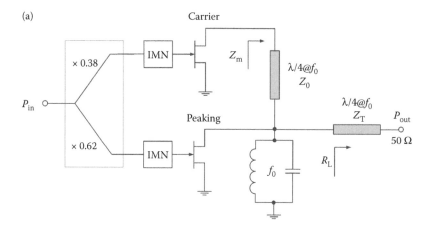

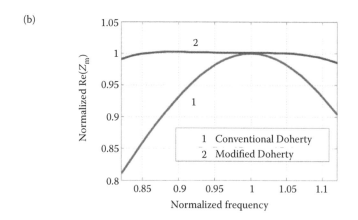

FIGURE 7.23
Doherty amplifier with extended bandwidth and its broadband property.

$Z_T < 50\ \Omega$, as shown in Figure 6.7c (Chapter 6) [68]. As a result, such an output combiner provides the impedance $\mathrm{Re}Z_m$ of 100 Ω seen by the carrier device at power backoff mode when the peaking device is turned off over more than 35% fractional bandwidth, compared with less than 10% bandwidth for the conventional output combiner, as shown in Figure 7.23b. In this case, the output capacitance of the peaking amplifier can be absorbed into the tank capacitance, and the tank inductor can act as the biasing feed for both the carrier and peaking devices, which in turn provides the small baseband impedance at the drain of the carrier and peaking transistors up to a few hundreds of megahertz. Based on 10-W and 25-W Cree GaN HEMT devices, a *PAE* of 42% was achieved at an average output power of 33.5 dBm when amplifying a concurrent dual-band 15-MHz WCDMA signal at 750 MHz and 15-MHz LTE signal at 900 MHz with a *PAR* of 9.4 dB.

Similarly, a wider bandwidth can be obtained for a three-stage Doherty amplifier in a backoff region when both peaking amplifiers are turned off, and the two quarterwave transmission lines connected in series compose a half-wave transmission line, which acts as a shunt resonant circuit. In this case, its input reactance at the fundamental varies in the opposite direction to that of the series transmission line connected to the carrier (main) amplifier, and variation of the inductive and capacitive parts of the reactances will depend on the load resistance. To control the second-harmonic impedance, three quarterwave short-circuit stubs (TL_2, TL_3, and TL_6), as shown in Figure 7.24, were connected to the drain line, and the characteristic impedances of these stubs were individually optimized to achieve high efficiency across the bandwidth [69]. As a result, based on two 25-W and one 10-W Cree GaN HEMT devices, a constant-wave (CW) drain efficiency from 49% to 64%

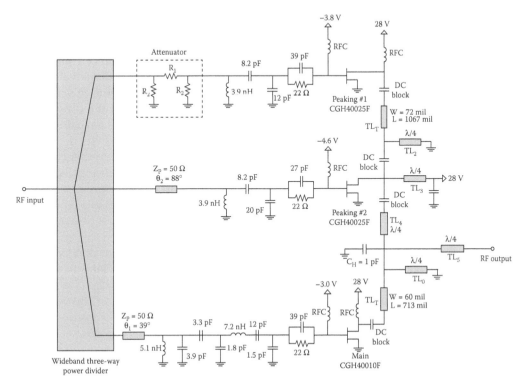

FIGURE 7.24
Circuit schematic of three-stage Doherty amplifier with extended bandwidth.

for output backoff power levels of up to 9 dB was achieved across the bandwidth from 730 to 950 MHz. For a clipped four-carrier 20-MHz WCDMA signal at 900 MHz with a *PAR* of 7.14 dB, an *ACLR* of −51.4 dBc was measured after applying DPD, with an output power of 35 dBm and a *PAE* of 53%.

7.5.3 Broadband Doherty Amplifier via Real Frequency Technique

The broadband properties of the Doherty amplifier significantly depend on the broadband capability of the output combining network. In conventional Doherty amplifier, this network includes two quarterwave transmission lines: one as an impedance transformer and the other as an output transformer. Therefore, optimizing the characteristic impedances of the quarterwave transmission lines may result in the bandwidth extension of the Doherty amplifier. The fractional bandwidth obtaining when it is necessary to transform the impedance from $Z_L = 25\ \Omega$ to the impedance $Z_m = 100\ \Omega$ seen by the carrier device at power backoff mode can be calculated from

$$\frac{\Delta f}{f_0} = 2 - \frac{4}{\pi}\cos^{-1}\left(\frac{2\Gamma_{max}}{\sqrt{1 - \Gamma_{max}^2}}\frac{\sqrt{Z_m Z_L}}{|Z_L - Z_m|}\right) \tag{7.14}$$

where f_0 is the center bandwidth frequency, $\Delta f/f_0$ is the fractional bandwidth of the quarterwave line, and Γ_{max} is the maximum reflection coefficient magnitude [70]. From Equation 7.14, it follows that the fractional bandwidths $\Delta f/f_0 = 17\%$, 35%, and 59% can be achieved for maximum reflection coefficients amplitudes $\Gamma_{max} = 0.10$, 0.20, and 0.32, respectively. For $\Gamma_{max} = 0.20$, the corresponding voltage standing wave ratio is $VSWR = 1.5$. Besides, the bandwidth of the quarterwave transmission line can be enhanced by bringing Z_m and Z_L near to each other, resulting in a reduced transformation ratio of the impedance transformer. In this case, if the common load impedance Z_L is increased from conventional value by 1.4 times, then the transformation ratio of the quarterwave transmission line that interconnects the carrier and peaking amplifiers will reduce to 1:2.85 instead of the conventional 1:4. In this case, the output quarterwave transmission line will also have a reduced transformation ratio, and hence, an enhanced frequency bandwidth can be provided.

As a result, the modified Doherty amplifier has a better efficiency-enhancement behavior than the conventional Doherty amplifier over frequency at 6-dB backoff efficiency of at least 11% up to 44% fractional bandwidth [57]. For example, if the quarterwave transmission lines have a center bandwidth frequency of 2.4 GHz, then the 6-dB backoff efficiency at 2.14 and 2.655 GHz will be of 68.1%, as shown in Figure 7.25a. In this case, a first peak efficiency of 71.2% at $f = f_0 \pm 0.11 f_0$ occurs at 6.4-dB output power backoff, as shown in Figure 7.25b, which is a small degradation from the conventional 6-dB backoff efficiency at f_0 [58].

An extended frequency bandwidth was achieved for the asymmetric Doherty amplifier with a 10-W Cree GaN HEMT device in the carrier amplifier and a 25-W Cree GaN HEMT device in the peaking amplifier [57]. In this case, the Class-AB and Class-C power amplifiers were designed to provide broadband performance using matching circuits with low-frequency response (comprising series microstrip lines with shunt capacitors). With an optimum load-line impedance of approximately 50 Ω for the carrier device, the modified output combining network was based on two quarterwave transmission lines with the characteristic impedances of 59.16 and 41.83 Ω. The 59.16-Ω line interconnects the carrier and peaking amplifiers and inverts the 100-Ω impedance seen by the carrier device in the low-power region to a common-load impedance of 35 Ω (transformation ratio 1:2.85). The

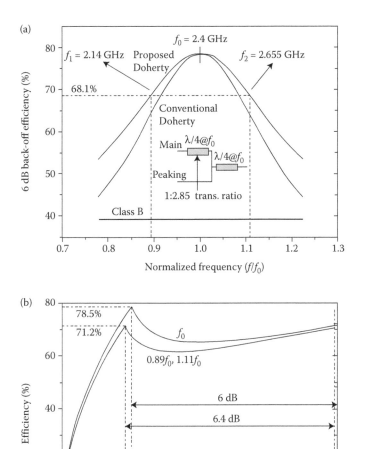

FIGURE 7.25
Efficiency of Doherty amplifier with conventional and modified combining network.

41.83-Ω line transforms the common-load impedance to the standard load of 50 Ω. As a result, the modified Doherty amplifier showed more wideband efficiency-enhancement performance than the conventional Doherty amplifier providing the drain efficiency from 41% to 55% at backoff of 5–6 dB from the maximum output power measured in the range of 42.1–45.3 dBm over the frequency bandwidth of 1.7–2.6 GHz, or 42% fractional bandwidth.

The broadband Doherty amplifier can be optimized at either the backoff or saturation power level. The optimization at the backoff power level implies that the optimum modulated impedance of the carrier power amplifier can be only achieved at the backoff power level, whereas the modulated impedance at the saturation power level assumes only a suboptimum value because of the nonideal load modulation. Figure 7.26 shows the block diagram demonstrating the design procedure associated with the backoff optimization method [71]. Here, the load network of the Doherty amplifier consists of three two-port networks for

(a)

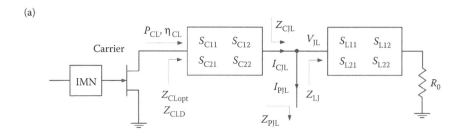

(b)

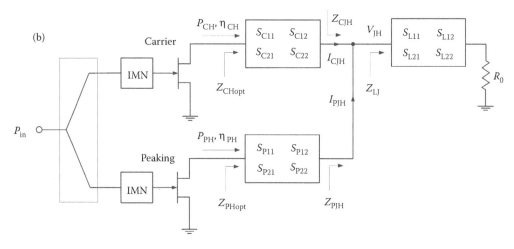

FIGURE 7.26
Block diagram for design procedure using real frequency technique.

carrier, peaking, and output matching paths, whose frequency properties are described by their associated scattering parameters S_{Cij}, S_{Pij}, and S_{Lij}. The impedances Z_{CHopt} and Z_{PHopt} denote the optimized impedances of the carrier and peaking power amplifiers at the saturation power level, the impedance Z_{CLopt} represents the optimized impedance of the carrier power amplifier at the given backoff power level, whereas the impedances Z_{CLD} is defined as the corresponding desired impedance and Z_{PJL} represents the impedance looking into the peaking power amplifier at the junction, when the peaking power amplifier is turned off. During the optimization procedure, both the frequency-dependent drain efficiency and resulting output power of the Doherty amplifier should be as flat and as high as possible.

Generally, the transducer power gain of a two-port network doubly terminated with the frequency-dependent source and load impedances Z_G and Z_L can be defined in terms of the scattering parameters as

$$T(j\omega) = \frac{\left(1 - S_G^2\right)|S_{21}|^2\left(1 - S_L^2\right)}{|1 - S_{11}S_G|^2 \left|1 - \left(S_{22} + \left(S_{21}^2 S_G / 1 - S_{11}S_G\right)\right)S_L\right|^2} \tag{7.15}$$

where S_G and S_L are the scattering parameters associated with the source and load reflections, respectively [71]. If either S_G or S_L is frequency independent, the double-matching solution degenerates into the simplified single-matching solution. In this case, the broadband matching is realized by applying the simplified real frequency technique with the desired frequency-dependent optimum impedances.

The design method for broadband Doherty amplifier at the backoff power level based on the real frequency technique is as follows:

- The scattering parameters S_{Lij} are constructed to transfer the load resistance R_0 to a low-impedance Z_{LJ} over the operating frequency bandwidth. The desired impedance Z_{CLD} associated with the maximum achievable drain efficiency is determined through the harmonic-balance simulation at each frequency point within the specified frequency range. Assuming the "quasi-open-circuit" impedance Z_{PJL}, the impedance Z_{CJL} is obtained based on the knowledge of Z_{LJ}. As shown in Figure 7.26a, the two-port network $[S_C]$ is optimized so that the transducer power gain, defined by Equation 7.15, and drain efficiency are as flat and as high as possible.

- The desired frequency-dependent load-modulated impedance Z_{CHopt} is subjectively selected at the saturation power level that enables the calculation of Z_{CJH} and Z_{PJH} shown in Figure 7.26b.

- The carrier power amplifier is simulated with the frequency-dependent complex load impedance Z_{CJH} at the saturation power level. The drain efficiency η_{CH}, power P_{CH}, and current I_{CJH} at the load termination Z_{CJH} are determined through the harmonic-balance simulation.

- The peaking power amplifier is simulated with the load termination Z_{PJH}, which is transferred to Z_{PHopt} via the two-port network $[S_P]$. The two-port network $[S_P]$ is optimized so that the transistor delivers flat output power around P_{PH} with the minimum drain efficiency η_{PH} over the given frequency range. The "quasi-open-circuit" requirement on Z_{PJL} is also included as an optimization boundary condition.

- The current I_{PJH} is simulated at the load termination Z_{PJH} of the peaking power amplifier. The phase difference between currents I_{CJH} and I_{PJH} is adjusted to equalize the phases in the carrier and peaking paths by tuning the phase-compensating lines in each IMN.

- All the circuit parameters are adjusted to achieve the best performance of the broadband Doherty amplifier.

The symmetric Doherty amplifier was designed using two Cree CGH40006P GaN HEMT devices with a center bandwidth frequency of 2.6 GHz. An output impedance transformation via optimized three-section ladder LC structure provides better broadband matching and results in less influence of the impedance inverter over broader frequency bandwidth. By using a multisection output transmission line with different characteristic impedances, the frequency range from 2.2 to 2.96 GHz can be covered with the drain efficiency over 40% at the output power backoff level of 5 to 6 dB, as shown in Figure 7.27a and b [71]. However, the measured power gain is sufficiently low, as shown in Figure 7.27c, varying between 5.5 and 8.8 dB across the entire frequency range at different output power levels.

7.5.4 Broadband Parallel Doherty Architecture

The classical two-stage Doherty amplifier has limited bandwidth capability in a low-power region since it is necessary to provide an impedance transformation from 25 to 100 Ω when the peaking amplifier is turned off, as shown in Figures 7.28a and 7.29a, thus resulting in a loaded quality factor $Q_L = \sqrt{100/25 - 1} = 1.73$ at 3-dB output-power backoff level, which is

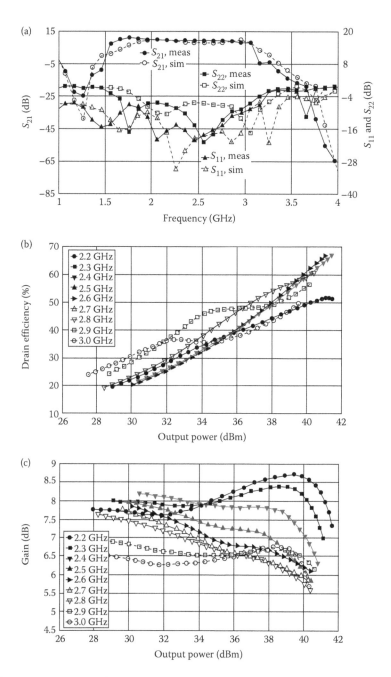

FIGURE 7.27
Simulated and measured performance of broadband Doherty amplifier.

sufficiently high for broadband operation. The parallel architecture of a two-stage Doherty amplifier with modified modulated load network, whose block schematic is shown in Figure 7.28b, can improve bandwidth properties in a low-power region by reducing the impedance transformation ratio by a factor of 2 [72]. Such an approach is very convenient for a high-power Doherty amplifier when the device optimum impedances for the carrier and peaking devices

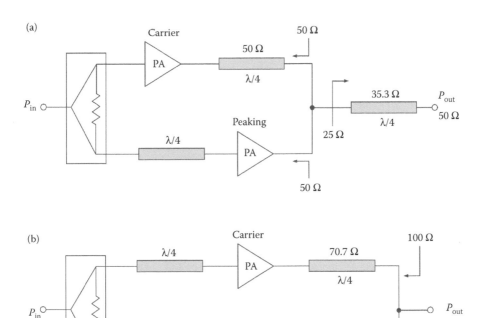

FIGURE 7.28
Block diagram of conventional and modified two-stage Doherty amplifiers.

are typically equal to a few ohms only. In this case, by providing the transmission-line characteristic impedances greater than the device optimum impedances, the drain efficiency of around 48% was achieved at the output powers above 49 dBm in a frequency bandwidth from 800 to 950 MHz for a 20-MHz WCDMA signal with a *PAR* of 7 dB after DPD linearization [73].

In this case, the load network for the carrier amplifier consists of a single quarterwave transmission line required for impedance transformation, the load network for the peaking amplifier consists of a 50-Ω quarterwave transmission line followed by another quarterwave transmission line required for impedance transformation, and the quarterwave transmission line at the input of the carrier amplifier is necessary for phase compensation. Both impedance-transforming quarterwave transmission lines, having a characteristic impedance of 70.7 Ω each, provide a parallel connection of the carrier and peaking amplifiers in a high-power region by parallel combining of the two 100-Ω impedances at their output into a 50-Ω load, with 50-Ω impedances at their inputs seen by each amplifier output. In a low-power region below output-power backoff point of −6 dB when the peaking amplifier is turned off, the required impedance of 100 Ω seen by the carrier amplifier output is achieved by using a single quarterwave transmission line with the characteristic impedance of 70.7 Ω to match with a 50-Ω load, as shown in Figure 7.29b.

This provides a loaded quality factor $Q_L = \sqrt{100/50 - 1} = 1$, resulting in a 1.73 times wider frequency bandwidth, as shown in Figure 7.29c by curve 1 compared with a conventional case (curve 2). Since the load network of the peaking amplifier contains two

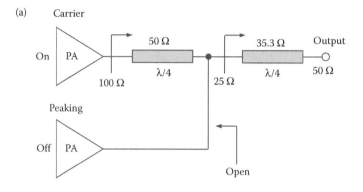

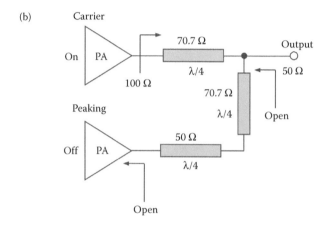

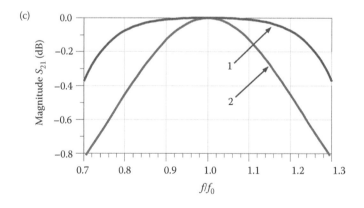

FIGURE 7.29
Load-network schematics and broadband properties.

quarterwave transmission lines connected in series, an overall half-wavelength transmission line is obtained, and an open circuit at the peaking-amplifier output directly translates to the load providing a significant isolation of the peaking-amplifier path from the carrier amplifier path in a wide frequency range. The input in-phase divider and phase-compensating transmission line can be replaced by a broadband coupled-line 90° hybrid coupler.

From Figure 7.29c, it follows that using a parallel Doherty architecture can provide a broadband operation within 25%–30% around center bandwidth frequency with minimum variation of the load-network transfer characteristic. As a result, a dual-band operation can be easily provided by this architecture, for example, in 1.8-GHz (1805–1880 MHz) and 2.1-GHz (2.11–2.17 GHz) or in 2.1-GHz and 2.6-GHz (2.62–2.69 GHz) WCDMA/LTE frequency bands, respectively.

Figure 7.30 shows the simulated circuit schematic of a dual-band parallel GaN HEMT Doherty architecture, where the carrier and peaking amplifiers are based on broadband transmission-line Class-E power amplifiers, whose circuit structure is shown in Figure 6.23b (Chapter 6). Here, the input matching circuits and output load network are based on microstrip lines with their parameters corresponding to a 20-mil RO4360 substrate. The ideal broadband 90° hybrid coupler is used at the input to split signals between the carrier and peaking amplifying paths. The electrical lengths of both offset and combining microstrip lines were optimized to maximize efficiency at saturated and backoff output power levels.

Figure 7.31 shows the simulation results for the small-signal S_{21}-parameters versus frequency demonstrating the bandwidth capability of a parallel transmission-line GaN HEMT Doherty amplifier covering a frequency range from 2.0 to 2.8 GHz with a power gain over 10 dB. In this case, an input return loss defined from the magnitude of S_{11} is less than 5 dB over the frequency bandwidth of 2.1–2.9 GHz.

Figure 7.32 demonstrates the broadband capability of a parallel Doherty structure, where the carrier and peaking amplifiers are based on a broadband transmission-line

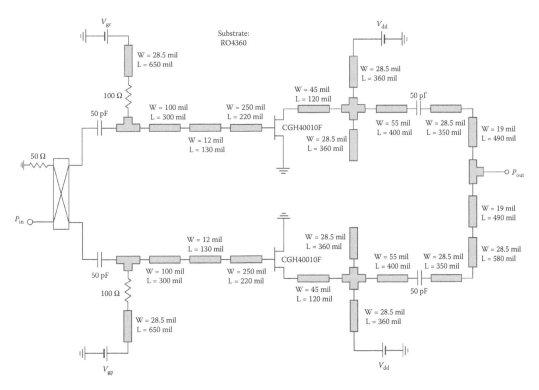

FIGURE 7.30
Circuit schematic of dual-band parallel GaN HEMT Doherty amplifier.

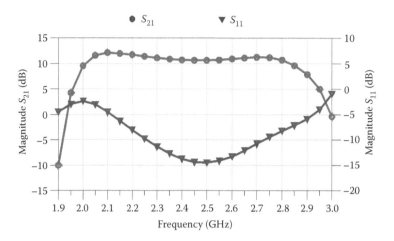

FIGURE 7.31
Simulated small-signal S_{11} and S_{21} versus frequency.

reactance compensation Class-E technique. In an amplifier saturation mode with an input power of 36 dBm, a drain efficiency of around 70% with an average output power of greater than 43 dBm and a gain variation of about 1 dB was simulated across the frequency range of 2.0–2.8 GHz, as shown in Figure 7.32a. At the same time, high-drain efficiency over 50% at backoff output powers of 5–6 dB from saturation can potentially be achieved across the frequency range from 2.1 to 2.7 GHz, as shown in Figure 7.32b. This means that the practical implementation of a parallel Doherty power amplifier, whose simulation setup is shown in Figure 7.30, can provide a highly efficient operation in two cellular bands of 2.11–2.17 GHz and 2.62–2.69 GHz without any tuning of the amplifier load-network parameters, either with separate or simultaneous dual-band transmission of WCDMA or LTE signals.

The large-signal simulations versus input power have been done at two center bandwidth frequencies of 2.14 and 2.655 GHz with optimized circuit parameters to achieve maximum performance. Figure 7.33 shows the simulated large-signal power gain and drain efficiencies of a dual-band parallel transmission-line GaN HEMT Doherty amplifier, with the carrier gate bias $V_{gc} = -2.45$ V, peaking gate bias $V_{gp} = -7$ V, and dc supply voltage $V_{dd} = 28$ V. In this case, a linear power gain of about 11 dB was achieved at an operating frequency of 2.655 GHz, whereas a slightly higher linear power gain of about 12 dB was achieved at lower operating frequency of 2.14 GHz [72]. At the same time, the drain efficiencies of 64% and 53% were simulated at backoff output powers of 39 dBm (4-dB backoff from saturated power of 43 dBm) and 37 dBm (6-dB backoff) at both center bandwidth frequencies, respectively.

The dual-band transmission-line GaN HEMT Doherty amplifier was fabricated on a 20-mil RO4360 substrate. An input splitter represents a broadband coupled-line coupler from Anaren, model 11306-3, which provides maximum phase balance of ±5° and amplitude balance of ±0.55 dB across the frequency range from 2 to 4 GHz. Figure 7.34 shows the test board of a dual-band parallel Doherty amplifier based on two 10-W Cree GaN HEMT power transistors CGH40010P in metal–ceramic pill packages [72]. The input matching circuit, output load network, and gate and drain bias circuits (having bypass capacitors on their ends) are fully based on microstrip lines of different electrical lengths and

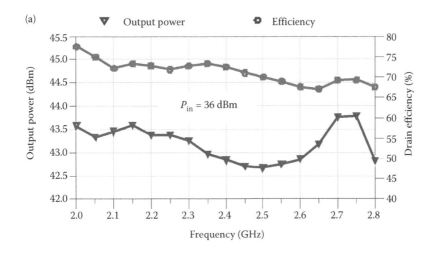

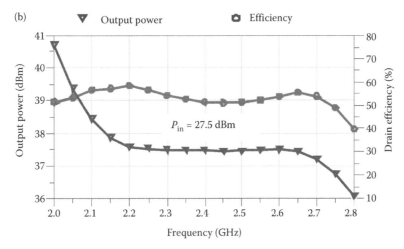

FIGURE 7.32
Broadband capability of parallel Doherty amplifier.

characteristic impedances according to the simulation setup shown in Figure 7.30. Special care should be taken in the device implementation process in order to minimize the input and output lead inductances of the packaged GaN HEMT device, which can significantly affect the amplifier performance such as power gain, output power, and drain efficiency.

For a single-carrier 5-MHz WCDMA signal with a *PAR* of 6.5 dB, a drain efficiency of 45% with a power gain of about 10 dB and *ACLR* (at 5-MHz offset) lower than −30 dBc at 2.14 GHz and a drain efficiency of 40% with a power gain of about 11 dB and *ACLR* around −30 dBc at 2.655 GHz were achieved at an average output power of 39 dBm. In both cases, optimization of the gate bias voltages for the carrier (Class-AB mode) and peaking (Class-C mode) amplifiers were provided.

7.5.5 Broadband Inverted Doherty Configuration

Figure 7.35a shows the modified broadband load network of an inverted Doherty amplifier, which consists of a two-section transmission-line output impedance transformer,

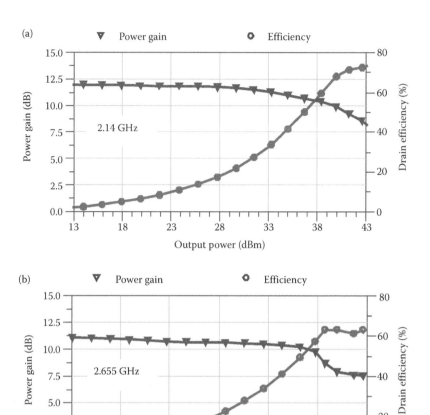

FIGURE 7.33
Simulated results of dual-band parallel Doherty amplifier at (a) 2.14 GHz and (b) 2.655 GHz.

FIGURE 7.34
Test board of dual-band GaN HEMT parallel Doherty amplifier.

(a)

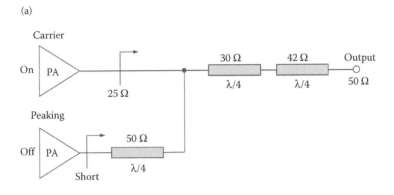

(b)

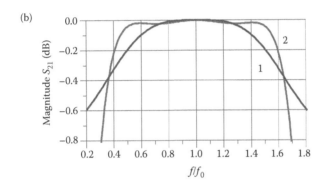

FIGURE 7.35
Load-network schematic and broadband properties.

where each quarterwave transmission line has a different characteristic impedance to match first the initial 25 Ω to intermediate 35.3 Ω and then to 50-Ω load. Such 25- to 50-Ω transformer provides a wide frequency range, as shown in Figure 7.35b by curve 1 [74]. However, broader frequency range with flatter frequency response can be achieved with a quarterwave open-circuit stub connected at the input of the two-line transformer when the peaking amplifier is turned off, as shown in Figure 7.35b by curve 2, resulting in greater than octave bandwidth in a low-power region at output power levels less than −6 dB backoff point. In this case, it is assumed that the output matching circuit of the carrier amplifier provides ideally a broadband impedance transformation from 25 to 100 Ω or close seen by the device multiharmonic current source. At the same time, broadband performance is also provided in a high-power region when both the carrier and peaking amplifiers are turned on.

Figure 7.36 shows the simulated circuit schematic of a tri-band inverted GaN HEMT Doherty amplifier configuration, where the carrier and peaking amplifiers using Cree CGH40010 GaN HEMT devices are based on the same broadband transmission-line Class-E power amplifiers, whose idealized circuit structure is shown in Figure 6.23a (Chapter 6), and the broadband load network corresponds to the impedance-transforming structure shown in Figure 7.35a. The input matching circuits and output load network are based on microstrip lines with their parameters corresponding to a 20-mil RO4360 substrate. In this

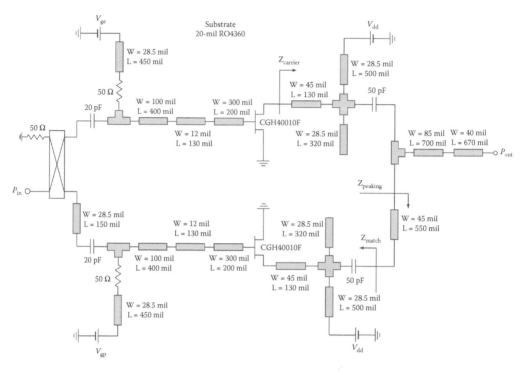

FIGURE 7.36
Circuit schematic of tri-band inverted GaN HEMT Doherty amplifier.

case, it was found that, when the broadband Class-E power amplifier as a peaking amplifier is turned off, a short-circuit condition is achieved at the input of a series 35-Ω transmission line shown in Figure 6.23b (Chapter 6), which has a quarter wavelength at high-bandwidth frequency to match with a 50-Ω load. Therefore, such a quarterwave transmission line was removed from the load networks of both the carrier and peaking amplifiers.

As a result, the overall combining load-network is significantly simplified, and only a small optimization of the electrical lengths of the short- and open-circuit stubs in the load networks of the carrier and peaking amplifiers is required to achieve broader frequency response. The broadband 90° hybrid coupler is used at the input to split signals between the carrier and peaking amplifying paths and to provide a 90° phase shift at the input of the carrier amplifier across the entire frequency bandwidth.

The impedance conditions at different points of the load network of the peaking amplifier when it is turned off are shown in Figure 7.37, where Z_{match} shown in Figure 7.37a indicates a low reactance at the output of a Class-E load network with short- and open-circuit stubs across the required frequency range from 1.8 to 2.7 GHz, having nearly zero reactance at high-bandwidth frequency of 2.7 GHz and increasing capacitive reactance when the operating frequency reduces to 1.8 GHz. At the same time, by using the series transmission line of a quarter-wavelength long at high-bandwidth frequency, an open-circuit condition is provided at higher bandwidth frequencies with sufficiently high-inductive reactances at lower bandwidth frequencies, indicating by $Z_{peaking}$ shown in Figure 7.37b. Hence, the broadband performance of such an inverted Doherty structure can potentially be achieved in a practical realization.

(a)

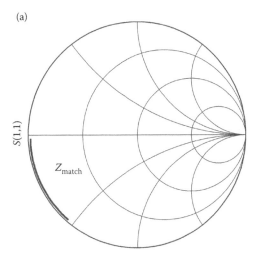

Frequency (1.800 GHz to 2.700 GHz)

(b)

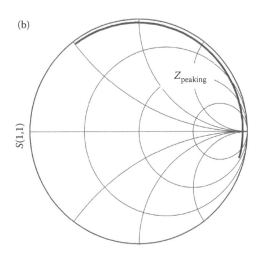

Frequency (1.800 GHz to 2.700 GHz)

FIGURE 7.37
Impedances for peaking amplifier.

Figure 7.38 shows the frequency behavior of the impedance $Z_{carrier}$ seen by the carrier device, as shown in Figure 7.36, which has an inductive reactive component required for a high-efficiency Class-E operation and its real component varies slightly between 17 and 22 Ω. This means that by taking into account the device output shunt capacitance of 1.3 pF and series bondwire inductor of about 1 nH, the impedances seen by the device multiharmonic current source at the fundamental across the entire frequency bandwidth of 1.8–2.7 GHz can be increased up to around 50 Ω, which is high enough to achieve high efficiency at backoff output power levels. In this case, the device output capacitance and bondwire inductor constitute a low-pass L-type matching section to increase the load impedance seen internally by the device multiharmonic current source at the fundamental.

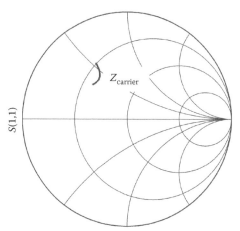

FIGURE 7.38
Impedance for carrier amplifier.

Figure 7.39 shows the simulation results for the small-signal S_{21}-parameters versus frequency, demonstrating the bandwidth capability of a modified inverted transmission-line GaN HEMT Doherty amplifier, which potentially can cover a wide frequency range of 1.6–3.0 GHz with a power gain over 11 dB.

Figure 7.40 shows the simulated large-signal power gain and drain efficiencies of a transmission-line tri-band inverted GaN HEMT Doherty amplifier, with the carrier gate bias $V_{gc} = -2.45$ V, peaking gate bias $V_{gp} = -7.45$ V, and dc supply voltage $V_{dd} = 28$ V [74]. In this case, a linear power gain of about 11.5 dB was achieved at higher bandwidth frequencies of 2.655 and 2.14 GHz, whereas a slightly higher linear power gain of about 13 dB was achieved at lower bandwidth frequency of 1842.5 MHz. At the same time, the drain efficiencies of 71.5%, 69.0%, and 64.0% at backoff output powers of 40 dBm (4-dB backoff from saturated power of 44 dBm) and 59.0%, 57.0%, and 53.5% at backoff output powers of

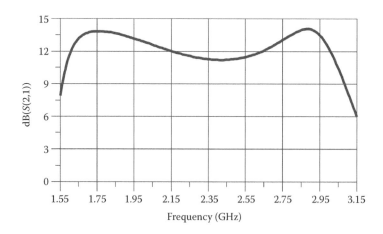

FIGURE 7.39
Simulated small-signal S_{21} versus frequency.

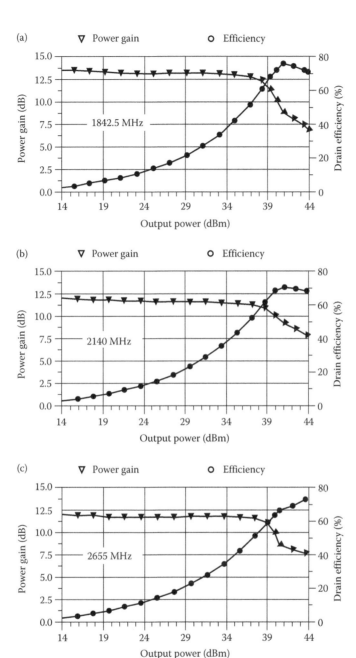

FIGURE 7.40
Simulated power gain and drain efficiencies of tri-band inverted Doherty amplifier at (a) 1842.5 MHz, (b) 2140 MHz, and (c) 2655 MHz.

38 dBm (6-dB backoff) were simulated at center bandwidth frequencies of 1842.5, 2140, and 2655 MHz, respectively. Here, the peak drain efficiency peaks near 4-dB backoff output power at low-bandwidth frequency of 1842.5 MHz and at medium bandwidth frequency of 2.14 GHz are clearly seen, whereas high efficiency remains almost constant at high-output powers at high-bandwidth frequency of 2.655 GHz.

The tri-band transmission-line GaN HEMT Doherty amplifier was fabricated on a 20-mil RO4360 substrate. An input splitter represents a broadband coupled-line coupler from Anaren, model X3C17A1-03WS, which provides maximum phase balance of ±5° and amplitude balance of ±0.5 dB across the frequency range of 690–2700 MHz.

Figure 7.41 shows the test board of a tri-band inverted Doherty amplifier based on two 10-W Cree GaN HEMT power transistors CGH40010P in metal–ceramic pill packages [74]. The input matching circuit, output load network, and gate and drain bias circuits (having bypass capacitors on their ends) are fully based on microstrip lines of different electrical lengths and characteristic impedances according to the simulation setup shown in Figure 7.36. Special care should be taken in the device implementation process in order to minimize the input and output lead inductances of the packaged GaN HEMT device. Additional tuning has been done in the input matching circuits to maximize power gain over the entire frequency range.

For a single-carrier 5-MHz WCDMA signal with a *PAR* of 6.5 dB, the drain efficiencies of 58%, 50%, and 42% at an average output power of 38 dBm with a power gain of more than 11 dB were achieved at the operating frequencies of 1.85, 2.15, and 2.65 GHz, respectively, with the *ACLR* (at 5-MHz offset) measured from −32 dBc at 1.85 GHz to −37 dBc at 2.65 GHz. The gate bias voltages for carrier (Class-AB mode with a quiescent current of 100 mA) and peaking (Class-C mode) amplifiers were the same for all three frequencies.

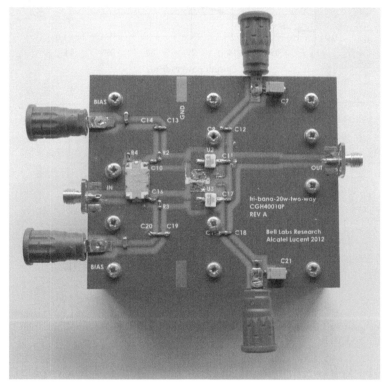

FIGURE 7.41
Test board of tri-band inverted GaN HEMT Doherty amplifier.

References

1. W. H. Doherty, Amplifier, U.S. Patent 2,210,028, August 1940 (filed April 1936).
2. W. H. Doherty, A new high efficiency power amplifier for modulated waves, *Proc. IRE*, 24, 1163–1182, 1936.
3. W. H. Doherty and O. W. Towner, A 50-kilowatt broadcast station utilizing the Doherty amplifier and designed for expansion to 500 kilowatts, *Proc. IRE*, 27, 531–534, 1939.
4. W. H. Doherty, *Proc. IRE*, 25, 922, 1937.
5. C. E. Smith, J. R. Hall, and J. O. Weldon, Very high power long-wave broadcast station, *Proc. IRE*, 42, 1222–1235, 1954.
6. J. B. Sainton, A 500 kilowatt medium frequency standard broadcast transmitter, Cathode Press (Machlett Company), vol. 22(4), pp. 22–29, 1965.
7. V. M. Rozov and V. F. Kuzmin, Use of the Doherty circuit in SSB transmitters, *Telecommun. Radio Eng.*, 1970–1971.
8. P. Y. Vinogradov, N. I. Vorobyev, E. P. Sokolov, and N. S. Fuzik, Amplification of a modulated signal by the Doherty method in a transistorized power amplifier, *Telecommun. Radio Eng., Part 1*, 31, 38–41, 1977.
9. Development of circuitry for multikilowatt transmitter for space communications satellites, Report No. CRI19803 (NASA N71-29212), General Electric Company, Space Systems Division, February 1971.
10. A. Mina and F. Parry, Broadcasting with megawatts of power: The modern era of efficient powerful transmitters, *IEEE Trans. Broadcast.*, BC-35, 121–130, 1989.
11. G. Clark, A comparison of current broadcast amplitude-modulation techniques, *IEEE Trans. Broadcast.*, BC-21, 25–31, 1975.
12. F. H. Raab, Efficiency of Doherty RF power-amplifier systems, *IEEE Trans. Broadcast.*, BC-33, 77–83, 1987.
13. B. Kim, J. Kim, I. Kim, and J. Cha, The Doherty power amplifier, *IEEE Microwave Mag.*, 7, 42–50, 2006.
14. R. J. McMorrow, D. M. Upton, and P. R. Maloney, The microwave Doherty amplifier, *1994 IEEE MTT-S Int. Microwave Symp. Dig.*, San Diego, CA, pp. 1653–1656, 1994.
15. C. F. Campbell, A full integrated *Ku*-band Doherty amplifier MMIC, *IEEE Microwave Guided Wave Lett.*, 9, 114–116, 1999.
16. K. W. Kobayashi, A. K. Oki, A. Gutierrez-Aitken, P. Chin, L. Yang, E. Kaneshiro, P. C. Grossman et al., An 18–21 GHz InP DHBT linear microwave Doherty amplifier, *2000 IEEE RFIC Symp. Dig.*, Boston, MA, pp. 179–182, 2000.
17. J. -H. Tsai and T. -W. Huang, A 38–46 GHz MMIC Doherty amplifier using post-distortion linearization, *IEEE Microwave Wireless Compon. Lett.*, 17, 388–390, 2007.
18. C. Steinberser, T. Landon, C. Suckling, J. Nelson, J. Delaney, J. Hitt, L. Witkowski, G. Burgin, R. Hajji, and O. Krutko, 250 W HVHBT Doherty with 57% WCDMA efficiency linearized to −55 dBc for 2c11 6.5 dB PAR, *IEEE J. Solid-State Circuits*, SC-43, 2218–2228, 2008.
19. H. Deguchi, N. Ui, K. Ebihara, K. Inoue, N. Yoshimura, and H. Takahashi, A 33 W GaN HEMT Doherty amplifier with 55% drain efficiency for 2.6 GHz base stations, *2009 IEEE MTT-S Int. Microwave Symp. Dig.*, Boston, MA, pp. 1273–1276, 2009.
20. Y. Yang, J. Yi, Y. Y. Woo, and B. Kim, Optimum design for linearity and efficiency of a microwave Doherty amplifier using a new load matching technique, *Microwave J.*, 44, 20–36, 2001.
21. G. K. Wong, T. R. Shah, and K. Titizer, Doherty power amplifier with phase compensation, U.S. Patent 7,295,074, November 2007 (filed March 2005).
22. S. -C. Jung, O. Hammi, and F. Ghannouchi, Design optimization and DPD linearization of GaN-based unsymmetrical Doherty power amplifiers for 3G multicarrier applications, *IEEE Trans. Microwave Theory Tech.*, MTT-57, 2105–2113, 2009.

23. M. Nick and A. Mortazawi, Adaptive input-power distribution in Doherty power amplifiers for linearity and efficiency enhancement, *IEEE Trans. Microwave Theory Tech.*, MTT-58, 2764–2771, 2010.

24. J. Moon, Ja. Kim, Ju. Kim, I. Kim, and B. Kim, Efficiency enhancement of Doherty amplifier through mitigation of the knee voltage effect, *IEEE Trans. Microwave Theory Tech.*, MTT-59, 143–152, 2011.

25. Y. Yang, J. Cha, B. Shin, and B. Kim, A microwave Doherty amplifier employing envelope tracking technique for high efficiency and linearity, *IEEE Microwave Wireless Compon. Lett.*, 13, 370–372, 2003.

26. Y. Zhao, M. Iwamoto, L. E. Larson, and P. M. Asbeck, Doherty amplifier with DSP control to improve performance in CDMA operation, *2003 IEEE MTT-S Int. Microwave Symp. Dig.*, Philadelphia, PA, pp. 687–690, 2003.

27. S. Bae, J. Kim, I. Nam, and Y. Kwon, Bias-switching quasi-Doherty-type amplifier for CDMA handset applications, *2003 IEEE RFIC Symp. Dig.*, Philadelphia, PA, pp. 137–140, 2003.

28. T. Kato, K. Yamaguchi, Y. Kuriyama, and H. Yoshida, An HPSK/OFDM 64-QAM dual-mode Doherty power amplifier module for mobile terminals, *IEICE Trans. Electron.*, E90-C, 1678–1684, 2007.

29. S. Kawai, Y. Takayama, R. Ishikawa, and K. Honjo, A GaN HEMT Doherty amplifier with a series connected load, *Proc. 2009 Asia-Pacific Microwave Conf.*, Singapore, pp. 325–328, 2009.

30. S. Watanabe, Y. Takayama, R. Ishikawa, and K. Honjo, A broadband Doherty amplifier without a quarter-wave impedance inverting network, *Proc. 2012 Asia-Pacific Microwave Conf.*, Kaohsiung, Taiwan, pp. 361–363, 2012.

31. R. F. Stengel, W.-C. A. Gu, G. D. Leizerovich, and L. F. Cygan, High efficiency power amplifier having reduced output matching networks for use in portable devices, U.S. Patent 6,262,629, July 2001 (filed July 1999).

32. G. Ahn, M. Kim, H. Park, S. Jung, J. Van, H. Cho, S. Kwon et al., Design of a high-efficiency and high-power inverted Doherty amplifier, *IEEE Trans. Microwave Theory Tech.*, MTT-55, 1105–1111, 2007.

33. S. Kwon, M. Kim, S. Jung, J. Jeong, K. Lim, J. Van, H. Cho, H. Kim, W. Nah, and Y. Yang, Inverted-load network for high-power Doherty amplifier, *IEEE Microwave Mag.*, 10, 93–98, 2009.

34. S. Jin, J. Zhou, and L. Zhang, A broadband inverted Doherty power amplifier for IEEE 802.11b/g WLAN applications, *Microwave Optical Technol. Lett.*, 53, 636–639, 2011.

35. J. Sirois, S. Boumaiza, M. Helaoui, G. Brassard, and F. M. Ghannouchi, A robust modeling and design approach for dynamically loaded and digitally linearized Doherty amplifiers, *IEEE Trans. Microwave Theory Tech.*, MTT-53, 2875–2883, 2005.

36. I. Takenaka, K. Ishikura, H. Takahashi, K. Hasegawa, T. Ueda, T. Kurihara, K. Asano, and N. Iwata, A distortion-cancelled Doherty high-power amplifier using 28-V GaAs heterojunction FETs for W-CDMA base stations, *IEEE Trans. Microwave Theory Tech.*, MTT-54, 4513–4521, 2006.

37. M.-W. Lee, S.-H. Kam, Y.-S. Lee, and Y.-H. Jeong, Design of highly efficient three-stage inverted Doherty power amplifier, *IEEE Microwave Wireless Compon. Lett.*, 21, 383–385, 2011.

38. C. P. McCarroll, G. D. Alley, S. Yates, and R. Matreci, A 20 GHz Doherty power amplifier MMIC with high efficiency and low distortion designed for broad band digital communication systems, *2000 IEEE MTT-S Int. Microwave Symp. Dig.*, Boston, MA, pp. 537–540, 2000.

39. D. Yu, Y. Kim, K. Han, J. Shin, and B. Kim, A 60-GHz fully integrated Doherty power amplifier based on 0.13-µm CMOS process, *2008 IEEE RFIC Symp. Dig.*, Atlanta, GA, pp. 69–72, 2008.

40. D. Yu, Y. Kim, K. Han, J. Shin, and B. Kim, Fully integrated Doherty power amplifiers for 5 GHz wireless-LANs, *2006 IEEE RFIC Symp. Dig.*, San Francisco, CA, pp. 177–180, 2006.

41. M. Elmala, J. Paramesh, and K. Soumyanath, A 90-nm CMOS Doherty power amplifier with minimum AM-PM distortion, *IEEE J. Solid-State Circuits*, SC-41, 1323–1332, 2006.

42. C. Tongchoi, M. Chongcheawchamnan, and A. Worapishet, Lumped element based Doherty power amplifier topology in CMOS process, *2003 IEEE Int. Circuits Syst. Symp. Dig.*, Bangkok, vol. 1, pp. I-445–I-448, 2003.

43. I. I. Blednov, High power Doherty amplifier, U.S. Patent 7,078,976, July 2006 (filed October 2005).

44. I. I. Blednov and J. van der Zanden, High power LDMOS integrated Doherty amplifier for W-CDMA, *2006 IEEE RFIC Symp. Dig.*, San Francisco, CA, pp. 1–4, 2006.
45. J. Jung, U. Kim, J. Jeon, J. Kim, K. Kang, and Y. Kwon, A new 'series-type' Doherty amplifier for miniaturization, *2005 IEEE RFIC Symp. Dig.*, Long Beach, CA, pp. 259–262, 2005.
46. C. Koo, U. Kim, J. Jeon, J. Kim, and Y. Kwon, A linearity-enhanced compact series-type Doherty amplifier suitable for CDMA handset applications, *2007 IEEE Radio Wireless Symp. Dig.*, Long Beach, CA, pp. 317–320, 2007.
47. D. Kang, J. Choi, D. Kim, and B. Kim, Design of Doherty power amplifiers for handset applications, *IEEE Trans. Microwave Theory Tech.*, MTT-58, 2134–2142, 2010.
48. D. Kang, D. Kim, and B. Kim, Broadband HBT Doherty power amplifiers for handset applications, *IEEE Trans. Microwave Theory Tech.*, MTT-58, 4031–4039, 2010.
49. D. Kang, D. Kim, Y. Cho, B. Park, J. Kim, and B. Kim, Design of bandwidth-enhanced Doherty power amplifiers for handset applications, *IEEE Trans. Microwave Theory Tech.*, MTT-59, 3473–3483, 2011.
50. D. Gustafsson, J. C. Cahuana, D. Kuylenstierna, I. Angelov, N. Rorsman, and C. Fager, A wideband and compact GaN MMIC Doherty amplifier for microwave link applications, *IEEE Trans. Microwave Theory Tech.*, MTT-61, 922–930, 2013.
51. J. S. Moon, H. Moyer, P. Macdonald, D. Wong, M. Antcliffe, M. Hu, P. Willadsen et al., High efficiency X-band Class-E GaN MMIC high-power amplifiers, *Proc. 2012 IEEE Topical Conf. Power Amplifiers Wireless Radio Appl.*, Santa Clara, CA, pp. 9–11, 2012.
52. R. Sperlich, G. C. Copeland, and R. Hoppenstein, Hybrid Doherty amplifier system and method, U.S. Patent Appl. 2008/0111622, May 2008 (filed November 2007).
53. D. R. Pehlke, Class E Doherty amplifier topology for high efficiency signal transmitters, U.S. Patent 6,396,341, May 2002 (filed December 2000).
54. R. Darraji, F. M. Ghannouchi, and H. Hammi, A dual-input digitally driven Doherty amplifier for performance enhancement of Doherty transmitters, *IEEE Trans. Microwave Theory Tech.*, MTT-59, 1284–1293, 2011.
55. R. Darraji and F. M. Ghannouchi, Digital Doherty amplifier with enhanced efficiency and extended range, *IEEE Trans. Microwave Theory Tech.*, MTT-59, 2898–2909, 2011.
56. Y. Suzuki and S. Narahashi, Multiband Doherty amplifier, U.S. Patent 7,602,241, October 2009 (filed June 2007).
57. K. Bathich, A. Z. Markos, and G. Boeck, Frequency response analysis and bandwidth extension of the Doherty amplifier, *IEEE Trans. Microwave Theory Tech.*, MTT-59, 934–944, 2011.
58. K. Bathich, D. Gruner, and G. Boeck, Analysis and design of dual-band GaN HEMT based Doherty amplifier, *Proc. 6th Eur. Microwave Integrated Circuits Conf.*, Manchester, UK, pp. 248–251, 2011.
59. L. Wu, Z. Sun, H. Yilmaz, and M. Berroth, A dual-frequency Wilkinson power divider, *IEEE Trans. Microwave Theory Tech.*, MTT-54, 278–284, 2006.
60. K. M. Cheng and F. Wong, A new Wilkinson power divider design for dual band applications, *IEEE Microwave Wireless Compon. Lett.*, 17, 664–666, 2007.
61. K. Rawat and F. Ghannouchi, A design methodology for miniaturized power dividers using periodically loaded slow wave structure with dual-band applications, *IEEE Trans. Microwave Theory Tech.*, MTT-57, 3380–3388, 2009.
62. W. Chen, S. A. Bassam, X. Li, Y. Liu, K. Rawat, M. Helaoui, F. M. Ghannouchi, and Z. Feng, Design and linearization of concurrent dual-band Doherty power amplifier with frequency-dependent power ranges, *IEEE Trans. Microwave Theory Tech.*, MTT-59, 2537–2546, 2011.
63. P. Colantonio, F. Feudo, F. Giannini, R. Giofre, and L. Piazzon, Design of a dual-band GaN Doherty amplifier, *Proc. 18th Int. Microwave Radar Wireless Commun. Conf.*, Vilnius, Lithuania, pp. 1–4, 2010.
64. X. A. Nghiem and R. Negra, Novel design of a concurrent tri-band GaN HEMT Doherty power amplifier, *Proc. 2012 Asia-Pacific Microwave Conf.*, Kaohsiung, Taiwan, pp. 364–366, 2012.
65. X. A. Nghiem, J. Guan, T. Hone, and R. Negra, Design of concurrent multiband Doherty power amplifiers for wireless applications, *IEEE Trans. Microwave Theory Tech.*, MTT-61, 4559–4568, 2013.

66. C. Monzon, A small dual-frequency transformer in two sections, *IEEE Trans. Microwave Theory Tech.*, MTT-51, 1157–1161, 2003.
67. N. Giovannelli, A. Cidronali, P. Singerl, S. Maddio, C. Schuberth, A. Del Chiaro, and G. Manes, A 250 W LDMOS Doherty PA with 31% of fractional bandwidth for DVB-T applications, *2014 IEEE MTT-S Int. Microwave Symp. Dig.*, Tampa, FL, pp. 1–4, 2014.
68. M. N. Ali Abadi, H. Golestaneh, H. Sarbishaei, and S. Boumaiza, An extended bandwidth Doherty amplifier using a novel output combiner, *2014 IEEE MTT-S Int. Microwave Symp. Dig.*, Tampa, FL, pp. 1–4, 2014.
69. H. Golestaneh, F. A. Malekzadeh, and S. Boumaiza, An extended-bandwidth three-way Doherty power amplifier, *IEEE Trans. Microwave Theory Tech.*, MTT-61, 3318–3328, 2013.
70. D. M. Pozar, *Microwave Engineering*, New York: John Wiley & Sons, 2004.
71. G. Sun and R. H. Jansen, Broadband Doherty power amplifier via real frequency technique, *IEEE Trans. Microwave Theory Tech.*, MTT-60, 99–111, 2012.
72. A. Grebennikov and J. Wong, A dual-band parallel Doherty power amplifier for wireless applications, *IEEE Trans. Microwave Theory Tech.*, MTT-60, 3214–3222, 2012.
73. D. Y. Wu, J. Annes, M. Bokatius, P. Hart, E. Krvavac, and G. Tucker, A 350 W, 790 to 960 MHz wideband LDMOS Doherty amplifier using a modified combining scheme, *2014 IEEE MTT-S Int. Microwave Symp. Dig.*, Tampa, FL, pp. 1–4.
74. A. Grebennikov, Multiband Doherty amplifiers for wireless applications, *High Freq. Electron.*, 13, 30–46, 2014.

8

Low-Noise Broadband Amplifiers

This chapter begins with the historical aspects and basic principles of the low-noise amplifier (LNA) design, including basic topologies, minimum noise figure, and linearization techniques. When it is necessary to achieve high-gain and low-noise performance over a sufficiently wide frequency range, the LNAs are designed using lossless matching circuits. However, the lossy feedback LNAs have been shown to be capable of the flat gain over a very wide bandwidth with a sufficiently low-noise figure, small size, and convenience in practical implementation. Several design techniques, including iterative optimization or interactive graphical design approach, are very useful in view of the many variables and conflicting objectives of high gain, flat and broadband gain, and low-noise figure. Finally, some practical circuit schematics of the broadband millimeter-wave LNAs are given and discussed.

8.1 Basic Principles of Low-Noise Amplifier Design

Over the past number of years, the advances made in LNA technology have been significant. From the early 1980s, the technological innovations achieved in this expanding field not only resulted in improved devices and systems but also enabled new markets to be developed such as the cellular market, direct TV, or satellite communications [1].

8.1.1 Historical Aspects

Figure 8.1 shows the state-of-the-art performance of the low- and ultra-low-noise amplifiers developed up to the early 1980s and based on the parametric amplifiers, microwave amplification by stimulated emission of radiation (MASER) technique, and bipolar or (field-effect transistor) FET technology [2]. Many LNAs were large, heavy, and consumed a lot of power at that time. The continuous development and improvement of three-terminal solid-state devices had led to better performance of the LNAs in noise temperature, gain, power dissipation, bandwidth, and frequency of operation, reaching the millimeter-wave frequency band and approaching the submillimeter-wave portion of frequency spectrum, with a significant reduction in size and weight. This improvement has been achieved moving from FETs to HEMTs, then to pHEMTs, and later to InP HEMTs having a transconductance greater than 1000 mS/mm and a maximum operating frequency above 400 GHz.

One of the first configurations of an LNA representing a vacuum-tube cascode configuration is shown in Figure 8.2 [3]. This amplifier arrangement consisting of a grounded-cathode triode followed by a grounded-grid cathode demonstrated a very good combination of noise factor, gain, and stability, with a noise factor averaging 0.25 dB at a carrier frequency of 6 MHz and 1.35 dB at 30 MHz. Here, the coil L_n in parallel with the grid-plate capacitance C_{gp} is a neutralizing coil, whose purpose is to achieve a low-noise factor. The combination

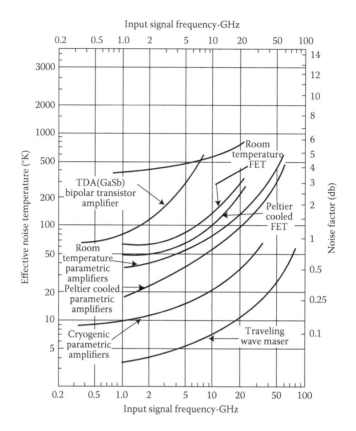

FIGURE 8.1
State-of-the-art performance of low- and ultra-low-noise amplifiers.

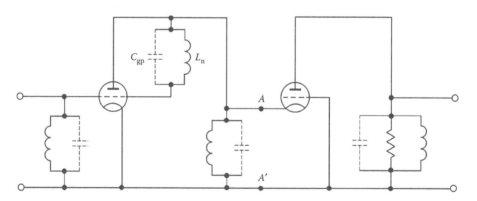

FIGURE 8.2
Circuit schematic of vacuum-tube low-noise amplifier.

of a very low resistance looking to the right at points AA' and a very high resistance to the left is a crucial characteristic of the grounded-cathode grounded-grid combination with regard to both stability and noise factor. Better noise figure can often be achieved if the double-tuned circuit is used in the inputs of the high-frequency amplifiers instead of a simpler single-tuned network [4].

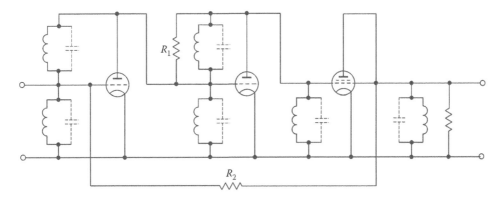

FIGURE 8.3
Circuit schematics of vacuum-tube three-stage low-noise amplifier.

In narrow-band vacuum-tube amplifiers, the optimum source impedance is generally lower than the input impedance of the amplifier. In this case, the use of a feedback network can not only provide a single-frequency impedance matching but can also equalize the response of the input circuit with sufficient accuracy over the desired frequency range. If the induced grid noise is significant so that the circuit to be equalized falls by less than 6 dB at the band limits, then the matching condition may evidently be satisfied with an even better overall response. Figure 8.3 shows the circuit schematic of a three-stage vacuum-tube LNA, where the feedback for compensation of the second-stage circuit is provided by the shunt resistor R_1 and feedback for the input circuit is over three stages, allowing the resistor R_2 to be large compared with the source impedance, having no measurable effect on the noise factor [5].

For a single-stage amplifier, the single-frequency noise figure depends on the coupling between the tube and the signal source, and there exists an optimum coupling network at the amplifier input, for which the noise figure assumes a minimum value. At the same time, it was proved that a multistage amplifying system cannot have a noise figure smaller than that of an optimum amplifying system using one amplifying device [6]. In many important cases, the best noise performance attainable with a particular type of amplifier is actually achieved by a simple cascade, in which the input of each stage is properly mismatched. However, the mismatch conditions for each stage do not in general coincide with those normally used to minimize its noise figure [7]. Noise measurements on silicon field-effect transistors indicated that the pinch-off mechanism is the predominant noise source and usually limits the device noise figure [8]. For a bipolar transistor amplifier, it was shown that its noise figure can be optimized by proper choice of emitter current with the source impedance remaining fixed [9]. In a balanced amplifier with the transistors having identical noise characteristics, it is possible to operate with minimum noise figure and matched input impedance simultaneously, with the reflection being absorbed by the ballast resistor in the input 3-dB directional coupler [10].

8.1.2 Basic LNA Topologies

The LNA is one of the most important building blocks of different communication and radar systems. As a key component in the receiver, the LNA should provide good input return loss, low-noise figure, high gain, low power consumption, and good linearity. Although excessive gain degrades the input dynamic range, it must be set high enough for the LNA noise figure to dominate the cascaded noise figure. Input dynamic range is particularly important when large interferences close in frequency are present. There are three basic LNA topologies with proper device configuration: with common source shown in Figure 8.4a, with common gate shown in Figure 8.4b, and cascode connection shown in Figure 8.4c.

Table 8.1 provides a concise performance comparison of three basic LNA topologies [11]. The cascode LNA is a good compromise between these three topologies as it provides the most stable signal gain over the widest bandwidth with only a slight sacrifice in noise figure performance and design complexity. The common-source transistor is sized to deliver the best possible noise figure, but that advantage often comes at the cost of greater sensitivity to bias, temperature, and component tolerances.

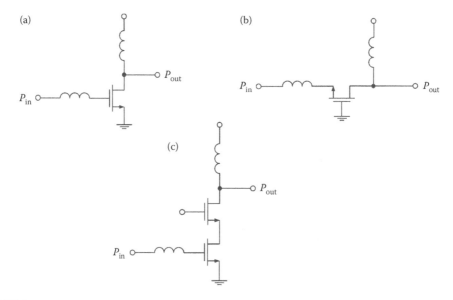

FIGURE 8.4
Three basic LNA topologies.

TABLE 8.1

Comparison of Three Basic LNA Topologies

Characteristic	Common-Source	Common-Gate	Cascode
Noise figure	Lowest	Rises rapidly with frequency	Slightly higher than CS
Gain	Moderate	Lowest	Highest
Linearity	Moderate	High	Potentially highest
Bandwidth	Narrow	Fairly broad	Broad
Stability	Often requires compensation	Higher	Higher
Reverse isolation	Low	High	High
Sensitivity to process variation, temperature, power supply, component tolerance	Greater	Lesser	Lesser

The most popular active devices used in LNAs are based on GaAs pHEMT and SiGe BiCMOS process technologies [11]. GaAs pHEMT devices generate very little noise due to the heterojunction between the doped AlGaAs layer and the extremely thin undoped GaAs layer, but the real advantage of GaAs is its capability to provide linear gain. Modern SiGe process technology is comparable to GaAs devices in terms of usable frequency range, but the relatively low breakdown voltage of SiGe devices limits dynamic range. GaAs HEMT demonstrates clear advantages over SiGe implementations in terms of noise figure and linearity performance, whereas SiGe has a cost advantage due to higher levels of integration. However, application of any technology seriously depends on specific requirements where technology advantages are much more important compared to disadvantages.

To minimize the need for external noise matching circuit components, it is necessary to properly choose the transistor size (gate finger dimension and number of gate fingers) and circuit topology, including package parasitics. In this case, careful insertion of the source degeneration feedback implemented with the series inductance L_s, as shown in Figure 8.5a, improves amplifier stability and linearity at the expense of gain, especially at higher frequencies. Figure 8.6a shows that generally the input impedance $Z_{in} = 1/Y_{in}$ of the common-source device with source degeneration feedback to provide a maximum

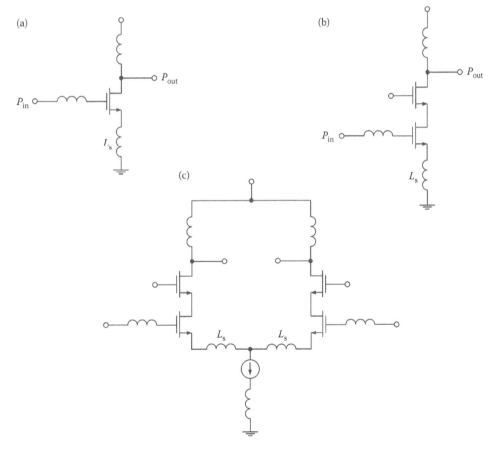

FIGURE 8.5
LNA topologies with source degeneration.

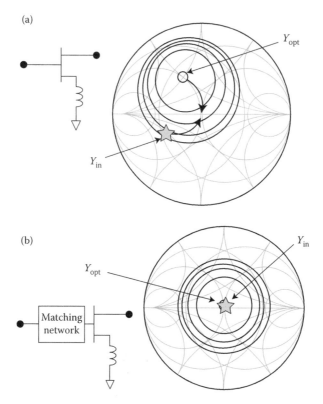

FIGURE 8.6
Optimum input matching for common-source configuration.

gain does not coincide with the optimum input impedance $Z_{opt} = 1/Y_{opt}$, corresponding to a minimum noise figure F_{min}, and the system characteristic impedance $Z_0 = 1/Y_0$ to minimize an input return loss. Consequently, the input matching network is required to achieve the desired maximum gain and optimal noise performance simultaneously when $Y_{in} = Y_{opt} = Y_0$, as shown on the Smith chart in Figure 8.6b using gain and noise circles [11,12]. The impedance looking into the gate of an inductively degenerated transistor can be written assuming zero gate–drain capacitance C_{gd} as

$$Z_{in}(j\omega) = j\omega L_s + \frac{1}{j\omega C_{gs}} + \omega_T L_s \tag{8.1}$$

where C_{gs} is the gate–source capacitance, $\omega_T = g_m/C_{gs}$ is the angular transition frequency, and g_m is the device transconductance. Adding an extra inductor in series to the gate as a matching element can resonate out the imaginary part of the input impedance at the specified frequency. In some cases, the real part of the input impedance $R_{in} = \text{Re}Z_{in}$ can be made equal to a source impedance R_s, usually of 50 Ω, by optimizing both inductances for a given bias voltage and device geometry [13]. Similar design approach can be applied to the LNAs based on bipolar devices with optimized transistor sizing [14]. At microwave frequencies, the source degeneration inductor can be replaced by a short-circuited trans-mission line [15]. Note that the lossless series inductive feedback adds no noise to the LNA circuit [16,17].

The input and output matching circuits of a microwave monolithic GaAs MESFET LNA can be constructed with microstrip lines and parallel tuning stubs, as shown in Figure 8.7a [18]. In this case, the characteristic impedance of the microstrip lines between the active device and the parallel tuning stubs were chosen to be 70 Ω by compromising the line length and the transmission loss when the high characteristic impedance reduces the line length needed for the same impedance transformation but gives a larger loss. The tuning stubs jB_S and jB_L are capacitive having the electrical length of $\lambda/8$, and the 70-Ω microstrip lines are bent rectangularly to reduce the chip size. The microstrip-line loss was about 0.22 dB per wavelength at 15 GHz or higher frequencies for a 50-Ω line (3-μm-thick gold) on a 0.22-mm-thick GaAs substrate. The source-ground reactance was about 4 Ω at 20 GHz, which corresponds to the inductance of about 32 pH. As it follows from Figure 8.7b, the power gain and noise figure strongly depend on the gate bias voltage.

The common-gate amplifier is also characterized by a low-noise figure, particularly at lower frequencies, but its noise figure increases rapidly with frequency. In this case, the high drain–source capacitance in a common-gate configuration requires inductive

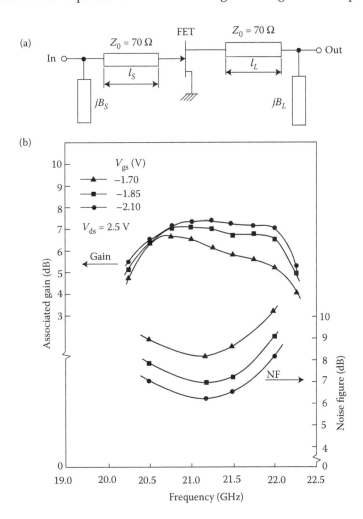

FIGURE 8.7
Schematic and performance of 20-GHz monolithic GaAs MESFET LNA.

feedback to improve noise figure, gain, and stability at higher frequencies. In a cascode configuration shown in Figure 8.5b with source degeneration feedback, the common-gate stage should be designed for maximum linearity. The common-source device also can be biased to predistort the AM/PM performance of the common-gate transistor. The differential LNA topology shown in Figure 8.5c is very useful when it is necessary to avoid the package parasitics (off-chip bondwires), which vary from part to part and require careful modeling. In this case, the package parasitics are only on the gate sides, and not on the common source of both transistors due to the effect of virtual grounding. The source degeneration inductors L_s are realized with on-chip inductors with tight process tolerances.

For the cascade two-stage design, the second stage can also employ inductive degeneration that is mainly used to improve the amplifier linearity rather than for matching. To lower the overall noise contribution of the input stage, it can be designed with a single transistor [19]. In a cascode second stage, the input impedance of a capacitively degenerated cascode transistor may have a negative real part. In this case, a high-Q parasitic inductance at the gate of the cascode device can form a Colpitts oscillator [20]. Therefore, to improve the stability of the cascode stage, a resistor can be added to the gate of the cascode device, whose value should be greater by magnitude to compensate for this negative resistance. However, this value should not be too large, as it degrades the noise figure and gain.

8.1.3 Minimum Noise Figure

The noise in a bipolar transistor is assumed to arise from three basic sources: diffusion fluctuations, recombination fluctuations in the base region, and thermal noise in the base resistance [21]. The noise behavior of the bipolar transistor can be described based on its equivalent circuit representation shown in Figure 8.8a, which includes the main elements responsible for the device electrical behavior and noise sources [22]. Since the process of the carrier drifting into the collector–base depletion region is a random process, the collector current I_c demonstrates white and shot noise contribution and is represented by a noise collector-current source $\overline{i_{nc}^2}$. The base current I_b is a result of the carrier injection from the base to the emitter and generation–recombination effect in the base and base–emitter depletion regions. Because all these components are independent, representing a random process, the base current also demonstrates a shot-noise behavior and is represented by a shot-noise base current source $\overline{i_{nb}^2}$. The series base, emitter, and collector resistances are represented by the voltage and current thermal noise sources $\overline{e_{nb}^2}$, $\overline{i_{ne}^2}$, and $\overline{e_{nc}^2}$, respectively.

The minimum noise figure F_{min} based on the simplified noise-free two-port network shown in Figure 8.8b can be calculated for a sufficiently high value of the low-frequency current gain $\beta = g_m r_\pi$ from

$$
\begin{aligned}
F_{min} = 1 + \frac{r_b}{r_\pi}\left[1 + \frac{1}{\beta}\left(\frac{f}{f_T}\right)^2\right] \\
+ \sqrt{\frac{2r_b}{r_\pi}\left(1 + \frac{r_b}{2r_\pi}\right) + \frac{2r_b}{\beta r_\pi}\left(1 + \frac{r_b}{r_\pi}\right)\left(\frac{f}{f_T}\right)^2}
\end{aligned}
\tag{8.2}
$$

where f is the operation frequency and $f_T = g_m/2\pi C_\pi$ is the bipolar transition frequency (the effect of the feedback collector capacitance C_c is not taken into account) [23]. A noise model for HBT device operated at very high frequencies should include the contribution

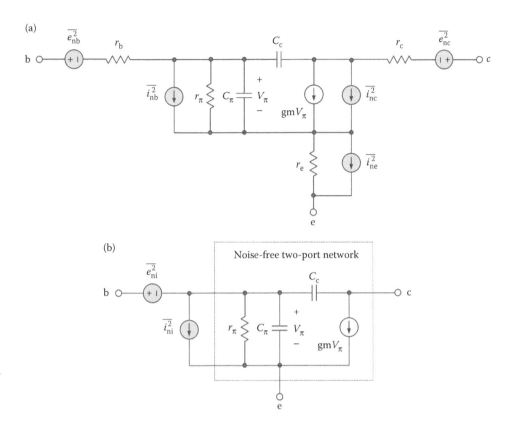

FIGURE 8.8
Bipolar equivalent circuits with noise sources.

of both space-charge layers (at the emitter–base junction and the base–collector junction) to the shot noise. These two noise sources related to the collector current I_c are the result of the same electrons, which are injected from the emitter into the base, cross this layer, and then reach the collector. Therefore, their correlation can be given by a time delay function $\exp(-j\omega\tau)$, where τ is the transit time through the base and the collector–base junction, which is τ_π for a π-type model [24–26]. Thus, in order to achieve minimum noise figure, the emitter length of the transistor has to be carefully scaled to get the optimum base resistance for the desired frequency band and for a 50-Ω source, with the collector bias current set for minimizing noise figure. Monolithic LNAs implemented in Si and SiGe bipolar technologies show the increasing advantage of SiGe over Si technology with rising frequency [27]. Using a series matching inductance at the input of the cascode bipolar LNA with degeneration feedback eliminates white-noise contribution due to the collector current I_c but enhances the contribution of the base resistance r_b and produces the frequency-dependent shot-noise component of the base current I_b [28].

The noise properties of a MESFET device can be described based on both its physical and equivalent circuit models. The dominant intrinsic noise of a microwave GaAs MESFET device is the diffusion noise introduced by electrons experiencing velocity saturation. In a device two-zone model, a portion of the channel near the source end is assumed to be in the constant mobility operation mode (zone I), while the remaining portion near the drain end is postulated to be in velocity saturation (zone II). The position of the boundary between these zones is a strong function of the source–drain bias with weak dependence on the gate–source bias. It is assumed that the noise in zone I is thermal enhanced by hot electron effects [29,30]. However, zone II cannot be treated as an ohmic conductor. Its contribution must be represented as a high-field diffusion noise, being dominant in microwave devices [31]. This diffusion noise is proportional to the high-field diffusion coefficient and is linearly dependent on drain current. On the other hand, the thermal noise of zone I decreases with increasing drain current. As a result, a strong correlation exists between the drain noise and the induced gate noise, which leads to a high degree of cancellation in the noise output of the GaAs MESFET [32].

The noise equivalent circuit of the MESFET device with both intrinsic and extrinsic noise sources is shown in Figure 8.9a [29,32]. The noise source $\overline{i_{ng}^2}$ represents the noise induced on the gate electrode by the passing thermal fluctuations in the drain current. The intrinsic drain noise source $\overline{i_{nd}^2}$ has a flat spectrum. The resistance R_{gs} represents the resistive charging path for the gate–source capacitance C_{gs}, and noise associated with this resistor is embedded in the gate noise source. The series gate, source, and drain resistances are represented by the voltage thermal noise sources $\overline{e_{ng}^2}$, $\overline{e_{ns}^2}$, and $\overline{e_{nd}^2}$, respectively. The noise current source $\overline{i_{ngl}^2}$ is responsible for the effect of the gate leakage current, which should

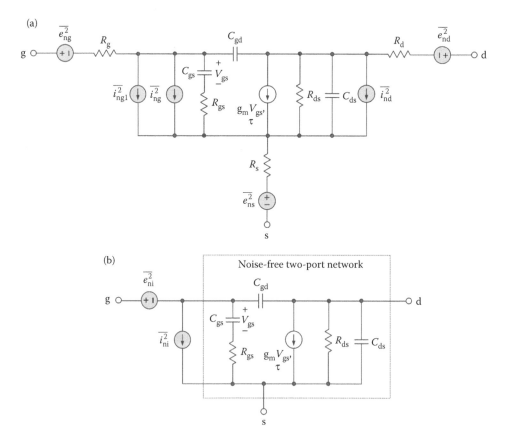

FIGURE 8.9
MESFET equivalent circuits with noise sources.

be taken into account when using a submicron gate-length HEMT device [33,34]. It should be noted that $\overline{i_{nd}^2}$ increases in the ohmic region and tends to saturate at high drain voltage, whereas $\overline{i_{ng}^2}$ increases with a near constant slope versus drain voltage.

In a first approximation, the gate noise source $\overline{i_{ng}^2}$, feedback capacitance C_{gd}, and series drain resistance R_d can be neglected. As a result, a simple approximate expression based on measurements can be obtained in terms of the parameters of the device equivalent circuit shown in Figure 8.9b as

$$F_{min} = 1 + 0.016 f \; C_{gs} \sqrt{\frac{R_g + R_s}{g_m}} \tag{8.3}$$

provided that R_g and R_s are in ohms, transconductance g_m is in mhos, capacitance C_{gs} is in picofarads, and operating frequency f is given in gigahertz [35,36].

The minimum noise figure F_{min} given by Equation 8.3 can also be expressed in terms of the device geometrical parameters as

$$F_{min} = 1 + 0.27 L \; f \sqrt{g_m(R_g + R_s)} \tag{8.4}$$

where the effective gate length L is in micrometers [37].

A comparison of the noise performance of both HEMT and conventional MESFET devices demonstrates the HEMT superiority, mainly related to its higher transition frequency and correlation coefficient [38]. The transition frequency of an HEMT device is greater for two main reasons: higher carrier mobility results in a higher average velocity and, therefore, a higher transconductance, whereas the small epilayer thickness yields higher transconductance and less effect of the parasitic capacitances.

In a MOSFET device, in view of the resistive material in a device channel, it exhibits thermal noise as a major source of noise, which can be represented by a noise current source $\overline{i_{nd}^2}$ connected between the drain and the source in the MOSFET small-signal equivalent circuit shown in Figure 8.10a [39]. The induced gate current noise is modeled by the gate noise current source $\overline{i_{ng}^2}$ connected across the gate–source capacitance C_{gs}. The series gate, source, and drain resistances are represented by the voltage and current thermal noise sources $\overline{e_{nb}^2}$, $\overline{i_{ne}^2}$, and $\overline{e_{nc}^2}$, respectively.

The minimum noise figure F_{min} as a function of the input-referred noise voltage $\overline{v_{ni}^2}$, noise current $\overline{i_{ni}^2}$, and can be approximately estimated through the bias conditions and parameters of the simplified noise-free two-port network shown in Figure 8.10b by

$$F_{min} = 1 + \left(\frac{f}{f_T}\right)\sqrt{\beta \; I_d \; R_g} \left[1 + \left(\frac{f}{f_T}\right)\sqrt{\beta \; I_d \; R_g} \right] \tag{8.5}$$

where

$$\beta = \frac{1}{V_{dsat}} + \frac{\alpha^2 V_{dsat}}{3(V_{gs} - V_{th})^2} \tag{8.6}$$

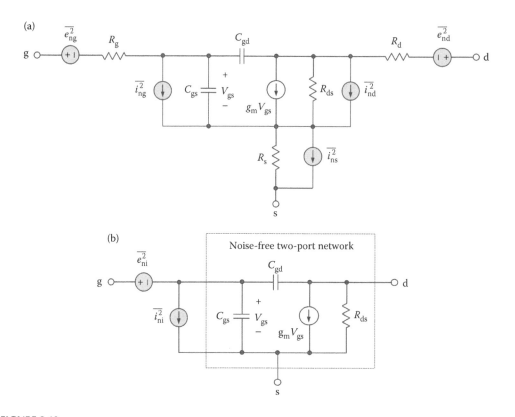

FIGURE 8.10
MOSFET equivalent circuits with noise sources.

V_{dsat} is the saturation drain–source voltage, α is the bulk-charge effect coefficient, f is the operation frequency, I_d is the drain bias current, and $f_T = g_m/2\pi(C_{gs} + C_{gd})$ [40]. Equation 8.5 suggests that devices with shorter channel length yield better noise figures because the transition frequency f_T is reversely proportional to the channel length. Besides, the better noise figure can be achieved with larger device width for the same bias conditions due to lower gate resistance R_g.

The noise figure F of a single-stage common-source CMOS LNA with the directly connected source resistance R_S to the device input (assuming that $R_S \gg R_g$) can be written as

$$F = 1 + \frac{R_g}{R_S} + \frac{\gamma}{\alpha}\left(\frac{f}{f_T}\right)^2 g_m R_S \qquad (8.7)$$

where g_m is the device transconductance and γ is the bias-dependent factor [41]. Here, α is equal to unity for long-channel devices and decreases as channel scales down, and γ equals to 2/3 in saturation mode for long-channel devices and can be greater than 2 in short-channel devices. The minimum noise factor F_{min} can be derived from Equation 8.7 by setting zero-derivative condition $\partial F/\partial R_S = 0$ that results in

$$F_{min} = 1 + 2\left(\frac{f}{f_T}\right)\sqrt{\left(\frac{\gamma}{\alpha}\right)g_m R_g} \qquad (8.8)$$

8.1.4 Filtering Multistage LNA Topology

For an LNA used in RF frontends, it is important not only to provide sufficient filtering of the harmonic components but also to reduce the out-of-band noise in receive (Rx) band. For example, this filtering can help to properly isolate the Rx band (27–31 GHz) from the transmit (Tx) band (17–21 GHz) in a satellite communication system. Figure 8.11a shows the block schematic of a three-stage monolithic LNA using a metamorphic HEMT (mHEMT) process to achieve a linear gain over 20 dB and a noise figure of less than 1.7 dB in Rx band of 27–31 GHz [42]. Since the requirements to out-of-band rejection specify 0 dB at 23 and 35 GHz, it is necessary to design a filtering three-stage LNA. In this case, the first stage is designed to minimize noise and provide input matching for specified return loss without specific filtering constraints. Then, the first interstage matching network provides a high-pass filtering, whereas the second interstage matching network allows a passband frequency behavior. Finally, the output matching network is used to complete the high-pass gain rejection. The high-pass matching structure using lumped elements with two shunt LC resonant circuits shown in Figure 8.11b can be used for the first interstage network, the bandpass matching structure with the series and shunt LC resonant circuits shown in Figure 8.11c can represent the second interstage network, and the desired low-pass frequency behavior can be achieved with open-ended transmission-line stubs.

To simplify a broadband input matching, the common-gate transistor configuration can be used. However, a single-stage common-gate LNA may not have enough gain, especially at higher frequencies, and cascaded common-source stages are needed to provide sufficient gain. To achieve a broadband LNA behavior, the interstage network can be designed in the form of a typical parallel-to-series reactance compensation circuit shown in Figure 8.12a that provides a bandpass filtering response. In this case, the bandpass frequency response is realized by selecting the value of inductors and capacitors to resonate the parallel and series LC networks at the same center bandwidth frequency, that is, to provide $L_1C_1 = L_2C_2$. Figure 8.12b shows the circuit schematic of a three-stage 0.18-μm CMOS LNA with a parallel-to-series reactance compensation circuit between the first and second amplifying

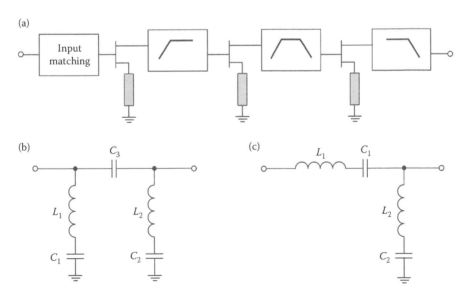

FIGURE 8.11
Principle of filtering LNA and circuit structures.

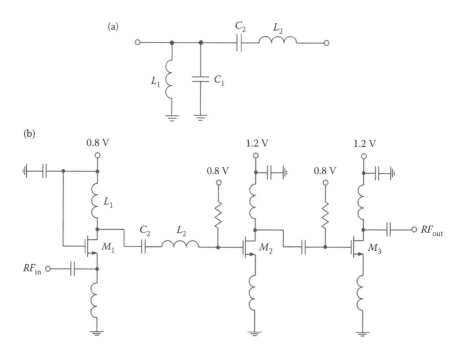

FIGURE 8.12
Parallel-to-series matching technique and broadband LNA schematic.

stages, where the parallel inductance L_1 resonates with the parasitic drain–gate capacitance C_{gd} of the transistor M_1 in a common-gate configuration at the midband resonant frequency $1/2\pi\sqrt{L_1 C_{gd}} = 1/2\pi\sqrt{L_2 C_2}$ [43]. As a result, a wide frequency range of 14.3–29.3 GHz was covered with a power gain of 8.25 ± 1.65 dB and a noise figure from 4.3 to 5.8 dB. In order to reduce the effective gate resistance, the transistor M_1 is implemented as two transistors in parallel with twice as many fingers compared to the second-stage device.

8.1.5 Linearization Techniques

Owing to the potential presence of large interference signals at the input of the LNAs in modern communication systems that support multiple radio standards across multiple frequency bands, the LNAs have to provide high linearity and low noise over a wide frequency range to suppress interference and maintain high sensitivity. The LNA linearization methods should be simple, should consume minimum power, and should preserve noise figure, gain, input matching [44]. Since the LNA typically has a low-amplitude and high-frequency input, it operates as a weakly nonlinear system with a few higher-order harmonics, typically only second and third. In this case, the main nonlinearity of the LNA based on a CMOS technology originates from the nonlinear transconductance g_m, which converts linear input gate voltage to nonlinear output drain current.

The output drain current i_d for an inductively source-degenerated LNA, whose simplified circuit is shown in Figure 8.13a, can be approximated by the first three Taylor-series terms as

$$i_d = g_{m1}(v_{in} - v_s) + g_{m2}(v_{in} - v_s)^2 + g_{m3}(v_{in} - v_s)^3 \tag{8.9}$$

where $g_{m1} = g_m$ is the linear device transconductance, and g_{m2} and g_{m3} represent its second- and third-order nonlinearities obtained by the second- and third-order derivatives of the

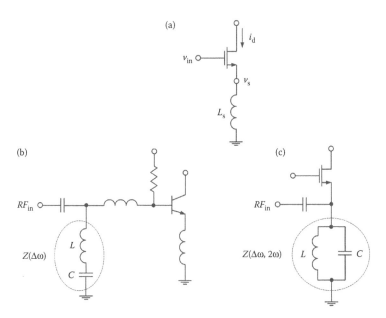

FIGURE 8.13
Linearization due to source degeneration and harmonic termination.

drain–source dc current amplitude I_{ds} with respect to the gate–source voltage amplitude V_{gs} at the dc bias point, respectively,

$$g_{m1} = \frac{\partial I_{ds}}{\partial V_{gs}} \quad g_{m2} = \frac{1}{2!}\frac{\partial^2 I_{ds}}{\partial V_{gs}^2} \quad g_{m3} = \frac{1}{3!}\frac{\partial^3 I_{ds}}{\partial V_{gs}^3} \tag{8.10}$$

The nonlinear properties are usually determined by a two-tone excitation test signal with individual components separated slightly in frequency, which can be represented in a common case of unequal amplitudes as

$$v_{in} - v_s = V_1 \cos\omega_1 t + V_2 \cos\omega_2 t \tag{8.11}$$

For the first three derivatives, the output signal can be represented by a Taylor-series expansion with the appropriate equating of the frequency component terms as

$$\begin{aligned}
i_d =\ & \frac{g_{m2}}{4}(V_1^2 + V_2^2) \\
& + \left[g_{m1} + \frac{g_{m3}}{4}\left(\frac{1}{2}V_1^2 + V_2^2\right)\right]V_1 \cos\omega_1 t \\
& + \left[g_{m1} + \frac{g_{m3}}{4}\left(V_1^2 + \frac{1}{2}V_2^2\right)\right]V_2 \cos\omega_2 t \\
& + \frac{g_{m2}}{4}(V_1^2 \cos2\omega_1 t + V_2^2 \cos2\omega_2 t) \\
& + \frac{g_{m3}}{24}(V_1^3 \cos3\omega_1 t + V_2^3 \cos3\omega_2 t) \\
& + \frac{g_{m2}}{2}V_1 V_2 \cos(\omega_1 \pm \omega_2)t \\
& + \frac{g_{m3}}{8}[V_1^2 V_2 \cos(2\omega_1 \pm \omega_2)t + V_1 V_2^2 \cos(\omega_1 \pm 2\omega_2)t]
\end{aligned} \tag{8.12}$$

The following conclusions can be drawn from the above Taylor-series expansion of the transistor transfer function:

- Variation of the device bias point is directly proportional to the second derivative of the transfer function.
- The device transfer function will be linear only if the third derivative is equal to zero.
- Second-harmonic components result from second derivatives of the device transfer function, whereas third-harmonic components result from the third derivative of the device transfer function.
- First-order mixing products (total and differential) are provided by the second derivative of the device transfer function.
- Mixing products of the third order are mainly determined by the third derivative of the device transfer function.
- Distortions, which are determined by the second derivative (second amplitude degree) or by the third derivative (third amplitude degree) of the device transfer function, are called the *second-order intermodulation distortions* (IMD_2) or the *third-order intermodulation distortions* (IMD_3), respectively.

Note that the *third-order intermodulation* (IM_3) *components* $2\omega_1 \pm \omega_2$ and $2\omega_2 \pm \omega_1$ exist in differential LNA because they are determined by the odd term and cannot be rejected by differential operation. Generally, an inductive degeneration in the LNA provides the linearization of its transfer characteristic depending on the feedback factor; however, special means should be provided to minimize the unwanted IMD_2 and IMD_3.

To improve the linearity, the resonant tanks can be added to optimally tune the terminal impedances to minimize the second-order mixing products. The terminations are commonly implemented with dedicated LC networks, which provide high impedance at operating frequency and low impedances at low modulation and high harmonic frequencies. Figure 8.13b shows the inductively degenerated common-emitter amplifier stage, where a series low-frequency LC trap is added to the base of the transistor to improve its third-order intercept point (IP_3). The third-order intercept point of a circuit is used to measure the linearity of the circuit with respect to its third-order intermodulation performance. The trap network has low-impedance at low-frequency due to the large capacitance, and appears open at RF due to large inductance [45]. In Figure 8.13c, the parallel LC network at the source terminal of a common-gate amplifier stage has its maximum impedance at operating frequency and negligible impedances at low frequencies due to low inductive reactance to ground and at the second harmonic due to low capacitive reactance to ground [46]. However, such an out-of-band termination technique only works well in narrowband systems when the modulation bandwidth is not very wide and the second harmonic does not vary considerably.

The derivative superposition (DS) method is a special case of the feedforward technique because it adds the third derivative (g_{m3}) of the drain current from the main and auxiliary transistors to cancel distortion. In its modified version with the CMOS circuit implementation shown in Figure 8.14a, instead of optimally scaling and rotating the second-order contribution by tuning the second-harmonic termination, the two source-degenerated inductors are connected in series with two transistor sources connected to different nodes of the inductor chain to adjust the magnitude and phase of the composite third-order contribution [47]. Here, the main transistor M_a is biased with a negative g_{m3a} and the auxiliary

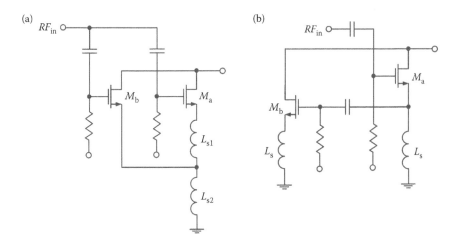

FIGURE 8.14
Circuit implementations of modified DS method.

transistor M_b is biased with positive g_{m3b}. The purpose of connecting the M_b source to the common node of the two inductors is to change the magnitude and phase of its g_{m3b} contribution to IMD_3 relative to the g_{m2a} and g_{m3a} contribution of M_a. As a result, g_{m3a} on the vector diagram is rotated properly such that the composite vector of g_{m3a} and g_{m3b} contribution is 180° out of phase with the g_{m2b} contribution, yielding zero net IM_3, and the choice of L_{s1} determines the angle of g_{m3a}. In an alternative implementation shown in Figure 8.14b, the gate of the auxiliary transistor M_b is connected to the source of the main transistor M_a instead of directly connecting to the input, thus minimizing the degradation in noise figure and input matching [48]. Note that the DS method is strongly dependent on the accuracy of the device models, and the transistor biased in the subthreshold region provides a very limited distortion cancellation range.

Figure 8.15a shows the circuit schematic of a CMOS LNA, which incorporates nMOS transistor M_1 and pMOS transistor M_2 in the common-gate stage to provide the cancellation of the second-order distortions [49]. At the same time, the transistors M_3 and M_4 are sized for the noise and intrinsic third-order distortion cancellation. The capacitor connected between two drain nodes of M_1 and M_2 shorts out two drain nodes at RF signal frequency and beyond. In this case, at higher IM_2 frequency $\omega_1 + \omega_2$, the distortion current from M_2 goes across the capacitor and loops back through M_1, resulting in no second-order distortion current if both transistors have the same g_m characteristics. On the other hand, for frequencies much lower than RF, this capacitor presents a high impedance path between two drain nodes and breaks the LNA into two standing-alone common-gate amplifiers. As a result, at lower IM_2 frequency $\omega_1 - \omega_2$, the M_1 and M_2 distortion current flowing through the separate common-gate and common-source paths are subtracted at the output, whereas the RF signal is added. Implemented in a 0.13-µm CMOS technology, this LNA achieved a peak input IP_3 of 16 dBm and a noise figure below 2.6 dB over a wide frequency range from 800 MHz to 2.1 GHz, consuming 11.6 mA from a 1.5-V supply voltage. To minimize the degradation of linearity caused by the output stage, the output matching can be achieved by using the series feedback resistors between the sources of M_3 and M_4 and ground, respectively. By using an additional broadband output buffer with the resistive shunt feedback, a noise figure below 3.6 dB was achieved over a wide frequency range from 2.5 to 5 GHz [50].

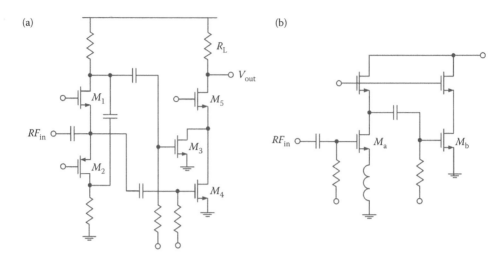

FIGURE 8.15
Circuit implementations of distortion cancellation techniques.

Similar to the DS method, the active postdistortion (APD) technique also includes an auxiliary transistor to cancel the nonlinearity from the main device. In this case, an auxiliary transistor M_b is connected to the output of the main device M_a instead of directly to the input, as shown in Figure 8.15b, thus minimizing the impact on input matching [51]. Besides, all transistors are operated in saturation, resulting in more robust distortion cancellation. The main transistor size and degeneration inductance are chosen to have optimum noise figure performance with high power gain. After that, the auxiliary device and cascade device are sized based on the transconductance ratio to reduce noise and gain loss. As a result, a power gain of 16.2 dB, a noise figure of 1.2 dB, and an input IP_3 of 8 dBm were obtained in a cellular band of 869–894 MHz, while consuming 12-mA current from 2.6-V power supply.

8.2 Lossless Matched Broadband Low-Noise Amplifiers

Usually, when it is necessary to achieve high-gain and low-noise performance over a sufficiently wide frequency range and to minimize the die size in monolithic implementation, the LNAs are designed using lossless matching circuits. Figure 8.16 shows the circuit schematic of a two-stage monolithic lossless matched MESFET, which provides a less than 2.8-dB noise figure with greater than a 16-dB power gain in the frequency band of 11.7–12.7 GHz [52]. Here, based on the parameters of the MESFET equivalent circuit, the amplifier circuit elements were optimized by a CAD program. The microstrip lines were folded in order to reduce the chip size, and spacings between adjacent lines were designed to be as large as possible to avoid parasitic couplings. In this case, the chip size, whose photo is shown in Figure 8.16b, was reduced to 1.5×0.9 mm^2, with a substrate thickness of 150 μm and a relative dielectric constant of 4.8. For a monolithic three-stage MESFET LNA with series feedback, a maximum noise figure of 2.0 was achieved from 8.5 to 11.5 GHz, with an output power of 10 dBm at 1-dB compression point and a maximum gain of 30 dB at 1.8-dB noise figure [53].

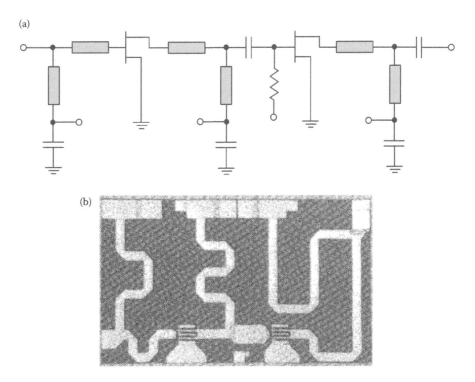

FIGURE 8.16
Schematic and implementation of monolithic two-stage GaAs MESFET LNA.

Much wider frequency bandwidth can be achieved by using the method to synthesize the interstage and output matching networks with prescribed gain versus frequency slopes. The initial value for the feedback inductance in the first transistor is determined by constructing noise circles for low-, mid-, and high-bandwidth frequencies. Then, the practical input matching network can be designed using the Smith chart. As a result, a typical noise figure of 2.5 dB, a power gain of around 15 dB, and an output power of 5 dBm were achieved in the frequency range of 6–18 GHz for a two-stage monolithic GaAs HEMT LNA for electronic warfare (EW) applications [54]. A resistive drain network (the series connection of a microstrip line and a resistor loaded by a 10-pF bypass capacitor) was used for each device to enhance the amplifier stability and to compensate for the transistor inherent gain slope. The LNAs can be combined by a pair of Lange couplers to provide better input return loss and higher output power over a wide frequency range. By using a single-stage MMIC HEMT LNA in each amplifying path, a noise figure of 1.8 dB and greater than 12 dB gain were obtained across the frequency range of 9–16 GHz, with an output power of about 9 dBm [55].

A monolithic *Ka*-band (26.5–40 GHz) reactively matched LNA, whose circuit schematic is shown in Figure 8.17, was designed using a 0.25-μm HEMT process [56]. The input and output matching networks representing the bandpass filters consist of the shunt-shorted stubs and series microstrip lines, with electrical lengths given at 40 GHz. In this case, the input matching network has an upward frequency slope of 4 dB/octave, a minimum insertion loss of 1 dB, and a 0.5-dB ripple from 20 to 40 GHz. The output matching network has an upward slope of 2 dB/octave and a minimum insertion loss of 1 dB with a 0.3-dB ripple. The gain slopes in the input and output matching networks compensate the 6 dB/octave

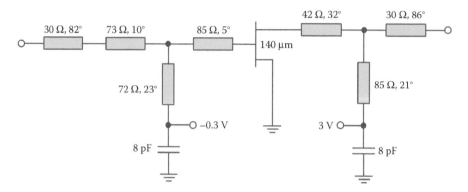

FIGURE 8.17
Circuit schematic of monolithic *Ka*-band 0.25-μm HEMT LNA.

gain roll-off of the HEMT device and result in a flat performance from 20 to 40 GHz. As a result, a power gain of 8 dB from 20 to 37 GHz and a noise figure up to 4 dB from 26 to 40 GHz were measured using a mushroom gate profile for a 0.25-μm HEMT device. By using a 0.1-μm InP HEMT technology, an average noise figure of 1.5 dB with a power gain of 21.9 ± 0.9 dB was achieved across the entire *Ka*-band for a three-stage LNA with source inductive stubs [57].

Figure 8.18a shows the circuit schematic of a highly survivable monolithic GaN HEMT LNA achieving a noise figure below 2.3 dB and a power gain about 20 dB across the bandwidth of 3.5–7 GHz [58]. The LNA consists of two stages, where both transistors have a gate width of 4 × 50 μm, and all dc bias networks are integrated on a chip. The source of the first device is inductively degenerated in order to improve the input return loss together with noise matching. It was shown that a series resistance in the gate-bias line can improve the LNA ruggedness due to voltage feedback, which reduces the gate current. This feedback effect results in a lower gate-bias voltage, thereby increasing the negative peak voltage. Wider operation frequency bandwidth can be obtained by using additional shunt and series inductors in the interstage matching network, as shown in Figure 8.18b for a two-stage 0.25-μm GaN HEMT LNA with inductive degeneration in both stages [59]. The amplifier is unconditionally stabilized from 1 to 12 GHz with the small-value series resistors on the drain lines. In this case, the first stage was designed for minimum noise figure while maintaining high gain, whereas the second stage was biased at higher drain voltage and current close to a Class-A operation to maximize the linearity. As a result, a noise figure lower than 1.8 dB (having a minimum value of 0.5) was measured across the frequency range from 1 to 3 GHz, with a 1-dB compression power of 29.5 dBm.

8.3 Lossy Feedback Broadband Low-Noise Amplifiers

The gain-bandwidth performance of the lossless matched LNAs, which are designed with reactive matching networks, is limited by the characteristics of the active devices such as maximum available gain and requirements of unconditional stability over a very wide frequency range. This makes it difficult to achieve multioctave bandwidth with

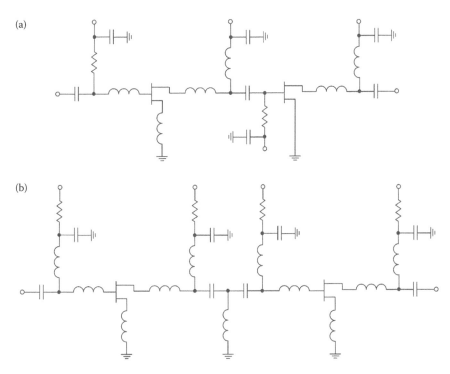

FIGURE 8.18
Circuit schematics of broadband two-stage GaN HEMT LNAs.

high gain and minimum gain flatness. As an alternative, the lossy feedback LNAs have been shown to be capable of the flat gain over a very wide bandwidth, good input matching, sufficiently low-noise figure, small size, and convenience in practical implementation. Such amplifier is cascadable for increased gain and easily made unconditionally stable.

8.3.1 Shunt Feedback

The circuit schematic of a resistive feedback two-stage monolithic 0.1-μm pHEMT LNA is shown in Figure 8.19a, where both transistors have a gate width of 4×50 μm yielding a transconductance g_m of nominally 160 mS [60]. Since the feedback resistance can be approximately estimated as $R_F = g_m Z_0^2$ according to Equation 4.25 in Chapter 4 for an idealized transistor consisting of only a current source with transconductance g_m and conditions of $S_{11} = S_{22} = 0$, where Z_0 is the system characteristic impedance, the feedback resistance was chosen as 500 Ω in series with a 2-pF capacitance. The second stage has an inductor in the feedback path in order to provide a gain boost at the high end of the frequency range. The lossy feedback LNA achieved a gain of more than 20 dB, a bandwidth from 2 to 20 GHz, and a noise figure of less than 2.7 dB with lower than 100-mW dc power consumption. Using a 50-nm InGaAs/InP pHEMT technology in a two-stage resistive feedback LNA had resulted in a very low-noise figure of lower than 1.1 dB and a minimum gain of 17 dB across 4–24 GHz, with a dc power consumption of 20 mW [61]. For a three-stage monolithic 0.1-μm InAs/AlSb HEMT LNA with a shunt *RLC* feedback in each stage, a 30-dB gain and a less than 2.6-dB noise figure were achieved across the frequency bandwidth of 2–11 GHz at a 7.5-mW dc power consumption [62].

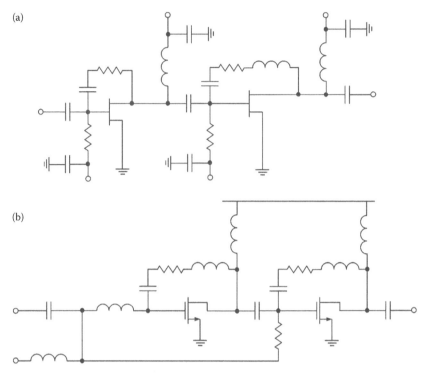

FIGURE 8.19
Circuit schematics of two-stage feedback amplifiers.

Assuming that a 10-dB power gain in a 50-Ω circuit environment is required, the device transconductance g_m of 83 mS should be provided according to Equation 4.26 in Chapter 4, with a feedback resistance of 208 Ω. To achieve the transconductance of 85 mS in a 0.9-μm CMOS process, a 100-μm wide transistor with 40 fingers was used. To compromise the instability and gain reduction at high bandwidth frequencies, the value of the feedback inductance was set to 1 nH. To further improve the input matching, a small inductor of 179 pH was placed in series with the input to the gate of the transistor to compensate for its gate–source capacitance. Figure 8.19b shows the final circuit schematic of a two-stage monolithic 0.9-μm CMOS LNA with identical stages, which provides both input and output return loss better than 10 dB and a power gain of greater than 12 dB from 5 to 26 GHz, with a noise figure below 7 dB [63].

Figure 8.20a shows the circuit schematic of a two-stage resistive feedback monolithic 0.15-μm GaAs HEMT LNA, where a two-finger 100-μm device is used in both stages [64]. The input matching circuit was designed to provide a trade-off between noise figure and input return loss, as shown in Figure 8.20b, by using the input matching network consisting of a shunt inductor and a series inductor, which transforms the 50-Ω source impedance to the device input impedance. The interstage matching network consisting of a series inductor and a shunt inductor provides an impedance transformation of the output impedance seen at the drain of the transistor M_1 to the input impedance seen at the gate of the transistor M_2, as shown in Figure 8.20c. Finally, the output matching network consisting of a series inductor and a shunt inductor transforms the output impedance seen at the drain of the transistor M_2 to the 50-Ω load impedance, as shown in Figure 8.20d. Each feedback

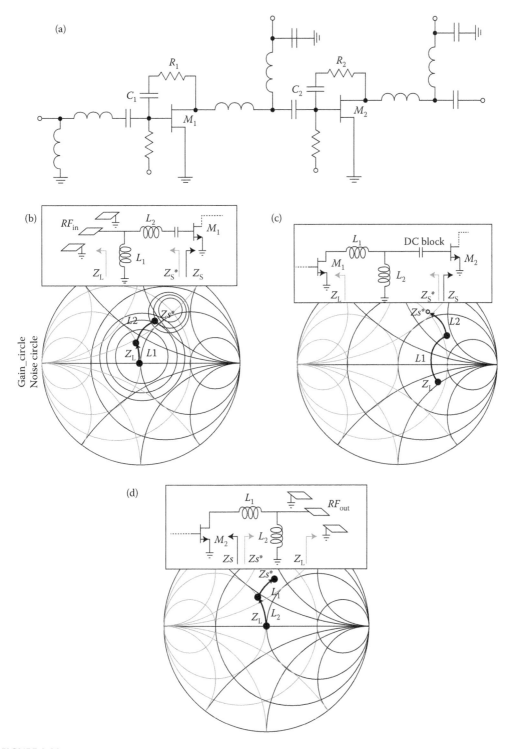

FIGURE 8.20
Schematics and traces of input, interstage, and output matching circuits.

network is composed of a resistor of 520 Ω and a capacitor of 0.11 pF. As a result, a noise figure of 3.5 ± 0.5 dB and a power gain around 20 dB were measured in the frequency range of 9–21 GHz, with a total dc power consumption of 88.8 mW from 2-V supply voltage.

8.3.2 Dual Feedback

The balanced configuration of the LNA shown in Figure 8.21 is well suited for octave-band designs and provides design flexibility in achieving optimum noise matching for the first-stage input impedance and broadband gain flatness without sacrificing *VSWR* [65]. Here, the series inductive feedback in the sources of the first transistors is used to improve the input broadband matching and stability of operation, whereas the second stages include shunt negative feedback designed for positive gain slope with frequency to compensate for the negative gain slope of the low-noise first stage. The shunt resistive feedback also helps to insure interstage stability. For a low broadband *VSWR*, these two-stage LNAs are embedded between Lange couplers. In this case, the noise figure is degraded only by the insertion loss in the input coupler. As a result, a noise figure of less than 2.2 dB with a flat gain of 23 ± 0.6 dB and input return loss better than 14 dB was achieved from 5 to 11 GHz for a balanced monolithic InGaAs HEMT LNA. For a two-stage single-ended monolithic AlGaN/GaN HEMT LNA with inductive degeneration in the first stage and shunt *RLC* feedback in the second stage, the measured noise figure was less than 3 dB from 4 to 11 GHz [66].

The five-stage LNA using a 0.13-μm InAlAs/InGaAs/InP HEMT technology and consisting of first two stages with high-pass reactive matching circuits and inductive degeneration and three final stages with shunt resistive feedback to achieve both low-noise and broadband high-gain characteristics achieved a power gain of greater than 40 dB and a minimum noise figure of 1.9 dB in the frequency band of 18–43 GHz, with a chip size of 1.8×0.9 mm^2 based on a thin-film microstrip-line technique [67].

GaN HEMT technology has an advantage of providing simultaneous low-noise and high-power capability due to high breakdown voltage and high power density. Figure 8.22a shows the circuit schematic of a single-stage LNA with dual feedback using a 0.2-μm AlGaN/GaN-SiC HEMT technology with f_T of about 75 GHz, achieving a noise

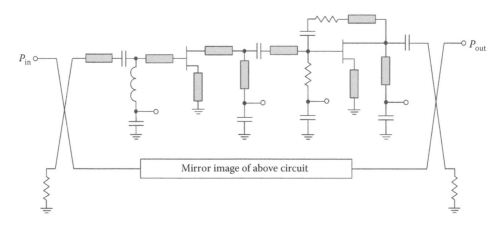

FIGURE 8.21
Circuit schematic of balanced two-stage monolithic InGaAs HEMT LNA.

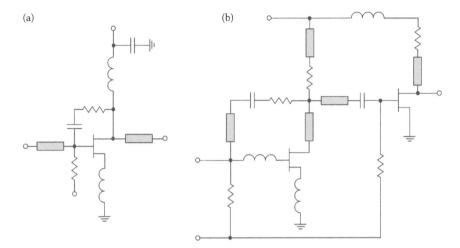

FIGURE 8.22
Circuit schematics of broadband GaN HEMT MMIC LNAs.

figure lower than 1 dB with a P_{1dB} of 32.8–33.2 dBm from 1 to 4 GHz at a high bias of 15 V and 400 mA [68]. Using an inductor in series to the transistor gate and transmission line connected in series to the drain terminal can significantly enhance the frequency bandwidth. As a result, for a 0.2-μm GaN HEMT LNA, whose circuit schematic is shown in Figure 8.22b, a maximum power gain of 13 dB and a minimum noise figure of 3.3 dB were achieved across the 3-dB bandwidth of 1–25 GHz [69]. Here, to simplify the external bias network, the gate biases of both stages are connected such that only one gate bias voltage is applied.

Figure 8.23a shows the circuit schematic of a dual-feedback 0.25-μm CMOS LNA for UHF applications, where the two transistors, nMOS M_1 and pMOS M_2, are used to boost the overall transconductance, resulting in a high gain and low-noise figure [70]. The feedback resistor R_F is used to flatten the gain over a wide bandwidth with small degradation in noise figure. Together with the source degeneration inductor L_s, these two feedback networks lead to simultaneous input impedance and noise matching. In this case, the effect of the feedback resistor R_F dominates the amplifier performance. The feedback resistor value of 500 Ω and source inductance of 1.1 nH were optimally chosen to achieve the best gain, bandwidth, noise figure, and impedance matching simultaneously, resulting in a gain over 13 dB and a noise figure lower than 3 dB in a wide bandwidth from 50 to 560 MHz.

The circuit schematic of a dual-feedback 90-nm CMOS LNA with extended bandwidth due to the series gate inductor L_g and drain peaking inductor L_d is shown in Figure 8.23b [71]. Here, the drain inductance L_d compensates for the gain degradation at high frequencies due to the degeneration inductance L_s. The input transistor M_1 is self-biased through the feedback resistor R_F, and no ac-coupling capacitors are added in series with the feedback resistor R_F or between two stages that results in no bandwidth degradation and no additional bias circuit. The inductor L_{sp} not only enhances the bandwidth but also suppresses the noise current of M_2, resulting from source degeneration at high frequencies, as shown in Figure 8.23c. As a result, a maximum power gain of 12.7 dB, a minimum noise figure of 3.3 dB, and a power consumption of 12.6 mW were measured across the bandwidth from 100 MHz to 20 GHz.

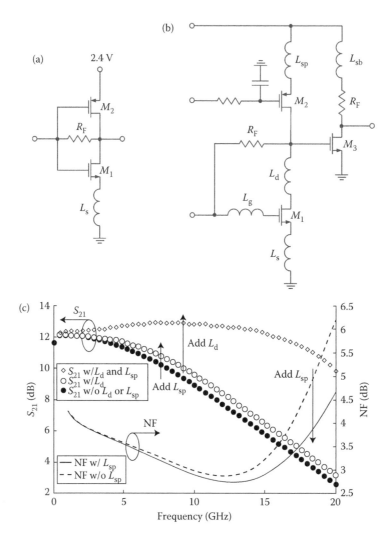

FIGURE 8.23
Circuit schematics and performance of broadband CMOS MMIC LNAs.

8.3.3 Active Feedback

An active feedback circuit included into the LNA can improve the linearity and gain-bandwidth performance without significantly impacting the noise figure. In this case, the use of the HBT active feedback provides several advantages over the FET active feedback such as smaller size, lower dc power consumption, active self-bias, and direct-coupled performance. Figure 8.24a shows the circuit schematic of a monolithic 0.2-μm InGaAs/GaAs HEMT LNA with an HBT active feedback, which is connected in series to the feedback resistor R_{fb} [72]. By adjusting V_{ss} that corresponds to the proper adjustment of an active-feedback current I_{fb}, various degrees of positive feedback can be induced by the resultant change in phase and amplitude characteristics of the active-feedback network. The compact size of the HBT active-feedback network requires no dc-blocking capacitors and is much smaller than a spiral inductor implementation. The HBT active feedback has resulted in a 50% improvement in gain-bandwidth performance (the measured 3-dB

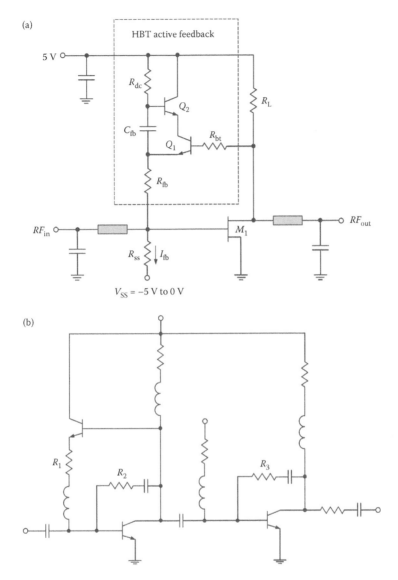

FIGURE 8.24
Circuit schematics of broadband MMIC LNAs with active feedback.

bandwidth was extended from 11 GHz to greater than 16 GHz) and an improvement in IP_3 by 4–10 dB without degradation in the noise figure of about 2.5 dB, compared to an equivalent resistive-feedback design.

Figure 8.24b shows the circuit schematic of a two-stage monolithic bipolar LNA with a shunt resistive feedback in both stages and an active feedback in the first stage [73]. In this case, the active feedback providing by a bias circuit can increase the input impedance in conjunction with the series resistor R_1. As a result, a noise figure of 1.5–2.7 dB and a gain of about 20 dB were measured over the frequency range of 0.5–5 GHz for a SiGe BiCMOS LNA, with a 1-dB gain-compression point of 6.3 dBm and a current consumption of 9 mA.

8.3.4 Noise Figure

A noisy amplifier with parallel negative feedback can be represented as an ideal noise-less amplifier with voltage and current noise sources connected at the input, as shown in Figure 8.25a [74]. In this case, the noise figure can be written as

$$
\begin{aligned}
F = 1 &+ \frac{G_n}{G_S} + \frac{R_n}{|Y_{21} - G_F|^2} \frac{|G_F(Y_{21} + Y_{11}) + G_S Y_{21}|^2}{G_S} \\
&+ \frac{G_F}{|Y_{21} - G_F|^2} \frac{|Y_{21} + Y_{11} + G_S|^2}{G_S}
\end{aligned}
\tag{8.13}
$$

where $G_S = 1/R_S$, $G_F = 1/R_F$, G_n and R_n are the equivalent input-referred noise conductance and noise resistance characterizing the noisy network, respectively, and Y_{ij} are the admittance parameters of the two-port noiseless network.

The optimum value of the source resistance R_{Sopt}, which yields the minimum noise figure F_{min} for a given feedback resistance R_F, is obtained by differentiating Equation 8.13 with respect to the source resistance R_S as

$$
R_{Sopt} = \sqrt{\frac{R_n\,|Y_{21}/Y_{21} - G_F|^2 + (G_F /|\,Y_{21} - G_F|^2)}{G_n + G_F(R_n G_F + 1)|Y_{21} + Y_{11}/Y_{21} - G_F|^2}}
\tag{8.14}
$$

Figure 8.25b shows the circuit schematic of a parallel-feedback MESFET amplifier [75]. Here, the gate lead inductance L_g changes the noise parameters and thereby the noise figure, but it does not alter the minimum noise figure. The optimum values of the feedback resistor R_F and electrical length θ of the drain transmission line are chosen to provide a practical compromise between noise figure, *VSWR*, gain, and stability of the amplifier. The noise parameters of the amplifier are strongly affected by the feedback resistor, especially at lower frequencies where the negative feedback is strongest. In this case, the difference

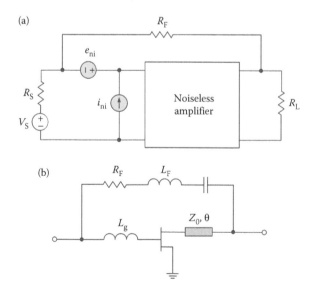

FIGURE 8.25
Basic schematics of negative-feedback LNA.

of the minimum noise figure for open-loop case and that of a parallel feedback may exceed 2 dB across the lower octave of the frequency band, with significant reduction over the second octave for optimum electrical length θ. As a result, a five-stage monolithic MESFET LNA consisting of the same single-ended parallel feedback stages achieved over 40-dB gain and a maximum noise figure of 4 dB between 2.4 and 8 GHz.

8.4 Cascode Broadband Low-Noise Amplifiers

Cascode configuration of the CMOS LNA allows better opportunity to provide simultaneously high gain, improved reliability, and wide frequency range with a reasonable noise figure. Figure 8.26a shows the circuit schematic of a two-stage K-band LNA implemented in a 45-nm semiconductor-on-insulator (SOI) CMOS process [76]. Owing to its lower coupling to the substrate, SOI transistors have inherently better performance at higher frequencies than their bulk CMOS counterparts. The common-source first stage is optimized for low noise, whereas the cascode second stage is designed to result in a higher gain with a sufficiently low-noise figure, being biased at the minimum noise figure point. The source degeneration in the first stage is mostly used for stability and input matching, whereas the source degeneration in the second stage serves for the interstage matching. In this case, the high-Q inductor is used in the source degeneration for lower noise contribution, whereas a low-Q inductor is used at the drain for better stability. The gate bias circuits are realized using high-value resistive dividers. As a result, a mean noise figure of 2.2 dB with a peak gain of 19.5 dB was measured across the frequency range of 16–24 GHz. An inclusion of the drain peaking inductors in the first and second stages and a series inductor between the cascode transistors in the second stage to resonate the gate–drain capacitance contributes to better gain-bandwidth performance when a flat gain of about 14.5 dB and a low-noise figure of about 3.1 dB were achieved from 21 to 27 GHz in a standard 0.18-µm CMOS process [77].

Figure 8.26b shows the circuit schematic of a two-stage single-ended 0.13-µm CMOS LNA, where the resistive feedback with a shunt resistor R_F is used in the first stage for wideband input matching and an inductive peaking with the series resistor R_L and inductor L_L as the peaking-load components is used in the second stage for gain compensation [78]. For the final differential LNA structure, a voltage gain of 14 dB with less than 1.7-dB variation and a noise figure from 4.0 to 4.7 dB were measured across the multioctave frequency bandwidth from 0.1 to 6.0 GHz. In this case, by using only one small-size inductor, the chip area is much smaller than that using filter-based multi-inductor designs, occupying 0.13 mm² and consuming 16 mW from a 1.2-V supply.

To provide a wideband impedance matching, the cascode CMOS LNA can be designed with a bandpass response at its input, where an input impedance of the common-source transistor is considered a part of the filter. However, an adoption of the filter at the input requires a number of reactive elements, thus increasing the overall die size and noise figure due to the finite quality factor of the passive components when implemented on a chip. Figure 8.27a shows the circuit schematic of a cascode 90-nm CMOS LNA, where the wideband input impedance matching, flat and high gain, flat and low-noise figure, and compact size were achieved with a shunt feedback resistor R_F in conjunction with a preceding π-type LC matching network with two shunt capacitances (external C_{in} and gate–source capacitance C_{gs1}) and series inductance L_g and a postcascode series-peaking

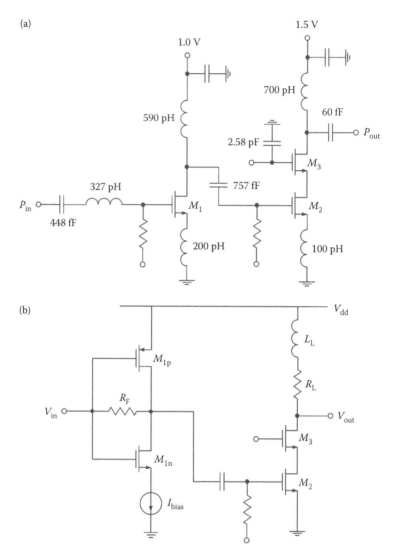

FIGURE 8.26
Circuit schematics of two-stage cascode CMOS LNAs.

inductance L_p [79]. As a result, a power gain of 9.6 ± 1.1 dB, an input return loss of greater than 10 dB, and a noise figure of 3.68 ± 0.72 dB were measured across the frequency range of 1.6–28 GHz, with the die size of only 0.139 mm². Similar wideband performance from 1.9 to 22.5 GHz with a low noise of 3.24 ± 0.5 dB was achieved for a cascode 90-nm CMOS LNA shown in Figure 8.27b with shunt resistive feedback for each transistor, where both input and output matching networks represent the second-order wideband bandpass filters, each equivalent to two parallel *RLC* branches with two notch frequencies, which are sufficiently far from each other [80].

Generally, the low-noise figure and low-power consumption can be hardly achieved simultaneously across a large frequency range. The noise cancellation technique can be used to relax this trade-off in the resistive-feedback LNAs. Figure 8.28 shows the circuit schematic of a broadband cascode 0.18-μm CMOS LNA, where the first stage is based

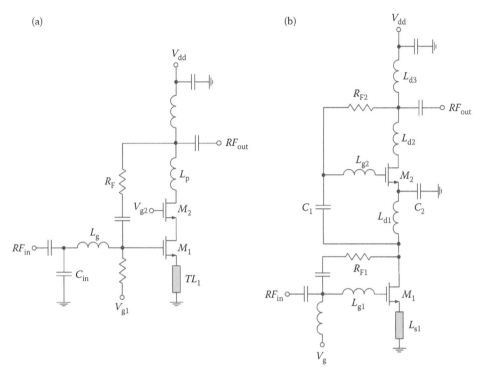

FIGURE 8.27
Circuit schematic of resistive-feedback cascode CMOS LNAs.

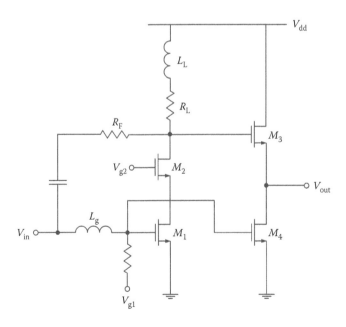

FIGURE 8.28
Circuit schematic of resistive-feedback noise-canceling CMOS LNA.

on the resistive feedback and inductive peaking, whereas the second stage is added for wideband output matching and partial noise cancellation by connecting the gate of the transistor M_4 to the gate of the transistor M_1 to provide the opposite phases of the noise signals at the source of the transistor M_3 and at the drain of the transistor M_4 [81]. As a result, a maximum power gain of 12.5 dB and a 3-dB bandwidth of 0.7–6.5 GHz with a noise figure from 3.5 to 4.2 dB were measured.

Figure 8.29 shows the simplified circuit schematic of a broadband cross-coupled resistive-feedback 90-nm CMOS LNA, which can be used in broadcasting and satellite communication systems [82]. This architecture is similar to the conventional broadband LNA with resistive matching; however, the overall noise figure is reduced by incorporating the transistor M_{p1} and connecting the gate of the transistor M_{p1} to the gate of the transistor M_{n1} in a cross-coupled fashion. In this case, the presence of the transistor M_{n1} reduces the output noise by one half, thus lowering the overall noise figure. The transistor M_{p2} is required to provide dc biasing and an additional gain to increase the overall gain of the LNA. The widths of the transistors are increased to reduce the flicker noise at lower frequencies. Finally, a measured gain of 20 dB across the frequency range from 2 to 1100 MHz and a minimum and maximum noise figure of 1.43 and 1.9 dB from 100 to 1100 MHz were obtained, respectively. The LNA consumes 18 mW from 1.8-V supply and occupies an area of 0.06 mm².

Figure 8.30a shows the circuit schematic of a cascode LNA using a 0.25-µm SiGe BiCMOS process, where the broadband operation of a cascode stage on the transistors T_1 and T_2 is achieved by using a shunt feedback network, which includes the resistor R_2 and a diode-connected transistor T_3 [83]. Here, the emitter-follower buffer stage is designed as a compromise between broadband frequency response, output matching, and linearity. Because of broadband matching and absence of lossy on-chip inductors, the noise figure varies between 2.5 and 2.9 dB from 2 GHz up to 11 GHz, with a gain exceeding 20 dB and a chip size of 270 × 370 µm. In a resistive-feedback cascode LNA with reactive input matching

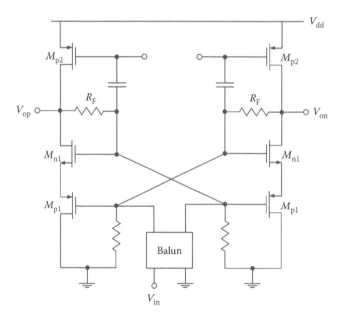

FIGURE 8.29
Circuit schematic of resistive-feedback cross-coupled CMOS LNA.

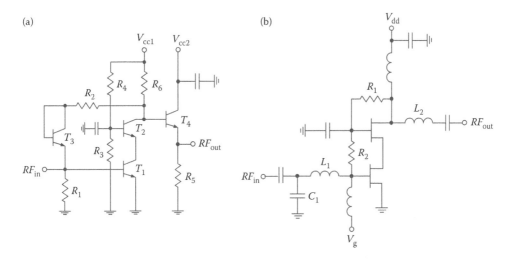

FIGURE 8.30
Circuit schematics of multioctave monolithic LNAs.

implemented in a 130-nm SiGe BiCMOS technology, a power gain of 9 dB with less than 1-dB variation from 3 to 26 GHz and a noise figure of less than 5 dB from 3 to 18 dB rising to 6.5 dB at 24 GHz were achieved [84]. By using a 200-GHz SiGe BiCMOS technology, a power gain greater than 23 dB from 1.1 to 2.0 GHz and a noise figure of 2.7–3.3 dB from 1.2 to 2.4 GHz were measured for a single-stage resistive-feedback cascode LNA with a fully integrated matching network [85].

A broadband low-noise performance can be achieved by using a standard 0.15-μm pHEMT technology. Figure 8.30b shows the circuit schematic of a monolithic cascode pHEMT LNA with two identical cells, where the positive drain bias is applied to the drain of the top cell and the negative gate bias is applied only to the gate of the bottom cell [86]. The shunt resistors R_1 and R_2 provide dc biasing and RF feedback simultaneously. Such arrangement ensures that the gate–source bias for two cells is the same so that each cell has the same drain–source voltage. The capacitor connected to the gate of the top cell is necessary to provide RF grounding at high frequencies. In a Class-A operation mode, the drain current of 80 mA at $V_{dd} = 4$ V results in an output impedance of 50 Ω, and the series inductor L_2 is used to compensate for the output capacitive reactance. As a result, a power gain of 12.5 ± 1 dB, a noise figure from 1.5 to 2.5 dB, and a 1-dB compression output power over 15-dBm covering the entire frequency band of 2–13 GHz with 25% peak efficiency.

For a single-stage resistive-feedback fully integrated cascode LNA based on a 0.35-μm enhancement-mode pHEMT technology to provide a positive threshold voltage and superior noise performance, a noise figure of less than 0.5 dB and a gain of around 20 dB were achieved over a wide band of 1.5–2.7 GHz [87]. A low noise of less than 3 dB and a high output power (P_{1dB}) up to 8 W with 20-dB flat gain can be achieved across a decade bandwidth from 250 MHz to 3.5 GHz for a resistive-feedback cascode 0.25-μm GaN HEMT-SiC LNA by applying a 40-V supply voltage and 500 mA bias current [88]. In this case, to provide unconditional two-port circuit stability with K-factor greater than 1, a resistor of 30 Ω was added in series to the gate of the common-gate transistor. For an optimum low-noise bias current density of ~100 mA/mm with a supply voltage of 20 V and dc current of 300 mA, much lower noise of less than 1.5 dB was achieved at the expense of the reduction in P_{1dB} by about 7 dB.

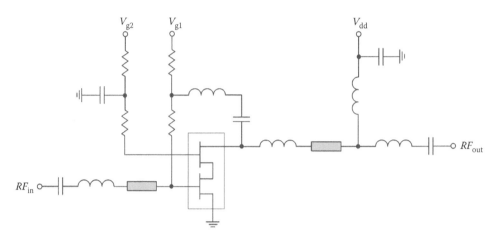

FIGURE 8.31
Circuit schematic of resistive-feedback dual-gate GaN HEMT LNA.

Instead of a cascode configuration with common-source and common-gate transistors, a single dual-gate MESFET can be used with attractive gain and noise performance, stability, and modulation capability [89]. For an RF-shorted second gate, the dual-gate MESFET represents a cascode circuit with a common-source input stage driving a common-gate output stage. In this case, to minimize the amplifier noise figure, an optimization of the second-gate termination and second-gate dc biasing is required. Two dual-gate devices can be combined in a balanced configuration with two Lange couplers at microwave frequencies. The Lange coupler is extremely useful for LNAs due to its property of showing matched input return loss for equal reflections from both amplifiers to the external circuit, whereas the internal circuit is matched for best noise performance. Three MMIC LNAs, each using two 0.1-μm dual-gate GaAs HEMTs in a balanced amplifier configuration, achieved less than 1.75-dB noise figure from 4 to 9 GHz, less than 2.75-dB noise figure from 9 to 20 GHz, and less than 2.5-dB noise figure from 20 to 40 GHz, respectively, with a high gain of around 20 dB [90]. Figure 8.31 shows the circuit schematic of a resistive-feedback 0.18-μm dual-gate GaN HEMT LNA [91]. Here, two cascaded spiral inductors are included in a feedback path to boost the gain at the high end. Spiral inductors are also used at the input and output for simple low-pass type impedance matching. As a result, a dual-gate LNA operating from 300 MHz to 3 GHz achieved over 17-dB flat gain and better than 2-dB noise figure, whereas a dual-gate LNA operating from 1.2 to 18 GHz achieved a gain of 13.3 ± 0.3 dB and a noise figure between 2 and 3 dB.

8.5 Graphical Design Technique

The design of low-noise broadband microwave integrated amplifiers is very well suited to the application of iterative optimization techniques in view of the many variables and conflicting objectives of high gain, flat and broadband gain, and low-noise figure. It may also include input and output return loss and requirements on phase shift and gain compression. An objective function, which includes both gain and noise figure, may be written as a sum over frequency of the weighted sum of the square of the noise figure and the squares of the difference between the calculated and the desired gain [92]. For example, the objective function can be chosen as the minimax kind written for a maximally flat 10-dB desired gain and minimum noise figure in the frequency range from 1 to 4 GHz [93]. When lumped lossy matching networks are used, the synthesis of the lumped matching network can incorporate the arbitrary loss factor associated with each reactive element and provide a very good initial design for practical MMIC circuit [94]. In this case, a solution from lossless matching network synthesis is used as an initial guess for the lossy matching network with certain gain reduction added to the gain function to compensate for the gain reduction due to the losses in the matching network. Analytical expressions giving the maximum available gain for a specified noise figure or the minimum noise figure for a specified available gain can be used to directly optimize the design of broadband low-noise microwave transistor amplifiers [95]. An interstage network of the sloped passband response to compensate for a 6-dB/octave roll-off so that the resultant amplifier gain-bandwidth characteristic is flat across the specified frequency range can be synthesized using a computer-optimization technique by finding poles and zeroes via a graphical user interface [96].

As an alternative for the designing of the broadband feedback LNAs, an interactive "visual" design technique is based on determining of acceptable regions (ARs) of the immittance at sample frequencies, from which a feedback two-port network is synthesized [97]. This technique allows the exact design directly from a simultaneous set of performance specifications, including gain and gain flatness, noise figure and stability, and input and output matching. Besides, this technique can be extended to determining the ARs of the source and load reflection coefficients in the complex planes [98]. To provide the amplifier stability with conditionally stable transistor, apart from ARs constructed at sample frequencies over an operating frequency band, the circular stability regions in the planes of the source and load reflection coefficients, Γ_S and Γ_L, should be used. These regions are only employed at frequencies where a transistor is potentially unstable, including frequencies outside the passband. Figure 8.32a shows the circuit schematic of a cascode feedback LNA with source degeneration for better input matching and circuit stability [98]. In this case, the first design step is to find elements of a parallel feedback network based on ARs on impedance (Z_p) plane in correspondence with amplifier requirements. The contours for the unit stability factor ($K = 1$) are plotted in Figure 8.32b, where arrows show the stability regions with noise figure lower than 1.1 dB and gain flatness within 18 and 19 dB. Using these ARs and stability regions, a series RC feedback circuit is synthesized using interactive "visual" design procedure. The feedback elements are finalized in such a way that full ARs in the source reflection coefficient plane include the origin, as shown in Figure 8.32c, and the output matching network is designed based on the source-terminated ARs in the load plane for $\Gamma_S = 0$, as shown in Figure 8.32d. As a result, the single-stage cascode feedback 0.18-μm pHEMT LNA provides a power gain of 17.6 ± 0.35 dB, a noise figure of less than 1.1 dB, and an input return loss better than 10.5 dB over the band of 1.5–2.5 GHz.

In a multistage feedback LNA, the first need is to assign requirements for each amplifier stage and then synthesize matching networks considering stability, gain, and noise figure circles mapped on the Smith chart. However, usually it is useful to start the design from the synthesis of the feedback networks in individual amplifier stages. At this stage, it is assumed that that the transistor with a feedback network is terminated in some optimum source and load impedances providing the minimum noise figure, maximum gain, etc. At the first design step, the ARs are constructed in the source Γ_S and load Γ_L termination planes. Then, at the second design step, each interstage matching network is designed so that its input and output reflection coefficients fall into the respective ARs. The synthesis

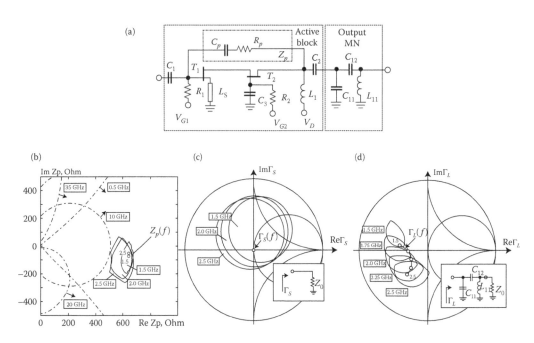

FIGURE 8.32
Circuit schematic and diagrams of broadband cascode LNA.

of lossless matching networks based on simultaneously prescribed ARs of input and output reflection coefficients can be performed with optimization routine using an appropriate goal function using the "visual" design tool as well.

Figure 8.33a shows the block diagram of a two-stage MMIC LNA using a 0.15-μm GaN HEMT technology that was design based on the following requirements: power gain of greater than 20 dB, gain flatness of less than ±1 dB, maximum noise figure of 1.5 dB, and input and output return loss of less than 9.63 dB [99]. Here, a shunt feedback network Z_{p2} is used to level gain response and maintain stability, whereas the series feedback inductors L_{s1} and L_{s2} are used to provide input matching and additionally improve stability. The first design step is to synthesize an input matching circuit MN1 using full ARs in $\Gamma_S^{(1)}$ plane at certain frequencies over 8–12 GHz, as shown in Figure 8.33b, when the power gain varies within 7–11 dB, the noise figure is less than 0.6 dB, and the input return loss is better than 10 dB. To guarantee the first stage stability, the circular stability regions were constructed in $\Gamma_S^{(1)}$ plane at some frequencies outside the passband, which are plotted in Figure 8.33b with dashed lines, and arrows show stability circle parts. In this case, the first stage with the input matching network and conjugately matched output exhibits 5 dB/octave gain roll-off from 8 to 12 GHz. Then, a parallel feedback network in the second stage is synthesized based on the ARs in Z_{p2} plane. At this step, the input and output ports of the second stage are assumed to be conjugately matched. In order to compensate for the first-stage gain roll-off, the gain is specified at 12 GHz by 2 dB greater than at 8 GHz, with $F \leq 2.5$ dB and $K > 1$. The inductances L_{g2}, L_{d2}, and L_{s2} are selected to obtain maximally wide ARs in Z_{p2} plane, as shown in Figure 8.33c. At the next design step, an interstage network MN2 is synthesized by constructing the source-terminated $\Gamma_L^{(1)}$ plane for the first stage, as shown in Figure 8.33d, and full ARs in $\Gamma_S^{(2)}$ plane for the second stage, as shown in Figure 8.34a. In this case, the input impedance of the second stage with feedback is close to 50 Ω.

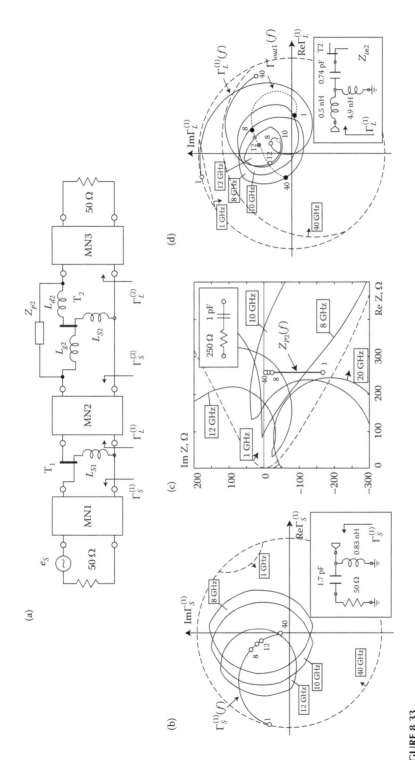

FIGURE 8.33
Circuit schematic and diagrams of broadband two-stage LNA.

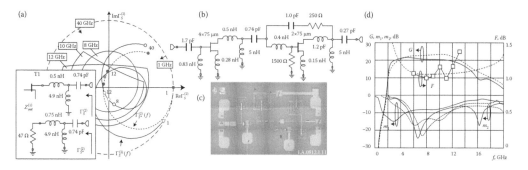

FIGURE 8.34
Diagram, layout, and performance of broadband two-stage LNA.

To design the output matching network MN3 at the last step, the requirements for an entire LNA are specified as follows: a power gain from 20 to 22 dB, a noise figure less than 1 dB, and input and output return loss of less than 9.63 dB. Considering a two-stage LNA with input and output matching networks as a two-port network, the ARs in $\Gamma_L^{(2)}$ plane were constructed. The circuit schematic of the designed stable two-stage feedback LNA with ideal lumped elements is shown in Figure 8.34b, with the simulated (without optimization) power gain of 20.5 ± 0.5 dB, noise figure $F \leq 0.8$ dB, input return loss of greater than 12 dB, and output return loss of greater than 8 dB in a frequency range of 8–12 GHz [99]. After replacing of all ideal elements by MMIC elements, a final LNA optimization was provided. Figure 8.34c shows the circuit layout of a two-stage MMIC LNA, with chip size of 1.4×1.2 mm². The measured results are given in Figure 8.34d, with a power gain of 20.0 ± 1.2 dB, a noise figure $F \leq 1.3$ dB, an input return loss of greater than 7 dB, and an output return loss of greater than 8 dB from 8 to 12 GHz.

8.6 Broadband Millimeter-Wave Low-Noise Amplifiers

The broadband millimeter-wave monolithic GaAs MESFET LNAs were first developed to operate over the entire U-band (40–60 GHz) at the beginning of the 1990s with typical performance for a two-stage LNA of at least 7-dB of gain with about 1.5-dB gain flatness and a maximum noise figure of 7.5 dB [100]. By using an InP-based HEMT MMIC process with a cutoff frequency of 300 GHz, a four-stage coplanar-waveguide MMIC LNA achieved a power gain of 20–25 dB and a noise figure of 3–4 dB in a frequency range of 85–110 GHz, whereas a three-stage microstrip MMIC LNA demonstrated a power gain of 14-dB and a noise figure of 7 dB from 165 to 190 GHz [101]. With an advanced 70-nm InAlAs/In GaAs mHEMT technology, a four-stage MMIC LNA exhibited a small-signal gain of greater than 18 dB between 216 and 238 GHz using airbridge-type transmission lines, whereas a four-stage MMIC LNA achieved a linear gain of greater than 12 dB over the bandwidth of 217–245 GHz using conductor-backed coplanar circuit technology [102]. Note that in all the above-mentioned implementations, the transistors were used in a common-source configuration. As opposed to a common-source LNA, a common-gate LNA with a matching output series inductor can be used to increase gain while maintaining broadband operation. Being fabricated in an 80-nm InP HEMT process, a three-stage common-gate MMIC

LNA obtained a gain of 18 dB and a noise figure of 3.5 dB from 68 to over 110 GHz, with a power consumption of 12 mW at 3-V supply voltage [103].

Figure 8.35a shows the circuit schematic of a two-stage 0.1-μm InP HEMT MMIC LNA with a coplanar waveguide structure [104]. Here, the first stage was designed for low-noise performance using source degeneration, whereas the second stage was designed for the high-gain performance. In this case, the input matching circuit is matched to two frequencies, the bottom and upper edges of the frequency range, whereas the output matching circuit is matched to the upper edge to achieve flat broadband operation. As a result, the LNA achieved greater than 10-dB gain over the frequency range of 44.6–67.2 GHz at the supply voltage of 0.4 V, with a noise figure of 2.86 dB at 60 GHz. For a four-stage 70-nm GaAs mHEMT MMIC LNA with source degeneration in each stage, whose schematic is shown in Figure 8.35b, a power gain of greater than 21 dB in a frequency range of 70–110 GHz and noise figure of 2.7 dB between 80 and 95 GHz and less than 3.2 dB up to 108 GHz were measured [105]. By using a 65-nm CMOS technology, a four-stage cascode MMIC LNA with a transmission line between the common-source and common-gate transistors in the final stage to achieve wideband output matching demonstrated a small-signal gain of greater than 20 dB from 75.5 to 120.5 GHz with a noise figure of 6.0–8.3 dB from 87

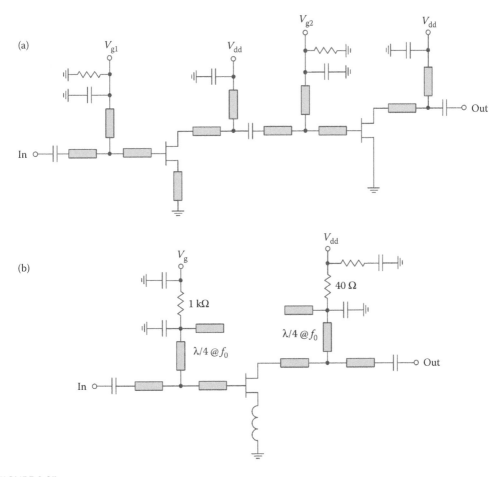

FIGURE 8.35
Circuit schematics of broadband common-source HEMT LNAs.

to 100 GHz [106]. Two differential LNAs realized in 0.25-μm and 0.13-μm SiGe BiCMOS technologies achieved a noise figure below 7.2 dB from 50 to 75 GHz and a noise figure below 7 dB from 78 to 110 GHz, respectively, with a measured maximum gain of 23 dB for both LNAs [107].

Figure 8.36a shows the circuit schematic of a single 35-nm InP HEMT MMIC LNA stage using a source degeneration and series grounded CPW transmission lines at the input and output [108]. The operation stability is provided by both inductive source feedback and resistor-capacitor networks at the bias lines. The final four-stage LNA based on this common-source amplifier stage achieved a noise figure of 7–8 dB over the frequency range from 220 to 252 GHz, with a gain of greater than 20 dB from 160 to 240 GHz. For a three-stage 35-nm InP HEMT MMIC LNA, the first two common-source stages to minimize noise figure and a final cascode stage for better reverse isolation were used [108]. The cascode stage is shown in Figure 8.36b, where the small-size MIM capacitors in a CPW environment for proper gate grounding were implemented. Besides, a high-impedance CPW transmission line between the common-source and common-gate transistors and a

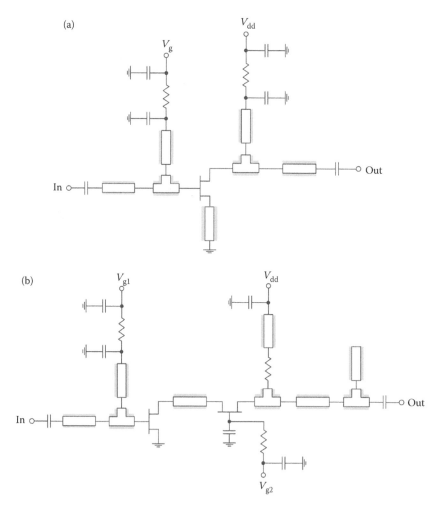

FIGURE 8.36
Circuit schematics of common-source and cascode InP HEMT LNA stages.

resistor connected in series to the output short-circuited shunt stub for stabilization were used. An open-circuited shunt stub was included in the output matching network to stabilize the cascode cell above the amplifier frequency band of operation. As a result, on-wafer measured gain of 17–22 dB was achieved from 160 to 270 GHz.

References

1. J. J. Whelehan, Low-noise amplifiers: Then and now, *IEEE Trans. Microwave Theory Tech.*, MTT-50, 806–813, 2002.
2. S. Okwit, An historical view of the evolution of low-noise concepts and techniques, *IEEE Trans. Microwave Theory Tech.*, MTT-32, 1068–1082, 1984.
3. H. Wallman, A. B. Macnee, and C. P. Gadsden, A low-noise amplifier, *Proc. IRE*, 36, 700–708, 1948.
4. M. T. Lebenbaum, Design factors in low-noise figure input circuits, *Proc. IRE*, 38, 75–80, 1950.
5. D. Weighton, Note on the design of wide-band low-noise amplifiers, *Proc. IRE*, 43, 1096–1101, 1955.
6. A. G. Bose and S. D. Pezaris, A theorem concerning noise figures, *IRE Trans. Circuit Theory*, CT-3, 190–196, 1956.
7. H. A. Haus and R. B. Adler, Optimum noise performance of linear amplifiers, *Proc. IRE*, 46, 1518–1533, 1958.
8. P. O. Lauritzen and O. Leistiko, Jr., Field-effect transistors as low-noise amplifiers, *1962 IEEE Int. Solid-State Circuits Conf. Dig.*, Philadelphia, PA, pp. 62–63, 1962.
9. F. M. Gardner, Optimum noise figure of transistor amplifiers, *IRE Trans. Circuit Theory*, CT-10, 45–48, 1963.
10. R. S. Engelbrecht and K. Kurokawa, A wide-band low noise *L*-band balanced transistor amplifiers, *Proc. IEEE*, 53, 237–247, 1965.
11. T. Das, Practical considerations for low noise amplifier design, Freescale Semiconductor, 2013.
12. L. Besser, Stability considerations of low-noise transistor amplifiers with simultaneous noise and power match, *1975 IEEE MTT-S Int. Microwave Symp. Dig.*, Palo Alto, CA, pp. 327–329, 1975.
13. Q. Liang, G. Niu, J. D. Cressler, S. Taylor, and D. L. Harame, Geometry and bias current optimization for SiGe HBT cascode low-noise amplifiers, *2002 IEEE RFIC Symp. Dig.*, Seattle, WA, vol. 1, pp. 407–410, 2002.
14. S. P. Voinigesku, M. C. Maliepaard, J. L. Showell, G. E. Babcock, D. Marchesan, M. Schroter, P. Schvan, and D. L. Harame, A scalable high-frequency noise model for bipolar transistors with application to optimal transistor sizing for low-noise amplifier design, *IEEE J. Solid-State Circuits*, SC-32, 1430–1439, 1997.
15. B. Jung, A. Gopinath, and R. Harjani, A novel noise optimization design technique for radio frequency low noise amplifiers, *2003 IEEE Int. Circuits Syst. Symp. Dig.*, Bangkok, vol. 1, pp. 209–212, 2003.
16. A. Anastassiou and M. J. O. Strutt, Effect of source lead inductance on the noise figure of a GaAs FET, *Proc. IEEE*, 62, 406–408, 1974.
17. J. Engberg, Simultaneous input power match and noise optimization using feedback, *Proc. 4th Eur. Microwave Conf.*, Montreux, Switzerland, pp. 385–389, 1974.
18. A. Higashisaka and T. Mizuta, 20-GHz band monolithic GaAs FET low-noise amplifier, *IEEE Trans. Microwave Theory Tech.*, MTT-29, 1–6, 1981.
19. B. Afshar and A. M. Niknejad, X/Ku band CMOS LNA design techniques, *Proc. 2006 IEEE Custom Integrated Circuits Conf.*, San Jose, CA, pp. 389–393, 2006.
20. A. Grebennikov, *RF and Microwave Transistor Oscillator Design*, John Wiley & Sons, Chichester, UK, 2007.
21. E. G. Nielsen, Behavior of noise figure in junction transistors, *Proc. IRE*, 45, 957–963, 1957.

22. H. Fukui, The noise performance of microwave transistors, *IEEE Trans. Electron Devices*, ED-13, 329–341, 1966.
23. L. Escotte, J. -P. Roux, R. Plana, J. Graffeuil, and A. Gruhle, Noise modeling of microwave heterojunction bipolar transistors, *IEEE Trans. Electron Devices*, ED-42, 883–889, 1995.
24. M. Rudolph, R. Doerner, L. Klapproth, and P. Heymann, An HBT noise model valid up to transit frequency, *IEEE Electron Device Lett.*, 20, 24–26, 1999.
25. G. Niu, J. D. Cressler, S. Zhang, W. E. Ansley, C. S. Webster, and D. L. Harame, A unified approach to RF and microwave noise parameter modeling in bipolar transistors, *IEEE Trans. Electron Devices*, ED-48, 2568–2574, 2001.
26. J. Gao, X. Li, H. Wang, and G. Boeck, Microwave noise modeling for InP-InGaAs HBTs, *IEEE Trans. Microwave Theory Tech.*, MTT-52, 1264–1272, 2004.
27. D. Zoeschg, W. Wilhelm, H. Knapp, K. Aufinger, J. Boeck, T. F. Meister, M. Wurzer, H. D. Wohlmuth, and A. L. Scholtz, Monolithic low-noise amplifiers up to 10 GHz in silicon and SiGe bipolar technologies, *Proc. 30th Eur. Microwave Conf.*, Paris, pp. 1–4, 2000.
28. G. Girlando and G. Palmisano, Noise figure and impedance matching in RF cascode amplifiers, *IEEE Trans. Circuits Syst.-II: Analog Digital Signal Processing*, CAS-II-46, 1388–1396, 1999.
29. H. Statz, H. A. Haus, and R. A. Pucel, Noise characteristics of gallium arsenide field-effect transistors, *IEEE Trans. Electron Devices*, ED-21, 549–562, 1974.
30. W. Baechtold, Noise behavior of GaAs field-effect transistors with short gate lengths, *IEEE Trans. Electron Devices*, ED-19, 674–680, 1972.
31. A. van der Ziel, Thermal noise in the hot electron regime in FET's, *IEEE Trans. Electron Devices*, ED-18, 977, 1971.
32. R. A. Pucel, D. F. Masse, and C. F. Krumm, Noise performance of gallium arsenide field-effect transistors, *IEEE Trans. Solid-State Circuits*, SC-11, 243–255, 1976.
33. P. Heymann and H. Prinzler, Improved noise model for MESFETs and HEMTs in lower gigahertz frequency range, *Electron. Lett.*, 28, 611–612, 1992.
34. D. -S. Shin, J. B. Lee, H. S. Min, J. -E. Oh, Y. -J. Park, W. Jung, and D. S. Ma, Analytical noise model with the influence of shot noise induced by the gate leakage current for submicrometer gate-length high-electron mobility transistors, *IEEE Trans. Electron Devices*, ED-44, 1883–1887, 1997.
35. H. Fukui, Design of microwave GaAs MESFET's for broad-band low-noise amplifiers, *IEEE Trans. Microwave Theory Tech.*, MTT-27, 643–650, 1979.
36. H. Fukui, Addendum to design of microwave GaAs MESFET's for broad-band low-noise amplifiers, *IEEE Trans. Microwave Theory Tech.*, MTT-29, 1119, 1981.
37. H. Fukui, Optimal noise figure of microwave GaAs MESFET's, *IEEE Trans. Electron Devices*, ED-26, 1032–1037, 1979.
38. A. Cappy, Noise modeling and measurement techniques, *IEEE Trans. Microwave Theory Tech.*, MTT-36, 1–9, 1988.
39. C. Enz, An MOS transistor model for RF IC design valid in all regions of operation, *IEEE Trans. Microwave Theory Tech.*, MTT-50, 342–359, 2002.
40. S. Asgaran, M. J. Deen, and C. -H. Chen, Analytical modeling of MOSFET's channel noise and noise parameters, *IEEE Trans. Electron Devices*, ED-51, 2109–2114, 2004.
41. D. K. Schaeffer and T. H. Lee, A 1.5-V, 1.5-GHz CMOS low noise amplifier, *IEEE J. Solid-State Circuits*, SC-32, 745–759, 1997.
42. V. Armengaud, J. Lintignat, B. Barelaud, B. Jarry, L. I. Babak, and C. Laporte, Design of a Ka-band MMIC filtering LNA with a metamorphic HEMT technology for a space application, *Proc. 38th Eur. Microwave Conf.*, Amsterdam, pp. 1358–1361, 2008.
43. Y. T. Lo and J. F. Kiang, Design of wideband LNAs using parallel-to-series resonant matching network between common-gate and common-source stages, *IEEE Trans. Microwave Theory Tech.*, MTT-59, 2285–2294, 2011.
44. H. Zheng and E. Sanchez-Sinencio, Linearization techniques for CMOS low noise amplifiers: A tutorial, *IEEE Trans. Circuits Syst.-I: Regular Papers*, CAS-I-58, 22–36, 2011.

45. K. L. Fong, High-frequency analysis of linearity improvement technique of common-emitter transconductance stage using a low-frequency trap network, *IEEE J. Solid-State Circuits*, SC-35, 1249–1252, 2000.

46. T. W. Kim, A common-gate amplifier with transconductance nonlinearity cancellation and its high-frequency analysis using the volterra series, *IEEE Trans. Microwave Theory Tech.*, MTT-57, 1461–1469, 2009.

47. V. Aparin and L. E. Larson, Modified derivative superposition method for linearizing FET low-noise amplifiers, *IEEE Trans. Microwave Theory Tech.*, MTT-53, 571–581, 2005.

48. S. Ganesan, E. Sanchez-Sinencio, and J. Silva-Martinez, A highly linear low-noise amplifier, *IEEE Trans. Microwave Theory Tech.*, MTT-54, 4079–4085, 2006.

49. W. H. Chen, G. Liu, Z. Boos, and A. M. Niknejad, A highly linear broadband CMOS LNA employing noise and distortion cancellation, *IEEE J. Solid-State Circuits*, SC-43, 1164–1176, 2008.

50. K. H. Chen and S. I. Liu, Inductorless wideband CMOS low-noise amplifiers using noise-canceling technique, *IEEE Trans. Circuits Syst.-I: Regular Papers*, CAS-I-59, 305–314, 2012.

51. N. Kim, V. Aparin, K. Barnett, and C. Persico, A cellular-band CDMA 0.25-μm CMOS LNA linearized using active post-distortion, *IEEE J. Solid-State Circuits*, SC-41, 1530–1534, 2006.

52. T. Sugiura, H. Itoh, T. Tsuji, and K. Honjo, 12-GHz-band low-noise GaAs monolithic amplifiers, *IEEE Trans. Microwave Theory Tech.*, MTT-31, 1083–1088, 1983.

53. R. E. Lehmann and D. D. Heston, X-band monolithic series feedback LNA, *IEEE Trans. Microwave Theory Tech.*, MTT-33, 1560–1566, 1985.

54. J. Panelli, N. Chiang, W. Ou, R. Chan, C. Shih, Y. C. Pao, and J. Archer, A 2.5 dB low noise 6 to 18 GHz HEMT MMIC amplifier, *1992 IEEE Microwave Millimeter-Wave Monolithic Circuits Symp. Dig.*, Albuquerque, NM, pp. 21–24, 1992.

55. P. Huang, W. L. Jones, A. Oki, D. Streit, W. Yamasaki, P. Liu, S. Bui, and B. Nelson, A 9–16 GHz monolithic HEMT low noise amplifier with embedded limiters, *1995 IEEE MTT-S Int. Microwave Symp. Dig.*, Orlando, FL, pp. 205–206, 1995.

56. C. Yuen, C. K. Nishimoto, M. W. Glenn, Y. C. Pao, R. A. LaRue, R. Norton, M. Day, I. Zubeck, S. G. Bandy, and G. A. Zdasiuk, A monolithic *Ka*-band HEMT low-noise amplifier, *IEEE Trans. Microwave Theory Tech.*, MTT-36, 1930–1937, 1988.

57. Y. L. Tang, N. Wadefalk, M. A. Morgan, and S. Weinreb, Full *Ka*-band high performance InP MMIC LNA modules, *2006 IEEE MTT-S Int. Microwave Symp. Dig.*, San Francisco, CA, pp. 81–84, 2006.

58. M. Rudolph, R. Behtash, R. Duerner, K. Hirche, J. Wuerfl, W. Heinrich, and G. Traenkle, Analysis of the survivability of GaN low-noise amplifiers, *IEEE Trans. Microwave Theory Tech.*, MTT-55, 37–43, 2007.

59. P. Chehrenegar, M. Abbasi, J. Grahn, and K. Andersson, Highly linear 1–3 GHz GaN HEMT low-noise amplifier, *2012 IEEE MTT-S Int. Microwave Symp. Dig.*, Montreal, Canada, pp. 1–3, 2012.

60. H. Zirath, P. Sakalas, and J. M. Miranda, A low noise 2–20 GHz feedback MMIC amplifier, *2000 IEEE RFIC Symp. Dig.*, Boston, MA, pp. 169–172, 2000.

61. P. S. Chen, D. H. Kim, J. Bergman, J. Hacker, and B. Brar, Wideband low-noise amplifier (LNA) with $L_g = 50$ nm InGaAs pHEMT and wideband RF chokes, *2011 IEEE MTT-S Int. Microwave Symp. Dig.*, Baltimore, MD, pp. 1–4, 2011.

62. B. Y. Ma, J. Bergman, P. Chen, J. B. Hacker, G. Sullivan, G. Nagy, and B. Brar, InAs/AlSb HEMT and its application to ultra-low-power wideband high-gain low-noise amplifiers, *IEEE Trans. Microwave Theory Tech.*, MTT-54, 4448–4455, 2006.

63. H. Jacobsson, L. Aspemyr, M. Bao, A. Mercha, and G. Carchon, A 5–25 GHz high linearity, low-noise CMOS amplifier, *Proc. 32nd Eur. Solid-State Circuits Conf.*, Montreux, Switzerland, pp. 396–399, 2006.

64. J. H. Tsai, J. Y. Lin, and K. Y. Ding, Design of a 9–25 GHz broadband low noise amplifier using 0.15-μm GaAs HEMT process, *Proc. 2012 Int. Microwave Millimeter Wave Technol. Conf.*, Shenzhen, China, vol. 5, pp. 1–4, 2012.

65. B. Nelson, W. Jones, E. Archer, B. Allen, M. Dufault, D. Streit, P. Liu, and F. Oshita, Octave band InGaAs HEMT LNA's to 40 GHz, *1990 IEEE GaAs IC Symp. Dig.*, New Orleans, LA, pp. 165–168, 1990.

66. G. A. Ellis, J. S. Moon, D. Wong, M. Micovic, A. Kurdoghlian, P. Hashimoto, and M. Hu, Wideband AlGaN/GaN HEMT MMIC low noise amplifier, *2004 IEEE MTT-S Int. Microwave Symp. Dig.*, Fort Worth, TX, pp. 153–156, 2004.

67. S. Masuda, T. Ohki, and T. Hirose, Very compact high-gain broadband low-noise amplifier in InP HEMT technology, *IEEE Trans. Microwave Theory Tech.*, MTT-54, 4565–4571, 2006.

68. K. W. Kobayashi, Y. C. Chen, I. Smorchkova, R. Tsai, M. Wojtowicz, and A. Oki, A 2 watt, sub-dB noise figure GaN MMIC LNA-PA amplifier with multi-octave bandwidth from 0.2–8 GHz, *2007 IEEE MTT-S Int. Microwave Symp. Dig.*, Honolulu, Hawaii, pp. 619–622, 2007.

69. M. Chen, W. Sutton, I. Smorchkova, B. Heying, W. B. Luo, V. Gambin, F. Ishita, R. Tsai, M. Wojtowicz, R. Kagiwada, A. Oki, and J. Lin, A 1–25 GHz GaN HEMT MMIC low-noise amplifier, *IEEE Microwave Wireless Comp. Lett.*, 20, 563–565, 2010.

70. I. Lo, O. Boric-Lubecke, and V. Lubecke, 0.25 μm CMOS dual feedback wide-band UHF low noise amplifier, *2007 IEEE MTT-S Int. Microwave Symp. Dig.*, Honolulu, Hawaii, pp. 1059–1062, 2007.

71. M. Chen and J. Lin, A 0.1–20 GHz low-power self-biased resistive-feedback LNA in 90 nm digital CMOS, *IEEE Microwave Wireless Comp. Lett.*, 19, 323–325, 2009.

72. K. W. Kobayashi, D. C. Streit, A. K. Oki, D. K. Umemoto, and T. R. Block, A novel monolithic HEMT LNA integrating HBT-tunable active-feedback linearization by selective MBE, *IEEE Trans. Microwave Theory Tech.*, MTT-44, 2384–2391, 1996.

73. M. Kawashima, Y. Yamaguchi, K. Nishikawa, and K. Uchara, Broadband low noise amplifier with high linearity for software-defined radios, *Proc. 10th Eur. Wireless Technol. Conf.*, Munich, Germany, pp. 323–326, 2007.

74. A. F. Bellomo, Gain and noise considerations in RF feedback amplifier, *IEEE J. Solid-State Circuits*, SC-3, 290–294, 1968.

75. K. B. Niclas, Noise in broad-band GaAs MESEFET amplifiers with parallel feedback, *IEEE Trans. Microwave Theory Tech.*, MTT-30, 63–70, 1982.

76. T. Kanar and G. M. Rebeiz, A 16–24 GHz CMOS SOI LNA with 2.2 dB mean noise figure, *2013 IEEE Compound Semiconductor Integrated Circuits Symp. Dig.*, Monterey, CA, pp. 1–4, 2013.

77. Y. S. Lin, J. H. Lee, S. L. Huang, C. H. Wang, C. C. Wang, and S. S. Lu, Design and analysis of a 21–29-GHz ultra-wideband receiver front-end in 0.18-μm CMOS technology, *IEEE Trans. Microwave Theory Tech.*, MTT-60, 2590–2604, 2012.

78. J. Wadatsumi, S. Kousai, D. Miyashita, and M. Hamada, A 1.2 V, 0.1–6.0 GHz, two-stage differential LNA using gain compensation scheme, *Proc. 2008 IEEE Topical Meeting Silicon Monolithic Integrated Circuits RF Syst. (SiRF)*, San Diego, pp. 175–178, 2008.

79. H. K. Chen, Y. S. Lin, and S. S. Lu, Analysis and design of a 1.6–28-GHz compact wideband LNA in 90-nm CMOS using a π-match input network, *IEEE Trans. Microwave Theory Tech.*, MTT-58, 2092–2104, 2010.

80. Y. S. Lin, C. C. Wang, and J. H. Lee, A 9.96 mW 3.24 ± 0.5 dB NF 1.9 ~ 22.5 GHz wideband low-noise amplifier using 90 nm CMOS technology, *2014 IEEE Radio Wireless Symp. Dig.*, Newport Beach, CA, pp. 208–210, 2014.

81. J. Hu, Y. Zhu, and H. Wu, An ultra-wideband resistive-feedback low-noise amplifier with noise cancellation in 0.18 μm digital CMOS, *Proc. 2008 IEEE Topical Meeting Silicon Monolithic Integrated Circuits RF Syst. (SiRF)*, San Diego, pp. 218–221, 2008.

82. M. El-Nozahi, A. A. Helmy, E. Sanchez-Sinencio, and K. Entesari, A 2–1100 MHz wideband low noise amplifier with 1.43 dB minimum noise figure, *2014 IEEE RFIC Symp. Dig.*, Tampa, FL, pp. 119–122, 2014.

83. S. Chartier, P. Lohmiller, J. Dederer, H. Schumacher, and M. Oppermann, SiGe BiCMOS wideband low noise amplifiers for applications in digital beam-forming receivers, *Proc. 40th Eur. Microwave. Conf.*, Manchester, UK, pp. 1070–1073, 2010.

84. P. K. Saha, S. Shankar, R. Schmid, R. Mills, and J. D. Cressler, Analysis and design of a 3–26 GHz low-noise amplifier in SiGe HBT technology, *2012 IEEE Radio Wireless Symp. Dig.*, Santa Clara, CA, pp. 203–206, 2012.

85. J. Chung, H. Poh, P. Cheng, T. K. Thrivikraman, and J. D. Cressler, High gain, high linearity, L-band SiGe low noise amplifier with fully-integrated matching network, *Proc. 2010 IEEE Topical Meeting Silicon Monolithic Integrated Circuits RF Syst. (SiRF)*, New Orleans, LA, pp. 69–72, 2010.

86. C. Y. Chiang and H. T. Hsu, Design and implementation of multioctave low-noise power amplifier (LNPA) using HIFET configuration, *Proc. 2012 Asia-Pacific Microwave Conf.*, Kaohsiung, Taiwa, pp. 956–958, 2012.

87. J. Yao, X. Sun, and B. Lin, Ultra low-noise highly integrated 1.5 to 2.7 GHz LNA, *2013 IEEE Int. Wireless Symp. Dig.*, Beijing, China, pp. 1–4, 2013.

88. K. Kobayashi, An 8-W 250-MHz to 3-GHz decade-bandwidth low-noise GaN MMIC feedback amplifier with >+ 51-dBm OIP3, *IEEE J. Solid-State Circuits*, SC-47, 2316–2326, 2012.

89. C. A. Liechti, Performance of dual-gate GaAs MESFET's as gain-controlled low-noise amplifiers and high-speed modulators, *IEEE Trans. Microwave Theory Tech.*, MTT-23, 461–469, 1975.

90. W. R. Deal, M. Biedenbender, P. Liu, J. Uyeda, M. Siddiqui, and R. Lai, Design and analysis of broadband dual-gate balanced low-noise amplifiers, *IEEE J. Solid-State Circuits*, SC-42, 2107–2115, 2007.

91. S. E. Shih, W. R. Deal, D. M. Yamauchi, W. E. Sutton, W. B. Luo, Y. Chen, I. P. Smorchkova, B. Heying, M. Wojtowicz, and M. Siddiqui, Design and analysis of ultra wideband GaN dual-gate HEMT low-noise amplifiers, *IEEE Trans. Microwave Theory Tech.*, MTT-57, 3270–3277, 2009.

92. T. W. Houston and L. W. Read, Computer-aided design of broad-band and low-noise microwave amplifiers, *IEEE Trans. Microwave Theory Tech.*, MTT-17, 612–614, 1969.

93. V. Andresciani and R. De Leo, Automated design of broad-band and low-noise microwave amplifiers, *Proc. 2nd Eur. Microwave Conf.*, Stockholm, vol. 2, pp. 1–4, 1971.

94. L. C. T. Liu and W. H. Ku, Computer-aided synthesis of lumped lossy matching networks for monolithic microwave integrated circuits (MMIC's), *IEEE Trans. Microwave Theory Tech.*, MTT-32, 282–290, 1984.

95. G. N. Link and V. S. Rao Gudimetla, Analytical expressions for simplifying the design of broadband low noise microwave transistor amplifiers, *IEEE Trans. Microwave Theory Tech.*, MTT-43, 2498–2501, 1995.

96. M. N. A. Halib, Z. Awang, and N. H. Baba, Computer-aided synthesis of matching networks for multi-stage broadband microwave amplifiers, *Proc. 2005 Asia-Pacific Appl. Electromagnetics Conf.*, Johor Bahru, Malaysia, pp. 245–249, 2005.

97. M. V. Cherkashin, D. Eyllier, L. I. Babak, L. Billonnet, B. Jarry, D. A. Zaitsev, and A. V. Dyagilev, Design of 2–10 GHz feedback MMIC LNA using visual technique, *Proc. 35th Eur. Microwave Conf.*, Paris, pp. 1–4, 2005.

98. L. I. Babak, M. V. Cherkashin, and A. Y. Polyakov, A new region technique for designing microwave transistor low-noise amplifiers with lossless equalizers, *Proc. 38th Eur. Microwave Conf.*, Amsterdam, pp. 1402–1405, 2008.

99. L. I. Babak, M. V. Cherkashin, F. I. Sheyerman, and Y. V. Fedorov, Design of multistage low-noise amplifiers using visual CAD tools, *2011 IEEE MTT-S Int. Microwave Symp. Dig.*, Baltimore, MD, pp. 1–4, 2011.

100. N. Camilleri, P. Chye, A. Lee, and P. Gregory, Monolithic 40 to 60 GHz LNA, *1990 IEEE MTT-S Int. Microwave Symp. Dig.*, Dallas, TX, pp. 599–602, 1990.

101. R. Lai, T. Gaier, M. Nishimoto, S. Weinreb, K. Lee, M. Barsky, R. Raja, M. Sholley, G. Barber, and D. Streit, MMIC low-noise amplifiers and applications above 100 GHz, *2000 IEEE GaAs IC Symp. Dig.*, Seattle, WA, pp. 139–141, 2000.

102. A. Tessmann, A. Leuther, H. Massler, W. Bronner, M. Schlechtweg, and G. Weimann, Metamorphic H-band low-noise amplifier MMICs, *2007 IEEE MTT-S Int. Microwave Symp. Dig.*, Honolulu, Hawaii, pp. 353–356, 2007.

103. M. Sato, T. Takahashi, and T. Hirose, 68–110-GHz-band low-noise amplifier using current reuse topology, *IEEE Trans. Microwave Theory Tech.*, MTT-58, 1910–1916, 2010.

104. K. Nishikawa, T. Enoki, S. Sugitani, and I. Toyoda, 0.4 V, 5.6 mW InP HEMT V-band low-noise amplifier MMIC, *2006 IEEE MTT-S Int. Microwave Symp. Dig.*, San Francisco, CA, pp. 810–813, 2006.

105. W. Ciccognani, F. Giannini, E. Limiti, and P. E. Longhi, Full W-band high-gain LNA in mHEMT MMIC technology, *Proc. 3rd Eur. Microwave Integrated Circuits Conf.*, Amsterdam, pp. 314–317, 2008.

106. D. R. Lu, Y. C. Hsu, J. C. Kao, J. J. Kuo, D. C. Niu, and K. Y. Lin, A 75.5-to-120.5-GHz, high-gain CMOS low-noise amplifier, *2012 IEEE MTT-S Int. Microwave Symp. Dig.*, Seattle, WA, pp. 1–4, 2012.

107. G. Liu and H. Schumacher, Broadband millimeter-wave LNAs (44–77 GHz and 70–140 GHz) using a T-type matching network, *IEEE J. Solid-State Circuits*, SC-48, 2022–2029, 2013.

108. M. Varonen, P. Larkoski, A. Fung, L. Samoska, P. Kangaslahti, T. Gaier, R. Lai, and S. Sarkozy, 160–270-GHz InP HEMT MMIC low-noise amplifiers, *2012 IEEE Compound Semiconductor Integrated Circuits Symp. Dig.*, La Jolla, CA, pp. 1–4, 2012.

9

Distributed Amplifiers

The potential of traveling-wave or distributed amplification for obtaining power gains over very wide frequency bands was recognized in the mid-1930s when it was found that the gain-bandwidth performance is greatly affected by the capacitance and transconductance of the conventional vacuum tube [1]. However, the first theoretical analysis and its practical verification were obtained for broadband vacuum-tube amplifiers more than a decade later [2,3]. The basic concept was based on the idea of combining the interelectrode capacitances of the amplifying vacuum tubes with series wire inductors to form two lumped-element artificial transmission lines coupled by the tube transconductances. As a result, the distributed amplifier overcomes the difficulty of a conventional amplifier, whose frequency limit is determined by the factor that is proportional to the ratio of the transconductance of the tube to the square root of the product of its input grid–cathode and output anode–cathode capacitances, by paralleling the tubes in a special way, in which the capacitances of the tubes can be separated while the transconductances may be added almost without limit and not affect the input and the output of the device. Since the grid–cathode and anode–cathode capacitances form part of low-pass filters that can be made to have a substantially uniform response up to filter cutoff frequencies, whose value can be conveniently set within a wide range by suitable choice of the values of the external inductor coils, it became possible to provide amplification over much wider bandwidths than was achievable with conventional amplifiers.

9.1 Basic Principles of Distributed Amplification

Figure 9.1 shows the basic circuit structure of a vacuum-tube distributed amplifier [2]. Here, an artificial transmission line consisting of the grid–cathode capacitances C_g and inductances between tubes L_g is connected between the input terminals 1-1 and 2-2, with the characteristic impedance of the grid line defined as $Z_{01} = \sqrt{L_g/C_g}$. If the proper terminating impedance is connected to terminals 2-2 and if this transmission line is assumed to be lossless, then it can be shown that the driving-point impedance at terminal 1-1 is independent of the number of tubes so connected. In a similar fashion, a second transmission line is formed by making use of the anode–cathode capacitances C_p to shunt another set of coil inductances L_p, resulting in the similar characteristic impedance of the anode (or plate) line independent of the number of tubes as $Z_{02} = \sqrt{L_p/C_p}$. Impedances connected to terminals 3-3 and 4-4 are intended to be equal to the characteristic impedance of the anode line. The impedance connected to terminals 2-2 is called the grid termination, the impedance connected to terminals 3-3 is called the reverse termination, and the impedance connected to the output terminals 4-4 is called the anode termination. These two artificial transmission lines are made to have identical velocities of propagation. The bandwidth of a distributed amplifier is determined by the cutoff frequency of the artificial line. In general,

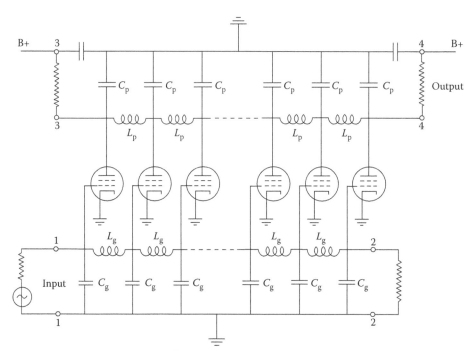

FIGURE 9.1
Basic structure of vacuum-tube distributed amplifier.

the higher this cutoff frequency, the lower will be the characteristic impedance of the line, and hence the less the voltage gain.

A signal generator connected to the input terminals 1-1 will cause a wave to travel along the grid line. As this wave reaches the grids of the distributed tubes, currents will flow in the anode circuits of the tubes. Each tube will then send waves in the anode line in both directions. If the reverse termination is perfect, the waves that travel to the left in the anode line will be completely absorbed, and will not contribute to the output signal. The waves that travel to the right in the anode line are added all in phase, and the output voltage is thus directly proportional to the number of tubes. Hence, the effective transconductance of such a distributed stage may be increased to any desired limit, no matter how low the gain of each tube (or section) is (even if it less than unity). As long as gain per section is greater than the transmission-line loss of the section, the signal in the anode line will increase and can be made large enough by using a sufficient number of tubes.

The total voltage gain A of the distributed amplifier consisting of n sections is written as

$$A = \frac{ng_{\mathrm{m}}}{2}\sqrt{Z_{01}Z_{02}} \tag{9.1}$$

where g_{m} is the tube transconductance, Z_{01} is the characteristic impedance of the grid line, and Z_{02} is the characteristic impedance of the anode line. Assuming that both transmission lines are identical,

$$Z_{0\pi} = Z_{01} = Z_{02} = \frac{R}{\sqrt{1 - x_{\mathrm{k}}^2}} \tag{9.2}$$

where $x_k = f/f_c$, $R = 1/\pi f_c C$ ($C = C_g = C_p$), f is the frequency, and f_c is the cutoff frequency of the transmission line. Equation 9.1 can be rewritten under conditions given by Equation 9.2 as

$$A = \frac{g_m R}{2} \frac{n}{\sqrt{1 - x_k^2}} \qquad (9.3)$$

where the second factor shows that the gain of the simple structure of a distributed amplifier shown in Figure 9.1 will be a function of frequency. This is due to the fact that the midshunt characteristic impedance of a low-pass constant-k filter section rises rapidly as the cutoff frequency is approached. This, in turn, causes the gain of the amplifier to increase sharply near cutoff, producing a large undesired peak.

There are several methods that can be used to eliminate this undesired peak and improve the frequency response, and one of them is to use the adjacent coils that are wound on the same form and in the same direction with large coupling coefficient M, as shown in Figure 9.2. In this case, each coupling section can be equated to the usual m-derived filter section. As a result, the total voltage gain A and phase shift ϕ for a distributed amplifier with n tubes can be written, respectively, as

$$A = \frac{g_m R_0}{2} \frac{nm^3}{\left[m^2 - (1 - m^2)x_k^2 \right] \sqrt{m^2 - x_k^2}} \qquad (9.4)$$

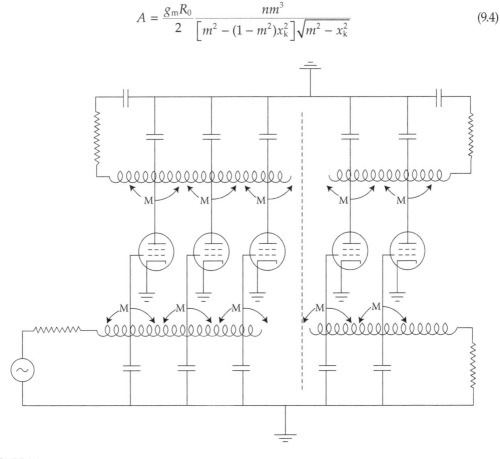

FIGURE 9.2
Vacuum-tube distributed amplifier with mutual coupling.

$$\varphi = 2n \tan^{-1} \frac{mx_k}{\sqrt{m^2 - x_k^2}} \qquad (9.5)$$

where $R_0 = 1/\pi f_0 C_g$, $x_k = f/f_0$, $f_0 = g_m/\pi\sqrt{C_g C_p}$ is the Wheeler's bandwidth-index frequency, and m is the design parameter selected for desired tolerance [2]. It should be noted that the presence of the parasitic capacitance distributed throughout the transmission-line coil windings results in lowering the amplifier cutoff frequency and in altering the impedance of the transmission lines, thus making it difficult to terminate properly [3]. An improvement of the gain/frequency response near cutoff frequency can be achieved by the insertion of extra sections into the grid or anode line, by the use of network whose image (or terminating) impedance (at a shunt-capacitance point) falls to zero at the cutoff frequency, or by the use of low-pass networks containing resistive elements [4,5]. In addition, the rise of gain can be eliminated by having different propagation functions for the sections in a distributed amplifier [6].

To improve both gain/frequency characteristic by making it flatter and phase-shift/frequency characteristic by making it more linear, a staggering principle can be applied when the lumped lines in the distributed amplifier are so arranged that the anode-line traveling wave and the grid-line traveling wave are not in phase at corresponding points along the lines [7]. In a distributed amplifier embodying the constant-k LC filter network as the elements of the lumped lines, the stagger is introduced by making the cutoff frequency of the grid line a little higher than that of anode line. At a given frequency, a line with a higher cutoff frequency produces a smaller phase than one with a lower cutoff frequency. The difference between the phase shifts produced by the two lines increases continuously as the frequency is increased.

The overall gain characteristic and phase shift of a staggered n-tube distributed amplifier with constant-k low-pass LC filter network is given, respectively, by

$$A = g_m \frac{Z_{0\pi}}{2} \frac{\sin n\psi/2}{\sin \psi/2} \qquad (9.6)$$

$$\varphi = (n-1)(\sin^{-1}x_k - \sin^{-1}qx_k) \qquad (9.7)$$

where $\psi = \theta_p - \theta_g$, $\theta_p = 2\sin^{-1}x_k$ is the phase shift introduced by the individual section of the anode line, $\theta_g = 2\sin^{-1}qx_k$ is the phase shift introduced by the individual section of the grid line, $x_k = f/f_{cp}$, $q = f_{cp}/f_{cg}$, f_{cp} is the anode-line cutoff frequency, and f_{cg} is the grid-line cutoff frequency. In order that the amplitude may fall to zero just below the cutoff frequency of the anode line, the last factor in Equation 9.6 has to vanish at f_{cp}, resulting in $(n\psi/2) = \pi$ at $x_k = 1$. Hence, the ratio between grid-line and anode-line cutoff frequencies can be obtained as

$$q = \sin\left(\frac{\pi}{2} - \frac{\pi}{n}\right) \qquad (9.8)$$

Similarly, the gain and phase-shift characteristics of a distributed amplifier with m-derived filter sections can be improved by staggering with different m-values of the anode and grid lines when the grid line has the larger m [7]. In practical implementation of a vacuum-tube distributed amplifier, the distributed cascode circuit can be used to minimize the degeneration at higher frequencies caused by the common lead inductances and feedback capacitances of the tubes [8]. However, if the attenuation in the grid line due to grid loading is substantial, the staggering of the lines may not necessarily provide much improvement of the gain characteristic, although the phase-shift characteristic is approximately the same as for the lossless case [9]. Generally, effects of staggering the lines in the distributed amplifiers based on vacuum tubes and transistors are different as their electrical behavior is characterized by different equivalent circuit representations [10].

If the attenuation in the grid line is neglected in order to obtain a general analysis of distributed amplifiers operating as large-signal devices, all tube currents add in phase as they progress toward the output end of the amplifier. Consequently, the output current is just n times the current that one tube produces in the output resistance and the output power varies as the square of the number of tubes. This is in contrast to the dc power input, which varies directly with the number of tubes. As a result, it is desirable to use as large a number of tubes as possible to increase efficiency, which increases directly with the number of tubes.

In a distributed amplifier designed for flat frequency response, the output power is constant with frequency as long as the driving power remains constant. Hence, the output power calculated for low frequencies should apply for any frequency in the operating region. Figure 9.3 shows the instantaneous values of anode voltage and anode current for

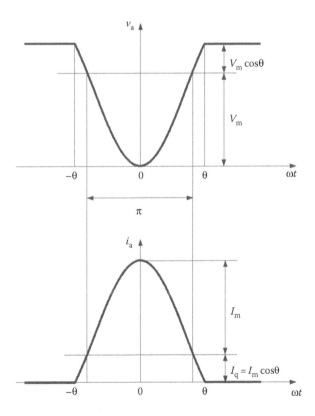

FIGURE 9.3
Instantaneous values of anode voltage and anode current in Class-AB operation.

idealized tube voltage–ampere characteristics in a Class-AB operation [11]. In this case, the instantaneous anode voltage $v_a(\omega t)$ varies according to

$$v_a(\omega t) = \begin{cases} V_m(1 - \cos \omega t) & -\theta \leq \omega t < \theta \\ V_m(1 - \cos \theta) & \theta \leq \omega t < 2\pi - \theta \end{cases} \qquad (9.9)$$

where the conduction angle 2θ indicates the part of the RF current cycle, during which a device conduction occurs.

The instantaneous anode current $i_a(\omega t)$ can be written as

$$i_a(\omega t) = \begin{cases} I_m(\cos \omega t - \cos \theta) & -\theta \leq \omega t < \theta \\ 0 & \theta \leq \omega t < 2\pi - \theta \end{cases} \qquad (9.10)$$

with a quiescent current $I_q = I_m \cos \theta$.

The dc value of the anode voltage must be equal to the anode supply voltage V_a as

$$V_a = \frac{V_m}{2\pi} \left[\int_{-\theta}^{\theta} (1 - \cos \omega t) \, d\omega t + (2\pi - 2\theta)(1 - \cos \theta) \right]$$

$$= \frac{V_m}{\pi} \left[\pi + \theta \cos \theta - \pi \cos \theta - \sin \theta \right]. \qquad (9.11)$$

The dc anode current I_0 is the average value of the instantaneous anode current given by

$$I_0 = \frac{I_m}{2\pi} \int_{-\theta}^{\theta} (\cos \omega t - \cos \theta) \, d\omega t = \frac{I_m}{\pi} (\sin \theta - \theta \cos \theta). \qquad (9.12)$$

The peak value of the fundamental component of the anode voltage is written as

$$V_1 = \frac{V_m}{\pi} \left[2 \int_{0}^{\theta} (1 - \cos \omega t) \cos \omega t \, d\omega t + \int_{\theta}^{2\pi - \theta} (1 - \cos \theta) \cos \omega t \, d\omega t \right]$$

$$= \frac{V_m}{\pi} (\theta - \sin \theta \cos \theta). \qquad (9.13)$$

The peak value of the fundamental component of the anode current is obtained by

$$I_1 = \frac{I_m}{2\pi} \int_{-\theta}^{\theta} (\cos \omega t - \cos \theta) \cos \omega t \, d\omega t = \frac{I_m}{\pi} (\theta - \sin \theta \cos \theta). \qquad (9.14)$$

The efficiency η is calculated as the ratio of the output fundamental power to the input dc power. At low frequencies, the anode circuit of the distributed amplifier with resistive terminations is only 50% efficient since half the output power from tubes is dissipated in the reverse termination. Consequently, the output power is one half the product of the fundamental current and fundamental voltage. As a result,

$$
\begin{aligned}
\eta &= \frac{1}{2}\left(\frac{V_1 I_1}{2}\right)\Big/(V_a I_0) \\
&= \frac{1}{4}\frac{(\theta - \sin\theta\cos\theta)^2}{(\pi + \theta\cos\theta - \pi\cos\theta - \sin\theta)(\sin\theta - \theta\cos\theta)}.
\end{aligned}
\tag{9.15}
$$

From Equation 9.15, it follows that the maximum theoretical efficiency of the distributed amplifier is derived to be about 30% and occurs with an anode current conduction angle of about 225° [11]. However, since the average value of the dc current drawn from the anode supply will decrease with frequency supply, the distributed amplifier will tend to become more efficient as the operating frequency increases. The change in efficiency is related to the change in the average plate current from no-signal conditions to full-signal conditions.

9.2 Microwave GaAs FET Distributed Amplifiers

The distributed amplifier using hybrid technology with lumped elements was first investigated based on silicon bipolar transistors in 1959 [12], MOSFETs in 1965 [13], and MESFETs in 1968–1969 [14,15]. In circuits that employ the field-effect transistors as active elements, the gate and drain loading plays a very significant role in operation and the high-frequency performance of the distributed amplifier. Therefore, it is only recently, with the availability of good-quality microwave GaAs FETs, the distributed amplifiers have again become popular at microwave frequencies [16]. In this case, GaAs FETs are used as the active devices, and the input and output lines can represent the periodically loaded microstrip transmission lines. With such an arrangement, the factors degrading the expected performance such as device input and output resistances and capacitances are either completely eliminated or their effect is included in the design. The resultant distributed amplifiers exhibit very low sensitivities to process variations and are relatively easy to design and simulate. In the early 1980s, the technology of distributed amplification was further improved by implementing the silicon and semi-insulating GaAs MMICs, which provide low-loss, small size, high reliability, circuit design flexibility, and a high level of integration. The first 0.5- to 14-GHz monolithic GaAs FET distributed (or traveling-wave) amplifier was designed in 1981 [17].

9.2.1 Basic Configuration with Microstrip Lines

The simplified schematic representation of a four-section GaAs FET distributed amplifier is shown in Figure 9.4, where the microstrip lines are periodically loaded with the complex gate and drain impedances of the devices, thus forming lossy transmission-line structures of different characteristic impedance and propagation constant [18]. An RF signal applied at the input end of the gate line travels down the line to the other end, where it is absorbed by the terminating impedance connecting at the end of the gate line, which includes the gate dummy resistor R_1. However, a significant portion of the signal is proportionally dissipated by the gate circuits of the individual FETs along the way. The input signal sampled

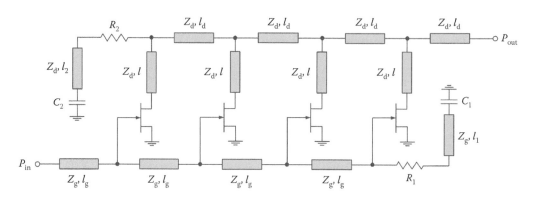

FIGURE 9.4
Schematic representation of four-section GaAs FET distributed amplifier.

by the gate circuits at different phases (and generally at different amplitudes) is transferred to the drain line through the FET transconductances. If the phase velocity of the
signal at the drain line is identical to the phase velocity of the gate line, then the signals
on the drain line add forming a traveling wave. The addition will be in phase only for the
forward-traveling signal. Any signal that travels backward, and is not fully canceled by
the out-of-phase additions, will be absorbed by the terminating impedance connecting at
the end of the drain line, which includes the drain dummy resistor R_2. The gate and drain
capacitances of the FET effectively become part of the gate and drain transmission lines,
while the gate and drain resistances introduce loss on these lines.

In conventional power amplifiers, it is impossible to increase the gain-bandwidth
product by just paralleling the FETs because the resulting increase in transconductance
g_m is compensated for by the corresponding increase in the input and output capacitances. The distributed power amplifier overcomes this problem by adding the individual device transconductances without adding their input and output capacitances,
which are now the parts of the artificial gate and drain transmission lines, respectively.
If the spacing between FETs is small compared to the wavelength, the characteristic
impedances of the gate and drain lines shown in Figures 9.5a and 9.5b, respectively, can
be approximated as

$$Z_g = \sqrt{\frac{L_g}{C_g + (C_{gs}/l_g)}} \tag{9.16}$$

$$Z_d = \sqrt{\frac{L_d}{C_d + (C_{ds}/l_d)}} \tag{9.17}$$

where C_{gs} is the gate–source capacitance and C_{ds} is the drain–source capacitances of the
unit FET cell, l_g and l_d are the lengths of the unit gate-line and drain-line sections, and
L_g, C_g and L_d, C_d are the per-unit-length inductance and capacitance of the gate and drain
lines, respectively. Here, the effect of the gate resistance R_{gs} and drain resistance R_{ds} is
neglected. It should be noted that the characteristic impedance expressions in Equations
9.16 and 9.17 are clearly independent of the number of FETs used in the circuit.

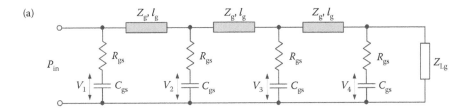

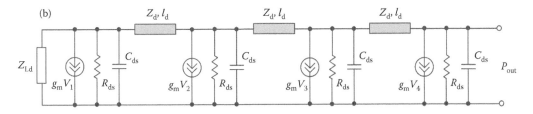

FIGURE 9.5
Simplified equivalent-circuit diagram of FET distributed amplifier.

As a result, the amplifier available gain G for n-section circuit by approximating the gate and drain lines as continuous structures can be written as

$$G = \frac{g_m^2 Z_g Z_d}{4} \left| \frac{\exp(-n\gamma_g l_g) - \exp(-n\gamma_d l_d)}{\exp(-\gamma_g l_g) - \exp(-\gamma_d l_d)} \right|^2 \tag{9.18}$$

where the propagation constants γ_g and γ_d are simplified using small-loss approximation as

$$\gamma_g = \frac{\omega^2 C_{gs}^2 R_{gs}}{2l_g} Z_g + j\omega \sqrt{L_g \left(C_g + \frac{C_{gs}}{l_g} \right)} \equiv \alpha_g + j\beta_g \tag{9.19}$$

$$\gamma_d = \frac{1}{2R_{ds}l_d} Z_d + j\omega \sqrt{L_d \left(C_d + \frac{C_{ds}}{l_d} \right)} \equiv \alpha_d + j\beta_d \tag{9.20}$$

Under normal operating conditions, the signals in the gate and drain lines are near synchronism when $\beta_g l_g \cong \beta_d l_d$, and Equation 9.18 can be simplified for $Z_g \cong Z_d \equiv Z_0$ and small losses to

$$G = \frac{g_m^2 Z_0^2}{4} \frac{[\exp(-n\alpha_g l_g) - \exp(-n\alpha_d l_d)]^2}{(\alpha_g l_g - \alpha_d l_d)^2} \tag{9.21}$$

from which it follows that, as the number of unit cells or sections n is increased, the available gain G does not increase monotonically and approaches zero in limit as n gets large.

For values of $n\alpha_g l_g \leq 1$ and when the drain-line losses are negligible compared to the gate-line losses, Equation 9.21 can be rewritten as

$$G \cong \frac{n^2 g_m^2 Z_0^2}{4} \left(1 - \frac{n\alpha_g l_g}{2} - \frac{n^2 \alpha_g^2 l_g^2}{6} \right)^2 \qquad (9.22)$$

which means that, in this operating conditions, the available gain G can be made proportional to n^2. In this case, by using the expression for the gate-line attenuation constant α_g given in Equation 9.19, one can find that

$$n Z_0 R_{gs} \omega^2 C_{gs}^2 \leq 2 \qquad (9.23)$$

which defines the upper limit to the total gate periphery that can be used in practical distributed amplifier or the maximum number of unit sections for a given FET device. Similar results can be obtained by applying a theoretical analysis based on matrix technique by employing a finite number of active and passive circuit elements [19].

In view of the gate-line and drain-line losses, from Equation 9.21, it follows that the power gain of a distributed amplifier approaches zero as $n \to \infty$. This happens due to the fact that the input voltage on the gate line decays exponentially, so the input signal does not reach FETs at the end of the amplifier gate line and, similarly, the amplified signals from FETs near the beginning are attenuated along the drain line. This implies that, for a given set of FET parameters, there will be an optimum value of n that maximizes the power gain of a distributed amplifier. Hence, by differentiating Equation 9.21 with respect to n and setting the result to zero, the optimum number of sections n_{opt} can be determined from

$$n_{opt} = \frac{1}{\alpha_g l_g - \alpha_d l_d} \ln\left(\frac{\alpha_g l_g}{\alpha_d l_d} \right) \qquad (9.24)$$

which depends on the device parameters, line lengths, and frequency through the attenuation constants given in Equations 9.19 and 9.20. For example, it is necessary to calculate the power gain of a distributed amplifier operated from 1 to 12 GHz, with a maximum gain at 10 GHz. Assuming that $\omega R_{gs} C_{gs} = 0.5$, $Z_0/R_{gs} = 4$, and $Z_0/R_{ds} = 0.2$ for the specified device parameters and $Z_g = Z_d = Z_0 = 50\ \Omega$, from Equations 9.19 and 9.20, it follows that $\alpha_g l_g = 0.5$ and $\alpha_d l_d = 0.1$, resulting in $n_{opt} = \ln(0.5/0.1)/(0.5 - 0.1) = 4.0$ or four sections.

Figure 9.6 shows the frequency dependence of a power gain for different numbers of FET unit cells or sections [18]. It is clearly seen that there is an optimum number n_{opt}, which provides maximum frequency bandwidth with minimum gain variations and reasonable power gain. For example, a power gain of 9 ± 1 dB over a bandwidth of 1–13 GHz was obtained for a four-cell distributed amplifier with a total GaAs FET gate periphery of $4 \times 300\ \mu m$. It should be mentioned that the resistive part of the gate loading typically results in a 3-dB gain reduction. Effect of the drain loading is not so significant; however, the power gain can be increased by about 1 dB for increased values of the drain loading resistance.

Replacing the conventional GaAs MESFETs with high-performance HEMT devices in a five-section monolithic distributed amplifier will result in significant improvements in

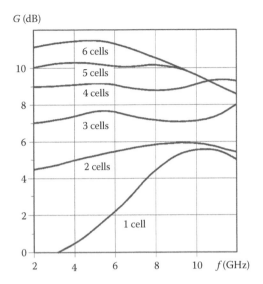

G (dB)

6 cells

5 cells

4 cells

3 cells

2 cells

1 cell

f (GHz)

FIGURE 9.6
Power gain versus frequency for different numbers of cells.

power gain and noise figure. For example, using 0.35-μm-gate-length HEMTs provides a low-noise figure by 2 dB lower and a power gain by 2.5 dB higher than achieved using 0.5-μm-gate-length MESFET devices in a frequency range from 2 to 20 GHz [20].

9.2.2 Basic Configuration with Lumped Elements

The equivalent gate and drain artificial transmission lines based on lumped inductors and capacitors are shown in Figures 9.7a and 9.7b, respectively [21,22]. For a constant-k-type transmission line, the phase velocity is a well-known function of the cutoff frequency f_c of the line. By requiring the phase shift between each gate-line and drain-line section to be equal, the cutoff frequency for the gate transmission line $f_{cg} = 1/2\pi R_{gs}C_{gs}$ and the cutoff frequency for the drain transmission line $f_{cd} = 1/2\pi R_{ds}C_{ds}$ must also be equal. As a result, the available gain G of the lumped distributed amplifier can be written as

$$G = \frac{g_m^2 Z_{01}Z_{02}\sinh^2[(n/2)(\alpha_d - \alpha_g)]\exp[-n\,(\alpha_d + \alpha_g)]}{4[1 + (f/f_{cg})^2][1 - (f/f_c)^2]\sinh^2[(1/2)(\alpha_d - \alpha_g)]} \qquad (9.25)$$

where α_g and α_d are the attenuations on the gate and drain lines per section, $Z_{01} = \sqrt{L_g/C_{gs}}$ and $Z_{02} = \sqrt{L_d/C_{ds}}$ are the characteristic impedances of the gate and drain line, respectively.

The attenuation on the gate and drain lines is the critical factor controlling the frequency response of the distributed amplifier. When attenuation per section is sufficiently small, the corresponding attenuations on the gate and drain lines can be given by

$$\alpha_g = \frac{x_k^2}{\sqrt{1 - [1 - (f_c/f_{cg})^2]x_k^2}}\frac{f_c}{f_{cg}} \qquad (9.26)$$

(a)

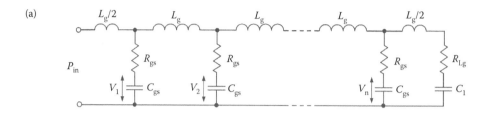

(b)

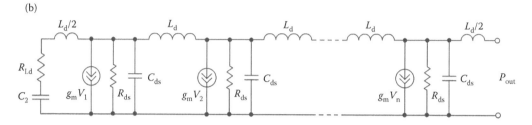

FIGURE 9.7
Simplified equivalent circuits of FET distributed amplifier with lumped inductors.

$$\alpha_d = \frac{1}{\sqrt{1 - x_k^2}} \frac{f_{cd}}{f_c} \tag{9.27}$$

where $x_k = f/f_c$ is the normalized frequency and $f_c = 1/\pi\sqrt{L_g C_{gs}} = 1/\pi\sqrt{L_d C_{ds}}$ [22]. From Equations 9.26 and 9.27, it follows that the gate-line attenuation is more sensitive to frequency than the drain-line attenuation, and the drain-line attenuation does not vanish in the low-frequency limit, unlike attenuation in the gate line. Therefore, the frequency response of the distributed amplifier can be expected to be predominantly controlled by the attenuation on the gate line. Generally, the attenuation on the gate and drain lines can be decreased by making f_c/f_{cg} and f_{cd}/f_c small when the transistor having high f_{cg} and low f_{cd} has to be chosen for a given f_c.

The maximum gain-bandwidth product of the distributed amplifier can be estimated by

$$\sqrt{G_0}\, f_{1dB} \approx 0.8\, f_{max} \tag{9.28}$$

where G_0 is low-frequency available gain of the amplifier, f_{1dB} is the frequency at which the power gain of the amplifier falls below G_0 by 1 dB, and

$$f_{max} = \frac{g_m}{4\pi C_{gs}} \sqrt{\frac{R_{ds}}{R_{gs}}} \tag{9.29}$$

is the frequency at which the maximum available gain (MAG) of the FET becomes unity [22].

9.2.3 Capacitive Coupling

The attenuation of the gate line α_g increases rapidly with frequency, as shown in Equation 9.19, resulting in a lower power gain at high bandwidth frequencies. In this case, if the gate-line attenuation can be made very small, the input signal is nearly evenly applied to all

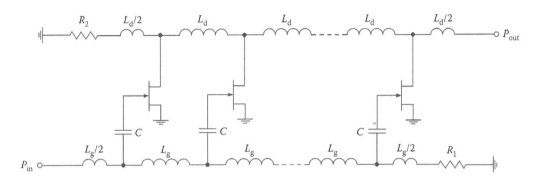

FIGURE 9.8
Schematic of distributed amplifier with series capacitors at FET gates.

FETs in the amplifier and the power gain will remain constant over a wide frequency band. However, since the gate-line attenuation is directly proportional to the gate–source capacitance C_{gs}, then it is possible to reduce its effect by adding a series capacitor C connected to each gate, as shown in Figure 9.8 [23,24]. As a result, since the effective gate capacitance is reduced by a factor of $q/(1+q)$ when $C = qC_{gs}$, the gate-line attenuation α_g decreases by a factor of $q/(1+q)$ and the gate-circuit cutoff frequency f_g is increased by a factor of $(1+q)/q$ at a fixed frequency. The gate voltage, however, divides between C and C_{gs}, and the FET can now be considered as a modified device having an effective gate–source capacitance of $C_g' = qC_{gs}/(1+q)$ and an effective transconductance of $g_m' = qg_m/(1+q)$ [25]. The series capacitor also reduces the gain per device, but the overall amplifier gain cannot be reduced if more devices are connected or a larger gate periphery is used. Moreover, the series capacitance and gate–source capacitance form a voltage divider, allowing for an increased signal level along the gate line, resulting in a significantly higher output power and efficiency for a distributed amplifier.

With a much larger total FET periphery, drain-line loading begins to limit output power, particularly at the upper end of operating band, resulting in a low or even negative real part of the impedance at the drains closer to the output. In this case, a capacitor can be inserted between the drain line and the drain of any FET with low real part of the impedance, thus decreasing drain-line loading and increasing the impedance at the drains [26]. As a result, a higher total FET periphery can be accommodated and higher output power can be achieved. Figure 9.9 shows an example of the three-cell GaAs FET distributed

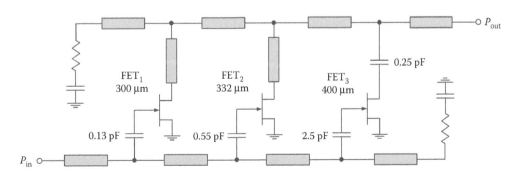

FIGURE 9.9
Schematic of distributed amplifier with series capacitor at FET drain.

amplifier using capacitive drain coupling. This circuit with a drain coupling capacitor connected to FET_3, which operates from 14 to 37 GHz, also features a varying gate periphery and capacitive gate coupling. Inserting the 0.25-pF capacitor between FET_3 and the drain line substantially increases the real part of the impedances at the drains of FET_2 and FET_3 over the frequency range from 18 to 38 GHz. This, in turn, results in higher and flatter output-power performance up to much higher frequencies, with a power increase by 1.5 dB at 18 GHz and by 5 dB at 27 GHz compared to the circuit without drain capacitive coupling.

9.2.4 Tapered Lines

As an alternative, to compensate for the attenuation due to the gate finite input resistance so that the FETs in the distributed network are not driven equally, equal drive to each transistor can be restored by increasing the characteristic impedance of the gate line in a systematically tapered manner from the input of the gate line to its end toward the gate load resistor [27]. In a manner analogous to the gate voltage equalization, the voltage at the drains of all of the FETs can be made the same by tapering the drain-line impedance, but with systematically decreasing impedance along the line toward the output load [2]. Improved performance in terms of the smaller gain flatness and wider frequency bandwidth can also be achieved by using a concept of the declining drain-line lengths when the lengths of the drain-line elements between the FET drains become shorter with optimized values the closer the drain line is located toward the output terminal [28].

The effect of a tapered drain line in terms of current distribution for a two-section distributed amplifier is shown in Figure 9.10, where the first FET device operates into a section of the drain line with a characteristic impedance Z_0 and entire drain current i_d flows to the next section [2]. If the next section has a lower characteristic impedance of $Z_0/2$, one-third of the incident drain current from the second FET device will cancel the reflected current from the first FET device at the junction of the second FET device. The remaining two-thirds of the drain current from the second FET device and four-thirds of the drain current from the first FET device add and propagate toward the end of the second section of the drain line into the new third section. At the next junction, the third section should have the characteristic impedance equal to $Z_0/3$. This process continues where each successive

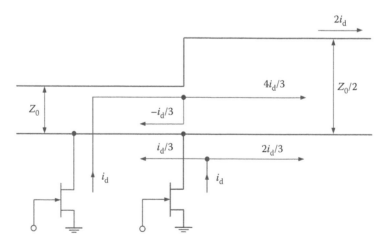

FIGURE 9.10
Current distribution in optimally tapered drain line.

transmission-line section has a characteristic impedance of Z_0/n, where n is the number of sections. The entire current of the FET devices may thus be effectively used in the load without the necessity of half the drain current flowing into the load and half the drain current flowing into the reverse termination. In this case, it should be noted that current equalization is difficult to achieve in practice due to unequal drive voltages on the gate line and FET process variation, and there exists a small range of useful realizable impedances for microstrip-line practical implementation.

Figure 9.11a shows the general structure of a distributed FET amplifier, where $Z_{g(i)}$ and $Z_{d(i)}$ are the optimum characteristic impedances of the gate- and drain-line ith sections,

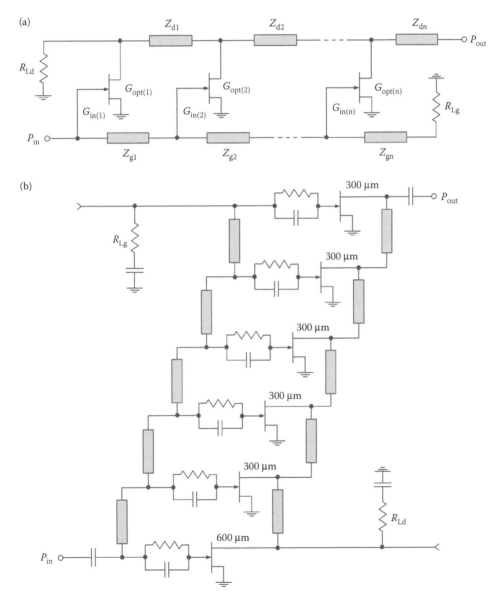

FIGURE 9.11
Schematics of nonuniform pHEMT distributed amplifiers.

while R_{Lg} and R_{Ld} are the gate and drain dumping resistors, respectively [29]. In this case, the optimum input and output capacitances of each FET device are absorbed into the artificial gate and drain lines to synthesize the optimum characteristic impedances $Z_{g(i)}$ and $Z_{d(i)}$. In the particular case of uniform distributed amplifiers, assuming an identical gate voltage amplitude on each transistor, the characteristic admittances of the drain-line sections can be given as

$$Y_{d(1)} = G_{opt} \tag{9.30}$$

$$Y_{d(i)} = G_{opt}\left(\frac{G_{opt}}{G_{opt} + G_{Ld}} + i - 1\right) \quad 2 \leq i \leq n \tag{9.31}$$

where G_{opt} is the optimum output conductance of each transistor, $Y_{d(1)} = 1/Z_{d(1)}$ is the optimum characteristic admittance of the first drain-line section, $Y_{d(i)} = 1/Z_{d(i)}$ is the optimum characteristic admittance of the ith drain-line section, and $G_{Ld} = 1/R_{Ld}$ is the drain dummy conductance.

The resulting optimum output power P_{out} of the uniform distributed amplifier is defined as

$$P_{out} = \left(\frac{G_{opt}}{G_{opt} + G_{Ld}} + n - 1\right)P_{max} \tag{9.32}$$

where P_{max} is the maximum power at 1-dB gain compression point and n is the total number of transistors within the amplifier.

In the case of nonuniform distributed amplifier, the generalized optimum power-matching structure can be analytically determined to add the individual power contribution in the direction of the output power as

$$Y_{d(1)} = G_{opt(1)} \tag{9.33}$$

$$Y_{d(i \geq 2)} = \left(\frac{G_{opt(1)}^2}{G_{opt(1)} + G_{Ld}} + \sum_{k=2}^{i} G_{opt(k)}\right) \quad 2 \leq i \leq n \tag{9.34}$$

$$P_{out} = \frac{G_{opt(1)}}{G_{opt(1)} + G_{Ld}} P_{max(1)} + \sum_{k=2}^{n} P_{max(k)} \tag{9.35}$$

where $P_{max(1)}$ is the maximum power of the first device, P_{out} is the amplifier output power, and $G_{opt(k)}$ and $P_{max(k)}$ are the optimum output conductance and output power of the kth transistor, respectively [29].

It should be noted that, in the case of moderate frequency bandwidth applications ($f_{max}/f_{min} < 3$), the drain dumping load R_{Ld} can be removed so that each transistor could be ideally matched and yield its maximum output power, resulting in

$$Y_{d(i)} = \sum_{k=1}^{i} G_{opt(k)} \quad 1 \leq i \leq n \tag{9.36}$$

$$P_{\text{out}} = \sum_{k=2}^{n} P_{\text{max}(k)} \tag{9.37}$$

To achieve the equal-gate voltage distribution, the characteristic impedances of the gate-line sections $Z_{g(i)} = 1/Y_{d(i)}$ are defined as

$$Y_{g(i)} = \sum_{k=i}^{n} G_{\text{in}(k)} \quad 1 \le i \le n \tag{9.38}$$

$$Z_{Lg} = 1/G_{\text{in}(n)} \tag{9.39}$$

and the electrical lengths $\theta_{g(i)}$ and $\theta_{d(i)}$ of the corresponding gate- and drain-line sections must always verify $\theta_{g(i)} = \theta_{d(i)}$.

Figure 9.11b shows the simplified circuit schematic of a monolithic nonuniform distributed amplifier composed of six amplifying cells and implemented in a 0.25-μm power pHEMT process, where the first transistor represents a 600-μm HEMT and the other transistors represent 300-μm HEMTs [29]. Here, discrete series capacitors couple each transistor to the gate line and act as voltage dividers to ensure equal drive levels on the transistor gates. Implanted GaAs resistors shunt the series MIM capacitors to supply gate bias. As a result, an output power of 30 dBm with a power gain of 7 dB and a *PAE* of greater than 20% was achieved across the frequency band from 4 to 19 GHz at a drain supply voltage of 8 V. By optimizing the nonuniform nature of the gate and drain lines and using the series capacitors at the device gates, the average 5.5-W output power and 25% *PAE* were achieved over 2–15 GHz for a five-cell monolithic distributed amplifier using a high-voltage 0.25-μm AlGaN/GaN HEMT on SiC technology with a total gate device periphery of 2 mm at a 20-V drain bias [30].

The efficiency can be further increased for the same multioctave frequency bandwidth when the resistive termination is neglected and optimum load can be presented to each active device by correspondingly tapering the drain-line characteristic impedance [31]. In addition, the transmission-line lengths can be adjusted such that the transistor currents add in-phase and the FET output capacitances can be absorbed into the transmission line. If the optimum load resistance $R_{\text{opt}(k)}$ for kth transistor is a known quantity for the process and is extracted from load-pull data, the unknown drain-line characteristic impedances $Z_{d(k)}$ can be calculated from

$$Z_{d(k)} = \frac{R_{\text{opt}}(\Omega \cdot \text{mm})}{\sum_{i=1}^{k} W_{Q_i}} \tag{9.40}$$

where $R_{\text{opt}} = R_{\text{opt}(k)} W_{Q_k}$ and W_{Q_k} is the gate width of kth transistor [32]. At low frequency, the individual FET optimum load resistances $R_{\text{opt}(k)}$ will combine in parallel and this parallel combination should be equal to the load impedance R_L to maximize the output power of the amplifier, where $Z_{d(k)} = R_L$.

Table 9.1 shows the device and transmission-line parameters corresponding to the 10-cell nonuniform distributed amplifiers with equal and unequal device gate sizes. In the case of equal gate widths for each cell, the maximum characteristic impedance of the transmission line is equal to 500 Ω, which is difficult to realize by microstrip line using normal

TABLE 9.1

Parameters of 10-Cell Nonuniform Distributed Amplifier

Parameters	FET Number	Equal FET Cells		Unequal FET Cells	
		W_Q (mm)	Z_d (Ohm)	W_Q (mm)	Z_d (Ohm)
FET R_{opt} (Ω-mm) = 120	1	0.24	500	0.60	200
R_L (Ω) = 50	2	0.24	250	0.20	150
Total FET width (mm) = 2.4	3	0.24	167	0.20	120
Number of cells = 10	4	0.24	125	0.20	100
Supply voltage (V) = 30	5	0.24	100	0.20	86
Maximum RF power	6	0.24	83	0.20	75
(W) = 9.0	7	0.24	71	0.20	67
	8	0.24	63	0.20	60
	9	0.24	56	0.20	55
	10	0.24	50	0.20	50

GaN on SiC process. However, by making the first FET cell thrice as large as that of the others, the maximum characteristic impedance of the first transmission line can be significantly reduced to an acceptable value. The estimated maximum RF output power shown in Table 9.1 is calculated assuming a sinusoidal output voltage across the load as $V_{dd}^2/2R_L$, where V_{dd} is the drain supply voltage.

The circuit schematic of a monolithic 10-cell nonuniform distributed amplifier designed to operate over a 10:1 bandwidth including as much of *Ku*-band as possible is shown in Figure 9.12a [32]. In this case, by using Equation 9.40 and assuming 35-Ω load impedance, the first FET cell was sized at 520 μm with the remaining nine cells sized at 320 μm each for a total periphery of 3.4 mm. To transform the standard 50-Ω load to 35-Ω impedance, a quarterwave microstrip line centered near the upper band edge was used, as shown in Figure 9.12b. As a result, the saturated output power greater than 8 W with a peak value of 13 W and *PAE* greater than 20% with a peak value of 38% were achieved over a frequency bandwidth of 1.5–17.0 GHz at a drain supply voltage of 30 V for an input power of 32 dBm using a 0.25-μm GaN HEMT on SiC technology. With some process modifications resulting in higher current handling capability for passive elements and increased gain and voltage handling at the device unit cell, a minimum output power of 11 W with a minimum *PAE* of 28% was achieved across 2–18 GHz at a drain supply voltage of 35 V [33].

The greater efficiency can be provided at lower frequencies when the power-added efficiencies of 30%–60% with an output power of around 40 dBm over the frequency bandwidth of 100 MHz to 2.2 GHz have been achieved for a monolithic four-cell distributed amplifier with a tapered drain line using a 0.5-μm GaN HEMT on Si process [34]. A low-temperature co-fired ceramic (LTCC) technology is an attractive choice for fabricating power amplifiers because it provides a sufficiently high circuit density integration, low RF loss, and good thermal performance. In this case, the discrete active devices can be placed directly on top of 400-μm silver-filled vias, which provide a good thermal dissipation to ground [35]. As a result, a five-cell distributed amplifier using pHEMT devices with a 2.1-mm gate periphery each provides a 1-W output power over the frequency bandwidth from 800 MHz to 2.1 GHz with a 3.2-V supply voltage. Here, a broadband multisection output impedance transformer (incorporating two 4:1 coupled-coil transformers in series with additional matching elements) is required to transform the optimum output impedance of 3.3 Ω to a standard 50-Ω load throughout the band. A three-cell design for a distributed amplifier using pHEMT devices with a 1.9-mm gate periphery each, whose circuit schematic is shown in Figure 9.13, can provide a 2-W output power over the frequency range

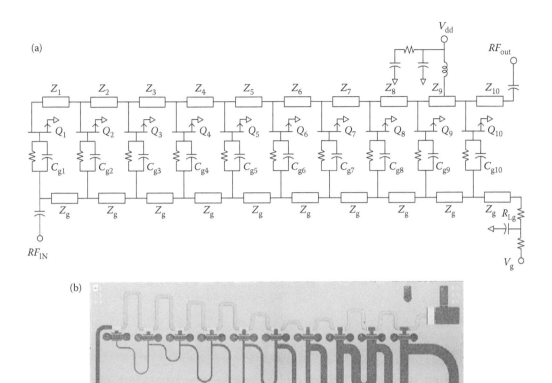

FIGURE 9.12
Circuit topology and MMIC of nonuniform GaN HEMT distributed amplifier.

from 0.6 to 2.2 GHz with a *PAE* greater than 30% at a 12-V supply voltage [36]. In this case, the drain-line termination was set well above 50 Ω, since the impedance at that point in the circuit is approximately 150 Ω, and it could be made even larger at the expense of gain flatness and stability with only 1%–2% efficiency drop due to the extensive drain-line impedance tapering. A 6-W distributed amplifier based on five LMOSFET devices having a 5-mm gate width each can achieve a *PAE* greater than 30% over the frequency bandwidth from 100 MHz to 1.8 GHz at a 28-V supply voltage [36]. With a hybrid implementation using Roger's RT5880 substrate and discrete transistors, a higher than 30% *PAE* over a frequency bandwidth of 20 MHz to 2.5 GHz was achieved for a 5-W three-cell distributed amplifier with an optimized tapered drain line using a 0.35-μm GaN HEMT on SiC process at a supply voltage of 28 V [37].

9.2.5 Power Combining

Since there is a strong demand for solid-state power amplifiers to provide high output power, high efficiency, and wide bandwidth in different microwave and millimeter-wave radar and communication systems, distributed amplifiers can be considered a

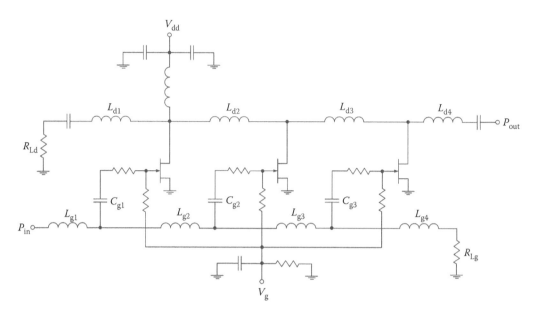

FIGURE 9.13
Schematic of three-cell pHEMT distributed amplifier using LTCC technology.

good candidate for ultra-wideband operation because their bandwidth performance is dominated by a high cutoff frequency for artificial input and output transmission lines. However, output power is generally limited by the drain-line termination and a maximum total gate periphery that can be included in a single-stage design. In this case, it is necessary to use power combining schemes using the transmission-line Wilkinson power divider or coupled-line Lange-type power dividers and combiners [38].

Figure 9.14 shows the circuit schematic of a monolithic two-stage distributed amplifier, where the input signal is equally divided into the gate lines, each employing four 150-μm FETs, using a Wilkinson power divider without the isolation resistor [39]. Such a configuration with a single-section Wilkinson divider had contributed in obtaining a

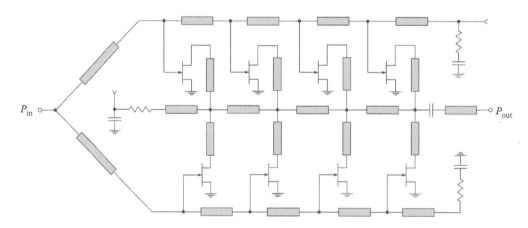

FIGURE 9.14
Distributed power amplifier with eight FET cells.

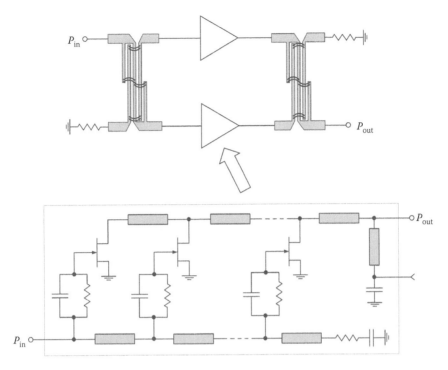

FIGURE 9.15
Schematic of balanced distributed amplifier with Lange couplers.

decade bandwidth performance because of the good input matching characteristics of the individual amplifiers. In this circuit, the FETs excited from the two separate gate lines are combined on a single drain line, effectively giving a $4 \times 300\text{-}\mu\text{m}$ drain periphery, thus doubling the output power over a frequency range of 2–20 GHz. Figure 9.15 shows the circuit diagram of a monolithic balanced nonuniform distributed amplifier, where an input power divider and an output power combiner represent the monolithic Lange couplers, which were designed to achieve operation over 6–18 GHz using highly over-coupled lines [40]. In this case, some output power degradation at upper operating frequencies because of the loss in artificial gate and drain lines due to the parasitic resistors in transistors was compensated by using a shunt short-circuited quarterwave microstrip line at the output of each distributed amplifier. As a result, the fabricated monolithic balanced nonuniform distributed amplifier using a 0.25-μm AlGaN/GaN HEMT technology was able to deliver an output power of more than 10 W over 6–18 GHz at a drain supply voltage of 40 V.

9.2.6 Bandpass Configuration

Similar to the conventional distributed amplifiers representing a low-pass configuration, it is possible to achieve sufficient power gain over extreme bandwidths with the bandpass distributed amplifiers limited only by the high-frequency effects of the amplifying devices [8]. In this case, additional inductive elements are placed in parallel with the shunt capacitance of the FETs, and this allows the gate capacitance to be effectively reduced, thus increasing the upper operating frequency for a given FET [41,42]. However, there has to be some drop in fractional bandwidth owing to the introduction of a lower cutoff frequency. Besides, better noise performance and linear phase response can be achieved with the bandpass distributed amplifiers. However, the basic bandpass distributed amplifier has an amplitude response that is inherently nonflat that may require additional parameter optimization procedure or inserting of additional compensating circuits.

The gate and drain lines in such a bandpass distributed amplifier have capacitors in the series arms of the gate and drain lines, thus preventing a direct biasing of the gate and drain terminals in the usual manner by means of two common dc power supplies connected to both lines. An alternative structure of the three-cell bandpass distributed amplifier with the series inductors and shunt series LC circuits in the gate and drain lines is shown in Figure 9.16 [43]. The value of the corresponding inductances in the gate and drain circuits are calculated from

$$L_{g1} = L_{d1} = \left(\frac{1-m}{1+m}\right)\frac{CR_L^2}{2} \tag{9.41}$$

$$L_{g2} = L_{d2} = \left(\frac{2}{m}\right)\frac{1}{\omega_0^2 C} \tag{9.42}$$

where $R_L = R_{Lg} = R_{Ld}$ is the load impedance, $\omega_0 = 2\pi\sqrt{f_1 f_2}$ is the center bandwidth frequency, f_1 is the lower bandwidth frequency, f_2 is the upper bandwidth frequency, $m = f_1/f_2$, and the effective capacitance $C = C_g = C_d$ is determined for particular device input and output capacitances, load impedance, and boundary frequencies. For example, $C = 3.18$ pF for $f_1 = 1$ GHz, $f_2 = 3$ GHz, $R_L = 50\ \Omega$, and GaAs FET NE72218 with $C_{in} = 0.96$ pF and $C_{out} = 0.64$ pF. In this case, the inductor values are calculated from Equations 9.41 and 9.42 as $L_{g1} = L_{d1} = 1.99$ nH and $L_{g2} = L_{d2} = 7.97$ nH. As a result, the passband of this

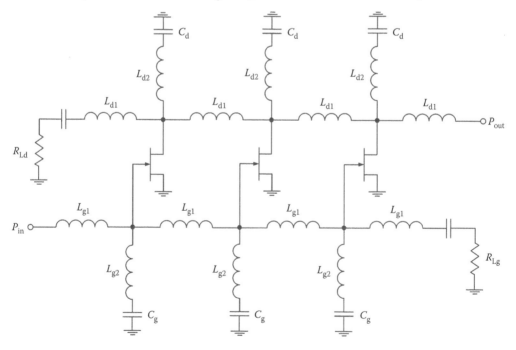

FIGURE 9.16
Schematic of three-cell bandpass distributed amplifier.

bidirectional and symmetric distributed amplifier with surface-mounted inductors covers a frequency bandwidth from 842 MHz to 3.17 GHz with a power gain of around 10 dB and linear phase response [43].

9.2.7 Parallel and Series Feedback

Among the various factors affecting the amplifier gain are the input and output loss factors of the active device. The input loss factor determines the rate at which the input signal decays along the gate line, and the output loss factor affects the growth rate of the output signal along the drain line. As a result, the optimum number of transistors to be used in the distributed amplifier is therefore determined by the input and output loss factors, which can be changed to achieve higher gain values through the use of feedback provided by the active element.

Figure 9.17 shows the small-signal equivalent circuit of a common-source FET device, where $Y_f = G_f + jB_f$ is a parallel feedback admittance and $Z_f = R_f + jX_f$ is a series feedback impedance. For studying the effects of various feedback elements on the maximum available gain (*MAG*) of the two-port network derived in Reference 44, a perturbation method can be used where the feedback is assumed small and only the first-order correction term ΔMAG is considered. As a result, for a series resistive feedback,

$$\Delta MAG = -R_f g_m \left[\frac{2\omega^2 R_{gs} R_{ds} C_{gs} C_{ds} (g_m R_{ds} C_{ds} - 2C_{gs})}{8\omega^2 C_{gs}^3 R_{gs}^2} \right.$$
$$\left. + \frac{2g_m C_{gs}(R_{gs} + R_{ds}) - (g_m R_{ds})^2 C_{ds}}{8\omega^2 C_{gs}^3 R_{gs}^2} \right] \tag{9.43}$$

and for a series reactive feedback,

$$\Delta MAG = X_f g_m \frac{4\omega^2 C_{gs}^2 R_{gs} - g_m^2 R_{ds}}{8\omega^3 C_{gs}^3 R_{gs}^2} \tag{9.44}$$

where $X_f = \omega L_f$, if the feedback element is an inductance, and $X_f = -1/\omega C_f$, if it is a capacitance [45]. From Equation 9.44, it follows that capacitive series feedback increases *MAG* at low frequencies acting as a positive feedback. On the contrary, inductive series feedback

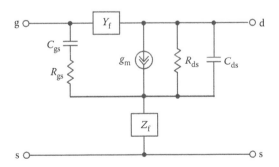

FIGURE 9.17
General FET model to evaluate effect of series and parallel feedback.

acts as a negative feedback at low frequencies. At the same time, at higher frequencies when the numerator in Equation 9.44 becomes positive, capacitive feedback is negative and inductive feedback is positive. The role of the resistor as a positive or negative feedback element is not only frequency dependent but also dependent on various FET parameters at a given frequency. Specifically, series source resistance acts as a negative feedback element only if the numerator in Equation 9.43 is positive. At low frequencies, the increase in gain due to positive capacitive feedback and the reduction in gain due to negative resistance feedback do not have the same frequency dependence. Therefore, a series RC feedback cannot be used for broadband loss compensation, unlike a parallel RC feedback, which can lead to the possibility of broadband loss compensation.

Similarly, for a parallel resistive feedback,

$$\Delta MAG = G_f g_m \left[\frac{(g_m R_{ds})^2 + 2 g_m R_{ds}(\omega^2 C_{gs}^2 R_{gs}^2 + \omega^2 C_{gs}^2 R_{gs} R_{ds} + 1)}{8(\omega^2 C_{gs}^2 R_{gs})^2} \right.$$
$$\left. + \frac{4 R_{gs} R_{ds} \omega^2 C_{gs}^2}{8(\omega^2 C_{gs}^2 R_{gs})^2} \right] \tag{9.45}$$

and for a parallel reactive feedback,

$$\Delta MAG = B_f g_m \frac{4\omega^2 C_{gs}^2 R_{gs} - g_m^2 R_{ds}}{8\omega^3 C_{gs}^3 R_{gs}^2} \tag{9.46}$$

where $B_f = \omega C_f$, if the feedback element is a capacitance, and $B_f = -1/\omega L_f$ for inductive feedback [45]. It can be seen from Equation 9.45 that a parallel resistive element always gives a negative feedback regardless of frequency or FET parameters. However, according to Equation 9.46, the role of capacitive and inductive elements in giving positive or negative parallel feedback is reversed compared to that in a series feedback.

It should be noted that the maximum level of positive feedback usable in a given distributed amplifier is limited by the requirement of unconditional stability of the amplifier over the whole frequency range. Generally, the risk of oscillations increases with greater device transconductance g_m, feedback gate–drain capacitance C_{gs}, or transmission-line cutoff frequency $f_c = 1/\pi\sqrt{L_g C_{gs}}$; however, the parasitic resistance R_{gs} and R_{ds} tend to moderate the oscillation phenomena. The detailed analysis of a two-cell distributed structure has shown that the oscillation conditions can be satisfied at high frequencies and is due to an internal loop formed by the transconductance and the feedback capacitance of the active devices, combined with the transmission lines [46]. The oscillations cannot occur due to reflections at the terminations (Z_{Lg} and Z_{Ld}) because the oscillation frequency is lower than the cutoff frequency of the transmission lines and an active feedback is necessary to provide gain in the loop to generate oscillations. The feedback gate–drain capacitance C_{gd} of the active devices strongly modifies the behavior of distributed amplifiers, but mainly when the frequency is high and transistors with high transconductance are used.

Figure 9.18 shows the circuit schematic of a monolithic three-cell distributed driving amplifier, where each active unit cell represents a cascade pHEMT amplifier with a self-biasing circuit through the parallel and series feedback resistors [47]. The feedback resistor values were determined for high stability and high gain in distributed amplifier using 200-μm pHEMTs as input devices and 480-μm pHEMTs as output devices (187 Ω

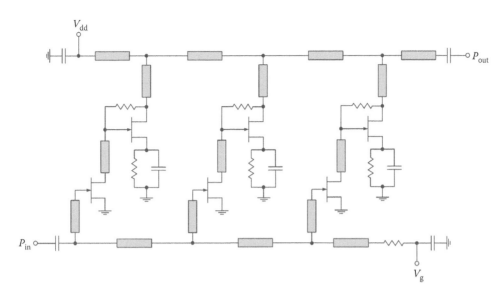

FIGURE 9.18
Schematic of three-cell distributed amplifier with parallel and series feedback.

for parallel feedback resistors and 53 Ω for series feedback resistors). The conventional drain termination resistor was eliminated for increasing output power and efficiency in this amplifier. As a result, a maximum available gain of more than 25 dB across 5–20 GHz and a stability factor of larger than unity over all frequencies were simulated, and a small-signal gain of 16 ± 0.6 dB over 5–21 GHz and an output power of more than 22 dB at 1-dB gain compression point over 6–18 GHz were achieved.

9.3 Cascode Distributed Amplifiers

In practical implementation of a vacuum-tube distributed amplifier, the distributed cascode circuit was used to minimize the degeneration at higher frequencies caused by the common lead inductances and feedback capacitances of the tubes [8]. By using a cascode connection of transistors, it is possible to significantly improve the isolation between the input and output artificial transmission lines. The cascode MESFET cell is characterized by a higher output shunt resistance, which reduces loading of the drain line, a lower gate–drain capacitance than the common-source cell, which reduces negative feedback at the high end, and provides a possibility of automatic gain control. As a result, a higher *MAG* can be achieved for the cascode cell over a very wide frequency range. For example, a cascode monolithic distributed amplifier based on a 0.25-μm InP HEMT technology was capable to produce gain as high as 12 ± 1 dB from 5 to 60 GHz and noise figure as low as 2.4–4.0 dB in the *Ka*-band [48].

Figure 9.19 shows the circuit schematic of a three-cell nonuniform cascode AlGaN/GaN HEMT distributed amplifier exhibiting an output power of 5–7.5 W and a *PAE* of 20%–33% over the operating frequency range from dc to 8 GHz [49]. In this case, three cascode-connected AlGaN/GaN HEMT cells each having a 0.3-μm gate length and a 1-mm

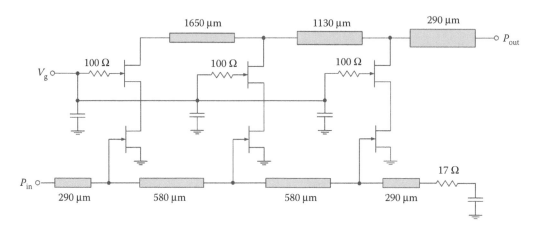

FIGURE 9.19
Schematic of three-cell nonuniform cascode GaN HEMT distributed amplifier.

gate periphery were employed to design and fabricate a high-power monolithic distributed amplifier. The drain-line dummy load was removed, and the gate- and drain-line sections were optimized to maximize an output power and efficiency and provide a flat gain throughout the operating frequency range. Note that the lines nearer the output are of lower characteristic impedance and, hence, are wider and more able to supply higher dc current in a Class-A bias mode. The corresponding center conductor widths of CPW drain lines are 27 μm (60 Ω), 40 μm (50 Ω), 66 μm (30 Ω), with ground-to-ground spacing of 80 μm. The nine-cell monolithic cascode distributed amplifier using a 0.2-μm AlGaN/GaN low-noise GaN HEMT technology with an $f_T \sim 75$ GHz was able to achieve 1–4 W from 100 MHz to over 20 GHz with noise figure of around 3 dB at a drain supply voltage of 30 V [50].

Figure 9.20a shows the MMIC of a five-cell cascode GaN HEMT distributed amplifier (Cree CMPA0060025F), which operates between 20 MHz and 6.0 GHz [51]. The amplifier typically provides a 17 dB of small-signal gain and an average 30 W (from 42 to 45 dBm) of saturated output power with a drain efficiency of better than 23% (better than 30% up to 4.0 GHz), as shown in Figure 9.20b for an input power of 32 dBm and two different drain supply voltages of 40 and 50 V. To achieve high efficiency, a nonuniform approach was used in the design of the drain line where the characteristic impedances changes cell by cell and the output reverse termination was eliminated. Proper design of the gate and drain lines and resizing of the individual cells provide a reasonable load-line impedance for each cell.

The dual-gate GaAs FET device, which is equivalent to cascode-connected single-gate devices, has input impedance comparable to that of a single-gate device but much higher isolation and output impedance. Note that high reverse isolation in the device is necessary for high amplifier isolation to achieve better operation stability, and high device output impedance improves gain flatness and output *VSWR*. Figure 9.21 shows the circuit schematic of a monolithic four-cell dual-gate MESFET distributed amplifier, which provides a power gain of 6.5 ± 0.5 dB with greater than 25-dB isolation across the frequency range of 2–18 GHz [52]. With a dual-gate AlGaN/GaN HEMT technology, the broadband performance can be improved when a small-signal gain of 12 ± 1 dB over the bandwidth of 2–32 GHz with a peak *PAE* of 16% and a peak output power of about 30 dBm is achieved for a five-cell monolithic dual-gate distributed amplifier [53]. Using a two-stage distributed

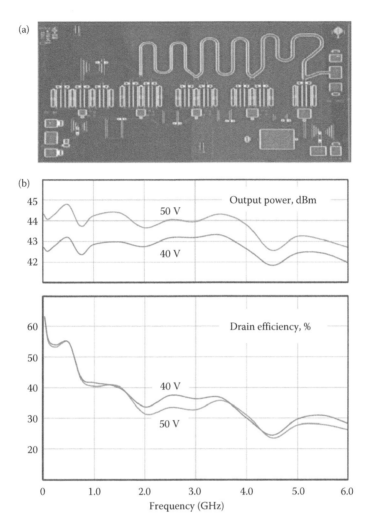

(a)

(b)

FIGURE 9.20
Schematic of five-cell nonuniform cascode GaN HEMT distributed amplifier.

amplifier when the first stage consists of six 4×50-μm dual-gate GaN HEMTs and the second stage consists of six 4×100-μm dual-gate GaN HEMTs allows the small-signal gain to be increased to more than 20 dB with a peak output power of 33 dBm over the bandwidth of 2–18 GHz [54].

When using a bipolar technology, superior gain-bandwidth performance of the HBT cascode cell compared to conventional common-emitter HBT shows that the cascode offers as much as 7 dB more maximum available gain, especially at higher bandwidth frequencies [55,56]. Figure 9.22a shows the circuit schematic of a three-cell cascode distributed amplifier based on SiGe HBT devices [57]. Here, the emitter degeneration resistor increases the device input impedance and helps to reduce the output distortion. Instead of a constant-k T-section consisting of a series lumped inductor L and a shunt capacitor C with the frequency-independent characteristic impedance $Z_0 = \sqrt{L/C}$ shown in Figure 9.22b, the m-derived T-section shown in Figure 9.22c can be used by adding a parallel inductor to

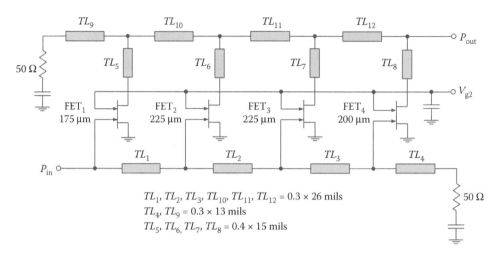

FIGURE 9.21
Schematic of four-cell dual-gate MESFET distributed amplifier.

provide an additional degree of freedom [58]. Both sections still maintain the same input and output impedances, but the *m*-derived *T*-section has an *LC* series resonance in its shunt arm. This resonance provides the ability to modify the passband attenuation. As a result, the *m*-derived *T*-section has a flatter passband and a better input reflection coefficient than its corresponding *T*-section with constant Z_0. The choice of a filter section is limited by the available die area, complexity of design, and the technology used. Being implemented in SiGe BiCMOS or HBT technology, such a cascode distributed power amplifier can achieve a measured passband from 100 MHz to 50 GHz with a 1-dB compression power gain varying from 6 to 8.5 dB with an output power of 4.2 ± 2 dBm [57]. In this case, the *m*-derived filter sections were used for the drain line, and the constant-*k* *T*-sections were used for the gate line.

To further improve the gain-bandwidth characteristics required for high-bit-rate telecommunication systems, the design approach based on the attenuation compensation technique when a common-collector stage is followed by a cascode transistor pair in each distributed cell can be used. In this case, a gain of 12.7 dB over a bandwidth of 50 MHz to 27.5 GHz was achieved for the fabricated monolithic three-cell "common-collector cascode" HBT distributed amplifier [59]. The measured midband saturate output power of 17.5 dBm with a peak *PAE* of 13.2% and the 3-dB output power bandwidth greater than 77 GHz were achieved with a monolithic eight-cell nonuniform cell-scaled cascode distributed amplifier using a 0.13-μm SiGe BiCMOS technology [60].

Monolithic integration of HEMT and HBT devices in a single chip can combine the advantages of both processes, such as low-noise figure and high input impedance from HEMTs and low 1/*f* noise, higher linearity, and high current driving capability from HBTs. In this case, the monolithic distributed amplifier can be designed using the stacked 2-μm InGaP/GaAs HBT and 0.5-μm AlGaAs/GaAs HEMT process for each cascode-connected cell [61]. To simplify the design procedure, the modified *m*-derived low-pass *T*-section can be used where the capacitance is fixed and only the inductance is varying with *m*. Figure 9.23 shows the circuit schematic of a monolithic six-cell HEMT-HBT cascode distributed amplifier where the peaking inductances L_{deg} and L_b are utilized to resonate with the parasitic capacitances of the common-source and common-base transistors, resulting

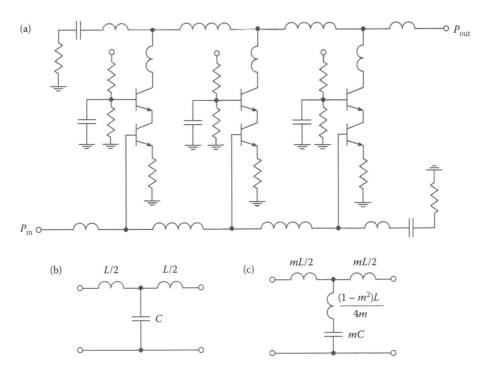

FIGURE 9.22
Schematic of three-cell bipolar cascode distributed amplifier.

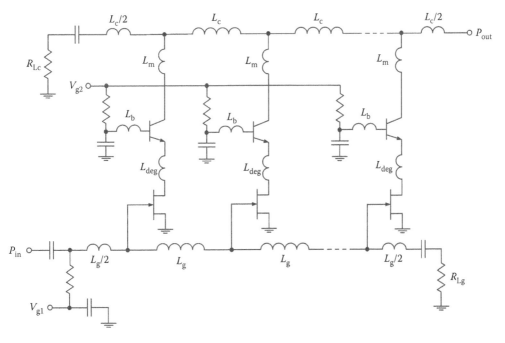

FIGURE 9.23
Schematic of six-cell HEMT-HBT cascode distributed amplifier.

in a gain peaking at the cutoff frequency. For a supply voltage of 4.5 V with a total dc current consumption of 50 mA, the distributed amplifier achieves an average small-signal gain of 8.5 dB and a 3-dB bandwidth of wider than 43.5 GHz.

9.4 Extended Resonance Technique

An extended-resonance power-combining technique can be used to space transistors in a distributed manner to proper distance from each other to form a resonant power combining/dividing structure where quarter- or half-wavelength spacing between transistors can be avoided [62]. Instead, the spacing is such that the input/output admittance of one transistor is converted to its conjugate value at the input/output of the next transistor. In this case, both gate- and drain-line termination resistors used in a traveling-wave structure are eliminated, resulting in a lower frequency bandwidth but higher efficiency. Since the magnitude of the voltage at each transistor is the same, the gain of the extended-resonance power-combining amplifier with equal-size n transistors is equal to the gain of a single-device amplifier, while its power capability is increased by n times. This approach enables a compact circuit to be designed that is particularly suitable for monolithic microwave integrated circuits.

Figure 9.24a shows the circuit schematic of a MESFET distributed amplifier using an extended-resonance technique to combine powers from n devices, where the gates and drains are sequentially linked with transmission lines [63]. Here, θ_{gk} and θ_{dk} are the electrical lengths of the transmission lines connecting each device for the gate and drain lines, respectively, where $1 \le k \le n-1$. The quarterwave transmission-line transformers are used at the input and output of the amplifier to match with the respective 50-Ω source and load impedances. The gate and drain extended resonance circuits can be designed separately after calculating the simultaneous conjugate-match admittances. It is assumed that each device has the same gate admittance $Y_g = G_g + jB_g$ and drain admittance $Y_d = G_d + jB_d$. The extra susceptance jB_g connected to the device input is provided by shunt capacitors or inductors, which can be realized in the form of open- or short-circuit stubs. To provide proper power combining, the voltage phase difference between successive drains should be equal to the voltage phase shift between successive gates. Example of a lumped four-device version of a distributed extended-resonance amplifier is shown in Figure 9.24b [64].

As shown in Figure 9.24a, the gate-line length $\theta_{g(n-1)}$ transforms the admittance $Y_g = G_g + jB_g$ from the Nth device to its conjugate value $Y_g{}^* = G_g - jB_g$ at the location of the next device. Adding the gate admittance of the next device, the imaginary components cancel, resulting in the gate admittance at $(n-1)$th device equal to $2G_g$. Then, a shunt susceptive element B_g is placed at this device so that line length θ_{g2} transforms the resulting admittance $Y_{i2} = 2G_g + jB_g$ from the second device to its conjugate value $Y_{i2}{}^* = 2G_g - jB_g$ at next device. This process continues all the way to the gate of the first device where the admittance $Y_{in} = nG_g$ can be matched to a given source impedance using a quarterwave transformer. Similarly, but only in the reverse direction, the drain-line length θ_{d1} transforms the admittance $Y_{o1} = Y_d + jB_1$, where the shunt susceptance jB_1 is connected to the device drain, from the first device to its conjugate value $Y_{o1}{}^* = G_d - j(B_d + B_1)$ at the location of the second device. Then, the drain-line length θ_{d2} transforms the resulting admittance

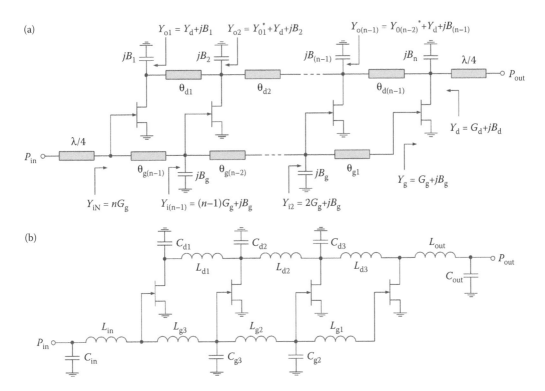

FIGURE 9.24
Schematics of MESFET distributed amplifiers using extended resonance technique.

$Y_{o2} = 2G_d + j(B_2 - B_1)$ at the second device to $Y_{o3} = 2G_d + j(B_2 - B_2 + B_1)$ at the third device. This continues to the drain of the nth device, where the resulting susceptance is canceled by jB_n. Consequently, an input signal applied to the gate of the first device will be divided equally among all devices. Then, each device amplifies $1/n$ of the input power and delivers it to the output combining circuit where the power from each device is recombined at the load. Assuming lossless transmission lines, it can be shown that the total output power is equal to n times the power generated by each device.

It should be noted that, due to the resonant nature of the dividing and combining circuits, their 3-dB frequency bandwidth is limited to about 5%. However, to maximize the bandwidth performance, the optimized low-pass ladder-type networks can be used in the dividing and combining circuits. Besides, the broadband matching circuit is required at the input to match low-impedance presented by the transistors at the gate to the 50-Ω source. As a result, the measured output power around 32 dBm with 1-dB flatness and a *PAE* of 20%–40% from 4 to 9 GHz were achieved for a hybrid four-device distributed extended-resonance amplifier using AlGaAs/InGaAs pHEMT transistors [65]. To increase an overall output power capability, a 40-device distributed multicell multistage amplifier based on extended resonance technique was designed where each multicell includes four 0.25-μm AlGaAs/InGaAs pHEMT devices with a 600-μm gate width and a total gate width of the MMIC of 24 mm, resulting in a small-signal gain of 15 dB and an output power of around 32 dBm at 1-dB compression point within the frequency bandwidths of 25–30 and 28–34 GHz [66].

9.5 Cascaded Distributed Amplifiers

High gain over wide frequency range can be achieved by cascading several stages of a single-stage amplifier, thus composing an artificial active transmission line, which includes active-device parameters. Figure 9.25 shows the schematic representation of a four-cascaded single-stage distributed amplifier (4-CSSDA), where each stage is based on equal-size transistors and equal-value characteristic impedance of all cascades. A feature of this arrangement is that the need to maintain the characteristic impedance Z_{0int} of the intermediate stages at the standard impedance of 50 Ω is eliminated. An increase in this intermediate impedance results in a higher available gain since the gate voltage at each intermediate stage is correspondingly increased.

The available gain for an *n*-cascaded single-stage distributed amplifier (CSSDA) can be calculated to be

$$G_{CSSDA} = \frac{g_m^{2n} Z_{0int}^{2(n-1)} Z_{0g} Z_{0d}}{4} \tag{9.47}$$

where g_m is the device transconductance, Z_{0g} is the characteristic impedance of the input gate line, and Z_{0d} is the characteristic impedance of the output drain line [67]. An RF signal from a matched generator will be coupled by the transconductance of the active device at each stage, and finally terminated by the matched output load port. The amplified signal is valid only up to the cutoff frequency, which is controlled by the gate circuits. Unlike the conventional distributed amplifiers, it is only necessary to equalize the characteristic impedances of the input gate and output drain ports of the active device involved at each stage, and this can be done by adding an extra capacitance with the output capacitance C_{ds} to equalize with the input capacitance C_{gs}. In this case, the CSSDA is characterized by two main features when the input and output stages are the only stages to match with the 50-Ω source and load, respectively, and the intermediate characteristic impedance Z_{0int} can be optimized to boost the overall available gain according to Equation 9.47.

In order for the CSSDA to produce available gain equal to or higher than the forward available gain for an ideal lossless *n*-stage conventional distributed amplifier (CDA) derived from Equation 9.1 as

$$G_{CDA} = \frac{n^2 g_m^2 Z_{0g} Z_{0d}}{4} \tag{9.48}$$

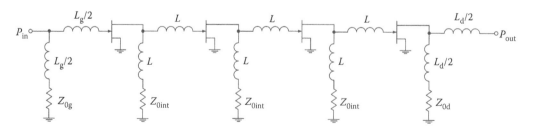

FIGURE 9.25
Schematic of four-cascaded single-stage distributed amplifier.

from Equations 9.47 and 9.48, it follows that

$$g_m Z_{0int} > \sqrt[(n-1)]{n} \qquad (9.49)$$

which shows that the required interstage characteristic impedance Z_{0int} decreases for the same devices as the stage number n increases [68]. Hence, a cascade of four single-stage FETs with intermediate characteristic impedance of 86.6 Ω and device transconductance $g_m = 28$ mS should yield a 50-Ω matched distributed amplifier configuration with a 20-dB available gain over wide frequency band, compared to only 9-dB available gain for a conventional distributed amplifier according to Equation 9.48. In a practical case of the 4-CSSDA using low-noise MESFET devices with $g_m = 55$ mS, a power gain of 39 ± 2 dB over a frequency bandwidth of 0.8–10.8 GHz was measured [68].

Owing to the typical second-order low-pass filter configurations, the bandwidth of the 2-CSSDA is band limited compared with the CDA. As the number of stages of the amplifier increases, the low-frequency gain also increases so it is not easy to design a flat gain performance for a multistage CSSDA. Therefore, to provide wider bandwidth with high-gain performance, the broadband distributed amplifier can combine the CDA and CSSDA with different number of stages [69]. As a result, the forward available gain of the distributed amplifier combining the n-stage CDA and n-stage CSSDA is given by

$$G = \frac{n^2 g_m^2 Z_{0g} Z_{0d}}{4} \frac{g_m^{2n} Z_{0int}^{2(n-1)} Z_{0g} Z_{0d}}{4} \qquad (9.50)$$

Figure 9.26a shows the circuit schematic of a monolithic two-stage CDA cascaded with a single-stage CSSDA using a 0.15-μm pHEMT technology, which provides a small-signal gain of 19 ± 1 dB over the frequency range of 0.5–27 GHz. To extend the amplifier gain-bandwidth performance for millimeter-wave applications, a monolithic seven-stage CDA cascaded with a two-stage CSSDA using the same technology with a die size of 1.5×2 mm^2 was designed, as shown in Figure 9.26b, achieving a small-signal gain of 22 ± 1 dB over the frequency range of 0.1–40 GHz with a total dc consumption of 484 mW [69]. The group delay of 30 ± 10 ps is sufficiently flat over whole bandwidth, which is very important for digital optical communications.

The gain response and efficiency of the CSSDA is increased if the intermediate impedance $Z_{0int} = R_{var} + j\omega L_{var}$ at the drain terminal of each active device is included, as shown in Figure 9.27 for a lumped three-stage cascaded reactively terminated single-stage distributed amplifier (CRTSSDA) [70]. Although the bandwidth of this amplifier is also limited by the gate and drain inductances L, it can be substantially improved by the inclusion of the inductance L_{var} and resistance R_{var}. The effect of the reactive termination is to enhance the voltage swing across the input gate–source capacitance C_{gs} of each active device, which results in an increased output drain current from each device. This consequently improves the amplifier overall gain performance over the multioctave bandwidth. In this case, to provide a flat gain response over the desired bandwidth, it is simply necessary to adjust the impedance Z_{0int}, because the effect of the inductance L_{var} is negligible at lower frequencies (in the range of 10 kHz to 1 GHz) when the intermediate impedance can be written as $Z_{0int} = R_{var}$. The selection of the bias components L_{bias} and C_{bias} also plays a critical role in optimizing the bandwidth and must have minimum intrinsic parasitics.

The initial value of the resistance R_{var} can be calculated from Equation 9.49, which is dependent on the device transconductance g_m and the number of stages n constituting

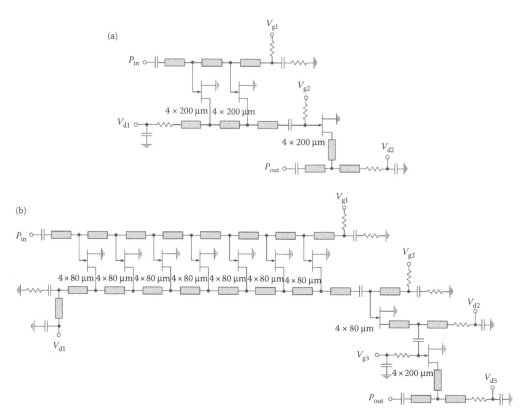

FIGURE 9.26
Schematics of cascaded conventional and single-stage distributed amplifiers.

the CRTSSDA. However, the calculated value of R_{var} will have to be optimized in order to achieve the required small-signal response. The inductive component L_{var} will have an effect on the small-signal gain at higher frequencies (over 2 GHz). The primary effect of this component is to alter the magnitude of the small-signal level at the input port of the respective device of the CRTSSDA chain to be amplified by its device transconductance. The initial value of the inductance can be calculated from

$$L_{var} \geq \frac{\sqrt[(n-1)]{n}}{g_m \omega} \tag{9.51}$$

The fabricated three-stage CRTSSDA based on a 0.25-μm double pHEMT technology with a gate periphery of 360 μm for each transistor with a self-biased mode of operation (gates are directly grounded through the inductances L_{bias}) providing a dc current of 120 mA achieved a gain of 26 ± 1.5 dB, the input and output return loss of better than 9.6 dB (*VSWR* of better than 2:1), and a *PAE* of greater than 12.6% across the frequency bandwidth of 2–18 GHz [71]. The output power of greater than 24.5 dBm with a *PAE* of greater than 27% across 2–18 GHz was achieved when a pHEMT device with a gate width of 720 μm was used in a final stage and a pHEMT device with a gate width of 200 μm was used in a first stage [72].

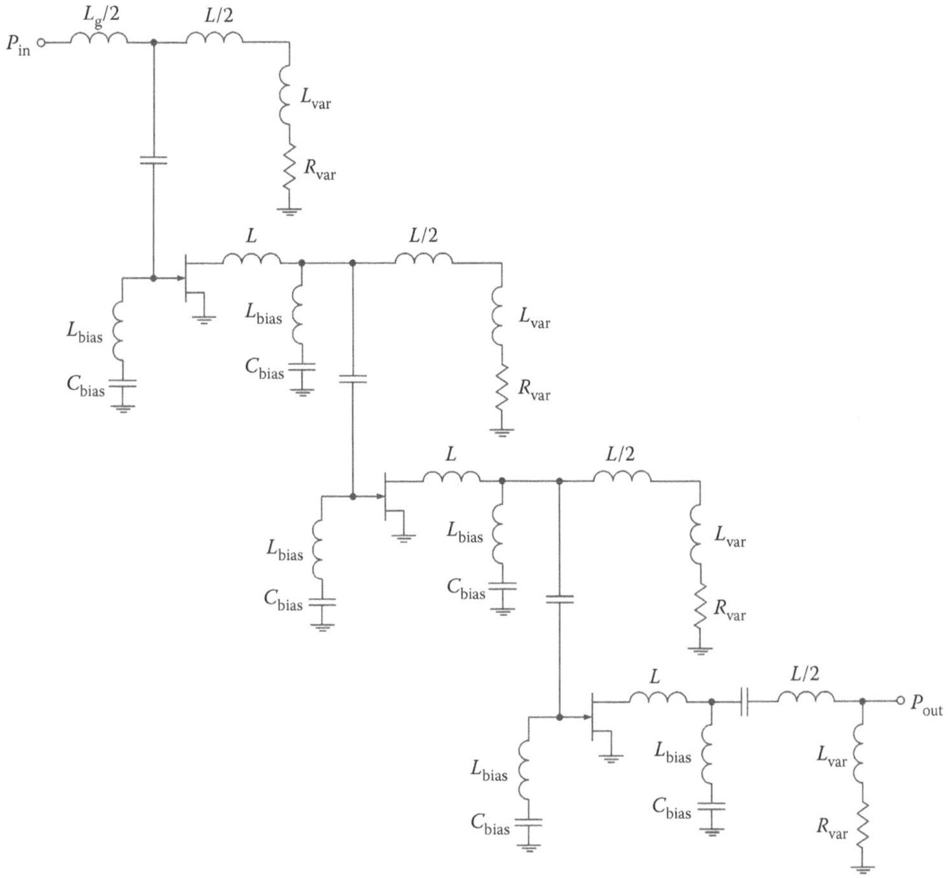

FIGURE 9.27
Schematic of three-cascaded reactively terminated single-stage distributed amplifier.

9.6 Matrix Distributed Amplifiers

The concept of the matrix amplifier combines the processes of additive and multiplicative amplification in one and the same module. Its purpose, therefore, is to combine the characteristic features of both principles, namely, to increase the gain of the additive amplifier concept and the bandwidth of the multiplicative amplifier concept. This can be accomplished in a module whose size is significantly reduced when compared with the traditional amplifier types of similar gain and bandwidth performance. In its most general form, the matrix amplifier consists of an array of m rows and n columns of active devices. Each column is linked to the next by inductors or transmission-line elements connected at the input and output terminals of each transistor, composing a lattice of circuit elements. For m active tiers, there are $2m$ idle ports that are terminated into power-dissipating loads. By adding the vertical dimension to the horizontal dimension of the distributed amplifier in the form of the $n \times m$ rectangular array, the multiplicative and additive process in one and the same module is achieved. The advantages of the matrix amplifier include significantly higher gain and reverse isolation over wide bandwidths at considerably reduced size.

Figure 9.28a shows the circuit schematic of a distributed matrix MESFET amplifier in the form of a 2×4 array representing a six-port flanked by the input and output four-ports [73,74]. The active six-port incorporates the transistor characterized their set of Y-parameters, the network of transmission-line elements represented by their respective characteristic impedances and electrical lengths, and the open-circuit shunt stubs capacitively loading the drain line. In contrast, the input and output four-ports contain only passive circuit elements, that is, the terminations of the amplifier idle networks and a simple input and output matching network. Each idle port is terminated into either a resistor or an impedance consisting of a resistor shunted by a short transmission line that allows

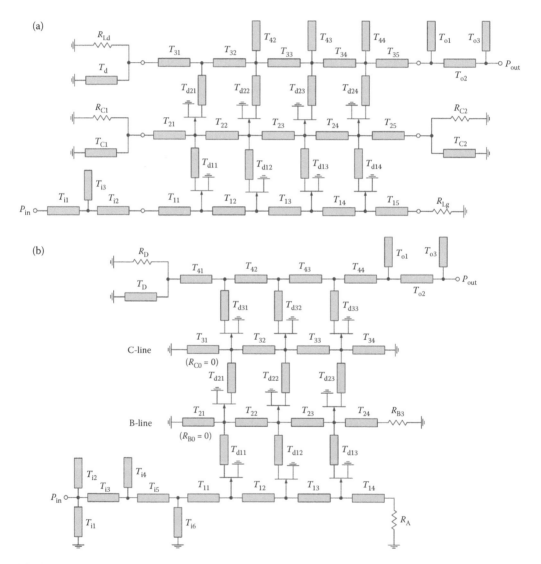

FIGURE 9.28
Schematics of matrix distributed amplifiers.

biasing of the active devices without any power dissipation in the termination resistors. The choice for the termination elements is critical for gain flatness, noise figure, gain slope, and operation stability. Based on a rigorous solution for voltages and currents involving GaAs MESFETs with 0.25×200-μm gate dimensions, the 2×4 matrix amplifier was fabricated with an overall size of 0.5×0.24 inches using a 10-mil-thick quartz substrate, achieving a large-signal gain of 11.6 ± 1.5 dB from 2 to 21 GHz with an output power of 100 mW [73]. In addition, it should be noted that the matrix amplifier can offer a most desirable compromise between its broadband maximum noise figure on the one hand and its gain and *VSWR* performance on the other. As a result, a computer-optimized two-tier (2×4) GaAs MESFET matrix amplifier could provide a noise figure $F = 3.5 \pm 0.7$ dB with an associated gain of 17.8 ± 1.6 dB across the frequency band of 2–18 GHz [75]. The monolithic 2×3 matrix amplifier using a 0.2-μm pseudomorphic InGaAs HEMT technology achieved a 20-dB gain and a 5.5-dB noise figure over the frequency band of 6–21 GHz [76].

The circuit schematic of a 3×3 matrix MESFET amplifier is shown in Figure 9.28b, where the left port of the artificial transmission line B and both ports of the artificial transmission line C are terminated into short circuits [77]. The input and output matching circuits are necessary to improve the reflection coefficients of the amplifier, and biasing of the active devices is easily provided through the short-circuited idle ports. The theoretical analysis of the amplifier circuit shows that a low-frequency gain of the matrix distributed amplifier with three tiers ($m = 3$) can be estimated by

$$G_{3\times n} \cong \left[\frac{2(ng_m)^3 Y_0}{(G_A + Y_0)(G_{B0} + G_{Bn} + nG_{ds})(G_{C0} + G_{Cn} + nG_{ds})(G_{D0} + Y_0 + nG_{ds})} \right]^2 \quad (9.52)$$

where $G_{ds} = 1/R_{ds}$ is the device drain–source conductance, $Y_0 = 1/Z_0$ is the characteristic admittance of the artificial transmission lines, and n is the number of MESFETs per tier. If one terminal of the artificial transmission line is short-circuited, the stability of the amplifier can only be maintained if the other terminal of the same line is terminated into a finite impedance or a short. By using GaAs MESFETs with 0.35×200-μm gate dimensions and termination resistors $R_A = 29$ Ω, $R_{B3} = 49$ Ω, and $R_D = 212$ Ω ($R_{B0} = R_{C0} = R_{C3} = 0$), a noise figure of 5.2 ± 1.2 dB and a gain of 27.7 ± 0.9 dB were achieved across the frequency band of 6–18 GHz. The low-frequency gain of the matrix distributed amplifier with two tiers ($m = 2$) can be calculated from

$$G_{2\times n} \cong \frac{1}{2} g_{m1} g_{m2} \; nZ_0 \; \frac{Z_{0c} R_{ds1}}{nZ_{0c} + 2R_{ds1}} \; \frac{\sinh(b)\exp(-b)}{\sinh(b/n)} \quad (9.53)$$

where Z_{0c} is the characteristic impedance of the central line and $b = (n/4)(Z_0/R_{ds2})$ [78].

The integration on the same chip of the active devices based on different technologies can merge the advantages inherent to these technologies. For example, using the HEMT and HBT devices simultaneously on a single chip allows high-gain performance in a multioctave frequency band together with low dc-power consumption and noise figure. Figure 9.29a shows the simplified circuit schematic of a 2×4 HEMT-HBT matrix amplifier, where the first tier consists of HEMTs and the second tier is replaced by HBT devices [79]. Here, the higher input capacitance of each HBT is absorbed in the central line, whose characteristic impedance can be different from 50 Ω without degradation of the input and output matching. As a result, a flat gain of 18 dB, which is only by 1 dB less than that for the HEMT matrix amplifier and 2 dB higher compared to the HBT matrix amplifier, a noise

(a)

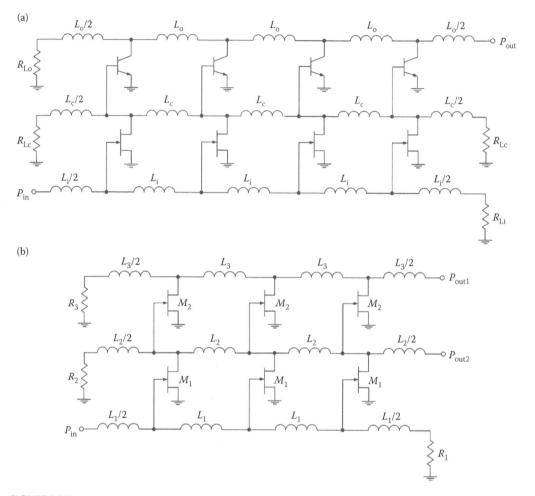

(b)

FIGURE 9.29
Schematics of lumped matrix amplifier and active balun.

figure of 5–6 dB, which is close to that for the HBT matrix amplifier and more than 2 dB better than the noise figure of the HEMT matrix amplifier, and more than 40% reduction in the dc-power consumption compared to the HEMT matrix amplifier were achieved across the frequency band up to 30 GHz.

The matrix balun that is based on the matrix amplifier concept can provide a decade bandwidth and a high gain, while having a small size, compared to the conventional active and passive baluns. Figure 9.29b shows the simplified circuit schematic of a 2×3 HEMT matrix balun, where the phase balance is achieved by utilizing the fact that the phase difference between two rows in a matrix amplifier with common-source transistors is 180° [80]. The analytical expression for the common-mode rejection ratio ($CMMR$) for this matrix balun with finite output conductances for zero normalized frequency Ω is defined as

$$CMRR = \left| \frac{S_{21} - S_{31}}{S_{21} + S_{31}} \right| = \left| \frac{3g_{m2}Z_0R_3 + R_3 + ((3R_3/R_{ds2}) + 1)Z_0}{3g_{m2}Z_0R_3 - R_3 - ((3R_3/R_{ds2}) + 1)Z_0} \right| \qquad (9.54)$$

where Z_0 is the characteristic impedance of the artificial transmission lines, R_3 is the idle-port termination resistance of the output line, and g_{m2} is the transconductance and R_{ds2} is the drain–source resistance of the transistors connected to the output line. As a result, a 2×3 matrix balun implemented in a 0.15-μm GaAs mHEMT technology with a chip size of 0.9×1.1 mm^2 ($R_1 = R_2 = 39$ Ω and $R_3 = 29$ Ω) achieved more than a decade bandwidth of 4–42 GHz with a *CMRR* higher than 15 dB, a gain of 2 ± 1 dB, and a maximum phase imbalance of $20°$ with a power consumption of 20 mW. The same matrix balun circuit may also be biased for amplification and used as a matrix amplifier. In this case, the circuit exhibited a 10.5-dB gain up to 63 GHz with a 1-dB ripple above 5.5 GHz and a power consumption of 67 W.

9.7 CMOS Distributed Amplifiers

Unlike the semi-insulating GaAs process that provides higher-quality lumped inductors and transmission lines, a CMOS-based implementation is advantageous in that it results in a lower cost and a higher level of integration. One of the first designs of a four-cell CMOS distributed amplifier was based on a 0.6-μm CMOS process with a three-layer Al-metal interconnect when a flat gain of 6.5 ± 1.2 dB over a bandwidth from 500 MHz to 4 GHz with approximately linear phase over the passband was achieved [81]. Figure 9.30a shows the basic circuit schematic of a four-cell lumped CMOS distributed amplifier. In this case, if the gate- and drain-line inductors are matched, and the drain capacitance is made equal to the gate capacitance for each transistor, then the input and output currents are phase synchronized. Another modification to the basic circuit relates to the proper gate- and drain-line termination. The impedance seen looking into the *LC* artificial transmission lines will exhibit a strong deviation from the nominal impedance near the cutoff frequency of the lines. Ideally, all four ports would be image-impedance matched to the lines to eliminate reflections. However, it is not practical to realize an image-impedance matching directly. Thus, the method used will be to insert the *m*-derived half-sections between the lines and the input port, output port, and terminations. These half-sections will greatly improve the impedance matching, while also allowing simple resistive terminations to be used. The modified circuit of a lumped four-cell CMOS distributed amplifier with matching sections is shown in Figure 9.30b [81]. Based on a 0.18-μm SiGe BiCMOS technology with six-layer metal interconnects and final tow thick-copper layers to realize low-loss inductors and using only nMOS transistors, a frequency bandwidth was extended from 500 MHz to 22 GHz with the flat gain of 7 ± 0.7 dB and input return loss better than 10 dB over most of the bandwidth [82].

Silicon-on-insulator (SOI) CMOS technologies can provide high-gain performance at millimeter-wave frequencies required for low-power broadband microwave and optical systems. For example, a 0.12-μm SOI CMOS process offers a low-parasitic nMOS transistor with a peak f_T in excess of 150 GHz for gate length smaller than 60 nm. Here, since the integration of the low-loss 50-Ω microstrip lines is difficult, the coplanar waveguide (CPW) structures were implemented on the last 1.2-μm-thick metal layers to reduce the parasitic capacitances to the substrate [83]. As a result, a gain of 4 ± 1.2 dB over the bandwidth of 4–91 GHz and an 18-GHz output 1-dB compression point of 10 dBm were measured for the cascode five-cell distributed amplifier with a power consumption of 90 mW. The cascode three-cell distributed amplifier implemented in a 45-nm SOI CMOS process with a peak f_T in excess of 230 GHz, whose value strongly depends on the layout parasitics and may

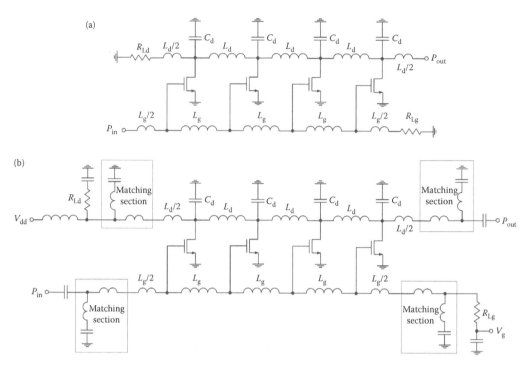

FIGURE 9.30
Schematics of lumped four-cell CMOS distributed amplifiers.

reach 380 GHz, achieved a 3-dB bandwidth of 92 GHz and a peak gain of 9 dB with a gain ripple of 1.5 dB and an input return loss better than 10 dB [84]. It should be noted that the noise behavior of a distributed amplifier over an entire frequency band depends significantly on the number of cells n. For example, the best low-frequency noise performance is achieved for larger values of n, whereas the lowest noise figures are reached at high frequencies for smaller values of n [85]. This noise behavior is attributed to the fact that the drain noise is inversely proportional to n, whereas the gate noise is proportional to n. Besides, for the same number of cells, a cascode CMOS distributed amplifier demonstrates better noise performance over most of the frequency bandwidth, especially at higher frequencies, compared to the conventional CMOS distributed amplifier with the transistors in a common-source configuration.

By employing a nonuniform architecture for the artificial input and output transmission lines, the CMOS distributed amplifier exhibits enhanced performance in terms of gain and bandwidth. Figure 9.31 shows the circuit schematic of a cascode nonuniform seven-cell distributed amplifier using a standard 0.18-μm CMOS technology where the transistor sizes and inductance values of the center cell is 2.5 times as large as those of the other cells [86]. In this design, the parameters of the common-source transistors are designed in consideration of the cutoff frequency of the input line and the transconductance of the gain stages. On the other hand, the common-gate transistors are designed to provide an output capacitance equivalent to the input capacitance of the gain stages such that matched input and output lines can be utilized to optimize the phase response of the distributed amplifier. The values of the inductive elements L_m between cascode transistors need to be adjusted for maximum bandwidth extension and better noise performance [87]. Finally, the high-impedance CPW structures were employed to realize the required gate and drain

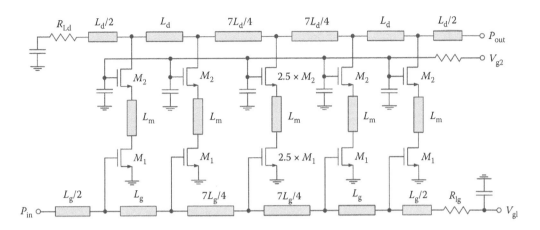

FIGURE 9.31
Schematic of cascode nonuniform seven-cell CMOS distributed amplifier.

inductances, resulting in a passband gain of 9.5 dB and a 3-dB bandwidth of 32 GHz for this distributed amplifier.

The main drawback of an integrated CMOS implementation of the single-ended common-source amplifiers including system-on-chip solutions is that parasitic interconnects, bondwires, and package inductors degenerately degrade their gain-bandwidth performance. Specifically, for a packaged single-ended distributed amplifier that exhibits a unity-gain-bandwidth of 4 GHz, there is a bandwidth degradation of 27% compared to its unpackaged performance [81]. Figure 9.32 shows the circuit schematic of a fully differential

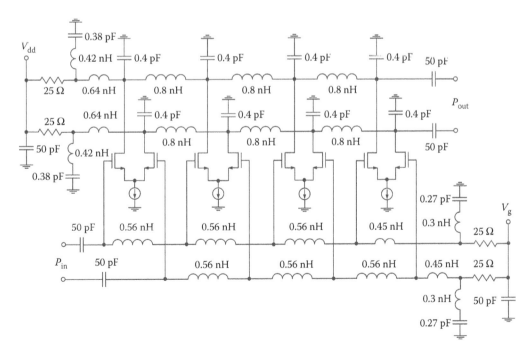

FIGURE 9.32
Circuit schematic of fully differential four-cell CMOS distributed amplifier.

four-cell CMOS distributed amplifier with the ideal passive components [88]. The characteristic impedances of the gate and drain lines are designed to be 25 Ω each to provide a load impedance of 50 Ω for fully differential signals. The highest achievable line cutoff frequency for the 0.6-μm CMOS process is about 10 GHz and is limited by the smallest practical values of the circuit inductances and capacitances. Since high-quality tail current sources are essential to achieving high common-mode and power-supply rejection ratios, a regulated cascode current source is employed in each stage. Such architecture also minimizes undesirable signal coupling between stages through the common substrate. To improve the impedance matching between the termination resistors and the artificial transmission line over a wide range of frequencies, two pairs of *m*-derived half-sections are used, and the drain and date bias voltages are supplied through the termination ports. As a result, the measured results for this fully differential CMOS distributed amplifier with transistor gate widths of 400 μm demonstrated a bandwidth of 1.5–7.5 GHz, which is about 50% greater than for a single-ended counterpart, but obtained at the expense of increased power consumption, die size, and noise figure.

Use of the bisected-*T* *m*-derived filter sections at the input and output of the distributed amplifiers, as shown in Figure 9.33a for $n = 4$, is widely used to improve matching and gain flatness near the amplifier cutoff frequency. In this case, a bisected-*T*-type *m*-section matches a *T*-type *k*-section (or cascade of them) on the one side and matches the constant real impedance on the other. Thus, the bisected-*T* *m*-section can couple power from a real source into a cascade of *T*-type *k*-sections over the full frequency range from dc to cutoff frequency. Similarly, the bisected-π *m*-section can couple power from a real source into a cascade of π-type *k*-sections over the same frequency range [89]. Here, the shunt 0.3*C* capacitance in the matching section is connected in parallel with the adjacent capacitance from the first *k*-section. This can be added together into a single transistor or gain cell that is 80% of the size of a full gain cell, thus resulting in a higher voltage gain by

$$\frac{A_\pi}{A_\mathrm{T}} = 1 + \frac{0.6}{n} \tag{9.55}$$

than its *T*-type equivalent. The second factor 0.6/*n* comes from extra transistor area in the matching sections at both the beginning and end of the artificial transmission line. As it follows from Equation 9.55, the gain boost is higher for smaller *n*, but gives appreciable improvement over the typical range of *n*, specifically of 1 dB for $n = 5$. Besides, since inductors are generally lossy and difficult to accurately model at microwave frequencies, the π-type topology reduces both the number and the size of inductors. Figure 9.33b shows the circuit schematic of a cascode five-cell CMOS distributed amplifier with bisected π-type *m*-sections. In a practical implementation, an overall area reduction of 17% (excluding pads) was achieved for a π-type topology [89]. To extend the flat bandwidth and improve the input matching of a cascode distributed amplifier, the gate artificial transmission line based on coupled inductors in conjunction with series-peaking inductors in cascode gain stages can be used [90].

The cascaded CMOS distributed structure is an alternative configuration to exhibit simultaneous high gain and wide operating bandwidth. When compared with the matrix CMOS amplifier topology, which has the same low-frequency gain characteristic, the cascaded structure offers robustness to high-frequency mismatches in the signal delays along the individual paths, thus compromising the overall gain [91]. It also has no loading effect at interstage artificial transmission lines, yielding a larger operating bandwidth, and the

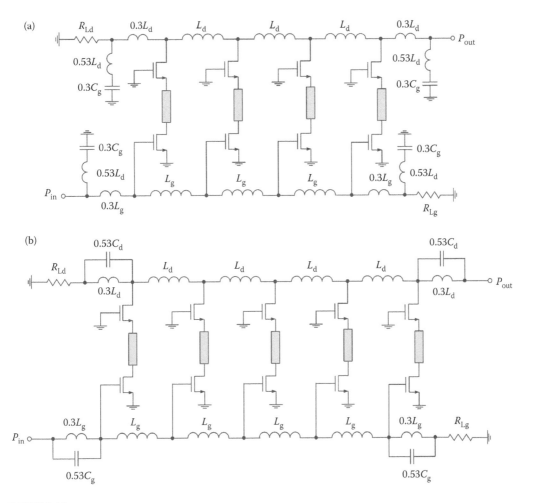

FIGURE 9.33
Schematics of CMOS distributed amplifier with bisected *T*-type and π-type *m*-sections.

omission of the idle drain terminations at intermediate cascaded stages or interstages results in a significant gain improvement. The basic structure of a generalized cascaded CMOS distributed amplifier includes the artificial transmission lines in the form of the constant-*k LC* network with simultaneous match to both *T*- and π-sections [92]. It consists of *m* cascading stages of *n*-cell distributed amplifiers with matched idle drain terminations. To achieve a gain improvement, the idle terminations are omitted, except for the output *m*th stage, so that the current and voltage waves along the interstage drain/gate lines are enhanced, and hence the total gain is increased. In this case, the tapered structure offers an additional bandwidth improvement as compared to the case of the uniform-drain cascaded distributed amplifier, with only a small sensitivity to impedance ratio variation. Figure 9.34 shows the circuit schematic of a cascaded tapered-drain double-stage cascode CMOS distributed amplifier along with the component parameters [92]. The size of the amplifier transistors M_1 and M_2 was selected at $W/L = 100\ \mu m/0.18\ \mu m$, whereas that of the cascoded transistors M_{C1} and M_{C2} was selected at a twice smaller gate periphery as $W/L = 50\ \mu m/0.18\ \mu m$, so that the drain/gate capacitance ratio was equal to 0.25 for $n = 2$.

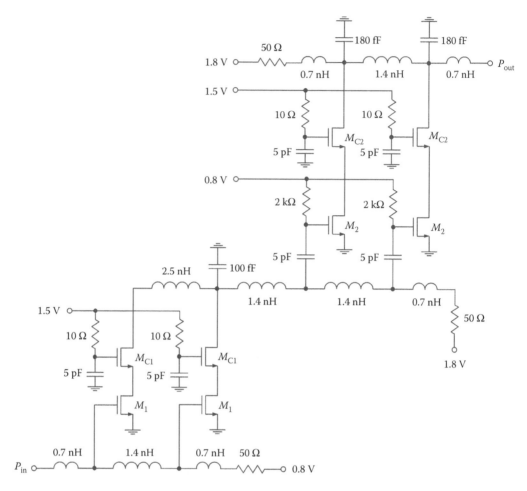

FIGURE 9.34
Schematic of cascaded double-cell cascode CMOS distributed amplifier.

The measured results demonstrated an output driving capability of −2.5 dBm and a gain of 14 dB with a noise figure of 5.5–7.5 dB over the bandwidth from 1.0 to 13.8 GHz.

9.8 Noise in Distributed Amplifiers

All four sources of noise of basic and avoidable nature that need to be considered in any high-frequency amplifier such as thermal noise in the input impedance, shot-effect noise due to the random emission of electrons from the cathode, and induced grid (or gate) noise, which is associated with transit-time effects at high frequencies, can also be taken into account when designing a distributed amplifier structure [2]. Since the gate line can be terminated with resistances on each end, both of them act as generators of thermal noise. The noise due to the gate termination produces a noise wave on the gate line, which is amplified by the active devices, and then the noise signals are added in the drain line in a way, which depends upon the phase shift per section. Note that the thermal noise due to the gate termination is usually small compared with the noise due to the input impedance.

The shot-effect noise source can be represented by a resistor in the gate circuit, which is assigned a value such that this virtual resistance generates as much noise as is actually observed in the drain circuit. At low frequencies and in narrow-band amplifiers, the input impedance can be made high and, consequently, the shot-effect noise can be made to be negligible, which is generally not the case for wideband amplifiers, including the distributed amplifier. However, in the case of the distributed amplifier, the shot-effect noise can be made negligibly small despite the fact that the gate-to-ground impedance is not high when compared to the equivalent noise resistance. Since the noise voltages caused by the noise currents to appear on the drain line are added in a random manner, the total noise power is proportional to the number of active devices. However, the signal voltage at the output terminals is proportional to the number of active devices, and the signal power is proportional to the number of active devices squared. Hence, the signal-to-noise ratio will be proportional to n, where n is the number of sections. Thus, by using a sufficient number of sections, it is possible to make the signal as large as one desires compared to the shot-effect noise.

The behavior of the high-frequency noise due to transit-time effects in the output of the distributed amplifier associated with induced gate noise is very complicated. On the one hand, the magnitude of the noise is a rapid function of frequency (the noise power per cycle is approximately proportional to frequency squared). On the other hand, each active device generates noise voltages that propagate in both directions from the device. Thus, noise generated by one active device is amplified by all the other active devices. Moreover, this amplification depends upon the particular position of the active device in the distributed amplifier. Thus, it can be seen that the noise power in the output due to gate loading effects is proportional to n^3, whereas the signal voltage is proportional to n^2 [2]. Hence, the increasing number of sections decreases the signal-to-noise ratio. As a result, the noise figure of the distributed amplifier, which is the sum of these three noises, depends on competing factors $1/n$ and n, and there should be an optimum value of n for minimum total noise.

For a noise analysis of a distributed amplifier, it is convenient to divide the entire structure into functional blocks such as input matching circuit, the elementary amplifiers, and, if necessary, the output matching circuit. The MESFET noise sources are characterized by the voltage and current at the input terminal of the transistor. The input and output links of the drain and gate lines may consist of either transmission line or lumped-circuit elements. All circuit elements as they are typically employed in a distributed amplifier can be represented by the four-port network. Figure 9.35 shows the four-port network to calculate noise figure of a distributed amplifier, where the idle ports are terminated with the admittances Y_{dp} on the drain and Y_{gp} on the gate side, and Y_0 is the load admittance [93]. Since all four terminations contain a finite conductance, they inject thermal noise, which in the case of Y_{gp} and Y_{dp} contributes to the noise power at the output terminal of a distributed amplifier. Furthermore, those components generated by the MESFETs have a strong influence on the amplifier noise behavior and may be represented by voltage sources at each of the four terminals. This is accomplished by transforming the transistor individual noise sources to the terminals of the four-port network and thereby making the four-port network itself free of noise, as shown in Figure 9.35 for a distributed amplifier with noise input power method where all noise sources are transformed to the amplifier input.

For a low-frequency model when the transforming characteristics of the linking elements may be neglected and the amplifier can be treated as a lossy match amplifier with

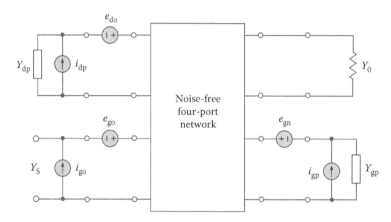

FIGURE 9.35
Four-port network to calculate noise figure of distributed amplifier.

n numbers of parallel transistors, the approximate minimum noise figure of an n-cell distributed amplifier with source admittance $Y_S = G_S + jB_S$ can be written as

$$F_{min} \equiv 1 + 2\Big[R_n((G_{gp}/n) + \text{Re}\,C)$$
$$+ \sqrt{R_n((G_{gp}/n) + G_n) + R_n^2((G_{gp}/n) + \text{Re}\,C)^2}\,\Big] \tag{9.56}$$

where R_n is the equivalent noise resistance, G_n is the equivalent noise conductance, and $C = \text{Re}\,C + j\text{Im}\,C$ is the correlation admittance [93]. From Equation 9.56, it follows that the minimum noise figure of a distributed amplifier at low frequencies may be reduced by increasing the number of sections. In this case, at low frequencies, the amplifier noise resistance R_n is a result of paralleling the noise resistances of individual transistors, while the noise conductance G_n is mainly dependent on the resistance $R_{gp} = 1/G_{gp}$ of the gate-line termination. Similarly, $\text{Re}\,C$ depends almost entirely on R_{gp} at low frequencies.

In order to calculate noise figure of a distributed amplifier through the active device parameters, consider the equivalent circuit of a MOSFET (or MESFET) device with noise sources, as shown in Figure 9.36 [94]. Generally, the intrinsic noise sources of a distributed amplifier can be identified as noise from gate source impedance R_S, noise from the gate load R_{Lg}, noise source from the left-hand drain load R_{Ld}, and noise associated with each of the n FETs. The FET noise behavior can be represented by a gate current generator placed in parallel to the gate–source capacitance and a drain current generator placed in parallel to the drain–source capacitance with their mean square values given by

$$\overline{i_{ng}^2} = \frac{4kT\omega^2 C_{gs}^2\, \delta \Delta f}{g_m} \tag{9.57}$$

$$\overline{i_{nd}^2} = 4kT g_m \gamma \Delta f \tag{9.58}$$

where δ and γ are the gate and drain noise coefficients depending upon the implementation technology and biasing conditions. In this case, noise from the nth-cell gate current

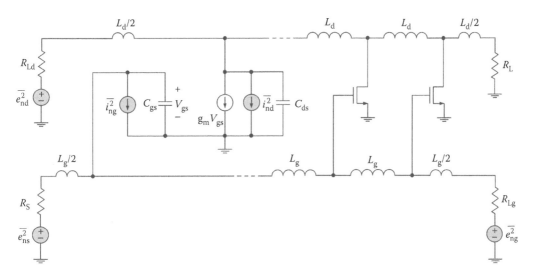

FIGURE 9.36
Schematic of MOSFET distributed amplifier with noise sources.

generator is dissipated in the right-hand drain load by both forward amplification in the succeeding cells and by reverse amplification occurring in earlier stages.

Assuming that the source resistance R_S is matched to the gate-termination resistance R_{Lg} ($R_S = R_{Lg}$) and the load resistance R_L is matched to the drain termination resistance R_{Ld} ($R_L = R_{Ld}$), the noise factor F of a n-cell distributed amplifier can be approximated by

$$F = 1 + \frac{4\gamma}{ng_mR_S} + \frac{\delta\omega^2 C_{gs}^2 nR_S}{3g_m} \tag{9.59}$$

where the second term describes the drain noise, which is dominated at low frequencies, whereas the third term represents the frequency-dependent gate noise determining the high-frequency performance [85,95]. Typically, values of $2/3 < \gamma < 1$ and $\delta = 4/3$ are used for long-channel MOSFET devices. However, owing to hot electron effects, significantly higher drain noise currents and coefficients γ are expected for short-channel devices.

There is an optimum value of n_{opt}, which minimizes noise figure F in Equation 9.59 to give a minimum noise figure F_{min} of

$$F_{min} = 1 + \frac{2\omega C_{gs}}{g_m}\sqrt{\frac{4\gamma\delta}{3}} \tag{9.60}$$

which can be derived for a number of cells of

$$n_{opt} = \frac{2}{\omega C_{gs}R_S}\sqrt{\frac{3\gamma}{\delta}} \tag{9.61}$$

From the noise analysis of a MESFET distributed preamplifier, it was concluded that the high-gain matching impedance and appropriate scaling of the MESFET gate width improve the noise performance [96]. However, the increase in gate-line characteristic impedance results in a reduction in the frequency bandwidth of the preamplifier. Similarly, the equivalent input noise current density of a common-collector-cascode HBT distributed preamplifier can be reduced by increasing the base-line characteristic impedance, which in turn leads to a lower cutoff frequency of the input distributed structure, thus limiting the frequency bandwidth [97]. In this case, the noise contribution of the base termination resistance dominates the amplifier equivalent input noise current density at low-frequency, while the gain cell noise contribution dominates at high frequency. Since the noise contribution of the collector termination resistance is relatively small, in order to improve high-frequency noise performance, the noise source of the gain cell should be reduced. This can be achieved by reducing the bias current of the HBTs.

References

1. W. S. Percival, Improvements in and relating to thermionic valve circuits, British Patent 460,562, January 1937.
2. E. L. Ginzton, W. R. Hewlett, J. H. Jasberg, and J. D. Noe, Distributed amplification, *Proc. IRE*, 36, 956–969, 1948.
3. W. H. Horton, J. H. Jasberg, and J. D. Noe, Distributed amplifiers: Practical considerations and experimental results, *Proc. IRE*, 38, 748–753, 1950.
4. H. G. Bassett and L. C. Kelly, Distributed amplifiers: Some new methods for controlling gain/frequency and transient responses of amplifiers having moderate bandwidths, *Proc. IEE, Part III: Radio Commun. Eng.*, 101(69), 5–14, 1954.
5. W. K. Chen, Distributed amplifiers: Survey of the effects of lumped-transmission-line design on performance, *Proc. IEE*, 114, 1065–1074, 1967.
6. W. K. Chen, Distributed amplification: A new approach, *IEEE Trans. Electron Devices*, ED-14, 215–221, 1967.
7. D. G. Sarma, On distributed amplification, *Proc. IEE, Part B: Radio Electron. Eng.*, 102, 689–697, 1955.
8. F. C. Thompson, Broad-band UHF distributed amplifiers using band-pass filter techniques, *IRE Trans. Circuit Theory*, CT-7, 8–17, 1960.
9. W. K. Chen, The effects of grid loading on the gain and phase-shift characteristics of a distributed amplifier, *IEEE Trans. Circuit Theory*, CT-16, 134–137, 1969.
10. W. K. Chen, Theory and design of transistor distributed amplifiers, *IEEE J. Solid-State Circuits*, SC-3, 165–179, 1968.
11. J. A. Gallagher, High-power wide-band RF amplifiers, *IRE Trans. Aerospace Electron. Syst.*, AES-1, 141–151, 1965.
12. L. H. Enloe and P. H. Rogers, Wideband transistor distributed amplifiers, *1959 IEEE Int. Solid-State Circuits Symp. Dig.*, Philadelphia, PA, pp. 44–45, 1959.
13. G. W. McIver, A travelling-wave transistor, *Proc. IEEE*, 53, 1747–1748, 1965.
14. G. Kohn and R. W. Landauer, Distributed field-effect amplifiers, *Proc. IEEE*, 56, 1136–1137, 1968.
15. W. Jutzi, A MESFET distributed amplifier with 2 GHz bandwidth, *Proc. IEEE*, 57, 1195–1196, 1969.
16. J. A. Archer, F. A. Petz, and H. P. Weidlich, GaAs FET distributed amplifier, *Electron. Lett.*, 17, 433, 1981.
17. Y. Ayasli, J. L. Vorhaus, R. L. Mozzi, and L. D. Reynolds, Monolithic GaAs travelling-wave amplifier, *Electron. Lett.*, 17, 413–414, 1981.
18. Y. Ayasli, R. L. Mozzi, J. L. Vorhaus, L. D. Reynolds, and R. A. Pucel, A monolithic GaAs 1–13 GHz traveling-wave amplifier, *IEEE Trans. Microwave Theory Tech.*, MTT-30, 976–981, 1982.

19. K. B. Niclas, W. T. Wilser, T. R. Kritzer, and R. R. Pereira, On theory and performance of solid-state microwave distributed amplifiers, *IEEE Trans. Microwave Theory Tech.*, MTT-31, 447–456, 1983.
20. S. G. Bandy, C. K. Nishimoto, C. Yuen, R. A. Larue, M. Day, J. Eckstein, Z. C. H. Tan et al., A 2–20 GHz high-gain monolithic HEMT distributed amplifiers, *IEEE Trans. Microwave Theory Tech.*, MTT-35, 1494–1500, 1987.
21. E. W. Strid and K. R. Gleeson, A DC-12 GHz monolithic GaAs FET distributed amplifier, *IEEE Trans. Microwave Theory Tech.*, MTT-30, 969–975, 1982.
22. J. B. Beyer, S. N. Prasad, R. C. Becker, J. E. Nordman, and G. K. Hohenwarter, MESFET distributed amplifier design guidelines, *IEEE Trans. Microwave Theory Tech.*, MTT-32, 268–275, 1984.
23. B. Kim and H. Q. Tserng, 0.5 W 2–21 GHz monolithic GaAs distributed amplifier, *Electron. Lett.*, 20, 288–289, 1984.
24. Y. Ayasli, S. W. Miller, R. L. Mozzi, and L. K. Hanes, Capacitively coupled traveling-wave power amplifier, *IEEE Trans. Microwave Theory Tech.*, MTT-32, 1704–1709, 1984.
25. S. N. Prasad, J. B. Beyer, and I. K. Chang, Power-bandwidth considerations in the design of MESFET distributed amplifiers, *IEEE Trans. Microwave Theory Tech.*, MTT-36, 1117–1123, 1988.
26. M. J. Schindler, J. P. Wendler, M. P. Zaitlin, M. E. Miller, and J. R. Dorman, A K/Ka-band distributed power amplifier with capacitive drain coupling, *IEEE Trans. Microwave Theory Tech.*, MTT-36, 1902–1907, 1988.
27. C. Z. Den Brinker and M. Parkyn, Amplifiers, British Patent 1,235,472, June 1971.
28. K. B. Niclas, R. D. Remba, R. R. Pereira, and B. D. Cantos, The declining drain line lenghts circuit—A computer derived design concept applied to a 2–28.5-GHz distributed amplifier, *IEEE Trans. Microwave Theory Tech.*, MTT-34, 427–435, 1986.
29. C. Duperrier, M. Campovecchio, L. Roussel, M. Lajugie, and R. Quere, New design method of uniform and nonuniform distributed power amplifiers, *IEEE Trans. Microwave Theory Tech.*, MTT-49, 2494–2500, 2001.
30. J. Gassmann, P. Watson, L. Kehias, and G. Henry, Wideband, high-efficiency GaN power amplifiers utilizing a non-uniform distributed topology, *2007 IEEE MTT-S Int. Microwave Symp. Dig.*, Honolulu, Hawaii, pp. 615–618, 2007.
31. B. M. Green, S. Lee, K. Chu, K. J. Webb, and L. F. Eastman, High efficiency monolithic gallium nitride distributed amplifier, *IEEE Microwave Guided Wave Lett.*, 10, 270–272, 2000.
32. C. Campbell, C. Lee, V. Williams, M. Y. Kao, H. Q. Tserng, P. Saunier, and T. Balisteri, A wideband power amplifier MMIC utilizing GaN on SiC HEMT technology, *IEEE J. Solid-State Circuits*, SC-44, 2640–2647, 2009.
33. E. Reese, D. Allen, C. Lee, and T. Nguyen, Wideband power amplifier MMICs utilizing GaN on SiC, *2010 IEEE MTT-S Int. Microwave Symp. Dig.*, Anaheim, CA, pp. 1230–1233, 2010.
34. C. Xie and J. Pavio, A high efficiency broadband monolithic gallium nitride distributed power amplifier, *2008 IEEE MTT-S Int. Microwave Symp. Dig.*, Atlanta, GA, pp. 307–310, 2008.
35. L. Zhao, A. Pavio, and W. Thompson, A 1 watt, 3.2 VDC, high efficiency distributed power PHEMT amplifier fabricated using LTCC technology, *2003 IEEE MTT-S Int. Microwave Symp. Dig.*, Philadelphia, PA, vol. 3, pp. 2201–2204, 2003.
36. L. Zhao, A. Pavio, B. Stengel, and B. Thompson, A 6 watt LDMOS broadband high efficiency distributed power amplifier fabricated using LTCC technology, *2002 IEEE MTT-S Int. Microwave Symp. Dig.*, Seattle, WA, pp. 897–900, 2002.
37. S. Lin, M. Eron, and A. E. Fathy, Development of ultra wideband, high efficiency, distributed power amplifiers using discrete GaN HEMTs, *IET Circuits Devices Syst.*, 3, 135–142, 2009.
38. D. E. Meharry, R. J. Lender, Jr., K. Chu, L. L. Gunter, and K. E. Beech, Multi-watt wideband MMICs in GaN and GaAs, *2007 IEEE MTT-S Int. Microwave Symp. Dig.*, Honolulu, Hawaii, pp. 631–634, 2007.
39. Y. Ayasli, L. D. Reynolds, R. L. Mozzi, and L. K. Hanes, 2–20-GHz GaAs traveling-wave power amplifier, *IEEE Trans. Microwave Theory Tech.*, MTT-32, 290–295, 1984.
40. S. Masuda, A. Akasegawa, T. Ohki, K. Makiyama, N. Okamoto, K. Imanishi, T. Kikkawa et al., Over 10 W C-Ku band GaN MMIC nonuniform distributed power amplifier with broadband couplers, *2010 IEEE MTT-S Int. Microwave Symp. Dig.*, Anaheim, CA, pp. 1388–1391, 2010.

41. P. N. Shastry and J. B. Beyer, Bandpass distributed amplifiers, *Microwave Optical Technol. Lett.,* 2, 349–354, 1989.

42. P. N. Shastry, A Kajjam, and Z. M. Li, Bandpass distributed amplifier design guidelines, *Microwave Opt. Technol. Lett.,* 10, 215–218, 1995.

43. N. P. Mehta and P. N. Shastry, Design guidelines for a novel bandpass distributed amplifier, *Proc. 35th Eur. Microwave Conf. Dig.,* Paris, vol. 1, pp. 1–4, 2005.

44. J. M. Rollett, Stability and power-gain invariants of linear twoports, *IRE Trans. Circuit Theory,* CT-9, 29–32, 1962.

45. M. Riaziat, S. Bandy, L. Y. Ching, and G. Li, Feedback in distributed amplifiers, *IEEE Trans. Microwave Theory Tech.,* MTT-38, 212–215, 1990.

46. P. Gamand, Analysis of the oscillation conditions distributed amplifiers, *IEEE Trans. Microwave Theory Tech.,* MTT-37, 637–640, 1989.

47. H. T. Kim, M. S. Jeon, K. W. Chung, and Y. Kwon, 6–18 GHz MMIC drive and power amplifiers, *J. Semiconductor Technology and Science,* 2, 125–131, 2002.

48. C. Yuen, Y. C. Pao, and N. G. Bechtel, 5–60-GHz high-gain distributed amplifier utilizing InP cascode HEMT's, *IEEE J. Solid-State Circuits,* SC-27, 1434–1438, 1992.

49. B. M. Green, V. Tilak, S. Lee, H. Kim, J. A. Smart, K. J. Webb, J. R. Shealy et al., High-power broad-band AlGaN/GaN HEMT MMICs on SiC substrates, *IEEE Trans. Microwave Theory Tech.,* MTT-49, 2486–2493, 2001.

50. K. W. Kobayashi, Y. C. Chen, I. Smorchkova, B. Heying, W. B. Luo, W. Sutton, M. Wojtowicz et al., Multi-decade GaN HEMT cascode-distributed power amplifier with baseband performance, *2009 IEEE RFIC Symp. Dig.,* Boston, MA, pp. 369–372, 2009.

51. R. S. Pengelly, S. M. Wood, J. W. Milligan, S. T. Sheppard, and W. L. Pribble, A review of GaN on SiC high electron mobility power transistors and MMICs, *IEEE Trans. Microwave Theory Tech.,* MTT-60, 1764–1783, 2012.

52. W. Kennan, T. Andrade, and C. C. Huang, A 2–18-GHz monolithic distributed amplifier using dual-gate GaAs FET's, *IEEE Trans. Microwave Theory Tech.,* MTT-32, 1693–1697, 1984.

53. R. Santhakumar, Y. Pei, U. K. Mishra, and R. A. York, Monolithic millimeter-wave distributed amplifiers using AlGaN/GaN HEMTs, *2008 IEEE MTT-S Int. Microwave Symp. Dig.,* Atlanta, GA, pp. 1063–1066, 2008.

54. R. Santhakumar, B. Thibeault, M. Higashiwaki, S. Keller, Z. Chen, U. K. Mishra, and R. A. York, Two-stage high-gain high-power distributed amplifier using dual-gate GaN HEMTs, *IEEE Trans. Microwave Theory Tech.,* MTT-59, 2059–2063, 2011.

55. K. W. Kobayashi, J. Cowles, L. Tran, T. Block, A. K. Oki, and D. C. Streit, A 2–32 GHz coplanar waveguide InAlAs/InGaAs-InP HBT cascode wave distributed amplifier, *1995 IEEE MTT-S Int. Microwave Symp. Dig.,* Orlando, FL, pp. 215–218, 1995.

56. J. P. Fraysse, J. P. Vlaud, M. Campovecchio, P. Auxemery, and R. Quere, A 2 W, high efficiency, 2–8 GHz, cascode HBT MMIC power distributed amplifier, *2000 IEEE MTT-S Int. Microwave Symp. Dig.,* Boston, MA, pp. 529–532, 2000.

57. J. Aguirre and C. Plett, 50-GHz SiGe HBT distributed amplifiers employing constant-*k* and *m*-derived filter sections, *IEEE Trans. Microwave Theory Tech.,* MTT-52, 1573–1579, 2004.

58. Y. Chen, J. B. Beyer, V. Sokolov, and J. P. Culp, A 11 GHz hybrid paraphase amplifier, *1986 IEEE Int. Solid-State Circuits Conf. Dig.,* Anaheim, CA, pp. 236–237, 1986.

59. S. Mohammadi, J. W. Park, D. Pavlidis, J. L. Guyaux, and J. C. Garcia, Design optimization and characterization of high-gain GaInP/GaAs HBT distributed amplifiers for high-bit-rate telecommunication, *IEEE Trans. Microwave Theory Tech.,* MTT-48, 1038–1044, 2000.

60. J. Chen and A. M. Niknejad, Design and analysis of a stage-scaled distributed power amplifier, *IEEE Trans. Microwave Theory Tech.,* MTT-59, 1274–1283, 2011.

61. H. Y. Chang, Y. C. Liu, S. H. Weng, C. H. Lin, Y. L. Yeh, and Y. C. Wang, Design and analysis of a DC–43.5-GHz fully integrated distributed amplifier using GaAs HEMT-HBT cascode gain stages, *IEEE Trans. Microwave Theory Tech.,* MTT-59, 443–455, 2011.

62. A. Martin, A. Mortazawi, and B. C. De Loach, A power amplifier based on an extended resonance technique, *IEEE Microwave Guided Wave Lett.,* 5, 329–331, 1995.

63. A. Martin, A. Mortazawi, and B. C. De Loach, An eight-device extended-resonance power-combining amplifier, *IEEE Trans. Microwave Theory Tech.*, MTT-46, 844–850, 1998.

64. A. Martin and A. Mortazawi, A new lumped-elements power-combining amplifier based on an extended resonance technique, *IEEE Trans. Microwave Theory Tech.*, MTT-48, 1505–1515, 2000.

65. X. Jiang and A. Mortazawi, A broadband power amplifier design based on the extended resonance power combining technique, *2005 IEEE MTT-S Int. Microwave Symp. Dig.*, Munich, Germany, pp. 835–838, 2005.

66. R. Lohrman, H. Gill, and S. Koch, A novel distributed multicell multistage amplifier structure, *Proc. 33rd Eur. Microwave Conf.*, Munich, Germany, pp. 379–382, 2003.

67. J. Y. Liang and C. S. Aitchison, Gain performance of cascade of single-stage distributed amplifier, *Electron. Lett.*, 31, 1260–1261, 1995.

68. B. Y. Banyamin and M. Berwick, Analysis of the performance of four-cascaded single-stage distributed amplifiers, *IEEE Trans. Microwave Theory Tech.*, MTT-48, 2657–2663, 2000.

69. K. L. Deng, T. W. Huang, and H. Wang, Design and analysis of novel high-gain and broadband GaAs pHEMT MMIC distributed amplifiers with travelling-wave gain stages, *IEEE Trans. Microwave Theory Tech.*, MTT-51, 2188–2196, 2003.

70. A. S. Virdee and B. S. Virdee, 2–18 GHz ultra-broadband amplifier design using a cascaded reactively terminated single stage distributed concept, *Electron. Lett.*, 35, 2122–2123, 1999.

71. A. S. Virdee and B. S. Virdee, Experimental performance of ultra-broadband amplifier design concept employing cascaded reactively terminated single-stage distributed amplifier configuration, *Electron. Lett.*, 36, 1554–1556, 2000.

72. A. S. Virdee and B. S. Virdee, A novel high efficiency multioctave amplifier using cascaded reactively terminated single-stage distributed amplifier configuration, *2001 IEEE MTT-S Int. Microwave Symp. Dig.*, Phoenix, AZ, vol. 1, pp. 519–522, 2001.

73. K. B. Niclas and R. R. Pereira, The matrix amplifier: A high-gain module for multioctave frequency bands, *IEEE Trans. Microwave Theory Tech.*, MTT-35, 296–306, 1987.

74. K. B. Niclas, R. R. Pereira, and A. P. Chang, On power distribution in additive amplifiers, *IEEE Trans. Microwave Theory Tech.*, MTT-38, 1692–1700, 1990.

75. K. B. Niclas, R. R. Pereira, and A. P. Chang, A 2–18 GHz low-noise/high-gain amplifier module, *IEEE Trans. Microwave Theory Tech.*, MTT-37, 198–207, 1989.

76. K. W. Kobayashi, R. Esfandiari, W. L. Jones, K. Minot, B. R. Allen, A. Freudenthal, and D. C. Streit, A 6–21-GHz monolithic HEMT 2×3 matrix distributed amplifier, *IEEE Microwave Guided Wave Lett.*, 3, 11–13, 1993.

77. K. B. Niclas and R. R. Pereira, On the design and performance of a 6–18 GHz three-tier matrix amplifier, *IEEE Trans. Microwave Theory Tech.*, MTT-37, 1069–1077, 1989.

78. C. Paoloni and S. D'Agostino, A design procedure for monolithic matrix amplifier, *IEEE Trans. Microwave Theory Tech.*, MTT-45, 135–139, 1997.

79. C. Paoloni, HEMT–HBT matrix amplifier, *IEEE Trans. Microwave Theory Tech.*, MTT-48, 1308–1312, 2000.

80. M. Ferndahl and H. O. Vickes, The matrix balun—A transistor-based module for broadband applications, *IEEE Trans. Microwave Theory Tech.*, MTT-57, 53–60, 2009.

81. B. M. Ballweber, R. Gupta, and D. J. Allstot, A fully integrated 0.5–5.5-GHz CMOS distributed amplifier, *IEEE J. Solid-State Circuits*, SC-35, 231–239, 2000.

82. G. A. Lee, H. Ko, and F. De Flaviis, Advanced design of broadband distributed amplifier using a SiGe BiCMOS technology, *2003 IEEE RFIC Symp. Dig.*, Philadelphia, PA, pp. 703–706, 2003.

83. J. O. Plouchart, J. Kim, N. Zamdmer, L. H. Lu, M. Sherony, Y. Tan, R. A. Groves et al., A 4–91-GHz travelling-wave amplifier in a standard 0.12-μm SOI CMOS microprocessor technology, *IEEE J. Solid-State Circuits*, SC-39, 1455–1461, 2004.

84. J. Kim and J. F. Buckwalter, A 92 GHz bandwidth distributed amplifier in a 45 nm SOI CMOS technology, *IEEE Microwave Wireless Comp. Lett.*, 21, 329–331, 2011.

85. F. Ellinger, 60-GHz SOI CMOS traveling-wave amplifier with NF below 3.8 dB from 0.1 to 40 GHz, *IEEE J. Solid-State Circuits*, SC-40, 553–558, 2005.

86. L. H. Lu, T. Y. Chen, and Y. J. Lin, A 32-GHz non-uniform distributed amplifier in a 0.18-μm CMOS, *IEEE Microwave Wireless Comp. Lett.*, 15, 745–747, 2005.

87. P. Heydari, Design and analysis of a performance-optimized CMOS UWB distributed LNA, *IEEE Trans. Solid-State Circuits*, SC-42, 1892–1905, 2007.

88. H. T. Ahn and D. J. Allstot, A 0.5–8.5-GHz fully differential CMOS distributed amplifier, *IEEE J. Solid-State Circuits*, SC-37, 985–993, 2002.

89. A. Kopa and A. B. Apsel, Alternative m-derived termination for distributed amplifiers, *2009 IEEE MTT-S Int. Microwave Symp. Dig.*, Boston, MA, pp. 921–924, 2009.

90. K. Entesari, A. R. Tavakoli, and A. Helmy, CMOS distributed amplifiers with extended flat bandwidth and improved input matching using gate line with coupled inductors, *IEEE Trans. Microwave Theory Tech.*, MTT-57, 2862–2871, 2009.

91. J. C. Chien and L. H. Lu, 40-Gb/s high-gain distributed amplifiers with cascaded gain stages in 0.18-μm CMOS, *IEEE J. Solid-State Circuits*, SC-42, 2715–2725, 2007.

92. A. Worapishet, I. Roopkom, and W. Surakampontorn, Theory and bandwidth enhancement of cascaded double-stage distributed amplifiers, *IEEE Trans. Circuits Systems I: Regular Papers*, CAS-57, 759–772, 2010.

93. K. B. Niclas and B. A. Tucker, On noise in distributed amplifiers at microwave frequencies, *IEEE Trans. Microwave Theory Tech.*, MTT-31, 661–668, 1983.

94. F. Zhang and P. R. Kinget, Low-power programmable gain CMOS distributed LNA, *IEEE J. Solid-State Circuits*, SC-41, 1333–1343, 2006.

95. C. S. Aitchison, The intrinsic noise figure of the MESFET distributed amplifier, *IEEE Trans. Microwave Theory Tech.*, MTT-33, 460–466, 1985.

96. A. P. Freundorfer and T. L. Nguyen, Noise in distributed MESFET preamplifiers, *IEEE J. Solid-State Circuits*, SC-31, 1100–1111, 1996.

97. X. Tien, A. P. Freundorfer, and L. Roy, Noise analysis of a photoreceiver using a P-I-N and GaAs HBT distributed amplifier combination, *IEEE Microwave Wireless Comp. Lett.*, 13, 208–210, 2003.

10

CMOS Amplifiers for UWB Applications

UWB transmission technology is very attractive for its low-cost and low-power communication applications, occupying a very wide frequency range, which was first proposed for communication systems yet in the 1940s [1]. Since then, it was mainly used for radar-based applications because of the wideband nature of the signal that results in very accurate-timing information. By the early 1970s, the system concept and basic components such as pulse train generators and modulators, detection receivers, and wideband antennas were available [2]. However, due to further development in high-speed switching and narrow-band pulse generation technology, UWB has become more attractive for low-cost communication applications, now representing any wireless transmission scheme that occupies a transmission frequency bandwidth of more than 20% of a center frequency, or more than 500 MHz [3]. Such large bandwidths are achieved by using very narrow time-duration baseband pulses of an appropriate shape and duration, including the family of Gaussian-shaped pulses and their derivatives. Larger-transmission bandwidths are preferred to achieve higher data rates without the need to increase transmitting power, resulting in the ability for increasingly fine resolution for multipath arrivals, which leads to reduced fading per resolved path since the impulsive nature of the transmitted waveforms prevents significant overlap and, hence, reduces the possibility of destructive combining. Market considerations require that UWB-based products be implemented in CMOSs to achieve low-power and low-cost integration.

10.1 UWB Transceiver Architectures

A key advantage of UWB designs is that highly linear power amplifiers are generally not required because the UWB pulse generator needs to only produce a peak-to-peak voltage swing on the order of 100 mV to meet spectral mask requirements that can be achieved by a suitable UWB waveform choice. The transmitting signal in a UWB system can be generally modulated by turning (keying) the pulse on and off (OOK), by providing the binary phase-shift keying (BPSK), or by dithering the pulse position (PPM), as well as the pulse amplitude modulation (PAM) can be used [4]. The pulse may have duration on the order of 200 ps and its shape can be designed to concentrate energy over the broad frequency range, for example, from 2 to 6 GHz. In this case, it is necessary to use a bandpass filter before the antenna to constrain the emissions within this desired frequency band when the filter would have a bandwidth on the order of 4 GHz. UWB systems can also be designed using spread-spectrum codes, which offer better coexistence with other UWB systems. The generation of spread-spectrum UWB signals can be achieved using time hopping or direct sequence methods. In multiband UWB systems, when the UWB band from 3.1 to 10.6 GHz is divided into 14 subbands, each with a bandwidth of 528 MHz, signal transmissions are staggered in time across the constituent subbands, and one of the several (typically 3–10) subbands are used sequentially for transmission. The requirement for maximum total power consumption of a UWB transceiver set by specification at 110 and 200 Mb/s is 100 and 250 mW, respectively.

A variety of UWB systems can be designed to use a 7.5-GHz-available UWB spectrum depending on the bandwidth and number of available bands, such as a WPAN, a wireless sensor network (WSN), or a wireless body area network (WBAN). If the system needs to provide 200 Mb/s, a single-band approach to instantaneously use the entire 7.5-GHz band can be used with a total signal power of 16 dBm [5]. For a multiband approach employing 15 500-MHz bands with a signal rate per band of 13.4 Mb/s, a total bandwidth signal power is 4 dBm. In this case, the transmit signal of a single-band UWB system is characterized by a much higher bandwidth than the multiband UWB system and requires very fast switching circuits. The multiband UWB system, on the other hand, requires a signal generator capable to quickly switch between frequencies. At the same time, multiband systems permit adaptive selection of the bands to provide good interference robustness and coexistence properties.

Generally, low-rate and high-rate-pulsed UWB transmitters are based on one of the two basic techniques to synthesize pulses in the 3.1–10.6-GHz band [6]. The upconversion pulse (or direct conversion) synthesis technique involves generating a pulse at the baseband and upconverting it into center frequency in the UWB band by mixing with a local oscillator (LO), whose output can be either enabled or disabled by a simple switch, thus effectively mixing the RF signal with a rectangular baseband pulse. The carrier-less pulse synthesis involves generating pulses that directly fall in the UWB band without requiring frequency translation where the pulse width is usually defined by delay elements that may be tunable or fixed. In this case, a baseband impulse may excite a filter that shapes the pulse or the pulse may be directly synthesized at RF with no additional filtering required. An advantage of carrier-less techniques over traditional mixed-based architectures is that the carrier frequency generation is inherently duty cycled when RF power is only generated when it is required. However, its main disadvantage is that an integrated downconverting receiver typically cannot share the RF generation circuits, and therefore must have its own LO. Pulse-based transmitters can additionally be categorized in terms of how pulses are delivered to an antenna, either by analog power amplification or digital buffering. However, in the latter case, linear amplitude modulation and pulse shaping are more difficult to achieve. The direct-conversion UWB transceiver is usually characterized by the large LO leakage to the RF front end and to the antenna because the LO is operated at the same frequency as the RF carrier. Therefore, a single-balanced mixer with LO cancellation can be used in a direct-conversion architecture or a dual-conversion zero-IF transceiver architecture with two-step upconversion can be used as a suitable alternative [7,8].

Figure 10.1a shows the generic UWB pulse-based transmitter architecture with a pulse generator followed by a modulator, both driven by a digital baseband. The modulator is then directly followed by a power amplifier to achieve the required power level, a bandpass filter to reduce spurious emission, and a wideband antenna to radiate the pulsed signal. A UWB pulse can be generated by multiplying a sinusoidal signal with an envelope close to zero-order Gaussian shape with a standard deviation of 341 ps and centered in 4.1 GHz to spread energy over a 2-GHz bandwidth [9]. The UWB transmitter based on a 0.18-μm CMOS technology can achieve the highest pulse repetition rate of 750 Mb/s with bi-phase-modulated pulses of 500-ps duration and 8-GHz center frequency, operating over the band of 6–10 GHz and complying with the standard spectrum mask [10].

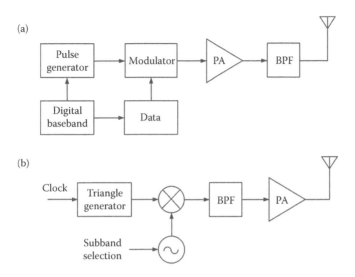

FIGURE 10.1
UWB transmitter architectures.

The UWB transmitters, the typical block schematic of which is shown in Figure 10.1b, are intended to operate in the 3.1–10.6-GHz band with 100 Mb/s data rate at a link distance of 30 ft using BPSK-modulated pulses at a maximum pulse repetition frequency of 100 MHz [4]. The transmitter implemented in a 0.18-μm SiGe BiCMOS process uses the direct-conversion architecture to upconvert a shaped baseband pulse train into one of the 14 subbands in the UWB band by modulating the bias current of the differential pair with a carrier frequency. The baseband UWB signal is generated using either a programmable arbitrary waveform generator (AWG) or a dedicated discrete pulse generator. Using an AWG enables a large amount of flexibility in the shape of the transmitted pulses, modulation scheme, and duration of transmission. The carrier frequency can be generated either by on-chip voltage-controlled oscillator (VCO) or from external LO. In the latter case, with a 0.13-μm CMOS technology, the frequency of the external differential input signal from 6 to 22 GHz is divided by the on-chip 2:1 frequency divider to obtain quadrature LO signals from 3 to 11 GHz to cover 14 bands of the UWB system at the same time [11]. With the LO input power of 0 dBm, the quadrature upconversion mixer and balanced power amplifier provide the conversion gain of 1.1 ± 1 dB and power gain of 12.4 ± 2 dB over the frequency range from 3 to 11 GHz, respectively.

In a typical UWB receiver, the received signal is first amplified by an LNA and correlated with the expected signal, and then sampled by the analog-to-digital converter (ADC) before it is demodulated [5]. The reference phase-locked loop (PLL) provides the required sequence to the correlator to detect the signals. The LNA and correlator usually operate across the whole-frequency spectrum from 3.1 to 10.6 GHz with a 30-dB dynamic range that guarantees operation in the required range from 1 to 30 ft. The reference PLL provides both transmit signals to the transmit driver and receive reference signals. It generates the sequence of short signals and is capable of fast switching between different center frequencies. Since the PLL for short UWB signals does not require low-phase noise, there are several alternatives to implement a reference PLL with these characteristics at low power.

The power amplifier in UWB systems should meet several stringent requirements simultaneously such as broadband matching, high gain, sufficient output power, and reasonable

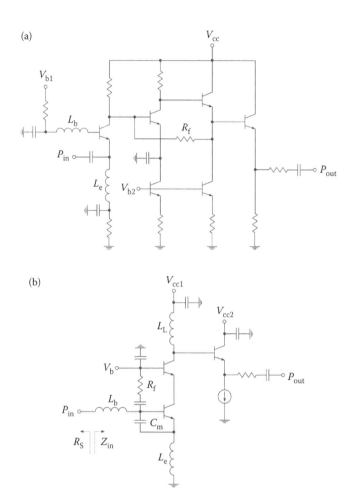

FIGURE 10.2
Schematics of HBT LNAs for UWB applications.

efficiency for low-power operation. Figure 10.2a shows the example of a simple broadband amplifier topology that enables an improved noise figure while keeping the power consumption relatively low using a 0.25-μm SiGe BiCMOS technology [12]. Here, to simplify the input-matching network and improve noise performance, an input stage is designed in an inductor-terminated common-base configuration, which is capable of good noise-matching performance over the UWB frequency ranges up to 10.6 GHz. The second stage represents a resistive-feedback amplifier, consisting of a common-emitter transistor, a feedback resistor R_f, two emitter followers, and two tail current sources. While the resistive-feedback topology is inherently suitable for broadband applications, it is characterized by a relatively large noise level due to the corresponding contribution of the feedback resistor. The role of the common-base input stage, with a gain of about 13 dB, is to reduce the noise contribution of the resistive-feedback amplifier, having a gain of 10 dB, and lower the noise figure at the LNA input. Besides, there is an optimum value of the inductor L_b in terms of balancing the noise figure and input impedance. As a result, while consuming only 13.2 mW, this SiGe HBT LNA obtained a flat gain characteristic within 1.1-dB variations around 22 dB and noise figure values ranging from 2.7 to 3.9 dB across the frequency range from 3.1 to 10.6 GHz.

In a conventional narrowband cascode LNA with inductive degeneration, the impedance looking into the input of the LNA can be written as

$$Z_{in}(\omega) = \frac{g_m}{C_\pi} L_e + j\omega \left(L_e + L_b - \frac{1}{\omega^2 C_\pi} \right) \qquad (10.1)$$

where L_e is the emitter degeneration inductance, L_b is the base series inductance, C_π is the device base–emitter capacitance, and g_m is the device transconductance. The input quality factor $Q_{in} = \omega_0/\Delta\omega$ determining the operating frequency bandwidth $\Delta\omega$ can be expressed through the circuit parameters as

$$Q_{in}(\omega) = \frac{1}{\omega_0 C_\pi \left(\dfrac{g_m L_e}{C_\pi} + R_S \right)} \qquad (10.2)$$

where

$$\omega_0 = \frac{1}{\sqrt{C_\pi (L_b + L_e)}} \qquad (10.3)$$

is the center bandwidth frequency and R_S is the source resistance usually equal to 50 Ω [13]. Consequently, as follows from Equation 10.2, to broaden the bandwidth for fixed ω_0 and R_S, it is possible to properly consider the device parameters C_π and g_m, since the degeneration inductance L_e must be very small to achieve high gain and low noise. In this case, the effective value of C_π can be increased by adding a shunt capacitor C_m between the base and emitter of the input common-emitter transistor, as shown in Figure 10.2b, to decrease the input quality factor Q_{in}. In this case, the transconductance g_m needs to be increased so that the factor $g_m L_e/(C_\pi + C_m)$ is still matched to 50 Ω, according to Equation 10.1, and the shift in the center frequency ω_0 can be eliminated by decreasing L_b, according to Equation 10.3. Since an increased value of the effective base–emitter capacitance degrades the gain and noise performance of the LNA at high frequencies, the value of this extra capacitance C_m can be carefully chosen. Besides, to further improve the LNA broadband performance, a large resistor R_f can be added between the base of the input transistor and the cascode output, as shown in Figure 10.2b. Thus, a broadband input matching can be achieved by using a small shunt capacitance C_m and a large feedback resistance R_f, which provide only a minimal effect on other LNA performance parameters. As a result, by using a 0.18-μm SiGe HBT BiCMOS technology, noise figures of 1.8–3.1 dB with a gain variation of 16.8–20.3 dB and a power consumption of 26 mW (without output buffer) were achieved across the frequency range of 3–10 GHz.

10.2 Distributed CMOS Amplifiers

A CMOS-based implementation of distributed amplifiers used in UWB transceivers has advantages of a low cost and full integration in a CMOS process with the capability to use arbitrary transmission-line impedances. One of the first fully integrated distributed amplifiers with a four-cell-distributed structure implemented in a standard 0.6-μm three-layer CMOS process achieved a power gain of 6.5 dB with a gain flatness of ±1.2 dB over the frequency range of 0.5–4.0 GHz, occupying an area of approximately 0.79 mm² [14]. For a two-stage low-noise-distributed amplifier with an applied inductive source degeneration technique implemented in a 0.18-μm CMOS process, a 10-dB gain with the 3-dB bandwidth, a power consumption of 7.8 mW, and a noise figure varying from 3.8 to 6.9 dB within the frequency band of 2.7–9.1 GHz were measured [15].

Figure 10.3a shows the circuit schematic of a cascode N-section CMOS-distributed amplifier with uniform gate and drain artificial LC transmission lines and identical cascode cells [16]. A cascode configuration of each cell improves the overall amplifier stability by reducing a negative feedback through the gate–drain capacitances. However, the common-gate transistors of each cascode cell contribute to increased noise to the output at

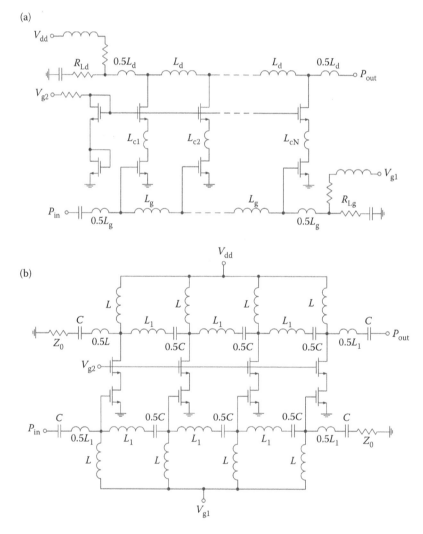

FIGURE 10.3
Schematics of cascode CMOS-distributed amplifiers.

high frequencies, thereby degrading the amplifier noise figure. In this case, to extend the frequency bandwidth to higher frequencies by absorbing the input and output parasitic capacitances and to reduce the noise level at high frequencies, each cascode cell incorporates a series inductor L_{Ck}, where $1 \leq k \leq N$, which are equal for a uniform-distributed structure. As a result, measurements for a three-cell cascode-distributed amplifier using bandwidth-enhancing inductors fabricated in a 0.18-μm SiGe BiCMOS process, where only nMOS transistors were used, had shown a 2.9-dB flat noise figure and a forward gain of 8 dB over the 7.5-GHz UWB bandwidth, with the overall power consumption of 12 mA at a 1.8-V supply voltage. For a low number of cascode sections, an additional improvement of the noise performance can be achieved by the replacement of the gate-terminating resistor as a significant contributor to the amplifier noise figure by an *RL* network with optimized parameters [17].

The combination of high-pass *T*-sections and low-pass π-sections can form a wide bandpass gate and drain artificial *LC* transmission lines, as shown in Figure 10.3b for a four-cell cascode CMOS-distributed amplifier, where $Z_0 = \sqrt{2L/C}$ is the termination characteristic impedance [18]. Here, the shunt inductors can be implemented using on-chip spiral inductors, package traces, or bondwires. For operation frequencies above 3 GHz, decoupling capacitance at each bias termination can be small enough allowing for integration. Such a bandpass-distributed amplifier can provide a flat gain and improved out-of-band suppression. The lower-band cutoff frequency of the amplifier is written as

$$f_{\text{lower}} = \frac{1}{2\pi\sqrt{LC}} \tag{10.4}$$

where the inductance *L* and capacitance *C* are determined based on the specified transmission-line characteristic impedance (usually equal to 50 Ω) for a given lower-band cutoff frequency. The upper-band cutoff frequency can be calculated from

$$f_{\text{upper}} = \frac{1}{2\pi\sqrt{L_1 C_{\text{eff}}}} \tag{10.5}$$

where the effective parallel capacitance C_{eff} includes parasitic capacitances from the inductor and associate transistor. The passband ripple will be minimal as long as the condition

$$\sqrt{\frac{2L}{C}} = \sqrt{\frac{L_1}{C_{\text{eff}}}} \tag{10.6}$$

is maintained, which also guarantees the impedance matching through the passband [19]. As a result, this bandpass cascode CMOS-distributed amplifier with identical transistors of a 144-μm gate width each, which was implemented in a 0.18-μm RF CMOS process, achieved an output power above 5.6 mW at 1-dB gain compression point with a power gain of about 10.5 dB and an input return loss better than 10 dB from 3 to 10 GHz [18].

A low-noise high-gain cascode CMOS-distributed amplifier can be designed by combining an *RL* gate termination with cascaded gain cells. Figure 10.4 shows the circuit schematic of a two-section cascode CMOS-distributed amplifier where high and flat gain is achieved by using the inductive-peaking cascaded gain cells, which can also contribute to an amplifier improved noise figure [20]. In this case, each cell constitutes a cascode stage with a

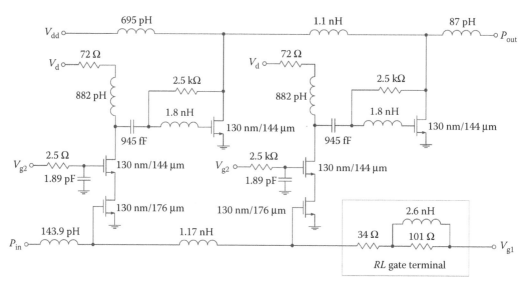

FIGURE 10.4
Circuit schematic of a two-cell high-gain-distributed CMOS amplifier.

low-Q (slightly >0.707) RLC load to extend the interstage bandwidth and a series-peaking inductance at the input of a common-source transistor to extend the upper 3-dB frequency. Instead of the conventional 50-Ω gate terminal, an optimized RL terminal network with a resistor connected in series with the parallel RL circuit is used. In a high-gain mode with the bias voltages $V_{dd} = V_d = 1.5$, $V_{g2} = 0.7$, and $V_{g1} = 0.51$ V, the two-section-distributed amplifier implemented in a 0.18-μm CMOS process consumes 37.8 mW and achieves a gain of 20.47 ± 0.72 dB with an average noise figure of 3.29 dB over the UWB frequency band of 3–10 GHz. With an additional inductive-peaking common-source stage in each amplifier section resulting in a dual inductive-peaking cascaded gain cell, the amplifier forward gain in a high-gain mode can be increased to 24.5 ± 1.5 dB with an average noise figure of 3.9 dB over the same frequency range of 3–10 GHz [21].

Since distributed amplifiers have significant disadvantages of high-power consumption and large dimensions, one of the possible solutions to significantly reduce amplifier area is to use multilayer structures. Figure 10.5a shows the circuit schematic of a three-cell CMOS-distributed amplifier with multilayer inductors implemented on different metal layers in a standard 0.18-μm RF/mixed-signal CMOS process occupying just 0.08 mm² of die area using transistors with a gate width of 136 μm [22]. The circuit exhibits a relatively flat gain of 6 dB from 3.1 to 10.6 GHz with less than 0.5-dB ripple, a noise figure less than 5 dB up to 14 GHz, and a power consumption of about 22 mW. Generally, the noise performance can be improved by using pMOS transistors instead of nMOS transistors because the lower mobility in pMOS transistors provides a lower flicker noise and less hot carrier effects. Since the pMOS transistor is characterized by lower gain due to lower transconductance, the pMOS-based distributed amplifier used as an input stage is followed by the common-source nMOS-amplifying stage as an output stage, as shown in Figure 10.5b, thus achieving a sufficient gain over the required frequency band [23]. As a result, a power gain of 9 ± 0.8 dB with a noise figure of 4.7 ± 0.6 dB was measured from 3.1 to 10.6 GHz with a total power consumption of 22.5 mW at a supply voltage of 1.5 V.

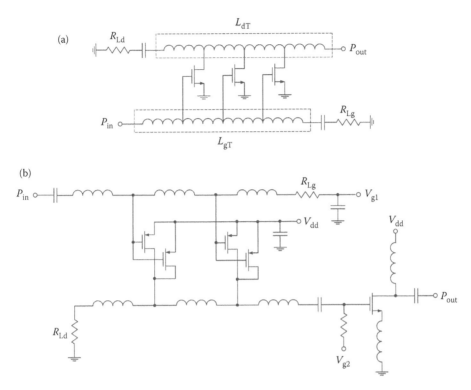

FIGURE 10.5
Schematics of small-size and low-noise CMOS-distributed amplifiers.

10.3 Common-Gate CMOS Amplifiers

Generally, distributed CMOS amplifiers are very popular circuit configurations to achieve broadband input matching and gain. However, their major drawbacks are the large silicon area, resulting from the presence of several on-chip inductors or transmission lines, and high-power consumption, which is related to the number of cells required to enhance the gain resulting in a poor efficiency. The common-gate CMOS amplifier occupies a small area, requires a low-power consumption, and provides a wideband input matching by setting the input-transistor transconductance equal to the reciprocal of the source resistance. In this case, if the common-gate transistor is adopted for a single-ended topology, it requires a wideband high-impedance biasing to avoid significant loss of the RF signal.

Figure 10.6 shows the circuit schematic of a two-stage broadband amplifier implemented in a 0.18-μm CMOS process, which consists of a common-gate input stage, a common-source cascode second stage, and an output buffer configured as a source follower [24]. In this case, the input common-gate stage provides the wideband noise and impedance matching. By neglecting the loading effect of the second stage and parasitic resistance of an input inductor L_s, the input impedance Z_{in} can be written in a simplified form as

$$Z_{in}(\omega) = \frac{j\omega L_s}{1 + j\omega(g_m + j\omega C_{gs})L_s}$$

(10.7)

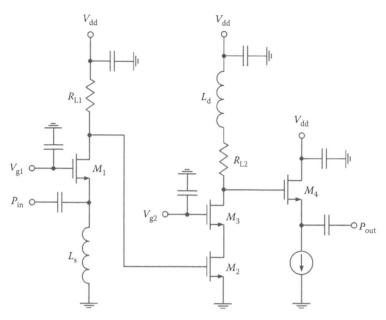

FIGURE 10.6
Schematic of a two-stage CMOS amplifier with input common-gate stage.

where g_m and C_{gs} are the transconductance and gate–source capacitance of the transistor M_1, respectively. From Equation 10.7, it follows that there is a zero near dc point determining the low 3-dB frequency. Since the input inductor L_s provides an extremely small reactance to the ground at lower frequency, the input impedance Z_{in} is dominated by L_s and its value approaches zero. As the operating frequency increases and $g_m \gg \omega C_{gs}$, the input impedance Z_{in} becomes close to $1/g_m$. Although the reactance of L_s changes the phase of Z_{in}, the magnitude of Z_{in} is still dominated by $1/g_m$ across the gigahertz operation frequencies. The value of L_s determines the input-matching range, in which the inductance varies from 3 to 11 nH, resulting in a frequency range for input return loss better than 10 dB of about 10 GHz. Optimum input matching close to 50 Ω can be achieved with $L_s = 5.3$ nH.

However, despite the input wideband matching, the transfer frequency response of the common-gate stage is not enough wideband, when a power gain of 8.5 dB is achieved with a 3-dB bandwidth of 0.4–3.5 GHz for $R_{L1} = 320$ Ω. In this case, the second stage representing a simple cascode common-source configuration provides a high-frequency gain and determines a higher 3-dB bandwidth of the CMOS amplifier. A series-peaking inductor L_d is resonant with the total parasitic capacitance at the drain node of the transistor M_3 around 10 GHz. The cascode transistor M_3 is chosen of a smaller size to have less parasitic capacitance, as well as the Q-factor of the inductor L_d is kept small for flat gain of the whole amplifier. To reduce the value of the Q-factor, an extra resistor $R_{L2} = 60$ Ω is added. As a result, a power gain of 11.2–12.4 dB and a noise figure of 4.4–6.5 dB were measured with a 3-dB bandwidth of 0.4–10 GHz, with an input return loss better than 10 dB across 2.2–12 GHz, a power consumption of 12 mW from 1.8-V supply, and a die size of 0.42 mm² [24]. In a broadband low-noise CMOS amplifier used in a UWB receiver front end, a tunable active notch-filter circuit can be connected between the cascode transistors M_2 and M_3 for interference rejection of a 5-GHz WLAN signal [25].

Figure 10.7 shows the circuit schematic of a two-stage broadband amplifier for a UWB receiver implemented in a 0.18-μm CMOS process, which includes an input common-gate transistor M_1 followed by a common-source transistor M_2 [26]. The first common-gate stage provides an input impedance matching of 50 Ω for RF signals from an antenna to LNA. The second common-source stage is operated as a gain stage to greatly amplify the weak RF signals. The first and second stages are constructed in a cascode configuration to lower power consumption, where the input impedance of the transistor M_2 becomes an output load for the transistor M_1. An output buffer configured as a source follower based on the transistor M_3 with a current source M_4 is cascaded with the second stage. Here, C_s and L_s are the elements of the high-pass input-matching circuit, C_1, L_1, and L_3 are the elements of the first interstage-matching circuit between the transistors M_1 and M_2, R_1 is added to supply a dc bias for the transistor M_2, and C_2 is a bypass capacitor to ground the source of the transistor M_2 for a high-frequency ac current. To provide an input impedance around 50 Ω, the inductance L_s and capacitance C_s generate a resonance at the center bandwidth frequency, with Q-factors of L_s larger than 5 across the full band.

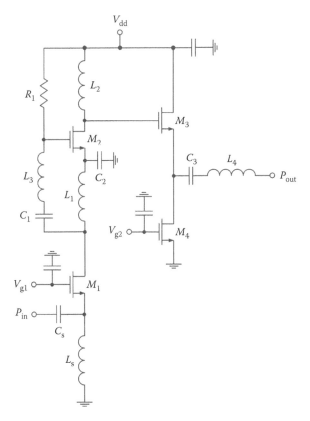

FIGURE 10.7
Schematic of a two-stage-cascoded CMOS amplifier.

Such a cascoded structure is transformed from a cascaded one without changing the amplifier type, where the interstage resonant circuits generate two different resonant frequencies, low-band frequency f_{low} and high-band frequency f_{high}, which provide the flat amplifier gain across the entire UWB bandwidth. The low-band frequency f_{low} can be derived as

$$f_{low} = \frac{1}{2\pi} \sqrt{\frac{1}{(L_1 + L_3) C_1}} \qquad (10.8)$$

from which it follows that the low-band frequency can be determined by adjusting C_1, L_1, and L_3 according to the design specification. At the same time, the high-band-resonant frequency f_{high} can be written as

$$f_{high} = \frac{1}{2\pi} \sqrt{\frac{1}{L_2 C_T}} \qquad (10.9)$$

where C_T represents the total parasitic capacitance at the drain node of the transistor M_2. It is clearly seen from Equation 10.9 that the high-band-resonant frequency can be determined by choosing the value of L_2 properly.

To achieve low-power consumption, the bias current of a two-stage-cascoded full-band LNA without the buffer was set to be 2.7 mA, and the width of the transistor M_1 was chosen to be 144 μm to provide the transconductance g_{m1} of approximately 20 mS. At the same time, the width of the transistor M_2 was chosen to be 80 μm to meet linearity requirements. The inductance L_1 resonates at 3 GHz with the parasitic capacitances, as well as the capacitance between the bottom plate of C_1 and ground at the drain node of M_1. Additionally, the inductance L_2 as an element of the second interstage-matching circuit is chosen to resonate at 11 GHz. Thus, by choosing $L_1 = 9.7$ nH, $L_2 = L_s = 3.3$ nH, $L_3 = 1.6$ nH, $C_1 = 2$ pF, $C_2 = 4$ pF, and $R_1 = 5.5$ kΩ, a maximum gain of 12 dB and a minimum noise figure of 5.27 dB with a total power consumption of 12 mW were measured from 3.1 to 10.6 GHz for a 0.18-μm CMOS two-stage-cascoded amplifier occupying a chip area of 1.17×0.88 mm², whose circuit schematic is shown in Figure 10.7 [26]. To enhance the input-matching conditions and improve noise figure at lower frequencies, an *RL*-type input-matching circuit can be used, with first a series inductor and a shunt circuit composed of the series-connected inductor and resistor located between the device source node and ground [27].

To improve the noise performance over a wide frequency bandwidth, it is preferable to use noiseless components in the input-matching circuit. Figure 10.8 shows the circuit schematic of a common-gate cascode broadband LNA implemented in a 0.18-μm CMOS technology, where the input-matching circuit represents the third-order bandpass Butterworth filter [28]. Here, the input gate–source capacitance C_{gs} of the transistor M_1 is considered as an element of this filter connected in parallel to the shunt inductor L_s. The capacitor C_3 is added to make the selection of the size of the transistor M_1 more flexible. Since the filter has a unit–gain transfer function in passband, the input impedance of the LNA is approximated as $1/g_m$, where g_m is the transconductance of the transistor M_1, which should be equal to 50 Ω to achieve the required impedance matching in a UWB band. To compromise the noise performance and isolation between the input and output, the size of the transistor M_2 was chosen to be 80 μm, which is half the size of the transistor M_1. The peaking inductor L_c between the cascode transistors M_1 and M_2 is necessary to compensate for

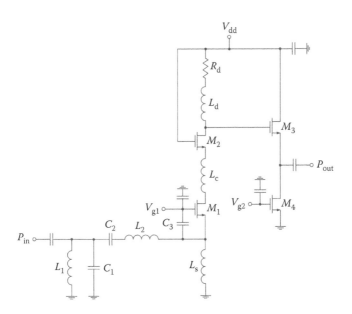

FIGURE 10.8
Schematic of a low-noise common-gate cascode CMOS amplifier.

the parasitic capacitive load reactance by forming a broadband π-type lowpass LC network with the parasitic gate–drain capacitance of the common-gate transistor M_1 and the parasitic gate–source capacitance of the common-gate transistor M_2. Similarly, the peaking inductor L_d is required to compensate for the parasitic gate–drain capacitance of the transistor M_2, as well as the resistor R_d is added to increase the low-frequency gain. The simulated results show that the flat gain of 14.5–15.3 dB, input return loss better than 8.27 dB, and noise figure of 3.57–4.27 dB in a frequency range from 3 to 10 GHz can be achieved. For a common-gate cascode LNA used in UWB receivers, a passive bandpass filter with three finite transmission zeros can be used at the input to achieve a low-frequency stop band (1.1 GHz) and a high-frequency stop band (14.6 and 25.7 GHz), with an additional 2.4-GHz-active notch filter connected at the output port of the LNA to provide a stop-band rejection in the WLAN band [29].

10.4 CMOS Amplifiers with Lossy Compensation Circuits

As a simple alternative, it was demonstrated that, by using the lossy input and output RLC-matching networks and two-stage cascode configuration, a full UWB band from 3.1 to 10.6 GHz can be covered with a power gain above 6 dB and an output power of around 0 dBm using a 0.18-μm CMOS technology [30]. Figure 10.9a shows the circuit schematic of a broadband UWB LNA implemented in a 0.18-μm CMOS process, where the input stage was designed with a series-inductive degeneration and a shunt lossy compensation circuit obtaining broadband gain and match simultaneously [31]. In this case, the device size is chosen to optimize the minimum noise figure at the proper bias condition. The second stage with interstage bandpass matching is necessary for high-gain operation.

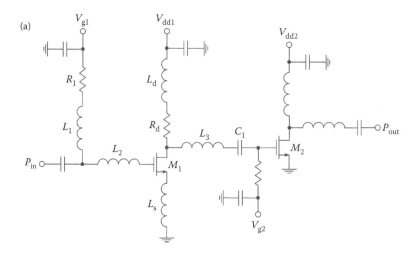

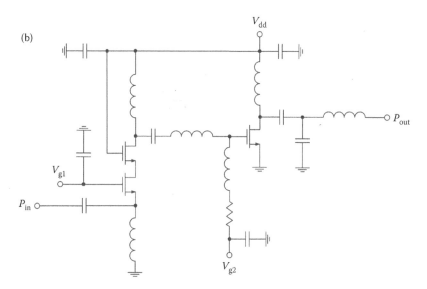

FIGURE 10.9
Schematics of broadband CMOS UWB power amplifiers with lossy matching.

By neglecting the feedback through the gate–drain capacitance C_{gd} of the transistor M_1, the input impedance Z_{in} can be written as

$$Z_{in}(\omega) = (R_1 + j\omega L_1)//\left[L_s \frac{g_m}{C_{gs}} + j\omega(L_s + L_2) + \frac{1}{j\omega C_{gs}}\right] \qquad (10.10)$$

with the corresponding real and imaginary parts of the input impedance Z_{in} derived and simplified from Equation 10.10 as

$$\mathrm{Re}\, Z_{in} = \frac{L_s R_1(g_m/C_{gs}) + (\omega L_1)^2}{L_s(g_m/C_{gs}) + R_1} \tag{10.11}$$

$$\mathrm{Im}\, Z_{in} = \omega L_1 \frac{L_s(g_m/C_{gs}) - R_1}{L_s(g_m/C_{gs}) + R_1} \tag{10.12}$$

From Equation 10.12, it follows that the imaginary part of Z_{in} can be canceled when $R_1 = (g_m/C_{gs})L_s$ and thus broadband matching is achieved. The real part of Z_{in} can be adjusted toward 50 Ω by selecting the proper L_1, L_2, L_s, and R_1. As a result, by choosing the gate width of 190 μm for M_1 and the gate width of 60 μm for M_2, a power gain of 10.8 dB with a 3-dB bandwidth from 1.6 to 13.2 GHz and a gain flatness within ±0.5 dB across the entire UWB band were achieved [31]. The input return loss better than 10 dB and a noise figure of 3.4–5.7 dB with a power consumption of 22 mW were measured from 3.1 to 10.6 GHz.

Figure 10.9b shows the circuit schematic of a 0.18-μm CMOS two-stage power amplifier with an interstage *RLC* broadband impedance matching (including the gate–drain capacitance of the first-stage common-gate device and the gate–source capacitance of the second-stage common-source device) to provide a flat power gain over a frequency band from 6 to 10 GHz [32]. In this case, a *PAE* of about 15% over the most bandwidth and a power consumption of only 18 mW were achieved with an output power of about 5 dBm at a supply voltage of 1.5 V. To simplify the input matching and provide significant interstage isolation, both first-stage nMOS devices are configured with a common gate. The overall die size was 0.82 × 1.32 mm².

10.5 Feedback CMOS Amplifiers

Among the feedback approaches applied to the CMOS amplifier design, the resistive feedback is an area-saving solution that proves to be appropriate to provide an input matching at lower UWB frequencies, especially when combined to a narrowband inductively degenerated common-source amplifier [33]. The additional noise coming from the source feedback resistor is one of the main limitations of this topology for the use in a whole UWB bandwidth. In this case, to cover the frequency range from 3 to 10 GHz, the feedback resistor value should be lowered or, as an alternative, the common-source transistor should be chosen with a larger gate width to provide higher transconductance. The reactive-feedback LNA with a feedback via a transformer provides broadband input matching with less noise [34]. However, the extra area required by the transformer-feedback network is a serious drawback for the overall chip size. The inductively degenerated common-source amplifier is one of the simplest and convenient approaches for narrowband applications in terms of gain and noise performance [35]. An extension of this circuit to broadband operation over the entire UWB bandwidth can be achieved by using an input multisection *LC*-matching network. A drawback of this approach is a significant group-delay variation due to its multiresonant nature, and it requires a large-area size and introduces a high-insertion loss that degrades the amplifier gain and noise figure.

10.5.1 Source Degeneration

Figure 10.10 shows the circuit schematic of a two-stage common-source high-efficiency power amplifier with inductive degeneration for UWB transmitters using a 0.18-µm CMOS process [36]. Here, to ensure a reasonable power gain and high-operating efficiency, the first stage is optimized for maximum gain using a cascode structure, while the second stage represents a simple common-source topology designed for maximum output power. Additionally, the source degeneration inductor L_{s1} in the first cascode stage contributes to better linearity and stability improvement, and a series inductor L_1 is necessary to complete the input impedance matching between the source impedance and the input of the transistor M_1. If the value of L_{s1} is sufficiently small (~0.5 nH), then, it can easily be replaced by a bondwire inductor to reduce the overall chip size. An inductor L_{d1} is placed as a shunt-peaking inductor resonating with the parasitic capacitances at the drain of the transistor M_2 at center bandwidth frequency. Generally, a larger transistor size of the transistor M_3 is needed to provide high gain and output power of the amplifier at high frequencies. However, the transistor with a large periphery is usually characterized by high-parasitic capacitances. Therefore, the transistor size should be optimized to provide broadband and high-efficiency performance.

The impedance looking into the input of the CMOS power amplifier shown in Figure 10.10 can be written as

$$Z_{in}(\omega) = \frac{g_m}{C_{gs}} L_{s1} + j\omega \left(L_{s1} + L_1 - \frac{1}{\omega^2 C_{gs}} \right) \tag{10.13}$$

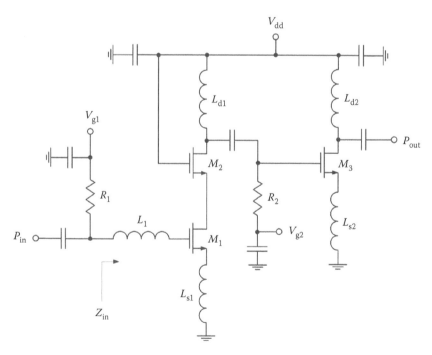

FIGURE 10.10
Schematic of a CMOS power amplifier with source degeneration.

where C_{gs} is the gate–source capacitance and g_m is the transconductance of the transistor M_1. Since the imaginary part of the input impedance can be fully compensated at center bandwidth frequency by L_1, the corresponding resonant frequency ω_0 is approximated as

$$\omega_0 = \frac{1}{\sqrt{C_{gs}(L_1 + L_{s1})}} \tag{10.14}$$

As a result, for this two-stage common-source CMOS power amplifier with inductive degeneration in both stages occupying a die area of 1.1×1.5 mm^2, an output power of more than 10 dBm with a maximum *PAE* of 34% and a power gain of 15.2 ± 0.8 dB at 1-dB compression point was achieved in a lower UWB bandwidth of 3–5 GHz, while consuming 25 mW from a 1.8-V supply [36].

For a broadband CMOS amplifier design, a three-section bandpass Chebyshev filter structure can be used to resonate the reactive part for the input impedance Z_{in} over the whole UWB band from 3.1 to 10.6 GHz, which is shown in Figure 10.11, where a series inductor L_3, a shunt capacitor C_p, a source degeneration inductor L_s, and an input device gate–source capacitance C_{gs} are also embedded in the filter structure [37]. In this case, the real part of Z_{in} is ideally chosen to be equal to the source resistance (filter termination) $R_S = (g_m/C_{gs})L_s$ by proper selection of the value of the degeneration inductor L_s for a given CMOS transistor. The gate width of the transistor M_1 (240 μm) is optimized for noise, while the cascode transistor M_2 (60 μm) is chosen to be as small as possible to reduce the parasitic capacitances. The load for the cascode stage is designed to achieve a flat gain, where an inductance L_d must resonate with the parasitic device output capacitance and the value of R_L is chosen to place the zero frequency $\omega_z = R_L/L_d$ as close as possible to the

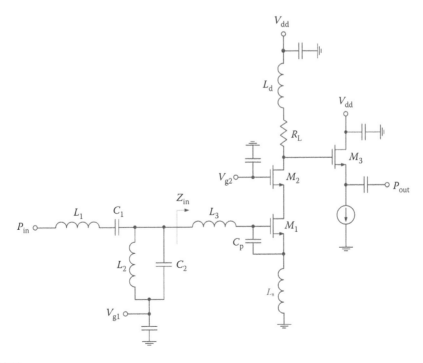

FIGURE 10.11
Schematic of a CMOS power amplifier with input band-pass filter.

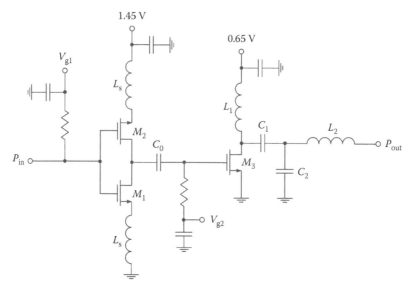

FIGURE 10.12
Schematic of a two-stage CMOS LNA with source degeneration.

lower-bandwidth edge to improve gain at lower frequencies. The buffer stage configured as a source follower must drive a 50-Ω external load. As a result, a two-stage cascode LNA with source degeneration for UWB receivers implemented in a 0.18-μm CMOS process achieved a power gain of 9.3 dB with a return loss better than 10 dB and a minimum noise figure of 4 dB over 3.1–10.6 GHz, while consuming 9 mW. Similar performance can be achieved with a slightly simplified structure of a three-section Chebyshev filter where the capacitors C_2 and C_p are removed and the remaining parameters are optimized [38]. With a two-section Chebyshev filter at the amplifier input, a 3.9-dB average noise figure with a 27-dB gain having a ripple less than 1.5 dB is achieved in an upper UWB bandwidth of 6–10 GHz [39].

To simplify the input-matching circuit and minimize group-delay variations with reduced current budget, the broadband CMOS LNA can be designed with two stages where the first stage adopts a complementary topology performance with inductive degeneration, while the second stage is implemented as a cascode structure with resonant load to improve gain and reverse isolation [40]. Figure 10.12 shows the circuit schematic of a two-stage LNA implemented in a 0.18-μm CMOS process, where the complementary topology of the first stage with inductive degeneration provides an input matching and low-noise performance [41]. The output conjugate impedance matching (the best matching point at 9.6 GHz) provided by the shunt capacitor C_2 and inductor L_2 accomplishes bandwidth extension to higher frequencies. The gate width of the second-stage transistor M_3 (100 μm) should be optimized for minimum noise. As a result, with a die area of 0.81 × 0.81 mm², the small-signal peak gain of 16.2 dB with a 3-dB bandwidth from 3 to 10 GHz and minimum noise figure of 2.3 dB at 3.5 GHz with a power consumption of 6.8 mW were measured.

10.5.2 Parallel Negative Feedback

Among the feedback configurations, the parallel resistive feedback represents an area-saving solution that proves to be appropriate for the implementation of the input-matching circuit in the UWB band of 3–5 GHz, especially when a narrowband inductively degenerated common-source amplifier is used. The additional noise coming from the feedback resistor is one of the main limitations of this topology for the use in the whole UWB bandwidth. In this case, to cover the full UWB range of 3–10 GHz, the value of the feedback resistor must be sufficiently low and the common-source transistor transconductance must be increased to provide a high-gain condition across the entire UWB band.

Figure 10.13a shows the circuit schematic of a shunt-series resistive-feedback LNA implemented in a 0.18-μm CMOS process that provides a proper design trade-off between the source impedance matching and noise figure performance [42]. To increase gain at higher

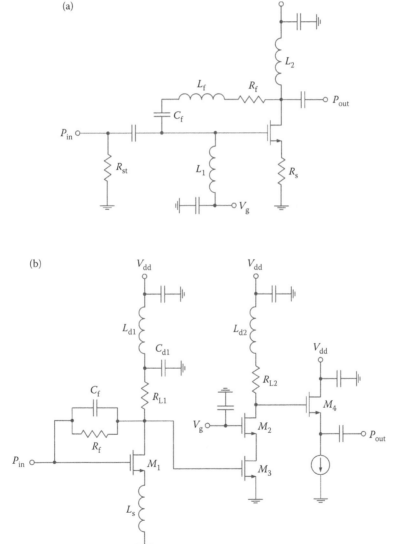

FIGURE 10.13
Schematics of parallel-feedback CMOS LNAs.

frequencies, the load inductance L_2 should be sufficiently high and an inductor L_f is added into the feedback circuit. When considering the ideal matching conditions of $S_{11} = S_{22} = 0$ at lower frequencies with infinite capacitance C_f and zero inductance L_f, the series resistance R_s can be derived for the circuit characteristic impedance $Z_0 = 50\ \Omega$ as

$$R_s = \frac{Z_0^2}{R_f} - \frac{1}{g_m} \tag{10.15}$$

where g_m is the transconductance of the idealized nMOS transistor, resulting in a low-frequency power gain of $(1 - R_f/Z_0)^2$. As a result, a broadband gain of 12 dB with a gain flatness of 0.27 dB at 1-dB compression point of better than −0.8 dBm, a minimum noise figure of 3.8 dB, and an input return loss of better than 9 dB were achieved across 3.1–10.6 GHz, with a low-power consumption of 9.8 mW. A parallel resistance–capacitance feedback with a source degeneration inductor L_s can also be used to obtain broadband input matching and to effectively reduce the noise level, as shown in Figure 10.13b for a two-stage UWB LNA with an output source-follower buffer implemented in a 0.18-μm CMOS process [43]. Measured results demonstrated a power gain of 10.9–13.9 dB and a noise figure of 2.5–4.7 dB over a full UWB bandwidth of 3.1–10.6 GHz, with a die size of 0.46 mm² and a power consumption of 14.4 mW from a 1.8-V supply.

On the design of broadband multistage LNAs, there are some specific requirements to the matching circuits to be optimized. For example, for a two-stage LNA, the input-matching circuit is optimized for both noise and flat gain characteristics simultaneously over a broad frequency band, as shown in Figure 10.14a. At the same time, the gain-sloped interstage-matching circuit should be applied to compensate for the transistor intrinsic frequency

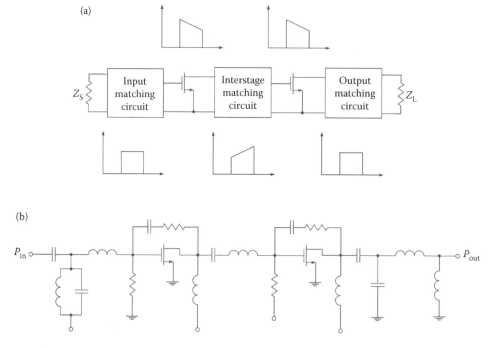

FIGURE 10.14
Two-stage CMOS LNA topology with sloped interstage matching.

roll-off to maintain gain flatness and extend frequency bandwidth of the amplifier. Figure 10.14b shows the circuit schematic of a fully integrated two-stage UWB LNA with a shunt-resistive feedback applied to both stages, which is implemented using a 0.18-μm CMOS technology and occupying the size of 1.37 × 1.19 mm² [44]. Operated on a 1.8-V supply, the LNA achieved a power gain of 19.1 dB within the frequency bandwidth of 2.8–7.2 GHz, a noise figure lower than 3.8 dB from 3.0 to 7.5 GHz, and a power consumption of 32.4 mW, with the group delay less than 0.5 ns within the UWB band.

A low-power consumption, full UWB band, and high gain can be achieved by using a cascaded CMOS structure with two common-source stages where a parallel resistive feedback is used in the first stage and output current of the first stage is reused in the input gate of the second stage [45]. In this case, an input return loss of better than 14 dB and a flat gain of about 8 dB were achieved across 3.1–10.6 GHz using a 0.13-μm CMOS technology, with a minimum noise figure of 2.5 dB at 10.5 GHz. Higher gain of around 18 dB across the entire UWB bandwidth is achieved using an additional source-follower buffer stage [46]. However, the noise figure variations are quite significant especially at lower frequencies. Figure 10.15 shows the circuit schematic of a cascaded current-reused UWB LNA implemented in a 0.18-μm CMOS process, where the source degeneration inductor L_s is added to the input stage to achieve wideband flat noise figure performance [47]. Here, the gate series inductor L_{g1} is used to achieve wideband input matching, the value of L_{d1} is chosen to be very large to provide high impedance to divert the signal to the gate of the second-stage

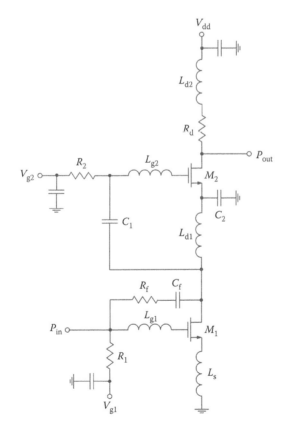

FIGURE 10.15
Schematic of a cascaded current-reused CMOS LNA with source degeneration.

transistor M_2, C_2 is used as a bypass capacitor to the ground, and the interstage circuit that consists of C_1 and L_{g2} provides a series resonance with the gate–source capacitance C_{gs} of the transistor M_2. The output of each stage is equivalently loaded with a low-Q RLC parallel resonant circuit to maximize the 3-dB gain bandwidth. As a result, a flat gain of 12.26 ± 0.63 dB, a flat noise figure of 4.24 ± 0.5 dB, and a group-delay variation of only ± 22 ps were measured over the 3.1–10.6 GHz. By appropriately selecting the values of L_s, L_{g1}, L_{g2}, L_{d1}, L_{d2}, C_f, R_f, C_1, and R_d, and the size and bias of the transistors M_1 and M_2, the noise figure can be improved to 2.87 ± 0.19 dB with a group-delay variation of less than 16 ps over the entire UWB band of 3.1–10.6 GHz [48].

The cascode feedback LNA topology can provide a stable low-noise operation with high gain and wide frequency bandwidth [49]. In this case, the common-source transistor is responsible for the noise performance, while the common-gate transistor increases the power gain and improves the reverse isolation. However, the noise figure cannot be optimized without sacrificing other important performances such as gain, input return loss, or frequency bandwidth. For example, a lower value of the feedback resistance results in a higher bandwidth with minimum gain flatness, but with a higher noise figure and lower gain at the same time. For a 0.18-μm CMOS cascode parallel-feedback LNA, the noise figure of less than 4 dB was simulated over the frequency bandwidth of 2–10 GHz [50]. Figure 10.16 shows the circuit schematic of a cascode parallel-feedback 0.18-μm CMOS LNA with source degeneration, where the feedback resistor $R_f = 300 \, \Omega$ is used to reduce the quality factor of the input series resonance circuit to achieve a broadband input matching [51]. Since the power gain is degraded at higher frequencies by the device parasitic capacitances, a series-peaking inductor L_d with an optimized value is added into the load circuit. The degeneration inductor L_s was chosen to be sufficiently small (0.12 nH) for matching at higher bandwidth frequencies and implemented by a high-impedance microstrip line,

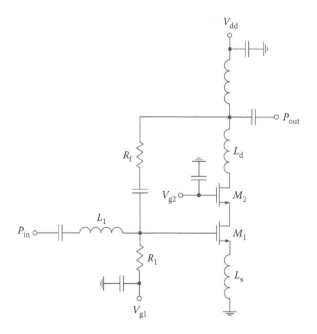

FIGURE 10.16
Schematic of a cascode parallel-feedback CMOS LNA with source degeneration.

which is less susceptible to process variations compared to the spiral inductor when the inductance is very small. As a result, the power gain greater than 10 dB and input return loss better than 10 dB were measured from 2 to 11.5 GHz, with a noise figure flatness of 3.58 ± 0.41 dB in a UWB bandwidth of 3.1–10.6 GHz. Generally, the parallel-feedback *RC* circuit can be applied to each stage in a two-stage cascode CMOS LNA to achieve higher gain with minimum variations of the noise figure over a very wide frequency range [52].

Figure 10.17 shows the circuit schematic of a transformer-feedback 0.18-μm CMOS LNA, where the first stage is self-biased by the high-value resistor R_{bias} representing a dc feedback and the second cascode stage provides a capacitive loading to the first stage [53,54]. By neglecting the effect of the gate–drain feedback capacitance and assuming that the value of the drain inductor L_d is chosen to resonate in series with the input capacitance of the transistor M_3, the impedance looking into the input series inductor L_g of the common-source CMOS stage can be written in a simplified form as

$$Z_{\text{in}}(\omega) = \frac{g_m}{C_{gs}} M + j\omega \left(L_g - \frac{1}{\omega^2 C_{gs}} \right) \tag{10.16}$$

where g_m is the transconductance and C_{gs} is the gate–source capacitance of the transistor M_1 and the real part of Z_{in} is fully determined by the transformer mutual coupling inductance M. The value of the gate inductor L_g can be optimally chosen to properly compensate for the input device capacitive reactance to provide minimum gain flatness over the required bandwidth. An input *LC* resonator formed by the inductor L_1 and capacitor C_1 is connected in parallel to suppress excessive gain at the out-of-band frequencies below 3 GHz. The cascode CMOS stage not only gives a sufficient reverse isolation, but also improves gain at higher frequencies due to the series-peaking inductor L_2 and shunt load inductor L_3. As a result, a power gain of about 11 dB with a gain variation of 1.2 dB, an input return loss better than 11 dB, and a noise figure of 4.7–5.6 dB across the frequency range of 3.1–10.6 GHz with a power consumption of 10.6 mW were achieved [53]. Fabricated in a 0.13-μm CMOS

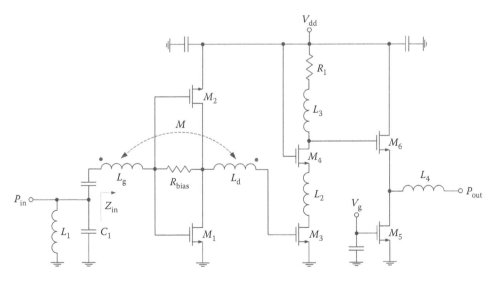

FIGURE 10.17
Schematic of a UWB CMOS LNA with transformer feedback.

technology, such a transformer-feedback LNA exhibits a noise figure of 2.2–3.1 dB and less than 5-mW power consumption from 2.4 to 5.4 GHz under 1-V power supply [55].

10.6 Noise-Canceling Technique

From consideration of different contributions to the drain current noise of a 0.18-μm *n*-channel MOSFET device, it was found that the major part is due to the channel thermal noise, which has a power spectral density of $4kT\gamma g_{d0}\Delta f$, where g_{d0} is the channel conductance for zero drain–source bias voltage and γ is the noise parameter called the *white-noise gamma factor*, while the relatively small contribution comes from the gate resistance, bulk resistance, or $1/f$ noise [56]. The theoretical long-channel value of γ is equal to 2/3, whereas shorter channels show a small enhancement of γ, which is partly due to thermal noise of parasitic resistances and partly due to short-channel effects such as velocity saturation and channel length modulation. For a deep-submicron MOSFET, the value of γ exceeds 1 in saturation and may become 2–3 under some biasing conditions [35]. For a common-gate CMOS LNA, whose simplified circuit schematic is shown in Figure 10.18a, the noise figure *F* is derived as

$$F = 1 + \frac{\gamma}{\alpha} + \frac{4R_S}{R_{L1}} \tag{10.17}$$

where $\alpha = g_m/g_{d0}$, and is often less than 1 in deep-submicron MOSFETs [35,57]. Assuming a conservative value of 1.33 for γ/α, the value of R_{L1} should be larger than 241 Ω for a noise figure *F* (in decibels) less than 5 dB. In this case, the bandwidth hardly exceeds 10 GHz with such a large R_{L1} in a 0.18-μm CMOS process since the LNA must properly drive the following receiver circuit such as a downconversion mixer. Besides, other noise sources such as gate-induced noise makes the noise figure even worse at higher frequencies since its power spectral density is proportional to the squared frequency. To obtain the best noise figure, the transistor should be biased to its peak transition frequency f_T [58]. Therefore, the common-gate CMOS LNA may be considered as an appropriate candidate for a UWB receiver operating only at lower frequencies of the UWB band.

The circuit schematic of a resistive-feedback LNA is shown in Figure 10.18b, where the input impedance is ideally equal to $(R_f + R_{L1})/(1 + g_{m1}R_{L1})$ that should be set to 50 Ω for matching. In this case, the voltage gain is calculated from $R_{L1}(1 - g_{m1}R_{L1})/(R_f + R_{L1})$. By neglecting the noise from the second transistor M_2, the noise figure *F* is derived as

$$F = 1 + \frac{R_f}{R_S}\left(\frac{1 + g_{m1}R_S}{1 - g_{m1}R_f}\right) + \frac{1}{R_S R_{L1}}\left(\frac{R_f + R_S}{1 - g_{m1}R_f}\right)^2$$
$$+ \frac{\gamma}{\alpha}\frac{g_{m1}}{R_S}\frac{(R_f + R_S)^2}{1 - g_{m1}R_f} \tag{10.18}$$

that also depends on γ/α [57]. As the frequency increases, the noise from M_2 is more pronounced, making the noise figure larger. Though inductive peaking can be applied to the input to tolerate the size of the transistor M_1 and hence a higher g_{m1}, this topology typically

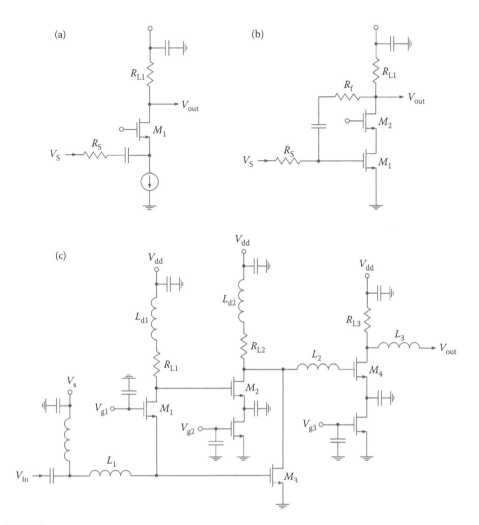

FIGURE 10.18
Circuit schematics of different CMOS LNAs.

requires larger power consumption or more advanced technologies to achieve an acceptable noise figure. This is primarily due to the inherently low transconductance of CMOS devices, which not only degrades the noise performance but also prohibits the use of a large feedback resistor.

To minimize a noise figure of the LNA over an entire UWB bandwidth, it is required to use a special broadband noise-canceling technique, which can be based on the decoupling of the input signal to create two fully correlated noise signals with opposite phases flowing through the different amplifying paths with their cancellation at the output [59]. Figure 10.18c shows the circuit schematic of a broadband noise-canceling CMOS LNA, where the first stage incorporates a common-gate topology to facilitate the ease of matching and signal splitting. In this case, the series inductor L_1 and parasitic capacitances of the transistors M_1 and M_3 form an LC ladder structure, the noise current due to the transistor M_1 flows in opposite directions, and then combine at the common drain of M_2 and M_3, while the signal currents will be added in phase. The noise cancellation will occur if the cancellation condition $g_{m2}R_{L1} = g_{m3}R_S$ is fulfilled, where g_{m2} is the transconductance of the transistor M_2 and g_{m3} is the transconductance of the transistor M_3. A current source with a bypass capacitor is used for M_2. The inductors L_1, L_2, and L_3 are added to extend the circuit bandwidth to higher frequencies, while the transistor M_4 with its source connected to a current source and shunt capacitor forms a high-pass stage to filter out the low-frequency components below 3 GHz, resulting in a bandpass frequency response of the entire LNA. By applying a noise-canceling technique, the noise figure of a broadband CMOS LNA dominated by R_{L1}, M_2, and M_3 is approximated as

$$F = 1 + \frac{R_S}{R_{L1}}\left(1 + \frac{\gamma}{\alpha}\frac{1}{g_{m2}R_{L1}}\right) + \frac{\gamma}{\alpha}\frac{1}{g_{m3}R_S} \tag{10.19}$$

which provides design insights for sizing circuit components [57]. As a result, the measured noise figure was of 4.5–5.1 dB over the 3.1–10.6 GHz, which is close to the simulated noise figure with $\gamma = 4/3$.

A similar circuit schematic of a broadband noise-canceling 0.13-μm CMOS LNA is shown in Figure 10.19, where the shunt-peaking inductor L_d is added to the drain circuit of the transistor M_2 to increase the frequency bandwidth to higher frequencies and an inductive degeneration is used for the transistor M_3 [60]. In this case, the gain can be increased by about 5 dB at high frequencies around 10 GHz. A second amplification stage based on the transistor M_4 is necessary to further increase the overall gain to more than 16 dB and provide an output impedance match from 3 to 10 GHz. A series inductor L_4 is used to compensate for the output capacitance of M_4. Inductors L_d and L_4 have a high-self-resonant frequency of more than 30 GHz for a constant inductance value from 3 to 10 GHz, which

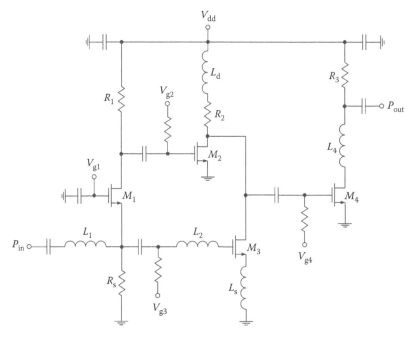

FIGURE 10.19
Schematic of a broadband noise-canceling CMOS LNA.

is important to obtain a flat gain over the entire UWB band. As a result, a noise figure of 3.9 ± 0.28 dB and a group delay of 105 ± 18 ps were measured in a frequency bandwidth of 3–10 GHz.

Figure 10.20 shows the circuit schematic of a two-stage noise-canceling 0.18-μm CMOS LNA, where the parallel resistive feedback is used in the first cascode stage and the noise cancellation is achieved at the output of the second stage [61]. The noise current at the output port will be successfully suppressed to zero by proper selection of the values of the feedback resistor R_f and transconductances g_{m1}, g_{m3}, and g_{m4} of the corresponding transistors M_1, M_3, and M_4. An inductor L_3 is added into the second-stage resistive feedback to construct an RLC feedback configuration to increase the total gain response and cancel the effect of the parasitic capacitances. The presence of this inductor allows the gain flatness above 7.2 GHz to be provided. However, the negative feedback generally does not improve the noise performance of the circuit. From simulation results, it follows that the noise figure below 3 dB can potentially be achieved in the frequency range from 3.1 to 10.6 GHz.

Figure 10.21 shows the basic schematic of a transformer noise-canceling CMOS LNA, which is suitable for low-voltage operation and achieves an input impedance matching and a low-noise performance across the UWB bandwidth of 3.1–10.6 GHz without additional circuits or increased power consumption [62]. In this case, the transistor is arranged in a common-gate configuration and the input and output shunt-peaking inductances L_s and L_d are magnetically coupled to form a transformer, which partly cancels the output noise voltage produced by the drain noise current. An output series inductor L_1 extends not only the gain bandwidth, but also improves the input bandwidth through the coupling

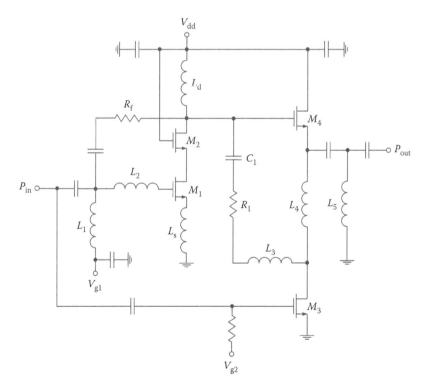

FIGURE 10.20
Schematic of a two-stage noise-canceling CMOS LNA.

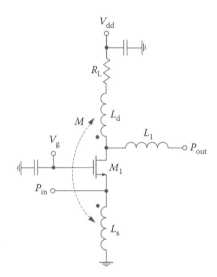

FIGURE 10.21
Basic schematic of a transformer noise-canceling CMOS LNA.

by reducing the effect of the device input capacitance over the entire UWB bandwidth. The noise factor of the transformer noise-canceling CMOS LNA is written as

$$F = 1 + F_{M_1} + F_{R_L} \tag{10.20}$$

with

$$F_{M_1} = \left| \frac{n^2 - n + ((1/R_S) + j\omega C_{in} + (1/j\omega L_s))R_L}{g_m R_L + n} \right|^2 \gamma g_{d0} R_S \tag{10.21}$$

$$F_{R_L} = \left| \frac{(1 - n)g_m R_L + ((1/R_S) + j\omega C_{in} + (1/j\omega L_s))R_L}{g_m R_L + n} \right|^2 \frac{R_S}{R_L} \tag{10.22}$$

where C_{in} represents the sum of the gate–source capacitance of the transistor M_1 and the parasitic capacitances of the input pad, $n = \sqrt{L_d/L_s}$ is the turns ratio of the transformer (magnetic coupling factor k is assumed to be one), R_S is the source resistance, g_m and g_{d0} are the transconductance and zero-bias drain conductance of M_1, and F_{M1} and F_{RL} represent noise factors contributed from M_1 and R_L, respectively. The gate-induced noise current from M_1 and the parasitic pad capacitances and inductor resistances are ignored. From Equations 10.20 to 10.22, it follows that the noise figure approaches its minimum value in the range of $1 < n < 1.2$. The transconductance of M_1 is determined by the input-matching conditions and the load resistor R_L is selected to achieve a gain of more than 10 dB at lower UWB frequencies. The transformer adopts a stacked configuration, which provides the greatest coupling factor and a small area. As a result, a noise figure of 2.7–3.3 dB, an input return loss of better than 10 dB, and a gain of more than 7.8 dB with a power consumption of 2.5 mW from a 1.0-V supply were measured across the 3.1–10.6 GHz for this transformer noise-canceling UWB LNA implemented in a 90-nm digital CMOS technology.

References

1. C. H. Hoeppner, Pulse communication system, U.S. Patent 2,999,128, September 1961 (filed November 1945).
2. G. F. Ross, Transmission and reception system for generating and receiving base-band duration pulse signals without distortion for short base-band pulse communication system, U.S. Patent 3,728,632, April 1973 (filed May 1971).
3. S. Roy, J. R. Foerster, V. S. Somayazulu, and D. G. Leeper, Ultrawideband radio design: The promise of high-speed, short-range wireless connectivity, *Proc. IEEE*, 92, 295–311, 2004.
4. D. D. Wentzloff, R. Blazquez, F. S. Lee, B. P. Ginsburg, J. Powell, and A. P. Chandrakasan, System design considerations for ultra-wideband communication, *IEEE Commun. Mag.*, 43, 114–121, 2005.
5. G. R. Aiello and G. D. Rogerson, Ultra-wideband wireless systems, *IEEE Microw. Mag.*, 4, 36–47, 2003.
6. A. P. Chandrakasan, F. S. Lee, D. D. Wentzloff, V. Sze, B. P. Ginsburg, P. P. Mercier, D. C. Daly, and R. Blazquez, Low-power impulse UWB architectures and circuits, *Proc. IEEE*, 97, 332–352, 2009.
7. H. Zheng, S. Lou, D. Lu, C. Shen, T. Chan, and H. S. Luong, A 3.1 GHz–8.0 GHz single-chip transceiver for MB-OFDM UWB in 0.18-μm CMOS process, *IEEE J. Solid-State Circuits*, SC-44, 414–426, 2009.
8. B. Park, K. Lee, S. Choi, and S. Hong, A 3.1–10.6 GHz RF receiver front-end in 0.18 μm CMOS for ultra-wideband applications, *IEEE MTT-S Int. Microw. Symp. Dig.*, Paris, 1616–1619, 2010.
9. D. Marchaland, M. Villegas, G. Baudoin, C. Tinella, and D. Belot, System concepts dedicated to UWB transmitter, *Proc. Eur. Wirel. Technol. Conf.*, Paris, France, pp. 141–144, 2005.
10. V. V. Kulkarni, M. Muqsith, K. Niitsu, H. Ishikuro, and T. Kuroda, A 750 Mb/s, 12 pJ/b, 6-to-10 GHz CMOS IR-UWB transmitter with embedded on-chip antenna, *IEEE J. Solid-State Circuits*, SC-44, 394–403, 2009.
11. W. C. Wang, C. P. Liao, Y. K. Lo, Z.-D. Huang, F. R. Shahroury, and C.-Y. Wu, The design of integrated 3-GHz to 11-GHz CMOS transmitter for full-band ultra-wideband (UWB) applications, *Int. Circuits Syst. Symp. Dig.*, 2709–2712, 2008.
12. N. Shiramizu, T. Masuda, M. Tanabe, and K. Washio, A 3–10 GHz bandwidth low-noise and low-power amplifier for full-band UWB communications in 0.25-μm SiGe BiCMOS technology, *IEEE RFIC Symp. Dig.*, 39–42, 2005.
13. Y. Lu, R. Krithivasan, W. L. Kuo, and J. D. Cressler, A 1.8–3.1 dB noise figure (3–10 GHz) SiGe HBT LNA for UWB applications, *IEEE RFIC Symp. Dig.*, 1–4, 2006.
14. B. M. Ballweber, R. Gupta, and D. J. Allstot, A fully integrated 0.5–5.5-GHz CMOS distributed amplifier, *IEEE J. Solid-State Circuits*, SC-35, 231–239, 2000.
15. Y. H. Yu, Y. E. Chen, and D. Heo, A 0.6-V low power UWB CMOS LNA, *IEEE Microw. Wirel. Compon. Lett.*, 17, 229–231, 2007.
16. P. Heydari, Design and analysis of a performance-optimized CMOS UWB distributed LNA, *IEEE J. Solid-State Circuits*, SC-42, 1892–1905, 2007.
17. K. Moez and M. I. Elmasry, A low-noise CMOS distributed amplifier for ultra-wide-band applications, *IEEE Trans. Circuits Syst.—II: Express Briefs*, CAS-II-55, 128–130, 2008.
18. C. Lu, A. V. Pham, and M. Shaw, A CMOS power amplifier for full-band UWB transmitters, *IEEE RFIC Symp. Dig.*, 397–400, 2006.
19. D. R. Webster, G. Ataei, and D. G. Haigh, High-pass lumped-element transmission lines, *IEEE Microw. Guid. Wave Lett.*, 8, 27–29, 1998.
20. J. F. Chang and Y. S. Lin, Low-power, high-gain and low-noise CMOS distributed amplifier for UWB systems, *Electron. Lett.*, 45, 634–636, 2009.
21. J. F. Chang and Y. S. Lin, A 3.9-dB NF, 24.5-dB gain 0.3–10-GHz distributed amplifier using dual-inductive-peaking cascade gain cell for UWB systems in 0.18-μm CMOS technology, *Microw. Opt. Technol. Lett.*, 53, 2228–2232, 2011.

22. M. K. Shirada, X. Guan, C. Huynh, and C. Nguyen, Design of an ultra-small distributed low-noise-amplifier for ultra-wideband applications, *IEEE Int. Antennas Propag. Symp. Dig.*, 3361–3364, 2011.

23. C. C. Wei, H. C. Chiu, and W. S. Feng, A low noise 3.1–10.6 GHz pMOS distributed amplifier for ultra-wideband applications, *Microw. Opt. Technol. Lett.*, 49, 1641–1644, 2007.

24. K. H. Chen, J. H. Lu, B. J. Chen, and S. I. Liu, An ultra-wide-band 0.4–10-GHz LNA in 0.18-μm CMOS, *IEEE Trans. Circuits Syst.—II: Express Briefs*, CAS-II-54, 217–221, 2007.

25. B. Park, K. Lee, S. Choi, and S. Hong, A 3.1–10.6 GHz RF receiver front-end in 0.18-μm CMOS for ultra-wideband applications, *IEEE MTT-S Int. Microw. Symp. Dig.*, 1616–1619, 2010.

26. R. M. Weng, C. Y. Liu, and P. C. Lin, A low-power full-band low-noise amplifier for ultra-wideband receivers, *IEEE Trans. Microw. Theory Tech.*, MTT-58, 2077–2083, 2010.

27. J. F. Chang and Y. S. Lin, 0.99 mW 3–10 GHz common-gate CMOS UWB LNA using T-match input network and self-body-bias technique, *Electron. Lett.*, 47, 658–659, 2011.

28. X. Fan, E. Sanchez-Sinencio, and J. Silva-Martinez, A 3 GHz–10 GHz common gate ultrawideband low noise amplifier, *Proc. 48th Midwest Circuits Syst. Symp.*, Covington, KY, pp. 631–634, 2005.

29. J. F. Chang and Y. S. Lin, A low-power 3.2–9.7 GHz ultra-wideband low-noise amplifier with excellent stop-band rejection using 0.18-μm CMOS technology, *Microw. Opt. Technol. Lett.*, 54, 1261–1263, 2012.

30. H. C. Hsu, Z. W. Wang, and G. K. Ma, A low power CMOS full-band UWB power amplifier using wideband RLC matching method, *Proc. IEEE Electron. Devices Solid-State Conf.*, Hong Kong, pp. 233–236, 2005.

31. H. Y. Liao, C. M. Tseng, and H. K. Chiou, Ultra-wideband CMOS low noise amplifier with simultaneous gain and noise matches, *Microw. Opt. Technol. Lett.*, 50, 158–160, 2008.

32. H. W. Chung, C. Y. Hsu, C. Y. Yang, K. F. Wei, and H. R. Chuang, A 6–10-GHz CMOS power amplifier with an interstage wideband impedance transformer for UWB transmitters, *Proc. 38th Eur. Microw. Conf.*, Amsterdam, pp. 305–308, 2008.

33. C. W. Kim, M. S. Kang, P. T. Ahn, H. T. Kim, and S. G. Lee, An ultra-wideband CMOS low noise amplifier for 3 – 5-GHz UWB system, *IEEE J. Solid-State Circuits*, SC-40, 544–547, 2005.

34. M. T. Reiha and J. R. Long, A 1.2-V reactive-feedback 3.1–10.6 GHz low-noise amplifier in 0.13-μm CMOS, *IEEE J. Solid-State Circuits*, SC-42, 1023–1033, 2007.

35. D. K. Schaeffer and T. H. Lee, A 1.5-V, 1.5-GHz CMOS low-noise amplifier, *IEEE J. Solid-State Circuits*, SC-32, 745–759, 1997.

36. S.-K. Wong, S. Maisurah, M. N. Osman, F. Kung, and J.-H. See, High efficiency CMOS power amplifier for 3 to 5 GHz ultra-wideband (UWB) application, *IEEE Trans. Consum. Electron.*, CE-55, 1546–1550, 2009.

37. A. Bevilacqua and A. M. Niknejad, An ultrawideband CMOS low-noise amplifier for 3.1–10.6-GHz wireless receivers, *IEEE J. Solid-State Circuits*, SC-39, 2259–2268, 2004.

38. B. Y. Chang and C. F. Jou, Design of a 3.1–10.6 GHz low-voltage, low-power CMOS low-noise amplifier for ultra-wideband receivers, *Proc. Asia-Pac. Microw. Conf.*, Suzhou, China, pp. 1–4, 2005.

39. M. Battista, J. Gaubert, M. Egels, S. Bourdel, and H. Barthelemy, 6–10 GHz ultra-wideband CMOS LNA, *Electron. Lett.*, 44, 343–344, 2008.

40. G. Sapone and G. Palmisano, A 3–10-GHz low-power CMOS low-noise amplifier for ultra-wideband communication, *IEEE Trans. Microw. Theory Tech.*, MTT-59, 678–686, 2011.

41. C. P. Liang, C. W. Huang, Y. K. Lin, and S. J. Chung, 3–10 GHz ultra-wideband low-noise amplifier with new matching technique, *Electron. Lett.*, 46, 1102–1103, 2010.

42. H. Xie, X. Wang, A. Wang, Z. Wang, C. Zhang, and B. Zhao, A fully-integrated low-power 3.1–10.6 GHz UWB LNA in 0.18 μm CMOS, *Proc. IEEE Radio Wirel. Symp.*, Long Beach, CA, pp. 197–200, 2007.

43. K. C. He, M. T. Li, C. M. Li, and J. H. Tarng, Parallel-RC feedback low-noise amplifier for UWB applications, *IEEE Trans. Circuits Syst.—II: Express Briefs*, CAS-II-57, 582–586, 2010.

44. Y. E. Chen and Y. I. Huang, Development of integrated broad-band CMOS low-noise amplifiers, *IEEE Trans. Circuits Syst.—I: Regul. Pap.*, CAS-I-54, 2120–2127, 2007.

45. J. H. Lee, C. C. Chen, H. Y. Yang, and Y. S. Lin, A 2.5-dB NF 3.1 – 10.6-GHz CMOS UWB LNA with small group-delay-variation, *IEEE RFIC Symp. Dig.*, 501–504, 2008.

46. A. I. Galal, R. Pokharel, H. Kanaya, and K. Yoshida, A low power UWB low noise amplifier using current reused and feedback techniques, *Microw. Opt. Technol. Lett.*, 54, 471–474, 2012.

47. Y. S. Lin, C. Z. Chen, H. Y. Yang, C. C. Chen, J. H. Lee, G. W. Huang, and S. S. Lu, Analysis and design of a CMOS UWB LNA with dual-RLC-branch wideband input matching network, *IEEE Trans. Microw. Theory Tech.*, MTT-58, 287–296, 2010.

48. C. H. Wu, Y. S. Lin, and C. C. Wang, 11.81 mW 3.1–10.6 GHz ultra-wideband low-noise amplifier with 2.87 ± 0.19 dB noise figure and 12.52 ± 0.81 dB gain using 0.18 μm CMOS technology, *Microw. Opt. Technol. Lett.*, 54, 1445–1450, 2012.

49. H. Doh, Y. Jeong, and S. Jung, Design of CMOS UWB low noise amplifier with cascode feedback, *Proc. 47th Midwest Circuits Syst. Symp.*, Hiroshima, Japan, vol. 2, pp. 641–644, 2004.

50. Y. C. Chen, W. K. Yeh, R. L. Wang, and H. D. Yen, 2~10 GHz UWB low noise amplifier using a cascode structure with resistive shunt feedback, *Proc. Asia-Pac. Microw. Conf.*, Bangkok, pp. 1–3, 2007.

51. H. K. Chen, D. C. Chang, Y. Z. Juang, and S. S. Luo, A compact wideband CMOS low noise amplifier using shunt resistive-feedback and series inductive-peaking techniques, *IEEE Microw. Wirel. Compon. Lett.*, 17, 616–618, 2007.

52. J. Jung, T. Yun, and J. Choi, Ultra-wideband low noise amplifier using a cascode feedback topology, *Microw. Opt. Technol. Lett.*, 48, 1102–1104, 2006.

53. C. T. Fu and C. N. Kuo, 3~11-GHz CMOS UWB LNA using dual feedback for broadband matching, *IEEE RFIC Symp. Dig.*, 67–70, 2006.

54. C. T. Fu, C. N. Kuo, and S. S. Taylor, Low-noise amplifier design with dual reactive feedback for broadband simultaneous noise and impedance matching, *IEEE Trans. Microw. Theory Tech.*, MTT-58, 795–806, 2010.

55. C. T. Fu, C. L. Ko, C. N. Kuo, and Y. Z. Juang, A 2.4 – 5.4-GHz wide tuning-range CMOS reconfigurable low-noise amplifier, *IEEE Trans. Microw. Theory Tech.*, MTT-56, 2754–2763, 2008.

56. A. J. Scholten, L. F. Tiemeijer, R. van Langevelde, R. J. Havens, A. T. A. Zegers-van Duijnhoven, and V. C. Venezia, Noise modeling for RF CMOS simulation, *IEEE Trans. Electron. Devices*, ED-50, 618–632, 2002.

57. C. F. Liao and S. I. Liu, A broadband noise-canceling CMOS LNA for 3.1–10.6-GHz UWB receivers, *IEEE J. Solid-State Circuits*, SC-42, 329–339, 2007.

58. A. Ismail and A. A. Abidi, A 3–10-GHz low-noise amplifier with wideband LC-ladder matching network, *IEEE J. Solid-State Circuits*, SC-39, 2269–2277, 2004.

59. F. Bruccoleri, E. A. M. Klumperink, and B. Nauta, Wide-band CMOS low-noise amplifier exploiting thermal noise canceling, *IEEE J. Solid-State Circuits*, SC-39, 275–282, 2004.

60. A. M. El-Gabaly and C. E. Saavedra, Broadband low-noise amplifier with fast power switching for 3.1–10.6-GHz ultra-wideband applications, *IEEE Trans. Microw. Theory Tech.*, MTT-59, 3146–3153, 2011.

61. S. C. Chen, R. L. Wang, H. C. Kuo, M. L. Kung, and C. S. Gao, The design of full-band (3.1–10.6 GHz) CMOS UWB low noise amplifier with thermal noise canceling, *Proc. Asia-Pac. Microw. Conf.*, Yokohama, Japan, pp. 409–412, 2006.

62. T. Kihara, T. Matsuoka, and K. Taniguchi, A 1.0 V, 2.5 mW, transformer noise-canceling UWB CMOS LNA, *IEEE RFIC Symp. Dig.*, 493–496, 2008.

Index